Tools and Techniques in Biomolecular

Tools and Techniques in Biomolecular Science

Edited by

Aysha Divan, University of Leeds, UK
Janice Royds, University of Otago, New Zealand

OXFORD
UNIVERSITY PRESS

OXFORD
UNIVERSITY PRESS

Great Clarendon Street, Oxford, OX2 6DP,
United Kingdom

Oxford University Press is a department of the University of Oxford.
It furthers the University's objective of excellence in research, scholarship,
and education by publishing worldwide. Oxford is a registered trade mark of
Oxford University Press in the UK and in certain other countries

Impression: 1

British Library Cataloguing in Publication Data
Data available

978–0–19–969556–0

Printed in Italy by
L.E.G.O. S.p.A.—Lavis TN

Dedication

Janice would like to dedicate this book to her three wonderful children,
Rachel, Martin, and Christopher.

Preface

What is this book about?

This textbook provides essential material on key technologies currently used to advance our understanding of the biomolecular sciences. A range of molecular techniques are covered, including working with DNA, RNA, and proteins, the role of cells and human tissues in research, and how mathematical modelling helps our understanding of biological systems. Each chapter focuses on the theoretical principles underpinning the technology covered, its applications, and its limitations. There are worked examples that illustrate the nature of the data generated and the types of problems that can be solved by each technique.

The book is designed to enable you to plan and conduct scientific research. Planning research requires an understanding of the scope and limitations of different methodologies so that the most appropriate technique can be selected to answer a particular research question. It also requires you to devise a workflow that incorporates all the relevant controls so that your research strategy and hence your data are robust. Once you have the results, it is necessary to appraise and interpret the data *critically.* The material covered in this book should assist you in this process.

The book does not provide step-by-step protocols, but highlights the types of question you should consider when using a particular technique so that you can justify the choices that you make.

Who is this book for?

This book is important for all students and researchers working in molecular biology or medical research who would like to understand how techniques work, the type of data generated, and how these data are interpreted. In other words, it is suitable for advanced undergraduate students, Masters and PhD students, research-active medical students and professionals, as well as research workers in industry. Educators will also find this book and the accompanying online material an essential aid in their work.

How has the content been structured and how should you use it?

The textbook is structured thematically into several parts. The first part (Chapters 1–6) covers topics that underpin use and analysis of DNA and RNA. Chapters 7–14 focus on proteins and the range of technologies used in their study. Techniques requiring cells and tissues are covered in Chapters 16–18. The textbook finishes with a chapter on the role of mouse models in scientific research and one on mathematical modelling in biological systems (Chapters 19 and 20, respectively).

Each chapter forms a review that illustrates theoretical principles, applications, and limitations. In addition, examples of the types of data the technique generates and the format in which the data are commonly presented are included. Each chapter includes worked examples of problem-based case studies or data-handling questions which illustrate the applications and help to contextualize the material covered.

There is extensive cross-referencing between chapters so that you can begin to build up an appreciation of how different techniques can be linked together to investigate particular research questions. For example, you may wish to explore the function of a particular gene in a series of mammalian cell lines. You may choose to silence its expression using RNAi technology and assess the downstream consequences such as changes to cell morphology or signalling pathways. This would require an understanding of the material covered in Chapter 1 (gene cloning essentials), Chapter 6 (RNAi technology), Chapter 15 (mammalian cell culture) and, depending on the types of validation and downstream assays performed, Chapter 2 (PCR), Chapter 16 (flow cytometry), and Chapter 17 (fluorescence microscopy). In addition, you may decide to extend your studies to animal systems, and hence Chapter 19 (mouse models in bioscience research) would also be relevant.

Alternatively, you may wish to express a particular protein and characterize its structure and function. This would require an understanding of the techniques covered in Chapter 1 (gene cloning essentials), Chapter 7 (recombinant protein expression), and Chapter 8 (protein purification). You would also need to be familiar with techniques that enable structural information to be determined such as those covered in Chapters 12 and 13 (X-ray crystallography, nuclear magnetic resonance, and mass spectrometry) and methods for generating mutant proteins (Chapter 3). Techniques that capture information on protein interaction partners may also be relevant (Chapters 10 and 11).

Online Resource Centre

The textbook is accompanied by an Online Resource Centre, which contains a bank of downloadable resources. These resources include:

- further case studies
- short problem-type questions
- multiple-choice type questions
- material that can be used to facilitate journal-club-type discussions.

You can access the Online Resource Centre at www.oxfordtextbooks.co.uk/orc/divan_royds/

Contributors

Luis Acevedo – *Active Motif Inc. Carlsbad, CA 92008, USA*

Gavin Allsop – *School of Biomedical Sciences, Faculty of Biological Sciences, University of Leeds, Woodhouse Lane, Leeds, LS2 9JT, UK*

Alison E. Ashcroft – *Astbury Centre for Structural Molecular Biology, Faculty of Biological Sciences, University of Leeds, Leeds, LS2 9JT, UK*

Josie A. Athens – *Department of Preventive and Social Medicine, University of Otago, PO Box 913, Dunedin 9054, New Zealand*

James R. Ault – *Astbury Centre for Structural Molecular Biology, Faculty of Biological Sciences, University of Leeds, Leeds, LS2 9JT, UK*

Johanna M. Avis – *Faculty of Life Sciences, Manchester Institute of Biotechnology, University of Manchester, 131 Princess Street, Manchester M1 7DN, UK*

Gavin Batman – *School of Medicine, University of Manchester, Oxford Road, Manchester, M13 9WL, UK*

Suzanne M. Birks – *Cellular and Molecular Neuro-Oncology, School of Pharmacy and Biomedical Sciences, University of Portsmouth, Portsmouth, UK*

David Brockwell – *School of Molecular and Cellular Biology, Faculty of Biological Sciences, University of Leeds, Woodhouse Lane, Leeds, LS2 9JT, UK*

Jeannie Chin – *Department of Neuroscience and Farber Institute for Neurosciences, Thomas Jefferson University, 900 Walnut Street, Philadelphia, PA 19107, USA*

John Colyer – *School of Biomedical Sciences, Faculty of Biological Sciences, University of Leeds, Woodhouse Lane, Leeds, LS2 9JT, UK*

Brian Corbett – *Department of Neuroscience and Farber Institute for Neurosciences, Thomas Jefferson University, 900 Walnut Street, Philadelphia, PA 19107, USA*

Sarah E. Deacon – *School of Molecular and Cellular Biology, Faculty of Biological Sciences, University of Leeds, Woodhouse Lane, Leeds, LS2 9JT, UK*

Kyle Dent – *School of Molecular and Cellular Biology, Faculty of Biological Sciences, University of Leeds, Woodhouse Lane, Leeds, LS2 9JT, UK*

Laura K. Donovan – *Cellular and Molecular Neuro-Oncology, School of Pharmacy and Biomedical Sciences, University of Portsmouth, Portsmouth, UK*

Thomas Edwards – *School of Molecular and Cellular Biology, Faculty of Biological Sciences, University of Leeds, Woodhouse Lane, Leeds, LS2 9JT, UK*

Martin Evison – *Northumbria University Centre for Forensic Science, Northumbria University, Newcastle upon Tyne, NE1 8ST, UK*

Oliver Farrance – *School of Molecular and Cellular Biology, Faculty of Biological Sciences, University of Leeds, Woodhouse Lane, Leeds, LS2 9JT, UK*

Alison Fitches – *Department of Pathology, University of Otago, PO Box 913, Dunedin, New Zealand*

Ian Hampson – *School of Medicine, University of Manchester, Oxford Road, Manchester, M13 9WL, UK*

Michael Harrison – *School of Biomedical Sciences, Faculty of Biological Sciences, University of Leeds, Woodhouse Lane, Leeds, LS2 9JT, UK*

David P. Hornby – *The Krebs Institute, Department of Molecular Biology and Biotechnology, Firth Court, Western Bank, University of Sheffield, Sheffield, S10 2TN, UK*

Gareth Howel – *School of Molecular and Cellular Biology, Faculty of Biological Sciences, University of Leeds, Woodhouse Lane, Leeds, LS2 9JT, UK*

Noelyn Hung – *Department of Pathology, University of Otago, PO Box 913, Dunedin, New Zealand*

Mary Anne Jelinek – *Active Motif Inc., Carlsbad, CA 92008, USA*

Arnout Kalverda – *School of Molecular and Cellular Biology, Faculty of Biological Sciences, University of Leeds, Woodhouse Lane, Leeds, LS2 9JT, UK*

Paul Labhart – *Active Motif Inc., Carlsbad, CA 92008, USA*

John Lawry – *BD Biosciences, Oxford, UK*

Iain Manfield – *Astbury Centre for Structural Molecular Biology, Faculty of Biological Sciences, University of Leeds, Leeds, LS2 9JT, UK*

Michael J. McPherson – *School of Molecular and Cellular Biology, Faculty of Biological Sciences, University of Leeds, Woodhouse Lane, Leeds, LS2 9JT, UK*

Gareth Morgan – *Scripps Research Institute, 10550 North Torrey Pines Road, La Jolla, CA 92037, USA*

Sanjay Nilapwar – *Faculty of Life Sciences, Manchester Institute of Biotechnology, University of Manchester, 131 Princess Street, Manchester M1 7DN, UK*

Arwen Pearson – *School of Molecular and Cellular Biology, Faculty of Biological Sciences, University of Leeds, Woodhouse Lane, Leeds, LS2 9JT, UK*

Geoffrey J. Pilkington – *Cellular & Molecular Neuro-Oncology, School of Pharmacy and Biomedical Sciences, University of Portsmouth, Portsmouth, UK*

Ana Sanz – *Active Motif Inc., Carlsbad, CA 92008, USA*

Qaiser I. Sheikh – *The Krebs Institute, Department of Molecular Biology and Biotechnology, Firth Court, Western Bank, University of Sheffield, Sheffield, S10 2TN, UK*

Tania Slatter – *Department of Pathology, University of Otago, PO Box 913, Dunedin, New Zealand*

Gary Thompson – *School of Molecular and Cellular Biology, Faculty of Biological Sciences, University of Leeds, Woodhouse Lane, Leeds, LS2 9JT, UK*

Richard Unwin – *Centre for Advanced Discovery and Experimental Therapeutics (CADET), Central Manchester University Hospitals NHS Foundation Trust, Oxford Road, Manchester, M13 9WL, UK*

Thomas Walker – *School of Medicine, University of Manchester, Oxford Road, Manchester, M13 9WL, UK*

Contents

1

Working with DNA and RNA

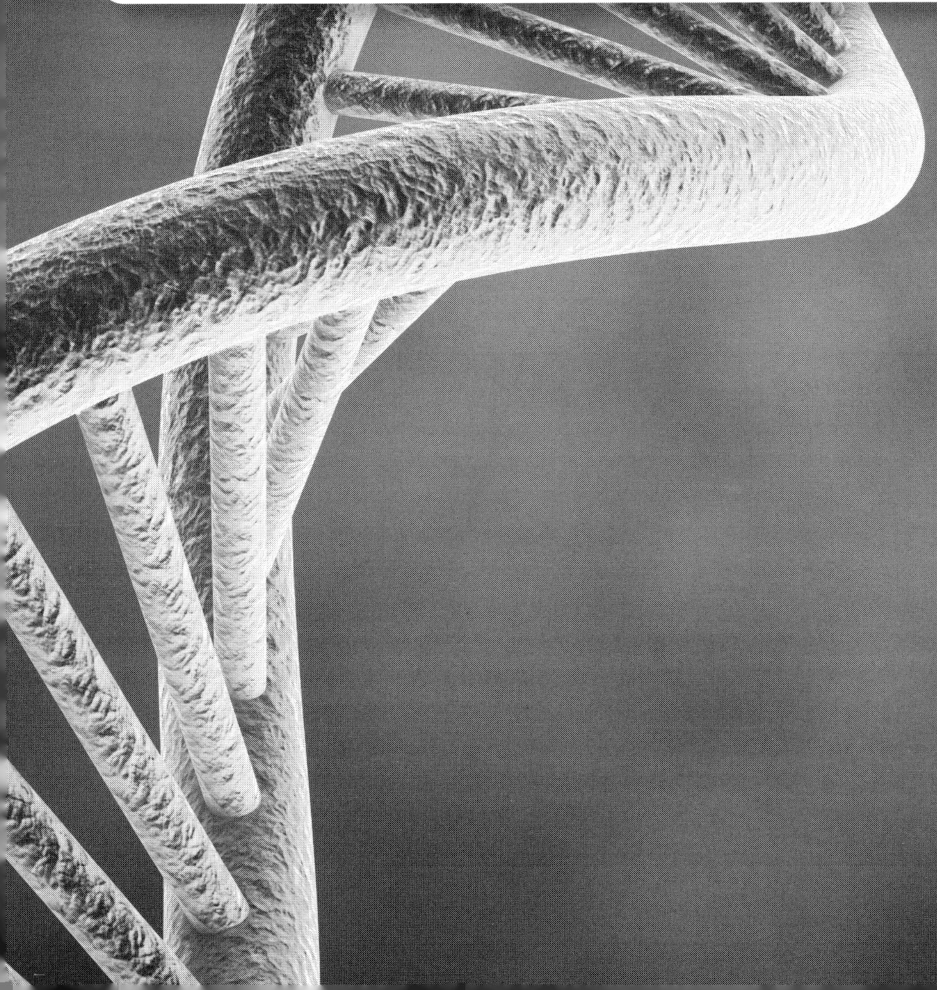

Gene cloning essentials

1

Qaiser I. Sheikh and David P. Hornby

Chapter overview

Gene cloning is a core part of contemporary molecular biology. It is generally defined as the insertion of a fragment of DNA into a cloning **vector** and the subsequent propagation of the **recombinant DNA** molecule in a host organism. Once a particular gene has been cloned and then copied, further analysis of its nucleotide sequence and function can occur. These approaches to manipulating gene sequences are invaluable for studying segments of DNA, including full-length genes. In this chapter we will describe the common tools—vectors and enzymes—used in gene cloning and provide an overview of some of the essential processes involved in gene cloning experiments, that is vector and target DNA preparation, ligation, **transformation**, and selection methods to identify recombinant clones. The content of this chapter is intended to provide introductory principles to gene cloning experiments to support the more detailed applications that are described in the later chapters of this textbook.

LEARNING OBJECTIVES

This chapter will enable the reader to:

- describe the different vectors that can be used in gene cloning experiments, including the advantages and disadvantages of each;
- describe the main applications of the different types of enzymes used in recombinant DNA research;
- describe the essential steps involved in gene cloning using restriction enzyme and PCR-based approaches;
- describe appropriate gene cloning strategies taking into account the starting information (for example, DNA sequence known or unknown, protein sequence, mRNA) and the desired endpoint of the experiment.

1.1 Introduction

In gene cloning, two core processes are employed by the molecular biologist for cloning a gene in the laboratory. The first, more traditional, method of cloning genes involves the use of restriction enzymes for cutting the DNA, followed by the use of the enzyme **DNA ligase** to join the DNA fragments prior to introduction into host cells. The **polymerase chain reaction** (PCR) has become an indispensable tool to the molecular biologist, since it combines the ability to amplify DNA with sequence specificity. In addition, PCR can be used to modify the target sequence during the amplification stage, thereby facilitating some of the downstream cloning steps. This will be discussed in more detail later in this chapter and in the next chapter dedicated to PCR. Alternative approaches to the more traditional gene cloning method described above are TA and TOPO TA cloning, site-specific recombination cloning using Gateway™ cloning technology, and ligation-independent cloning. Some of these alternative approaches will be briefly summarized at the end of this chapter.

Nearly 50 years ago, with the discovery of the enzymes DNA ligase (Zimmerman et al. 1967) and **restriction endonucleases** (Arber and Linn 1969), scientists began cutting DNA molecules at specific locations and recombining the fragments in different combinations using these enzymes (Mertz and Davis 1972). Subsequently, **vectors** were used (Cohen et al. 1973) to efficiently replicate and transfer genes into bacterial cells. This technique, described as 'recombinant DNA technology', has dominated research in biology and increased our fundamental knowledge and understanding of life processes. Although the technique was once considered catastrophic and seriously hazardous (Berg et al. 1974), after decades of serious debate and risk assessment this fear has now disappeared, although the implications are still being debated today, for example for the creation of genetically modified foods.

1.2 Gene cloning applications

The advent of molecular cloning and the subsequent developments in high-level protein expression now underpin many areas of applied life sciences including the following.

1. Isolation of genes from any organism, which may then be characterized further by nucleotide sequencing. See Chapter 4
2. The expression of recombinant proteins and purification for structural analysis and a range of pharmaceutical and biotechnological applications. See Chapter 7
3. The identification of the genetic basis of certain inherited diseases.
4. The analysis of evolution at the molecular level.
5. The engineering of whole organisms for specific purposes, such as tolerance of plants to different stress conditions and insect resistance (this is the emerging field of 'synthetic biology').

1.3 Gene cloning in the laboratory

In gene cloning experiments, a gene is generally isolated from host DNA and then joined to a self-replicating **plasmid** DNA to yield a recombinant DNA molecule. In some protocols, the vector is **bacteriophage** DNA, such as engineered forms of bacteriophage lambda (λ). Irrespective of the vectors used, the basic steps involved in gene cloning (and shown schematically in Figure 1.1) are:

i) isolation and purification of target DNA (the insert) to be cloned (e.g. from genomic DNA, **cDNA**, or PCR-amplified DNA);

Figure 1.1 Outline of the key steps in a basic cloning experiment using plasmid DNA as vector and *E. coli* as host. In this example, insert DNA is derived from isolated genomic DNA by PCR amplification of the region of interest. Extraction of mRNA followed by conversion to cDNA for cloning into a vector is also commonly used.

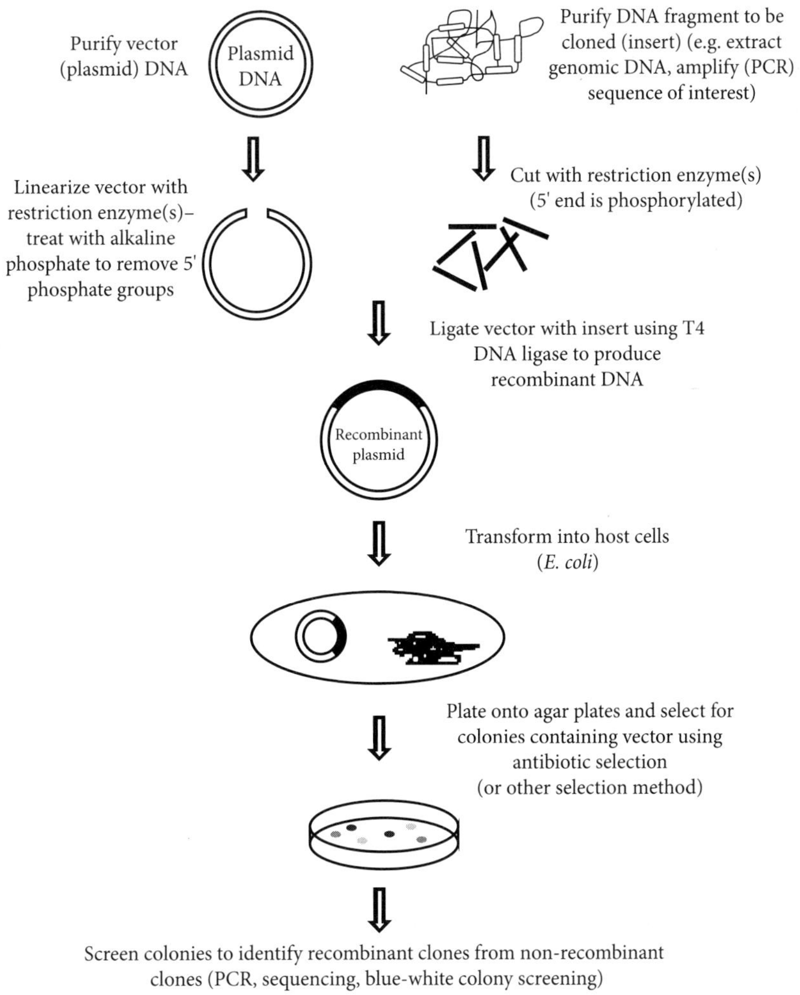

ii) isolation and purification of the vector DNA;

iii) enzymatic restriction digestion of the vector and target DNA to prepare them for ligation;

iv) ligation of the vector and target DNA;

v) transformation of the ligation mix into competent *E. coli* cells (or other host);

vi) plating of recombinant DNA molecules on agar plates and selection of colonies carrying the vector using selectable markers (such as antibiotic resistance);

vii) screening to identify colonies that are positive for the recombinant DNA.

Before moving on to describe each of the steps outlined above, we will describe the tools used in cloning experiments—vectors and enzymes.

1.3.1 Cloning vectors

Cloning vectors are DNA molecules that are used as a vehicle to transfer genetic material into host cells. Once inside the host cell, the vector is replicated to make many copies of itself and hence the gene it carries. Many types of vectors for cloning are available, but all facilitate the amplification of a cloned gene within a host cell. Common cloning vectors are bacterial plasmid, yeast plasmids, **yeast artificial chromosome** (YACs), **bacterial artificial chromosome** (BACs), **cosmid**, bacteriophage, retroviral vectors (RNA viruses), or viral genomes. These are described briefly below and summarized in Table 1.1.

Bacterial plasmids are extra-chromosomal circular double-stranded DNA molecules that are able to self-replicate and maintain themselves in a living cell independent of the host chromosome. They are used for cloning DNA fragments ranging in size from several base pairs to several thousand base pairs (<10kbp). However, they are less useful for cloning very large DNA fragments (>10kbp). Many different types of plasmids have been developed for gene cloning over the last 40 years and are available commercially. Examples include pBR322 (one of the first widely used plasmids), pACYC and pUN plasmids which are derivatives of pBR322, the pUC series, and the pLITMUS series. These plasmids contain specific sequences referred to as origins of replication, or Ori sites, where DNA replication is initiated, a multiple cloning site (MCS) carrying a variety of restriction sites to enable insertion of the target gene, and a genetic marker for selection of cells containing the vector (Figure 1.2(a)). The latter are usually genes coding for antibiotic resistance (e.g. ampicillin and tetracycline resistance in the case of pBR322) or containing the *LacZα* gene which facilitates identification of transformed colonies using blue–white screening (see section 1.4.5). Some plasmid vectors carry two origins of replications, such as bacterial as well as phage, and are called phagemids, an example of which is pBluescript II KS. The bacterial origin of replication enables double-stranded DNA replication, and the phage f1 origin of replication (from a filamentous bacteriophage—see below) enables single-stranded replication. **Expression plasmids** allow a cloned segment of DNA to be translated into protein inside a bacterial or eukaryotic cell, and the pET series plasmids are widely accepted as the most efficient vectors for induction and expression of recombinant proteins (Figure 1.2(b)). Production of functional proteins is still a major challenge and to overcome this problem **shuttle vectors** have been designed. These are generally plasmids (e.g. pBR322 and pBIN19), that can

Table 1.1 Vectors used in gene cloning experiments

Vector	Description	Functions
Bacterial cloning plasmids	Derived from double-stranded circular DNA naturally found in bacterial cells, e.g. pUC series and pLITMUS series.	Widely used for cloning small gene inserts, sub-cloning, and downstream manipulation. Easy to use owing to the presence of selection feature for isolation of recombinant clones. Only small DNA fragments can be cloned (<10kb).
Bacterial expression plasmids	Allow a cloned segment of DNA to be translated into protein inside a bacterial, e.g. pET series plasmids.	Large amounts of protein can be produced. Bacteria cannot carry out post-translational modification of eukaryotic recombinant proteins.
Shuttle vectors	These can replicate in two different organisms (e.g. bacteria and yeast) or mammalian cells and bacteria (e.g. pBR322).	Multifunctional usage in prokaryotic and eukaryotic cells. Enables, for example, cloning a gene in bacteria, to modify or mutate it, and then testing its function by introducing it into yeast or animal cells.
Yeast plasmids	Plasmids used for the transmission of DNA sequences in yeast, e.g. YCp, Yep, and Yip.	Capable of replicating in both bacteria and yeast. Carry dual-selection markers, i.e. an antibiotic resistance gene for bacteria and nutritional markers such as *LEU2* or *HIS3* for yeast. Expression systems based on yeast plasmids can enable post-translational modification of recombinant eukaryotic protein.
Yeast artificial chromosomes (YACs)	YACs are linear DNA vectors that resembles yeast chromosomes. Contain telomere, centromere, and selective marker gene sequences. Also contain bacterial Ori. Can be used in bacteria and yeast host cells.	Can carry 100kb to 1Mb of DNA. Useful for the analysis of larger genomes (genome mapping and genome sequencing projects).
Bacterial artificial chromosomes (BACs)	Based on F plasmids. Contain bacterial Ori and an antibiotic resistance gene for plasmid selection in *E. coli*.	Can carry large DNA sequences of up to 300kb. *E. coli* transformation is efficiently achieved by electroporation. Disadvantages include low yield of DNA (only 1–2 copies per cell). Used in the analysis of larger genomes/genomic DNA library construction.
P1-derived artificial chromosomes (PACs)	Derived from the P1 phage	Can be used for cloning of DNA insert size of up to 300kb.
Cosmids	Plasmids with λ Cos site that permits *in vitro* packaging of DNA into λ particles. Cosmids do not form plaques but rather cloning proceeds via *E. coli* colony formation.	Used to clone up to ~40kb inserts of DNA. Infection process rather than transformation for entry of recombinant DNA into *E. coli* host. Commonly used to generate large insert libraries.
Bacteriophage	Phage or virus particles are used to infect *E. coli* with recombinant DNA, e.g. λgt10, λgt11, EMBL3, EMBL4, or M13.	Phage-based vectors have proved successful for cloning up to 25kb DNA sequences. Useful in genomic and cDNA cloning and production of expression libraries. The transfer of phage-containing recombinant DNA is easy by transduction, and storage of clones is also easy.
Viral vectors (mammalian)	Engineered to enable introduction (and expression) of new or altered genes into mammalian cells and hence carry a number of viral elements that enable this.	Analysis of the function of oncogenes and other human genes. Used in mammalian cell lines and organisms.

propagate and express genes in different host species. These vectors have been designed to carry different origins of replication and hence one can clone a gene in bacteria, perhaps modify it or mutate it in bacteria, and then test its function by introducing it into yeast or animal cells. These plasmids also carry two selectable markers, such as an antibiotic resistance gene for selection in bacteria and a yeast selectable marker such as *URA3* (a gene that codes for the synthesis of uracil) for selection in yeast.

Figure 1.2 Key features of a bacterial cloning vector and a bacterial expression vector. (a) Key features of a cloning vector include an origin of replication (Ori), an antibiotic resistance gene, and a multiple cloning site (MCS) carrying a variety of restriction sites. The MCS is also called the polylinker. In this example, the MCS is inserted within the *LacZα* gene to enable selection of recombinant clones by blue–white colony screening (see section 1.4.5). (b) Key features of an expression vector include an origin of replication (Ori) an antibiotic resistance gene, a promoter upstream of the MCS, and a ribosome binding site (RBS). An affinity tag is also included at either the N or C terminus (or at both). A terminator sequence is positioned downstream of the MCS.

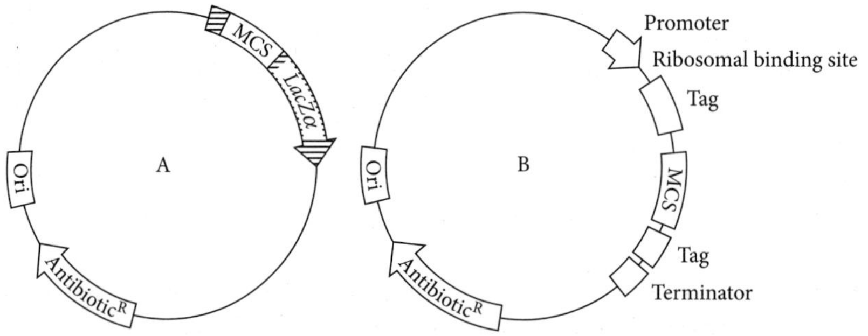

Yeast plasmids are mostly derived from bacterial plasmids and are capable of replicating in both bacteria and yeast because of the presence of two origins of replication (derived from bacterial and yeast plasmids) and dual selection markers, i.e. an antibiotic resistance gene for selection in bacteria and nutritional markers such as *LEU2* or *HIS3* for selection in yeast. Yeast plasmid classification is based on their replication and transmission once inside the yeast cell; examples include YCp, YEp, Ylp, and YRp. Yeast plasmids may replicate independently inside the host cell or may integrate into the yeast chromosome. YEp (yeast episomal plasmids) and YRp (yeast replicative plasmids) are able to replicate independently within the host cell, but others such as Yip (yeast integrative plasmids) integrate into the yeast genome, for example through homologous recombination between similar sequences carried within the plasmid vector and the yeast chromosome. The use of yeast vectors in recombinant protein expression is discussed further in Chapter 7.

Yeast artificial chromosomes (YACs) are linear DNA vectors that resemble yeast chromosomes and contain specific yeast DNA sequences encoding telomere and centromere functions, together with a selective marker gene. They also carry a bacterial Ori sequence and yeast replication sequences and hence can replicate in both bacteria and yeast host cells. YACs can be used for cloning large DNA fragments (100kb–1Mb) and are therefore useful for cloning genes that are too large to clone in bacterial plasmids; hence YACs are used in, for example, genome sequencing projects. pYAC is an example of this type of vector that has been used extensively.

See Chapter 4

Bacterial artificial chromosomes (BACs) are based on the F (fertility) plasmids and are analogous to YAC vectors developed for cloning in *E. coli*. They also contain a bacterial origin of replication and an antibiotic resistance gene for plasmid selection. An example of a BAC is pBeloBac11, which can be used efficiently for *E. coli* transformation. A large DNA sequence insertion of up to 300kb is achievable and is often used for the analysis of large genomes. Stable clones can be generated, but selection of recombinants versus non-recombinants is difficult if no screening feature, such as blue–white screening or auxotrophic selection, is incorporated (see section 1.4.5). In addition, one major disadvantage of using BACs is low yield of DNA during purification.

P1-derived phage vectors and artificial chromosomes (PACs) are vectors that combine the features of P1 bacteriophage vectors and BACs and are used for cloning large DNA fragments in *E. coli* cells. The pAD series of vectors, which are approximately 30kbp, can accommodate up to 100kb of DNA and includes the P1 phage head determining P1 *pac* site sequence in addition to an origin of replication and an antibiotic resistance gene. PACs can be used for cloning very large DNA inserts of up to 300kb.

Cosmids are bacterial plasmids with a pair of lambda (λ) cohesive (Cos) end-sites. The presence of the Cos site permits *in vitro* packaging of cosmid DNA into λ phage particles (see bacteriophage below). Cosmids contain a bacterial origin of replication and carry a selectable marker. Cosmids can be used to clone DNA inserts of up to 40–45kb which are then introduced into the host cells using *in vitro* packaging methods. Upon introduction into *E. coli*, transformants are produced as colonies instead of plaques as the cosmids lack all the λ genes.

Bacteriophage vectors (DNA viruses) are the phage or virus particles that can infect *E. coli*. The bacteriophage vectors are designed and constructed from these phages such as bacteriophage lambda (λ) derivatives e.g. λgt10, λgt11, λEMBL3, λEMBL4 or M13 (the latter is a 6.4kb long single stranded filamentous phage). The major components of λ include the head, tail, and double-stranded DNA packaged inside the head. λ is able to transfect bacterial cells and establish either a lytic or a lysogenic infection cycle. The λ genome comprises broadly three regions: genes coding for the capsid proteins (head) at one end, genes involved in controlling integration into the host chromosome in the middle, and genes coding for DNA synthesis, early and late regulation, and host cell lysis at the other end. The ends of the linear molecule are flanked by cohesive end (Cos) sites which promote circularization. These features have been exploited to develop two types of cloning vectors: **replacement vectors** (e.g. λEMBL4) or **insertion vectors** (e.g. λgt10). The λ-phage-based vectors have proved successful for cloning large DNA sequences (8–20kb fragments are possible depending on the vector type). Entry of phage-containing recombinant DNA into *E. coli* cells by **transduction** is easy. The phenomenon of transduction is a process by which DNA is transferred from one bacterium to another. However, this term is also used to describe the transfer of recombinant viral DNA into a cell. Figure 1.3 shows a schematic diagram of the cloning steps involved when phage replacement vectors are used.

Viral vectors for mammalian cells are derived from the genomes of viruses including SV40 (simian virus 40), adenoviruses, papilloma viruses, and retroviruses (RNA viruses). Viral vectors are engineered to enable introduction (and expression) of new or altered genes into mammalian cells and hence carry a number of viral elements that allow this. Many different viral vectors are available commercially. Viral vectors enable the study of gene function, for example by knocking out the expression of a particular gene or over-expressing a particular gene. The target gene may incorporate into the host chromosome and reside there practically indefinitely (e.g. lentiviruses) or it may remain episomal (does not integrate into the host chromosome, e.g adenoviruses). Viral vectors carrying the target gene are usually packaged into a viral packaging cell line such as HEK 293 along with a helper plasmid so that copes of recombinant viral particles can be generated. These particles are released from the cells, harvested, and then used to infect the target host cell (Thomas et al. 2003).

Figure 1.3 Typical cloning steps using phage replacement vectors. The phage genome comprises three regions encoding the capsid proteins at one end (left arm), genes involved in controlling integration and excision in the centre (non-essential genes that can be replaced), and genes involved in DNA synthesis, regulation, and lysis at the other end (right arm). The genome is flanked by cohesive (cos) sequences at either end. The purified vector is digested with restriction enzymes to remove the central portion of the genome (restriction sites are denoted by small triangles) to yield the two arms (left and right). The DNA fragment to be cloned (insert) is ligated to the arms to form a recombinant molecule. Note: Although not shown here, recombinant concatemer is formed, i.e. repeating units of left arm-DNA-right arm joined together. This construct is packaged into phage heads by mixing with an *in vitro* packaging extract (empty pre-heads, tails, and packaging proteins). The packaged phage is used to infect bacterial cells plated onto agar. Recombinant phage is replicated inside *E. coli cells*, the cells lyse, and plaques develop on the plate. Each plaque represents progeny from a single recombinant phage.

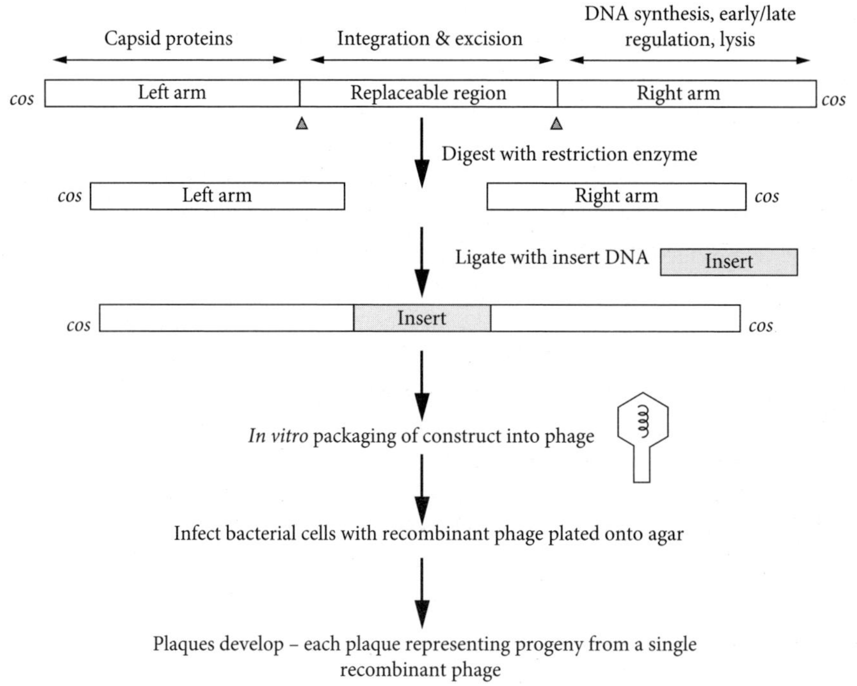

Selecting the appropriate vector

Vectors are generally chosen according to the size and type of DNA to be cloned. Larger DNA fragments are more easily cloned in viral vectors such as bacteriophage, retrovirus, or cosmids. Linear vector systems also have certain advantages over circular supercoiled plasmid vectors, for example the ability to clone sequences such as inverted repeats and AT-rich regions (which may exhibit instability in circular plasmid vectors).

One of the most important considerations in any cloning vector is the compatibility of the origins of replication sequences between vector and host cells. These Ori sequences vary somewhat between organisms, but all provide the information required to assemble the DNA replication machinery. However, there are many significant differences between prokaryotic and eukaryotic origins of replication. For example, shuttle vectors carry sequences that enable replication in different host species whilst other vectors may enable replication in a single host species only. In addition, some vectors are maintained at low copy numbers within the cell and some at high copy numbers. A higher copy number generally means that larger amounts of the cloned gene can be obtained. You may also want to consider whether you wish to use a vector that is maintained in the cell independently of the chromosome or a vector that integrates into the host genome and therefore produces a more stable recombinant.

The choice of cloning vector is also based on the final requirements of a specific gene cloning experiment for which the commercial manufacturers generally give guidance. For example, if you wish to clone a gene of interest and express recombinant protein, you will need to select a vector that is appropriate for your expression system. Phage can be used for generating expression libraries, and retroviral vectors are commonly used to study the function of oncogenes and other human genes.

See Chapter 10

Expression vectors

Expression vectors (many are commercially available) can be plasmid-based or engineered viruses which combine the features required for gene insertion with features that enable transcription and translation of the cloned gene. MCS are usually flanked with various features suitable for appropriate gene expression, such as a start codon, a **promoter** site, a ribosome binding site, and **fusion tags** that are used for the downstream affinity purification of the encoded protein product (Figure 1.2(b)). Promoter sequences are located upstream of the translational initiation start codon and support gene expression. Many of the promoter sequences found in plasmids are derived from *lac*, *gal*, or *trp* operons of *E. coli* and T7 or Sp6 from bacteriophage. Those plasmids that include the T7 promoter often yield very high levels of expression of the encoded protein in appropriate *E. coli* strains such as BL21 (DE3). These strains have been genetically modified to carry the T7 RNA polymerase gene which itself has been modified by addition of a lactose (IPTG (isopropyl β-d-1thiogalactopyranoside)) inducible genetic element which ultimately drives the promoter system for the over-expression of the encoded recombinant protein. This and other expression vectors are described further in Chapter 7.

Fusion tags

Fusion tags (also called **affinity tags**) are sequences that encode short peptides or additional proteins fused to the coding sequence of the protein of interest. When translated, a **fusion protein** is produced. Fusion tags can be placed at the N- or C-terminus of the protein (or both) and are used to detect and purify the expressed protein. Common examples of fusion tags are the polyhistidine (His) tag (6 or 8 amino acids), the strep II tag (9 amino acids), the glutathione-S-transferase (GST) tag (211 amino acids), haemaglutinin (HA) (9 amino acids) the c-myc tag (11 amino acids), and the maltose binding protein (MBP) (396 amino acids). Examples of the use of these tags can be found in Chapter 7 and Chapter 10. Purification of proteins using affinity tags is described in Chapter 8. For a review of the use of fusion tags, see Brizzard (2008) and Malhotra (2009).

1.3.2 Enzymes used in gene cloning

A number of enzymes are commonly used in gene cloning experiments. These include nucleases, polymerases, and ligases. Nucleases cleave phosphodiester bonds between the nucleotide subunits of DNA and are further subgrouped into **endonucleases** and **exonucleases**. Endonucleases cleave phosphodiester bonds within a DNA sequence, whereas exonucleases act on the 3′ or 5′ termini (or both) of the nucleic acid chain to sequentially remove one nucleotide at a time. Endonucleases can cleave DNA at specific

Table 1.2 Enzymes used in gene cloning experiments

Enzymes	Example(s)	Function
Restriction endonucleases	See Table 1.3	Sequence-specific cleavage of double-stranded DNA.
DNA polymerases	*Taq* polymerase	Directs 5′ → 3′ synthesis of DNA from a DNA template. Requires a primer and dNTPs. Does not have 3′ → 5′ exonuclease (proof-reading) activity. Used in PCR reactions.
	Pfu, Phusion, Kod HiFi	Directs 5′ → 3′ synthesis of DNA from a DNA template. Requires a primer and dNTPs. Has 3′ → 5′ exonuclease (proof-reading) activity. Used in reactions where a high degree of accuracy in DNA sequence is required after amplification.
	T7 and T4 phage DNA polymerase Klenow enzyme	Has 5′ → 3′ DNA synthesis activity and 3′ → 5′ exonuclease activity, but no 5′ → 3′ exonuclease activity (repair). Can be used to fill 5′ overhangs with dNTPs or for the generation of blunt ends from DNA molecules with 3′ overhangs.
	Reverse transcriptase (RNA-dependent DNA polymerase)	Synthesizes cDNA from an RNA template using a primer and dNTPs.
DNA ligases	T4 DNA ligase	Ligates DNA fragments by forming a phosphodiester bond between 3′-hydroxyl and 5′-phosphate termini. Can also repair single-stranded breaks.
Alkaline phosphatase	Calf alkaline phosphatase (CAP), bovine alkaline phosphatase (BAP), shrimp alkaline phosphatase	Removes phosphate groups from 5′ DNA ends. Can be used to treat DNA fragments to prevent self-ligation.
Polynucleotide kinase	T4 polynucleotide kinase	Adds phosphates to the 5′-OH groups at the ends of single- or double-stranded DNA.

sequences as in the case of restriction enzymes (see section 1.3.3) or non-specifically, for example DNase I.

See Chapter 2

DNA polymerases synthesize DNA using a template (either DNA or RNA), a primer, and dNTPs mixture. For example, thermostable DNA polymerases are used in PCR at temperatures in excess of 60°C. Reverse transcriptase can be used to generate cDNA, using RNA as a template. DNA ligases join a 5′-phosphate end to a 3′-hydroxyl end of DNA to create a new phosphodiester bond. They are used for joining different DNA pieces if both DNA strands have compatible cohesive or blunt ends. DNA ligases can also repair single-stranded breaks. These and other enzymes used in gene cloning are summarized in Table 1.2.

1.3.3 Restriction enzymes

Sequence-specific endonucleases are called restriction enzymes and cut double-stranded DNA molecules at specific recognition sites (sequences) within DNA. Table 1.3 shows the recognition and cleavage sites of several commonly used restriction enzymes. Some

Table 1.3 Examples of restriction enzymes and their recognition sequences

Enzyme	Recognition sequence	DNA after cleavage		Type of cleavage
*Bam*HI	5′ G*GATCC 3′ 3′ CCTAG*G 5′	5′—G 3′—CCTAG	GATCC—3′ G—5′	Sticky end with 5′ overhang
*Eco*RI	5′ G*AATTC 3′ 3′ CTTAA*G 5′	5′—G 3′—CTTAA	AATTC—3′ G—5′	Sticky end with 5′ overhang
*Eco*RV	5′ GAT*ATC 3′ 3′ CTA*TAG 5′	5′—GAT 3′—CTA	ATC—3′ TAG—5′	Blunt end
*Hin*dIII	5′ A*AGCTT 3′ 3′ TTCGA*A 5′	5′—A 3′—TTCGA	AGCTT—3′ A—5′	Sticky end with 5′ overhang
*Kpn*I	5′ GGTAC*C 3′ 3′ C*CATGG 5′	5′—GGTAC 3′—C	C—3′ CATGG—5′	Sticky end with 3′ overhang
*Pst*I	5′ CTGCA*G 3′ 3′ G*ACGTC 5′	5′—CTGCA 3′—G	G—3′ ACGTC—5′	Sticky end with 3′ overhang
*Sac*I	5′ GAGCT*C 3′ 3′ C*TCGAG 5′	5′—GAGCT 3′—C	C—3′ TCGAG—5′	Sticky end with 3′ overhang
*Stu*I	5′ AGG*CCT 3′ 3′ TCC*GGA 5′	5′—AGG 3′—TCC	CCT—3′ GGA—5′	Blunt end
*Sma*I	5′ CCC*GGG 3′ 3′ GGG*CCC 5′	5′—CCC 3′—GGG	GGG—3′ CCC—5′	Blunt end
*Sal*I	5′ G*TCGAC 3′ 3′ CAGCT*G 5′	5′—G 3′—CAGCT	TCGAC—3′ G—5′	Sticky end with 5′ overhang
*Xho*I	5′ C*TCGAG 3′ 3′ GAGCT*C 5′	5′—C 3′—GAGCT	TCGAG—3′ C—5′	Sticky end with 5′ overhang

* indicates site of enzyme cleavage

restriction enzymes are able to cleave more than one different but related (degenerate) sequence. For example, the restriction enzyme *Hinf*I is able to recognize the sequence GANTC where N represents any of the four nucleotides.

Three major classes of restriction endonuclease (type I, type II, and type III) are found in nature, but the type II class is the one commonly employed in molecular cloning. The sequences recognized by most type II restriction endonucleases are palindromic, i.e. the sequence is the same when read 5′ to 3′ and 3′ to 5′, and vary between four and eight base pairs.

Digestion with restriction enzymes can generate either blunt or sticky (cohesive) ends. Sticky ends carry an overhang at either the 5′ or the 3′ end, whereas there are no overhangs with blunt-ended DNA. Examples of the types of ends produced by different restriction enzymes are listed in Table 1.3. Sticky ends generated by the same enzymes on different DNA can normally be ligated more easily and therefore are complementary to each other. For example, DNA duplexes cut with a given restriction enzyme can be recombined with plasmids which have been cut with the same restriction enzyme. Some enzymes, such as *Sal*I and *Xho*I, can create similar overhangs that are also compatible for cloning.

Many thousands of restriction endonucleases are commercially available. New England Biolabs is one of the longest established molecular biology reagent suppliers and also provides additional information on restriction endonucleases and other DNA cloning enzymes.

1.4 Gene cloning processes

In gene cloning experiments, the DNA sequence or the gene has to be transferred from the source organism to a vector and then amplified in a prokaryotic or eukaryotic host cell. The sources of DNA for cloning maybe genomic DNA, cDNA, or a PCR-amplified DNA fragment. A **genomic DNA library** can be constructed by extracting genomic DNA from the cells followed by partial digestion with a restriction enzyme to generate ~20kb fragments which are then joined to a suitable vector for identification, analysis, and further processing. The clones in a genomic library collectively represent the entire genome of the organism. Alternatively, **cDNA libraries** can be constructed in which total mRNA is extracted, copied into single-stranded DNA using a reverse transcriptase enzyme, and amplified into double-stranded DNA prior to cloning into a cloning vector. The libraries are then screened for identification of the clone carrying the gene of interest and further analysed. In a cDNA library, the clones collectively represent a subset of the genes that are actively expressed in the cell or tissue from which the mRNA was extracted. Construction of cDNA libraries requires more processing steps than the synthesis of genomic DNA libraries. However, screening genomic DNA libraries is a significant task. It may be more convenient to purchase a particular library or clone from commercial suppliers.

Various vectors, such as plasmids, cosmids, or bacteriophage λ, can be used for the construction of these libraries, and the choice is based on the overall size of the DNA pieces to be cloned. For example, as plasmids can accommodate small DNA inserts, they are used for cDNA cloning, whereas cosmids and bacteriophage λ, which can accept very large DNA sequences, are used for genomic DNA cloning.

If the sequence information for a particular gene is available (you can find this out by conducting a search on a database such as GenBank) *in vitro* amplification of regions of genomic DNA (instead of selecting a desired clone from a library) is another option. Enzymatic replication of selective genomic DNA regions by PCR has facilitated the cloning of desired DNA sequences into suitable cloning vectors. This process, described in Chapter 2, requires genomic DNA, two short oligonucleotide primers flanking the DNA sequence to be amplified, dNTPs, and polymerase enzyme. As a result many millions of copies of the desired blunt-ended DNA are synthesized and can be cloned into cloning vectors. PCR can also be carried out using mRNA or cDNA as the template. Some typical processes involved in cloning experiments are described below. The focus is mainly on plasmid vectors, as these are frequently used. However, there are very many excellent books on gene cloning, such as Brown (2010), that you can refer to for more detailed information than it is possible to provide in this chapter.

See Chapter 2

1.4.1 Extraction of DNA and mRNA

Various methods can be used to isolate total genomic DNA and commercial kits are available, but all follow common steps. These include disruption and lysis of the cells or tissues from which the DNA is to be extracted, followed by removal of proteins and contaminating RNA and recovery of the DNA by ethanol (or isopropanol) precipitation. The procedure for isolating plasmid vector DNA is similar, but the plasmid DNA is selectively recovered from the contaminating genomic DNA, typically using the alkaline lysis method (Birnboim and Doly 1979). If your starting source for cloning experiments is cDNA, then you will need to extract mRNA and then convert it to cDNA using the enzyme reverse transcriptase. As mRNA represents the genes that are actively expressed and as only a subset of genes are expressed in different cells (and at any given time), your starting cell/tissue type will need to be selected carefully. mRNA isolation is based on the use of oligodT beads to which the mRNA binds via its poly A tail. The bound mRNA is subsequently eluted. Again, mRNA purification kits are commercially available. DNA and mRNA can then be quantitated by UV spectrophotometry and/or **agarose gel electrophoresis** (Gallagher and Desjardins 2008).

1.4.2 Preparation of vector and insert DNA

Circular plasmid vectors are first linearized using restriction enzymes. The choice of restriction enzyme generally depends upon the cloning strategy. PCR-amplified DNA fragments can be cloned directly into plasmid that has been linearized using a restriction enzyme that generates blunt-ended linear plasmid vector. For blunt-end cloning when a single enzyme is used, dephosphorylation of the vector is usually carried out (to prevent self-ligation of vector) using calf or bovine intestinal alkaline phosphatase enzymes. In blunt-end cloning, the insert DNA ends are 5′-phosphorylated to enable ligation by T4 DNA ligase. Blunt-end DNA cloning is used either to generate a cDNA library or where insertional inactivation of any gene is the end goal of the experiment.

Different DNA pieces cut with the same enzyme (or by enzymes that generate similar overhangs) will generate compatible cohesive ends that can join or recombine. Often two different enzymes are used to cut a DNA fragment, thereby generating an 'asymmetric DNA duplex'. This method is generally used for directional cloning and enables the insert to be ligated into the vector in a defined orientation. Self-ligation of vector is also less likely.

In this method the desired DNA molecules are cut from the genome using restriction enzymes or the desired sequence is amplified using PCR. The PCR-amplified DNA fragment can be modified at the termini by the introduction of new restriction sites which match those to be used to restrict the vector to facilitate cloning. Designing and synthesizing primers with flanking extra sequences corresponding to the restriction enzyme sites can achieve this. The PCR-amplified DNA is then cleaved using restriction enzymes which generate cohesive ends compatible with the cleaved vector DNA.

When restricting DNA inserts, it is important to check for the presence of endogenous recognition sequences so that the desired sequence is not cleaved. A number of programs to locate restriction sites, such as Webcutter and NEBcutter, are available online. Internal restriction sites can be modified using site-directed mutagenesis if any are found that match the restriction enzymes you plan to use in your cloning experiments.

See Chapter 3

It has now become more common for laboratories to synthesize long DNA sequences (in excess of 2000bp), and synthetically prepared oligodeoxynucleotides (linker inserts) are sometimes preferred for ligation into linear plasmid vectors using the reaction catalysed by a bacteriophage T4 DNA ligase enzyme.

1.4.3 Ligation of vector and insert

Once the vector and target DNA are prepared, the next stage is to ligate the two together to form a recombinant DNA molecule. This reaction is catalysed by the enzyme T4 DNA ligase which forms a phosphodiester bond between the 3′-OH and 5′-phosphate termini. In a ligation reaction, different ligation products are likely. These could include vector–vector and insert–insert ligation products as well as the desired vector–insert recombinant ligation product. When setting up a ligation reaction, the goal is to increase the frequency of producing a recombinant DNA molecule and minimize the frequency of vector–vector and insert–insert ligation products. Insert-to-vector ratio, the total concentration of DNA molecules, and the purity of the DNA preparation are all important factors which determine the success of a ligation reaction in producing the desired circular recombinant DNA molecule.

Ligation efficiency can be improved by, for example, treating the vectors with alkaline phosphatase (as described in section 1.4.2), using pure preparations of vector and insert (free from unwanted digested products and other contaminants/enzymes), and optimizing the experimental conditions. Ligations are typically performed in small volumes (e.g 10 or 20µl) using an approximate molar ratio of plasmid DNA to insert DNA (1:1, 3:1, and 1:3) in the presence of buffer (supplied by the manufacturer) and T4 DNA ligase (Sambrook et al. 1989). However, it will be necessary to optimize the molar ratio of vector to insert for your experiment as the optimal ratio can vary. The optimal total quantity of vector and insert DNA is 20–100ng per reaction, since higher concentrations can have an inhibitory effect on the transformation efficiency of competent cells. Ligation reactions can be carried out at room temperature (for 1–3 hours prior to transformation) or at 16°C overnight or according to the supplier's instructions.

1.4.4 Introduction of DNA into host cells

Once the DNA of interest is ligated into the cloning vector, the next stage is to introduce it into the host cell so that it can be replicated. Host cells can be bacterial or eukaryotic and each requires a different method of introducing DNA into the cell.

DNA is introduced into bacterial cells by a process called transformation. First the cells are made **competent** chemically by treating the cells with a solution of chloride salts such as calcium chloride or rubidium chloride or made electrocompetent by washing the cells in a buffer to keep the medium polar. Aliquots of competent cells are then combined with the ligation reaction mixtures and subjected to a transient heat shock or alternatively a brief electric field is applied (**electroporation**) to stimulate uptake by the cells. After incubation, cells are plated out onto appropriate antibiotic-containing LB agar plates and incubated at 37°C for colonies to develop. Cells that have taken up the vector DNA harbouring the antibiotic resistance gene will grow, whilst cells that have taken up vector or linear DNA will not grow. The **transformants** can then be picked off and grown in appropriate media, and the plasmid can be isolated from the host cell for further analysis or manipulations.

Introduction of λ phage particles into bacterial cells is termed transduction and simply involves incubating bacterial cells with packaged λ phage particles and plating on an appropriate growth medium to generate recombinant plaques. When setting up a viral growth (bacteriophage or eukaryotic virus) it is important to calculate the multiplicity of infection (MOI), i.e. the ratio of infectious virus to the number of host cells, as this will determine the productivity of infection. In some cases lower MOIs may be suitable (e.g.

for cells that are easy to transduce or when establishing a stable cell line), whilst higher MOIs may be required for cells that are hard to transduce.

The process of introducing DNA into eukaryotic cells is called **transfection** and a number of different methods can be used. For example, yeast cells can be transformed with recombinant plasmid by electroporation or by using polyethylene glycol–lithium salt mixtures followed by heat shock using **sphaeroplasts**. Electroporation or **lipofection** can be used to introduce DNA into cultured mammalian cells. In lipofection, DNA is packaged into liposomes using a lipofection reagent. The liposome then fuses with the lipid bilayer of the host cell membrane and releases the DNA into the cell. Alternately, **microinjection** can be used, whereby the DNA is injected directly into the nucleus of the cell.

See Chapter 6

See Chapter 19

1.4.5 Screening for recombinant molecules

The appearance of colonies on appropriate antibiotic-containing agar plates (or other selection media) is not always a sign that the cells carry the recombinant clone (i.e. the DNA insert). Although recombinant DNA molecules are likely to be made, some 'positives' due to either self-ligation (empty vector re-ligated) or the presence of uncut plasmid DNA may also appear. Generally the latter will be low in directional cloning experiments (if different restriction enzymes are used for each termini) or if alkaline phosphatase treatment has been carried out. Hence, it is imperative to confirm the authenticity of the recombinant clones. For accuracy, this is generally carried out by more than one technique. The common methods used are PCR screening, analytical restriction digestion, and DNA sequencing.

See Chapters 2 and 4

PCR screening

Colony PCR is often used to test a large number of bacterial colonies to check that the colony is carrying the vector with the desired DNA insert. Forward and reverse PCR primers flanking the target region are used to amplify the insert. The PCR products are then run on an agarose gel to identify whether a band corresponding to the size of the insert has been amplified.

Analytical restriction digestion

Purified plasmid DNA is analysed via a series of diagnostic restriction digests using an appropriate restriction endonuclease(s). Digests are incubated at 37°C for 2–4 hours and analysed on an agarose electrophoresis gel for confirmation of a recombinant clone.

DNA sequencing

A definitive identification of a recombinant plasmid is only possible via nucleotide sequencing. Various methods and techniques are available for DNA sequencing but the dideoxy chain termination method, as described by Sanger in 1970s, has now become the method of choice. The basic principle of this method is based on the use of dideoxy nucleotide phosphates (ddNTPs) as DNA chain terminators. This requires a single primer that is used for sequencing PCR, usually 50–100bp upstream from the target area under question, to determine the order of nucleotide sequence. The recombinant plasmids are sequenced to identify whether the target DNA has successfully ligated. DNA sequencing methods are described in Chapter 4.

Blue–white colony screening

Cells carrying the recombinant plasmid can also be distinguished from those carrying vector only using blue–white colony screening. This method utilizes the *LacZ* gene which encodes the enzyme β-galactosidase. This enzyme can break down the substrate X-gal in the presence of IPTG to produce a blue-coloured product. Colonies appear blue in the presence and white in the absence of functional β-galactosidase when plated on media supplemented with X-gal and IPTG. This enzyme comprises two parts, one of which is the N-terminal α fragment and is coded for by the plasmid. An MCS is also inserted within this region to enable insertion of insert DNA. The remaining part of the enzyme is coded on the host chromosome. Functional β-galactosidase is produced in cells that carry the empty vector and hence blue colonies are observed. In cells that harbour a recombinant plasmid, the *LacZ* is disrupted; hence functional β-galactosidase is not produced and these colonies appear white.

Auxotrophic selection

An alternative system to antibiotic selection is commonly used in yeast cells. In this system, the vector carries a gene coding for an enzyme involved in the synthesis of an amino acid such as *TRP1*, *LEU2*, or *HIS3* (for tryptophan, leucine, or histidine, respectively) or the bases of nucleic acids (*ADE2* and *URA3* for adenine and uracil, respectively). The vector is transformed into host yeast strains that are defective for that particular gene (i.e. an auxotrophic mutant). Hence, when cells are grown on minimal media lacking the particular amino acid or base, uptake of vector will favour the growth of the transformants only. The authenticity of the recombinant clone in this selection method will need to be determined by sequencing as for antibiotic selection, for example, or even by blue–white colony screening.

1.5 Further types of gene cloning

In addition to the traditional cloning approach described above, which involves ligating the insert with a vector using T4 DNA ligase after digestion of both vector and insert with restriction enzymes, several alternative cloning methods have also been developed. These include TA and TOPO TA cloning, site-specific recombination cloning using Gateway™ cloning technology, and ligation-independent cloning. Here we briefly describe two of these additional methods of gene cloning: site-specific recombination and TOPO TA cloning. For a review of these and other alternative methods, see Wang et al. (2012) and the references cited therein.

1.5.1 Recombination cloning/enzyme-free cloning

Homlogous recombination is a naturally occurring universal mechanism of genetic exchange that can take place between two genes (or parts thereof) carrying identical or similar DNA sequences (reviewed by Filippo et al. 2008). The phenomenon of homologous recombination is widely adapted by most organisms to replace and repair damaged DNA and to introduce genetic diversity. In its practical applications, a DNA sequence to be cloned by this approach is flanked at both ends by DNA sequences similar to the target DNA sequence as described in Figure 1.4. Therefore it is essential to know the precise sequence of the target DNA at the position where the recombinant segment of DNA is to be inserted. This process has been exploited by molecular biologists for gene targeting, gene knockout, and inserting DNA segments into specific cloning vectors.

Recombineering has its origins in work such as that described by Jones and colleagues (Jones and Howard 1991; Jones and Winistorfer 1992). These authors have developed a powerful and relatively simple cloning technique based on homologous recombination. Two simple methods for site-specific mutagenesis are described here. In each method (Jones and Howard 1991) PCR is used in two separate amplifications to mutate the sequence of interest and to add 'ends' to one PCR product that are homologous to the ends of the other PCR product. In the first method (Jones and Winistorfer 1992), the two PCR products are combined, denatured, and re-annealed prior to transformation of *E. coli* in order to form recombinant circles *in vitro*. In the second method, the two linear products are co-transformed directly into *E. coli* without prior manipulation,

Figure 1.4 Explanation of homologous recombination. The insertion of a DNA sequence via site-specific homologous recombination proceeds by the initial alignment of the homologous ends of a pair of DNA duplexes, with one duplex (the target) acting as recipient of the sequence to be cloned. The steps are illustrated schematically. (a) DNA with a flanking sequence similar to target DNA will integrate into the target DNA to generate recombinant DNA. Gene targeting, knockout, insertions of DNA segments for DNA repair, and DNA tagging are carried out routinely using this method. (b) Insertion/joining of fragments of DNA using homologous recombination. PCR-amplified DNA with similar sequences to the termini of the linearized target vector can be joined in this way.

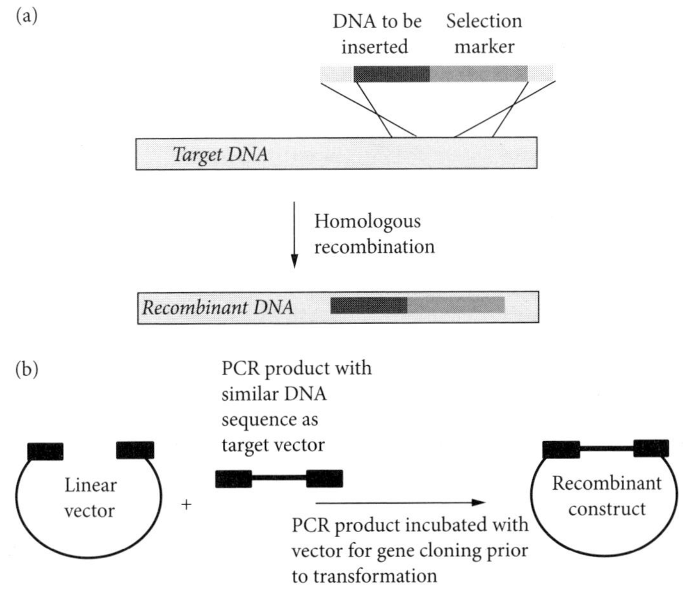

resulting in transformation of *E. coli* with the recombinant of interest by recombination *in vitro*. Each PCR amplification uses a plasmid template that has been linearized by restriction enzyme digestion outside the region to be amplified. This permits use of unpurified PCR products in these two protocols and generation of the mutant of interest with no other enzymatic manipulation *in vitro* apart from PCR amplification. Both methods work equally well without any detectable errors.

More recently, several alternative methods which build on the original protocols have been developed (Copeland et al. 2001; Datta et al. 2006). Gibson et al (2008) have also reported an *in vitro* recombination method to assemble the *Mycoplasma genitalium* genome by transformation-associated recombination cloning in the yeast *Saccharomyces cerevisiae*. Gibson's group have also reported another successful molecular engineering tool (Gibson et al. 2009). Both methods remove the need for compatible restriction sites. The idea is to generate the fragments that overlap by 15–30bp using PCR (melting temperature >50°C). The simplest use of this method lies in directional cloning of a fragment into a vector (the vector simply needs to contain a unique restriction site for linearization). This method is particularly useful for assembling multiple fragments, such as PCR primers with a homology region for the vector added to the 5′ end of fragment 1, and homology to fragment 1 3′ added to fragment 2 5′ and so on. Based on this principle, In-Fusion® cloning kits (Clontech) are now commercially available to facilitate the process of gene cloning efficiently without the need for restriction, phosphatase, and ligase enzymes. This method has also been used for site-directed mutagenesis by positioning the base to be changed in the overlap between two primers and for cassette in-and-out sequences that are not surrounded by restriction sites.

Prokaryotes have become popular choices as hosts for recombinant DNA technology in view of their rapid growth, simple media requirements, and general ease of manipulation for the production of proteins. However, sometimes it is necessary to move a cloned gene into different *E. coli* vectors encoding different promoter system for optimized protein expression or to overcome differences in the availability of post-translation modifications, which generally requires the use of a eukaryotic expression host–vector system. Biotechnology companies have adapted the modified techniques described above to simplify cloning and to facilitate the movement of sequences of interest into variety of expression vectors. Invitrogen has introduced the Gateway™ technology which has enormously facilitated the transfer of cloned genes between cloning vectors. This technology is based on the λ phage site-specific recombination system (*att*B × *att*P → *att*L × *att*R) which eliminates the requirements to work with restriction enzymes and DNA ligase after the initial entry clone is constructed, since recombination via strategic recombinogenic flanking sequences mediates the transfer of target sequences into different destination vectors.

1.5.2 TOPO TA cloning

TOPO TA cloning also provides a highly efficient cloning method for the insertion of *Taq* polymerase amplified PCR products into a plasmid vector. In this approach the ability of a single enzyme, vaccinia DNA topoisomerase, to both cleave and rejoin DNA strands with extreme specificity at each stage is exploited in the synthesis of recombinant DNAs (Shuman 1994). Topoisomerase I from vaccinia virus binds to duplex DNA at specific sites and cleaves the phosphodiester backbone after 5′-CCCTT in one strand (Shuman 1994). In this way, the linearized or 'activated' vector contains CCCTT cleavage sites at both ends which facilitate the cloning of PCR-amplified DNAs using *Taq* polymerase that adds deoxyadenosine (A) to the 3′ ends of a small but significant percentage of PCR product. Invitrogen provide these 'activated' vector backbones (covalently bound Topoisomerase I) with single 3′-thymidine (T) overhangs for TA cloning that do not require ligase, post-PCR manipulation, or PCR primers containing specific sequences.

1.6 Chapter summary

- The basic elements of molecular cloning are described in a way that combines practical aspects with theoretical explanations.
- DNA (insert) for cloning can be obtained from restriction or PCR amplification from a genomic DNA source, reverse transcriptase from mRNA to form cDNA, or by direct chemical synthesis.
- Once a suitable starting DNA has been obtained, it is cloned into a vector for propagation in either a prokaryotic or eukaryotic host. A number of different vectors are available and the type selected will depend on a number of factors including insert size to be cloned, copy number, compatibility, and the desired endpoint of the experiment.
- These steps require the use of various enzymes such as restriction enzymes for sequence-specific cutting of DNA, polymerases, and ligases.
- More specialized cloning methods are currently emerging which incorporate greater use of chemical synthesis of DNA fragments together with recombination phenomena to eliminate inefficient steps such as restriction digestion and ligation.
- The fundamental principles and technologies that were developed in the early phase of molecular biology are likely to remain intact, but will undoubtedly be modified by the addition of technologies that improve efficiency and lead to opportunities for automation.

✓ References

Arber, W. and Linn, S. (1969) DNA modification and restriction. *Annu Rev Biochem* **38**: 467–500.

Berg, P., Baltimore, D., Boyer, H.W., et al. (1974) Letter: Potential biohazards of recombinant DNA molecules. *Science* **185**: 303.

Birnboim, H.C. and Doly, J. (1979) A rapid alkaline extraction procedure for screening recombinant plasmid DNA. *Nucleic Acids Res* **7**: 1513–23.

Brizzard, B. (2008) Epitope tagging. *BioTechniques* **44**: 693–5.

Brown, T.A. (2010) *Gene Cloning and DNA Analysis: An Introduction* (6th edn). Chichester: Wiley–Blackwell.

Cohen, S.N., Chang, A.C., Boyer, H.W., and Helling, R.B. (1973) Construction of biologically functional bacterial plasmids in vitro. *Proc Natl Acad Sci USA* **70**: 3240–4.

Copeland, N.G., Jenkins, N.A., and Court, D.L. (2001) Recombineering: a powerful new tool for mouse functional genomics. *Nat Rev Genet* **2**: 769–79.

Datta, S., Costantino, N., and Court, D.L. (2006) A set of recombineering plasmids for gram-negative bacteria. *Gene* **379**: 109–15.

Filippo, J., Sung, P., and Klein, H. (2008). Mechanism of eukaryotic homologous recombination. *Annu Rev Biochem* **77**: 229–57.

Gallagher, R. and Desjardins, P.R. (2008) *Quantitation of DNA and RNA with Absorption and Fluorescence Spectroscopy. Current Protocols in Protein Science*. New York: Wiley-Interscience.

Gibson, D.G., Benders, G.A., Andrews-Pfannkoch, C., et al. (2008) Complete chemical synthesis, assembly, and cloning of a *Mycoplasma genitalium* genome. *Science* **319**: 1215–20.

Gibson, D.G., Young, L., Chuang, R.Y., Venter, J.C., Hutchison, C.A., 3rd, and Smith, H.O. (2009) Enzymatic assembly of DNA molecules up to several hundred kilobases. *Nat Methods* **6**: 343–5.

Jones, D.H. and Howard, B.H. (1991) A rapid method for recombination and site-specific mutagenesis by placing homologous ends on DNA using polymerase chain reaction. *BioTechniques* **10**: 62–6.

Jones, D.H. and Winistorfer, S.C. (1992) Recombinant circle PCR and recombination PCR for site-specific mutagenesis without PCR product purification. *BioTechniques* **12**: 528–30, 532, 534–5.

Malhotra, A. (2009). Tagging for protein expression. *Methods Enzymol* **43**: 239–58.

Mertz, J.E. and Davis, R.W. (1972) Cleavage of DNA by R_1 restriction endonuclease generates cohesive ends. *Proc Natl Acad Sci USA* **69**: 3370–4.

Sambrook, J., Fritsch, E.F., and Maniatis, T. (1989) *Molecular Cloning : A Laboratory Manual*. Cold Spring Harbor, NY: Cold Spring Harbor Laboratory.

Shuman, S. (1994) Novel approach to molecular cloning and polynucleotide synthesis using vaccinia DNA topoisomerase. *J Biol Chem* **269**: 32678–84.

Thomas, C.E., Ehrhardt, A., Kay, M.A. (2003) Progress and problems with the use of viral vectors for gene therapy. *Nat Rev Genet* **4**: 346–58.

Wang, T., Ma, X., Zhu, H., Li, A., Du, G., and Chen, J. (2012). Available methods for assembling expression cassettes for synthetic biology. *Appl Microbiol Biotechnol* **93**: 1853–63.

Zimmerman, S.B., Little, J.W., Oshinsky, C.K., and Gellert, M. (1967) Enzymatic joining of DNA strands: a novel reaction of diphosphopyridine nucleotide. *Proc Natl Acad Sci USA* **57**: 1841–8.

2 Polymerase chain reaction

Martin Evison

Chapter overview

In this chapter, we will learn about the polymerase chain reaction (PCR), a phenomenally powerful technique that has applications in many areas of medicine and biology. PCR is central to the development of our understanding of what genes are and how they work. The discovery of DNA was only a prelude to these exciting developments. PCR allows us to target and copy almost any gene in the vast quantities needed to analyse, sequence, or visualize the DNA itself. As well as being an invaluable tool in research into gene expression, genome analysis, and gene cloning, PCR is used routinely in clinical diagnosis and has revolutionized forensic science.

LEARNING OBJECTIVES

This chapter will enable the reader to:

- understand how the discovery of PCR is central to the development of molecular biology;
- describe how PCR works;
- consider the main factors that affect PCR precision, efficiency, and yield when designing PCR-based experiments;
- select appropriate PCR-based approaches to answer defined research questions in medicine, biology, and forensic science.

2.1 Introduction

When considering the significance of PCR, it is worth understanding how quickly molecular biology developed in the last quarter of the twentieth century. In 1944, Avery, MacLeod, and McCarty found that DNA could transform the properties of pneumococcal bacteria and identified it as the genetic material (Avery et al. 1944). In 1953, Watson and Crick used X-ray crystallography to discover the double-helix structure of the DNA molecule (Watson and Crick 1953), allowing the mechanisms of DNA replication and recombination to be understood, along with the basis of how DNA is able to code for proteins and, in essence, regulate the structure and metabolism of the organism. However, techniques permitting the isolation and sequencing of individual genes were slow and cumbersome.

It was a series of exciting developments occurring in a very short period of time during the 1970s and early 1980s that led to an explosion of knowledge that really allowed us to begin to understand the mechanisms and significance of gene expression in species of all kinds. There were advances in **DNA sequencing** and the discovery of **restriction endonucleases** that cleave DNA molecules only at specific sites permitted **recombinant DNA** molecules to be constructed. A region of interest could be inserted into the DNA contained in a host organism, such as *E. coli*—and copied simply by culturing-up the transformed host. This technique—**molecular cloning**—could also be used with the aim of expressing the cloned gene. For example, Martial et al. (1979) described the cloning and expression of a DNA sequence coding for human growth hormone in *E. coli*.

However, molecular cloning is a time-consuming and tedious process, and the alternative, attempting DNA replication *in vitro*, was much the same. The process required a series of steps at different temperatures involving moving the reaction between water baths and microwave ovens. A short sequence of DNA, an oligonucleotide **primer**, was required to initiate synthesis of a new DNA strand by **DNA polymerase** enzymes. However, these enzymes denatured at the temperatures (typically, 94°C) required to separate the new strand from the template, which meant that they had to be added freshly with each cycle. As only a single primer was used, each reaction involved the unidirectional synthesis of only one of the two strands of a double-stranded DNA duplex.

The final phase in the development of PCR occurred in the 1980s and was pioneered by Kary Mullis. Mullis pursued the concept of using *two* primers on opposite strands of the DNA, flanking the sequence intended for copying and allowing both strands of the DNA duplex to be copied in opposite directions. The need to add fresh enzyme with each cycle was overcome by using a *thermostable* DNA polymerase originally isolated from *Thermus aquaticus*, a thermophile bacterium found in geothermal hot springs. **Taq polymerase** is not denatured at the temperature required to separate DNA strands, allowing DNA replication to proceed without the need to reintroduce reagents during the reaction (Saiki et al. 1988). The polymerase chain reaction was patented, and in 1993 Mullis was awarded the Nobel Prize in Chemistry for his work.

2.2 How PCR works

PCR remains an *in vitro* adaptation of component processes of DNA replication occurring *in vivo*. This is a reaction governed by the properties of DNA polymerase enzymes, including *Taq*, and fundamentally involves three steps.

1. Denaturation or **strand separation**

   The reaction temperature is raised to a point at which the strands of the DNA duplex separate—typically 94°C.

2. **Annealing**

   The reaction temperature is lowered to a point at which the short oligonucleotide primers flanking the sequence intended for copying anneal to their respective targets on opposite strands of the single-stranded DNA. This temperature is tailored according to the specific requirements discussed in section 2.3.1, but temperatures of 60–65°C are common.

3. **Extension** or elongation

   The reaction temperature is raised to a point at which synthesis of new DNA strands can occur under the control of *Taq* or another suitable DNA polymerase. A typical elongation temperature for *Taq* is 72°C.

Strand separation, annealing, and extension constitute one cycle of the PCR (Figure 2.1). When extension is completed at the end of each cycle, the synthesis of two new strands

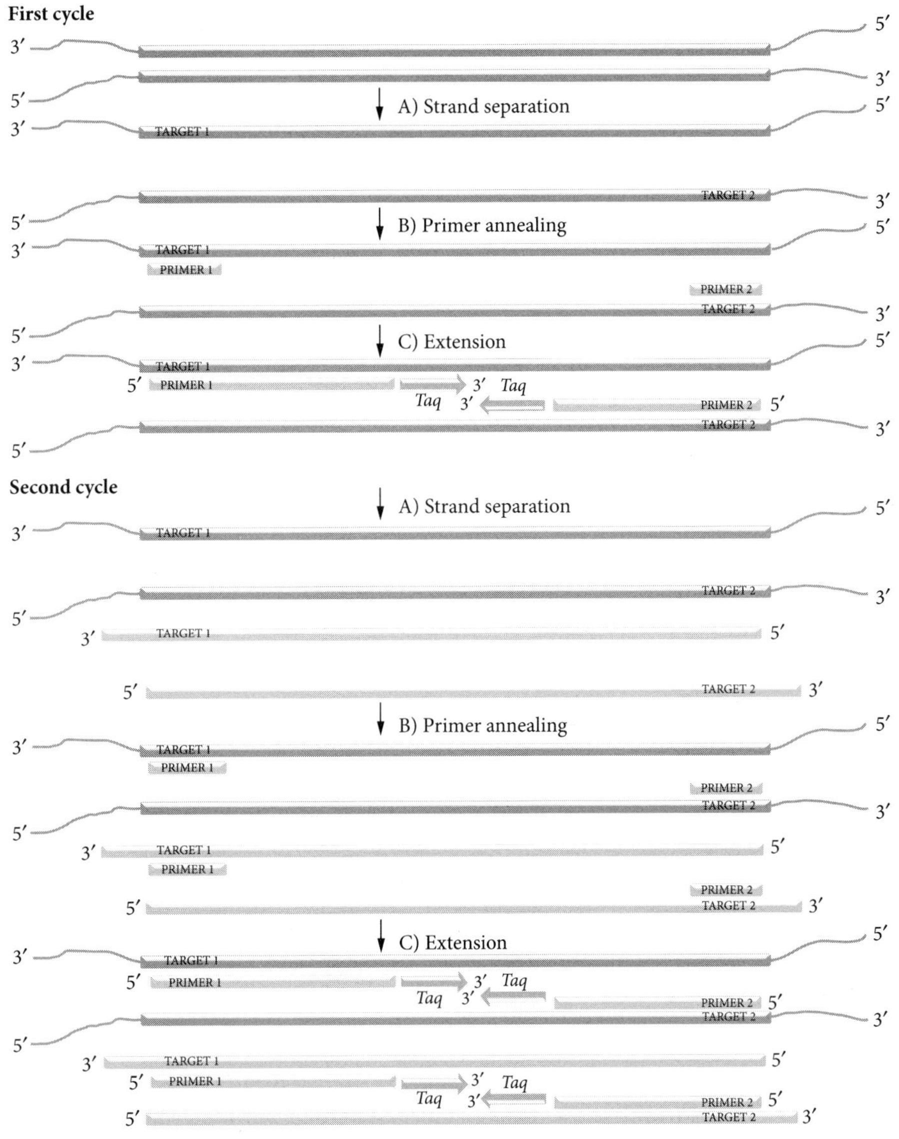

Figure 2.1 The polymerase chain reaction. During the each cycle the reaction temperature is raised until (A) strand separation of double-stranded DNA occurs. The reaction is then (B) cooled to permit primer annealing to specific target sites flanking the sequence of interest on the template DNA molecule. The reaction temperature is then raised (C) to permit extension of newly synthesized strands by *Taq* polymerase. At the end of each cycle the number of strands of the DNA target sequence increases exponentially. Two cycles are shown for illustration.

or **amplicons** is complete and the quantity of the target sequence is doubled. Doubling the number of template molecules with each PCR cycle allows the synthesis of new molecules to continue exponentially. Therefore the quantity of PCR product generated during the reaction is roughly proportional to the initial quantity of the target. As new molecules can only be initiated at each primer binding site, PCR product of the length and sequence flanked by the primer pair quickly proliferates.

2.3 The PCR protocol

The PCR procedure involves preparation of the reaction mixture and its components, and of the PCR itself, followed by analysis and visualization of the PCR product.

A basic PCR protocol is as follows.

HLA-B27 amplification by PCR (10 × 10µl)

1. Dispense 69.0µl master mix into a 1.5ml polypropylene tube. Add 1.0µl *Taq* polymerase (5U/µl) (final concentration 0.05U/µl) prior to PCR.
2. Dispense 7.0µl reaction mix into each of 10 microtubes in cycler rack.
3. Add 3µl DNA substrate (10ng/µl) to each tube. Mix by gently drawing solution in and out of the pipette tip.
4. Centrifuge rack briefly to collect mix at the bottom of tube.
5. Carry out PCR amplification using 30 cycles of the following programme:
   (i) 94°C, 20s (strand separation)
   (ii) 63°C, 50s (primer annealing)
   (iii) 72°C, 20s (primer extension).
6. Ramp to 4°C on completion.

The corresponding recipe for the master mix is as follows:

HLA-B27 master mix for 10 × 10µl reactions

Quantity	Ingredient	[Final]
10.0µl	X10 PCR Buffer (No Mg)	–
10.0µl	2.0mM dNTPs	0.2mM
10.0µl	10.0mM $MgCl_2$	1.0mM
9.5µl	2.5µM C3 control primer	0.24µM
9.5µl	2.5µM C5 control primer	0.24µM
10.0µl	5.0µM E91 B27 primer (sense)	0.5µM
10.0µl	5.0µM E136 B27 primer (antisense)	0.5µM

This master mix includes an internal positive control used to demonstrate that the PCR reaction has been successful in HLA-B27-negative samples. Similarly, it is usual to include a negative or 'blank' control of purified (DNA-free) water instead of a DNA sample in one of the tubes included with each reaction.

The PCR protocol is carefully designed to optimize the reaction according to the often competing demands of *efficiency*, *fidelity*, and *yield*. Efficiency refers to the speed of the reaction, fidelity to the precision with which the target sequence is copied, and yield to the quantity of PCR product produced at the end of the reaction. These in turn may depend on the source and condition of the template DNA and the overall aims of the experiment or test. PCR protocols may vary considerably depending on the specific technique being used and application requirement being addressed (see section 2.4).

2.3.1 Components of the PCR reaction

Template DNA

Normally, the **template** intended for amplification is double-stranded DNA. The DNA must be sufficiently pure. The quantity of DNA and protein in a sample, for instance, can be compared spectroscopically using the ratio of A260 (DNA) to A280 (protein) absorbance. The proportion of protein in the reaction should be at a minimum and as a guideline certainly lower than the quantity of DNA.

DNA damaged prior to or during the PCR reaction may result in base **mis-incorporation**, **partial products**, or a variety of artefacts affecting PCR fidelity, efficiency, and yield. PCR can be susceptible to a variety of organic and inorganic **inhibitors** which may reduce yield or prevent amplification occurring altogether. Intrusive DNA from an unintended source can easily lead to **contamination** of the reaction and hence confound the results.

Purification protocols must be optimized to ensure that the amount of co-purifying proteins and other molecules, which may inhibit PCR, are kept to a minimum, and stringent steps are taken to minimize the likelihood of contamination (see section 2.3.5).

Primers

Oligonucleotide primers must be chosen carefully. A variety of considerations affect primer choice and design. The key consideration is **stringency**. The primers must anneal only to *specific sequences* flanking the target template. If they anneal with any other region, amplification of unwanted template may occur. This applies to regions of the DNA expected to occur in the target sample, as well as to DNA that may originate from wildly different and unexpected sources such as microbial or other cross-species contamination. As strand extension proceeds only in a 5′ to 3′ direction, primers must be chosen for a specific upstream target on their complementary template strand. As PCR tends to deteriorate in efficiency with the length of the sequence being amplified, primers that *flank closely to the target template* or region of interest are desirable.

The **melting temperatures (T_m)** of primers will vary according to the number of A–T and G–C bonds involved in hybridization. Primers with similar melting temperatures (T_m) must be chosen. If not, a primer with a T_m unsuited to the annealing step may mis-hybridize leading to the amplification of unwanted DNA sequences. The sequences of the primers themselves should be such that they cannot form artefacts such as **loops** or **primer dimers**.

There are certain circumstances in which the strictest primer stringency is unimportant or even undesirable, such as when the aim of the experiment is to amplify a template whose sequence is insufficiently known to specify the primer sequence precisely or even to amplify the *whole genome*. In these cases, one or more **degenerate primers** may be chosen. These are so named because they are assumed not to match the target sequence perfectly, as their sequences only presuppose or anticipate their probable target.

Primers can be modified in a variety of ways: for example, to add an 'adapter sequence' that can act as a target for a further primer (see section 2.4.4) and to incorporate a fluorophore that can be used to measure the progress of the reaction by its luminescence (see section 2.4.5).

Software applications to assist in primer design are available. These include programs that search for known homologues in human or other species, and programs which measure the potential for primer-dimer or loop formation for given oligonucleotide sequences. Software can also be used to predict the melting temperature of the primer from its nucleotide sequence, using simple and complex formulae. The sequences of the primer pair can be optimized accordingly to arrive at a consensus annealing temperature.

Ultimately, the PCR reaction must be tested in practice and, if necessary, further optimized empirically (see section 2.3.5).

DNA polymerase enzyme

Several *Taq* and other DNA polymerase enzymes are available with a variety of properties that can be tailored toward a particular application. Factors affecting the choice of polymerase include *rate of synthesis* of the new strand, *fidelity* or *error* in replication, and requirements of particular techniques and applications (see section 2.4). For example, some protocols require a high-temperature initialization phase (see section 2.4.1). Proof-reading polymerases may be favoured where precision is of overriding importance or where the target sequence is particularly long (see Section 2.4.10). Original DNA polymerases derived from native *Thermus aquaticus* are relatively efficient, generating rapid strand elongation, but have poor proof-reading capabilities. More recent *Taq* alternatives rely on greater 3′–5′ exonuclease activity which excises mis-incorporated bases, offering greater proof-reading properties.

dNTPs

Dinucleotide triphosphates (dNTPs), corresponding to the four bases adenine, guanine, cytosine, and thymine, are incorporated to form the newly synthesized DNA strands. Adjustment of the required dNTP concentrations is based on the base composition of the target sequence. Enough is required to ensure sufficient dNTPs to allow the reaction to progress efficiently, but too large an excess of any one dNTP may lead to base mis-incorporation.

Other reagents

Other reagents can be chosen in order to optimize or tailor the reaction. DNA polymerase efficiency is particularly dependent on the concentration of **divalent cations** such as calcium (Ca^{2+}) and magnesium (Mg^{2+}). While divalent cations are generally essential for PCR, excess magnesium has been shown to lead to an increase in the rate of nucleotide mis-incorporation. The reaction mix includes a suitable PCR **buffer** (see the basic PCR protocol). Typically, the optimum pH for PCR is 8.4. Most PCR buffers are supplied commercially. Again, base mis-incorporation has been shown to increase in conditions of increased pH.

Figure 2.2 Diagrammatic representation of the simple thermal cycler programme for HLA-B27 PCR described in the text. An initialization phase is intended to prevent mispriming prior to commencement of the PCR proper. The three phases of a 30 cycle PCR are represented in the central section of the diagram. Upon completion, the reaction temperature ramps down to 4°C prior to post-PCR analysis or storage.

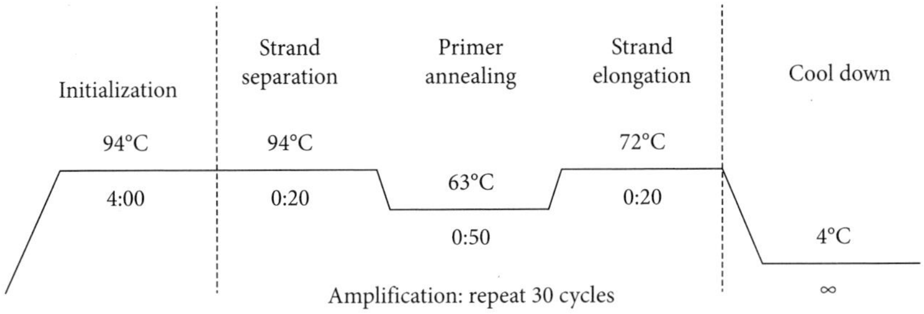

2.3.3 Instrumentation and consumables

The standard instrument used is the 'PCR machine' or **thermal cycler** which is capable of robust cycles of heating and of cooling. Thermal cyclers are programmable and can be linked together in array systems controlled by a single 'master' instrument. Figure 2.2 shows the phases of the PCR cycle in diagrammatic form for the HLAB 27 PCR protocol outlined in section 2.3. Thermal cyclers vary in the extent and accuracy with which PCR parameters may be controlled, in heating and cooling (and hence reaction) efficiency, and in the number and volume of reactions they can hold in individual wells or tubes.

Plastic-ware tailored to particular thermal cycler formats is readily available. Typical formats include 96-well reaction blocks and racks, holding 0.2ml polypropylene tubes, or blocks accepting 384 microwell plates with volumes of up to 120μl per well.

2.3.4 Optimization

Optimization of the PCR reaction is critical to its effectiveness. Standardized protocols are often the result of considerable experimentation in which the many parameters of the reaction and other **PCR additives** are carefully adjusted in order to arrive at a compromise driven by the often competing demands for fidelity, yield, and efficiency (Kennedy and Oswald 2011).

The key variables are the *length of time* and the *temperature* of each of the annealing, extension, and separation steps of each cycle, the *number of cycles*, and the design of the **initialization phase** before the main reaction. The temperature and duration of each phase of the PCR reaction must be chosen with considerable care.

If the strand separation step occurs at too low a temperature or is too short, primer **misannealing** or **mis-hybridization** to non-specific targets may result in the generation of unwanted products. Misannealing is particularly significant in the early cycles of the reaction as product begins to amplify and may overwhelm the subsequent cycles of the reaction (see also sections 2.4.1 and 2.4.2).

A high annealing temperature will reduce the time taken for the thermal cycler to 'ramp' between maximum and minimum temperatures, increasing efficiency, but annealing phases that are too short or are at temperatures that are too high will prevent primer binding and reduce efficiency and yield.

Similarly, an extension phase of incorrect temperature or too short a duration may lead to the generation of partial sequences, reducing the efficiency of the reaction and quantity of product. Too long a duration of the extension phase may lead to amplification of non-specific product.

Generally, if the duration of any of the phases is longer than necessary, efficiency is reduced. Similarly, increasing the number of cycles will increase the yield, but an excess of yield will reduce efficiency. On the other hand, insufficient yield will interfere with post-PCR analysis.

The length of the target sequence is also important. PCR tends to decrease in efficiency and fidelity with the length of the PCR product. Therefore short sequences are preferable and primer template sequences should be identified accordingly, wherever possible.

Optimization can be tailored to a specific priority according to the required outcome or application (see section 2.4).

2.3.5 Preparation

It is essential that the areas where PCR is carried out are clean, in order to minimize exposure to **intrusive DNA** and other contaminants that may affect PCR efficiency. **Amplified DNA** is a particularly powerful source of contaminating intrusive DNA and should never be introduced into DNA extraction or PCR preparation areas: *pre-PCR* and *post-PCR* activities should be carried out in separate laboratories. Laboratory surfaces should be routinely cleaned using a bleach solution (e.g. 10% sodium hypochlorite) or other reagent that destroys contaminating DNA. Any recycled consumables, such as plastic racks used to place tubes on thermal cycler blocks, should also be thoroughly cleaned in this way. Liquid-handling equipment, particularly pipettes, should also be routinely cleaned and checked for contamination. Dedicated sets should be used for handling reagents, purified DNA, and PCR product. Aerosol-resistant tips should be used for general purposes, and positive displacement pipettes should be used for handling the reagents most sensitive to contamination, such as *Taq*. DNA-free consumables and reagents may be required for low-template work, and the PCR process should be routinely checked for background contamination—for example, by assessing the frequency of false-positive reactions in 'blank control' amplifications where water replaces the DNA template.

2.3.6 Analysis

Completion of the PCR is followed by analysis of the PCR product. This may be undertaken using a number of different approaches: **agarose gel electrophoresis**, such as the ubiquitous 'minigel' (Figure 2.3), **polyacrylamide gel electrophoresis** (PAGE), **capillary electrophoresis**, dye-based detection methods, and **fluorescence-based detection methods**.

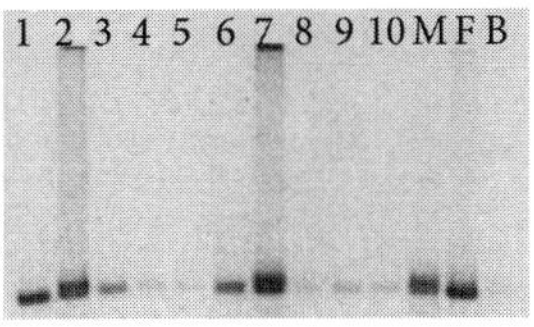

Figure 2.3 Agarose 'minigel' showing the results of PCR amplification of the amelogenin gene in a low-template DNA experiment. A 6bp deletion in the X-chromosome homologue allows males (two bands) and females (one band) to be distinguished. Lanes 1–10 show the results of low-template amelogenin PCR from dried blood spots up to 26 years old. Lanes M and F are male and female controls, respectively, and lane B is an extraction blank control (sterile filtered distilled water).

2.4 PCR techniques and applications

PCR has become a complex technology supported by advanced instrumentation and a plethora of alternative enzymes, reagents, and reaction protocols that can be tailored to different application areas. The essential utility of PCR lies in its value in copying DNA (or RNA, see below) from a tiny starting amount to the quantities needed to enable further analysis, such as PCR-based cloning, DNA sequencing, and genotyping, including microarray analysis or visualization by even the most simple and robust methods as shown in Figure 2.3.

See Chapters 1 and 4

The Barcode of Life Project (see Box 2.1) is a example of how DNA extraction, PCR amplification, and sequencing can be used to distinguish, identify, and catalogue many of the vast and diverse number species on the planet.

A brief review of a range of further PCR techniques with a description of the work in which they can be usefully applied is given in the following sections.

2.4.1 Hot-start PCR

Hot-start PCR is used to counter primer mis-hybridization or misannealing in which primers anneal to an unintended target usually similar in sequence to their intended

PCR and the Barcode of Life Project BOX 2.1

The Barcode of Life Project (Hajibabaei et al. 2007) is an international project aimed at using short universal standard sequences of DNA to distinguish and catalogue as many of the world's different species of plants and animals as possible. In animals, a 648bp region of the mitochondrial cytochrome c oxidase 1 gene (CO1) has proved effective in species identification. This region does not exhibit sufficient variation in plants, where the chloroplast maturaseK (matK) and ribulose bisphosphate carboxylase (rbcL) genes have been used.

Specimens have been collected from a variety of sources including aquariums, zoos, botanical gardens, seed banks, and so forth. Standard protocols for DNA extraction, PCR amplification, and sequencing are offered by the Barcode of Life project. Once amplified and sequenced, the DNA barcodes are entered into international databases—the International Nucleotide Sequence Database Collaborative and the Barcode of Life Database—which act as permanent repositories of the species, identifying sequences and offering a new resource for biological scientists.

target (Chou 1992). This problem is overcome by using a high-temperature initialization phase (D'Aquila et al. 1991). Misannealing commonly occurs at temperatures below the optimal, and a high-temperature initialization phase means that no primers can anneal until the reaction temperature is lowered, ideally to the optimum annealing temperature.

Using hot-start PCR with a **high-temperature-activated DNA polymerase** also prevents any potential extension from mis-hybridized primers, as the enzyme is initially inactive but is activated when the high start temperature is reached.

2.4.2 Touchdown PCR

Touchdown PCR uses high-temperature initialization and annealing phases in the earliest PCR cycles to improve fidelity. Using an annealing temperature slightly higher than the optimum reduces the likelihood of misannealing, and ensures that PCR amplicons generated in the first cycles are more likely to be accurate copies of the target sequence. As these products proliferate, the annealing temperature can be sequentially lowered to the optimum value so that the remaining cycles of the PCR reaction can proceed more efficiently. Figure 2.4 shows a thermal cycler programme for touchdown PCR.

2.4.3 Nested PCR

Nested PCR is a technique that can be employed when the primers most closely flanking the region of interest are known to hybridize elsewhere on the template DNA, leading to amplification of unwanted targets. To minimize the effect, a second set of primers of high specificity is used initially, flanking the region more broadly and generating longer amplicons containing the target sequence. Once these sequences proliferate sufficiently, the original closely flanking primers can be used as the region of interest will be amplified preferentially and more efficiently. Nested PCR (Figure 2.5) has been used in highly sensitive methods of pre-implantation genetic diagnosis (see Box 2.2).

2.4.4 RT-PCR and RACE

While traditional PCR is used to amplify DNA sequences, a variant known as RT-PCR (reverse transcriptase PCR) is used to amplify RNA, allowing the sequences transcribed during gene expression to be investigated. During RT-PCR, **reverse transcriptase** is used

Figure 2.4 Diagrammatic representation of a touchdown PCR programme. After an initialization phase the touchdown stage commences. The annealing phase begins at 60°C and is lowered by 1°C in each of 10 cycles. The routine PCR programme that follows is able to use a reduced number of cycles because of the availability of template following the touchdown stage.

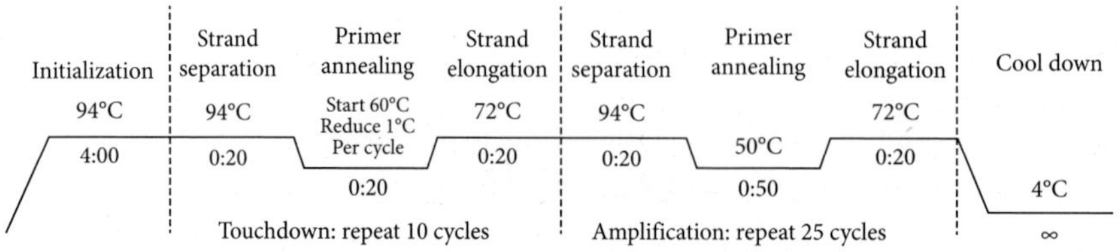

Figure 2.5 Nested PCR reaction. The initial stage involves amplification of a longer sequence for which primers highly stringent to their target are available. The second phase uses less stringent primers amplifying a shorter sequence whose primer binding sites are located within the longer amplicons. The availability of an excess of the desired template means that amplification of the less stringent primers to unwanted template is insignificant.

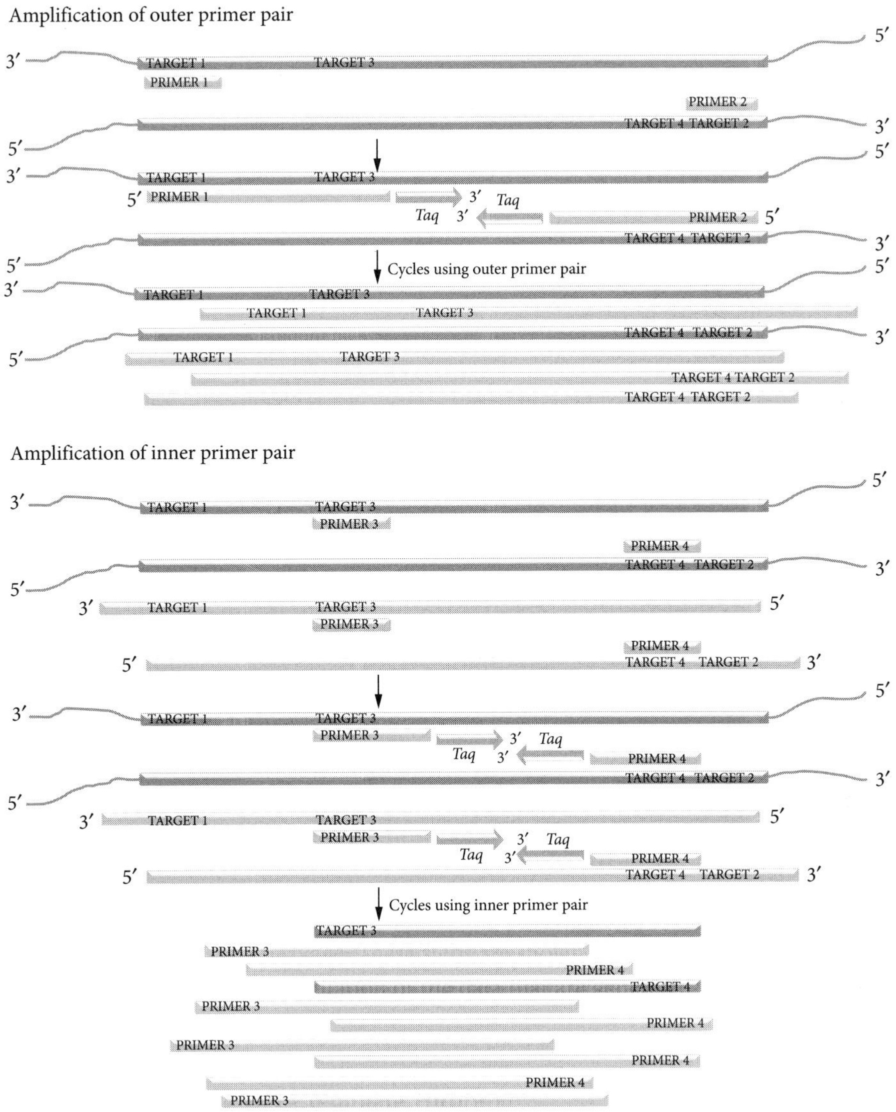

to reverse transcribe RNA sequences into **cDNA**. A primer targeting a known sequence on this strand can be used to initiate DNA polymerase synthesis of the opposite strand of the duplex. The cDNA may then be subject to analysis by a range of PCR methods.

An approach to RT-PCR called rapid amplification of cDNA ends (RACE) is important in the amplification of the unknown sequences at the ends of complete RNA transcripts. The reverse transcriptase step is again used to generate a cDNA copy (see Figure 2.6). RACE can be 5′ or 3′ depending on whether there is a known sequence available as a primer target site at the 5′ end of the RNA or whether the 3′ polyA tail is to be used.

In 5′ RACE (Edwards et al. 1991) an antisense primer recognizing a known target sequence is used to synthesize a single cDNA strand in a 3′ to 5′ direction.

BOX 2.2

Nested PCR in pre-implantation genetic screening and diagnosis

Pre-implantation genetic diagnosis and pre-implantation genetic screening are techniques that permit embryos to be tested for disease prior to implantation following *in vitro* fertilization (IVF) treatment (Harper et al. 2001). Two specific examples of pre-implantation genetic diagnosis are the use of PCR to amplify a sequence of the ΔF508 locus of the cystic fibrosis transmembrane regulator (CFTR) gene and a nullified *Xho*I restriction site in the hypoxanthine phosphoribosyl transferase (HPRT) gene occurring in Lesch–Nyhan syndrome. Diagnosis can be performed on biopsies of one or two cells from eight to ten cell embryos. Nested PCR can be used to amplify the region of interest sufficiently to avoid significant mis-hybridization by the more closely flanking primer set. Amplification of as few as one or two copies of the target present in a single haploid or diploid cell has been shown to be possible. Pre-implantation genetic screening and diagnosis are powerful scientific techniques. In view of the associated questions of termination of pregnancy and genetic selection of offspring, they also raise important social, moral, and ethical questions.

A homopolymeric tail or known 'anchor' oligonucleotide is then added to the 3′ end of the cDNA strand. A PCR reaction using a second forward primer targeting a known sequence paired with a universal primer targeting the homopolymeric tail or anchor is then used to copy the cDNA strand in the 5′ to 3′ direction.

In 3′ RACE a special primer is used that targets the polyA tail found at the 3′ end of eukaryotic mRNAs to synthesize a single cDNA strand incorporating an 'adapter sequence' at the 5′ end as part of the primer molecule. A PCR reaction using a primer targeting a known sequence paired with a primer targeting the adapter sequence is then used to copy the cDNA strand in the 3′ to 5′ direction.

2.4.5 Quantitative PCR

As PCR proceeds exponentially, the quantity of PCR product generated during the reaction is roughly proportional to the initial quantity of the target. Therefore measuring the amount of product during or at the **endpoint** of the reaction (by comparison with a reliable internal control) offers an approximate means of quantifying the number of copies of the region of interest initially present in the reaction (VanGuilder et al. 2008). Internal controls are frequently offered as purposely designed template and primer sets, whose PCR properties are well characterized.

Of immense value to both research and application, quantitative real-time PCR relies on specialist instrumentation that uses signals from fluorescent dyes or probes to detect and measure the quantity of PCR product being generated during the PCR reaction. Fluorescent dyes are frequently used to measure PCR output. Figure 2.7 shows a commonly used approach—the TaqMan® method—where a specially designed probe targeting a known sequence on the template strand is used. Reporter and quencher fluorescent dyes are attached to the probe. As strand synthesis progresses during PCR, the *Taq* polymerase exonuclease activity cleaves the probe, releasing the fluorophore and allowing it to be detected fluorometrically.

Quantitative PCR (qPCR) can be combined with real-time PCR (in RT-qPCR) in order to measure the level of messenger RNA (mRNA) molecules being produced in various tissues, allowing us to improve our understanding of gene expression (see Box 2.3).

Figure 2.6 Illustration of 5′ and 3′ RACE. Both 5′ and 3′ RACE commence with reverse transcription. In 5′ RACE an oligonucleotide (RT primer) targeting a known region at the 3′ end of the mRNA molecule is used to initiate reverse transcription. PCR synthesis from the single cDNA strand can be initiated via two routes. Either (a) a polynucleotide tail or (b) an anchor of known sequence is attached to the newly synthesized cDNA molecule. Synthesis of the sense strand can then be initiated using a primer (PRIMER 1) targeting one or other sequence. Synthesis of the opposite strand uses a primer (PRIMER 2) for DNA corresponding to the known target sequence on the mRNA molecule. In 3′ RACE, reverse transcription is primed with an oligonucleotide targeting the polyA tail of the mRNA to which a known adapter sequence is attached. PCR synthesis from the cDNA molecule commences using one oligonucleotide (PRIMER 1) targeting a known sequence at the 3′ end of the cDNA molecule to prime synthesis of the sense strand and a second oligonucleotide (PRIMER 2) targeting the known adapter sequence.

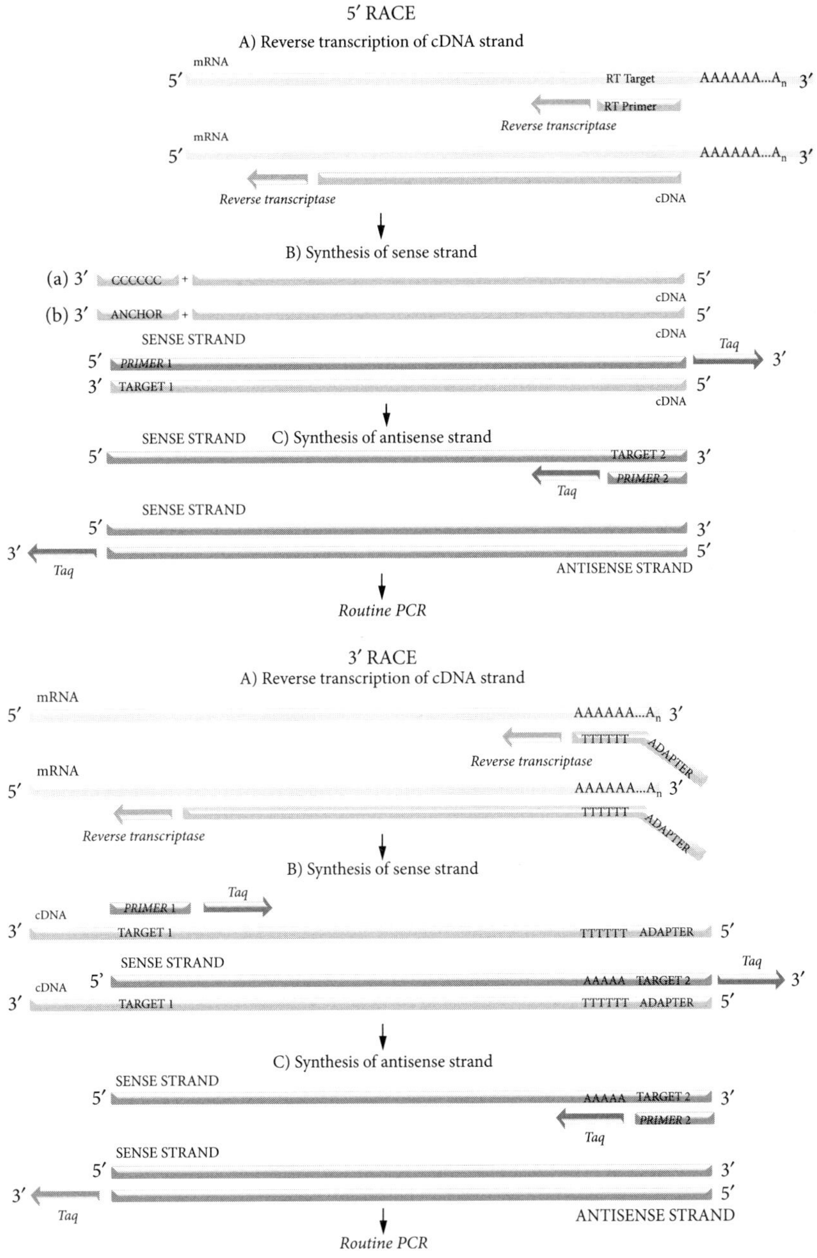

Figure 2.7 Illustration of the TaqMan® method (a) for fluorescence-based detection of PCR product in real time. The TaqMan® probe targets a known sequence on the template strand (A). Reporter (light grey) and quencher (dark grey) fluorescent dyes are attached to the probe. An oligonucleotide primer initiates strand synthesis by *Taq* polymerase (B). As strand synthesis progresses, the *Taq* polymerase's exonuclease activity cleaves the TaqMan® probe, releasing the light grey fluorophore which, no longer quenched by its proximity to the dark grey fluorophore, can be detected (C). Fluorescence from experimental and control samples can be measured in real time (graph (b)), permitting PCR product to be quantified.

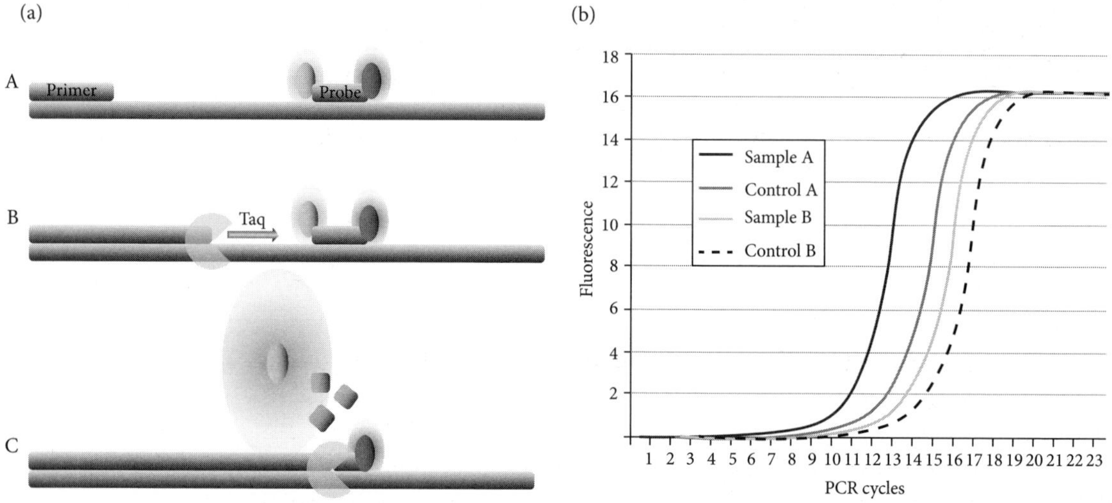

BOX 2.3

Using PCR to study gene expression

Our knowledge of the development and regulation of many processes in biology and medicine relies on our understanding of gene expression, which means that we need to measure and quantify the level of mRNA produced by different genes in different tissues at various stages of development and during different physiological processes.

RT-PCR is first used to reverse transcribe mRNA molecules into cDNA and then qPCR is used to quantify them. As the range of mRNA molecules potentially expressed in many tissues is large, microarrays which permit 'expression profiling' of entire genomes have been developed. Lombardino et al. (2006) used expression profiling to investigate neurons. They used laser dissection microscopy to physically excise intermingled long-range projection neurons prior to expression profiling, elucidating mRNAs implicated in neuronal development. As microarrays do not usually offer the quantitative sensitivity of qPCR, interesting findings can be verified using RT-PCR and qPCR in the usual way.

CASE STUDY 2.1

Identifying the allele responsible for a genetic syndrome

You are researching an extremely debilitating genetic syndrome with a low life expectancy. Neuronal cell lines have been developed, which strongly suggest that an unusual allele in one of a complex of differentially expressed homologous genes may be responsible. Partial DNA sequences have been established for some of the homologues and a robust cell culture model is available for the investigation of differential gene expression relevant to the disease.

? QUESTION
Which PCR techniques could you use in an effort to identify the allele responsible for the disease?

The cell culture model looks like a promising means for isolating relevant mRNAs, and RT-PCR could allow cDNA sequences to be amplified. However, the indication that known sequences are only partial suggests that 5′ and 3′ RACE may be required to properly establish the mRNA sequences in order to maximize the prospects of confident identification from DNA of the genetic locus responsible. If differential gene expression is implicated in the disease pathway, then RT-qPCR may establish which relevant mRNAs are expressed when. Assuming that the relevant RNAs can be isolated and their cDNA sequenced, it may be possible to identify the specific DNA sequence and gene responsible from a reference database, and to compare the sequence in healthy individuals with those possessing the disease.

2.4.6 Digital PCR

Digital PCR (Vogelstein and Kinzler 1999) goes one step beyond quantitative PCR in that it relies on an approach based on the separation of the reaction into numerous partitions on a microarray, each of which may contain few or no target molecules. The presence or absence of a target allows a binary or digital quantification of the number of target molecules originally present to be estimated according to standard statistical assumptions. Digital PCR is more precise than quantitative PCR, which relies on the assumption that the PCR reaction proceeds exponentially to arrive at quantification and relies on standardized controls to calculate the amount of target substrate present. Although the amount of product is roughly proportional to the initial quantity of target, this relationship is not exact and may be subject to various extraneous influences, such as the quality and quantity of initial template, the optimality of the reaction, and stochastic variation, particularly with low copy numbers of template. Digital PCR avoids this uncertainty by making a direct calculation of the amount of target template based on the number of positive and negative partitions, and without reference to standardized controls.

2.4.7 Multiplex PCR

A multiplex PCR is simply one in which more than one target DNA sequence is amplified in a single reaction, typically by using multiple sets of primers to amplify multiple targets. Multiplex PCR is very useful, as it avoids the need to laboriously set up and process individual reactions for each target sequence. It also means that when the quantity of template is restricted to a small and precious amount, it does not necessarily need to be diluted or separated for multiple analyses to be carried out.

However, optimization of multiplex PCR reactions is particularly challenging in that the overall reaction must accommodate parameters satisfactory for each and every individual PCR reaction included in the multiplex. Failure to optimize the reaction correctly can lead to artefacts and poor amplification, typically of longer PCR products. Artefacts include heterozygote imbalance, where one allele in a heterozygote pair is preferentially amplified, and allelic drop-out, where one allele fails to amplify altogether.

Allelic drop-in, by contrast, occurs as a consequence of amplification of a single contaminating allele sequence, typically in the form of a PCR product of that allele from an earlier reaction.

Forensic DNA profiling uses multiplex PCR (see Box 2.4) and typically relies on multiplexes of over a dozen sets of primer pairs.

2.4.9 ARMS PCR

Amplification Refractory Mutation System PCR (**ARMS PCR**) uses a panel of reactions, each containing primer pairs aimed at different targets, to differentiate between similar but polymorphic DNA sequences, typically allelic variants at a given locus. For example, in ARMS PCR for HLA DR-DQ loci, the results from a panel of 24 PCR reactions can be used to differentiate between major alleles that are important in tissue typing (Figure 2.8).

2.4.10 Long PCR

PCR efficiency typically diminishes in relation to the length of target sequence being amplified. Long PCR attempts to optimize PCR reactions intended to amplify longer templates, typically several kilobases long, by using longer extension times, adjusting a variety of reaction parameters (including pH, using additives that permit lowering of strand separation and annealing temperatures), and, as *Taq* polymerase is known to have a relatively high error rate, introducing a second proof-reading DNA polymerase into the reaction (Cheng et al. 1995). Long PCR and long-range PCR are valuable in improving the efficiency of PCR-based next generation sequencing (NGS) methods.

See Chapter 4

BOX 2.4

PCR forensic DNA profiling

DNA fingerprinting, which was discovered by Alec Jeffreys (Jeffreys et al. 1985), has been developed into a profiling technique central to modern forensic science. DNA profiling targets short tandem repeats (**STRs**), which are short stretches of DNA containing repeated pairs of bases. The number of pairs at each STR locus may vary, yielding STR alleles of different lengths. By using primers flanking each side of a locus it is possible to amplify the STR, and STRs of different lengths are distinguished by capillary electrophoresis. While allelic variation at a single locus is not sufficient to distinguish individuals, variation at ten or more loci increases the chance of distinguishing individuals astronomically. Forensic DNA profiling kits are designed to generate an STR profile for an individual at ten loci such that the chances of finding another individual with the same profile—the **random match probability**—is over one in a billion. The kits also amplify a short sequence of the **amelogenin** gene. This gene resides on the X and Y chromosomes and has a 6bp deletion on the X chromosome version, allowing the sex of the individual to be determined as well. **Low-template DNA analysis** is used for trace evidence, including evidence from remains of forensic interest (Evison et al. 1997), where the number of cycles in the PCR reaction is increased in order to detect the tiny quantities of template that may be present.

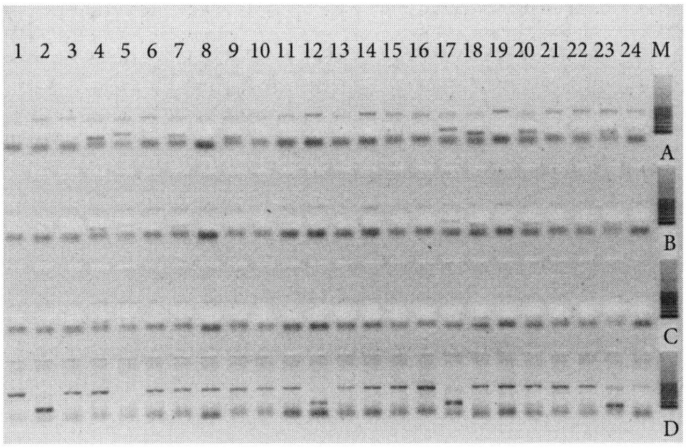

Figure 2.8 Polyacrylamide gel showing the results of a PCR-ARMS experiment determining the HLA DR-DQ type of four individuals (horizontal rows). Lanes 1–24 contain the results of amplification of a panel of PCR reactions targeting different allelic variants at regions of HLA DR and DQ loci. Each reaction contains an internal PCR positive control and is evident as a high molecular weight band. Low molecular weight products (e.g. sample D, rows 2, 12, 17, and 23) permit the HLA DR-DQ type to be assigned according to which of the 24 reactions are positive for the polymorphic sequences and which are not.

2.4.11 High-fidelity PCR

The quality and validity of a PCR-based assay may depend on the precision or purity of the PCR product generated at each phase of the reaction. High-fidelity PCR is directed at reducing the risk of potential mis-incorporation of nucleotides by optimizing the reaction for a minimum error rate (the concentration of nucleotides and the pH can each affect the rate of mis-incorporation, for example) again by using a proof-reading DNA polymerase such as the Pfu DNA polymerase extracted from *Pyrococcus furiosus* (Lundberg et al. 1991).

A potential new genetic diagnostic test

CASE STUDY 2.2

You have established that changes occurring in a 20bp sequence of DNA are responsible for all the known cases of an extremely debilitating genetic syndrome with low life expectancy. The syndrome is relatively frequent in some communities, and some groups are calling for a diagnostic test to be developed. Unfortunately, the 20bp region is flanked by long sequences of several hundred base pairs that are homogenous in sequence and have analogues in other parts of the genome.

QUESTION

You may be able to develop a diagnostic test, but it is essential that the PCR design problems be overcome and that the method can be applied to early-stage embryos. Which PCR protocols would you choose to employ?

The test must be precise and issues of PCR fidelity seem to be compounded by the implication that long sequences may need to be amplified first in a nested PCR. Primers designed to flank the long sequences would permit a long PCR to be undertaken as a first stage in a nested PCR. The second stage could use primers more closely flanking the 20bp region, given that an excess of the relevant template would then be available. Nevertheless, high precision, involving hot-start, touchdown, or high-fidelity PCR, would be essential in both cases. The low quantity of template may mean that an extended-cycle low-template DNA protocol may also be required.

2.4.12 *In situ* PCR

In situ PCR (Martínez et al. 1995) is a method designed to allow the position of target substrate molecules to be localized and visualized *in situ* in a fixed tissue preparation on a microscope slide. For example, the site of a viral infection could potentially be localized using *in situ* PCR or *in situ* RT-PCR. *In situ* PCR requires specialized instruments (a thermal cycler capable of accepting purpose-designed microscope slides holding both histological section and PCR reagents) and consumables. Reaction efficiency and yield are typically much reduced in *in situ* PCR reactions.

2.4.13 Whole genome amplification

Whole genome amplification (WGA) is intended to generate large quantities of product of the entire genome from a small quantity of substrate—potentially even a single cell (Zhang et al. 1992). A PCR-based approach can be taken which relies on 'degenerate' primers (a mixture of primers similar, but not identical, in sequence) anticipated to flank targets representative of the entire genome. Whole genome PCR is frequently a precursor to genome-wide association studies where PCR product is hybridized to microarrays to which are fixed an enormous range of DNA oligonucleotide probes.

2.5 Forensic DNA analysis

Forensic science has been revolutionized by the discovery of DNA fingerprinting (Jeffreys et al. 1985) and the application of PCR in the investigation of crime (Butler 2005). Routine **forensic DNA profiling** relies on the analysis of highly polymorphic STR alleles whose loci are situated on different chromosomes. As a consequence, STR alleles arising at different loci are not genetically linked and the product rule can be applied to calculate the random match probability of the STR profile. If ten or more STR loci are used, this value can easily exceed one in a billion. Forensic DNA profiles are readily amenable to electronic storage on a **forensic DNA database**, searching of which may allow suspects to be identified from biological material recovered from a crime scene. Forensic applications relying on PCR amplification of maternally inherited mitochondrial DNA (mtDNA) and paternally inherited Y-chromosome STR (**Y-STR**) analysis are also valuable, with the latter method being used extensively in **paternity testing** as STR alleles tend to be shared between close biological kin. Forensic STRs have also been used in **familial searching** of forensic DNA databases for relatives of unidentified offenders where a precise database match cannot be achieved. While STR applications predominate, **single-nucleotide polymorphism** (SNP) analyses are being developed for forensic applications, including the detection of externally visible characteristics (**EVCs**) like hair colour or eye colour (Kayser and Schneider 2009). RT-PCR methods have been applied to analysis of mRNA and micro-RNA (**miRNA**), in efforts to establish the source tissues of biological material deposited at crime scenes (Hanson et al. 2012).

Atrocity victim identification

CASE STUDY 2.3

You are a forensic scientist investigating skeletal remains from a mass grave. Forensic anthropologists have established the minimum number of individuals present, and established the probable sex and likely age of the victims. The human rights abuse allegations hinge on a dispute regarding kin relationships between victims, survivors, and alleged perpetrators. Some reference samples may be available from living relatives.

QUESTION

What PCR analyses would you employ to resolve the supposed kin relationships of the victims and living relatives?

Routine PCR can be used to amplify DNA from buccal swabs from living subjects, but only extended-cycle low-template DNA analysis is likely be effective on skeletal material. Forensic STR analysis may be sufficient to resolve most kin relationships, but further information may be usefully gained from mtDNA and Y-STR analysis regarding maternal and paternal lineages, respectively. In some circumstances, such as complex close kin relationships, statistical estimates rather than precise determinations may be the best that can be achieved, but some ambiguities might be resolved using the biological profiles offered by the forensic anthropologist's analysis of the skeletal remains.

2.6 Future prospects

Novel approaches to the rapid heating and cooling necessary in PCR have the potential to substantially reduce reaction times (Wheeler et al. 2011). The prospect of many exciting developments in PCR is offered by microfluidic chemistry and lab-on-a-chip technology. While microarrays offer high throughput analysis of hundreds of thousands of targets, other microfluidic devices have been developed that permit PCR to be carried out in small portable instruments that can be used in point-of-care diagnosis or, in a forensic context, at a crime scene. Like digital PCR, both **emulsion PCR** (Williams et al. 2006) and **droplet PCR** (Markey et al. 2010) use microfluidic methods of compartmentalizing numerous complex reactions into individual assays that can be carried out in parallel.

2.7 Chapter summary

- PCR emerged as a consequence of developments in molecular biology which included:
  - the discovery of DNA and its structure and function
  - the discovery of restriction endonucleases
  - the development of recombinant DNA techniques
  - the discovery of thermostable DNA polymerases.

- **PCR is a method for copying DNA sequences exponentially.**
- **PCR protocols must be optimized for precision, efficiency, and yield.**
- **PCR may be subject to inhibition and the DNA template must be pure.**
- **PCR may be subject to contamination with intrusive DNA and requires dedicated clean pre-PCR and post-PCR facilities.**
- **PCR parameters can be adjusted to tailor PCR protocols for fidelity, efficiency, and yield according to the requirements of particular applications.**
- **PCR can be applied to RNA molecules via RT-PCR.**
- **PCR is central to research and applications in biology, medicine, and forensic science.**

✓ References

Avery, O.T., MacLeod, C.M., and McCarty, M. (1944) Studies on the chemical nature of the substance inducing transformation of pneumococcal types: induction of transformation by a deoxyribonucleic acid fraction isolated from Pneumococcus type III. *J Exp Med* **79**: 137–59.

Butler, J.M. (2005) *Forensic DNA Typing: Biology Technology and Genetics of STR Markers* (2nd edn). New York: Elsevier.

Cheng, S., Chen, Y., Monforte, J.A., Higuchi, R., and Van Houten, B. (1995) Template integrity is essential for PCR amplification of 20- to 30-kb sequences from genomic DNA. *PCR Methods Appl* **4**: 294–8.

Chou, Q. (1992) Prevention of pre-PCR mis-priming and primer dimerization improves low-copy-number amplifications. *Nucleic Acids Res* **20**: 1717–23.

D'Aquila, R.T., Bechtel, L.J., Videler, J.A., Eron, J.J., Gorczyca, P., and Kaplan, J.C. (1991) Maximizing sensitivity and specificity of PCR by pre-amplification heating. *Nucleic Acids Res* **19**: 3749.

Edwards, J.B., Delort, J., and Mallet, J. (1991) Oligodeoxyribonucleotide ligation to single-stranded cDNAs: a new tool for cloning 5′ ends of mRNAs and for constructing cDNA libraries by in vitro amplification. *Nucleic Acids Res* **19**: 5227–32.

Evison, M.P., Smillie, D.M., and Chamberlain, A.T. (1997) Extraction of single-copy nuclear DNA from forensic specimens with a variety of postmortem histories. *J Forensic Sci* **42**: 1032–8.

Hajibabaei, M., Singer, G.A.C., Hebert, P.D.N., and Hickey, D.A. (2007) DNA barcoding: how it complements taxonomy, molecular phylogenetics and population genetics. *Trends Genet* **32**: 167–72.

Harper, J.C., Delhanty, J.D.A., and Handyside, A.H. (eds) (2001) *Preimplantation Genetic Diagnosis*. Chichester: John Wiley.

Hanson, E., Haas, C., Jucker, R., and Ballantyne, J. (2012) Specific and sensitive mRNA biomarkers for the identification of skin in 'touch DNA' evidence. *Forensic Sci Int Genet* **6**: 548–58.

Jeffreys, A.J., Wilson, V., and Thein, S.L. (1985) Hypervariable 'minisatellite' regions in human DNA. *Nature* **314**: 67–73.

Kayser, M. and Schneider, P.M. (2009) DNA-based prediction of human externally visible characteristics in forensics: motivations, scientific challenges, and ethical considerations. *Forensic Sci Int Genet* **3**: 154–61.

Kennedy, S. and Oswald, N. (2011) *PCR Troubleshooting and Optimization: The Essential Guide.* Caister: Caister Academic Press.

Lombardino, A.J., Hertel, M., Li, X.C., et al. (2006) Expression profiling of intermingled long range projection neurons harvested by laser capture microdissection. *J Neurosci Methods* **157**: 195–207.

Lundberg, K.S., Shoemaker, D.D., Adams, M.W.W., Short, J.M., Sorge, J.A., and Mathur, E.J. (1991) High-fidelity amplification using a thermostable DNA polymerase isolated from *Pyrococcus furiosus*. *Gene* **108**: 1–6.

Markey, A.L., Mohr, S., and Day, P.J.R. (2010) High-throughput droplet PCR. *Methods* **50**: 277–81.

Martial, J.A., Hallewell, R.A., Baxter, J.D., and Goodman, H.M. (1979) Human growth hormone: complementary DNA cloning and expression in bacteria. *Science* **205**: 602–7.

Martínez, A., Miller, J.M., Quinn, K., Unsworth, E.J., Ebina, M., and Cuttitta, F. (1995) Non-radioactive localization of nucleic acids by direct in situ PCR and in situ RT-PCR in paraffin-embedded sections. *J Histochem Cytochem* **43**: 739–47.

Saiki, R., Gelfand, D., Stoffel, S., et al. (1988) Primer-directed enzymatic amplification of DNA with a thermostable DNA polymerase. *Science* **239**: 487–91.

VanGuilder, H.D, Vrana, K.E., and Freeman, W.M. (2008) Twenty-five years of quantitative PCR for gene expression analysis. *BioTechniques* **44**: 619–26.

Vogelstein, B. and Kinzler, K.W. (1999) Digital PCR. *Proc Natl Acad Sci USA* **96**: 9236–41.

Watson, J.D. and Crick, F.H.C. (1953) Molecular structure of nucleic acids: a structure for deoxyribose nucleic acid. *Nature* **171**: 737–8.

Wheeler, E.K., Hara, C.A., Deotte, J.F.J., et al. (2011) Under-three minute PCR: probing the limits of fast amplification. *Analyst* **136**: 3707–12.

Williams, R., Peisajovich, S.G., Miller, O.J., Magdassi, S., Tawfik, D.S., and Griffiths, A.D. (2006) Amplification of complex gene libraries by emulsion PCR. *Nature Methods* **3**: 545–50.

Zhang, L., Cui, X., Schmitt, K., Hubert, R., Navidi, W., and Arnheim, N. (1992) Whole genome amplification from a single cell: Implications for genetic analysis. *Proc Natl Acad Sci USA* **89**: 5847.

3 DNA mutagenesis

Sarah E. Deacon and Michael J. McPherson

Chapter overview

In order to understand how a gene or its encoded protein functions, we usually make mutant versions by introducing changes into the DNA sequence in order to explore the functional consequences. For example, to identify amino acid residues that might be important for a protein's structure or function, we can change the DNA coding sequence to encode a different amino acid. The **mutated protein** can be produced and its structure and function examined. Changes may be **point mutations** where only a single base pair is altered, or multiple base pairs can be changed in parallel, or **insertions** or **deletions** of DNA sequences can be made. Alternatively, and of increasing importance, random mutations can be introduced into a DNA sequence to generate **libraries of variants** from which molecules with new and often improved properties can be selected. Thus **mutagenesis** of DNA is an essential tool in the modern molecular biology laboratory and allows significant insights into features such as:

- regulation of gene expression;
- the structure–function relationship of DNA, RNA, and proteins;
- molecular interactions, including those between an enzyme and its substrate or a receptor and its ligand which may be important in drug design;
- improving structural and functional properties such as stability and activity of proteins.

Selection of the correct mutagenesis method can be critical to the success of a project. In this chapter, we will describe a range of *in vitro* methods for introducing mutations into a known DNA sequence and will discuss advantages and disadvantages, allowing the reader to select the most appropriate approach to answer their biological research question.

LEARNING OBJECTIVES

This chapter will enable the reader to:

- understand the principles of mutagenesis and how they can be applied;
- gain an appreciation of the important features of various mutagenesis methods;
- select an appropriate method for introducing DNA mutations for a particular project.

3.1 Introduction

DNA can be mutated in a number of different ways, either spontaneously through replication errors in nature or through exposure to physical, chemical, or biological agents. Mutations in DNA can have significant impacts on gene expression and on the structure and function of encoded proteins and peptides. Some mutations can be lethal or may cause serious diseases, such as cancer, cystic fibrosis, or Huntington's disease. Others may have a minor effect or be phenotypically silent, either because they fall within non-essential regions of the DNA or because the mutation results in a conservative amino acid change (for example Leu to Ile) or occurs in a position that does not alter the amino acid sequence. Naturally occurring mutations also provide a mechanism by which new species evolve.

The ability to introduce mutations in the laboratory dates back to 1927 when Hermann Muller demonstrated that X-rays could cause genetic mutation in fruit flies (Muller 1927), and in 1933 his collaborator Edgar Altenburg showed that UV radiation could also cause mutation (Altenburg 1933). In the early 1940s Charlotte Auerbach and John Robson demonstrated that, in addition to physical sources, the chemical mustard gas could cause mutation (Auerbach et al. 1947). At this time, there was no control over the location or types of mutations introduced, and indeed no understanding of their effects at the molecular level, with only the phenotypic consequence being observed.

Today, as we can clone essentially any gene, we are able to rapidly and efficiently introduce site-specific mutations into any cloned DNA molecule to produce mutant genes, protein variants, or even altered strains of simple laboratory organisms. This allows us to study the effect of an amino acid change and the role of a mutation in a gene regulatory element in a disease process, and allows us the capability to engineer new molecules with novel functions.

There are two major classes of *in vitro* mutagenesis:

- **site-directed mutagenesis**, in which specific nucleotides within a DNA sequence are targeted—this is essentially a rational design strategy;
- **random mutagenesis**, in which mutations are usually introduced either through inducing errors during DNA replication or by recombination of related DNA sequences.

Mutagenesis approaches from both classes are described in the sections that follow.

3.2 Rational or site-directed mutagenesis

Key tools allowing us to introduce site-directed mutations into DNA are chemically synthesized **oligonucleotides** that are used as DNA synthesis **primers**. These are cheap and readily available from a number of commercial companies. The critical issue when ordering such oligonculeotides is to ensure that you have defined the 5′ and 3′ ends of your sequences so that you order the correct sequences. The principal steps in a site-directed mutagenesis experiment are outlined in Figure 3.1.

Figure 3.1 Outline of the steps in a basic site-directed mutagenesis experiment.

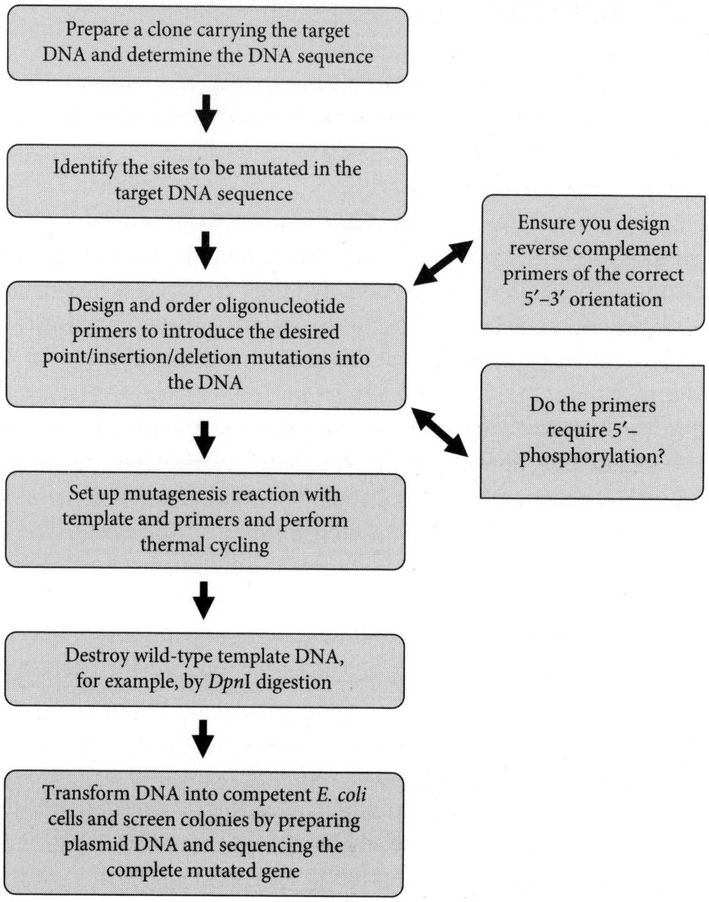

The original method for site-directed mutagenesis of a DNA sequence, developed by Mark Zoller and Michael Smith, required a single-stranded DNA **template**, isolated as the genome of **bacteriophage M13** (Zoller and Smith 1982), or subsequently M13-derived phagemids containing the cloned gene. A single 5′-phosphorylated primer containing the mutation(s) was used to generate the complementary strand from this single-strand template and **DNA ligase** was added to create closed circular DNA. The resulting **heteroduplex** circular double-stranded DNA molecule comprising the original **wild-type** template strand and the newly synthesized mutant strand was transformed into *E. coli* cells. However, the vast majority of **transformants** contained the original wild-type gene. This is because the wild-type DNA was methylated during its production in the original bacterial cells. In contrast, the *in vitro* synthesized mutant strand was not methylated, and hence the wild-type DNA was preferentially recognized by the new host cells and used as the template strand for correction of the mismatch between the wild-type and mutated strands. Mutants occurred at very low frequency, typically around 1 in 1000 or lower, and so had to be detected by a plaque or colony hybridization experiment using radiolabelled mutant oligonucleotide primer. Therefore these early experiments were difficult and time consuming.

As the sole purpose of wild-type DNA is to act as a template for the synthesis of the complementary primer-directed mutant DNA strand, destruction of the wild-type DNA once the mutant DNA has been synthesized should enhance the recovery of clones carrying the mutant DNA strand. Thus the underlying principle of **high-efficiency mutagenesis** approaches is to destroy or dramatically dilute the wild-type DNA.

Several such approaches have been developed, including the incorporation of uracil into the wild-type DNA template (Kunkel 1985) (see section 3.3) and restriction selection involving selective restriction enzyme cleavage of methylated DNA (Carter et al. 1985) (see section 3.4), with some significant improvements over the last 25 years. A number of PCR-based methods which result in the production of a large excess (>10^5 copies) of mutated DNA over the original wild-type (see section 3.7) are also used. Other approaches, such as cassette mutagenesis (see section 3.6), rely on replacing a section of wild-type DNA with the mutated DNA. It quickly became clear that single-stranded DNA isolated from phage such as M13, which involved additional cloning steps, was not necessary. The single-stranded template can simply be generated by heat denaturing essentially any double-stranded plasmid DNA, and this has simplified the process of carrying out mutagenesis reactions.

3.3 Uracil-containing DNA

An early approach to select against the wild-type template strand was developed by Thomas Kunkel and involved the incorporation of uracil bases into the DNA template (Kunkel 1985; Kunkel et al. 1991). The DNA to be mutated is purified from a *dut*$^-$, *ung*$^-$ *E. coli* strain which introduces dUTP into DNA at some positions where dTTP should be incorporated. In standard *E. coli* strains the mis-incorporated dUMP is removed by the enzyme uracil-*N*-glycosylase encoded by the *ung* gene. However, as this enzyme is knocked out in the *ung*$^-$ strain, the vector DNA contains several dUMPs. After mutagenic primer-directed DNA synthesis the new mutant strand does not contain dUMP. When the resulting heteroduplex DNA is transformed into a normal *ung*$^+$ strain of *E. coli* the original wild-type dUMP-containing template DNA is repaired by the uracil-*N*-glycosylase using the mutated strand as the template, and thus leads to efficient selection of mutational variants.

Whilst this provides one approach for selection against the wild-type template, another method called QuikChange® mutagenesis, marketed by the company Stratagene (now part of Agilent Technologies), is now predominantly used. This is based on a restriction–selection approach, involving the use of a restriction enzyme to selectively digest *in vivo* produced plasmid DNA but not *in vitro* produced mutant DNA.

3.4 QuikChange® mutagenesis

QuikChange® mutagenesis can be used to introduce single- or multiple-point mutations, or to delete or insert sequences cloned within a vector. The strategy requires two oligonucleotide primers that are complementary to the two strands of the target region of the gene, and which are therefore complementary to each other (Papworth et al. 1996).

They are designed to contain the required nucleotide changes to be introduced into the gene, and these should be located centrally. The primers should ideally be between 25 and 45 bases in length, have a melting temperature ≥78°C, and have a G or C at the 3′ end, and are not 5′-phosphorylated. Figure 3.2 shows an example of primer design for a QuikChange® mutagenesis reaction. A useful site for checking the melting temperature of primers and for generating the sequence of the reverse primer when you provide the forward primer is http://depts.washington.edu/bakerpg/primertemp/primertemp.html.

The primers are used to amplify the remainder of the vector strands, resulting in overlapping single-stranded DNA regions corresponding to the primer sequences, which can anneal to create a nicked plasmid A schematic diagram of the QuickChange® approach is shown in Figure 3.3. The QuikChange® reaction is carried out using a **proof-reading DNA polymerase**, such as *Pfu* Turbo, in a PCR machine. The reaction steps are controlled by heating to denature the plasmid template (95°C for 5min, then at each cycle 95°C for 30sec), cooling to anneal the primers to the template (55°C for 30sec), and then performing the DNA synthesis reaction (65°C typically for 1min per kilobase of template DNA being copied). This cycling process is repeated around 20–30 times. However, unlike a PCR, the primers can only productively anneal to the original template DNA and not to the newly synthesized mutated DNA, and hence QuikChange® is a **linear amplification** reaction. This means that each cycle results in the creation of one new DNA strand per original template strand, so after, say, 20 cycles the number of mutated

Figure 3.2 Primer design for a QuikChange® mutagenesis experiment. (a) Section of a wild-type protein coding region. (b) The design of complementary forward and reverse oligonucleotide primers designed to alter a tryptophan (Trp, W) codon (TGG) to an alanine (Ala, A) codon (GCC). The codon to be changed is shown in bold. (c) Sequence of the mutated DNA after the primer extension. The primers overlap, allowing circularization of the plasmid, but with phosphodiester backbone gaps indicated by the arrow heads. These gaps are repaired by the bacterial host following transformation.

(a) **Wild type DNA sequence**

N A K P L S W F C P S N N
■ ■ ■ ■ AATGGCAAGCCCTTGTCTTGGTTCTGCCCGTCTAACAAT ■ ■ ■ ■
■ ■ ■ ■ TTACCGTTCGGGAACAGAACCAAGACGGGCAGATTGTTA ■ ■ ■ ■

(b) **Forward primer**

Trp to Ala mutation

5′-GCAAGCCCTTGTCT**GCC**TTCTGCCCGTCTAAC-3′

N A K P L S **W** F C P S N N
■ ■ ■ ■ AATGGCAAGCCCTTGTCT**TGG**TTCTGCCCGTCTAACAAT ■ ■ ■ ■
■ ■ ■ ■ TTACCGTTCGGGAACAGA**ACC**AAGACGGGCAGATTGTTA ■ ■ ■ ■

3′-CGTTCGGGAACAGA**CGG**AAGACGGGCAGATTG-5′

Reverse primer

(c) **Mutated DNA sequence**

N A K P L S **A** F C P S N N
■ ■ ■ ■ AATG▼GCAAGCCCTTGTCT**GCC**TTCTGCCCGTCTAACAAT ■ ■ ■ ■
■ ■ ■ ■ TTACCGTTCGGGAACAGA**CGG**AAGACGGGCAGATTG▲TTA ■ ■ ■ ■

Figure 3.3 Schematic diagram of the QuikChange® muatgenesis process. Following denaturation of the plasmid DNA, the primers anneal to their complementary sequences and act as primers for DNA synthesis; mutations are indicated by X. In subsequent cycles the resulting strands denature from the template and new primers anneal; up to 25 cycles of this denaturation, annealing, and DNA synthesis are normally performed. The newly synthesized mutated complementary DNA strands, which are not methylated, anneal together and form circular plasmid molecules by annealing the primer sequences. The restriction enzyme *Dpn*I, which only digests DNA at sites that are methylated, will cleave the methylated wild-type parental template DNA but will not digest the mutated DNA, which is then transformed into *E. coli* cells.

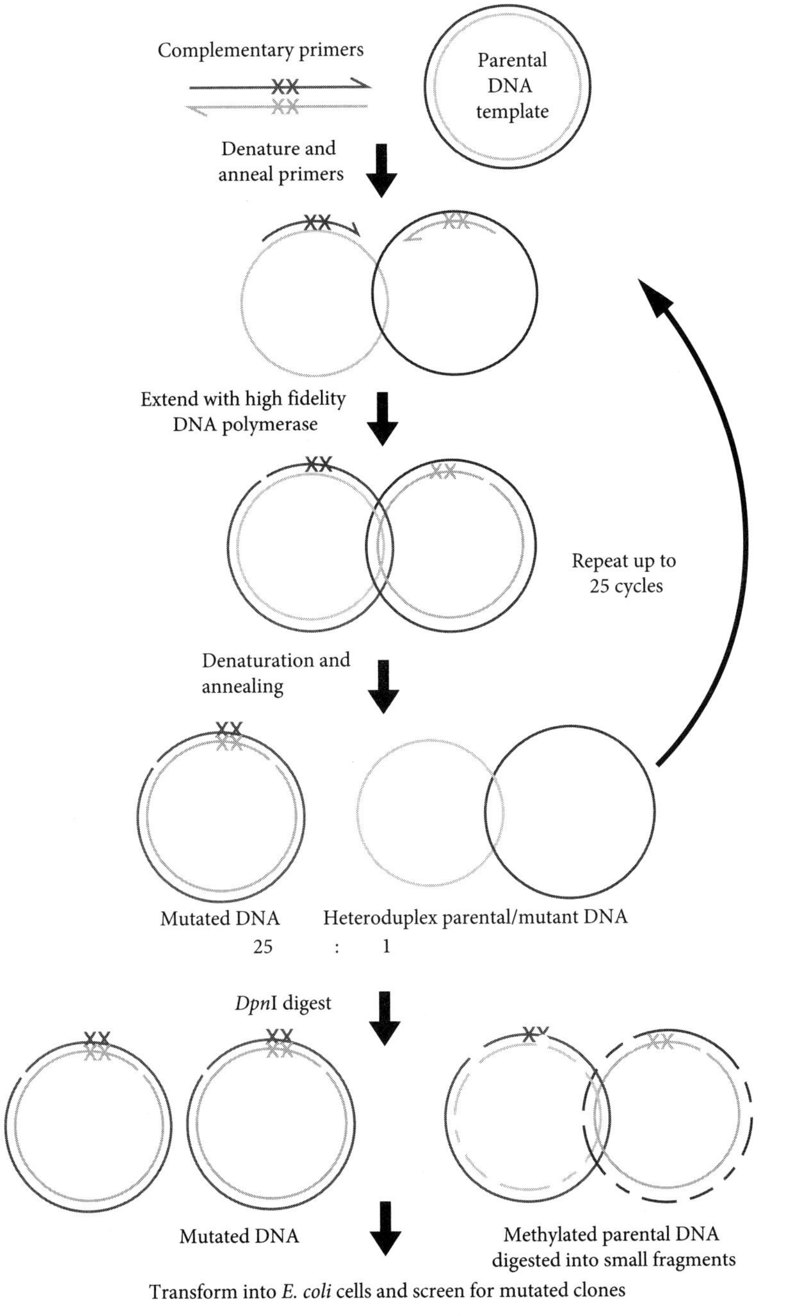

strands will be 20 times the number of the original wild-type strands. By comparison, an exponential PCR would have generated 10^5–10^6-fold more product than template.

After the reaction is complete, parental template DNA, but not the newly synthesized mutated DNA, is destroyed by digestion with the methylation-dependent restriction enzyme ***DpnI*** which targets the sequence 5′-Gm^{6}ATC-3′ (i.e. the GATC sequence where the N6 adenine residue is methylated). An aliquot of the resulting product, comprising mainly mutated DNA, is transformed into *E. coli* cells. Typically four colonies are picked from the transformation plate, cultured, and plasmid DNA prepared for DNA sequencing. In the vast majority of experiments one or more mutant plasmids will be identified in this way. A sample of the mutagenesis reaction can also be removed before and after *Dpn*I digestion for analysis by agarose gel electrophoresis. Since the majority of the DNA in the sample should be newly synthesized DNA that is resistant to *Dpn*I digestion, the intensity of the *Dpn*I digested and undigested bands on the gel should be essentially identical. If the digested DNA band is significantly less intense than the undigested sample, it is unlikely that the mutagenesis reaction has worked. As an example, QuikChange® mutagenesis was used to investigate substrate binding residues and to alter substrate specificity in the enzyme galactose oxidase (Deacon et al. 2004).

It is possible to purchase kits for QuikChange® mutagenesis from Agilent Technologies, or to use other reagents, such as a different DNA polymerase and an alternative source of *Dpn*I, to perform your own reactions. An improved kit, the QuikChange® Lightning Kit, contains a DNA polymerase blend and improved *Dpn*I that are reported to increase the speed of the DNA extension reaction to 30sec per kilobase of DNA template. The QuikChange® approach is a form of ligation-independent mutagenesis and a range of variations have been described; one example is site-directed ligase-independent mutagenesis (Chiu et al. 2004).

3.4.1 Single-primer reactions in parallel (SPRINP)

One issue with QuikChange® mutagenesis is that tandem repeats of the primers can sometimes be inserted. This occurs through the formation of 'primer dimers' as a result of partial annealing of the forward and reverse primers in the reaction. One approach to overcome this is to use the SPRINP method (Edelheit et al. 2009). In this technique QuikChange® primers are used but two reactions are set up in parallel, one with the forward primer only and one with the reverse primer only. On completion of DNA synthesis, the two reactions are mixed, heated to 95°C, and allowed to cool slowly so that the newly synthesized strands anneal, with the overlapping primer sequences allowing circularization of the DNA. The parental DNA template is then destroyed with *Dpn*I prior to transformation into competent *E. coli* cells.

3.4.2 Insertions

Insertion of short stretches of DNA within another DNA sequence can be achieved using a modified version of QuikChange® mutagenesis. The primers are designed such that the DNA to be inserted is added to the 5′-end of both the forward and reverse primers, creating overlapping complementary sequences, the 3′ segment (c. 15–20nt) of the primer sequence is complementary to the vector adjacent to the site of insertion (see Figure 3.4).

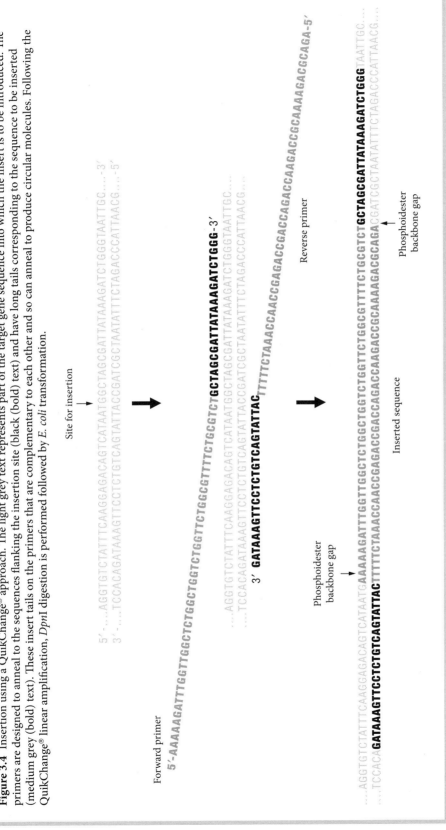

Figure 3.4 Insertion using a QuikChange® approach. The light grey text represents part of the target gene sequence into which the insert is to be introduced. The primers are designed to anneal to the sequences flanking the insertion site (black (bold) text) and have long tails corresponding to the sequence to be inserted (medium grey (bold) text). These insert tails on the primers that are complementary to each other and so can anneal to produce circular molecules. Following the QuikChange® linear amplification, *Dpn*I digestion is performed followed by *E. coli* transformation.

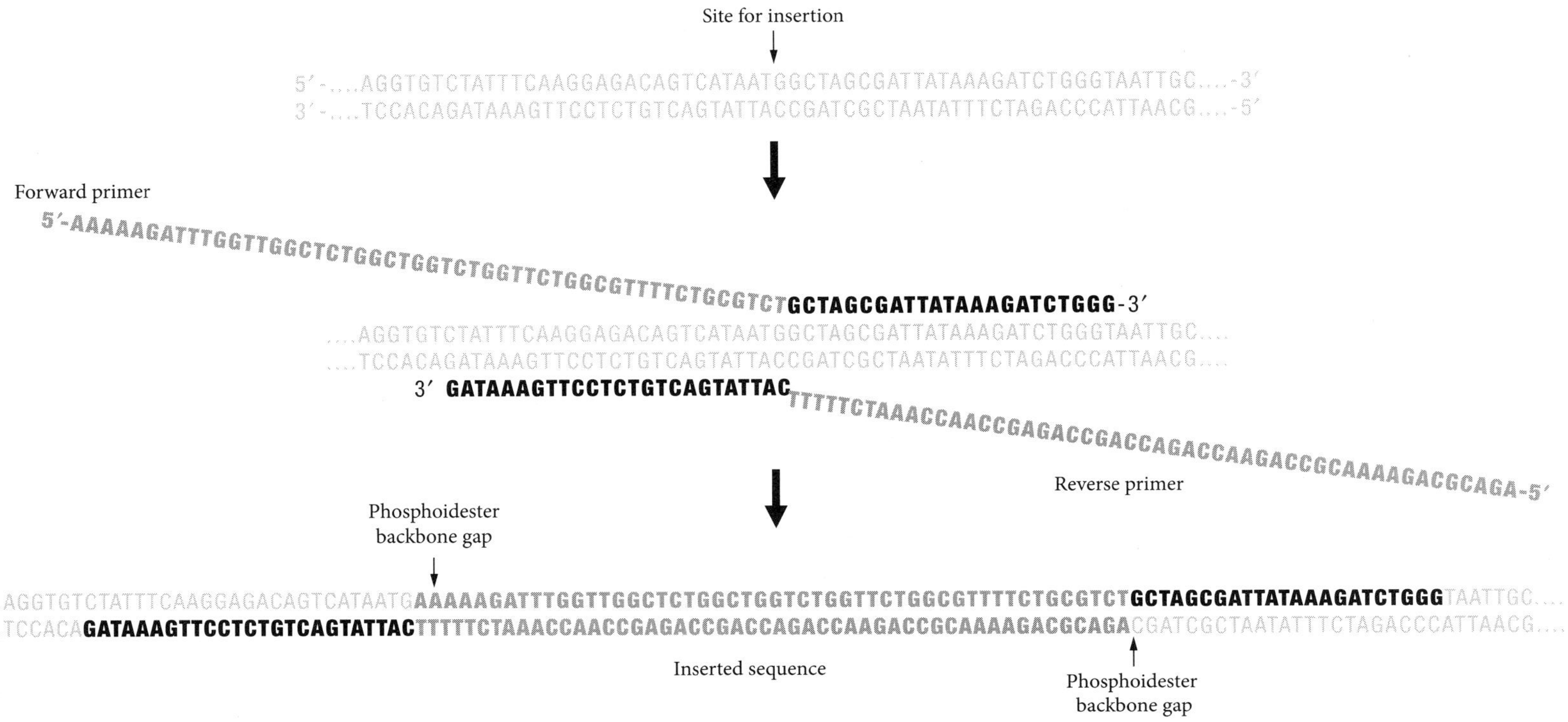

After several rounds of thermal cycling the newly synthesized strands anneal via the overlapping insertion sequences. The products are *Dpn*I digested and transformed into competent *E. coli* before screening by DNA sequencing.

The size of the insert is limited by the length of the synthetic oligonucleotide sequences which cannot be greater than ~100nt. This means that the maximum insertion size is around 120nt, with c. 60nt from each primer. For longer insertions the sticky-feet approach (see section 3.7.2) can be used to amplify a region of essentially any length, from one DNA molecule, with 5′-extensions that are complementary to the insertion sites on the new template.

3.4.3 Deletions

The simplest method of deleting a section of DNA is to use a modified QuikChange® approach. Two primers are designed to anneal to sequences flanking the region to be deleted. The 3′ region of the primer anneals to one deletion junction, while the 5′ region anneals to the other deletion junction. The remainder of the gene and vector are therefore replicated without the section of DNA to be deleted (Figure 3.5). Once the amplified products anneal, the parental template DNA is destroyed by *Dpn*I digestion and, following transformation into competent *E. coli* cells, clones can be isolated and sequenced.

3.5 Multi-site mutagenesis

On occasion it may be desirable to simultaneously introduce multiple mutations within a gene. If the mutations lie within about 20–30nt of each other, they can be introduced by using one pair of QuikChange® primers. However, if they are further apart, various modified QuikChange® approaches may be used.

First, if two mutation sites are required, it is possible to use the SPRINP method (see section 3.4.1) with the forward primer containing one mutation in one reaction and the reverse primer containing the other mutation in a separate reaction. Once the mutated DNA is synthesized, the reactions are combined and the methylated template DNA is digested using *Dpn*I before transformation into *E. coli*.

The second method involves the use of a pair of QuikChange® primers which have been 5′-phosphorylated. The reaction products are treated with DNA ligase to create a closed circular DNA and with *Dpn*I to remove the original template DNA. The *Dpn*I is heat inactivated and the DNA is treated with *dam* methylase in the presence of *S*-adenosylmethionine, which results in DNA methylation. Following a clean-up step, the methylated mutated DNA can act as a template for the next set of mutagenic primers and the whole procedure can be repeated.

The third method uses the Stratagene QuikChange® Multi Site-Directed kit (or Lightning Multi Site-Directed kit), in which a primer is designed for each mutation and added to a single reaction simultaneously. The primers all anneal to the same strand of denatured template DNA. A proof-reading high-fidelity DNA polymerase is used to extend the mutagenic primers, resulting in a double-stranded DNA containing multiple mutations and phosphodiester backbone breaks adjacent to each primer. The primers are 5′-phosphorylated so that these breaks can be sealed by DNA ligase before treatment

Figure 3.5 Introducing a deletion using a QuikChange® approach. The target gene is shown in light grey and the region to be deleted (medium (bold) grey) is boxed. The regions to be used in the production of the primers are shown in black (bold) text. These regions flank the deletion site, but in the primers these sequences are contiguous. The primers are annealed to the template strands and following QuikChange® amplification and *Dpn*I digestion the DNA is transformed into *E. coli* cells.

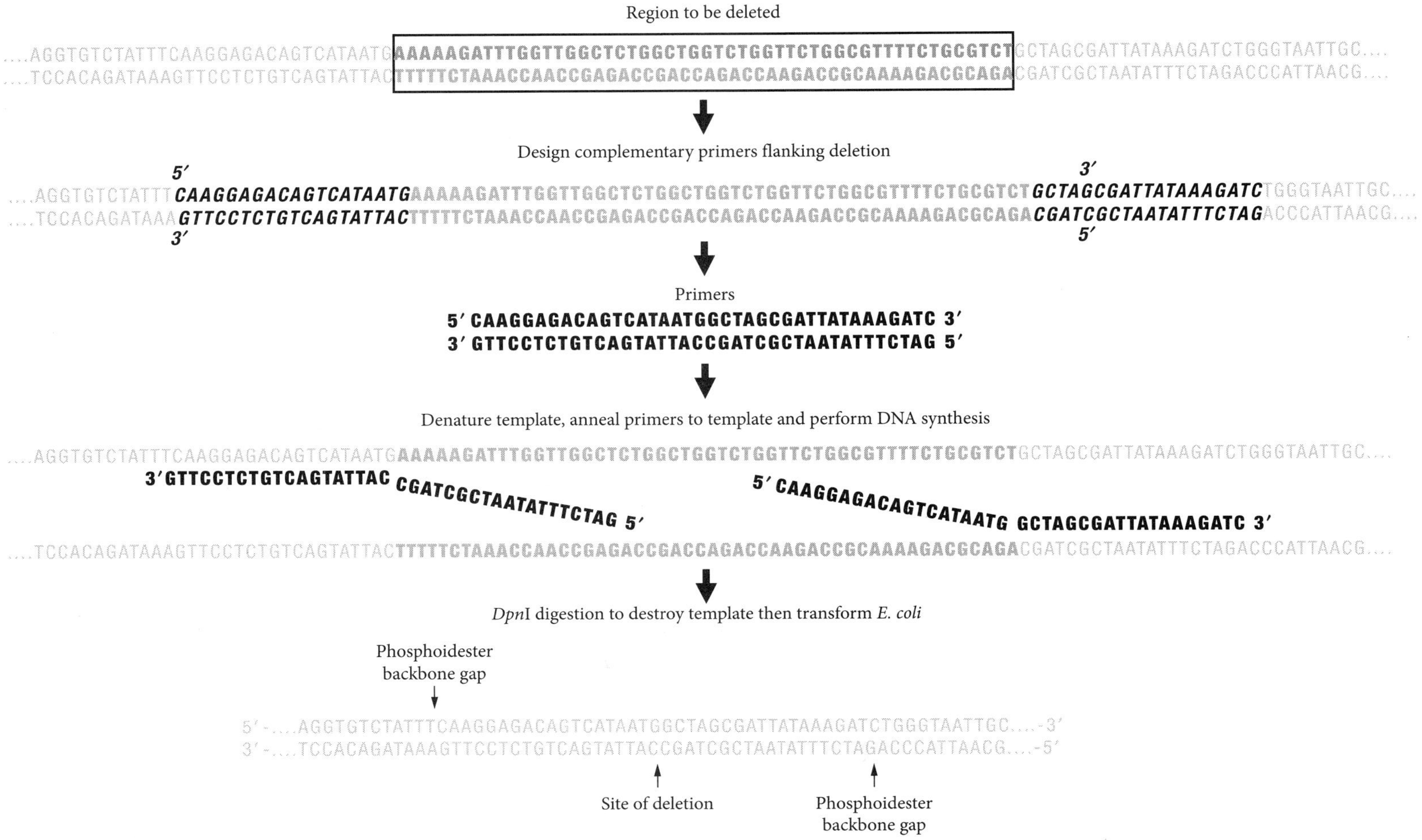

with *Dpn*I to digest the DNA template. The resulting product is single-stranded multiply mutated DNA which is transformed into *E. coli* competent cells where it is converted into double-stranded DNA and replicated. Preparing plasmid DNA from a number of colonies and screening by DNA sequencing should confirm the presence of the desired mutations. It is reported that the QuikChange® Multi Site-Directed kit has an efficiency of 55% for generating three mutations, and can be used to generate five mutations simultaneously. An outline of this procedure is shown in Figure 3.6.

See Chapter 2

The fourth approach involves the design of two PCR primers to incorporate the desired mutations. The primers are used in a PCR reaction to create a DNA fragment containing the mutations. The product is then purified by gel extraction and used in a QuikChange®-type reaction with the original template. Following denaturation and annealing, the product acts as a primer for synthesis of the remaining plasmid. The products of the reaction can anneal, forming double-stranded molecules with single-stranded DNA (ss-DNA) tails, corresponding to the primer regions, which can also anneal to circularize the plasmid. The reaction is treated with *Dpn*I to remove any parental DNA template. An example of this type of approach is given by Tian et al. (2010).

3.6 Cassette mutagenesis

If you are interested in studying a particular region of a gene in detail and wish to introduce a wide variety of different combinations of mutations in a series of different experiments, a **cassette mutagenesis** approach may be appropriate (Wells et al. 1985; Reidhaar-Olson and Sauer 1988). A synthetic DNA fragment is generated from complementary oligonucleotides containing the desired mutated sequence. The two oligonucleotides are heated to 95°C and then allowed to cool slowly so that they anneal to generate a double-stranded DNA cassette. The complimentary oligonucleotides are designed so that when they anneal they generate overhangs that are complementary to restriction sites in the selected vector. This cassette is used to replace the corresponding wild-type region of the gene. The vector is digested with appropriate restriction enzymes and ligated with the cassette, followed by transformation into *E. coli* cells. Transformants are then screened for the presence of the mutations by DNA sequencing.

A limitation of this method is that the cassette must be flanked by unique restriction enzyme sites to allow directional cloning of the cassette into the gene backbone. This can require initial mutagenesis of the cloned gene to create unique restriction sites by using an approach such as QuikChange® (see section 3.4) or a PCR approach (see section 3.7). To improve ligation efficiency, the oligonucleotide primers can be 5′-phosphorylated.

Two examples of a cassette mutagenesis approach are shown in Figure 3.7. The first shows insertion of a hexa-histidine tag between *Nde*I and *Nco*I restriction sites, and the second shows the mutagenesis of two specific amino acid codons. The size of cassette that can be generated is limited by the length of oligonucleotides that can be chemically synthesized; hence it cannot exceed around 100nt and in most cases will be significantly shorter. The efficiency of this mutagenesis approach stems from the fact that the wild-type sequence is removed from the vector before ligation of the vector backbone with the mutated DNA fragments, and so the transformants should be almost exclusively mutants.

Figure 3.6 Multi-site QuikChange® mutagenesis. In this case mutagenic oligonucleotides, with X representing the mutations, are 5′-phospohorylated and annealed to only one strand of the template plasmid. Following DNA synthesis, a DNA ligase repairs the phosphodiester backbone gaps to generate a closed circular single-strand molecule. Following several cycles of primer-directed DNA synthesis and ligation, *Dpn*I digestion destroys the parental methylated DNA. The mutant circular DNA is used to transform *E. coli* cells.

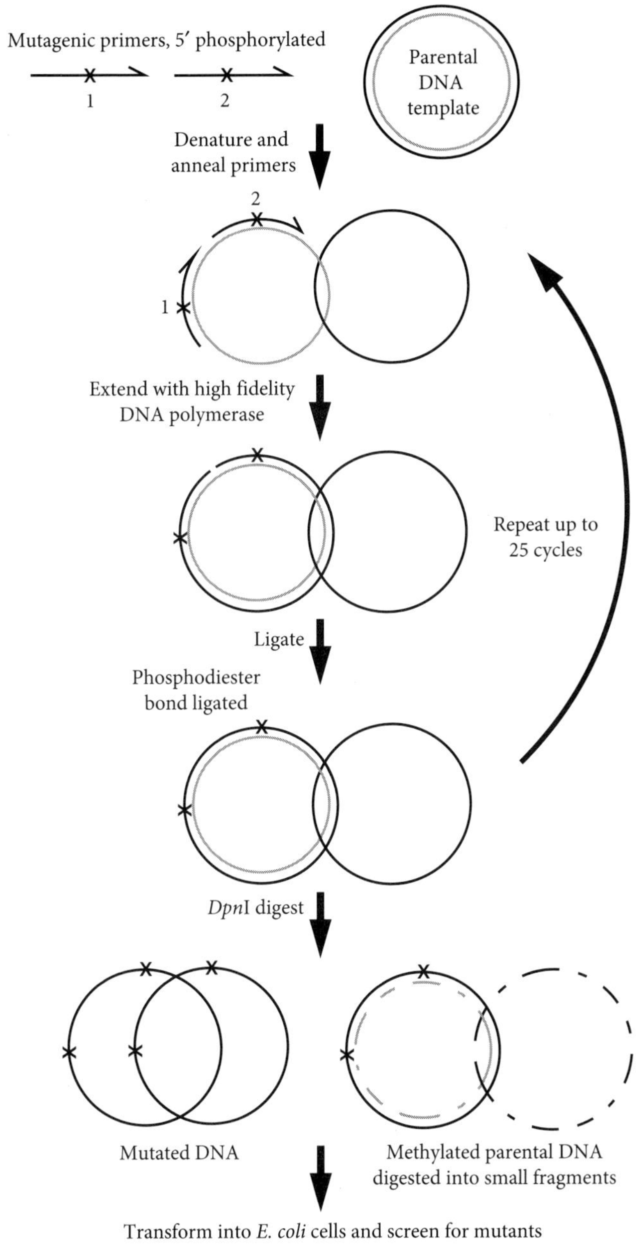

Figure 3.7 Cassette mutagenesis. (a) Two complementary oligonucleotides are designed to insert a six-histidine affinity purification tag. They are heated and allowed to cool to anneal with one another to generate the synthetic cassette (grey) with overhanging single-strand ends that are complementary to the restriction site ends of the sequence within the vector (black). The cassette is used to replace the corresponding wild-type region by ligating into the vector at the restriction site complementary ends (boxed) before transformation into *E. coli* cells. (b) The wild-type sequence is being changed at two positions: the Tyr codon at position 3 to a Phe codon, and the Ser codon at position 8 to a Cys codon. The positions of the changes are shown in italic. Two oligonucleotides are designed to generate a cassette to introduce the changes (Phe = TTC; Cys = TGC shown in italic), again with overhangs complementary to the vector restriction site ends for ligation to the vector.

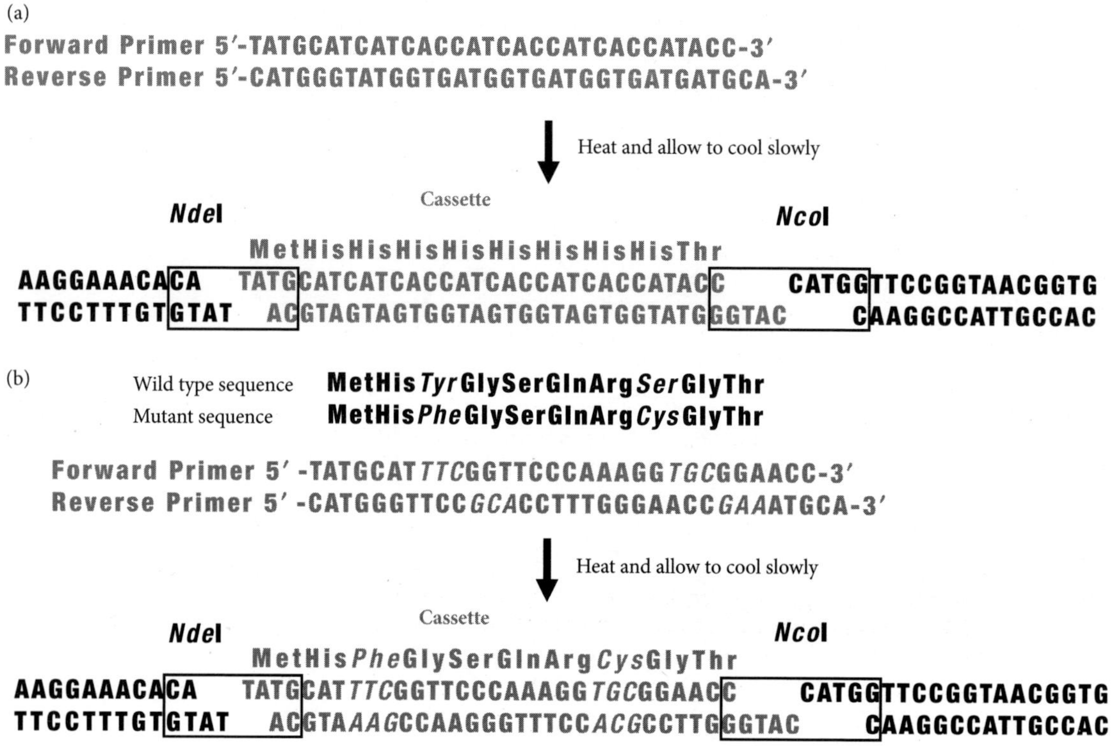

3.7 PCR mutagenesis

See Chapter 2

PCR is a powerful tool for manipulating DNA sequences and has a wide range of applications in modern biology (McPherson and Møller 2006). PCR can be used to introduce nucleotide changes into a DNA sequence by incorporating the changes within a PCR primer such that it anneals with the template except at the sites of mutation. It is important that the 3′-end of a PCR primer matches the template perfectly, and so mismatches should be located as centrally as possible. It is also possible to insert a DNA sequence using a PCR approach, for example to incorporate the coding sequence for a protein purification tag. Deletion of sequences is also straightforward by PCR. Whether point mutations, insertions, or deletions, the resulting PCR fragment and the vector are digested with appropriate restriction enzymes and are ligated by a DNA ligase before transformation into *E. coli* cells.

In PCR mutagenesis, the wild-type template represents a miniscule amount of the DNA present in the final reaction and so it does not normally affect the recovery of mutant clones. However, where a direct transformation step is to be performed, it is sensible to perform a *Dpn*I digest to destroy the template DNA. There are a variety of approaches for PCR mutagenesis of DNA (McPherson and Møller 2006).

3.7.1 Megaprimer mutagenesis

The PCR megaprimer approach involves the design of a reverse primer which contains the mutagenic sequence (Sarkar and Sommer 1990; Tyagi et al. 2004). In a first PCR the reverse mutagenic primer is used together with a forward primer to amplify a region of DNA, resulting in incorporation of the mutation within the 3′ end (Figure 3.8). This PCR product is then used in a second PCR reaction as a 'megaprimer' with a downstream reverse primer to produce a larger DNA fragment now containing the

Figure 3.8 Megaprimer PCR approach to mutagenesis. (a) A region of the target gene is PCR amplified by a forward primer and a mutagenic reverse primer that carries two mutations (XX) in this case. The product (b) is then purified and used in a second PCR (c) as a megaprimer along with a reverse primer. The resulting fragment (d) now has the mutations more centrally located and will normally contain restriction sites (RE1/RE2) that allow the product to be digested and used as a cassette to replace the corresponding region of the wild-type gene by ligation and then transformation.

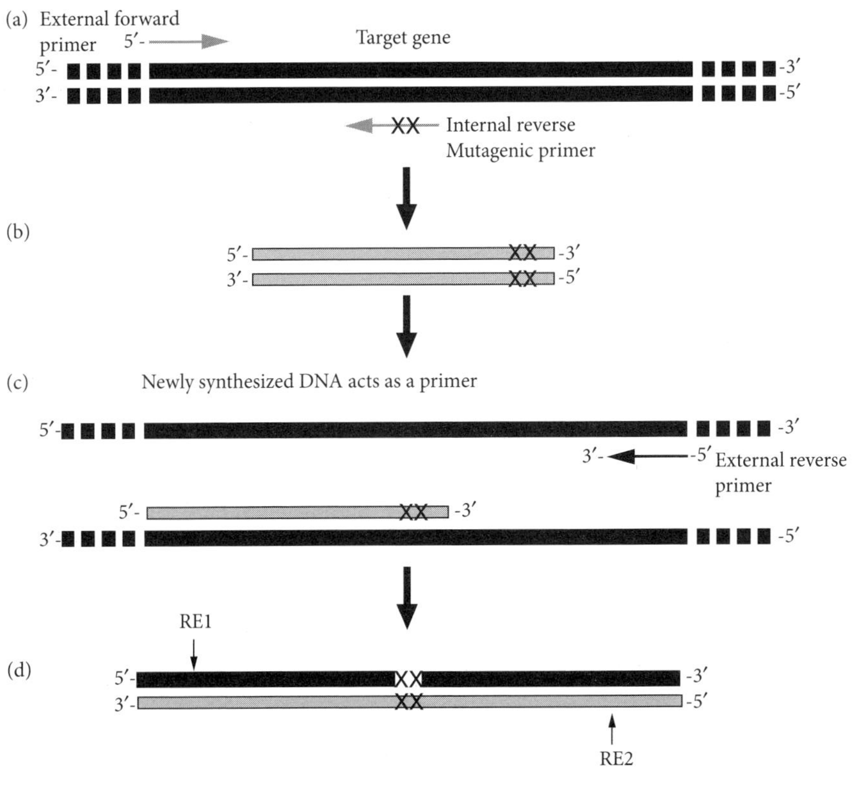

mutation(s) centrally. This fragment is then cloned into the desired suitable vector by exploiting restriction sites flanking the mutation site or incorporated into the flanking forward and reverse primers. The fragment could also be cloned using the sticky-feet mutagenesis approach outlined in section 3.7.2. These approaches represent an alternative approach to cassette mutagenesis (see section 3.6) for the incorporation of longer mutated DNA sequences.

3.7.2 Sticky-feet mutagenesis

'Sticky-feet' mutagenesis (Clackson and Winter 1989) can be used to replace sections of DNA sequence without the need for restriction sites. A region of DNA is PCR amplified using two oligonucleotide primers that each have 5′-extensions complementary to the sequences at which the insertion will be made. Once the DNA fragment has been amplified, these 5′-extensions act as 'sticky feet' to anneal the fragment to the target gene and as primers to synthesize the remainder of the gene and vector strand. The original method used a single-stranded M13 genome as the template to synthesize a second strand via the uracil incorporation approach (see section 3.3) to improve efficiency. A more generally useful approach is to use a *Dpn*I digestion strategy to destroy the original template DNA prior to *E. coli* transformation (Figure 3.9).

3.8 Saturation mutagenesis

Saturation mutagenesis involves changing one amino acid codon to the codons for all or a subset of the other 19 amino acids. This can be achieved by using most of the mutagenesis approaches described in this chapter. Traditionally, the oligonucleotide primers for saturation mutagenesis experiments have been based on the degeneracy of the codon table. Thus a codon designation NNN describes a mixture of sequences which can have A,G,C, or T at any of the three positions of the codon. Therefore this mixture would encode all 64 codons, but would result in mutational bias because some amino acids are encoded by different numbers of codons (between one and six). For example, methionine has only one codon (ATG), while serine has six (TCT, TCA, TCG, TCC, AGT, and AGC); therefore you would be six times more likely to introduce a serine codon than a methionine codon. Equally, there are three termination codons and so many products would result in truncated protein. This bias can be reduced to a modest extent by using the degenerate sequence NN(G/C) which encodes 32 amino acids with only one stop codon (TAG). The complexity of the degenerate oligonucleotide mixture and the different distribution of codons for different amino acids means that many clones (~100 for a single codon change) would need to be screened to be 95% confident of representing all possible amino acids in the saturation library.

Although the initial cost of oligonucleotides is higher, an efficient approach to generating a set of saturation substitutions of an amino acid is to generate a family of 19 QuikChange® primers, each encoding a different amino acid. Although this requires a set of parallel mutagenesis reactions and transformations, it increases the chance of isolating all possible mutational variants and reduces the downstream screening required.

Figure 3.9 Sticky-feet mutagenesis. This approach is commonly used to insert a sequence from one gene into another gene to create a fusion. The approach does not rely upon restriction and ligation. A region of the source gene (black) is PCR amplified using primers which contain 5′ tails corresponding to the sequences flanking the insertion site in the target gene (grey). These tails act as 'sticky feet' to anchor the amplified DNA to the strands of the source gene where they act as primers for DNA synthesis to generate mutated strands. Following *Dpn*I digestion, the mutants are recovered after *E. coli* transformation.

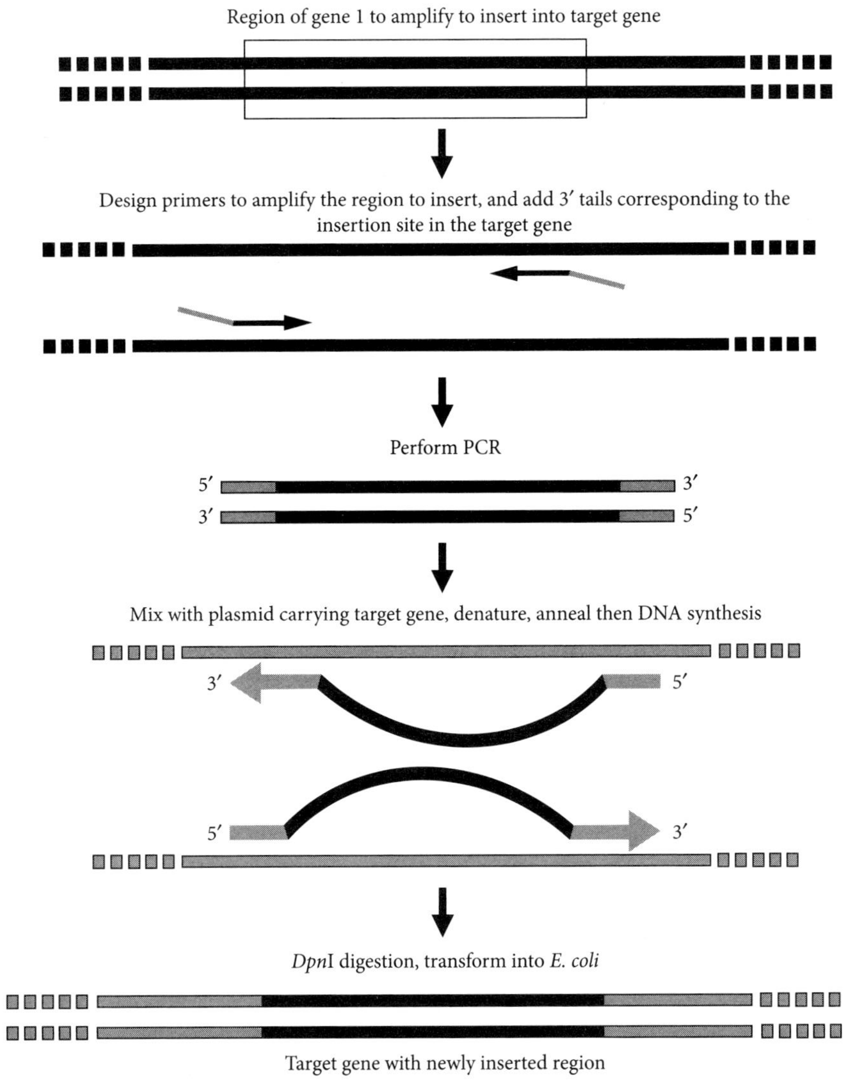

An alternative is to use primers synthesized using trinucleotides at the site of the mutations. These are available from some specialist suppliers (e.g. Ella Biotech). Their use reduces the mutational bias in the library by using a single codon per amino acid, allows a single mutagenesis reaction to be performed, and reduces the number of clones that need to be screened compared with a degenerate codon strategy.

Saturation mutagenesis is often used to explore the function of a specific amino acid or to discover new functional proteins by analysing a range of combinations of saturation mutants at different sites in a protein (Reetz et al. 2010). It is also often used in combination with random mutagenesis approaches (see section 3.9) to determine whether a particular mutational variant contains the best possible amino acid substitution. For example, saturation mutagenesis has been used to investigate an active site residue in the enzyme galactose oxidase (Deacon and McPherson 2011).

3.9 Random mutagenesis

Site-directed mutagenesis has increased our understanding of how particular DNA changes affect the function of a particular gene or encoded protein. However, despite excellent structural and functional data, we often do not fully understand why a particular mutation results in a particular phenotype. We are also poor at predicting which mutations or combination of mutations is likely to alter the function of a protein in a predictable manner. Therefore, in order to identify mutations that result in improvement of function of a protein, we often resort to introducing random mutations into the DNA sequence. This involves generating libraries of mutated DNA molecules that are screened for the desired characteristics, such as encoding higher catalytic activity or altered substrate specificity in the case of an enzyme. The process of generating such random mutation libraries and iteratively screening to identify improved variants is commonly known as directed evolution, and is often applied to enzyme systems (Jaeckel and Hilvert 2010). Approaches that are used for random mutagenesis, such as those described below, are usually distinct from those described for site-directed mutagenesis.

3.9.1 Error-prone PCR

See Chapter 2

Normally in a PCR we are trying to ensure that DNA is copied with high fidelity so that errors are not introduced. In contrast, in error-prone PCR, we attempt to achieve the opposite—the deliberate introduction of mutations during DNA replication. This can be achieved by altering the PCR reaction conditions, such as increasing the $MgCl_2$ concentration, adding $MnCl_2$, or using unequal ratios of nucleotides. Such conditions can result in the mis-incorporation of nucleotides during DNA synthesis by *Taq* DNA polymerase. An alternative and more controllable approach uses DNA polymerases that have been engineered for increased error rates during DNA replication. However, many of these have a bias favouring mutation of particular nucleotides. The GeneMorph® II Random Mutagenesis kit (Agilent Technologies) contains an enzyme mix of Mutazyme I and *Taq* DNA polymerase designed to reduce this mutational bias. Using this method it is possible to adjust the mutation frequency (the number of mutations per kilobase) by altering (a) the levels of DNA added into the reaction and/or (b) the number of amplification cycles performed.

3.9.2 Mutator strains

An apparently simple approach to the random incorporation of mutations is to introduce a plasmid carrying the target gene into a mutator strain of *E. coli*. An example is XL1

Red from Agilent Technologies, which is deficient in three DNA repair pathways: the error-prone mismatch repair pathway (*mutS*), the 3-′ to 5′-exonuclease activity of DNA polymerase III (*mutD*), and the ability to hydrolyse 8-oxo dGTP (*mutT*). As a consequence, the mutation rate in this strain is 5000 times higher than in the wild-type strain from which it is derived. The advantages of using this approach are that it requires little genetic manipulation and it is not necessary to have a selectable or screenable phenotype. However, there are a number of disadvantages: the cells have a slow doubling time and the introduction of mutations is not controlled. In addition, other genes within the cell, such as the antibiotic resistance gene used for plasmid selection or genes that affect other growth parameters, are often mutated. Therefore when clones are being characterized it is necessary to reclone the mutated gene into a clean vector backbone and to transform fresh non-mutator *E. coli* cells to determine whether the phenotype observed is actually associated with any mutations in the target genes, rather than in the original vector or cells.

3.9.3 Recombination strategies

After identifying clones from random mutagenesis variant libraries you may wish to explore the consequences of combining such mutations for enhancing protein function. Alternatively, you may have a number of naturally occurring homologous genes that encode the same protein but from different organisms. It can be useful to recombine the naturally occurring mutations within these genes to generate improved proteins. Various approaches have been used to recombine such genetic variants. Stemmer (1994) described a **DNA shuffling** method which has been widely applied to a range of systems. This involves mixing DNA fragments from a number of mutational variants and subjecting them to DNase I digestion or physical fragmentation by sonication. Short DNA fragments (50–100bp) are then isolated by gel extraction and combined in a PCR-type reaction, but with no flanking primers. DNA synthesis occurs through self-priming of the annealed fragments which show sequence homology, but which can be derived from different mutational variants or different sources of naturally occurring genes. If a number of cycles of this self-replication are undertaken, the various mutations become recombined, or shuffled, as full-length copies of the genes are generated. The addition of flanking primers then allows the amplification of full-length copies of re-shuffled genes which can be cloned into a suitable vector for transformation and screening for the desired function (Figure 3.10). An example of a landmark paper describing a gene family shuffling experiment involves the selection of improved subtilisin-like serine proteases (Ness et al. 1999). A number of other approaches to recombining mutations such as StEP (Zhao et al. 1998) and RACHITT (Coco et al. 2001) have also been developed.

3.10 Chapter summary

- It is possible to introduce nucleotide changes, insertions, and deletions into essentially any cloned gene.
- Using saturation mutagenesis all possible amino acid changes can be introduced at a specific codon within a target protein coding sequence.

Figure 3.10 DNA shuffling. Two copies of a gene, or two similar genes from distinct species (black and grey), are shown for simplicity. These are combined, and the DNA is fragmented and 50–100bp regions are purified. These small fragments are subjected to a series of denaturation, self-annealing, and self-primed DNA synthesis steps. During this process, it is possible to incorporate point mutations (white X) by using a *Taq* DNA polymerase. The introduction of mutations in this way is important for random mutagenesis when starting with a single gene. When starting with different members of a family of genes, such mutations are not usually introduced by using an efficient proof-reading DNA polymerase rather than *Taq*. During a number of cycles, typically 40 or more, the various DNA fragments become shuffled and recombined, as do any mutations introduced. Eventually, full-length copies of the gene can be recovered by a PCR amplification with flanking primers. This library of randomly mutated and recombined sequences is then cloned into a suitable vector, normally by restriction digestion and ligation. Following transformation into *E. coli* the library of clones can be screened for the desired property of any random mutational variants.

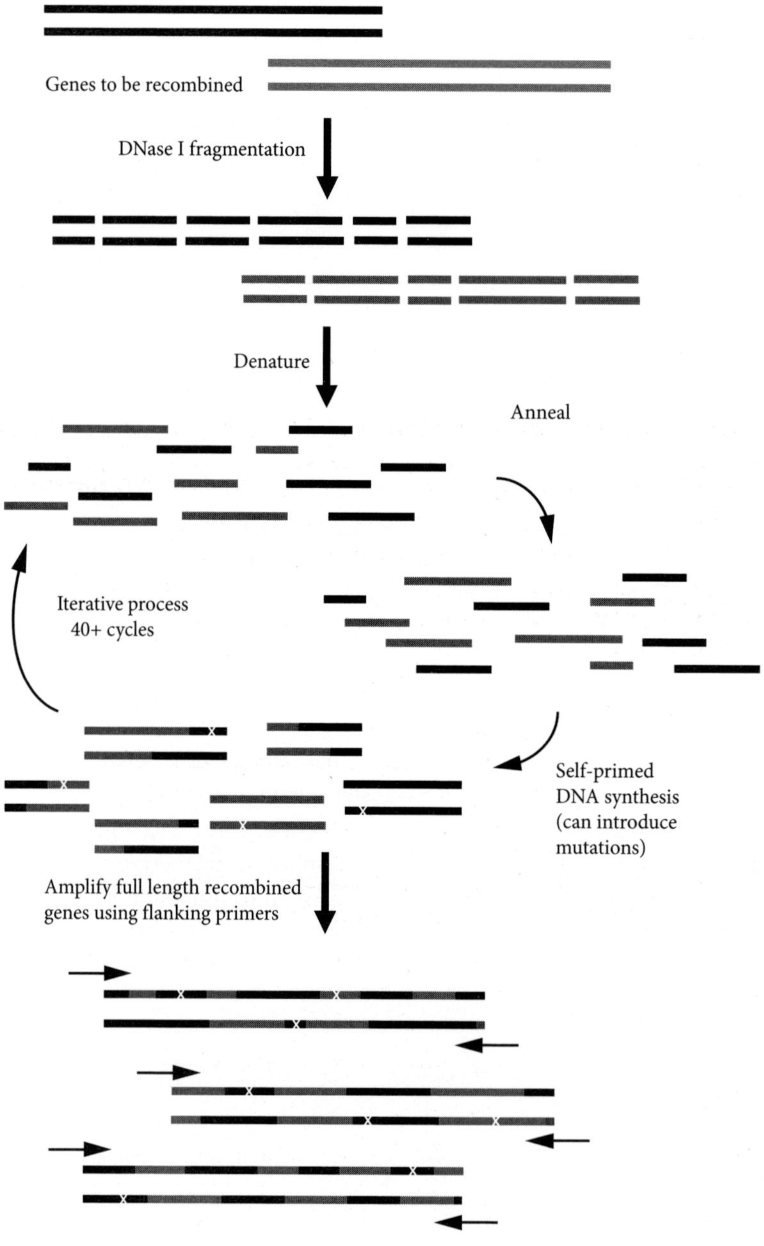

- A critical feature for ensuring high-efficiency recovery of mutant genes is the destruction of the wild-type template DNA which is methylated and otherwise preferentially replicated.
- A commonly used mutagenesis approach is based on the QuikChange® method, a restriction–selection approach involving the use of the restriction enzyme *Dpn*I to selectively destroy wild-type template DNA.
- PCR is an efficient approach for generating mutations using two separate reactions and usually cloning the resulting cassette in place of the corresponding wild-type region of a gene.
- Cassette mutagenesis can also be used to clone a series of oligonucleotide-derived cassettes to explore a short region of a target gene.
- Random mutagenesis approaches are increasingly being used for the discovery of new functional properties and the selection of improved variant proteins. These are *in vitro* approaches most usually used to identify new proteins within large libraries of random variants by a process called directed evolution.
- For any given project you can use more than one form of mutagenesis to investigate and/or improve your target gene or encoded protein.

CASE STUDY 3.1

Site-directed mutagenesis of a gene encoding a reporter protein

Green Fluorescent Protein (GFP) is a reporter used in cell and molecular biology experiments. Its fluorophore is excited by light at 395 nm and emits light at 509 nm. You are studying two proteins (X and Y) that you suspect are co-located in the cell. You have produced each protein fused to GFP and have shown that Protein X is located at the nuclear envelope. To determine whether Protein Y co-localises with protein X you need a fluorescent protein with a different emission wavelength to GFP. It is known that a single amino acid change from tyrosine (Y) at position 66 to histidine (H) produces a protein with a blue chromophore that emits at 448 nm. To introduce this Y66H mutation into your Protein Y- GFP fusion vector you decide to use QuikChange® mutagenesis to make this mutation by changing the tyrosine codon TAT to a histidine codon (CAC or CAT).

QUESTION 1

What factors should you consider when designing the QuikChange® primers?

The melting temperature should be ≥78°C; the primer lengths should be 25–45 nt; the mutation should be centrally located within the primers; the primers should be complementary; ideally there should be G or C at the 5′ and 3′ end.

QUESTION 2

A section of the GFP gene is shown below. Can you design forward and reverse primers of around 30 nucleotides in length to generate the desired mutation, the Y66 codon is in bold.

P W P T L V T T F S **Y** G V Q C F S R Y P D

CCGTGGCCAACACTTGTCACTACTTTCTCT**TAT**GGTGTTCAATGCTTTTCCCGTTATCCGGAT

GGCACCGGTTGTGAACAGTGATGAAAGAGA**ATA**CCACAAGTTACGAAAAGGGCAATAGGCCTA

The exact primer sequence may vary slightly. Select a region of around 13–15 nt on either side of the tyrosine codon (TAT); you could choose the sequence

5′GTC ACT ACT TTC TCT **TAT** GGT GTT CAA TGC3′

Make sure you include the 5′ and 3′ orientation. Then make the changes TAT to CAC:

5′–GTC ACT ACT TTC TCT **CAC** GGT GTT CAA TGC–3′

The reverse primer should be complementary to this

Forward Primer 5′–GTC ACT ACT TTC TCT **CAC** GGT GTT CAA TGC–3′

Reverse Primer 3′–CAG TGA TGA AAG AGA **GTG** CCA CAA GTT ACG–5′

But, the reverse primer must also be written in the 5′–3′ orientation

Reverse Primer 5′–GCA TTG AAC ACC **GTG** AGA GAA AGT AGT GAC–3′

? QUESTION 3

After performing the linear amplification reaction with your primers you treat the reaction products with the restriction enzyme *DpnI*. Why?

*Dpn*I only digests methylated DNA and so will cleave the original wild-type DNA template which was methylated in the bacterial cells from which it was isolated. The newly synthesized, mutated, DNA is unmethylated and so will remain intact allowing efficient transformation of competent cells and high efficiency recovery of the mutated gene.

✓ References

Altenburg, E. (1933) The production of mutations by ultra-violet light. *Science* **78**: 587.

Auerbach, C., Robson, J.M., and Carr, J. G. (1947) The chemical production of mutations. *Science* **105**: 243–7.

Carter, P., Bedouelle, H., and Winter, G. (1985). Improved oligonucleotide site-directed mutagenesis using M13 vectors. *Nucleic Acids Res* **13**: 4431–3.

Chiu, J., March, P.E., Lee, R., and Tillett, D. (2004) Site-directed, Ligase-Independent Mutagenesis (SLIM): a single-tube methodology approaching 100% efficiency in 4 h. *Nucleic Acids Res* **32**, e174.

Clackson, T. and Winter, G. (1989) 'Sticky feet'-directed mutagenesis and its application to swapping antibody domains. *Nucleic Acids Res* **17**: 163–70.

Coco, W.M., Levinson, W.E., Crist, M.J., et al. (2001) DNA shuffling method for generating highly recombined genes and evolved enzymes. *Nat Biotechnol* **19**: 354–9.

Deacon, S.E. and McPherson, M.J. (2011) Enhanced expression and purification of fungal galactose oxidase in *Escherichia coli* and use for analysis of a saturation mutagenesis library. *Chembiochem* **12**: 593–601.

Deacon, S.E., Mahmoud, K., Spooner, R.K., et al. (2004) Enhanced fructose oxidase activity in a galactose oxidase variant. *Chembiochem* **5**: 972–9.

Edelheit, O., Hanukoglu, A., and Hanukoglu, I. (2009) Simple and efficient site-directed mutagenesis using two single-primer reactions in parallel to generate mutants for protein structure-function studies. *BMC Biotechnology* **9**: 61.

Jaeckel, C. and Hilvert, D. (2010) Biocatalysts by evolution. *Current Opinion in Biotechnology* **21**: 753–9.

Kunkel, T.A. (1985) Rapid and efficient site-specific mutagenesis without phenotypic selection. *Proc Nat Acad Sci USA* **82**: 488–92.

Kunkel, T.A., Bebenek, K., and McClary, J. (1991) Efficient site-directed mutagenesis using uracil-containing DNA. *Methods Enzymol* **204**: 125–39.

McPherson, M.J. and Møller, S.G. (2006) *PCR: The Basics.* Abingdon, BIOS at Taylor & Francis.

Muller, H.J. (1927) Artificial transmutation of the gene. *Science* **66**: 84–7.

Ness, J.E., Welch, M., Giver, L., et al. (1999) DNA shuffling of subgenomic sequences of subtilisin. *Nat Biotechnol* **17**: 893–6.

Papworth, C., Bauer, J.C., Braman, J., and Wright, D.A. (1996) QuikChange® site-directed mutagenesis. *Strategies* **9**: 3–4.

Reetz, M.T., Prasad, S., Carballeira, J.D., Gumulya, Y., and Bocola, M. (2010) Iterative saturation mutagenesis accelerates laboratory evolution of enzyme stereoselectivity: rigorous comparison with traditional methods. *J Am Chem Soc* **132**: 9144–52.

Reidhaar-Olson, J. and Sauer, R. (1988) Combinatorial cassette mutagenesis as a probe of the informational content of protein sequences. *Science* **241**: 53–7.

Sarkar, G. and Sommer, S.S. (1990) The megaprimer method of site-directed mutagenesis. *BioTechniques* **8**: 404–7.

Stemmer, W.P. (1994) DNA shuffling by random fragmentation and reassembly: in vitro recombination for molecular evolution. *Proc Natl Acad Sci USA* **91**: 10 747–51.

Tian, J., Liu, Q., Dong, S., Qiao, X., and Ni, J. (2010) A new method for multi-site-directed mutagenesis. *Anal Biochem* **406**: 83–5.

Tyagi, R., Lai, R., Duggleby, R.G. (2004) A new approach to 'megaprimer' polymerase chain reaction mutagenesis without an intermediate gel purification step. *BMC Biotechnology* **4**: 2.

Wells, J.A., Vasser, M., and Powers, D.B. (1985) Cassette mutagenesis: an efficient method for generation of multiple mutations at defined sites. *Gene* **34**: 315–23.

Zhao, H.M., Giver, L., Shao, Z.X., Affholter, J.A., and Arnold, F.H. (1998) Molecular evolution by staggered extension process (StEP) in vitro recombination. *Nat Biotechnol* **16**: 258–61.

Zoller, M.J. and Smith, M. (1982) Oligonucleotide-directed mutagenesis using M13-derived vectors—an efficient and general procedure for the production of point mutations in any fragment of DNA. *Nucleic Acids Res* **10**: 6487–6500.

4 DNA sequencing

Tania Slatter and Alison Fitches

Chapter overview

DNA sequencing determines the precise order of the nucleotide bases—adenine, guanine, cytosine, and thymine—in a DNA fragment, and is a technique widely used in the molecular diagnostic and scientific laboratory. Sequencing allows us to build a gene map of a variety of life-forms from animals to microbes, to identify a cause of disease, or in forensics to find a suspect. This chapter describes the traditional sequencing methods, 'first-generation sequencing', and the newer technologies, 'next-generation sequencing'. First-generation sequencing methods were developed in the 1970s; modifications of these early methods are still widely used. When carried out correctly, first-generation sequencing is an accurate and robust technique, which is easy to perform. However, if your project requires giga base-pairs of sequence data in days, next-generation sequencing is ideal. This fast-moving technology is increasingly used with many more advances expected in the near future.

The underlying principle of the two sequencing types is similar. The DNA requiring sequencing is isolated, and in many applications specific regions are amplified. A starting point from which sequencing will commence is then designated. A pool of labelled DNA fragments is created, capturing the order of nucleotides from the original DNA fragment. Each fragment is separated by size, the label is recognized, and the order of nucleotides revealed.

LEARNING OBJECTIVES

This chapter will enable the reader to:

- determine if cycle or next-generation sequencing is the most appropriate technique to use for their research question;
- decide on an appropriate type of **template** for their sequencing requirements;
- carry out a cycle sequencing reaction and analyse the data generated;
- explain the key steps of next-generation sequencing, and the different **platforms** available;
- become acquainted with the steps involved in the preparation of templates for next-generation sequencing;
- identify the limitations of next-generation sequencing.

4.1 First-generation sequencing

Three approaches encompass first-generation sequencing: the chemical method (Maxam and Gilbert 1977), the Sanger dideoxy-chain-termination sequencing method, and the chain-termination cycle sequencing method (Sanger modification). These sequencing methodologies have accelerated biological research and as a consequence we now know the sequence of thousands of human genes as part of the Human Genome Project. Indeed, the NCBI contains genome data for over 1000 organisms. Each sequencing method is described in turn below.

4.1.1 The Maxam and Gilbert sequencing method

The chemical cleavage sequencing technique was introduced by Maxam and Gilbert (1977). In this method, the 5′ end of a DNA double-stranded template is radiolabelled with γ^{32}P-ATP, usually by means of a kinase reaction. The labelled DNA strand is cleaved by a series of base-specific chemical reagents to produce radiolabelled fragments of different lengths. These fragments are then separated by denaturing acrylamide gel electrophoresis and visualized by autoradiography. This method is only suitable for sequencing small DNA fragments (up to 100bp). The chemicals used for the cleavage reactions, which include formic acid and dimethyl sulphate, are extremely hazardous and hence this method is no longer commonly used.

4.1.2 Sanger dideoxy-chain-termination sequencing method

The enzymatic dideoxy-chain-termination sequencing method was developed by Sanger et al. (1977). This is a reliable and efficient sequencing method, which is easy to perform and requires fewer toxic chemicals. Sanger sequencing is the basis for the automated cycle sequencing widely used today that is outlined in section 4.1.3. The basic principle behind Sanger sequencing is the use of chemically modified deoxynucleotides (dNTPs) known as dideoxynucleotides (ddNTPs). The ddNTPs are introduced into the growing DNA chain in the same manner as the dNTPs, but they lack the 3′ hydroxyl group that in dNTPs allows the addition of the next nucleotide. Therefore incorporation of a ddNTP prevents the addition of any further dNTPs, stopping the extension of that specific fragment.

The original Sanger chain-termination method required a single-stranded DNA template, a sequencing primer, **DNA polymerase**, a reaction buffer, dNTPs, and ddNTPs. The primer and the dNTP or ddNTP can be radiolabelled (usually ^{35}S-dATP) or chemiluminescent labelled. Four separate reaction mixes are set up and one of the four specific ddNTPs (ddATP, ddCTP, ddTTP, or ddGTP) is added. DNA polymerase extends the 3′ end of the short primer annealed to the template by incorporating nucleotides complementary to the template sequence. This chain extension is terminated when a ddNTP is added instead of the corresponding dNTP. Hence in the reaction to which ddATP is added, for example, all the fragments generated will terminate with an adenine. It can be seen in Figure 4.1 that ten fragments are produced by Sanger sequencing a 10nt piece of the template DNA. Once synthesis is complete, the four sample reactions are

Figure 4.1 Overview of the Sanger sequencing method. Four separate reactions are set up. Each tube contains identical components, apart from the labelled ddNTP. The fragments generated in each reaction are indicated; for example, the various lengths of fragment in the A tube are show as ending in green because of the presence of a green fluorophore attached to the ddNTP. Ten fragments are produced in sequencing a 10nt piece of the template DNA. Fragments are separated by size during the electrophoresis process (smallest migrating the furthest) (middle panel) and the fragments are visualized. The DNA sequence is obtained from the gel by reading the position of each band from the bottom to the top, corresponding to the 5′ to 3′ direction. For a full colour reproduction of this figure, please see Plate 1.

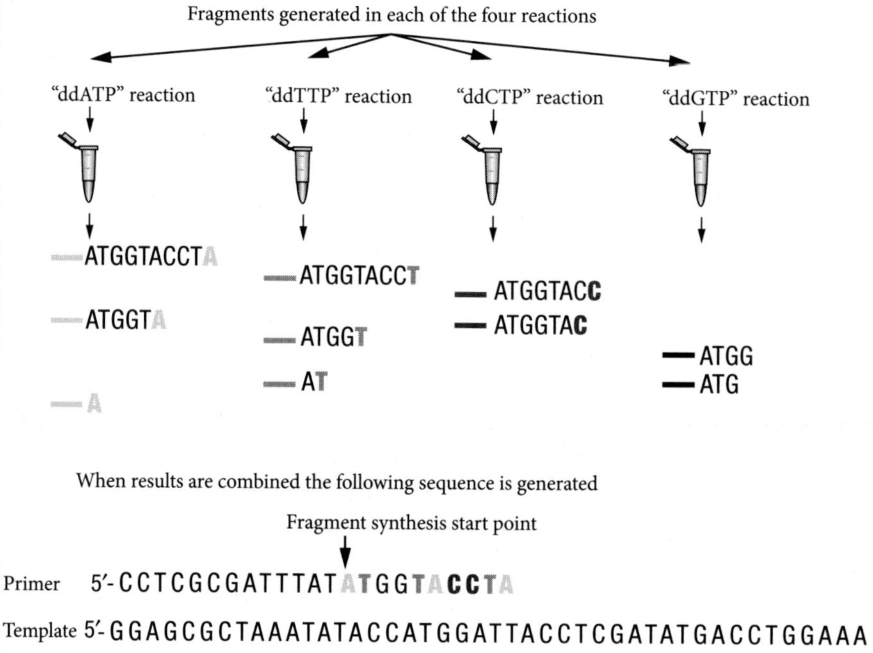

loaded into one of four lanes and the fragments separated by size (smallest migrating the furthest) using denaturing polyacrylamide-urea gel electrophoresis. Visualization of the fragments depends on the method of labelling; for example, X-ray film and autoradiography are used for detecting radiolabelled fragments. The DNA sequence is read from the gel by reading the position of each band from the bottom to the top corresponding to the 5′ to 3′ direction.

4.1.3 The chain-termination cycle sequencing method (Sanger modification)

A number of refinements to the original Sanger method have been introduced, but the most common modification is the use of a thermostable DNA polymerase. This polymerase allows the sequencing reaction to be repeated many times (cycles) in a single tube.

Cycle sequencing requires a DNA template, either single- or double-stranded, a sequencing primer, thermostable DNA polymerase, reaction buffer, dNTPs, and ddNTPs. As with the original Sanger method outlined earlier, the extension products are terminated by the addition of a ddNTP. In cycle sequencing a primer annealed to the template is extended by the DNA polymerase and amplified in a linear manner using cycles of denaturation, annealing, and extension.This procedure is outlined in Figure 4.2(a). As illustrated, the purified template to be sequenced is denatured into single stands. Only one primer is used—a short string of synthetic nucleotides with a complementary sequence to one of the template strands. The primer anneals to one template strand, and the strand is extended from the primer with dNTPs until the addition of the terminating ddNTP. In cycle sequencing each ddNTP is labelled with a different fluorescent dye, each with a different excitation and emission wavelength. As can be seen in Figure 4.2(b), the use of different dyes allows all sequencing components to be added to a single tube, and each cycle produces a fragment that will elucidate one base in the final sequence. Subsequently, the sample is transferred to an automated DNA sequencing machine (such as the ABI Prism 3100 Genetic Analyser) which separates the fragments by size using capillary electrophoresis and detects the fluorescence emitted by each dye. The data are then presented electronically in the form of a chromatogram (Figure 4.3(a) and section 4.1.6) with four coloured peaks. Each colour represents a different nucleotide; for example, G is black, T is red, A is green, and C is blue.

4.1.4 Uses of cycle sequencing

Cycle sequencing has a multitude of uses that fall into two broad categories: de novo sequencing and re-sequencing. The term de novo sequencing is best used to define sequencing of the unknown, meaning there will be no sequence data with which you can compare the data that you generate. De novo sequencing may involve sequencing a whole genome or just parts of it, and this will subsequently be used as a reference sequence for that genome. While de novo sequencing can be performed using the cycle sequencing method, it is far more suited to next-generation sequencing, as described in section 4.2.

If de novo sequencing of a genome is to be undertaken using cycle sequencing, it will be necessary to fragment and clone the target DNA into a plasmid or viral vector, sequence the fragments individually, and then assemble the sequencing data to create the partial or whole genome sequence. Because of the size and complexity of the whole or partial genome, the coverage of sequencing required, and the data analysis and assembly requirements, a high-throughput genetic analyser will be necessary to cope with the workload associated with this.

In contrast to de novo sequencing, re-sequencing is a broad term that covers all uses of sequencing that compare the data generated with a known reference sequence. The most common uses of re-sequencing are as follows.

- *Identification of genetic variants* Genomes show a huge degree of intra-species variability. Sequencing allows the identification, classification, and population screening of variants: polymorphisms (common genetic variants identified in at least 1% of the population) and rare variants that may be associated with a genetic disorder.
- *Identification of DNA* ***methylation status*** Methylation is the addition of a methyl group to a cytosine, and this modification is pivotal to a number of biological processes including imprinting and the development of cancer. Treatment of a sample with sodium bisulphite will preserve its methylation status by converting all

Figure 4.2 (a) Overview of the one-tube cycle sequencing procedure. The diagram shows the cycles of strand separation, annealing, and extension that are integral to this method. (b) Overview of the one-tube cycle sequencing fragment generation. All the components are added to one tube and multiple transcripts of varying lengths are produced. The transcript length will be determined by when the chain termination labelled ddNTP is incorporated into the growing fragment. Each ddNTP is labelled with a specific fluorophore, allowing visualization of the sequence. The fragments are separated by size, with the smallest transcripts moving past the laser first.

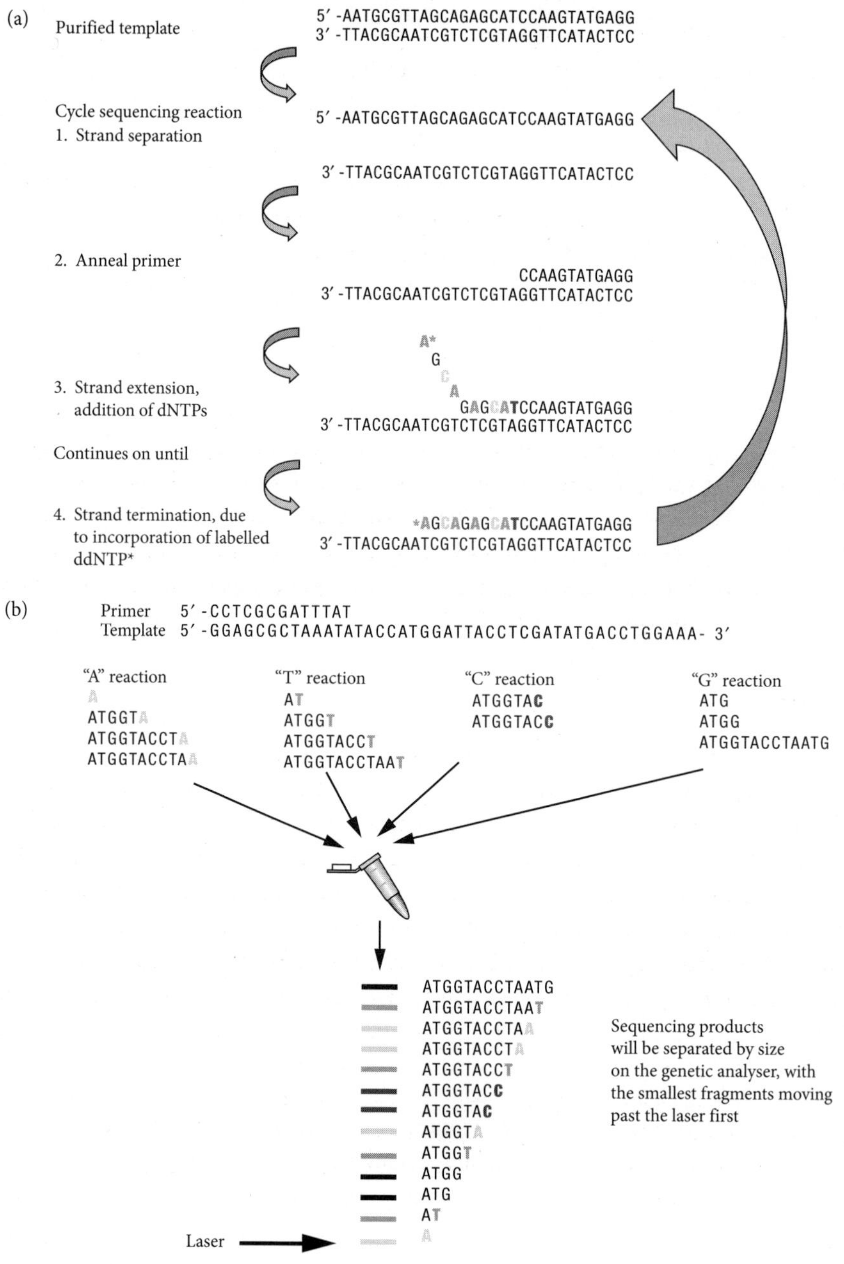

unmethylated cytosine residues to uracil, and is followed by sequencing to identify methylation sites.

- *Confirmation of the integrity of cloned constructs* The cloning of DNA fragments into a plasmid is an integral part of research and sequencing is the ideal method of confirming the veracity of the construct.
- *Species identification* Multi-locus sequence typing of house-keeping genes and the sequencing of ribosomal genes are both fast and efficient methods of species identification and classification, which is important in both research and clinical diagnostic laboratories.
- ***Gene expression** studies.* Sequencing can analyse alterations in gene expression by studying the presence or absence of a transcript, the methods by which a transcript can be modulated, and transcript variants.
- *Confirmation of next-generation sequencing results* Next-generation sequencing produces vast quantities of data, but it is not always completely accurate, and so verification of results by cycle sequencing is necessary.

4.1.5 Performing a cycle sequencing experiment

As described in section 4.1.3, there are four main steps in a cycle sequencing experiment: first, the template to be sequenced is prepared; second, the sequencing fragments are generated; third, the fragments are separated by size and the bases are called; and fourth the data are analysed. Each step is outlined in more detail in the remainder of this section. Sequencing is often performed in centralized facilities. In this case the sequencing fragment generation and separation may be done for you, and all you need to do is supply the template and primer.

Template preparation

A good template is essential for a successful sequencing reaction, i.e. a sequence in which the majority (>90%) of the bases have been identified. A number of factors will affect this, including the source of the template, its size, structural complexity, and quality, and the quantity available. A variety of templates can be used. The starting material for large-scale de novo sequencing will be plasmid or **cosmid** DNA, **bacterial artificial chromosome** (BAC), or **yeast artificial chromosome** (YAC). BAC, which has the capacity to contain large fragments (50–350kb), is often used for de novo sequencing.

Templates for re-sequencing are usually derived from extracted DNA, with the region of interest amplified using PCR or plasmid DNA. If you wish to screen for **single-nucleotide polymorphisms** (SNPs) or rare genomic variants that result in a genetic disorder, the most appropriate starting material will be a PCR product. Direct sequencing of human genomic DNA is not recommended because its size and complexity is not compatible with successful cycle sequencing under standard conditions.

A variety of methods of extracting DNA for PCR and subsequent sequencing are available, including simple boiling methods (Holmes and Quigley 1981), alkaline lysis (Hattori and Sakaki 1986), caesium chloride gradients (Garger et al. 1983), and commercial extraction kits. To produce high-quality sequencing data the extraction process should remove residual quantities of proteins, RNA, detergents, salts, or organic compounds that can inhibit the cycle sequencing reaction or interfere with sample injection electrokinetics of the capillary-based electrophoresis system of the DNA sequencing machines. In addition, the PCR product should be free of primer–dimers and non-specific amplification products, as well as any excess dNTPs and primers, as these will also interfere with the cycle sequencing reaction, resulting in poor-quality sequencing data. Again, a number of methods can be used to purify the PCR product, including ethanol precipitation, gel purification, ultrafiltration, and enzymatic purification using shrimp alkaline phosphatase and exonuclease I, as well as commercially available PCR product purification kits.

Sequencing reaction mix preparation

The quantity of template used in the cycle sequencing reaction is also crucial. An excess of template will significantly reduce the quality and length of the sequencing read, and too little template will result in either no readable data or poor-quality data that results in a low level of accurate base calling. The quantity of template required depends on the source of the template, and the type of **sequencing chemistry** that you are using. Follow the manufacturer's instructions for the sequencing equipment you are using. However, some general guidelines are as follows:

- PCR product (the longer the PCR, the more product required) 2–50ng
- Bisulphite-modified PCR product 3–10ng
- Single-stranded DNA 50–150ng
- Double-stranded DNA 200–300ng
- BAC, YAC, cosmid 200ng–3mg
- Plasmid DNA 2000ng–3mg

In the cycle sequencing reaction, all the components are added to a single tube which includes the template, the single primer, the dNTP/ddNTP mix, thermostable DNA polymerase, and the reaction buffer. The reaction mix is then placed into a thermal cycler for successive rounds of denaturating, annealing, and extension as illustrated in Figure 4.2(b). Cycle sequencing kits containing thermostable DNA polymerase, reaction buffer, and nucleotide mix are available from a number of companies including, but not limited to, Illumina, GE Healthcare, Applied Biosystems, Affymetrix, and Invitrogen. In these reaction mixes, the specific ratio of dNTP to ddNTP will have been carefully calculated and formulated to allow the DNA chains to elongate but at the same time ensuring that they are terminated at every possible position in the fragment. Manufacturers offer a variety of sequencing reaction kits with a selection of sequencing chemistries. The choice of sequencing chemistry will depend on the sequencing application, the type of template you use, and whether you are using labelled primers or labelled ddNTPs. Cycle sequencing protocols are always supplied by the kit manufacturer and these should be followed in order to obtain good-quality sequencing data.

A post-cycle sequencing purification step is necessary to remove all the unused reaction components, particularly the dye component, as these will have a deleterious effect on the subsequent analysis of the sequence data. The two most commonly used methods for post-sequencing purification are commercial kits, which are generally based on spin-columns and size-exclusion membranes and ethanol precipitation. Commercial kits are quick and easy to use, whereas ethanol precipitation, although a cheaper method, is more labour intensive.

Sequencing electrophoresis and instrumentation

Initially, the cycle sequencing products were separated using large denaturing polyacrylamide slab gel electrophoresis and the sequence was read manually from the associated autoradiogram. Now the separation and detection of labelled fragments is automated using genetic analysers. The introduction of automated sequencing has significantly improved the speed and accuracy of analysing DNA sequences. The original automated machines analysed up to ~600bp in 2 hours; this would have required 12–16 hours on a slab gel.

Genetic analysers use capillary-based electrophoresis systems to separate the fragments. The fluorescent labelled fragments are injected into a capillary and are fractionated by size, with the shortest fragments moving the fastest. Each fragment is excited by a laser beam as it moves past it, and the fluorescent signal emitted is detected by the genetic analyser and converted into digital data to allow analysis. A number of genetic analysers are available; these include Applied Biosystems (range of machines from 1 to 96 capillaries), GE Healthcare (MegaBACE systems, 16–96 capillaries), and Beckman Coulter (eight array capillary). Single-capillary machines are still available, but it is now far more common to use multi-capillary machines, which vary from 1 to 96 capillaries, allowing analysis of up to 348 samples per run.

Data analysis

Each company has proprietary software that will convert the signal into base calls to allow analysis of the data. The software, known as Phred (http://www.phrap.com/phred/), is free to academic users, and it can be used with data produced from GE Healthcare, Applied Biosystems, Beckman Coulter, and LI-COR Life Sciences analysers and can produce a variety of output files (Ewing et al. 1998).

Once converted into base calls, the data can be visualized in the form of a chromatogram using proprietary software. There are also a number free software downloads available that can be used to visualize and edit your data, for example 4Peaks (http://www.4peaks.en.softonic.com) and FinchTV (http://www.geospiza.com/Products/finchtv. shtml). A full range of **bioinformatics** tools are available to analyse your sequencing data. These tools allows data mining, sequence analysis and alignment, short read assembly, mirRNA assembly, spliced map reading, transcript quantification, and sequence variance calling.

4.1.6 What should your chromatogram look like?

Figure 4.3(a) shows an example of a high-quality DNA sequencing chromatogram. In a perfect chromatogram all the peaks will have an equal height and there will be no base-line pull-up of nucleotides. Peak height correlates with the amount of sequencing product produced, whilst base-line pull-up refers to a 'dirty sequence' in which more than one peak is seen at multiple bases along the sequence, although this does not always prevent correct reading of the sequence. However, it is more likely that your data will look like that presented in Figure 4.3(b); the peaks heights are not equal, but the data are still perfectly readable and the variation in the base line is not enough to interfere with data analysis. When sequencing to study genetic variants you may be looking for variations from the norm on your chromatogram. For example, a heterozygous SNP or **substitution** will present with two peaks of equal height in the same position, but each peak will be lower than those surrounding it (Figure 4.3(c)). The presence of a heterozygous **insertion**, **deletion**, or duplication will produce a very distinct pattern—the sequence will appear normal until the first nucleotide that is part of the variation and from that point on it becomes unreadable as illustrated in Figure 4.3(d).

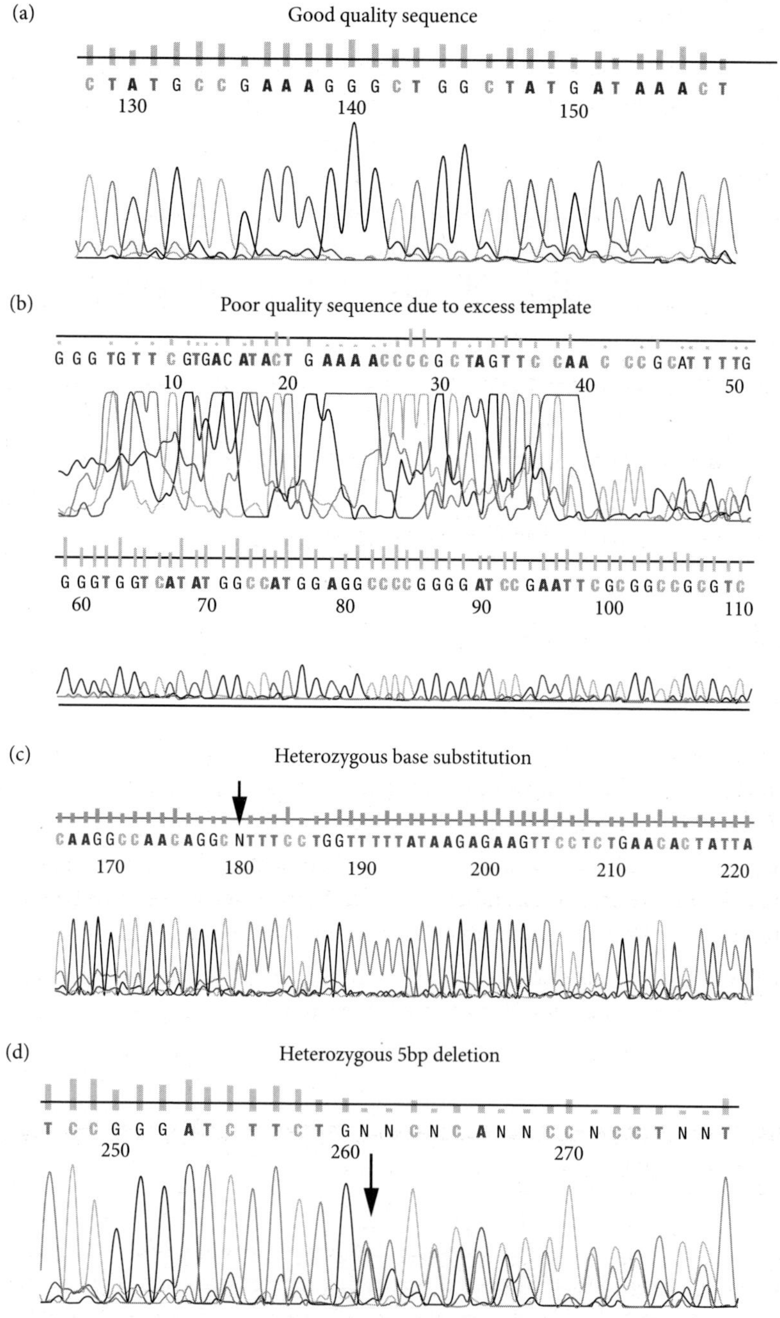

Figure 4.3 Examples of sequence chromatographs. (a) An example of a good-quality sequence from the TES testis-derived transcript (3 LIM domain) gene. Each base is represented by a clear peak with little background noise. (b) An example of a poor-quality sequence. Excess sequencing template results in short poor-quality data. If excess template is added, all the consumables are used up early in the reaction, resulting in a very large number of short fragments (the high peaks at the beginning), and not enough consumables remain for the successful sequencing of the rest of the fragment (peaks rapidly become lower and tail off much earlier than they should). (c) An example of a heterozygous substitution in exon 6 of the SERPINC1 gene. The substitution of a C to T base (arrow) results in two peaks of similar height. This base change results in the conversion from a proline to a theronine amino acid residue at codon 407. (d) An example of a heterozygous deletion mutation in exon 2 of the SERPINC1 gene. A good-quality sequence is present at the start of the 5bp deletion of nucleotides CAGAA (arrow) followed by a poor-quality sequence due to two different base calls at each position. For a colour reproduction of this figure, please see Plate 2.

It is always advisable to confirm sequencing variations by repeating the sequencing reaction using a primer that will allow you to read the alternative template strand. If the variation does not appear in both strands, the anomaly is a sequencing artefact and not real.

4.1.7 Advantages of cycle sequencing

- Easy to perform, and very reliable.
- The same protocol can be used for both single- and double-stranded products and a variety of different templates.
- As thermal cycling produces linear amplification of synthesized product, the typical reaction will result in a 20–30-fold amplification which means that limited quantities of starting material are less of an issue.
- High temperatures in the cycle sequencing reaction:
  - result in the reduction of secondary structure issues that could prevent successful sequencing;
  - allow precise annealing and extension, improving sequencing quality.
- The introduction of modified thermostable DNA polymerase:
  - allows improved extension of product with GC-rich regions and other difficult templates, such as BACs.;
  - improves dye incorporation, reducing issues of variable peak heights.
- The use of high throughput with automated analysis has increased the speed, efficiency, and accuracy of sequence analysis.

4.1.8 Limitations of cycle sequencing

- Poor quality in the first 15–40 bases of a sequence read.
- Deterioration of quality in a long sequencing trace. This depends on the quality and quantity of the template and the type of kit and analyser used, but as you reach 700–900 bases the quality of calls will start to reduce.
- Current methods still only allow relatively short reads (up to 1000 bases) in a single reaction. There are still a number of technical challenges to overcome before we can resolve large DNA fragments that differ in length by only one nucleotide.
- Although the method allows for high-throughput automated analysis, the speed and depth of sequencing is not sufficient for a number of today's applications, hence the advent of next-generation sequencing.

Cycle sequencing application **CASE STUDY 4.1**

A haematological screen has just identified antithrombin deficiency in a patient and you have been asked to confirm the diagnosis by identifying the underlying genetic variation that has caused the disorder. Antithrombin deficiency is caused by mutation in the *SERPINC1* gene and is highly heterogeneous. Mutations have been identified in six of the seven exons, and there are no common mutations that can be screened for using another technique. Design a sequencing strategy to identify the mutation.

QUESTION 1
What is your starting material?

You will be supplied with a sample from the patient, probably blood from which you need to generate a template. You could isolate DNA or RNA. DNA is easier to isolate, but because of intronic sequence more PCR products are required to screen the full coding region. If RNA is used, intronic sequence variants will be missed.

QUESTION 2
Having chosen your starting material you still need templates for sequencing. How will you go about generating a template and how many templates will be required? Also, is it easier to design primers or create plasmid clones? Is the time frame important, which method would be quicker?

The number of templates will depend on the make-up of the gene, how many exons need to be screened, and the size of the exons. For example, will one template cover an exon or is more than one required? Designing primers is generally easier and quicker, and most likely the preferred choice if the genome sequence is known. The inclusion of intron–exon boundaries allows potential splice-site mutations to be identified. Remember, the number of templates you decide to generate will influence the overall cost of the assay.

QUESTION 3
How will the templates be prepared for cycle sequencing?

There are many protocols for template preparation. Mostly likely there is an established protocol in a given research laboratory. If not, many commercial kits are available that are easy to use and are optimized for low-abundance templates (e.g. if your template has been difficult to generate and hence is in short supply).

QUESTION 4
Which commercial kit and labelling method should be used?

Consider whether this set of sequencing reactions will be carried out regularly, which method is in general use in your laboratory, and whether colleagues are experienced in the method you have chosen and therefore are able to help you if problems arise. Numerous commercial kits are available that are easy to use and are optimized for difficult templates. You will need to consider which kit best suits the type of template you have. For example, are your templates AT- or GC-rich? Do they contain GT regions or homopolymers of A or T?

QUESTION 5
How should the cycle-sequencing results be analysed, and what else would need to be done to show that the result is relevant?

Coding sequences for many genomes, and for the entire genome for many organisms, are available on the internet, and can be used as a reference sample for direct comparison. You will need to inspect your sequence first, remove any poor-quality data (usually at the start or end of the sequence), and change any ambiguous base calls (only where you can clearly identify the correct base on the chromatogram). To validate your result you need to repeat the sequencing of the exon using a new PCR product.

4.2 Next-generation sequencing

In this section, methods for sequencing using newer technologies, commonly referred to as 'next-generation sequencing' or 'second generation sequencing', are outlined. The major advantage of next-generation sequencing is its high throughput at a reduced cost compared with the traditional capillary-based sequencing outlined earlier in the previous section. Instead of generating a single sequence in one run and up to one megabase of sequence per day, next-generation sequencing generates millions of sequences by sequencing DNA fragments simultaneously. Although there are different methods for next-generation sequencing, the overall procedure is similar. The template of interest, DNA or RNA, is fragmented into many smaller pieces to provide a very large number of smaller templates and added to a sequencing **platform**. The three dominant next-generation sequencing platforms systems are available from: Roche (Roche 454) (http://www.my454.com), Illumina (http://www.illumina.com), and Applied Biosystems (SOLiD system) (http://www.appliedbiosystems.com/absite/us/en/home.html). As constant improvements are made to this technology, visiting the vendor's website is a good starting point for any project.

4.2.1 Uses of next-generation sequencing

Next-generation sequencing is used in an ever-increasing number of projects. The reduced cost allows sequencing to be used to characterize whole genomes and samples from a large number of individuals. As short templates are analysed, poorer-quality DNA samples can also be sequenced. Applications of next-generation sequencing are as follows.

- *Analysis of poorer-quality DNA samples*, such as ancient DNA from a Neanderthal or a 28 000-year-old mammoth (Poinar et al. 2006). Many modern samples also contain poorer-quality DNA (e.g. tumours with apoptotic or necrotic cells, and DNA from formalin-fixed paraffin-embedded tissues).
- *Re-sequencing regions of interest*, i.e. sequencing the whole genome or a particular region of interest from a species with a reference genome. This approach is used for the analysis of single nucleotide polymorphisms and **copy number variation**, and to identify sequence variants associated with rare disorders.
- *Sequencing of whole genomes* Next-generation sequencing is used to generate a reference genome for a particular organism.
- *Analysis of complex genomes*, for example mapping structural rearrangements (substitutions, gains, losses, and rearrangements) in cancerous cells.
- *Metagenomics*, i.e. the characterization of genomes in complex environmental niches without the need for prior culturing (e.g analysis of microbial genomes associated with the human body).
- *Analysis of gene expression (transcriptomes)*, i.e. sequencing mRNA to characterize new transcriptomes or differences in transcriptomes in development and disease, to map the start and stop sites of genes with greater accuracy, and

to characterize different transcripts that originate from the same gene, such as splice variants.

- *Analysis of non-coding RNAs* Non-coding RNAs are native RNA molecules that are not translated into protein and can affect gene regulation. Next-generation sequencing can identify novel non-coding RNAs. The quantitative feature of this technology allows the levels of non-coding RNAs to be compared between samples (e.g. a comparison of normal and tumour tissue).
- *Analysis of epigenetic variation* Methylation sites involved in cancer have been identified by next-generation sequencing. Here, modified DNA that has been treated with sodium bisulphite is used as the template for sequencing.
- *Analysis of DNA–protein interactions* Chromatin immunoprecipitation (**ChIP**) is performed first to create a library containing the protein of interest and its bound DNA fragments, which will be identified by **ChIP sequencing** (ChIP-seq). This has been used to map nucleosome positioning and transcription factor binding sites.

See Chapter 5

4.2.2 Overview of the next-generation sequencing procedure

The general principles of next-generation sequencing used with the Roche 454, Illumina, and SOLiD platforms can be divided up into a number of key steps. These are shown in Figure 4.4 and summarized briefly here. There are two key factors to consider before embarking on a next-generation sequencing project. The first is whether to create a **fragment library** or **mate pair library**, and the second is the choice of sequencing platform that will be used.

To create a fragment library, DNA is extracted from its source material, broken up into smaller pieces, and special **adapters** added to each DNA fragment end. Fragment libraries are usually used to assemble new genomes.

To create a mate pair library, DNA is extracted from its source material and broken up into fragments of a known size (typically 2–5kb) produced by nebulization or hydrodynamic shearing (described in more detail later in this chapter). These fragments are then ligated together either side of an internal adapter, and finally special adapters are added to each end of the DNA fragment–internal adapter–DNA fragment units. Mate pair libraries are useful for assembling complex genomes.

The special adapters allow the DNA library to be added to a next-generation sequencing platform by tethering each single-stranded adapter-flanked DNA fragment to a single bead or a glass support. Each DNA fragment is then amplified by PCR to produce many copies of each DNA fragment to record sequence events accurately.

To sequence DNA fragments, a priming string of nucleotides along with labelled single nucleotides (Roche 454, Illumini) or probes (SOLiD) are used to bind to the complementary sequence of the DNA fragment. Incorporated nucleotides are detected upon pyrophosphate release (Roche 454) or excitation of fluorescent labels (Illumina and SOLiD). The analysis of sequence data uses bioinformatics software to align to a reference genome, and to identify and quantify differences between samples. The final step is to validate the next-generation sequencing data. Here, other techniques are used to test specific results identified from the next-generation sequencing. Each step is described in turn in the following sections.

Figure 4.4 The general strategy for next-generation sequencing with the Roche 454, Illumina, and SOLiD platforms. *Library construction* The DNA for sequencing is formed into a library. DNA isolated from its source material is broken up into small fragments to provide an enormous number of templates for sequencing. To create a fragment library, adapters are added to each fragment end. To create a mate pair library, DNA fragments on either side of a fragment of known size are ligated either side of an internal adapter, and additional adapters are added to each end of the DNA fragment–internal adapter moiety. Adapter-flanked DNA is added to a solid support, which is bead-based (Roche 454, SOLiD) or glass-based (Illumina). *Library amplification* DNA fragments are denatured to single-stranded fragments, PCR reagents are added, and each DNA fragment is amplified to produce multiple copies to permit detection of sequencing events. *Library sequencing* Each platform uses a slightly different method of sequencing. The Roche 454 system uses pyro-sequencing. In each round of sequencing a single type of labelled nucleotide is added. An incorporated base is detected upon pyrophosphate release and luciferase emitted light detection. In the Illumina system each type of nucleotide is labelled with a different fluorescent label. An incorporated base is detected upon fluorescence emittance. The SOLiD system uses a ligation rather than a polymerase method. Probes consisting of eight nucleotides labelled with fluorescent tags are added. An incorporated probe is detected upon fluorescence emittance, with each round of sequencing identifying two adjacent bases in the DNA template. For a colour reproduction of this figure, please see Plate 3.

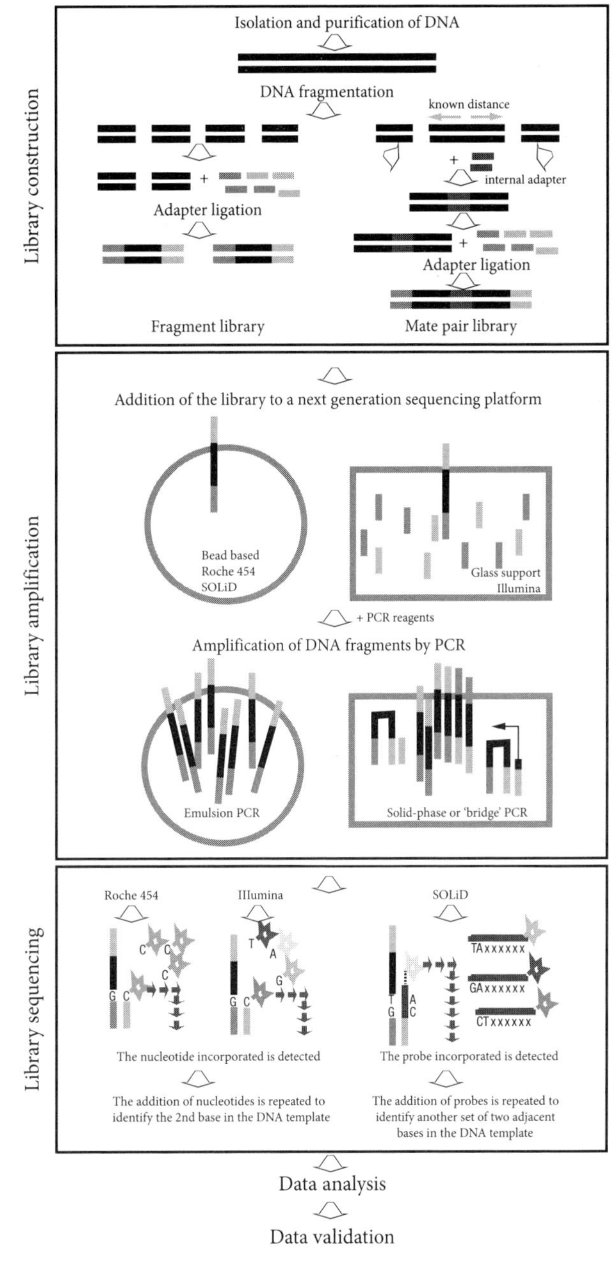

4.2.3 Template preparation for next-generation sequencing

Successful next-generation sequencing begins with preparation of the nucleic acid sample that is to be sequenced. An overview of the steps involved in preparing a DNA library from DNA, RNA, or ChIP starting materials is shown in Figure 4.5. Preparing a library for each type of starting material is slightly different, with that for DNA being the most straightforward. Preparing RNA and ChIP samples requires additional steps prior to the adapter ligation step: RNA requires conversion to cDNA, and ChIP requires

Figure 4.5 The workflow for creating DNA libraries for next-generation sequencing. The sample material, DNA, RNA, or chromatin immunoprecipitation (ChIP) is isolated and purified. Large nucleic acids are broken down into smaller fragments. For RNA and ChIP, further steps are required to isolate or convert the nucleic acid for sequencing. All samples are checked to make sure they are of suitable quantity and quality to continue with library construction. If insufficient sample is available, the sample can be enriched further by PCR, or selection of specific fragments. The library is constructed, which involves the addition of adapter sequences to the ends of each nucleic acid fragment. Finally, the library is added to the sequencing platform.

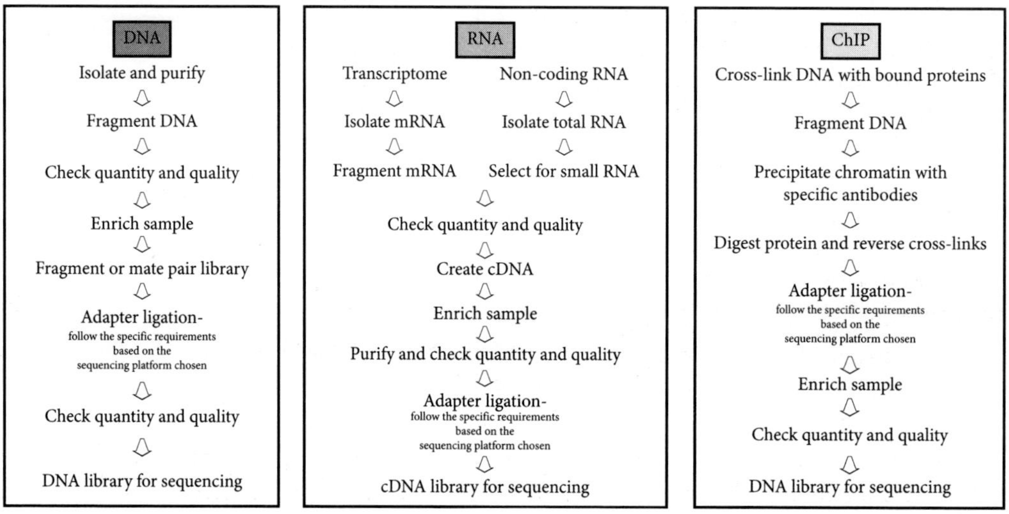

precipitation of specific proteins cross-linked to DNA. There are many protocols and commercially available kits for isolating nucleic acids from a range of source materials. You must choose the correct protocol/kit for the type of nucleic acid you would like to isolate, i.e. DNA, RNA, or, more specifically, non-coding RNA or coding RNA (mRNA), and ensure that it is one that removes any contaminant from the source material (such as RNA from DNA preparations) and any particles that may affect subsequent PCR and sequencing steps. For example, if DNA is required from paraffin-embedded tissue, the protocol will need to include a step to remove paraffin from the tissue sections.

DNA fragmentation

Once isolated, each nucleic acid source is broken up into smaller fragments. There are various methods for fragmentation which utilize physical or enzymatic means.

- Nebulization, in which DNA is broken up into 500–800bp fragments by compressed nitrogen or air.
- Sonication, in which DNA is broken up into fragments using ultrasound. A range of fragments can be obtained by this method, including very small fragments (100–300bp).
- Covaris' Adaptive Focused Acoustics Technology (AFA): DNA fragments as small as 200bp are achieved using focused acoustic energy. The strength of this method is a reduction in the amount of DNA that is wasted and a greater control on the length of fragments produced. A protocol for using this method can be found on the manufacturer's website.

- Hydrodynamic shearing, in which hydrodynamic shear stress is used for DNA fragmentation. The DNA fragments are larger (2–12kb) than those obtained with the other fragmentation methods, which is beneficial if larger DNA libraries are required. This method is often used to create mate pair libraries.
- Enzymatic, where enzymes are used to cut DNA up into smaller fragments. Restriction enzymes that cut the genome at regular intervals are used, and are useful if a particular sequence of interest is located between two restriction enzyme cutting sites. Many specialized commercial enzymes have been developed specifically for this purpose. These may be beneficial for some applications (e.g. AT-rich templates). Some enzyme preparations include subsequent steps in preparing libraries for sequencing (i.e. DNA end preparation for adapter ligation).

Adapter ligation

Once the DNA for sequencing has been fragmented, specialized adapter sequences are ligated to the end of each fragment. Adapters are key for addition to the solid support and for priming subsequent PCR and sequencing reactions. A list of adapter sequences compatible with the Illumina system for different applications is available from the manufacturer's website. Consult the manufacturer's instructions for how to add their (or compatible) adapters to your DNA fragments. In most cases the DNA fragment ends need to be modified, and additionally appropriate-sized fragments may need to be selected before the adapters are added. This can be achieved by separating the DNA fragments out onto agarose gels, and cutting out a gel slice corresponding to fragments within a recommended size range. The size and quality of the DNA library can be assessed using gel electrophoresis and spectrophotometry. Microfluidics chip-based equipment, such as Agilent Technologies' Bioanalyzer, offer an ideal easy assessment of nucleic acid fragment size, quality, and quantity.

Using a mate pair library instead of a fragment library allows sequence information to be more easily aligned to an existing genome. To create a mate pair library, following DNA fragmentation, fragments are circularized with an internal adapter, circular DNA is digested, and additional adapters are added to each linear fragment end (as for creating a fragment library). The net result is a DNA library composed of DNA fragment–internal adapter–DNA fragment units. Commercial kits are available which enable mate pair libraries to be created easily.

4.2.4 Sequencing platforms

The choice of sequencing platform may be based on what is optimal for a particular application. Since the equipment and analysis software is fairly expensive and specialized, the choice of platform may be based on the proximity, assistance available, and cost to the end user.

Roche 454

This platform was produced in 2004 and was the first to be available commercially (Margulies et al. 2005). A typical read length for this platform is 400bp, with read lengths of up to 1kb. Approximately 0.5Gb of data are collected per run, and the typical run time is 0.4–1 day. This platform produces longer read lengths than others and therefore may be advantageous in applications such as metagenomic and whole genome sequencing projects. In this technology, the adapter-flanked DNA fragments for sequencing are added to special magnetic beads, amplified, and sequenced by the pyrosequencing method (Ronaghi et al. 1996, 1998). Upon addition to the sequencing platform, the DNA library is denatured to produce single-stranded DNA fragments. Only one of the adapters contains a biotin label for binding to the streptavidin-coated beads. The net result is a unique single-stranded DNA fragment bound to each bead. Amplification of each DNA fragment by PCR enriches each bead with many copies of a specific DNA fragment. The type of PCR used is referred to as **emulsion PCR** as each bead is present with the amplification reagents in a water–oil mixture (Dressman et al. 2003). Following amplification and subsequent denaturation to produce single-stranded DNA templates for sequencing, each bead is loaded into a single well within a plate. For sequencing, a single type of nucleotide with a pyrophosphate tag is added to the plate with the other nucleotide types added individually in sequential steps. If a nucleotide binds to the template, light is produced via pyrophosphate release and the firefly enzyme luciferase. The emitted light is detected by a CCD camera on the platform and the specific nucleotide added is recorded.

Illumina system

This is currently the most widely used platform, partly because of the relative ease of creating the DNA library. Eight separate DNA libraries can be analysed per run, with a typical read length of 75bp. Approximately 18–35Gb of data are produced per run, with a typical run time of 4–9 days. In this technology, the adapter-flanked DNA fragments for sequencing are added to a sealed glass microfabricated device, amplified, and sequenced by detection of emitted fluorescent signals. The solid support contains a multitude of attached adapters that hybridize to the complementary adapter upon addition of the single-stranded DNA library. On the solid support surface each DNA fragment is amplified by PCR, often referred to as **solid phase PCR** or bridge PCR, to produce many copies in close proximity to the original DNA fragment (a cluster) (Adessi et al. 2000). For sequencing, all four nucleotides, each with a unique fluorescent label, are added at the same time. If a nucleotide binds to the DNA fragment template, a fluorescent signal is emitted upon laser excitation. The base incorporated is identified, the fluorochrome added is removed, and the addition of nucleotides is repeated to identify the next base in the template.

SOLiD system

This platform is generally cheaper per run compared with the others. A typical read length for this platform is 50bp. Approximately 30–50Gb of data are collected per run, with a typical run time of 7–14 days. In this technology, like the Roche 454 system, the adapter-flanked DNA fragments for sequencing are added to special magnetic beads and amplified, but are sequenced using a slightly different (ligase-mediated) method. Again, the net result following attachment of a single-stranded DNA fragment to the bead and emulsion PCR is a bead with many copies of the original DNA fragment. DNA-fragment-coated beads are added to a glass support for sequencing. The principle of sequencing is based on a two-base encoding method, where each round of sequencing interprets two adjacent bases in the template (McKernan et al. 2009). Following the addition of the universal primer, a library of fluorescently labelled probes which can hybridize and ligate to their complementary sequence on the DNA fragment template is added. Each probe is a nucleotide polymer (eight bases) with two nucleotides that are interpreted upon detection of the fluorescent tag. Following detection of the first probe, the fluorescent tag and part of the probe are removed and another round of sequencing, which interprets another two bases in the template, is initiated. The two-base encoding system allows each nucleotide in the template to be read twice, and the improved accuracy is an advantage of this platform. The two-base coding system produces data in a code format called Colour Space Code where four colours (0, 1, 2, 3, and 4) each represent a dinucleotide. To analysis this data you will need to choose software that can read Colour Space Code.

4.2.5 Data analysis for next-generation sequencing

Once the sequence reads have been obtained, they must be analysed using a bioinformatic approach. The bioinformatic software used depends on the aims of the project and can be chosen to do the following:

- align sequence reads to a reference genome;
- assemble sequence reads to create a reference genome;
- identify sequence variants;
- identify structural variants
- copy number analysis.

A large number of software tools are available for the analysis of next-generation sequencing data. A list of some of the software tools available for alignment, assembly, variant detection, and copy number variation analyses is given in Table 4.1. Some of the software tools available are produced by the sequencing platform vendors. When selecting the appropriate software, check that it is appropriate for the sequencing platform used. Software called SEAL (http://compbio.case.edu/seal/) can help select the best alignment software to use specifically for your next-generation sequencing data (Ruffalo et al. 2011).

4.2.6 Validation of next-generation sequencing data

Other techniques are used to confirm the results of interest from next-generation sequencing. For example, to validate mRNA results, reverse transcriptase PCR is used to identify, or quantitative reverse transcriptase PCR is used to quantify, a particular mRNA moiety. If the validation tests use the same nucleic acid template as that used for next-generation sequencing, the validation is limited to a test of the technology. It is best to use a different sample of extracted nucleic acid from that used for the next-generation sequencing, thus providing a biological validation of the next-generation sequencing results.

See Chapter 2

4.2.7 Limitations of next-generation sequencing

- Currently, the read lengths obtained with next-generation sequencing are shorter and less accurate than those obtained with Sanger sequencing. Therefore Sanger sequencing may be more appropriate for small-scale projects.

Table 4.1 Bioinformatic tools for the analysis of next-generation sequencing data

Use	Program	Website
Alignment	BFAST	http://bfast.sourceforge.net
	Bowtie 2*	http://bowtie-bio.sourceforge.net/index.shtml
	BWA (newer version of MAQ)*	http://bio-bwa.sourceforge.net
	Corona Lite (Applied Biosystems)	http://solidsoftwaretools.com/gf/project/corona
	ELAND (Illumina)	http://www.illumina.com
	Mosaik	http://code.google.com/p/mosaik-aligner/
	mrFAST and mrsFAST*	http://mrfast.sourceforge.net/
	Novoalign*	http://biowulf.nih.gov/apps/novocraft.html
	SHRiMP*	http://compbio.cs.toronto.edu/shrimp/
	SOAPv2*†	http://soap.genomics.org.cn
	SSAHA2	http://www.sanger.ac.uk/resources/software/ssaha2
	SXOligoSearch	http://www.synamatix.com/secondGenSoftware.html
Assembly	ALLPATHS-LG	http://www.broadinstitute.org/science/programs/genome-biology/crd
	Edena	http://www.genomic.ch/edena
	Euler-SR	http://euler-assembler.ucsd.edu
	GS De novo Assembler (Roche Diagnostics)	http://my454.com/products/analysis-software/index.asp
	SHARCGS	http://sharcgs.molgen.mpg.de/
	SSAKE	http://www.bcgsc.ca/platform/bioinfo/software/ssake
	VCAKE	http://sourceforge.net/projects/vcake
	Velvet	http://www.ebi.ac.uk/%7Ezerbino/velvet
Variant detection	CASAVA (from Illumina)	http://www.illumina.com/software/genome_analyzer_software.ilmn
	GS Amplicon Variant Analyzer (Roche Diagnostics)	http://my454.com/products/analysis-software/index.asp
	PbShort	http://bioinformatics.bc.edu/marthlan/PbShort
	Pindel (for large deletions and medium-sized insertions)	http://www.ebi.ac.uk/%7Ekye/pindel/
	SNVMix	http://www.bcgsc.ca/platform/bioinfo/software/SNVMix
	ssahaSNP	http://www.sanger.ac.uk/resources/software/ssahasnp/
	Unified genotyper	http://www.broadinstitute.org/gsa/gatkdocs/release/org_broadinstitute_sting_gatk_walkers_genotyper_UnifiedGenotyper.html
	Var-Scan	http://varscan.sourceforge.net
Copy number variation	seqCBS	http://cran.r-project.org/web/packages/seqCBS/
	CNV-Seq	http://tiger.dbs.nus.edu.sg/cnv-seq/

*Used in SEAL to help select the most suitable alignment programme for a specific project.

†Various software packages are available for specific applications such as SOAPsnp, SOAPsplice, SOAPdenovo, and SOAPindel.

- Each platform has specific limitations (Metzker 2010). For example, the Roche 454 platform has difficulty in deciphering repeats of the same nucleotide (e.g. TTTTT), while the current Illumina platform under-represents AT- and GC-rich templates.
- Insertion/deletion errors are more prominent with the Roche 454 system, while substitution errors occur more frequently with the Illumina system.
- The bottle-neck in next-generation sequencing is the analysis of the enormous amount of data acquired.
- Data analysis requires considerable bioinformatic expertise and data storage facilities.
- Many sequence analysis tools are not optimal for shorter read lengths. The demand for this technology will lead to continuous improvements in the analytical tools available.

4.3 Third-generation sequencing

Another limitation of the next-generation sequencing platforms is the use of a PCR amplification step to detect sequencing events. This potential source of sequencing bias is removed in the latest high-throughput sequencing platforms—third-generation sequencing. One such platform is the the Helicos Genetic Analysis System from Helicos BioSciences Corporation (http://www.helicosbio.com). In this technology DNA libraries are created by fragmenting the DNA source in 100–200bp pieces, adding adapters to each fragment end, and addition to a solid support. No amplification step is required; the sequencing reagents and imaging equipment are sensitive enough to detect individual sequencing events. Hence, this technology offers single-molecule sequencing.

Next-generation sequencing application

CASE STUDY 4.2

The aim of your project is to determine, using next-generation sequencing, which regions of the genome, if any, show copy number variation in a malignant tumour (grade 4 astrocytoma).

QUESTION 1
What next-generation sequencing platform will be used?

The cost of the equipment required for next-generation sequencing is considerable; therefore it is likely that the sequencing will be performed by a centralized facility and that only one next-generation sequencing platform will be available at a particular location. However, if a different platform is preferred, samples could be sent to another centre or commercial provider. You will also need to consider the cost of the sequencing run and the amount of sequence per fragment that is obtained. The SOLiD system generally costs the least per run. The Roche 454 platform provides longer read lengths.

? QUESTION 2
What samples will be used, and how will the samples be prepared for next-generation sequencing?

Will you need to isolate DNA or RNA?

Copy number variation is usually studied using DNA. Different protocols are used for extracting DNA. A number of commercial kits are available that may be ideal if your starting material provides a low yield or poor quality of nucleic acid.

Would one tumour sample be adequate?

If you are trying to identify the underlying cause of disease for one individual, one tumour sample would be adequate. However, if you are trying to find sequences common to a particular tumour type, analysing tumours from different individuals provides a better representation. Bear in mind that next-generation sequencing is relatively expensive, which may limit the number of samples that can be analysed.

Are control tissues, such as normal tissue from the same organ, available?

Control tissues can aid in identifying disease-specific sequences. A nucleic acid library will need to be made to add to the sequencing platform. A choice between a fragment and a mate pair library will need to be made. The nucleic acid source will need to be broken up in smaller fragments, so a method of doing this will need to be selected; you may be limited to equipment readily available in your laboratory. Adapter sequences will need to be added to each fragment; a commercially available kit may be a good option, as a number of steps are required to prepare the fragment ends.

? QUESTION 3
How could a result of interest from the next-generation sequencing be identified, and what else would need to be done to show that the result is biologically relevant?

Bioinformatics software will need to be used to analyse the large amount of data obtained. A choice of software will need to be made, i.e. that provided by the sequencing platform vendor, another commercially available source, or a freely available source. Analysis of the data requires considerable expertise; therefore you may wish to consult with someone experienced and use their recommended software. Results of interest will need to be validated using another technique. Here, multiple tumours could be tested to show that a particular part of the genome is amplified or deleted.

4.4 Chapter summary

- First-generation and next-generation technologies are used for DNA sequencing.
- Of the first-generation sequencing methods, chain-termination cycle sequencing is widely used for DNA sequencing. It is a robust method that is easy to perform.
- Cycle sequencing has many data analysis tools available, which adds to its usability.
- Next-generation sequencing is increasingly being used for high-throughput sequencing.

- **Next-generation sequencing is used to assemble new genomes and identify causes of disease.**
- **The analysis of next-generation sequencing data is complex and therefore assistance from someone familiar with the bioinformatics software will make data analysis easier.**

References

Adessi, C., Matton, G., Ayala, G., et al. (2000) Solid phase DNA amplification: characterization of primer attachment and amplification mechanisms. *Nucleic Acids Res* **28**: e87.

Dressman, D., Yan, H., Traverso, G., Kinzler, K.W., and Vogelstein, B. (2003) Transforming single DNA molecules into fluorescent magnetic particles for detection and enumeration of genetic variations. *Proc Natl Acad Sci USA* **100**: 8817–22.

Ewing, B., Hillier, L., Wendl, M.C., and Green, P. (1998) Base-calling of automated sequencer traces using phred. I: Accuracy assessment. *Genome Res* **8**: 175–85.

Garger, S.J., Griffith, O.M., and Grill, L.K. (1983) Rapid purification of plasmid DNA by a single centrifugation in a two-step cesium chloride–ethidium bromide gradient. *Biochem Biophys Res Commun* **117**: 835–42.

Hattori, M. and Sakaki, Y. (1986) Dideoxy sequencing method using denatured plasmid templates. *Anal Biochem* **152**: 232–8.

Holmes, D. S. and M. Quigley (1981) A rapid boiling method for the preparation of bacterial plasmids. *Anal Biochem* **114**: 193–7.

Margulies, M., Egholm, M., Altman, W.E., et al. (2005) Genome sequencing in microfabricated high-density picolitre reactors. *Nature* **437**: 376–80.

Maxam, A.M. and Gilbert, W. (1977) A new method for sequencing DNA. *Proc Natl Acad Sci USA* **74**(2): 560–4.

McKernan, K.J., Peckham, H.E., Costa, G.L., et al. (2009) Sequence and structural variation in a human genome uncovered by short-read, massively parallel ligation sequencing using two-base encoding. *Genome Res* **19**: 1527–41.

Metzker, M.L. (2010) Sequencing technologies—the next generation. *Nat Rev Genet* **11**: 31–46.

Poinar, H.N., Schwarz, C., Qi, J., et al. (2006) Metagenomics to paleogenomics: large-scale sequencing of mammoth DNA. *Science* **311**: 392–4.

Ronaghi, M., Karamohamed, S., Pettersson, B., Uhlén, M., and Nyrén, P. (1996) Real-time DNA sequencing using detection of pyrophosphate release. *Anal Biochem* **242**: 84–9.

Ronaghi, M., M. Uhlen, and Nyrén, P. (1998) A sequencing method based on real-time pyrophosphate. *Science* **281**: 363, 365.

Ruffalo, M., LaFramboise, T., and Koyutürk, M. (2011). Comparative analysis of algorithms for next-generation sequencing read alignment. *Bioinformatics* **27**: 2790–6.

Sanger, F., Nicklen, S., and Coulson, A.R. (1977). DNA sequencing with chain-terminating inhibitors. *Proc Natl Acad Sci USA* **74**: 5463–7.

5 Measuring DNA–protein interactions

Luis Acevedo, Ana Sanz, Paul Labhart, and Mary Anne Jelinek

Chapter overview

DNA can be viewed as the genetic blueprint or hardware of the cell, and DNA-binding proteins as the functional readers, or software, decoding the genetic information and providing the interface between heredity and cellular identity. DNA binding proteins regulate and control a variety of cellular processes including gene expression, DNA replication, recombination, and repair. Therefore identifying the DNA sequences that bind proteins, which proteins bind to DNA, and how they bind is critical to understanding the roles that DNA binding proteins play in regulating these processes.

In this chapter, we will look at commonly used methods for analysing DNA–protein interactions. These methods can be divided into two general categories. The first enables the identification or characterization of the proteins that interact with a DNA fragment of uniform length and sequence. These methods can be used in a variety of applications, such as identifying previously unknown DNA binding proteins in crude cell extracts, determining the strength and specificity of DNA binding with purified protein preparations, or mapping the nucleotides and amino acids involved in the interaction. The second category of methods identify the DNA sequences associated with a protein of interest, whether or not the protein recognizes a specific DNA sequence, or whether the protein is only indirectly associated with DNA as a component of a multi-protein complex. The latter methods require a means of capturing the protein of interest, either through the use of a specific antibody or through the ectopic expression of a recombinant form of the protein engineered with a **tag** or label to facilitate its isolation. The methods covered will include analysis of protein binding distribution spanning from single-gene interrogation through to genome-wide analysis.

LEARNING OBJECTIVES

This chapter will enable the reader to:

- **understand the importance of studying DNA–protein interaction methods;**
- **summarize the major workflow steps in the methods used to study DNA–protein interactions;**
- **interpret data that are produced and understand its significance;**
- **select an appropriate workflow to answer a defined biological problem and provide reasons for the selection.**

5.1 Introduction

The genome of all cells is organized by architectural proteins into higher-order structures, which are necessary to compact large genomes into the spatial confines of the cell. In prokaryotes the genome is associated with a variety of small basic sequence-independent DNA-binding proteins, and is called the nucleoid or nucleoid body. These proteins create bends and bridges between DNA tracts that result in the compaction of the nucleoid into interconnected ~10kb loops (Dame 2005). In the case of *E. coli*, the nucleoid containing 4.6Mb of DNA occupies about a quarter of the intracellular volume of the 2 × 1μm cell. In eukaryotic cells, DNA exists in the form of **chromatin**, whose fundamental unit structure is the **nucleosome**, consisting of 146bp of DNA wrapped around a histone core comprising two copies each of H3, H4, H2A, and H2B histone proteins (Kornberg 1977). During **interphase**, the 11nm 'beads on a string' chromatin structure is found in regions of transcriptionally active unfolded DNA (**euchromatin**). The more compacted 30nm fibre formed by the insertion of the linker histone H1 and architectural chromatin-associated proteins is associated with transcriptionally repressed heterochromatin that is further compacted into larger looped chromatin domains of 300–700nm by nuclear scaffold proteins. The most condensed chromatin structures occur during mitosis and meiosis, when a 10 000-fold compaction results in the formation of **metaphase** chromosomes of size 1.5μm to permit accurate segregation of chromosomes during cell division. To put all this in context, the human genome is 3400Mb of DNA, roughly 2m in length, and is compacted into the confines of the nucleus, which has an average diameter of only ~6μm (10% of the volume of a cell).

Historically, histones were considered to play only a structural role in eukaryotic cells. However, recent developments stemming from **epigenetic** research have revealed functional roles for individual histones and the nucleosome (Jenuwein and Allis 2001). These findings have created renewed interest in the analysis of DNA–protein interactions, and have spurred the development of an expanding number of techniques geared towards studying how proteins interact with chromatin.

5.1.1 DNA binding proteins

DNA binding proteins regulate a variety of cellular processes including gene expression, replication, and recombination, and maintenance of genome integrity. Examples of DNA binding proteins are **transcription factors**, polymerases, nucleases, and structural proteins. Some DNA binding proteins, such as transcription factors and restriction endonucleases, recognize specific DNA sequences, while others, such as DNA and RNA polymerases and histones, are sequence-independent DNA binders. Whether these proteins interact directly with either single-stranded (ss) DNA or double-stranded (ds) DNA, all do so via a protein domain called the **DNA binding domain** (DBD). DBDs that bind specific DNA sequences often bind in the DNA **major groove**, where amino acids in the DBD can readily interact with specific purine or pyrimidine bases on the DNA. In contrast, many of the DNA sequence-independent DBDs bind within the **minor groove** and have generalized affinity with the sugar–phosphate backbone.

DNA binding proteins are often multi-domain, comprising not only a DNA binding domain but also other domains with a wide variety of biological functions, such as regulating DNA binding (through post-translational modifications) or association with other proteins (transcription factor–RNA polymerase interactions), and/or enzymatic functions (histone or DNA methylation). For example, the p53 tumour suppressor protein, involved in preserving genome stability (Suzuki and Matsubara 2011), exists as a tetramer *in vivo*. It comprises three domains: an N-terminal transactivation domain, a central DNA binding domain, and a C-terminal regulatory domain. It is the single central DNA binding domain that has the ability to recognize and interact with several forms of DNA. For example, p53 is capable of DNA interactions that are both sequence-specific and non-sequence-specific (Kern et al. 1991a; el-Deiry et al. 1992; Foord et al. 1993). p53 also has the ability to bind both ds- and ssDNA (Kern et al. 1991b), including 3′ termini of mismatched DNA generated as a result of DNA damage (Bakhanashvili et al. 2010).

Some DNA binding proteins contain more than one DBD. For example, MBD1, one of the members of the methyl-CpG binding domain (MBD) family of proteins involved in the transcriptional repression of genes with methylated promoter DNA sequences, contains an amino-terminal MBD domain followed by three zinc-coordinating CXXC domains. The MBD domain targets the protein to methylated CpG sites, while the CXXC domains target the protein to non-methylated CpG sites in heterochromatic regions (Jorgensen et al. 2004). The repressive function of MDB1 at both methylated and non-methylated CpG sites makes it unique among the MDB protein family. Figure 5.1 shows the helix-turn-helix structure, the leucine zipper, and the homeodomain as examples of the structural variety of DBDs capable of binding DNA.

Figure 5.1 Schematic representation of the interaction of different DNA binding domains with the DNA double helix. Alpha-helical secondary structures contacting the DNA are depicted as cylinders. The helix-turn-helix domain is about 20 amino acid residues long and comprises two α-helices (depicted as cylinders), which bind in the major groove of DNA, separated by an unstructured region called the 'turn'. The homeodomain is a 60 amino acid long domain comprised of three α-helices connected by short unstructured regions. The first two helices are anti-parallel. The longer C-terminal helix, roughly perpendicular to the first two helices, makes direct contact with DNA. The leucine zipper is a basic α-helix domain with a leucine residue spaced every seven amino acids. These leucines form a zipper enabling dimerization with a second protein containing such a domain. The basic residues of the domain interact with the sugar–phosphate backbone while the α-helices bind within the major groove.

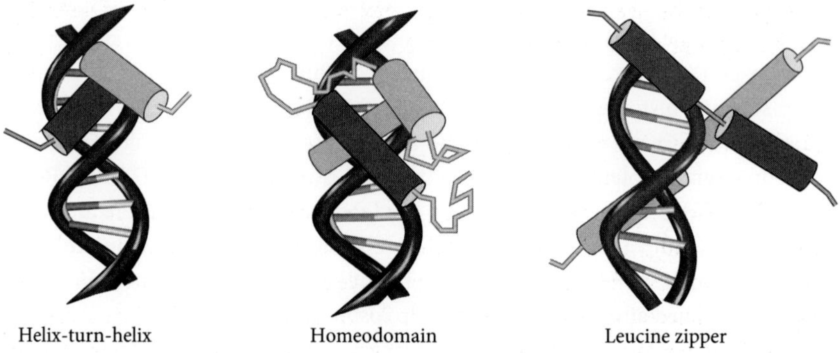

5.1.2 Methods for studying DNA–protein interactions

DNA–protein interaction studies can be directed towards identifying which amino acid residues are involved in making direct contact with the DNA or can be directed towards analysing the DNA regions which make contact with the protein. In the former case, these typically involve structural studies such as X-ray crystallography, with either the DBD or the full-length protein, or genetic studies, which require the generation of mutant proteins. Nested sets of amino-terminal and carboxyl-terminal truncation mutations (deletion mutants) are subsequently used to roughly delineate the limits of the DBDs. Point mutation strategies, in which discrete base-pair changes are introduced into the DNA sequence encoding a DBD to achieve the alteration of a single amino acid residue, are used to identify or confirm the amino acid residues involved in making direct contact with DNA. Often, both genetic and structural approaches are used in concert to reinforce the findings. Thus DNA-contacting amino acid residues identified in structural studies should abolish DNA binding when mutated.

See Chapter 3

The methods used to analyse the DNA regions which make contact with the protein are more diverse and will be the focus of this chapter. Several are listed in Table 5.1 and include:

- **footprinting** and gel mobility assays which require the use of cloned purified DNA or synthetic oligonucleotides for the mapping of protein binding sites on DNA;
- **ch**romatin **i**mmuno**p**recipitation (**ChIP**) which is a more recently developed technique that enables a snapshot analysis of the genome-wide distribution of proteins that bind DNA directly or indirectly through associations with DNA-bound proteins.

Each of these methods is described in turn in the following sections.

5.2 Footprinting

DNA footprinting enables the identification of the nucleotides that contact a DNA binding protein. It is based on the principle that the DNA binding protein will protect the underlying DNA sequences from DNA cleaving agents such as nucleases or chemicals. Footprinting experiments can be performed with purified proteins or with crude cell extracts, and can be used to monitor the partitioning of an unidentified DNA binding protein during a physical purification scheme.

5.2.1 The DNase I footprinting technique

Footprinting reactions are conducted in four stages. First, the DNA template (also called the probe) containing a protein binding site is prepared and end-labelled. This template is then incubated with the purified protein or crude cell extract followed by the addition of DNase I at concentrations such that each DNA template is nicked on average only once. The digestion products are then separated by high-resolution denaturing polyacrylamide electrophoresis to reveal the 'footprint'—a series of fragments missing in the

Table 5.1 Comparison of methods of analysis of DNA–protein interactions

Method			Protein		DNA			Applications	Limitations
	In vitro	*In vivo*	Crude	Pure	Chromatin	Synthetic oligonucleotides	Cloned Fragments		
Footprinting	√		√	√		√	√	• Provides base-pair resolution of protein binding sites • Can be used to determine DNA–protein binding affinity • Chemical or enzymatic agents can be used for DNA cleavage	• Labour intensive and technically challenging • Requires Mg^{2+} and Ca^{2+} for DNase I activity which are also cofactors for nucleases that may be present in crude protein preparations and will affect probe integrity
Gel mobility	√		√	√		√	√	• Provides some information on relative binding affinities • Reveals multiple protein–DNA complexes • Widely used because of its technical simplicity • Can be used with antibodies to confirm identity of protein of interest	• Cannot be used for weak or transient DNA–protein interactions • Does not localize DNA regions interacting with protein binding site
ChIP (PCR readout)		√	√		√			• Enables analysis of *in vivo* protein binding in the context of native chromatin • Can be used to assess the functional effects of mutations in DNA binding proteins • Single-gene to genome-wide interrogation is possible • Can be quantitative	• Requires ChIP-qualified antibodies • High background signals • Does not localize binding site • Number of genomic regions that can be analysed is limited
ChIP-on-chip		√	√		√			• Hundreds of thousands of genomic regions can be analysed at once • Different dyes can be used for competition experiments (several samples in the same array) • Faster experimental time than ChIP-seq	• Analysis is restricted to the probes present on the chip array • Repetitive regions are eliminated from the design (repeat-mask) • Indirect quantification between dye intensity and peak-finding (normalization required)
ChIP-seq		√	√		√			• Enables genome-wide mapping of DNA interaction sites • Coverage through some repetitive regions (not eliminated by repeat-mask) • Direct quantification of binding sites (peak finding)	• Costly • Requires bioinformatics expertise • Longer experimental work and analysis time

Check marks denote whether assays are performed *in vitro* or *in vivo*, and the DNA and protein formulation requirements.

Figure 5.2 Flowchart of the DNase I footprinting method. The DNA template (also called the probe) containing a protein binding site is incubated with DNase I in (a) the presence or (b) the absence of a DNA binding protein (grey oval). Predetermined concentrations of DNase I are used to produce a population of probes containing an average of one nick per probe. The digestion products are then separated by high-resolution denaturing polyacrylamide electrophoresis to reveal the 'footprint'—the region of the probe protected by protein binding.

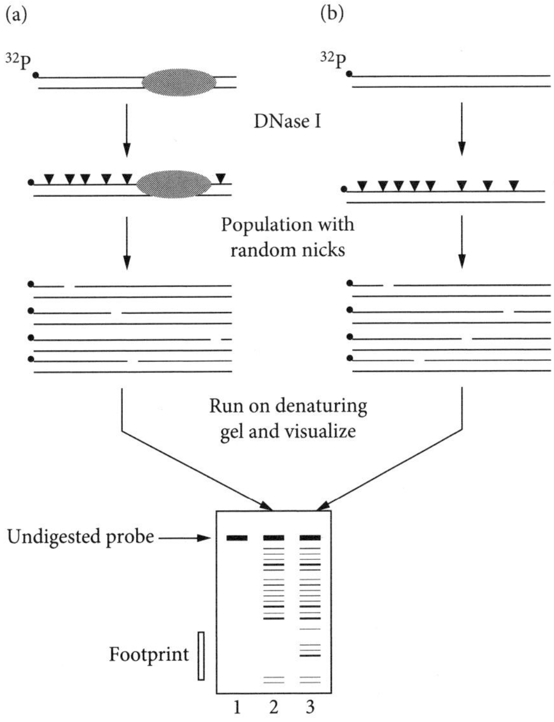

ladder of DNA fragments. This corresponds to the region of the DNA template protected by protein binding and hence DNase I cleavage. These steps are shown as a flowchart in Figure 5.2 together with a schematic representation of the data that are generated.

5.2.2 Step 1: Preparation of the DNA template (probe)

Footprinting requires that both strands of the DNA template are of uniform length and sequence and that only one strand is labelled. The probes are generated by chemical synthesis, PCR, or cleavage with restriction endonucleases, and then end-labelled using one of the probe-labelling strategies listed in Box 5.1. Since only a single strand of DNA can be analysed, experiments are often repeated with the same DNA fragment but labelled on the second strand. The **DNA probe** should be intact, as the presence of nicks will give false-positive signals indistinguishable from genuine DNase I cleavage and could mask an otherwise good footprint. If a radionuclide is used as the label, the probe must be prepared with high specific activity (>1000cpm/pmol DNA) so that very little probe can be used in each reaction. Lower probe amounts translate into lower amounts of DNA

BOX 5.1

DNA probe-labelling strategies

Label type: radionuclide

- Either ^{32}P- or ^{33}P-labelled dNTPs can be used.
- The radionuclide is incorporated enzymatically:
- at the 5′ DNA end with T4 kinase or **Klenow fragment**;
- at the 3′ DNA end with T4 DNA polymerase or terminal transferase.

Advantages

- High sensitivity—very low amounts of probe are needed for each reaction.
- Ideal when protein is limiting.

Disadvantages

- Short probe shelf-life because of radionuclide decay.
- Decay can also cause nicking of the DNA template which may complicate data analysis.
- Extra precautions are required for personal safety and waste disposal.

Label type: fluorophore

- A fluorophore compatible with available equipment can be selected.
- Stoichiometric incorporation of the fluorophore during chemical synthesis is achievable.
- Indirect labelling is also possible by incorporating biotin and any one of a large number of streptavidin–fluorophore conjugates. Biotin incorporation can be achieved:
  - during chemical synthesis of the template;
  - by enzymatic end-labelling as described for radionuclide labelling.

Advantages

- Ease of use and safety.
- Large selection of fluorophores.
- Long probe shelf-life.

Disadvantages

- Lower sensitivity requiring higher mass of probe per reaction.
- Use may not be feasible if protein is limiting.

binding protein needed to produce a footprint. Probes must be used within a few days, and their integrity should be evaluated on a denaturing gel in advance of their use since radioisotope decay can cause nicking.

5.2.3 Step 2: Incubation of DNA template with protein

Once labelled, the probe is incubated with the DNA binding protein (pure protein or protein mixtures). As the DNA–protein interactions are dynamic (i.e. the protein–DNA complex can dissociate and re-form), the DNA binding protein must be present in excess of the probe to produce a successful footprint. Typically 2–5ng of probe (200–800bp in length) is required in reaction volumes of 50–200µL. The amount of protein required depends on DNA binding affinity and can range from 20 to 200nM. Clear footprints are seen when a high percentage of the probe is protein bound. Thus preliminary rounds of experiments are necessary in which the DNase I (see below) and probe concentrations are titrated to identify the extract conditions that will produce a clear footprint.

The formation of DNA–protein complexes is performed in the presence of a binding buffer comprising a buffer (HEPES or Tris), a salt (KCl, NaCl), a reducing agent such as DTT, and either EDTA or EGTA to inhibit any nuclease contaminants. An excess of a non-specific competitor (such as poly dI-dC, poly dA-dT, plasmid DNA, calf thymus DNA, or *E. coli* DNA) is also needed to block non-specific DNA–protein interactions if protein mixtures (lysates or crude fractions) are used.

5.2.4 Step 3: Partial digestion of DNA template

Chemicals such as hydrogen peroxide (a hydroxy radical), dimethylsulphate (DMS), or diethyl pyrocarbonate (DEPC) can be used to digest the DNA. However, footprinting with the nuclease DNase I is the preferred method for studying transcription factor binding of DNA. DNase I binds in the minor groove of DNA and cleaves the phosphodiester backbone of the DNA strand, producing single-strand breaks or nicks. Due to its large size, digestion of DNA by DNase I is prevented by steric hindrance of the DNA-bound protein. Consequently, DNase I footprints are more likely to be specific than those produced by chemical agents.

It is necessary to optimize the concentration of DNase I since DNA competitor concentrations and protein amount and purity affect how much of the nuclease is needed to produce a footprint. Once optimized, footprinting reactions are performed with varying amounts of the protein preparation or fraction. Digestion is typically performed for 1–5 min at room temperature in the presence of 10mM $MgCl_2$ and 5mM $CaCl_2$, essential cofactors for the nuclease.

5.2.5 Step 4: Analysing the digestion products

Once the digestion stage is complete, the DNA is purified and concentrated by precipitation, and the digestion products are analysed. Radiolabelled reaction products are analysed by denaturing electrophoresis on a polyacrylamide gel, followed by autoradiography and quantitation with a densitometer. If you look at Figure 5.3(a), you will immediately notice the footprint—the region of DNA protected by the binding protein. Denaturing slab polyacrylamide gel electrophoresis, which is capable of achieving single-base resolution, was the sole means of analysis in the past when only radioisotope-based methods for labelling DNA were available. Now, with the development of fluorophore-based DNA sequencing and automated capillary DNA analysis instruments, electropherograms are produced for fluorophore-labelled probes. These two analytical methods are compared and discussed further in Box 5.2.

See Chapter 4

By performing footprinting experiments with a dilution series of the DNA binding protein, binding data can be obtained that can then be used to generate **binding curves** and **equilibrium constants** for the interactions at each of the footprinted regions. In the case of the electropherograms, this involves integration of the area under the peaks corresponding to the binding site. For autoradiograms, the films are scanned by a

Figure 5.3 Direct comparison of DNA footprinting methods: 'old' versus 'new' technologies. Comparison of the CbbR protected regions in the *cbb*$_I$ promoter as determined by densitometer traces of autoradiograms of ^{32}P-labelled DNA fragments resolved on a denaturing gel (panels (a) and (c)) or electropherograms of fluorescent dye-labelled fragments (panels (b) and (d)). Binding reactions with CbbR (panels (a) and (b)) or controls with bovine serum albumin (BSA) (panels (c) and (d)) were digested with DNase I as described in the literature (Dubbs and Tabita 2003; Zianni et al. 2006). The open reading frames of the *cbb*$_I$ operon (panel (e)) are depicted as shaded boxes. The arrows indicate the direction of transcription and the two triangles correspond to the DNA regions to which the CbbR protein binds. *Journal of Biomolecular Techniques*, **17**(2): 103–13, reprinted by permission of the Association of Biomolecular Resource Facilities.

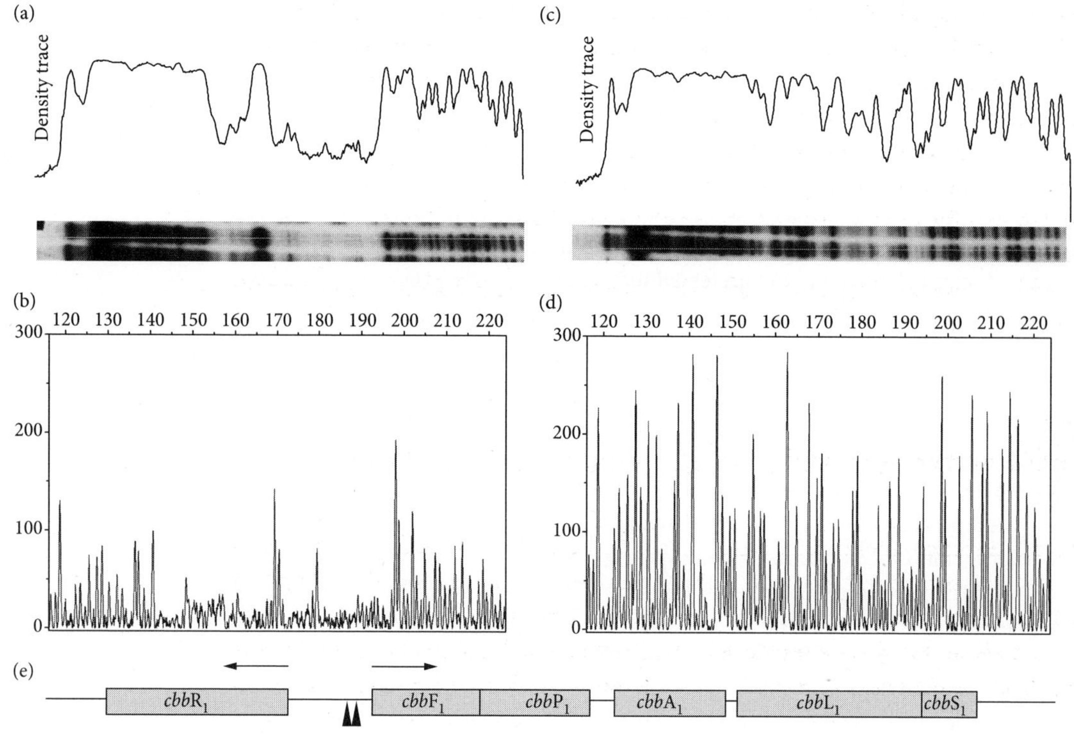

BOX 5.2 Comparison of DNA footprinting methods: old versus new technologies

Figure 5.3 depicts a footprinting experiment comparing two analytical methods, the traditional radionuclide-based method (panels (a) and (c)) and dye-primer sequencing on an automated capillary DNA sequencing instrument (panels (b) and (d)), for the CbbR protein which binds to the promoter of the *cbbI* operon of *Rhodobacter sphaeriode*. This bi-directional operon encodes enzymes involved in carbon dioxide fixation (Gibson et al. 1991; Dubbs et al. 2000). Motivated to generate an alternative non-radionuclide method for DNase I footprinting, Zianni et al. (2006) reproduced the footprint of the CbbR protein published by Dubbs et al. (2000) using an identical DNA fragment for the probe but labelled with a fluorophore. The original footprint was produced using a ^{32}P-labelled probe and visualized by autoradiography of the DNA fragments resolved on a denaturing polyacrylamide gel (Dubbs and Tabita 2003).

Figure 5.3 shows that the regions absent in the autoradiogram in panel (a) line up exactly with the peaks of reduced height in the electropherogram in panel (b), corresponding to DNA protected regions. No protected regions are discernible in panels (c) and (d) when only BSA was present in the binding reaction.

densitometer (preferably a two-dimensional optical scanning device) to enable digitization of the bands. Following normalization of the total amount of DNA loaded onto each lane, the fractional protection of the site, which is proportional to fractional saturation, can be calculated and plotted against protein concentration to produce a binding curve.

When analysing the digestion products, molecular weight markers and undigested probes are usually run in parallel to the footprinting samples to assist in the identification of the residues within the footprinted region. The undigested probe is included as a quality control measure to verify that no internal nicks were present. One limitation of the footprinting method is that DNase I does not cut DNA uniformly, with some sequences being rapidly cleaved whereas others remain intact even after extensive digestion. This results in gaps and bands of varying intensity, producing an uneven ladder that limits the resolution of this technique. The formation of the DNA–protein complex can result in structural changes in the DNA, such as bending, which may also result in the creation of some DNase I hypersensitive sites. These are identified as bands of greater intensity in protein-containing lanes relative to non-protein or non-specific-protein control lanes. A further limitation of the DNA footprinting is that it is labour intensive and technically challenging. However, it can be used to analyse low-affinity DNA binding proteins whose binding is too weak to be detected by the simpler gel mobility shift method described in section 5.3.

5.3 Electrophoretic mobility shift assay (EMSA)

The electrophoretic mobility shift assay (**EMSA**), also known as gel shift or gel retardation, is another *in vitro* technique that can be used to analyse protein binding to a defined DNA sequence of interest. First described in 1981 in two independent research articles (Fried and Crothers 1981; Garner and Revzin 1981), the method relies on the principle that the electrophoretic mobility of a protein–nucleic acid complex is reduced relative to the free nucleic acid in a native polyacrylamide or agarose gel. As the method is technically simple, EMSA is often selected for preliminary experiments designed to identify regions in a stretch of DNA where DNA–protein interactions occur.

5.3.1 The EMSA technique

There are four stages in the EMSA method. A double-stranded DNA fragment containing a protein binding site is prepared and end-labelled. The DNA probe is then incubated with a protein fraction (such as purified protein, protein complex, or nuclear extract) and protein–DNA complexes are allowed to form. The samples are then subjected to electrophoresis in a polyacrylamide or agarose gel under native conditions to separate free probe from protein-bound probe. Finally, the gel is dried and the position of the probe is detected. These steps are shown as a flowchart in Figure 5.4 together with various control reactions and a schematic representation of the data that are generated. Lane 2 shows that a fraction of the probe bound by protein or protein complexes migrates with an apparently higher molecular weight (being retarded in the gel) than the free unbound probe run in lane 1. The ability to separate free and bound probes makes EMSA assays useful in the determination of the relative affinities of a protein or

Figure 5.4 EMSA and supershift assay set-up. This schematic diagram shows a typical experimental set-up for an EMSA or supershift assay including the combination of experimental and control reactions that are required. The various binding reactions and their components are illustrated in the top two panels. Schematic diagrams of the protein–probe complexes that are formed in the binding reaction are depicted in the middle (third) panel and their relative migration on a native polyacrylamide gel is shown in the bottom panel. Lane 1, DNA probe only; lane 2, DNA probe with DNA binding protein (Ø, no competitor); lane 3, competition reaction with an excess of an unlabelled DNA fragment identical in sequence (wild-type) to the labelled probe; lane 4, competition reaction as in lane 3 but the excess of unlabelled DNA fragment contains point mutations which abolish protein binding; lane 5, supershift reaction produced by the binding of a protein specific antibody to the DNA–protein complex; lane 6, antibody competition reaction where the addition of excess neutralizing peptide blocks antibody binding to the DNA-protein complex, demonstrating the specificity of the supershift detected in lane 5.

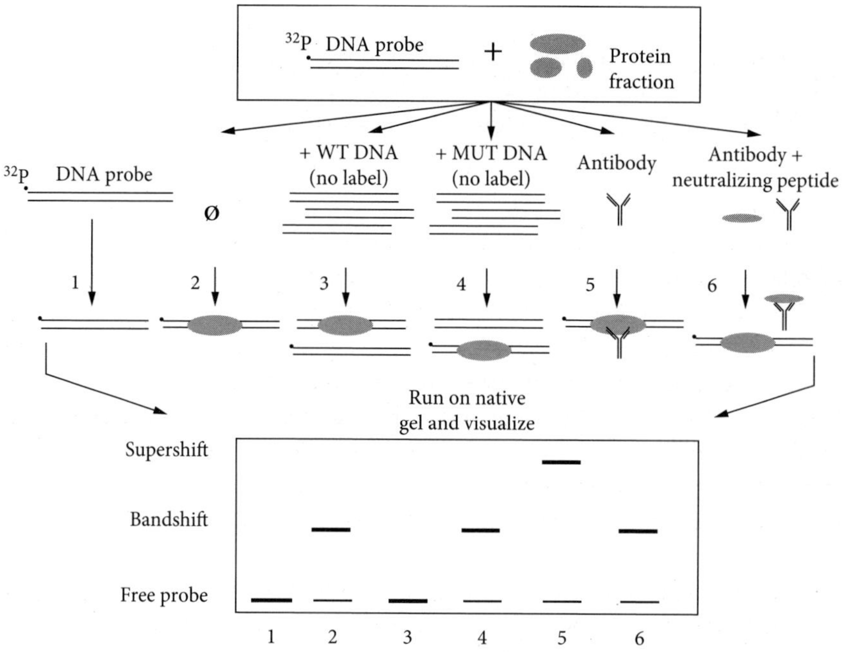

transcription factor for different DNA sequences or to compare the affinities of different proteins for the same site. Antibodies specific for a protein of interest can also be used in EMSA experiments to confirm the presence of protein. Migration of complexes formed as a result of antibody binding is slowed, resulting in a **supershift**. This is depicted in Figure 5.4 and described further in section 5.3.3.

5.3.2 Step 1: Preparation of the DNA template (probe)

DNA template or probe preparation for use in EMSA is identical to that for the foot-printing method (see section 5.2). The probe used in EMSA is usually a double-stranded DNA fragment; both strands are of uniform size and sequence and are generated by either annealing of complementary synthetic oligonucleotides or gel purification of a cloned restriction fragment bearing the binding sequence of interest. A typical target duplex DNA has a length of 20–35bp. Longer probes can be used, but the longer the fragment, the more likely it is that multiple binding sites are present, increasing the likelihood of the formation of multiple protein–probe complexes and non-specific binding. If the use of a long DNA probe is necessary, EMSA results will not identify the nucleotides or the location of the sequences that are contacted by the protein(s). In this case, complementary assays, such as DNA footprinting (see section 5.2), can be used to identify specific DNA binding sites.

EMSA probes can be labelled using the methods described in Box 5.1. Experiments using radionuclide-labelled probes provide higher sensitivity and enable use of smaller quantities of protein, and are more quantifiable than biotin or fluorophore-based EMSAs which are nevertheless increasingly popular.

5.3.3 Step 2: Incubation of the probe with protein fraction

The second step is the incubation of the probe with either a cellular extract or purified protein. As with footprinting, a non-specific competitor (such as poly dI-dC, poly dA-dT, plasmid DNA, calf thymus DNA, or *E. coli* DNA) is also added to eliminate non-specific interactions. The amount of protein or extract needed to produce a gel shift should be determined experimentally. Depending on the amount of active protein in the sample, the amount needed could range from 0.25μg for an abundant protein to 10μg if the protein is less abundant. However, an excess of protein extract might result in generation of an artefact through non-specific interactions with the DNA probe.

Negative controls in an EMSA experiment include reactions where the labelled probe is incubated without any protein extract and/or with an irrelevant control protein. Optimally, a positive control reaction with a known DNA binding protein should be included, although this protein might not always be available. Competition assays using a molar excess of unlabelled DNA probe of identical sequence to that of the labelled probe (Figure 5.4, lane 3) are performed to demonstrate that the observed mobility shift is exclusively due to binding of the probe to components of the extracts and is not an artefact. This competition should result in the partial or total disappearance of the shifted band owing to binding of the protein components to the excess of unlabelled probe. Using this principle of competition, binding specificity can be demonstrated by using an excess of unlabelled variants of the labelled probe which contain altered sequences not recognized by the DNA binding protein. Look at lane 4 in Figure 5.4, where the addition of excess of a mutated DNA does not compete for protein binding and so the bandshift remains as in lane 1. On the other hand, if irrelevant sequences can efficiently compete with the original probe, the DNA binding is not sequence-specific and the corresponding bandshift should be ignored. It is also possible to determine the relative apparent affinity of the different DNAs by performing competition assays with increasing concentrations of competitor.

Antibodies specific for a protein of interest can also be used in EMSA experiments to confirm the presence of the protein of interest in the gel-shifted band. The antibody is usually incubated with the protein or nuclear extracts before DNA probe addition. The extra mass of the antibody results in a complex whose electrophoretic mobility is slower than the related no-antibody protein–DNA complex (Figure 5.4, lane 5), producing what is called a 'supershift'. Alternatively, if antibody binding blocks DNA–protein interaction, the observed DNA–protein complex is diminished, yielding essentially the same information. The specificity of the antibody–protein interaction can be demonstrated by including a reaction with an antibody-neutralizing peptide. As this peptide will compete with the protein for antibody binding, the supershift will be diminished or abolished (Figure 5.4, lane 6).

The use of antibodies in EMSA experiments requires some prior knowledge about which proteins bind the DNA probe, and the availability of antibodies capable of recognizing native protein. Consequently, this technique is not suitable for identifying novel proteins. The nature of the antibody–protein interaction must be stable and strong enough to withstand electrophoresis. A preliminary survey of a panel of antibodies is often required to identify a supershifting antibody.

5.3.4 Steps 3 and 4: Analysing the protein–probe complexes

The resulting protein–DNA complexes are subjected to electrophoresis under native conditions. Although both native polyacrylamide and agarose gels can be used for EMSA, resolution is usually better with polyacrylamide gels.

Following electrophoresis, handling of the gel differs depending on how the probe was labelled. If a radionuclide-labelled probe was used, the gel is imaged using an X-ray film or a phospho-imager (Fried and Crothers 1981). With fluorophore-labelled probes, the gel is imaged directly with an appropriate instrument (Ruscher et al. 2000). For biotinylated probes, electrophoretic transfer of the probe to a nylon membrane followed by UV cross-linking to immobilize the transferred material is required. The membrane is then incubated in a solution containing streptavidin conjugated to either a fluorophore or an enzyme such as horseradish peroxidase (HRP). While fluorophore-containing probes can be directly imaged, when HRP is used the membrane is incubated in an appropriate chemiluminescent substrate and imaged with a digital CCD camera or X-ray film (Rosenau et al. 2004).

CASE STUDY 5.1 **Interpreting EMSA results**

Treatment of HeLa cells with the potent tumour-promoting phorbol ester 12-*O*-tetradecanoylphorbol-13-acetate (TPA) leads to various biological changes that mimic those observed in cells transformed by chemical carcinogens or viruses.

The aim of the experiment is to identify if the transcription factor NF-kB p50 is activated in TPA-treated U-937 cells—a human histolytic lymphoma cell line. Activation of NF-kB results in its translocation into the nucleus from the cytoplasm where it was sequestered in an inactive state by an associated inhibitory protein. For this purpose, a mobility shift and supershift assay was performed using a radiolabelled double-stranded oligonucleotide including a NF-kB p50 binding site and nuclear extracts derived from TPA-treated U-937 cells. The results are shown in Figure 5.5. Competition experiments with a 100-fold molar excess of unlabelled double-stranded wild-type or mutant oligonucleotide were also performed (lanes 2 and 3, respectively). Polyclonal antibody against NF-kB p50 factor was included in lane 4 and the same antibody plus a neutralizing peptide was included in lane 5.

Figure 5.5 EMSA analysis of DNA binding by NF-kB p50. A ^{32}P-labelled double-stranded DNA oligonucleotide probe containing a consensus NF-kB binding site was incubated with nuclear extracts from TPA-treated U-937 cells. The various reaction components are listed above the gel. The positions of the free probe, three shifted probe complexes (denoted by a, b, and c), and a super-shifted probe are indicated. Reprinted by permission of Active Motif Inc.

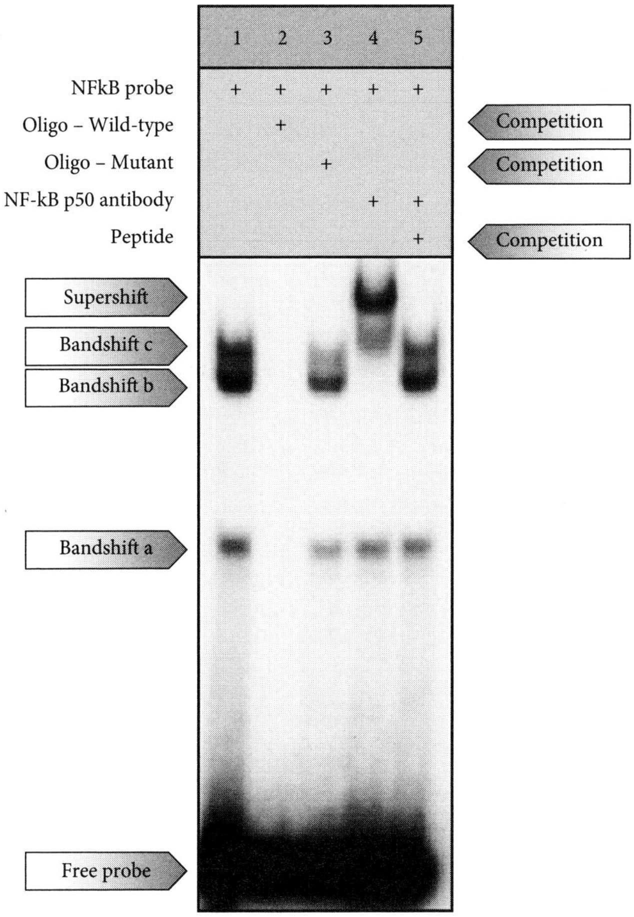

QUESTION
What do the different bands that appear in the gel mean?

Free labelled DNA from each reaction that is not bound to any protein appears as a band at the bottom of the gel. It migrates the furthest because it is free DNA without any protein and hence has the lowest molecular weight. DNA–protein complex formation can be seen as different bands that run more slowly on the gel in different lanes (1, 3, 4, 5). The addition of excess unlabelled competitor wild-type oligonucleotide (lane 2) was able to abolish the three bandshifts in lane 1, meaning that all those bands contain specific DNA–NF-kB complexes. This is supported by the inability of the mutant oligonucleotide added to lane 3 to compete effectively with the labelled DNA for binding to NF–kB proteins. You may have noticed that the intensity of the bands in lane 3 is slightly diminished with respect to those in lane 1. There are two possible interpretations. Either there is a little competition of the mutant oligonucleotide for NF-kB binding, or, for an unknown reason, this lane contained less radioactivity probe, perhaps due to loss of sample during the assay. In lane 4, the anti-p50 antibody generated a supershift of bands labelled b and c, so we can conclude that these complexes contain p50 protein. The third bandshift c does not move, so it does not include p50 protein. The supershift is specific for p50 protein because the neutralizing peptide from lane 6 is able to abolish the supershifting of the protein–DNA complexes.

The major advantages of EMSA as a technique for studying DNA–protein interactions are its simplicity and sensitivity. Due to its high sensitivity, it requires very low protein and nucleic acid concentrations (≤0.1nM) and small sample volumes (≤20μL). In addition, it is compatible with both highly purified proteins and crude cell extracts, and mobility shifts can be detected for a wide range of protein sizes, from small peptides to large multi-protein transcription complexes with Mr $\geq 10^6$. However, since EMSA requires that the DNA–protein interactions are strong and stable enough to withstand electrophoresis, weak or short-lived DNA–protein interactions may not be detected or may be altered when this method is used. The electrophoretic mobility of a protein–DNA complex depends on many factors besides the size of the protein. Consequently, the magnitude of the mobility shift does not provide much information about the molecular weight or identity of the proteins present in the complex. In situations when several proteins are known to recognize the same DNA sequence, this limitation can be redressed to some extent by using protein-specific antibodies in supershift assays (Figures 5.4 and 5.5) or in antibody-based EMSA variant assays, as described in section 5.3.5.

5.3.5 Alternative non-gel methods for the classical EMSA

The simple principle behind the EMSA technique has prompted development of alternative approaches to achieve the separation of protein–probe complexes from unbound probe. One of these methods is a variation of ELISA, where instead of using antibodies the probe is immobilized in the wells of a microtitre plate. When nuclear or whole-cell extract is added to the well, the protein of interest binds the oligonucleotide at its consensus binding site. The presence of bound protein is detected by protein-specific antibodies using detection methods commonly used in ELISA. For these assays to be quantitative a standard curve with purified protein is required (Renard et al. 2001; Shen et al. 2002; Brand et al. 2010).

The ELISA variant of EMSA provides several advantages over the original electrophoretic method. It is 10- to 100-fold more sensitive and requires only nanograms of purified protein or few micrograms of nuclear or whole-cell extracts. It also eliminates the use of radioactivity and the need to run gels, and therefore is less time consuming. Results can be obtained in less than 5 hours. Luminescent or colorimetric readout enables easy analysis that can be quantitative, and the assays are performed in a 96-well plate so that high throughput of samples is possible. As with EMSA, the specificity of the protein–DNA interaction can be demonstrated in competition experiments with oligonucleotides in solution. However, the method needs a highly specific antibody to prevent off-target antibody binding which would otherwise produce a false-positive result arising from the detection of a cross-reacting protein.

Other variants of classical EMSA include use of RNA probes to study RNA–protein interactions, and 'reverse EMSA' where the protein is radiolabelled instead of the DNA and bound nucleic acid is analysed (Worthington et al. 2002). EMSA assays can also be coupled to other techniques, such as Western blotting to identify the binding proteins or mass spectrometry to identify protein size (Granger-Schnarr et al. 1988; Woo et al. 2002), or electron microscopy to analyse the conformation of the binding protein (Jett and Bear 1994).

See Chapters 9, 13, and 17

5.4 Chromatin immunoprecipitation

In contrast with the *in vitro* footprinting and EMSA methods described in the preceding sections, the **Ch**romatin **I**mmuno**P**recipitation (ChIP) assay allows *in vivo* identification of DNA sequences associated with proteins when the DNA is in its native chromatin state. The technique is based on the principle of immunoprecipitation that is described in Chapter 9. A protein-specific antibody is used for immunoselection of chromatin fragments containing (bound to) the protein of interest. The associated DNA is then purified and analysed by PCR, sequencing, or hybridization to a microarray. The technique was first developed in the 1980s (Solomon and Varshavsky 1985) and has been instrumental in advancing our knowledge of how post-translational modifications of histones at their N-terminal histone tails are involved in cellular processes such as regulation of gene expression, repair of DNA damage, apoptosis, and mitosis (Cheung et al. 2000; Jenuwein and Allis 2001; Wells and Farnham 2002; Kouzarides 2007).

ChIP is unique amongst the various methods used to analyse DNA–protein interactions because it can be used to study the distribution not only of proteins which make direct physical contact with DNA (like histones and transcription factors) but also those which do not actually bind DNA but are associated with the chromatin through protein–protein interactions. In this section, we will discuss the traditional ChIP method followed by the **ChIP-on-chip** and **ChIP-sequencing** approaches.

5.4.1 Traditional ChIP technique

The ChIP assay can provide an in-cell snapshot of the physical location or distribution of the protein of interest, either throughout the genome or at specific loci. The key steps involved in the ChIP assay are represented as a flowchart in Figure 5.6. First, the cell culture or tissue samples are treated with a reversible cross-linking agent and the chromatin is fragmented. This stage is followed by selective immunoprecipitation of the protein(s) of interest and their associated DNA fragments. The chemical cross-linking is then reversed, the enriched DNA purified and the purified DNA analysed by PCR, hybridization, or sequencing.

See Chapters 2 and 4

5.4.2 Step 1: Chemical cross-linking and chromatin extraction

For ChIP-based analysis of transcription factors and chromatin-associated proteins, a reversible cross-linker is used to create temporary physical bonds between DNA–protein

Figure 5.6 Schematic diagram of the chromatin immunoprecipitation (ChIP) technique. Cells are treated with formaldehyde to cross-link chromatin with its associated proteins. Next, the cells are lysed and chromatin is extracted, sheared, and precipitated using an antibody specific to the protein of interest. Antibody–chromatin complexes are captured using beads coated with either protein A or protein G. Use of paramagnetic beads (depicted) improves chromatin recovery because bead loss in the subsequent wash steps is significantly reduced. Following cross-link reversal, the DNA is purified and can be analysed (using PCR microarray or sequencing) to determine which gene(s) is/are bound by the protein of interest. Reprinted by permission of Active Motif Inc.

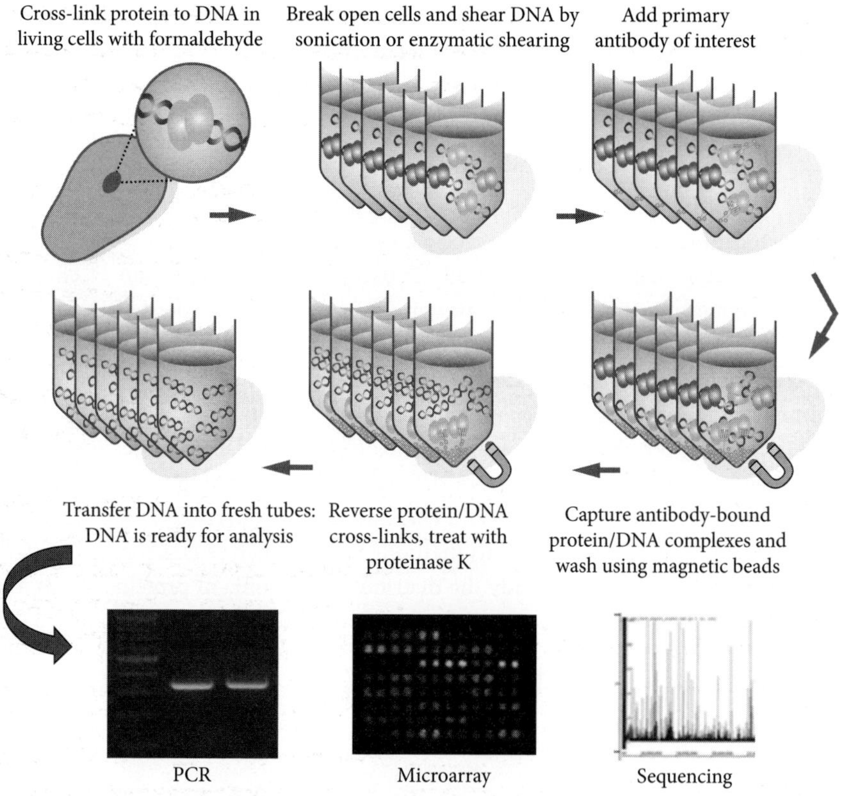

and chromatin–protein complexes. However, when analysing the distribution of histone modifications, native ChIP (no cross-linking) is used.

While formaldehyde is the most widely used cross-linking agent, other reversible chemical agents can be used, as can UV light (Gilmour and Lis 1985). One of two approaches is used for lysate preparation. Either the entire sample is lysed (whole-cell lysate) using strong ionic detergents, such as SDS and nuclease inhibitors, to maximize chromatin extraction efficiency and preserve DNA integrity, or the nuclei are first recovered in a hypotonic solution to reduce background interference from cytoplasmic proteins.

5.4.3 Step 2: Chromatin fragmentation

The next step is to fragment the chromatin, which not only reduces sample viscosity, but is also critical for efficient formation of antibody–chromatin or antibody–protein–chromatin complexes. Fragmentation is achieved either mechanically through sonication, or enzymatically using either a restriction endonuclease, with four-base recognition specificity, or micrococcal nuclease, which cleaves DNA between nucleosomes in a sequence-independent manner. Each fragmentation method has its advantages and limitations. Sonication is more prone to variations in nuclear lysis or over-cross-linking than enzymatic digestion and therefore is more suited to tissues or cells which are hard to lyse. By contrast, enzymatic digestion has the advantage that it can be performed under controlled conditions to ensure reproducibility among samples once all the experimental conditions are optimized. The readout method of the ChIP experiment dictates the desired fragment length. If endpoint or quantitative PCR is used, chromatin fragments of length 200–1000bp are preferred to ensure that both PCR primers can co-localize to the same DNA fragment. However, for hybridization (ChIP-on-chip) and sequencing (ChIP-seq) readouts, smaller fragment of length <200bp are preferred to improve read resolution. ChIP-on-chip and ChIP-seq specific adaptations of the traditional ChIP method are described in sections 5.4.7 and 5.4.8.

5.4.4 Step 3: Immunoprecipitation

The next step in ChIP is the immunoprecipitation of the protein of interest and its associated chromatin fragments using an antibody specific to the protein of interest. Since ChIP is an enrichment technique dependent on the specificity and capture efficiency of the antibody, a reference sample is needed for comparison to assess the level of enrichment. In addition, control immunoprecipitations are needed to monitor antibody specificity. To measure enrichment, a fraction of the starting material is saved and used as the reference 'input chromatin' in the assay's readout steps. This material can be used to normalize for the amount of DNA present across samples, thereby enabling inter-sample comparisons of the relative abundance of the DNA region of interest in the enriched fraction. To demonstrate antibody specificity, control immunoprecipitations are performed with a non-specific antibody or pre-immune serum. Antibodies previously validated for ChIP can also be used as controls to confirm correct execution of the technique.

A major challenge to a successful ChIP experiment is the identification of ChIP-compatible antibodies. Not all antibodies that function in immunoprecipitation will work in ChIP. The cross-linking steps used to preserve protein–chromatin interactions in ChIP may adversely affect antibody recognition of its antigen, making these antibodies useless in ChIP. In other cases, the antibody–antigen affinity may not be sufficiently strong to withstand the mechanical forces associated with the immunocapture of a 100–1000bp chromatin fragment, and their enrichment efficiency is too poor. Weak enrichment translates into higher background relative to input, and therefore interpretation of the results becomes more difficult.

5.4.5 Step 4: DNA capture and isolation

Antibody–chromatin complexes are collected using paramagnetic, agarose, or sepharose beads coated with either protein A or G or a secondary anti-host antibody which reacts with the immunoprecipitating (primary) antibody. Next, a series of wash steps of increasing stringency are performed to reduce non-specific associations. To isolate the immunopreciptiated DNA from the protein to which it is bound, the formaldehyde cross-links are first reversed by heating the samples. Antibody and chromatin proteins are then digested with proteinase K and the DNA is purified by either phenol–chloroform extraction or **affinity chromatography** using silica-based DNA purification spin columns.

This isolated DNA is now enriched for the fragments that were associated with the protein of interest and is ready for analysis. The amount of enriched DNA isolated after a ChIP experiment is orders of magnitude less than the starting material. Usually a ChIP assay starts with micrograms of chromatin and results in nanograms of enriched material (depending on the antibody affinity and abundance of the protein in the sample). In general, histone proteins are very abundant and result in ChIP enrichments in the hundreds of nanograms. However, transcription factors are not as abundant and therefore may result in sub-nanogram amounts of enriched material.

5.4.6 Step 5: Analysing the captured DNA

The methods of analysis of ChIP-enriched DNA have evolved in parallel with the technological advances made in DNA analysis. Initially, the enriched chromatin was analysed by immobilization on nitrocellulose and probing with labelled DNA complementary to the loci of interest. However, this approach limited the sensitivity of the technique and hence PCR-based readouts were developed. At first endpoint PCR assays and gel electrophoresis were used to visualize the amplified fragments, but these have now been replaced by real-time quantitative PCR (qPCR). Figure 5.7 shows both these readout methods. Case study 5.2 shows that while there is very good correlation between the two methods in the observed enrichment of the genomic target analysed, quantitative information can be extracted from a quantitative PCR experiment. This includes percentage recovery (DNA yield from the ChIP relative to Input DNA) and percentage enrichment (DNA yield from the ChIP relative to DNA yield from the IgG immunoprecipitation).

See Chapter 2

5.4.7 ChIP-on-chip

ChIP-on-chip is a technique in which ChIP-enriched DNA is analysed by **microarray technologies** instead of qPCR. It enables the identification of DNA-binding proteins not from a few regions, as is the case for ChIP, but from tens of thousands of DNA regions, i.e. on a genome-wide scale. However, the amount of DNA isolated by ChIP is sufficient for the analysis of only a few genomic regions by PCR. To overcome this limitation, the ChIP-enriched DNA has to be amplified by one of two approaches: ligation-mediated PCR or random priming amplification, with the latter being the preferred method.

Amplification of DNA by random priming amplification uses degenerate oligonucleotide sequences that are followed by a linker of known and conserved sequence. The amplification is performed with a defined number of cycles to achieve a linear rather than an exponential amplification of DNA. This linear amplification ensures that all the DNA fragments in the enriched material will be amplified equally to minimize any bias, although some sequences may amplify better than others. Amplification results in a two- to threefold increase in the amount of DNA, producing micrograms of DNA which is sufficient for whole-genome analysis with DNA microarrays. The DNA is then labelled

Figure 5.7 Endpoint and real-time PCR analysis of a typical ChIP experiment. (a) Images from ethidium bromide stained 2% agarose gels showing products of endpoint PCR analysis of chromatin captured by various antibodies: RNA polymerase II (RNAP) without (lane 1) or with a bridge (rabbit anti-mouse) antibody (lane 2), histone H3K4me3 (lane 3), H3K27me3 (lane 4), and non-specific IgG (lane 5) as a negative control. Water is used as a PCR negative control (lane 6). Input DNA (lane 7) is an aliquot of original chromatin reserved for use as a positive control for the PCR reaction. The genomic regions analysed by PCR (listed on the right) are the transcriptionally active promoters for RNA polymerase II (RNAP, top panel) and glyceraldehyde 3-phosphate dehydrogenase (GAPDH, middle panel), and the transcriptionally repressed HoxB13 promoter (bottom panel). H3K4me3 and RNA polymerase II are associated with transcriptionally active open chromatin while H3K27me3 is a well-known silencing marker that identifies transcriptionally repressed genes. (b) Graph illustrating quantitative PCR for the GAPDH promoter. Quantitation of GAPDH promoter sequences isolated by ChIP from HeLa cells using the same antibodies as in (a). PCR primers corresponding to the GAPDH using the Input DNA as the standard curve for quantification.

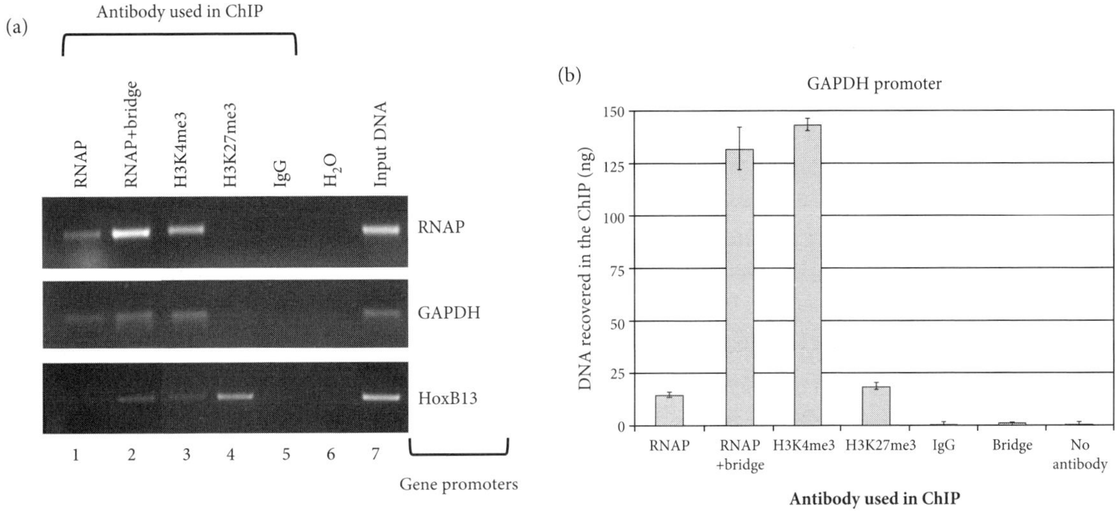

Antibody considerations for ChIP

CASE STUDY 5.2

Figure 5.7 shows endpoint and real-time PCR analysis of a typical ChIP experiment.

QUESTION 1
What do the data in Figure 5.7 show?

By comparing band intensities in the RNA polymerase II and H3K4me3 ChIPs (lanes 1–3) with that of the H3K27me3 ChIP (lane 4) you can see that that RNA polymerase II (top panel) and GAPDH genes (middle panel) are both transcriptionally active in HeLa cells, while HOXB13 is transcriptionally repressed. RNAP with bridge (secondary) antibody and histone H3K4me3 antibody show the highest enrichment levels, whereas the IgG, bridge alone, or no antibody negative controls depict the low background in this experiment. Overall, there is good correlation between endpoint and real-time PCR analysis. However, real-time PCR is a more robust method since it is able to reflect the enrichment levels with higher precision.

Antibodies can be generated in different animal hosts and prepared in a variety of formulations ranging from whole-serum immunoglobulin fractions to affinity purified over immunogen-containing columns. The primary immunoprecipitating RNA polymerase II antibody is a mouse monoclonal antibody. The other antibodies used in the experiment are rabbit polyclonal antibodies. In Figure 5.7(b) you can see that a significant enrichment difference was obtained in the RNAP ChIP when a bridge antibody was used (compare the RNAP bar with the RNAP + bridge bar).

QUESTION 2

Why and how can a bridging antibody improve chromatin enrichment?

A bridging (secondary) antibody functions as a high-affinity linker between the beads and the immunoprecipitating antibody (primary antibody).The ChIP technique depends on the efficient capture of antibody–chromatin complexes. Typically this is achieved with either protein-A- or protein-G-coated beads. Antibody affinity for binding protein A or G varies between immunoglobulin subtypes and between species of the animal host, and cannot always be predicted. In this example, the primary immunoprecipitating RNA polymerase II antibody is a mouse monoclonal antibody of an isotype with a modest to low affinity for protein G. Addition of a rabbit anti-mouse bridging antibody results in a tenfold improvement in the capture efficiency of the RNA polymerase II antibody. The other antibodies used in the experiment are rabbit polyclonal antibodies capable of high-affinity binding to the protein G beads.

with fluorescent dyes such as Cy5 or Cy3 fluorescently labelled nucleotides (Figure 5.8(a)) and the DNA sample is deposited into the array. The DNA microarray comprises single-stranded DNA fragments attached to a solid surface, such as glass, to which complementary fluorescently labelled fragments hybridize. Hybridization is carried out under denaturing conditions in the presence of non-specific DNA (such as salmon or herring sperm DNA) to block the sample from binding to the array surface. The length of the DNA fragments varies from 20 to a few hundred bases and the selection of the length depends on the resolution and the hybridization stringency requirements. The main advantage of DNA microarrays is the density with which the spots can be placed, ranging from thousands to millions of unique DNA probes in a single glass slide. A small genome such as the yeast organism can be placed into a single slide, whereas several slides may be required to represent the entire genome of a more complex organism, such as the human.

After incubation of the ChIP material, the array is washed to remove any non-binding DNA sequences. Finally, the array is scanned at a specific wavelength to detect the labelled nucleotides bound to the array and the enrichment levels are plotted relative to the levels observed in the reference input sample. Figure 5.8(b) shows a typical ChIP-on-chip experiment at two distinct chromosomes (chromosome 19 (top panel) and chromosome 18 (bottom panel)) and two different resolutions (megabases and kilobases).

5.4.8 ChIP-sequencing (ChIP-seq)

See Chapter 4

With the recent development of high-throughput parallel DNA sequencing technologies capable of generating tens to hundreds of millions of sequence reads from samples containing complex mixtures of DNA fragments, direct sequencing of ChIP DNA can provide genome-wide analysis of protein–DNA interactions. Chapter 4 describes the sequencing technology platforms available, the most relevant of which are from

Figure 5.8 ChIP-on-chip experiment. (a) Schematic flowthrough of a ChIP-on-chip experiment: (1) cells are cross-linked and DNA is sonicated; (2) an antibody is used to pull down and enrich for DNA sequences interacting with the specific protein of interest; (3) DNA material is purified (the left side is the ChIP-enriched material and the right side is the Input DNA (not enriched, and used as a reference)); (4) each DNA is amplified separately by random priming; (5) each DNA is labelled with fluorescent dyes; (6) both DNA samples (ChIP-enriched and Input) are combined in a DNA microarray. (b) Typical results from a ChIP-on-chip experiment. The top panel depicts the enrichment values for an entire human chromosome in three independent H3K9me3 ChIP-on-chip experiments (A, B, C). Enrichment values relative to the Input DNA are shown on the *y*-axis. The *x*-axis represents the chromosomal coordinate. Note that the centre region (around 30Mb) is devoid of signal; this region is the centromere and is not represented on the DNA microarray because of the presence of repetitive elements. the bottom panel shows the binding patterns for an 80 000nt region in three independent H3K9me3 ChIP-on-chip experiments (A, B, C). The *x*-axis represents the chromosomal coordinate, and the *y*-axis represents the enrichment levels. Note the difference in scale (megabases and kilobases) which allows identification of peaks and binding patterns on the DNA. For a colour reproduction of Figure 5.8, please see Plate 4.

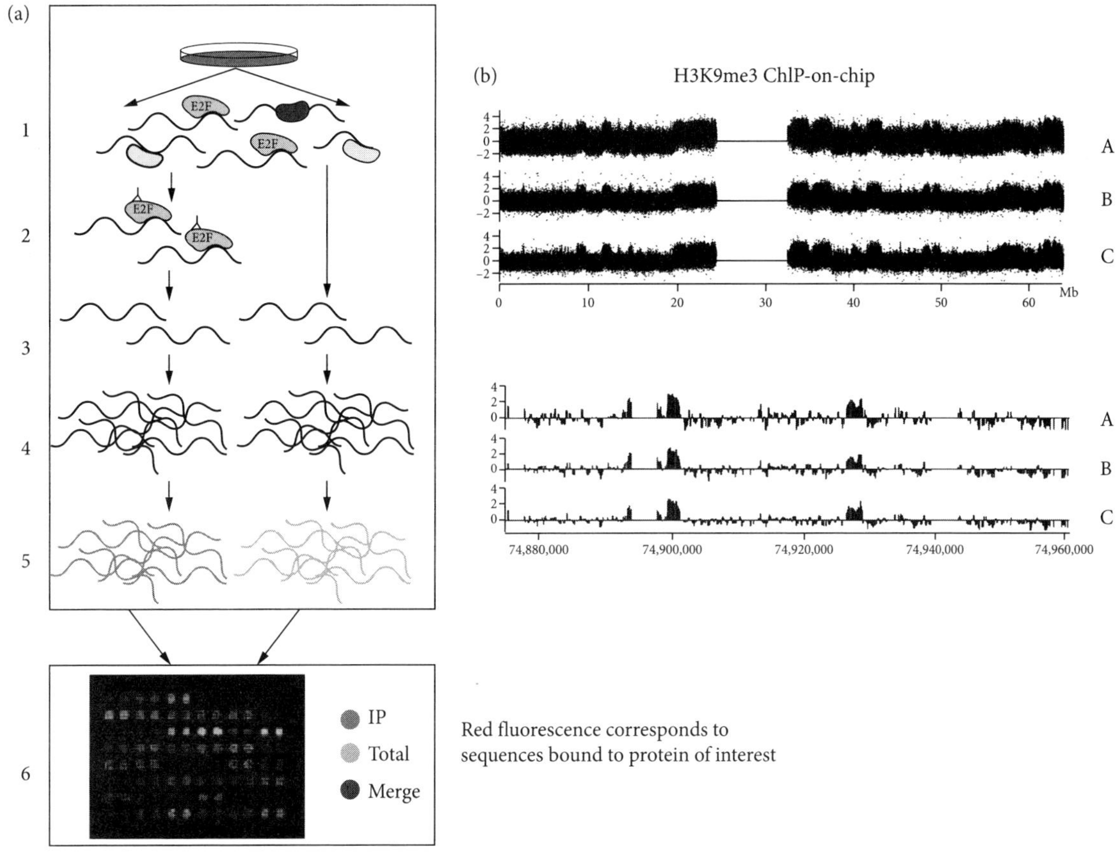

Illumina, Life Technologies, and Roche (Morozova and Marra 2008). Sample preparation for high-throughput sequencing is generally referred to as 'library construction' (Bentley et al. 2008; Quail et al. 2008) and is described in Chapter 4. These sequencing platforms generate short-sequence reads (e.g. 36nt) which correspond to the ends of the DNA fragments isolated by ChIP. For most ChIP-seq experiments, these short reads are sufficient to map the ChIP-isolated DNA fragments to their appropriate genomic locations. These read alignments are called tags, and their abundance in the sequenced sample is translated into local enrichments. Tag frequencies in the sample are plotted as peaks and represent candidate protein-binding sites.

A single sequencing reaction (e.g. one lane on an Illumina genome analyser) generates 25–30 million reads, of which 50–80% will map uniquely to the reference genome. Eight to fifteen million tags produce sufficient data for genome-wide mapping of binding sites for most DNA binding proteins. The sequence and alignment information is stored in a huge table containing detailed information for each sequencing attempt (such as location on flow cell, sequence quality, alignment quality, and of course chromosomal location for those reads where a match to the genome was identified). Analysis of this sequencing output file requires some expertise in bioinformatics and computing. Although the details of these post-sequencing data processing steps may differ between investigators and laboratories, the following basic steps will be performed in most studies.

1. Generation of a 'signal map' showing tag metrics along the genome. Different parameters can be used to make such signal maps. In the example shown in Figure 5.9(a), each alignment was extended *in silico* to the fragment length in the library (e.g. 140bp) and the number of fragments counted for all 32bp segments across the genome. The resulting histogram was visualized in the Integrated Genome Browser (http://bioviz.org/igb/). Figure 5.9(a) shows a 30Mbp view of p53 ChIP-seq data (bottom track). Note the many peaks in the ChIP sample (bottom track), and almost no peaks in the control Input sample (top track), which demonstrates the enrichment of the p53-bound DNA fragments by the ChIP procedure.

2. Peak finding—the identification of genomic regions with local enrichments of sequence tags, indicating DNA binding of the protein of interest. Many open-source programs are available for this step (Pepke et al. 2009). Figure 5.9(b) shows the ~24kb region containing the CDKN1A gene and five peaks identified by the peak caller MACS (Zhang et al. 2008). The peak intervals are shown as solid bars under each peak. The top track again shows the signal map of the Input control. This control is highly recommended since it will allow the peak-finding algorithms to calculate more precise p-values (statistical significance of peaks).

3. Gene annotation—the determination of the position of the binding sites relative to annotated genes. Figure 5.9(c) shows the tabular output for the five peaks in the CDKN1A gene region. In addition to the distance of the peaks to the CDKN1A transcription start site (TSS), the table also shows various peak metrics, such as peak signal height, number of tags, p-value, and false discovery rate (FDR), i.e. the probability that the peak is false. These peak metrics can be used to compare two or more samples and to quantify changes in protein–DNA binding between different samples. More tags in an enriched region, and thus higher peaks, mean that more cells in the examined sample have the protein bound, and therefore cross-linked, to this genomic site. However, since ChIP-seq involves a number of steps, many of which are not very quantitative (ChIP reaction, PCR amplification, sequencing), it is very important to confirm any quantitative changes of interest, especially smaller ones (e.g. <2-fold), by validation using real-time PCR of ChIP DNA and/or performing replicate assays.

5.4.9 Other variations of ChIP

Since ChIP was first described, the method has evolved and been refined from a procedure taking 4–5 days and requring 1–10 million cells to a single-day experiment requiring

Figure 5.9 ChIP-Seq data visualized in the Integrated Genome Browser. (a)The top track shows the essentially random distribution of the read alignments ('tags') for Input DNA in a 30Mbp region of chromosome 1. In contrast, aligned reads of p53-immunoprecipitated chromatin show a non-random distribution with hundreds of peaks, i.e. sites with significant tag enrichments. Many of the peaks are higher than the *y*-axis scale shown here, indicating that they contain >100 tags at the site of highest tag density. (b) Close-up view of the same p53 data but for a different chromosome (chromosome 6) than that in (a) (24kb). In addition to the Input and p53-ChIP tracks, the Browser shows a track with identified peak locations (boxes below the peaks) and the presence of the CDKN1A gene in the plus-strand of the genomic DNA. The three exons and two introns of the gene are indicated by bars and connecting thin lines, respectively (TSS, transcription start site). (c) The table shows the annotation of the five peaks with precise chromosome coordinates (Peak centre), name of gene (CDKN1A), distance to the start site (negative numbers indicate that the peak is located upstream from the start site), and five different peak metrics and statistics. A detailed description of these peak statistics can be found in Zhang et al. (2008).

(a)

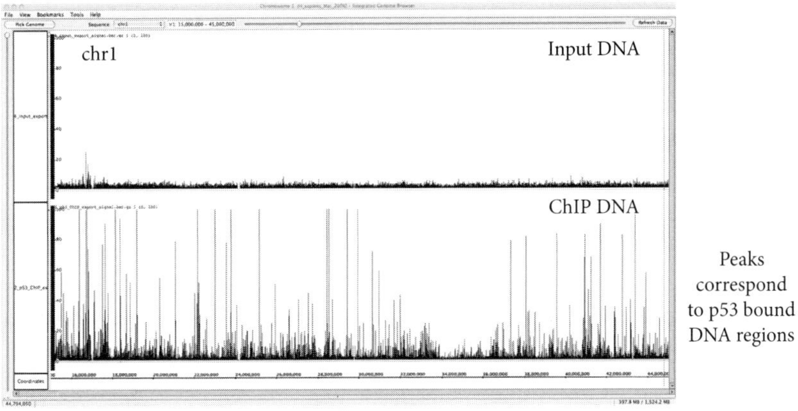

(b)

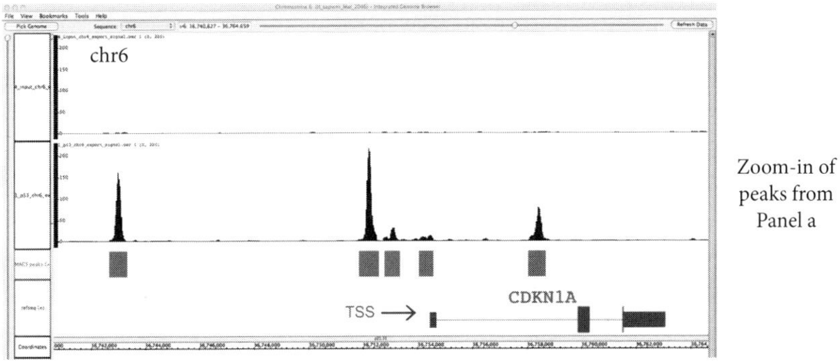

(c)

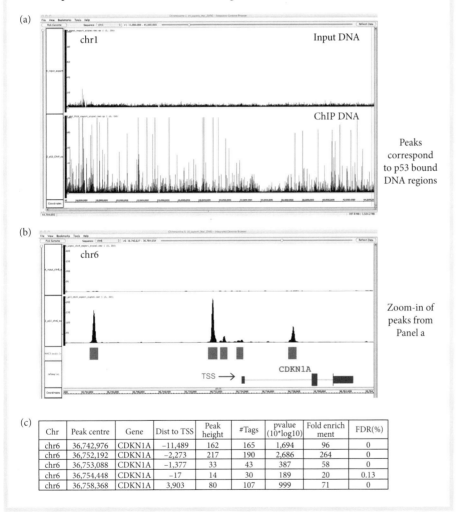

Chr	Peak centre	Gene	Dist to TSS	Peak height	#Tags	pvalue (10*log10)	Fold enrichment	FDR(%)
chr6	36,742,976	CDKN1A	−11,489	162	165	1,694	96	0
chr6	36,752,192	CDKN1A	−2,273	217	190	2,686	264	0
chr6	36,753,088	CDKN1A	−1,377	33	43	387	58	0
chr6	36,754,448	CDKN1A	−17	14	30	189	20	0.13
chr6	36,758,368	CDKN1A	3,903	80	107	999	71	0

merely 100–1000 cells (Acevedo et al. 2007). One adaptation is Carrier ChIP (CChIP) which uses chromatin from as few as 100 cells. Chromatin loss during the procedure is minimized by the addition of chromatin extracted from an evolutionarily divergent species such as the fruit fly *Drosophila*, but PCR primers specific for the target cell chromatin are required (O'Neill et al. 2006). The ChIP method has been adapted to microplates for the simultaneous processing of up to 96 samples (Flanagin et al. 2008), and for the analysis of RNA binding proteins (RIP-ChIP) (Baroni et al. 2008). Methods for mapping the genomic distribution of proteins that do not depend on protein-specific antibodies have also been developed. One approach is to express an **epitope-tagged** version of the protein of interest and using the anti-Tag antibody for ChIP. Another approach, named DamID, involves expression of the DNA binding protein as a fusion protein with the *E. coli* DNA adenine methyltransferase (DAM) (Greil et al. 2006). Binding the protein of interest to DNA localizes the methyltransferase in the region of the binding site. Adenosine methylation does not occur naturally in eukaryotes, so it can be inferred that any region of adenine methylation is located near a binding site. Digestion of the genome by *Dpn*I, which cuts only methylated GATCs, releases these regions. Double-stranded adapters with a known sequence are then ligated to the ends generated by *Dpn*I, which can then be amplified by PCR and sequenced to give a rough map of binding sites.

5.5 Chapter Summary

- Identifying the DNA sequences to which proteins bind, which proteins bind, and how they bind are critical to understanding a variety of important cellular processes such as DNA replication, repair, and gene expression.
- DNA binding proteins may bind through sequence-specific or sequence-independent DNA binding. The former DNA binding proteins make physical contacts with DNA via the purine and pyrimidine bases in the major groove. The latter types of DNA binding proteins make contact with DNA via the sugar–phosphate backbone of the DNA duplex.
- Footprinting methods identify the DNA bases that are contacted by DNA binding proteins. In this method, DNA residues in the probe bound by protein are protected from nicking by DNase I. Sites of protection appear as a gap or 'footprint' when the digestion products are run on a denaturing polyacrylamide gel.
- Mobility shift assays provide a quick and simple means of monitoring DNA–protein interactions. In this method, native electrophoresis is used to separate slow migrating DNA–protein complexes from fast migrating protein-free DNA probes.
- The development of ChIP methods has enabled analysis of *in vivo* DNA–protein interactions with chromatin on scales that range up to whole-genome interrogation.
- In ChiP methods, a protein-specific antibody is used to selectively precipitate DNA fragments bound to the protein of interest. The associated DNA is then purified and analysed by PCR, sequencing, or microarray technology.

References

Acevedo, L.G., Iniguez, A.L. Holster, H.L., Zhang, X., Green, R., and Farnham, P.J. (2007) Genome-scale ChIP-chip analysis using 10,000 human cells. *BioTechniques* **43**: 791–7.

Bakhanashvili, M., Hizi, A., and Rahav, G. (2010) The interaction of p53 with 3′-terminal mismatched DNA. *Cell Cycle* **9**: 1380–9.

Baroni, T.E., Chittur, S.V., George, A.D., and Tenenbaum, S.A. (2008) Advances in RIP-chip analysis: RNA-binding protein immunoprecipitation–microarray profiling. *Methods Mol Biol* **419**: 93–108.

Bentley, D.R., Balasubramanian, S., Swerdlow, H.P., et al. (2008) Accurate whole human genome sequencing using reversible terminator chemistry. *Nature* **456**: 53–9.

Brand, L.H., Kirchler, T., Hummel, S., Chaban, C., and Wanke, D. (2010) DPI-ELISA: a fast and versatile method to specify the binding of plant transcription factors to DNA in vitro. *Plant Methods* **6**: 25.

Cheung, P., Allis, C.D., and Sassone-Corsi, P. (2000) Signaling to chromatin through histone modifications. *Cell* **103**: 263–71.

Dame, R.T. (2005) The role of nucleoid-associated proteins in the organization and compaction of bacterial chromatin. *Mol Microbiol* **56**: 858–70.

Dubbs, J.M., Bird, T.H., Bauer, C.E., and Tabita, F.R. (2000) Interaction of CbbR and RegA* transcription regulators with the *Rhodobacter sphaeroides* cbbIPromoter-operator region. *J Biol Chem* **275**: 19 224–30.

Dubbs, J.M. and Tabita, F.R. (2003) Interactions of the cbbII promoter–operator region with CbbR and RegA (PrrA) regulators indicate distinct mechanisms to control expression of the two cbb operons of *Rhodobacter sphaeroides*. *J Biol Chem* **278**: 16 443–50.

el-Deiry, W.S., Kern, S.E., Pietenpol, J.A., Kinzler, K.W., and Vogelstein, B. (1992) Definition of a consensus binding site for p53. *Nat Genet* **1**: 45–9.

Flanagin, S., Nelson, J.D., Castner, D.G., Denisenko, O., and Bomsztyk, K. (2008) Microplate-based chromatin immunoprecipitation method, Matrix ChIP: a platform to study signaling of complex genomic events. *Nucleic Acids Res* **36**: e17.

Foord, O., Navot, N., and Rotter, V. (1993) Isolation and characterization of DNA sequences that are specifically bound by wild-type p53 protein. *Mol Cell Biol* **13**: 1378–84.

Fried, M. and Crothers, D.M. (1981) Equilibria and kinetics of lac repressor–operator interactions by polyacrylamide gel electrophoresis. *Nucleic Acids Res* **9**: 6505–25.

Garner, M.M. and Revzin, A. (1981) A gel electrophoresis method for quantifying the binding of proteins to specific DNA regions: application to components of the *Escherichia coli* lactose operon regulatory system. *Nucleic Acids Res* **9**: 3047–60.

Gibson, J.L., Falcone, D.L., and Tabita, F.R. (1991) Nucleotide sequence, transcriptional analysis, and expression of genes encoded within the form I CO_2 fixation operon of *Rhodobacter sphaeroides*. *J Biol Chem* **266**: 14 646–53.

Gilmour, D.S. and Lis, J.T. (1985) In vivo interactions of RNA polymerase II with genes of *Drosophila melanogaster*. *Mol Cell Biol* **5**: 2009–18.

Granger-Schnarr, M., Lloubes, R. de Murcia, G., and Schnarr, M. (1988) Specific protein–DNA complexes: immunodetection of the protein component after gel electrophoresis and Western blotting. *Anal Biochem* **174**: 235–8.

Greil, F., Moorman, C., and van Steensel, B. (2006) DamID: mapping of in vivo protein–genome interactions using tethered DNA adenine methyltransferase. *Methods Enzymol* **410**: 342–59.

Jenuwein, T. and Allis, C.D. (2001) Translating the histone code. *Science* **293**: 1074–80.

Jett, S.D. and Bear, D.G. (1994) Snapshot blotting: transfer of nucleic acids and nucleoprotein complexes from electrophoresis gels to grids for electron microscopy. *Proc Natl Acad Sci USA* **91**: 6870–4.

Jorgensen, H.F., Ben-Porath, I., and Bird, A.P. (2004) Mbd1 is recruited to both methylated and nonmethylated CpGs via distinct DNA binding domains. *Mol Cell Biol* **24**: 3387–95.

Kern, S.E., Kinzler, K.W., Baker, S.J., *et al.* (1991a) Mutant p53 proteins bind DNA abnormally in vitro. *Oncogene* **6**: 131–6.

Kern, S.E., Kinzler, K.W., Bruskin, A., et al. (1991b) Identification of p53 as a sequence-specific DNA-binding protein. *Science* **252**: 1708–11.

Kornberg, R.D. (1977) Structure of chromatin. *Annu Rev Biochem* **46**: 931–54.

Kouzarides, T. (2007) Chromatin modifications and their function. *Cell* **128**: 693–705.

Morozova, O. and Marra, M.A. (2008) Applications of next-generation sequencing technologies in functional genomics. *Genomics* **92**: 255–64.

O'Neill, L.P., VerMilyea, M.D., and Turner, B.M. (2006) Epigenetic characterization of the early embryo with a chromatin immunoprecipitation protocol applicable to small cell populations. *Nat Genet* **38**: 835–41.

Pepke, S., Wold, B., and Mortazavi, A. (2009) Computation for ChIP-seq and RNA-seq studies. *Nat Methods* **6**(Suppl): S22–32.

Quail, M.A., Kozarewa, I., Smith, F., et al. (2008) A large genome center's improvements to the Illumina sequencing system. *Nat Methods* **5**: 1005–10.

Renard, P., Ernest, I., Houbion, A., et al. (2001) Development of a sensitive multi-well colorimetric assay for active NFkappaB. *Nucleic Acids Res* **29**: E21.

Rosenau, C., Emery, D., Kaboord, B., and Qoronfleh, M.W. (2004) Development of a high-throughput plate-based chemiluminescent transcription factor assay. *J Biomol Screen* **9**: 334–42.

Ruscher, K., Reuter, M., Kupper, D., Trendelenburg, G., Dirnagl, U., and Meisel, A. (2000) A fluorescence based non-radioactive electrophoretic mobility shift assay. *J Biotechnol* **78**: 163–70.

Shen, Z., Peedikayil, J., Olson, G.K., Siebert, P.D., and Fang, Y. (2002) Multiple transcription factor profiling by enzyme-linked immunoassay. *BioTechniques* **32**: 1168, 1170–62, 1174.

Solomon, M.J. and Varshavsky, A. (1985) Formaldehyde-mediated DNA–protein crosslinking: a probe for in vivo chromatin structures. *Proc Natl Acad Sci USA* **82**: 6470–4.

Suzuki, K. and Matsubara, H. (2011) Recent advances in p53 research and cancer treatment. *J Biomed Biotechnol* **2011**: 978312.

Wells, J. and Farnham, P.J. (2002) Characterizing transcription factor binding sites using formaldehyde crosslinking and immunoprecipitation. *Methods* **26**: 48–56.

Woo, A.J., Dods, J.S., Susanto, E., Ulgiati, D., and Abraham, L.J. (2002) A proteomics approach for the identification of DNA binding activities observed in the electrophoretic mobility shift assay. *Mol Cell Proteomics* **1**: 472–8.

Worthington, M.T., Pelo, J.W., Sachedina, M.A., Applegate, J.L., Arseneau, K.O., and Pizarro, T.T. (2002) RNA binding properties of the AU-rich element-binding recombinant Nup475/TIS11/tristetraprolin protein. *J Biol Chem* **277**: 48 558–64.

Zhang, Y., Liu, T., Meyer, C.A., et al. (2008) Model-based analysis of ChIP-seq (MACS). *Genome Biol* **9**: R137.

Zianni, M., Tessanne, K., Merighi, M., Laguna, R., and Tabita, F.R. (2006) Identification of the DNA bases of a DNase I footprint by the use of dye primer sequencing on an automated capillary DNA analysis instrument. *J Biomol Tech* **17**: 103–13.

6 RNA interference technology

Ian Hampson, Gavin Batman, and Thomas Walker

Chapter overview

The RNA interference (**RNAi**) machinery regulates post-transcriptional gene silencing. In this process, non-coding double-stranded RNA sequences, approximately 20–30nt in length, bind to complementary sequences in mRNA transcripts to suppress (i.e. silence) their expression. Three main classes of small regulatory RNAs have been discovered so far: short interfering RNA (**siRNA**), microRNA (**miRNA**), and PIWI-interacting RNA (**piRNA**). These small RNAs are derived from longer dsRNA precursors which can originate from endogenous sources within the cell, such as genome regulation mechanisms, or can be exogenous sources that are introduced into the cell through viral infection and other foreign genetic material. The specificity of small RNAs towards the targeted mRNA transcript is principally a function of nucleotide complementarity. This property can be exploited in the laboratory to design and introduce small RNAs directed at particular mRNA transcripts to prevent the expression of that gene.

The manipulation of gene regulation by RNAi technology is a vital tool in molecular and cellular research. It is used for gene expression studies, for elucidating cellular and molecular pathways, for forcing specific phenotypes within a cell population, for array-based and high-throughput screening for purposes such as drug responses, and as an emerging therapeutic agent.

In this chapter we will provide an overview of the RNAi pathway and describe the two main approaches currently used to silence gene expression *in vitro*: short interfering RNA (siRNA) technology and short hairpin RNA (**shRNA**) technology. The former exploits our knowledge of siRNAs and the latter exploits our knowledge of miRNAs.

LEARNING OBJECTIVES

This chapter will enable the reader to:

- describe the RNAi pathway and explain the key similarities and differences between siRNA and miRNA gene-silencing mechanisms;
- summarize the major steps involved in using siRNA principles within siRNA technology and miRNA principles within shRNA technology for *in vitro* silencing of target cellular transcripts;
- select the appropriate silencing technology (siRNA or shRNA) to answer a defined research question and provide reasons for making that selection.

6.1 Regulation of gene expression: the RNA interference pathway

RNA interference (RNAi) was first described in the nematode *Caenorhabditis elegans* in the 1990s by Fire and colleagues (Fire et al. 1991; Nellen and Lichtenstein 1993). They reported that injected dsRNA with sequence complementarity to the chosen mRNA transcript was able to specifically silence the expression of that gene (Fire et al. 1998). This provided the first insights into the RNAi pathways. Since then, the molecular pathways of RNAi have been studied extensively within several model organisms, and although different proteins and mechanisms are known to modulate RNAi in the nematode *C.elegans*, the fruitfly *Drosophila melanogaster*, the yeast *Schizosaccharomyces pombe*, the plant *Arabidopsis thaliana*, and humans, remarkable similarities across the pathways are apparent (Siomi and Siomi 2009). These studies have provided new insights into molecular and cellular systems and evolutionary biology, and have been integral to establishing a framework for successfully designing, synthesizing, and delivering potent small RNAs into cells.

Before considering how the RNAi pathway can be exploited in the laboratory, it is necessary to understand how the small regulatory RNAs are generated and how they silence gene expression. To date, two main classes of small regulatory RNAs have been described: siRNAs and miRNAs. A third class, piRNAs, are not currently established within RNA interference technologies as the pathways they participate in are yet to be fully established. Therefore they are not included in this chapter.

6.1.1 siRNA generation

siRNAs were the first small regulatory RNAs to be discovered and characterized (Fire et al. 1998; reviewed by Zamore et al. 2000). Figure 6.1(a) shows how siRNAs are generated. They are derived almost exclusively from exogenous dsRNA precursors introduced into the cytoplasm through, for example, viral infection or transfection during siRNA technology procedures. These longer dsRNAs are cleaved by the enzyme Dicer, a member of the RNaseIII family endonucleases, to produce shorter siRNA duplexes 21–25nt long containing a 5′ monophosphate group and a 3′ dinucleotide overhang at each end of the duplex, characteristic of all dsRNAs cleaved by RNaseIII (Jinek and Doudna 2009; Siomi and Siomi 2009). The mammalian orthologue **DICER1** contains one dsRNA-binding domain (dsRBD) and two RNaseIII domains. Their spatial arrangement ensures that the dsRNA is cleaved into siRNA duplexes of a consistent length (Jinek and Doudna 2009). DICER1 activity is aided by the dsRBD cofactor TRBP.

6.1.2 miRNA generation

miRNAs were discovered after siRNAs (Grishok et al. 2001; Hutvágner et al. 2001; Ketting et al. 2001). The generation of miRNAs is illustrated in Figure 6.1(b). miRNA genes are found within the genome and are transcribed within the nucleus by RNA polymerase II to generate a single-strand RNA transcript. The nucleotide sequence towards the 3′ region

Figure 6.1 Small regulatory RNA biogenesis and RISC-directed silencing. (a) siRNA generation. Long dsRNA precursors are processed into siRNA duplexes 21–25nt long with 5′ monophosphate groups and 3′ dinucleotide overhangs by the enzyme DICER1 (in mammals) and cofactor TRBP. (b) miRNA generation. RNA polymerase II transcribes a single-strand RNA transcript from the genome. The strand self-anneals to create a dsRNA stem–loop structure with a terminal loop, called pri-miRNA. Excess ssRNA and dsRNA upstream of the terminal loop are removed by the RNaseIII enzyme Drosha and cofactor DGCR8 to generate a precursor miRNA (pre-miRNA) The pre-miRNA is exported to the cytoplasm and processed by DICER1 into an approximately 22nt miRNA duplex with a 2nt overhang at both 3′ ends. (c) RISC assembly. siRNA duplexes are loaded into the RNA-inducing silencing complex (RISC) comprised of DICER1, cofactor TRBP, and AGO2. miRNA is loaded into a trimeric complex comprised of DICER1, cofactor TRBP, and AGO1, AGO2, or AGO3. The RNA strand whose 5′ end is the least thermodynamically stable of the duplex is preferentially selected as the guide strand. The passenger strand is discarded. (d) RISC-mediated transcript silencing. RISC machineries target mRNA transcripts with complementarity to the RISC guide strand. AGO2 RISCs slice mRNA transcripts, whilst AGO1/3/4 RISCs catalyse mRNA translational inhibition or mRNA decay through deadenylation of the poly A tail and removal of the 5′ cap.

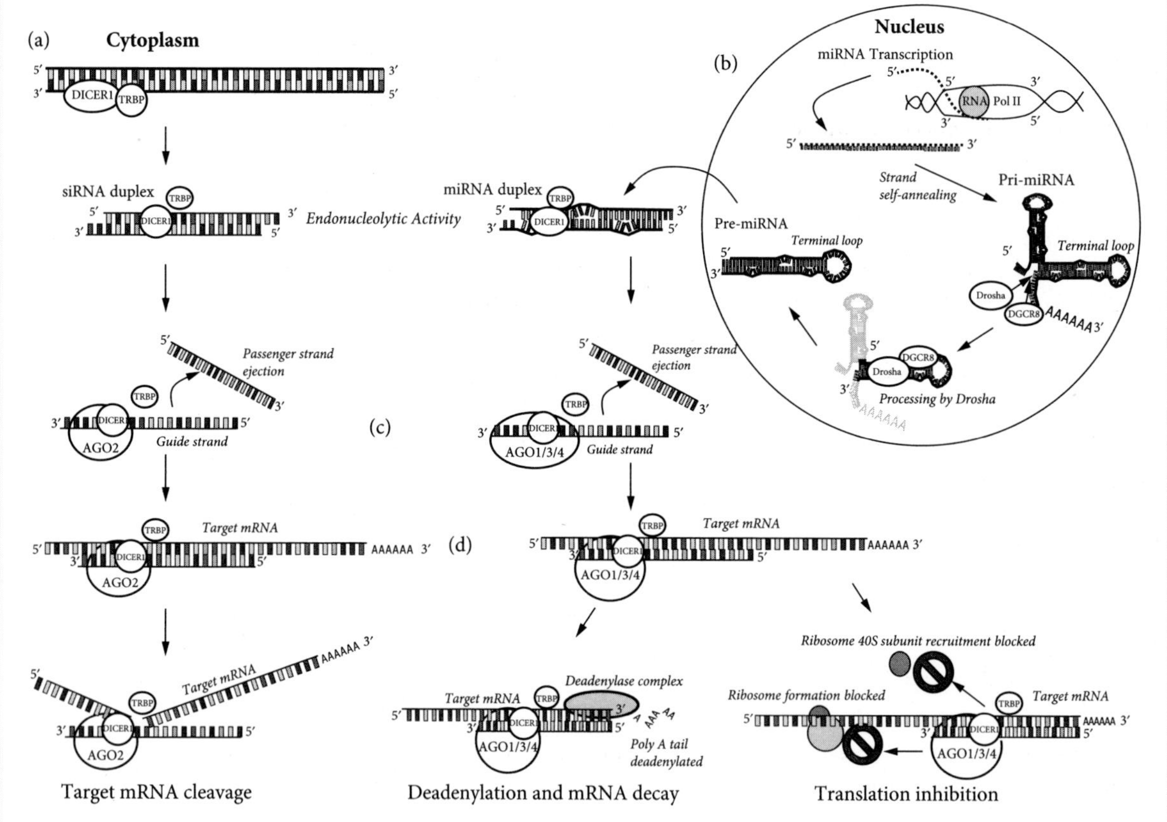

of the transcript is a reverse complement with respect to the sequence towards the 5′ region and so the strand self-anneals to create a dsRNA sequence. Crucially, complementarity between the 5′ and 3′ regions of the ssRNA transcript is not 100%, and this creates a dsRNA with a series of stem–loop structures. The most significant stem–loop is the 'terminal loop'—the central portion of the ssRNA transcript where the transcript folds upon itself to form the double-stranded molecule (Jinek and Doudna 2009). This structure is called the primary miRNA (**pri-miRNA**) and is subjected to endonucleolytic processing within the nucleus by an RNaseIII enzyme called Drosha (Zeng et al. 2005). Drosha, with the aid of a dsRBD cofactor (DGCR8 in mammals), removes the excess ssRNA and dsRNA upstream of the terminal loop to generate a precursor miRNA (**pre-miRNA**) approximately 70nt long containing a characteristic 2nt overhang at the 3′ end. The pre-miRNA is exported to the cytoplasm and subjected to further endonucleolytic processing by DICER1 to remove the terminal loop and generate a mature miRNA comprised of an approximately 22nt duplex with characteristic 2nt overhangs at both 3′ ends (Paddison 2008).

6.1.3 RNA-induced silencing complex (RISC) assembly and transcript silencing

Once the small siRNAs or miRNAs are generated, they are loaded into the RNA-induced silencing complex (**RISC**). This trimeric complex is comprised of DICER1, its cofactor TRBP, and an **Argonaute** protein—AGO2 in the case of siRNA, or AGO1, AGO3, or AGO4 in the case of miRNA. The RNA strand whose 5′ end resides at the least thermodynamically stable end of the duplex is preferentially selected to be the **guide strand** for the Argonaute protein and is loaded into the RISC complex. The remaining strand, the **passenger strand**, is discarded (Figure 6.1(c)). The guide strand is then directed to bind the mRNA transcripts with a complementary nucleotide sequence and induce gene silencing. An siRNA guide strand is 100% complementary in sequence to the target mRNA, whilst an miRNA guide strand contains mismatches in sequence complementarity with the target mRNA. The Argonaute (AGO2) recruited into an siRNA RISC complex catalyses mRNA transcript cleavage, whereas the Argonaute (AGO1, AGO3, or AGO4) recruited into an miRNA RISC complex catalyses mRNA translation inhibition or mRNA decay through de-adenylation of the poly A tail and removal of the 5′ cap (Figure 6.1(d)). Therefore the method by which the target mRNA is silenced depends on the degree of sequence complementarity between the guide strand and the target mRNA strand. Complete sequence match leads to transcript cleavage whilst mismatch between the sequences will lead to inhibition of translation, or de-adenylation and mRNA decay (Fabian and Sonenberg 2012).

6.2 Applications of RNAi

Two main approaches are currently available to silence genes of interest at the post-transcriptional level in the laboratory:

- short interfering RNA (siRNA) technology which is derived from our understanding of the siRNA pathway;
- short hairpin RNA (shRNA) technology which is derived from our understanding of the miRNA pathways.

An overview of the gene-silencing pathway mediated by siRNA and shRNA *in vitro* is shown in Figure 6.2.

In short interfering RNA (siRNA) technology, pre-designed synthetic siRNA duplexes with sequence specificity to the mRNA transcript of interest are introduced into the cell cytoplasm. Once in the cytoplasm, they are loaded into the endogenous siRNA/AGO2 RNAi machinery to cleave the target mRNA transcript. This results in a decrease in the transcript levels and a corresponding decrease in the target protein levels. The effects of

Figure 6.2 *In vitro* siRNA- and shRNA-mediated gene silencing overview. (a) siRNA duplexes are introduced into the cell cytoplasm by lipofection or electroporation. They are loaded into the RISC machinery and subsequently the target mRNA transcript is degraded. (b) DNA oligonucleotides encoding the shRNA sequence are cloned into a vector (plasmid-based or viral) and integrated into the cell genome. The sequence is expressed from the cassette either as pri-miRNAs (termed pri-shRNAs) or as pre-miRNAs (termed pre-miRNAs) depending on the type of RNA polymerase promoter used. These are exported to the cytoplasm, cleaved by DICER1 to produce the shorter duplex, and loaded into the RISC machinery. Subsequently the target mRNA transcript is degraded.

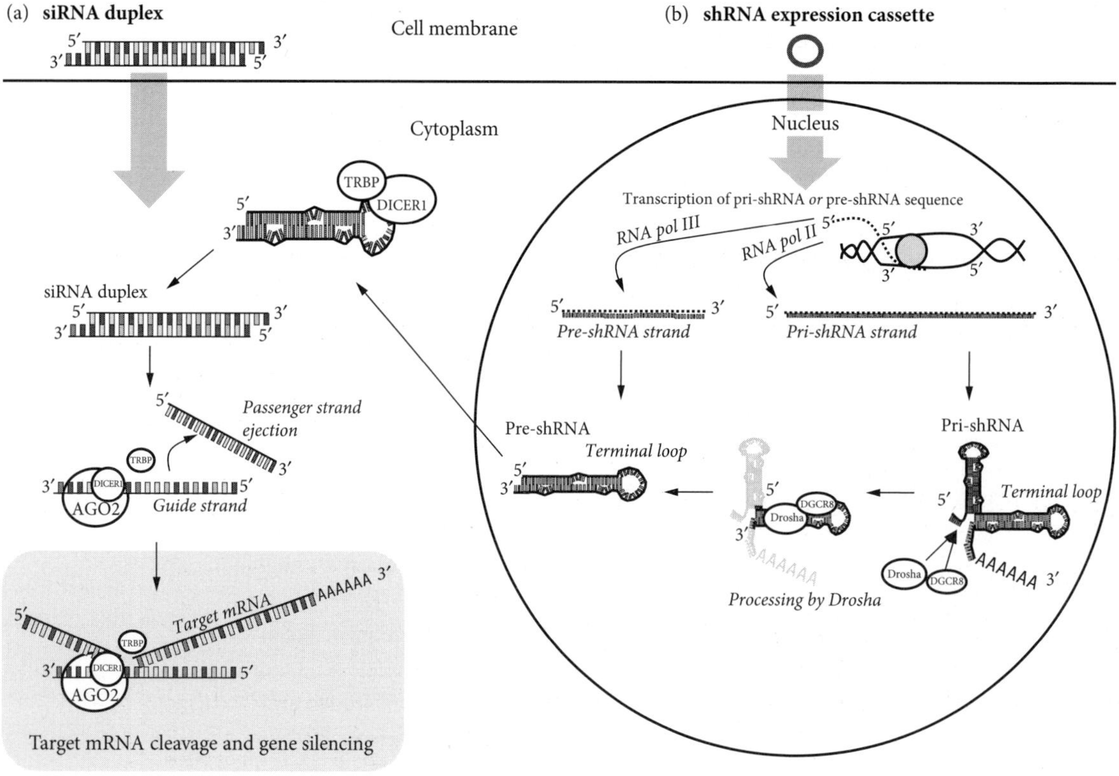

siRNA technology are transient, with peak activity typically occurring 24–48 hours following the introduction of the siRNA duplex. Activity then progressively diminishes and lasts little longer than 72 hours.

Short hairpin RNA (shRNA) technology is used to achieve **stable knockdown** (indefinite knockdown) of the gene of interest, although transient silencing is also possible with this system. In shRNA technology, the shRNA sequence is initially cloned into an **expression vector** under the control of an RNA polymerase promoter and the vector is then introduced into the cell where it permanently resides. shRNA sequences can be expressed under the control of an RNA polymerase II or RNA polymerase III promoter. If a polymerase II promoter is used, **pri-shRNAs** will be expressed within the nucleus (akin to pri-miRNAs (see section 6.1.2)) and then processed into **pre-shRNAs** by the endogenous miRNA regulation machinery. If a polymerase III promoter is used, pre-shRNAs (akin to pre-mRNA) will be directly expressed. The pre-shRNAs are then exported to the cytoplasm where they are further processed by DICER1 (in mammalian cells) to create the shorter 22nt mature duplex. shRNAs are designed so that there is a high sequence complementarity with the target mRNA and therefore are loaded into the siRNA/AGO2 mRNA-slicing RISC instead of the miRNA/AGO1/3/4 translation-inhibiting RISC. As with siRNA-mediated silencing, this leads to the degradation of the target mRNA transcript and corresponding reduction in target protein levels.

Typically, siRNA systems are used to assess the short-term effects of silencing gene expression. They are easy to use; data can be introduced into a wide range of cell lines and data is collated quickly. However, siRNA cannot be used to measure the long-term effects of gene silencing and some cell types, such as primary cell lines, non-dividing cells (e.g. neuronal cell lines), and stem cells, are difficult to transfect using siRNA. In contrast, shRNA allows gene-silencing changes to be assessed on a long-term basis and can be used with cell lines which are hard to transfect. However, shRNA systems are more expensive to use, take longer to collate the data, and require extensive optimization of protocols. The advantages and disadvantages of the two systems are summarized in Table 6.1 and should be considered before starting your experiments.

Table 6.1 Advantages and limitations of siRNA and shRNA

	Advantages	Limitations
siRNA	• siRNA sequences and protocols are typically cheaper than shRNAs • siRNA requires only between 24–72 hours to collate data following introduction into the cell • Successful knockdowns with siRNA can often be achieved across various cell lines with minimal optimization of the procedure. Significant calibration with shRNA may be required on a per-cell-line basis to achieve successful silencing	• The silencing effect of siRNA on mRNA transcripts is transient and thus not appropriate for experimental designs which require phenotypic variables to be studied on a long-term basis, where high yields of silenced cells are required, or when cell lines to be silenced exhibit very high division rates • Data can be difficult to reproduce if repeat transfections are required because of the number of variables that can influence the experimental outcome (precise cell density at siRNA delivery stage) • Not suitable for silencing mRNA transcripts for proteins that are required for cell viability (e.g. cell cycle or apoptosis pathways) • The percentage of cells which take up siRNAs during transfection of a population is routinely very low; therefore the observed silencing effect of the siRNAs is substantially lower than the actual effect as the heterogeneous cell cohort of both silenced and non-silenced cell populations are assayed • This masking can be exacerbated if the gene of interest is expressed at only basal levels, since any actual phenotypic change may be too low to be detected
shRNA	• Stable knockdowns can be generated which allow longer-term study of gene silencing. • Stable knockdowns are appropriate for cell lines with fast division times, or when the gene of interest is only expressed at basal levels • High yields of silenced cells are produced since the expression cassette is replicated during cell division • The observed knockdown of the cell population is much greater than that of an siRNA population since a homogenous cell cohort is assayed; either only silenced cells are viable (due to antibiotic selection marker) or only silenced cells are detected (e.g. via fluorescent protein markers) • shRNA sequence expression can be regulated by placing it under the control of an inducible promoter	• Successful silencing requires careful design and synthesis of shRNA sequences and optimization of protocols • Careful selection (and design) of vector is required to achieve delivery into the cell, and successful integration and expression • Stable shRNA systems take substantially more time to achieve data (weeks to months) as a consequence of the subcloning required to establish monoclone and polyclone cell lines • shRNA is more expensive than siRNA and can incur increased health and safety risks if virus vectors are used

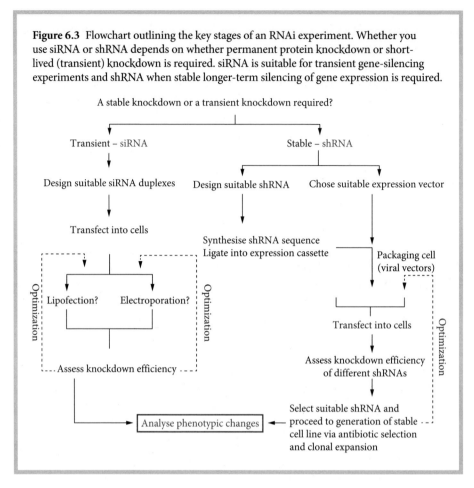

Figure 6.3 Flowchart outlining the key stages of an RNAi experiment. Whether you use siRNA or shRNA depends on whether permanent protein knockdown or short-lived (transient) knockdown is required. siRNA is suitable for transient gene-silencing experiments and shRNA when stable longer-term silencing of gene expression is required.

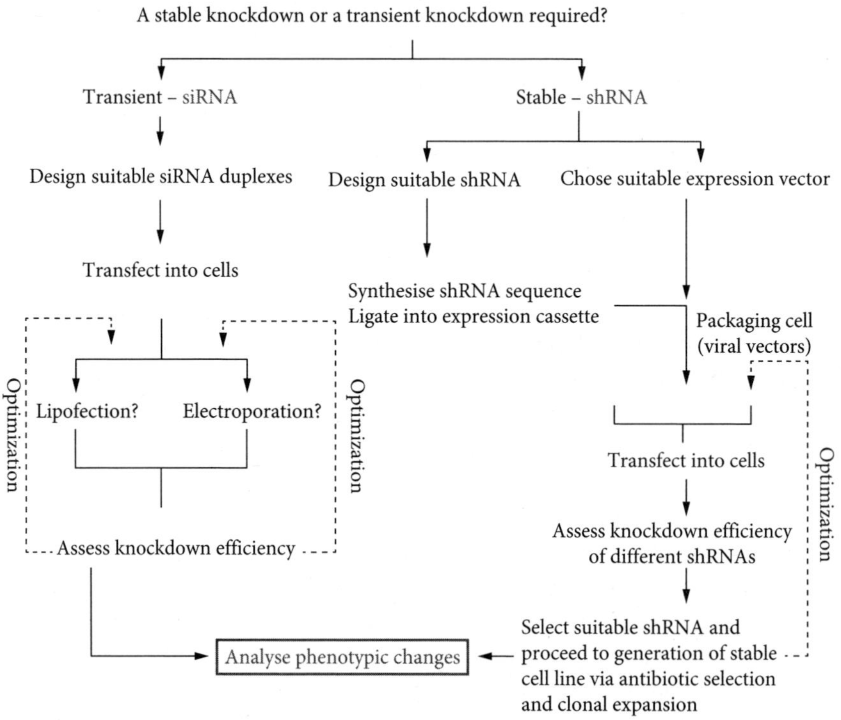

Both siRNA and shRNA *in vitro* protocols incorporate some standard steps:

(i) design of a suitable siRNA or shRNA sequence targeting the mRNA of interest;

(ii) introduction of the sequence into the cells that you are working with;

(iii) preliminary assessment of knockdown efficiency, i.e. confirmation that the target mRNA has been silenced;

(iv) conducting the actual experiment to analyse the phenotypic changes associated with the silencing of the mRNA transcript.

Figure 6.3 shows a flowchart of the key steps involved in an RNAi-based experiment. You may find it useful to work through the flow chart to determine the experimental approach best suited to your research question.

6.3 Short interfering RNA (siRNA)

6.3.1 Designing siRNA duplexes

When designing siRNA duplexes, there are two key goals to accomplish: one is to achieve *efficient* knockdown of the gene of interest and the other is to ensure that the knockdown is *specific* so that the siRNA duplex silences only its intended gene target. **Off-target effects** can arise due to two factors: the binding of siRNA duplexes to non-target sequences or induction of an innate immune response within the host cell (Hajeri and Singh 2009). siRNAs with as few as 11 contiguous nucleotides can result in elimination of non-targeted mRNA transcripts (Jackson et al. 2003). Hence, when designing siRNA duplexes a number of key criteria should be applied. These include taking into account factors such as the length of the siRNA duplex, the GC content of the sequence, the structural and thermodynamic stability of both siRNA and the intended target sequences and similarity with non-target sequences (Elbashir et al. 2001; Hajeri and Singh 2009; Samuel-Abraham and Leonard 2010; Walton et al. 2010).

The guidelines you should use when designing siRNA duplexes follow.

Identifying the siRNA target region within the mRNA sequence

The most efficient siRNA duplexes are 21–23nt long, comprising a **sense strand** and an **antisense strand**, where the antisense strand harbours 100% complementary to the mRNA target sequence (the sense sequence), and possess a 2nt overhang at the 3′ termini and AA bases at the 5′ end. Thus, when selecting suitable regions within the target mRNA sequence, use the general formula 5′-AA(N_{17-19})TT-3′ (where N represents any nucleotide). In situations where there are no 5′-AA(N_{17-19})TT-3′ motifs in the target mRNA sequence, apply the formula 5′-AA(N_{19-21})-3′ (Elbashir et al. 2001). An accurate target mRNA sequence can be obtained from the NCBI's Reference Sequence (RefSeq) database (http://www.ncbi.nlm.nih.gov). Consideration should be given to any single-nucleotide polymorphisms (SNPs) and mRNA splice variants of the gene of interest. If your experimental rationale dictates that all expressed transcripts for your gene of interest must be silenced, a homologous region within the DNA sequence, which is common to all splice variants and excludes all SNPs, should identified. Alternatively, if your experimental rationale requires that any number of splice variants or specific SNPs must be omitted from RNAi-mediated gene silencing, the chosen region of the DNA sequence must not be homologous with the sequence of those splice variants or SNP transcripts to be omitted.

Although the 3′ UTR of the mRNA can be targeted, siRNA duplexes should preferably not target regions within the 5′ UTR or the first 50–100 nucleotides downstream of the ATG start codon, as the binding of regulatory proteins and translation initiation complexes within these regions may interfere with siRNA/RISC-target mRNA binding (Elbashir et al. 2002).

The nucleotide composition of the siRNA duplex is also important. Extremely GC-rich sequences should be avoided as these can potentially form either Watson–Crick hairpins or parallel-stranded quadruplex structures and thus interfere with siRNA function (Hardin et al. 1992; Cui et al. 2004). In addition, AT-rich sequences exhibit thermodynamic instability (Cui et al. 2004) and therefore should not be targeted as this will reduce silencing efficiency. Ideally, sequences should contain a GC content of 30–70% and avoid any extended nucleotide repeats of four or more nucleotides.

siRNA duplex target specificity

siRNA duplexes should not exhibit any significant sequence homology with off-target regions, particularly within the **seed region** of the siRNA duplex guide strand. Therefore use the NCBI **BLAST** tool (http://blast.ncbi.nlm.nih.gov/Blast.cgi) to search for regions of similarity between proposed siRNA sequences and off-target regions. Proposed siRNA duplexes demonstrating off-target homology should be eliminated from the design process at this stage.

The design parameters described above are by no means exhaustive; siRNA design is a complex process and several studies have demonstrated that specific nucleotides at specific locations within the duplex, chemical modifications of the 3′ and 5′ regions, variations in T_m values, and modification of siRNA backbone chemistry can all influence target gene knockdown efficiency and target specificity (Reynolds et al. 2004; Ui-Tei et al. 2004; Jagla et al. 2005; Shabalina et al. 2006). Although it is possible to design siRNA duplexes manually, many web-based siRNA design tools are available that utilize algorithms based on the successful siRNA design principles outlined above.

Frequently used siRNA design web resources are as follows (although others do exist):

- siRNA Target Finder—Ambion (http://www.lifetechnologies.com/uk/en/home.html);
- BlOCK-iT^{tm} RNAi Designer—Invitrogen (http://www.lifetechnologies.com/uk/en/home.html);
- Dharmacon siDESIGN centre (http://www.dharmacon.com/DesignCenter/DesignCenterPage.aspx).

In addition, several biotechnology companies now offer pre-designed fully validated siRNA duplexes created with proprietary design processes, again using algorithms based on the known design parameters to ensure efficient and specific target knockdown.

6.3.2 Introducing siRNA into the cell cytoplasm

There are two main factors that can limit the effectiveness of any RNAi-based cell culture investigation: poorly selected siRNA duplexes and inefficient delivery of duplexes to the target cell. The phospholipids that form the structural components of the cell membrane of virtually all cells are largely impervious to negatively charged molecules, including siRNA which carries a net negative charge as a consequence of the sugar–phosphate backbone structure. Therefore efficient siRNA-mediated gene silencing depends on the ability of siRNA to traverse the lipid bilayer of the cell membrane. Two common methods for siRNA delivery can be used: lipid-based transfection (lipofection) or electroporation.

In **lipofection** siRNA duplexes are incubated with a lipofection reagent such as Lipfectamine-2000 (Invitrogen). These reagents contain cationic lipids which encapsulate the negatively charged siRNA duplexes within a lipid bilayer, forming a packaging vesicle referred to as a liposome. The liposome then fuses with the lipid bilayer of the host cell membrane. Following liposome–host cell fusion, the siRNA duplex is released into the cell, where it combines with the cellular RNAi machinery and brings about target mRNA silencing.

In **electroporation** an electric field is briefly applied to the cell, causing the lipid bilayer to temporarily lose its highly structured arrangement and form pores through which the siRNA duplex can freely pass. When the voltage is removed, the lipid bilayer reforms. As with lipofection, once in the cytoplasm, the siRNA duplex combines with the cellular RNAi machinery and initiates siRNA-mediated silencing.

Selecting the most appropriate siRNA delivery method

The siRNA delivery method used should achieve maximum delivery (transfection) efficiency and at the same time minimal cytotoxicity. Transfection efficiencies can vary widely between cell types, but lipofection is generally the most favourable siRNA delivery route. A wide range of lipofection reagents specifically developed for siRNA delivery and formulated to provide high transfection efficiency whilst minimizing cytotoxicity are commercially available, and are generally considered effective across a wide range of cell types.

However, some cell types, such as suspension cells, primary cell lines, non-dividing cells, and stem cells, are notoriously hard to transfect using lipofection. For these cell types, electroporation may be required to achieve sufficiently high transfection rates. Electroporation may also be the preferred method when extremely large volumes of lipofection reagent or a high concentration of siRNA duplexes are required to produce a measurable siRNA-mediated phenotype.

Optimizing lipofection-based protocols

The efficiency of siRNA uptake by cells depends on the quality of the cells in the culture, siRNA quality and concentration, the choice of lipofection reagent, and the duration of exposure to siRNA. Each is described in turn below.

Cell quality, cell confluency, and culture media Prior to siRNA lipofection, it is essential that cells are maintained under optimal culture conditions (percentage confluency, culture temperature, CO_2 content, pH, nutritional supplements) and used at an early passage number. Suboptimal cell confluency at the time of transfection will result in poor siRNA uptake and hence poor gene silencing. When optimizing a silencing protocol using either a new siRNA duplex sequence or a new cell line it is essential to assess transfection efficiencies at a range of cell densities. Growth-medium-lacking antibiotics should be used when conducting siRNA-based experiments as they may also be taken up during lipofection and accumulate at toxic levels within cells. In addition, some lipofection reagents require transfection to be carried out in a serum-free or reduced-serum medium, regardless of the serum requirements of individual cell types, so it is essential that manufacturers' guidelines are followed.

siRNA quality and concentration If you are manually designing and producing siRNA duplexes, care should be taken to ensure that they are free from contamination with synthesis reagents, including ethanol, proteins, and salts, which may impair the lipofection process. Increasing concentrations of siRNA duplexes can result in off-target effects, and therefore an initial titration of siRNA concentration should be carried out to identify the minimum effective concentration of siRNA (see section 6.3.3).

Choice of lipofection reagent A wide range of lipofection reagents which have been specifically formulated for efficient siRNA delivery, such as Lipofectamine-2000 (Invitrogen), are commercially available. Control transfections using a well-validated positive-control siRNA (see section 6.3.3) should also be carried out in order to assess both transfection efficiency and cytotoxic effects at a range of concentrations and using a range of lipofection reagents (following manufacturers' recommendations). If too little reagent is used, the siRNA will fail to penetrate target cells, whilst too much reagent may produce toxic effects.

Duration of exposure to siRNA siRNA expression is transient, peaking within 24 hours, and is generally lost within 48 hours as cells divide (Järve et al. 2007). However, certain transfection reagents require serum-free culture medium for successful delivery of siRNA into cells and, whilst this can boost transfection efficiency, the absence of serum within the culture medium can also impair the growth of some cell lines. Therefore it is necessary to titrate the duration of exposure to siRNA against the possible deleterious effects of serum-free growth medium to determine the optimum length of exposure of the siRNA to the cells. Cell culture growth medium can be replaced with serum-containing growth medium (where required) approximately 6–8 hours after the addition of the siRNA with no noticeable loss of siRNA-mediated silencing activity. Harvesting cells at a range of time points post-transfection (e.g. 0, 6, 12, 24, 48, and 72 hours) and assessing target mRNA and protein knockdown will provide a clearer understanding of the kinetics of the siRNA silencing potential of individual siRNA duplexes in any given cell line.

Optimizing electroporation-based protocols

As with lipofection-based transfection protocols, siRNA quality, concentration, cell health, and cell confluency all influence transfection efficiency when using electroporation-based methods. However, a number of additional variables should be considered with standard electroporation methods. These are waveform, voltage, capacitance, pulse length, pulse number, and electroporation buffer. Identifying the optimal combination of these variables can dramatically enhance siRNA delivery efficiency and limit cell mortality. Again, these parameters vary between cell types, and therefore it is essential that the protocol is optimized. For further details on optimizing electroporation protocols, see Jordan et al. (2008).

6.3.3 Validation of successful siRNA-mediated gene silencing

It is essential to confirm that the target mRNA has been silenced successfully before using a silenced cell line in any downstream assay. Confirmation studies should include analysis of mRNA transcript levels and the levels of the protein translated from the targeted transcript. mRNA transcript levels are typically assayed using reverse transcriptase PCR (RT-PCR), either endpoint or quantitative real-time PCR. Suitable assays for measuring protein levels include Western blot analysis, flow cytometry, and immunofluorescence using target-specific antibodies.

See Chapter 2

See Chapters 9, 16 and 17

A series of controls are necessary to ensure the validity of the experimental data.

(i) Two negative controls should be included—untreated cells and mock transfected cells—within both of which the siRNAs are omitted. Comparisons may then be drawn between the transfected cells and both controls, thus enabling identification of any toxic effects that might arise as a result of the transfection procedure.

(ii) A scrambled siRNA duplex—essentially a duplex with the same nucleotide composition as the test siRNA but with no sequence homology to any known mRNA transcript—should be included to control for non-specific effects.

(iii) A positive control, comprising a fully validated siRNA targeting an alternative sequence (e.g. MAP kinase), should be included to monitor siRNA delivery efficiencies and for optimizing experimental protocols.

6.4 Short hairpin RNA (shRNA)

The short hairpin RNA (shRNA) technology introduced in section 6.2 uses an **expression cassette** to produce an RNA strand with a nucleotide sequence that will self-anneal into a dsRNA hairpin (stem–loop) structure. The expression cassette is comprised of a DNA oligonucleotide sequence that encodes the shRNA sequence flanked by promoter

CASE STUDY 6.1

Lopinavir upregulates the expression of the anti-viral protein ribonuclease L and selectively kills HPV-positive cervical carcinoma cells: Does ribonuclease L play a key role in mediating lopinavir toxicity?

Human papillomavirus (HPV) related cervical cancer is the most common gynaecological malignancy in developing countries. Two virally encoded proteins, E6 and E7, 'hijack' the ubiquitin–proteasome system in the host cell to inappropriately degrade the p53 and Rb tumour suppressor proteins, respectively, resulting in the loss of regulated cell cycle control. The HIV protease inhibitor lopinavir can selectively inhibit the chymotryptic-like activity of the 26S proteasome resulting in the stabilization of p53 and induction of apoptosis of HPV-positive cervical carcinoma cells. Lopinavir also stabilizes a protein known as ribonuclease L (RNase L), a major cellular antiviral protein capable of initiating apoptosis in virally infected cells (Batman et al. 2011). This potentially indicates a previously unknown antiviral activity of lopinavir mediated by RNase L.

QUESTION 1

How can siRNA be used to demonstrate that RNase L plays an active role in regulating the toxicity of lopinavir against HPV-positive SiHa cells in culture?

Preliminary assessment of RNase L knockdown

Four siRNA duplexes (designated 1–4) specifically targeting the mRNA coding for the antiviral protein RNase L were purchased from Qiagen. A scrambled siRNA duplex (labelled AllStar) was used as a negative control and a fully validated alternative siRNA duplex (targeting MAPK) was used as a positive control. Lipofection was carried out using the Lipofectamine-2000 reagent (Invitrogen). Twenty-four and forty-eight hours post-transfection, and following a brief exposure to 25μM lopinavir, RNase L protein knockdown was assessed by SDS-PAGE separation of proteins and Western blotting for RNase L protein, followed by densitometric quantification. You see from Figure 6.4(a) that siRNA-1 produces the greatest level of RNase L protein knockdown (>98% knockdown compared with control cells).

Experimental testing

SiHa cells were cultured in 96-well plates for 24 hours prior to Lipofectamine-2000-based transfection using the siRNA-1 duplex. Twenty-four hours post-transfection, the growth medium was aspirated and replaced with fresh medium containing either a dose range of lopinavir (5–30μM) or DMSO control. Cell viability was assessed using an MTT-based colorimetric growth assay at 0, 24, 48, and 72 hours after addition of lopinavir. Figure 6.4(b) shows the data obtained with 25μM lopinavir. In the presence of siRNA, and hence reduced RNase L expression, lopinavir is significantly less toxic ($p < 0.05$), outgrowing those cells containing fully functioning RNase L (Batman et al. 2011).

Figure 6.4 (a) Confirmation of RNase L protein knockdown. Top panel: Western blot analysis of RNase L protein levels 24 hours after exposure of cells to siRNA. Four siRNA duplexes (labelled 1–4) specifically targeting the mRNA coding for the antiviral protein RNase L, a scrambled negative siRNA duplex (labelled AllStar), and a fully validated alternative siRNA duplex (targeting MAPK) were used. Bottom panel: quantitation of protein knockdown by densitometric analysis using Image J densitometric software. (b) Seventy-two hour cell proliferation assay following transient siRNA mediated silencing. HPV16-positive SiHA cells treated with siRNA-1 and grown in the presence of either DMSO control or lopinavir. Data shown represent mean values ± standard deviation based on triplicate observations. Reproduced with permission from Batman, G., Oliver, A.W., Zehbe, I., Richard, C., Hampson, L., and Hampson, I.N. (2011) *Antivir Ther* **16**: 515–25.

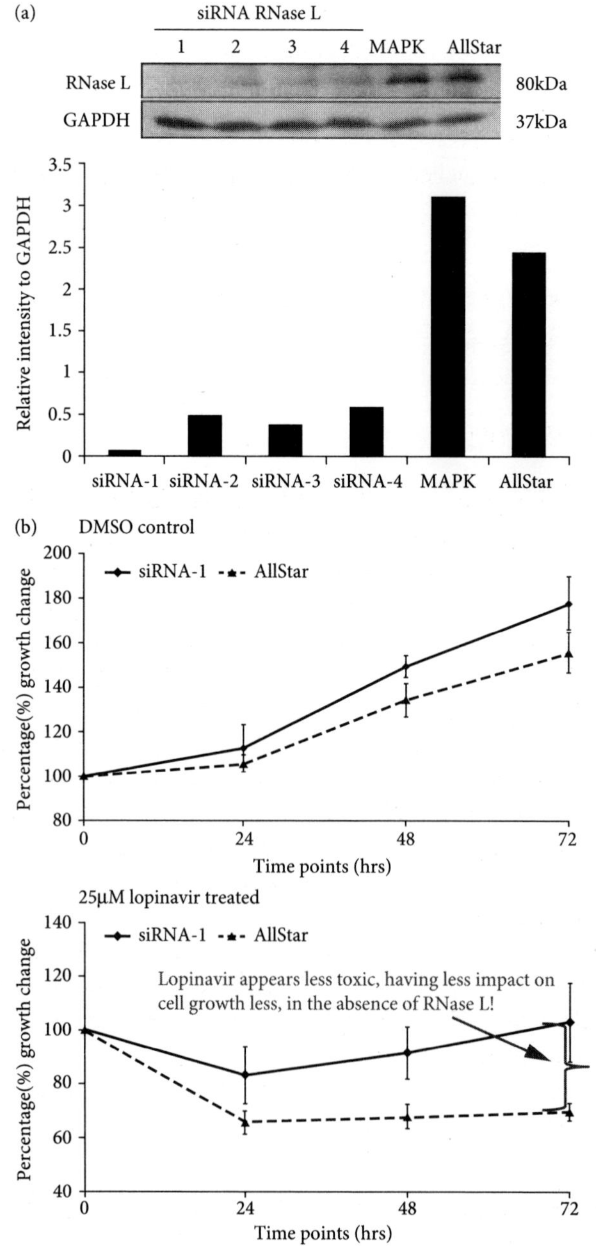

and terminator sequences. During transcription this DNA sequence transcribes an RNA transcript that folds into a stem–loop structure. This transcript is processed by the endogenous miRNA pathway to silence the expression of the target mRNA sequence (as described in section 6.2).

Although transient transfection of shRNA expression cassettes in target cells is possible, the shRNA expression cassette can be cloned into a vector which is subsequently used to transfect (or infect if a viral vector is used) a target cell line to produce a stable transfection. This is a major advantage of shRNA technology as it enables longer-term silencing effects to be studied. For a summary of the benefits and limitations of shRNA, see Table 6.1.

The steps involved in a shRNA experiment are similar to that of a siRNA protocol but include additional stages ((ii) and (v)):

(i) design of a suitable shRNA sequence targeting the mRNA of interest;

(ii) cloning of the sequence into an expression vector;

(iii) introduction of the vector into your target cells;

(iv) preliminary assessment of knockdown efficiency, i.e. confirmation that the target mRNA has been silenced;

(v) generation of stable shRNA cell lines;

(vi) conducting the actual experiment to analyse the phenotypic changes arising as a result of silencing the mRNA transcript.

6.4.1 Designing shRNA sequences

In shRNA technology, the starting sequence is a DNA oligonucleotide that encodes the shRNA sequence. The DNA oligonucleotide is typically 19–29nt long comprising a sense sequence (which therefore has sequence homology to the target mRNA sequence) separated from the antisense sequence (which will become the guide strand for targeting the mRNA sequence) by a spacer region. This spacer region is typically 7–9nt long. The sequence is carefully designed so that the transcribed shRNA can fold to form a stem–loop structure with a 3′ overhang. Figure 6.5 shows the construction of a shRNA sequence to silence expressed mRNA transcripts of the *Homo sapiens* gene *YWHAZ*. In the example shown, a 23nt region of the sense sequence is identified through a review of the accession sequences for all *YWHAZ* mRNA transcripts (NCBI RefSeq accession number: NM_003406.3 for transcript variant 1). The primary structure of the complete shRNA sequence is constructed, comprising a 5′ restriction site (*Bam H1*), the sense strand sequence (to become the passenger strand), a 7nt terminal loop, a 1nt stem–loop; the antisense sequence (reverse complement to the sense sequence, to become the guide strand), an RNA pol III termination sequence, and finally a 3′ restriction site (*Hind III*). This sequence is ligated into an expression vector and, upon expression, folds into the secondary structure of the shRNA as illustrated.

The key criteria that should be considered when designing shRNA sequences (and shown in Figure 6.5(d)) are summarized below.

Identifying the shRNA target region within the mRNA sequence

As stated, the shRNA sequence should be 19–29nt long ('19–29mer'), preferably within the lower range of these boundaries as this favours higher silencing rates (Taxman

Figure 6.5 Construction of a shRNA sequence to silence expressed mRNA transcripts of the *Homo sapiens* gene *YWHAZ*. (a) A 23nt region of the sense sequence is identified through a review of the accession sequences for all tyrosine 3-mono-oxygenase/tryptophan 5-mono-oxygenase activation protein, zeta polypeptide (*YWHAZ)* mRNA transcripts (NCBI RefSeq accession number: NM_003406.3 for transcript variant 1). The design parameters selected are shown adjacent to the sequence. (b) The primary structure of the complete shRNA sequence comprising of a 5′ restriction site (*Bam H1*), the sense strand sequence (to become the passenger strand), a 7nt terminal loop, a 1nt stem–loop, the antisense sequence (reverse complement to the sense sequence, to become the guide strand), an RNA pol III termination sequence, and a 3′ restriction site (*Hind III*). (c) This constructed sequence is ligated into an expression vector and upon expression (d) folds into the secondary structure of the shRNA. Number annotations in (c) and (d) are explained in text.

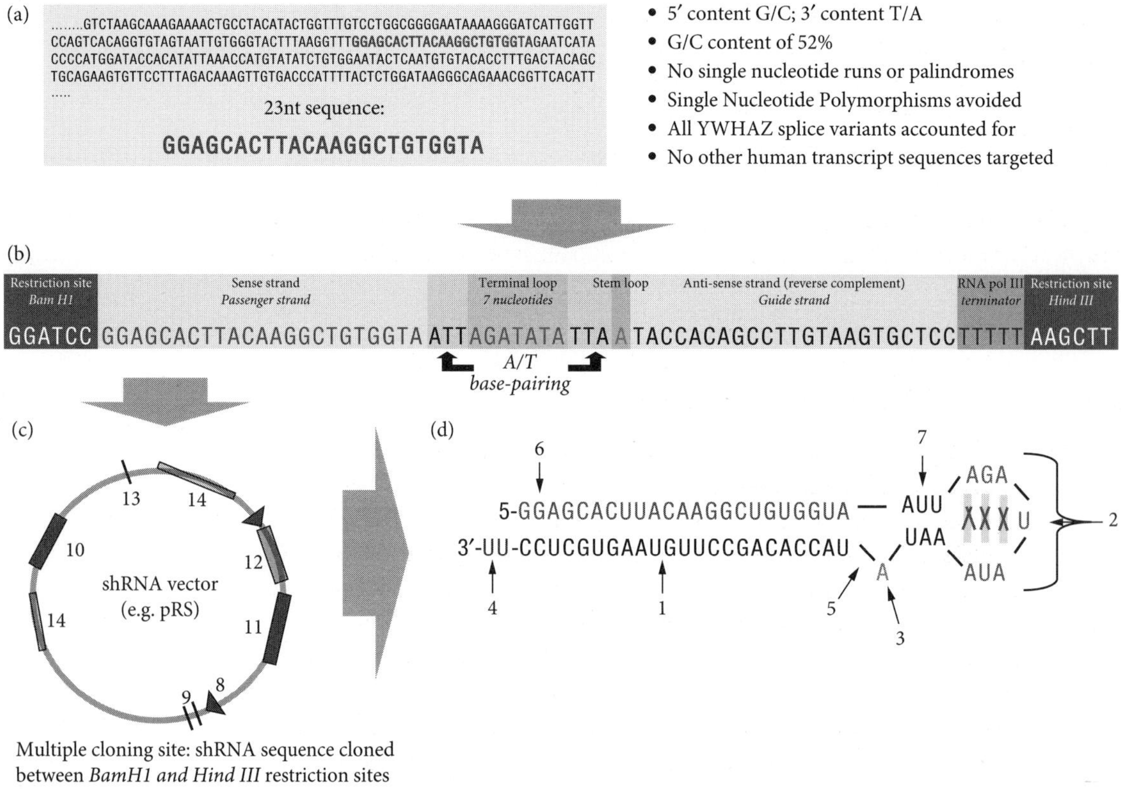

et al. 2006; Rao et al. 2009). In addition, the GC content of the shRNA sequence should be between 45% and 60%. Runs of ≥4nt repeats should be avoided to prevent off-target annealing. Nucleotide palindromes, i.e. a sequence which reads the same from 5′ to 3′ on the sense *and* the antisense strands (e.g. GAATTC) should also be avoided to prevent self-annealing. As with siRNA design (see section 6.3.1) consideration should be given to any SNPs and mRNA splice variants of the gene of interest. For example, if you wish to silence all splice variants for your gene of interest, a homologous region within the DNA sequence that is common to all splice variants and which excludes all SNPs should be identified. Alternatively, if you wish to exclude the silencing of any number of splice variants or particular SNPs, the chosen region of the DNA sequence must not be homologous with the sequence of those splice variants or SNP transcripts to be omitted. As with siRNA, a homology search of all possible shRNA sequence templates should be undertaken prior to use to eliminate any templates that may produce off-target effects.

Whether the target mRNA is degraded or translation is repressed will depend on the degree of complementarity between the shRNA sequence and the target mRNA sequence (see section 6.1.3). shRNA sequences are designed so that upon expression and processing (from a pri-shRNA into a pre-shRNA and ultimately a mature shRNA duplex), the sequence of the guide strand has 100% complementarity to the target mRNA sequence (annotated 1 in Figure 6.5(d)). This ensures that the guide strand is loaded onto an AG02 Argonaute RISC mRNA slicing complex and elicits silencing by cleaving the mRNA transcript.

Length and sequence of the spacer region

The spacer region separating the sense and antisense strands is typically 7–9nt long (annotated 2 in Figure 6.5(d)). To become the terminal loop, this space region must not be able to self-anneal. Consequently, an odd number of base pairs is routinely used (7 or 9) and base pairing is avoided by placing purines opposite purines or pyrimidines opposite pyrimidines. A lower G–C content is used, if possible, to maintain thermodynamic instability.

To further promote thermodynamic instability so that the antisense strand is taken up as the guide strand, a single-nucleotide mismatch can be included within the sequence so that a single-base stem–loop upstream of the DICER1 cleavage site is introduced when the hairpin structure forms (annotated 3 in Figure 6.5(d)).

Designing the 5′ and 3′ ends of the shRNA target sequence

shRNA sequences can be constructed so that the transcribed RNA sequence carries a 3′ terminal UU overhang, characteristic of pre-miRNAs. This is achieved by using the RNA polymerase III promoter termination sequence TTTTT at the 3′ end of the expression cassette. This results in an RNA transcript with a 3′ UU overhang (Cullen 2006) (annotated 4 in Figure 6.5(d)). This 3′ overhang is recognized by the enzyme DICER which, once bound, cleaves the terminal hairpin structure to form an siRNA duplex.

It is critical that the *antisense sequence* to the target mRNA is incorporated into the RISC (i.e. it becomes the guide strand instead of the sense strand). As thermodynamic instability directs which 5′ strand is taken up, less stable A–T residues can be used at the 5′ region of the *antisense sequence to the mRNA* and the more stable G–C residues can be used at the 5′ end of the *sense sequence to the mRNA* (annotated 5 and 6, respectively, in Figure 6.5(d). This will reduce the thermodynamic stability of the antisense strand and increase the thermodynamic stability of the sense strand. The outcome will be that the antisense strand is incorporated into the RISC complex.

Acceptance of the antisense strand as the guide strand can be further enforced with the use of A–T base pairing to connect the single-nucleotide mismatch stem–loop and the terminal stem–loop since A–T base pairing is less thermodynamically stable (annotated 7 in Figure 6.5(d)). Once DICER cleaves the terminal loop, the resulting A–T 3′ overhang and the base-pair mismatch will force the 5′ strand at this end of the duplex to become the guide strand.

6.4.2 DNA vector choice and shRNA sequence synthesis

shRNA expression vectors

For stable expression, the DNA oligonucleotide encoding the shRNA sequence is ligated into a DNA expression vector under the control of an RNA polymerase promoter and

See Chapter 7

then transfected into the host cell. A schematic diagram of an expression vector is shown in Figure 6.5(c).

The vector for expressing shRNA carries the following essential features.

- An RNA polymerase dependent promoter. Promoters may be RNA polymerase II dependent or RNA polymerase III dependent (such as U6 or H1) (annotated 8 in Figure 6.5(c)).
- A multiple cloning site which offers a variety of restriction sites suitable for cloning the oligonucleotide sequence in a defined orientation (annotated 9 in Figure 6.5(c)).
- A prokaryotic selectable marker which allows propagation of the shRNA expression vector in bacterial cells. Prokaryotic selectable markers commonly encode enzymes that are resistant to antibiotics such as ampicillin, chloramphenicol, and kanamycin (annotated 10 in Figure 6.5(c)).
- A eukaryotic selectable marker which is required to distinguish successfully transfected cells from non-transfected cells, and specifically allow selection of vector-transformed cells for stable cell-line generation (annotated 11 in Figure 6.5(c)). Examples of eukaryotic selectable markers include neomycin and puromycin antibiotic resistant genes.

Optional features within the vector include the following.

- Expression of the shRNA sequence can be **constitutive** or **inducible**. In constitutive expression the shRNA sequence is continually expressed. However, it may not be desirable (or indeed possible) to continually express your shRNA sequence within the generated clones, for example if the target mRNA is required for cell growth or survival and hence silencing would cause loss of cell viability. In such cells, expression of the shRNA cassette could be induced when required using inducible promoter systems. Popular inducible systems are the tetracycline-controlled transcription-based 'Tet-Off' (Gossen and Bujard 1992) and 'Tet-On' systems (Gossen et al. 1995). The 'Tet-Off' system achieves expression of the cassette in the absence of tetracycline (expressed when 'Tet Off') whilst the newer 'Tet-On' system achieves expression of the cassette in the presence of tetracycline (expressed when 'Tet On'). Inducible systems and their use in shRNA technology are reviewed by Lee and Kumar (2009).
- Reporter gene sequences such as GFP (green fluorescent protein) can also be included in the vector (annotated 12 in Figure 6.5(c)). GFP fluorescence can be utilized to visualize stable shRNA-expressing cells within a population via flow cytometry or microscopy.

See Chapters 16 and 17

- Before any DNA sequence is integrated into the host genome it must be linearized. Linearization of a transfected circular DNA vector by the host cell enzymes would cut the sequence at a random location and potentially shear within the expression cassette, rendering it useless. To circumvent this, a single restriction site is included in a region of the vector outside the expression cassette so that the circular vector can be linearized prior to transfection (annotated 13 in Figure 6.5(c)).
- Viral vectors are also frequently used for shRNA expression as they can easily transduct primary cells and cells *in vivo* in addition to cell lines (Cullen 2006). The most widely used viral vectors for shRNA delivery are adenoviruses, retroviruses, and the retroviral subfamily lentiviruses. Adenoviruses can be used to transiently express shRNA in a wide range of cell types. Retroviral vectors can integrate into the host

genome and produce stable silencing but are only able to infect dividing cells (Cullen 2006; Li et al. 2006; Yamashita and Emerman 2006), except for vectors which employ the retroviral subfamily lentiviruses which are able to infect non-dividing cells (Li et al. 2006; Zimmermann et al. 2011). Viral vectors contain additional features such as 3′ and 5′ long terminal repeats (LTRs) which flank the expression cassette and the antibiotic resistance genes (annotated 14 in Figure 6.5(c)). These signals enable viral vectors to be transfected into packaging cell lines to produce virions which can then be used to infect the target cell line.

shRNA sequence synthesis

The shRNA sequence can be fully synthesized as a double-stranded DNA oligonucleotide encoding the chosen shRNA with the sense and antisense sequences separated by the loop sequence and terminating in restriction sites at the 5′ and 3′ ends. These restriction sites match the restriction sites selected within the expression vector multiple cloning site so that it can be ligated in a defined orientation downstream of the promoter sequence. Alternatively, the shRNA sequence without the restriction sites is synthesized. The restriction sites are subsequently added during PCR amplification, using

Figure 6.6 shRNA sequence synthesis. shRNA sequences can be fully synthesized as a double stranded DNA oligonucleotide incorporating the restriction sites at each end or can be synthesized without the restriction enzyme sites. In the latter case, the restriction sites are added by PCR using primers which incorporate the restriction site sequences. Following PCR amplification and purification, the fragment is digested and ligated into the vector in a defined orientation.

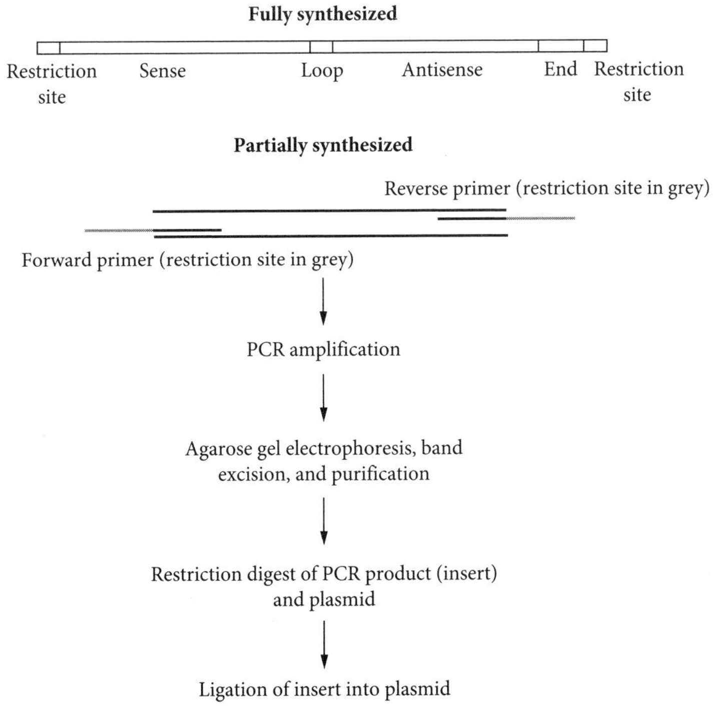

primers that incorporate restriction sites that match the sites selected within the vector. The amplified shRNA DNA fragment generated is then purified and ligated into the linearized expression vector downstream of the promoter sequence, again in a defined orientation. Synthesis of the shRNA sequence is shown in Figure 6.6.

shRNA kits targeting a vast number of mammalian genes are now available commercially and include pre-validated shRNA sequences and vectors. This saves significant design, calibration, and validation time.

shRNA stock generation and sequencing

See Chapter 4

Following ligation of the shRNA sequence into the expression vector, a stock of shRNA recombinant plasmid clones is generated by transforming the plasmid into a bacterial cell host. Transformed colonies are selected by applying antibiotic selection pressure. The shRNA sequence is then checked by DNA sequencing. Sequencing is necessary as the generation of synthetic oligonucleotides carries a high error rate, as does generating the shRNA sequence template by *Taq* polymerase activity during PCR. However, the high degree of secondary structure (hairpin) within shRNAs can be problematic for successful automated sequencing, resulting in partial sequence data when the hairpin is encountered. One approach is to incorporate a restriction site within the hairpin and then sequence both stands that result following a restriction digest of the shRNA (McIntyre and Fanning 2006). Alternatively, a number of modifications can successfully relax the shRNA secondary structure (e.g. addition of DMSO or betaine, or the use of the enzyme ThermoFidelase I) within the reaction and achieve usable sequencing data (Taxman et al. 2006).

6.4.3 Introducing shRNA into the cell cytoplasm

Transfection of the shRNA recombinant plasmid into target cell lines can be achieved by lipofection or electroporation (see section 6.3.2). If using viral vectors, the viral vector is first transfected into a packaging cell line. The resulting viral stock is harvested from the packaging cell line supernatant by centrifugation and filtration and then used to infect the target cells (Hannon 2002; Cullen 2006).

6.4.4 Validation of transient shRNA-mediated gene silencing

See Chapters 2, 9, 16, and 17

As with siRNA, it is necessary to confirm that the target mRNA has been silenced successfully. mRNA transcript levels can be assayed by endpoint PCR or quantitative real-time PCR, and protein levels can be analysed using Western blot analysis, flow cytometry, or immunofluorescence. As the shRNA silencing effect at this stage is transient, the validation assays should be carried out at suitable time points (24, 48, and 72 hours) to capture the most effective phenotypic changes.

A series of controls should be included in any shRNA experiment.

- A positive control shRNA known to elicit a high degree of silencing against a control transcript is used to ensure that that the experimental set-up is working. Positive controls target transcripts for an integral cell maintenance gene which is highly expressed. Examples include MAP kinase, GAPDH, actin, and cyclophilin.

- A number of negative controls are also required.
  - Non-transfected cells should be cultured to provide a matched wild-type phenotype.
  - A 'scrambled shRNA' negative control should be included. This is an shRNA with a guide strand sequence that does not target any known mRNA transcript and therefore should not produce any specific phenotypic change.
  - A 'vector only' negative control should be included. This is the parent vector without a shRNA insert. This control is used to identify any non-specific effects that could occur as a consequence of the vector entering the cell.
  - A 'delivery reagent only' negative control (e.g. lipofection reagent) should be included. This controls for non-specific effects which could occur as a consequence of the transfection procedure.

6.4.5 Generation of stable shRNA cell lines

To generate stable shRNA cell lines from a transiently silenced cell culture, the cells are transfected with the vector construct. Twenty-four hours after transfection, the cells are passaged into new vessels (typically plates or dishes) and incubated with a pre-determined concentration of antibiotic that will kill 100% of the parental cell line. This procedure will select for cells that carry the expression cassette and therefore are positive for the antibiotic resistance gene carried by the vector. After several rounds of passaging a large proportion of cells will have died, indicating that they did not take up, or have lost, the expression cassette. Isolated cell colonies, each derived from a single clonal ancestor, can be transferred to a new plate and grown to generate monoclonal stocks. PCR can be used to check that the shRNA expression cassette portion of the shRNA vector has integrated into the genome. The effects of gene silencing can then be analysed by assessing target mRNA and protein levels.

CASE STUDY 6.2

Generating stable cell lines expressing shRNA targeting the CSMD1 mRNA in head and neck squamous cell carcinoma (HNSCC) cell lines

Head and neck cancer is the sixth most common cancer in the UK and comprises 6% of all cancers. The *Cub and Sushi multiple domains 1* (*CSMD1*) gene is expressed at low levels in human adult epithelial cells. However, the gene is frequently deleted in tumours of the head and neck. The role that the CSMD1 protein plays in tumour suppression is currently unknown. Elucidating the cell signalling pathway(s) which CSMD1 participates in and/or governs could aid our understanding of the head and neck cancer phenotype.

QUESTION

Two HNSCC cell lines express CMSD1 and are suitable candidates for CMSD1 gene silencing. How would you go about generating CSMD1-silenced cell lines which can then be used to further our understanding of the molecular role(s) of the CSMD1 protein within the cell?

Candidate shRNA vector validation

Both the low transcript expression level of *CSMD1* in squamous epithelial cells and the requirement for stable (long-term) silencing of *CSMD1* negate the use of siRNA technology as the silencing method. Consequently, four pre-validated shRNA plasmid vectors (designated targets 1–4) which target *CSMD1* were acquired (HuSH plasmids, Origene) and stocks of each were established by bacterial transformation and antibiotic selection. To determine which of the four shRNA vectors elicited the greatest silencing of the *CSMD1 transcript*, cultures of one of the two HNSCC cell lines were transiently transfected with the shRNA vectors using Lipofectamine-2000 (Invitrogen). mRNA samples were harvested 24 and 48 hours post-transfection and *CSMD1* expression levels analysed by RT-PCR. Figure 6.7(a) shows that the greatest reduction in *CSMD1* mRNA was achieved using shRNA vector expression Target 2 at 24 hours (1 and 2 in Figure 6.7(a)). This vector construct was subsequently used to generate *CSMD1*-silenced stable shRNA cell lines.

Generation of CSMD1-silenced stable cell lines

The minimum concentration of the eukaryotic selection marker (puromycin) which will kill 100% of cells in culture was determined for both cell lines. The shRNA 'target plasmid 2' vector was transfected into the HNSCC cell lines. Twenty-four hours post-transfection the cultures were passaged into a series of new vessels at varying cell densities. Puromycin was applied after 16 hours to select for those cells which successfully integrated the shRNA plasmid into their genomes. Subsequent passaging for 15 weeks established a series of *CSMD1*-silenced stable monoclone and polyclone shRNA cell lines. Stable cell lines were assessed for genomic integration of the shRNA expression cassette portion of the shRNA plasmid by endpoint PCR. Primers were designed to generate an amplicon which flanked the U6 promoter of the shRNA expression cassette and an amplicon which flanked the 3′ region of the shRNA sequence. The parent cell line was used as a negative control. The data are shown in Figure 6.7(b)). All but one of the generated cell lines (arrow 3) were positive for both amplicons.

CSMD1 silencing was validated by both quantitative real-time RT-PCR (Figure 6.7(c)) and flow cytometry (Figure 6.7(d)) which showed successful silencing of the mRNA transcript and the protein, respectively.

6.5 Targeted gene silencing: therapeutic possibilities

RNAi is a powerful tool that can be exploited in the laboratory to knock out the expression of not only a single gene of interest but also multiple genes simultaneously to identify the functions of these genes. RNAi technology can also be used for low-cost genome-wide screening experiments which can identify the role(s) of specific genes (Iorns et al. 2007). Examples of such experiments are as follows.

- By screening a human cell line model using a pool of 6650 siRNA targeting a range of known druggable targets, Yang et al. (2012) identified several potential novel pharmacological targets for the treatment of type 2 diabetes.
- Triple negative breast cancer (TNBC), referring to any breast cancer in which cells are negative for oestrogen, progesterone, and human epidermal growth factor-2 (HER2) receptors, generally results in a high rate of relapse. By screening the TNBC model cell line SUM149 with a genome-wide siRNA library targeting 691 kinases, Hu et al. (2012) identified that inhibition of polo-like kinase 1 (PLK1) blocked the growth of TNBC cells and therefore may be a potential therapeutic drug target for triple negative breast cancer.

Figure 6.7(a–c) (a) Validation of transient shRNA-mediated gene silencing. Four shRNA plasmids (Targets 1–4) were tested for their silencing efficacy against *CSMD1* mRNA transcripts. Negative controls comprise vector backbone (Target vector), a scrambled shRNA sequence (Scrambled), and non-transfected cells (No transfection). Duplicate transfections were prepared and cells were harvested at both 24 and 48 hours. Endpoint RT-PCR was used to determine the degree of silencing. A reduction in *CSMD1* expression due to shRNA Target 2 can be observed at the 24-hour time point (arrow 1) and is lost at 48 hours (arrow 2). (b) Validation of genomic integration of the shRNA expression cassette. Endpoint PCR was used to establish whether the shRNA expression cassette had successfully integrated into the genome of the cell lines. Primers were designed to generate an amplicon which flanked the U6 promoter of the shRNA expression cassette and an amplicon which flanked the 3′ region of the shRNA sequence. The parent cell line was used as a negative control (arrow 3 depicts cell line that was not positive for both amplicons). (c) Validation of silenced *CSMD1* transcript by quantitative real-time RT-PCR. Hydrolysis probes were employed against transcripts for *CSMD1* and an endogenous control in a multiplex assay. A linear ΔR amplification plot is shown. Duplicate data points are shown for one *CSMD1*-silenced stable shRNA cell line, the corresponding parent cell line, and the non-template control (NTC). The endogenous control probe was detected for both cell lines. The *CSDM1* probe was detected for the parent cell line but not for the shRNA clone, indicating successful *CSMD1* silencing. The NTC samples did not amplify.

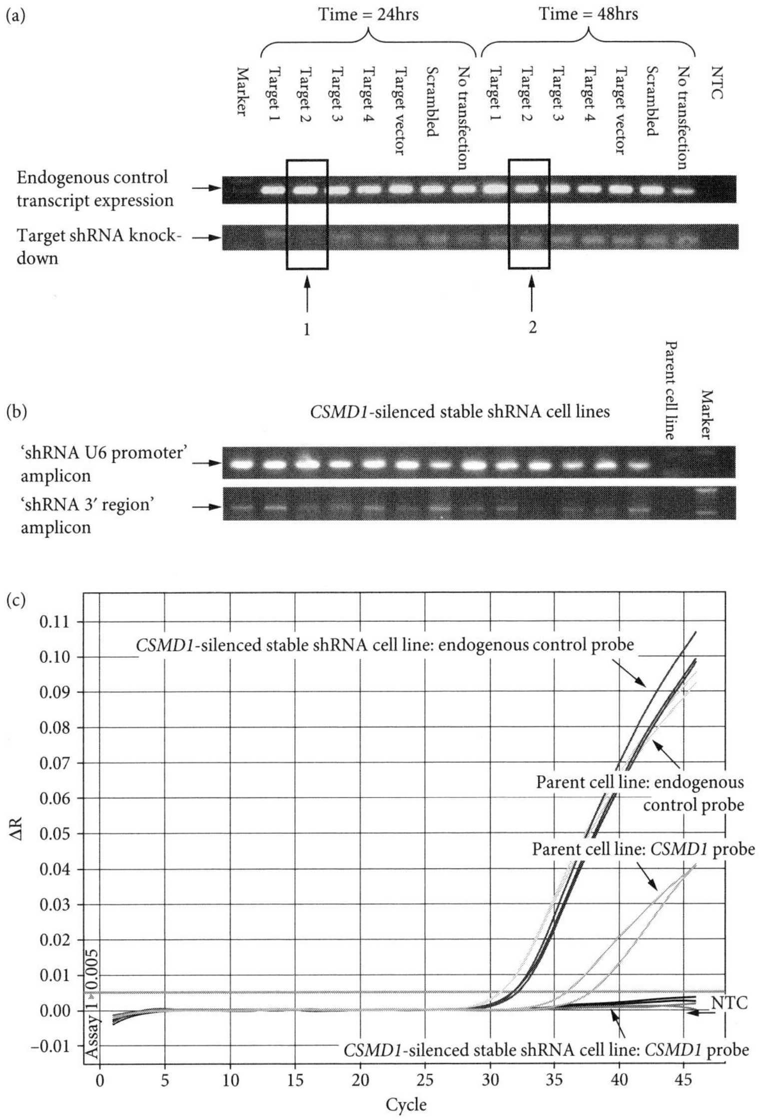

Figure 6.7(d) (d) Validation of silenced CSMD1 protein by flow cytometry. Goat anti-CSMD1 primary antibody validated for native conformation (SantaCruz Biotechnology) and a donkey anti-goat Alexa Fluor 488 fluorescent secondary antibody were used. One CSMD1-silenced stable shRNA cell line is depicted (top panel) and the corresponding parent cell line (bottom panel). There is decreased protein expression in the top panel, evidenced by less fluorescence intensity compared with the parent cell line control. The 530/40 log fluorescence mean and median values of the silenced cell line (top panel) are both significantly lower than the corresponding 530/40 log fluorescence values of the parent cell line (bottom panel). This reduction in protein expression is further illustrated by the percentage of the histograms which occupy the region labelled 'R1' (silenced cell line: 9.33% c.f. parent cell line: 91.77%).

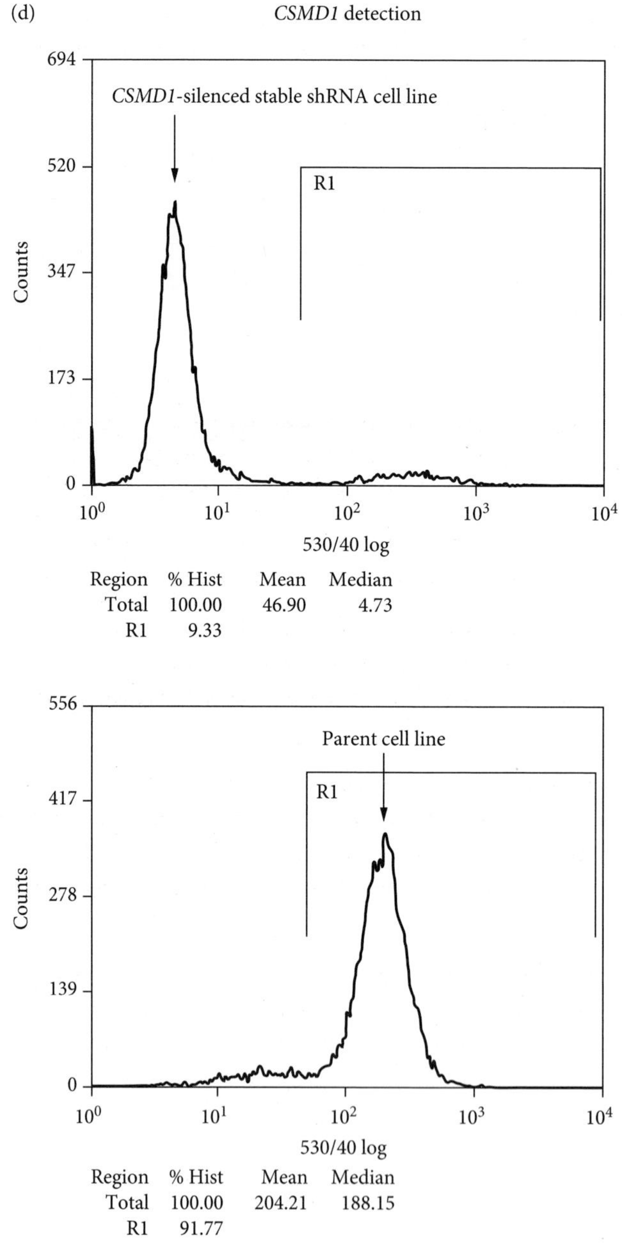

An additional application of RNAi technology is to use it as a potential therapeutic approach *in vivo*. The expression of specific disease-causing/promoting genes could be silenced by delivering nucleotide sequences complementary to the target mRNA sequence. For example, the expression of the hypoxia-inducible gene *RTP801* as a result of hypoxia or oxidative stress leads to the induction of neuronal cell apoptosis. This is a key stage in the development of acute macular degeneration (AMD). Consequently, blocking the function of *RTP801* using siRNA reduces neuronal cell apoptosis. A phase 1 clinical trial investigating the actions of an intravitreal injection of an *RTP801*-specific siRNA for the treatment of AMD is under way and is showing promising results (Nguyen et al. 2012).

As of June 2012 there were 25 clinical trials utilising RNAi registered with ClinicalTrials.gov, with conditions ranging from solid cancers through to acute and chronic viral infections.

A full discussion of RNAi-based therapeutics is beyond the scope of this chapter, but you may want to consult the following references: Pecot et al. (2011), Blake et al. (2012), and Kole et al. (2012).

6.6 Chapter summary

- The RNA interference (RNAi) machinery regulates post-transcriptional gene silencing. In this process, non-coding double-stranded RNA sequences, approximately 20–30nt in length, bind to complementary sequences in mRNA transcripts to silence their expression.
- Synthetic siRNA duplexes can be introduced directly into a wide range of cells *in vitro* to silence the expression of target genes. However, the gene-silencing effect is transient.
- shRNA sequences can be expressed from either plasmid-based or viral-based vectors. Once introduced into cells, these constructs can integrate into the host genome to achieve long-term silencing (stable knockdown) of gene expression. Depending on the promoter type, expression can be constitutive or inducible.
- A number of key parameters must be considered when designing an RNAi experiment. These include good design of siRNA or shRNA sequences, whether stable or transient expression of the silencing duplex is required, which cell line is being used, methods of introducing the synthetic nucleotide sequences into the cell, and the use of appropriate controls.
- This manipulation of gene regulation by RNAi technology has become a vital tool in basic molecular and cellular genetic research, functional genomics, gene expression profiling, drug discovery, prospective disease targeting, and therapies (both in vivo and ex vivo).

References

Batman, G., Oliver, A.W., Zehbe, I., Richard, C., Hampson, L., and Hampson, I.N. (2011) Lopinavir up-regulates expression of the antiviral protein ribonuclease L in human papillomavirus-positive cervical carcinoma cells. *Antivir Ther* **16**: 515–25.

Blake, S.J., Bokhari, F.F., and McMillan, N.A. (2012) RNA interference for viral infections. *Curr Drug Targets* **14**: 1311–20.

Cui, W., Ning, J., Naik, U.P., and Duncan, M.K. (2004) OptiRNAi, an RNAi design tool. *Comput Methods Programs Biomed* **75**: 67–73.

Cullen, B.R. (2006) Induction of stable RNA interference in mammalian cells. *Gene Ther* **13**: 503–8.

Elbashir, S.M., Martinez, J., Patkaniowska, A., Lendeckel, W., and Tuschl, T. (2001) Functional anatomy of siRNAs for mediating efficient RNAi in *Drosophila melanogaster* embryo lysate. *EMBO J* **20**: 6877–88.

Elbashir, S.M., Harborth, J., Weber, K., and Tuschl, T. (2002) Analysis of gene function in somatic mammalian cells using small interfering RNAs. *Methods* **26**: 199–213.

Fabian, M.R. and Sonenberg, N. (2012) The mechanics of miRNA-mediated gene silencing: a look under the hood of miRISC. *Nat Struct Mol Biol* **19**: 586–93.

Fire, A., Albertson, D., Harrison, S.W., and Moerman, D.G. (1991) Production of antisense RNA leads to effective and specific-inhibition of gene-expression in *C.elegans* muscle. *Development* **113**: 503–14.

Fire, A., Xu, S. Montgomery, M.K., Kostas, S.A., Driver, S.E., and Mello, C.C. (1998) Potent and specific genetic interference by double-stranded RNA in *Caenorhabditis elegans*. *Nature* **391**: 806–11.

Gossen, M. and Bujard, H. (1992) Tight control of gene expression in mammalian cells by tetracycline-responsive promoters. *Proc Natl Acad Sci USA* **89**: 5547–51.

Gossen, M., Freundlieb, S., Bender, G., Müller, G., Hillen, W., and Bujard, H. (1995) Transcriptional activation by tetracyclines in mammalian cells. *Science* **268**: 1766–9.

Grishok, A., Pasquinelli, A.E., Conte, D., et al. (2001). Genes and mechanisms related to RNA interference regulate expression of the small temporal RNAs that control *C.elegans* developmental timing. *Cell* **106**: 23–34.

Hajeri, P.B. and Singh, S.K. (2009) siRNAs: their potential as therapeutic agents–Part I. Designing of siRNAs. *Drug Discov Today* **14**: 851–8.

Hannon, G.J. (2002) RNA interference. *Nature* **418**: 244–51.

Hardin, C.C., Watson, T., Corregan, M., and Bailey, C. (1992) Cation-dependent transition between the quadruplex and Watson–Crick hairpin forms of d(CGCG3GCG). *Biochemistry* **31**: 833–41.

Hu, K., Law, J.H., Fotovati, A., and Dunn, S.E. (2012) Small interfering RNA library screen identified polo-like kinase-1 (PLK1) as a potential therapeutic target for breast cancer that uniquely eliminates tumor-initiating cells. *Breast Cancer Res* **14**: R22.

Hutvágner, G., McLachlan, J., Pasquinelli, A.E., Bálint, E., Tuschl, T., and Zamore, P.D. (2001) A cellular function for the RNA-interference enzyme Dicer in the maturation of the let-7 small temporal RNA. *Science* **293**: 834–8.

Iorns, E., Lord, C.J., Turner, N., and Ashwood, A. (2007) Utilizing RNA interference to enhance cancer drug discovery. *Nat Rev Drug Discov* **6**: 556–8.

Jackson, A.L., Bartz, S.R., Schelter, J., et al. (2003) Expression profiling reveals off-target gene regulation by RNAi. *Nat Biotechnol* **21**: 635–7.

Jagla, B., Aulner, N., Kelly, P.D., et al. (2005). Sequence characteristics of functional siRNAs. *RNA* **11**: 864–72.

Järve, A., Müller, J., Kim, I.H., et al. (2007) Surveillance of siRNA integrity by FRET imaging. *Nucleic Acids Res* **35**: e124.

Jinek, M. and Doudna, J.A. (2009) A three-dimensional view of the molecular machinery of RNA interference. *Nature* **457**: 405–12.

Jordan, E.T., Collins, M., Terefe, J., Ugozzoli, L., and Rubio, T. (2008) Optimizing electroporation conditions in primary and other difficult-to-transfect cells. *J Biomol Tech* **19**: 328–34.

Ketting, R.F., Fischer, S.E.J., Bernstein, E., Sijen, T., Hannon, G.J., and Plasterk, R.H. (2001) Dicer functions in RNA interference and in synthesis of small RNA involved in developmental timing in *C.elegans*. *Genes Dev* **15**: 2654–9.

Kole, R., Krainer, A.R., and Altman, S. (2012) RNA therapeutics: beyond RNA interference and antisense oligonucleotides. *Nat Rev Drug Discov* **11**: 125–40.

Lee, S.-K. and Kumar, P. (2009) Conditional RNAi: towards a silent gene therapy. *Adv Drug Deliv Rev* **61**: 650–64.

Li, C.X., Parker, A., Menocal, E., Xiang, S., Borodyansky, L., and Fruehauf, J.H. (2006) Delivery of RNA interference. *Cell Cycle* **5**: 2103–9.

McIntyre, G.J. and Fanning, G.C. (2006) Design and cloning strategies for constructing shRNA expression vectors. *BMC Biotechnol* **6**: 1.

Nellen, W. and Lichtenstein, C. (1993) What makes an mRNA anti-sensitive. *Trends Biochem Sci* **18**: 419–23.

Nguyen, Q.D., Schachar, R.A., Nduaka, C.I., et al. (2012) Phase 1 dose-escalation study of a siRNA targeting the RTP801 gene in age-related macular degeneration patients. *Eye (Lond)* **26**: 1099–1105.

Paddison, P.J. (2008) RNA interference in mammalian cell systems. *Curr Top Microbiol Immunol* **320**: 1–19.

Pecot, C.V., Calin, G.A., Coleman, R.L., Lopez-Berestein, G., and Sood, A.K. (2011) RNA interference in the clinic: challenges and future directions. *Nat Rev Cancer* **11**: 59–67.

Rao, D.D., Vorhies, J.S., Senzer, N., and Nemunaitis, J. (2009) siRNA vs. shRNA: similarities and differences. *Adv Drug Deliv Rev* **61**: 746–59.

Reynolds, A., Leake, D., Boese, Q., Scaringe, S., Marshall, W.S., and Khvorova, A. (2004). Rational siRNA design for RNA interference. *Nat Biotechnol* **22**: 326–30.

Samuel-Abraham, S. and Leonard, J.N. (2010). Staying on message: design principles for controlling nonspecific responses to siRNA. *FEBS J* **277**: 4828–36.

Shabalina, S.A., Spiridonov, A.N., and Oqurtsov, A.Y. (2006) Computational models with thermodynamic and composition features improve siRNA design. *BMC Bioinformatics* **7**: 65.

Siomi, H. and Siomi, M.C. (2009) On the road to reading the RNA-interference code. *Nature* **457**: 396–404.

Taxman, D.J., Livingstone, L.R., Zhang, J., et al. (2006) Criteria for effective design, construction, and gene knockdown by shRNA vectors. *BMC Biotechnol* **6**: 7.

Ui-Tei, K., Naito, Y., Takahashi, F., et al. (2004) Guidelines for the selection of highly effective siRNA sequences for mammalian and chick RNA interference. *Nucleic Acids Res* **32**: 936–48.

Walton, S.P., Wu, M., Gredell, J.A., and Chan, C. (2010) Designing highly active siRNAs for therapeutic applications. *FEBS J* **277**: 4806–13.

Yamashita, M. and Emerman, M. (2006) Retroviral infection of non-dividing cells: old and new perspectives. *Virology* **344**: 88–93.

Yang, R., Lacson, R.G., Castriota, G., et al. (2012) A genome-wide siRNA screen to identify modulators of insulin sensitivity and gluconeogenesis. *PloS One* **7**: e36384.

Zamore, P.D., Tuschl, T., Sharp, P.A., and Bartel, D.P. (2000) RNAi: double-stranded RNA directs the ATP-dependent cleavage of mRNA at 21 to 23 nucleotide intervals. *Cell* **101**: 25–33.

Zeng, Y., Yi, R., and Cullen, B.R. (2005) Recognition and cleavage of primary microRNA precursors by the nuclear processing enzyme Drosha. *EMBO J* **24**: 138–48.

Zimmermann, K., Scheibe, O., Kocourek, A., Muelich, J., Jurkiewicz, E., and Pfeifer, A. (2011) Highly efficient concentration of lenti- and retroviral vector preparations by membrane adsorbers and ultrafiltration. *BMC Biotechnol* **11**: 55.

2

Working with protein

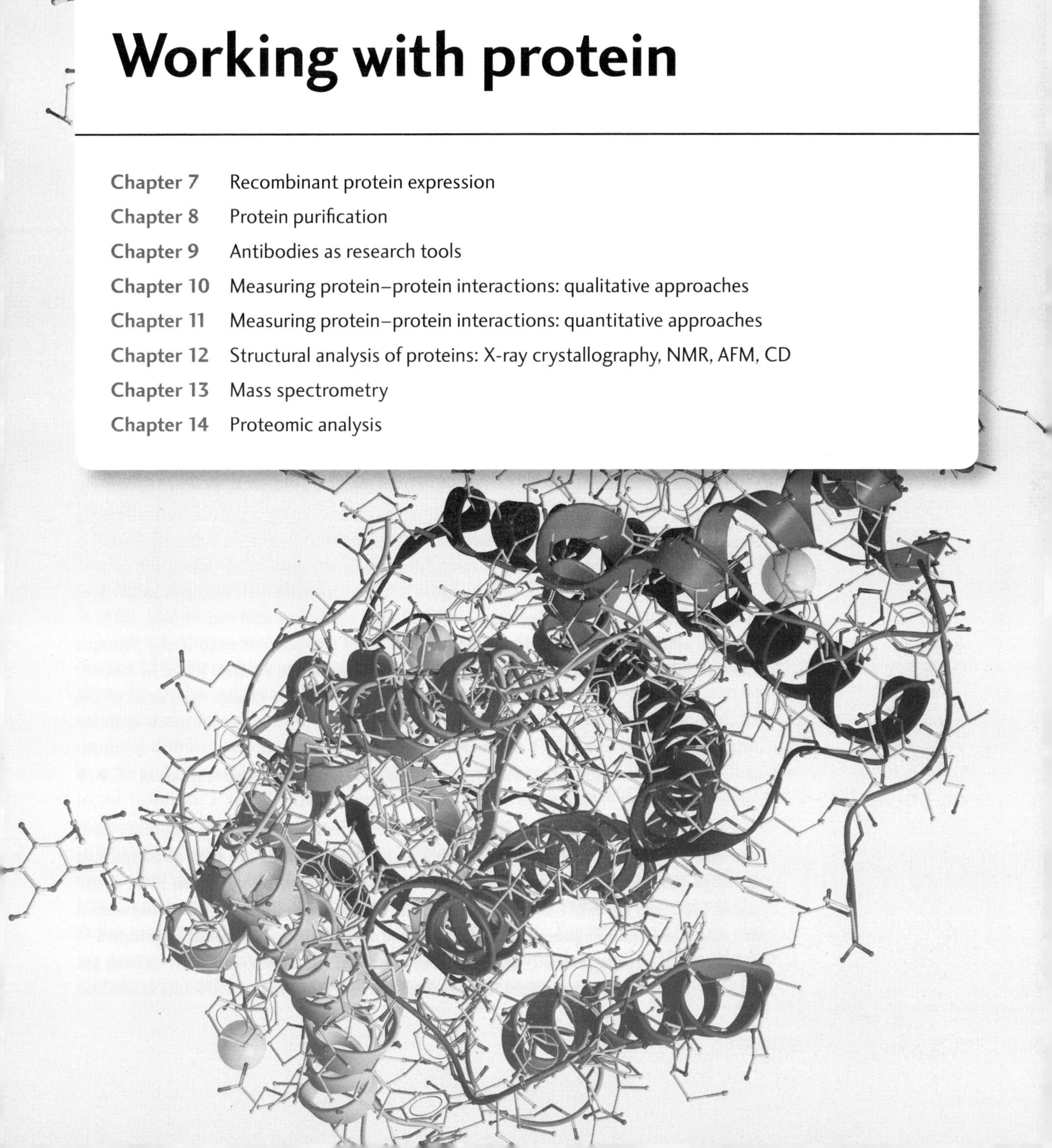

Recombinant protein expression

7

Michael Harrison and Michael J. McPherson

Chapter overview

Before the advent of recombinant DNA technology, purifying a protein involved laborious and often inefficient extraction from sources such as cells or tissues. We now have access to complementary DNAs (cDNAs) encoding virtually any protein, and this has revolutionized the way proteins are made in the research laboratory. To obtain high levels of expression of a protein, its cDNA can be introduced into a host cell, typically of bacterial, yeast, or animal origin, which then acts as a 'biofactory' to produce the 'foreign' experimental protein. Since production of the protein depends on **recombinant DNA**, the product is referred to as a **recombinant protein**. The expressed protein can become the most abundant protein in the host cell, making its purification much simpler.

Different forms of cDNA delivery systems (termed **expression vectors**) that maximize production of protein in a specific host cell have been developed. These can be simple plasmid DNA molecules or much more complicated recombinant viruses. In this chapter, we will guide you through some of the technical issues relating to the systems commonly used for recombinant protein production, and explore the advantages and disadvantages of each. This should enable you to make the correct choices at the start of your study.

LEARNING OBJECTIVES

This chapter will enable the reader to:

- understand how recombinant proteins are made in the research laboratory using different prokaryotic and eukaryotic host cells;
- gain a working knowledge of currently available vector technologies for protein over-expression in bacterial, yeast, and insect larval cells;
- appreciate the experimental variables that can influence the yield of expressed protein in each host–vector system;
- evaluate the advantages and disadvantages of different host–vector combinations and make an informed choice of a system with which to express your 'protein of interest'.

7.1 Introduction

Why might you want to produce a recombinant protein? There are many possible reasons. You may wish to study the protein, but cannot purify enough from its natural source and hence you clone the gene and express the protein in a simpler cell system. This recombinant protein could then be used to study the activity or function of the protein, or to determine its 3D shape and structure. You may want pure protein in order to produce antibodies that recognize the protein so that you can study its location and natural expression pattern in the original organism, or to use the protein for therapy such as insulin for diabetics or human growth hormone for dwarfism or as a vaccine to protect against the original organism. If you work out the 3D structure using nuclear magnetic resonance (NMR) or X-ray crystallography, you may want to study how the protein works by changing specific amino acids in the protein to examine the effect of the mutation on function. The 3D structure is usually important if you want to try and design a drug that might interfere with the activity of the protein. These are just some of the ways in which expressing a recombinant protein can be of value

See Chapter 12

See Chapter 3

Recombinant protein expression can be achieved in a wide variety of prokaryotic or eukaryotic cells ('hosts') using essentially only two types of cDNA delivery system ('vector')—plasmids or engineered viruses. Prokaryotic host cells commonly used for protein expression are the non-pathogenic strains of the Gram-negative bacterium *Escherichia coli* (Vaillancourt 2003; Balbás and Lorence 2004; Zerbs et al. 2009) (see section 7.3). Prokaryotic hosts have many advantages for expression which exploit their natural capacity to take up and replicate plasmids and to grow to very high cell densities under laboratory conditions. Eukaryotic hosts include the yeasts *Pichia pastoris* and *Saccharomyces cerevisiae* (see section 7.4) and larval insect cells (see section 7.5). These offer many of the advantages of cost and convenience of bacterial hosts, but as true eukaryotes are more likely to produce recombinant eukaryotic proteins in a correctly folded and hence fully active form (Romanos 1995; Cereghino and Cregg 2000; Macauley-Patrick et al. 2005; Murhammer 2007; Cregg et al. 2009; Jarvis 2009). Cultured mammalian cells can also be used as expression hosts, but these systems are generally unsuitable for large-scale production of purified protein because of the scale-up costs or the need for specialized laboratory containment facilities if viral vectors are used. More recently developed cell-free expression systems using bacterial, wheat germ, or insect cell lysates have the potential to produce soluble proteins in milligram amounts, but still at relatively high cost (Katzen et al. 2005). Hence in this chapter we will focus on the systems commonly used for protein production—*Escherichia coli*, *Pichia pastoris*, *Saccharomyces cerevisiae*, and larval insect cells. More detailed coverage of the most widely used systems than is possible here can be found in the literature (Vaillancourt 2003; Balbás and Lorence 2004; Murhammer 2007; Cregg et al. 2009; Jarvis 2009).

Regardless of the system used, production of recombinant protein includes the following steps (outlined in Figure 7.1).

(i) Obtain the coding sequence (cDNA) for the 'protein of interest'.

Figure 7.1 Outline scheme for setting up protein expression.

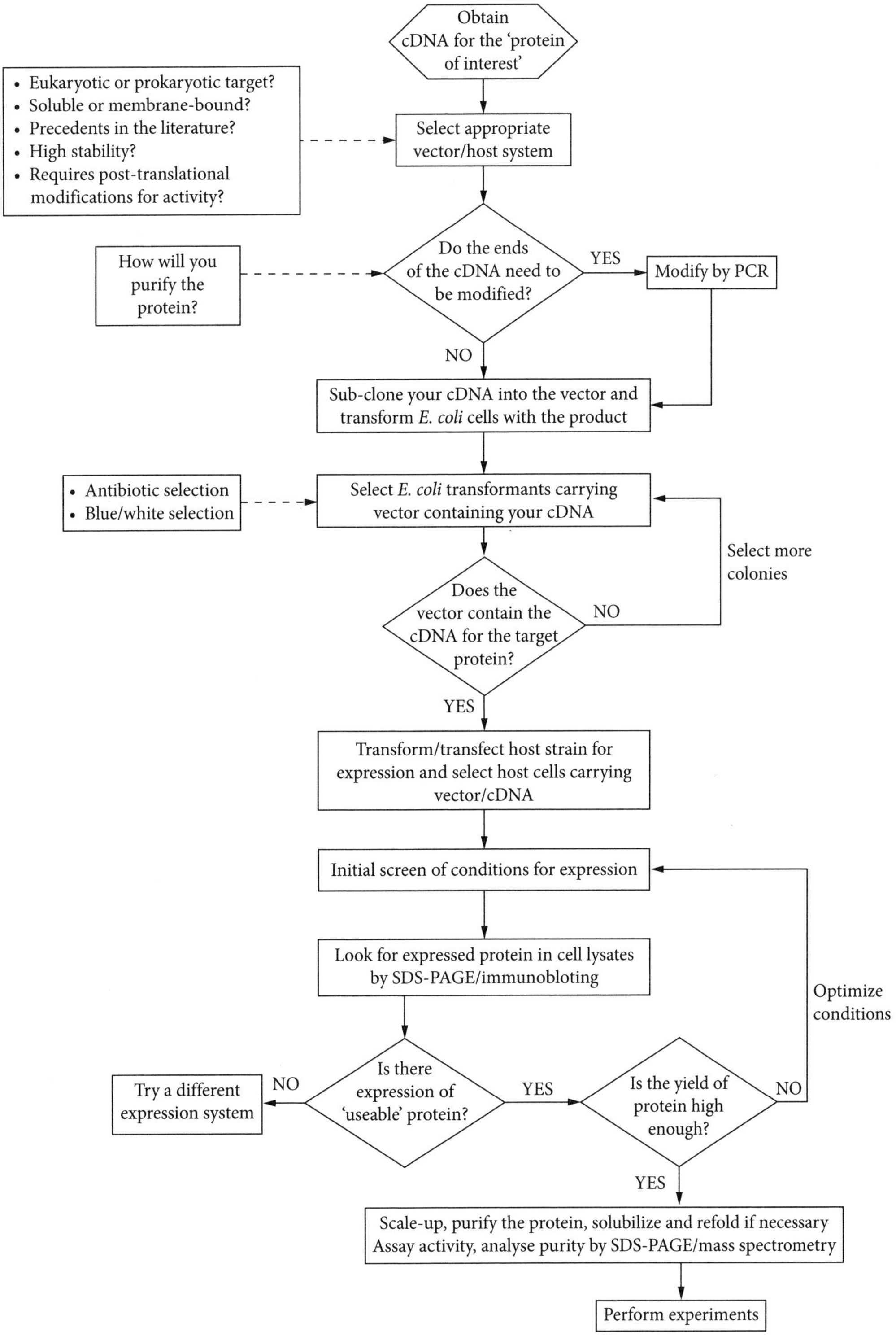

(ii) Select an expression system that has the greatest chance of producing the protein needed for your experiments. The criteria that can be applied to help with this decision are outlined in section 7.2.

(iii) The cDNA is then **sub-cloned** into the vector DNA by ligation *in vitro* after digestion with restriction endonucleases.

See Chapter 1

(iv) *E. coli* cells are transformed with the vector-cDNA construct, and efficient replication allows large amounts of it to be purified.

(v) After ensuring that the replicated vector contains the cDNA, it is purified and introduced into the host cell strain customised for protein expression. Transformed *E. coli*, yeast, or **transfected** insect cells are initially screened for expression and the growth/induction conditions are optimized to maximize protein yield. Once established, these conditions can be scaled up for expression before purification of the protein.

Each of these steps is described further below for the *E. coli*, yeast, and larval insect expression systems.

7.2 Choosing the right expression system

The following simple questions can be applied to select the most appropriate expression system for your 'protein of interest' (Balbás and Lorence 2004; Brondyk 2009; Bernaudat et al. 2011).

1. **What do you want to do with the protein?**

   This most important question is essentially asking how much protein you need. Structural biology techniques such as X-ray crystallography and NMR spectroscopy will require large amounts (>10mg) of pure protein, and production of protein as antigen for antibody generation can also require milligram amounts. For this scale of protein production, most investigators will initially test for expression in *E. coli* because of the speed and ease with which it can be set up. If this is unsuccessful, fungal cells may be assessed next because of their dual benefits of relatively low growth costs and potential for very high yield. An alternative to these is insect cell expression which, although relatively expensive and time-consuming to set up, has proved very successful at producing both soluble and membrane-bound proteins of eukaryotic origin (Murhammer 2007; Jarvis 2009). Large-scale manufacture of recombinant protein from cultured mammalian cells is possible, of course, but the high costs and technical challenges associated with scale-up of this type of system make it impractical for large-scale protein production in most laboratories.

2. **How are you going to purify your protein?**

   You must consider how you are going to purify the recombinant protein that you express. It is usually difficult and time-consuming to purify a protein based on its inherent properties, such as size, charge, and hydrophobicity, even from a recombinant source where it may be expressed at a reasonably high level. Such purification strategies are likely to involve many steps, and quite large amounts of your protein can be lost at each step. However, when we create a DNA clone to make a recombinant protein we can modify the protein sequence to add on a sequence called a **tag**. This tag can allow us to purify the protein very simply in large yields in only one or two steps. You *must* decide how you want to purify your protein, and decide which tag you are going to use and whether it will be at the N- or C-terminus of your protein, *before* you decide how you will clone the protein coding region from the DNA or cDNA of the original organism. This is because you can then decide how you might need to modify the protein coding sequence before cloning it into an expression vector.

3. **Is your target protein from a prokaryotic or eukaryotic organism?**

   Expression of recombinant protein in a cell similar to its natural host is more likely to produce recombinant proteins in a correctly folded and hence fully active form. Although there are many examples of fully functional eukaryotic (including human) proteins being successfully made by *E. coli*, the general rule is that bacterial cells are less competent at making proteins of eukaryotic origin. This can be due to differences in **codon usage** and protein folding mechanism between *E. coli* and eukaryotes. Eukaryotic integral membrane proteins, in particular, are notoriously difficult (but not impossible) to express in bacteria. However, because of their convenience and utility, *E. coli* systems are still likely to be the first to be tested when attempting to make a eukaryotic target protein for the first time. In many cases this can result in expression of misfolded protein in **inclusion bodies** that can be purified and solubilized using standard protocols before refolding to yield functional protein.

4. **Are there any essential post-translational modifications or required cofactors?**

   If a cofactor such as a metal ion is required for full protein activity, it may be necessary to add it to the culture medium since it is otherwise likely to be limiting, particularly where high levels of protein expression are achieved. If post-translation modification of the protein is necessary for it to gain full activity or stability, expression in *E. coli* is unsuitable as bacteria lack the enzymes to perform these modifications, although they can be engineered to contain them (Wacker et al. 2002). Yeast and insect cells will perform glycosylation, but the types of sugar and pattern of branching may differ from those in mammalian cells, affecting function. In some cases, yeasts may hyperglycosylate expressed proteins, making them inactive or difficult to purify. Yeast and insect cells are also capable of forming disulphide bonds in proteins targeted for secretion, a structural property that can only be achieved in specifically engineered *E. coli* strains (see section 7.3).

5. **Is the expressed protein toxic to the host cell?**

   If expression results in retarded growth or cell death, protein yields are likely to be very low. This is a particular problem where expression is constant and uncontrolled (**constitutive expression**). It can be overcome by using **inducible expression** systems in which expression can be switched on once a high host cell biomass is reached. Such systems are available for *E. coli*, yeast, and mammalian cell expression.

6. **Can the process of evaluating different expression systems be streamlined?**

   If you anticipate needing to introduce new restriction sites at the ends of the cDNA to facilitate sub-cloning into a vector, by careful forward planning you may find sites compatible with all the vectors you intend to test. This can make it easier to shuttle a cDNA for the 'protein of interest' between vectors, speeding up the screening process. Alternatively, you could use a **combinatorial cloning** system, such as Invitrogen's Gateway™ (Life Technologies Inc. 2010a) or IBA's Stargate™ (IBA 2011), which allows easy transfer of a cDNA between different expression vectors using a recombination mechanism.

See Chapter 1

The advantages and disadvantages of each system are summarized in Table 7.1.

Table 7.1 Comparison of commonly used protein expression systems

Expression system	Advantages	Disadvantages
Escherichia coli (prokaryotic)	Well-described genetics Wide variety of vectors available Rapid growth to high cell density Low-cost media and fast set-up High yield and easy scale-up	Over-expressed proteins may be misfolded and insoluble (especially proteins of eukaryotic origin) No post-translational modifications
Lactobacillus lactis (prokaryotic)	Commercially available kit Fast set-up Rapid growth to high cell density Inducible expression allows tight control 'Food grade' culture requires no special media or facilities Easy scale-up	Limited number of vectors Eukaryotic proteins may not be folded/active
Saccharomyces cerevisiae (yeast)	Well-described genetics Commercially available vectors Low-cost media High cell densities Easy scale-up Eukaryotic post-translational modifications	Plasmids may be unstable Low yields Non-mammalian-type glycosylation and potential for hyperglycosylation Uncertain targeting of expressed membrane proteins Insufficient aeration can lead to toxic build-up of ethanol
Pichia pastoris (yeast)	Well-described genetics Commercially available vectors Low-cost media Extremely high cell densities Very high yield and easy scale-up Eukaryotic post-translational modifications	Transformation can be difficult Cells do not support episomal DNA (plasmids) Long lead time to set up and optimize Non-mammalian-type glycosylation Safety risk due to methanol usage
Insect larval cells/ baculovirus	Variety of commercially available systems Successful history of expressing eukaryotic proteins, including membrane proteins Post-translational modifications High yield	Time-consuming to set up and optimize High cost of media Requires specialized culture facilities Limited potential for scale-up Non-mammalian-type glycosylation
Mammalian cells	Variety of commercially available vectors Wide range of cells can be transfected Post-translational modifications Eukaryotic proteins highly likely to be biologically active	Some cells difficult to transfect Transfection efficiencies can be low Over-expression may cause 'non-natural' cellular effects High cost of media Requires specialized culture facilities Limited potential for scale-up

Table 7.1 (Continued)

Expression system	Advantages	Disadvantages
Cell-free systems	Fast (2 hour reaction typical) High yields (up to 1mg/mL reaction possible) Lysates for *in vitro* production available from both prokaryotic and eukaryotic sources Coupled transcription/translation reaction requires only addition of plasmid Allows incorporation of labelled or modified amino acids	Cost Many of the problems of expressing eukaryotic proteins in prokaryotic systems remain Limited scope for making membrane proteins

7.3 *E. coli*-based expression systems

The steps involved in an *E. coli* expression experiment are as follows.

- A small plasmid (expression) vector carrying the cDNA for the 'protein of interest' is constructed, replicated in an *E. coli* cloning strain, and purified.
- A suitable *E. coli* expression strain such as BL21(DE3) is transformed with the recombinant plasmid.
- The cells are grown to a specific cell density and expression of the recombinant protein is induced.
- The cells are harvested and lysed, and the recombinant protein is purified usually using an associated affinity tag.

A wide variety of both expression vectors and host strains are commercially available and different combinations of these may need to be tested before successful expression is achieved. Fortunately, the ease of the molecular biology and speed of growth of *E. coli* mean that at least a preliminary test for expression can be set up within 2–3 weeks.

7.3.1 Expression vectors for *E. coli*

The key elements of all bacterial expression vectors are shown in Figure 7.2 and include the following.

- An origin of replication (ORI) that allows the plasmid to be maintained at high copy number in the host cell.
- An antibiotic resistance gene (typically for ampicillin, kanamycin, or chloramphenicol) to allow selection of transformed cells that carry the plasmid.
- A multiple cloning site (MCS) with a range of restriction sites that allow insertion of the cDNA encoding the 'protein of interest' in the correct orientation with respect to the promoter.
- A promoter sequence upstream of the MCS that drives transcription of mRNA for the 'protein of interest'. The promoter can be constitutive (always active) or

Figure 7.2 Example of an *E. coli* expression vector. Key elements of all bacterial vectors are the origin of replication (ORI), an antibiotic resistance gene, a promoter upstream of a multiple cloning site (MCS), and a ribosome binding site (RBS). The cDNA sequence encoding the protein of interest (POI) is ligated downstream of the promoter. The RBS is optimally placed 8bp from the transcriptional start site (ATG). Typically, expressed proteins will be fused to an affinity tag to facilitate purification. In this example, a hexahistidine tag is added to the C-terminus-encoding part of the POI cDNA. The fusion must maintain the correct reading frame. In the example shown, the 'stop' codon at the end of the POI cDNA has been removed and a Bam H1 restriction site introduced that is compatible with a Bam H1 site in the vector. Ligation of POI cDNA and vector at this point maintains the reading frame through the end of the POI cDNA, across the join, and into the hexahistidine tag. The Bam H1 site introduces new glycine and serine residues at the POI–tag fusion, and the 'stop' codon (TAA) is at the end of the polyhistidine tag.

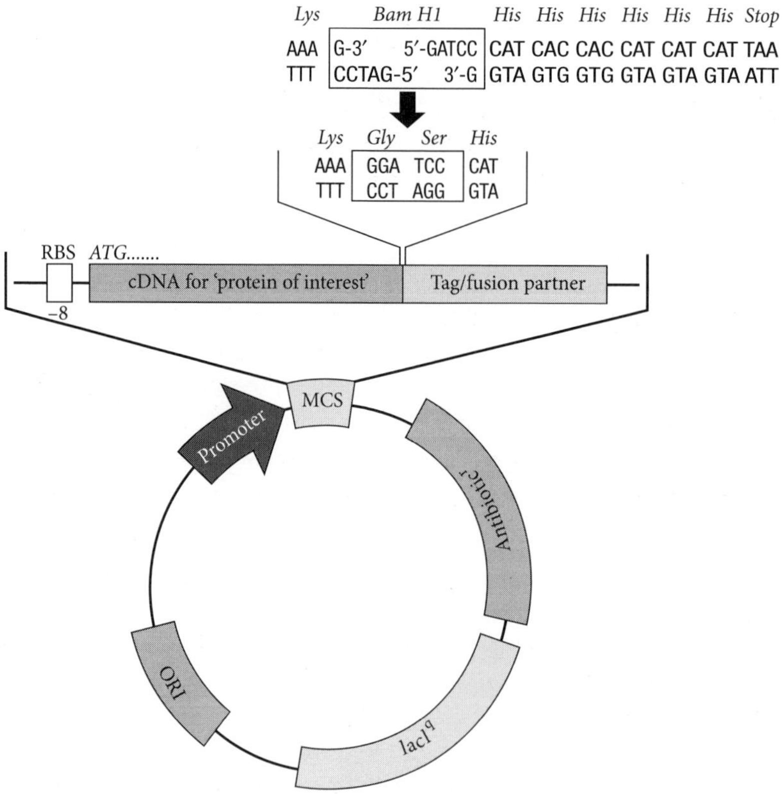

inducible (switched on by changes to growth conditions). Commercially available vectors use inducible promoters because of the level of control they allow.

- A terminator sequence downstream of the MCS.
- A ribosome binding site (RBS) which is optimally placed 8bp from the transcriptional start site.

The cDNA is sub-cloned into the MCS, oriented such that the transcriptional start is immediately downstream of the promoter in the vector and upstream of the termination site. Typically, expressed proteins undergo **in-frame fusion** to an affinity tag to facilitate protein identification and purification. The coding sequences for these affinity tags are

See Chapter 8

built into the plasmid vectors. In the example shown in Fig. 7.2, the cDNA sequence coding for the C-terminus of the 'protein of interest' is fused during the sub-cloning process to the sequence for a polyhistidine tag.

The promoters used in E. coli *expression vectors*

Several different promoters are used in *E. coli* vectors to drive protein expression.

The *lac* **promoter**–Typically, elements of the tightly regulated *E. coli lac* operon are used in the promoters of bacterial expression vectors. This operon encodes genes for lactose metabolism and is only active when lactose or synthetic mimics are included in the growth medium. The promoter/operator of the *lac* operon is repressed by a constitutively expressed **repressor** protein, LacI, ensuring that transcription is switched off. When analogues, such as isopropylthiogalactoside (IPTG) are added, the *lac* promoter/operator is de-repressed and transcription occurs. The *lac* operon is also subject to **catabolite repression**, as it is inactive in the presence of glucose even if galactose is available. This *lac* promoter/operator system has been exploited in two ways to control expression in the laboratory. First, components are built into plasmids to drive transcription only when the inducer IPTG is added to the culture medium (and glucose is excluded). Vectors, such as pTAC (Sigma) and pTrcHis (Invitrogen), have hybrid *tac* or *trc* promoters which combine the ~10-fold greater transcriptional activity of the promoter from the *trp* operon (for tryptophan biosynthesis) with the sugar control of the *lac* promoter (de Boer et al. 1983). In addition, these vectors also carry a constitutively active *lacI*^{q} mutant gene to provide sufficient repressor to maintain repression before IPTG is added.

Although high levels of expression are easily achieved using these *lac*-based systems, they tend to be leaky, with some expression even in the absence of inducer. This can be a major problem if the expressed protein is toxic to the host cell. The second use of the *lac* system is for control of the T7 system.

The T7 promoter–The T7–pET system devised by William Studier and colleagues uses the bacteriophage T7 RNA polymerase to drive transcription of the cDNA for the 'protein of interest'. T7 polymerase expression itself is under the control of the inducible *lac* promoter (Studier and Cregg 1986; Studier et al. 1990; Novagen 2011). Key features of the pET system are shown in Figure 7.3. The first component is an engineered *E. coli* expression strain in which a bacteriophage DE3 lysogen has been inserted into the genome. This contains the coding region for T7 RNA polymerase under the control of a hybrid *lacUV*5 promoter that is repressed by LacI from the host cell's *lacI*^{q} gene. The second component of the system is the plasmid vector that contains the cDNA for the 'protein of interest' under the control of a T7 promoter that is recognized only by the T7 polymerase. If inducer is not present, T7 polymerase is not expressed and the target protein is not expressed. Downstream of the T7 promoter on the vector is a *lac* operator element that allows additional repression by LacI, providing a further degree of control.

When inducer (such as IPTG) is added to the culture medium, LacI repressor binding is relieved and transcription and translation of the T7 polymerase can occur. T7 polymerase binds to the pET T7 promoter and initiates transcription of the target protein cDNA. As the T7 promoter is very powerful, expressed protein can account for 30–40% of total cell protein. Expression from the T7 promoter can also be leaky, with low levels in the absence of inducer having potentially harmful effects if the expressed protein

Figure 7.3 T7-based expression in *E. coli*. *E. coli* strains engineered for protein expression by the pET system have a DE3 phage lysogen integrated into their genome that includes the coding region for the phage T7 RNA polymerase (T7 pol) under the control of a hybrid *lacUV5* promoter. In expression vectors of the pET type, transcription of the cDNA for the target protein is under the control of a T7 promoter that requires the *E. coli* cell to express T7 polymerase. (a) In the presence of glucose/absence of galactose, expression of T7 polymerase is repressed by binding of the LacI repressor to three *lacO* sites in the *lacUV5* promoter region. The repressor is provided by a mutant *lacI*^{q} gene in the host cell genome, but this is often supplemented by expression from the vector plasmid. The repressor also binds to a site on the plasmid vector to minimize leaky expression of the target protein. (b) In the presence of IPTG, an analogue of galactose, LacI binding is blocked, the *lacUV5* promoter is de-repressed, and T7 polymerase expression can proceed. Thus the T7 promoter on the expression vector becomes active, driving mRNA transcription and its translation into the 'protein of interest'.

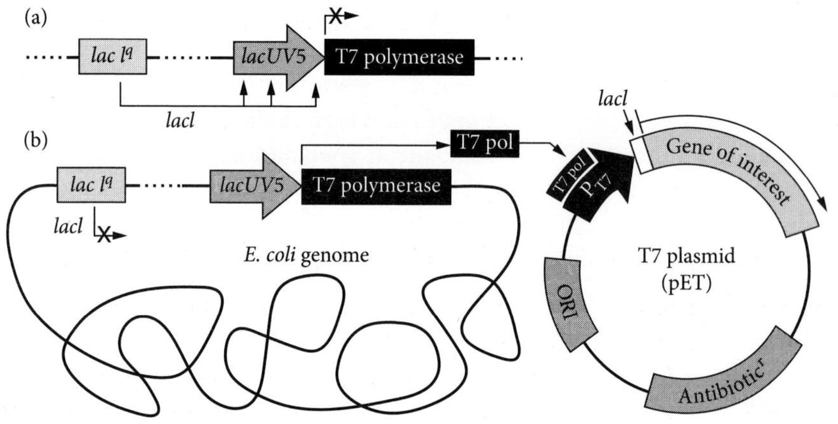

is toxic. More stringent control can be exerted by using host strains that also express T7 lysozyme, a natural inhibitor of the T7 polymerase. Host strains expressing this protein harbour additional plasmids pLysE and pLysS.

Many commercially available vectors use the T7 system, ranging from the basic pET design (Novagen) to more sophisticated versions with additional features:

- the four vectors in the Duet series from Novagen contain two oppositely oriented T7*lac* promoters that allow simultaneous co-expression of two proteins, but can be used in combination to allow expression of up to eight targets;
- the pDEST T7-based vector is a destination vector for Invitrogen's Gateway™ cloning system that enables swapping of cDNA of interest between multiple vectors;
- the pRSET vector, also from Invitrogen, incorporates a phage gene 10 transcript stabilizing sequence which can enhance target protein expression levels.

The adaptability of some T7 vectors is increased by building in additional promoters that drive expression in cell systems other than *E. coli*, hence allowing screening of multiple expression systems using the same recombinant plasmid construct which can cut down on set-up time. These include pBK-CMV (Stratagene), with a cytomegalovirus (CMV) promoter and a neomycin resistance cassette for expression in eukaryotic cells, and pTriEx (Novagen) with CMV- and baculovirus-specific sequences for both eukaryotic and insect cell expression.

It is often useful to tune down expression levels rather than force the host cell to manufacture as much protein as possible. Lower expression can minimize toxic effects on the cell and enhance the production of correctly folded protein. While absolute yields may be lower, a higher yield of useful active protein can be achieved. Various vector–host systems have been developed that give a strict linear correlation between inducer concentration and expression. For example, in the pBAD system (Invitrogen), expression of the target protein is under the control of the arabinose-inducible promoter of the *ara* operon. Arabinose concentrations across two to three orders of magnitude can be tested to determine the optimal concentration for production of active protein. Lowering growth temperature can increase the efficiency of folding of translation products, improving yield of active protein. The cold-responsive cspA promoter of the pCOLD vector (Takara) allows the start of transcription and translation to be synchronized with a decrease in culture temperature (to ~15°C).

7.3.2 Host strains

The majority of host expression cells, including the basic BL21(DE3) strain that drives T7 RNA polymerase expression, are based on the parental protease-deficient BL21 strain (Table 7.2). Versions of this strain have been engineered to provide higher levels of expression by deleting enzymes that degrade messenger RNA (BL21(DE3) Star). Codon usage by bacteria can differ significantly from that in eukaryotes, and can be a major contributing factor to low yields of eukaryotic proteins in *E. coli*. You can overcome this by using strains such as BL21(DE3) Codon Plus and Rosetta that have enhanced expression of a subset of tRNAs that are common in eukaryotes but rare in *E. coli* (AUA (Ile), AGG (Arg), AGA (Arg), CUA (Leu), CCC (Pro), GGA (Gly), and CGG (Arg)). Other strains are designed to give greater control over T7 RNA polymerase expression in conjunction with pET-type vectors. These include Tuner cells that show tight correlation between inducer concentration and expression, BL21-SI and BL21-AI strains in which T7 polymerase is salt-inducible and arabinose-inducible, respectively, and Lemo21(DE3) cells where control is by the sugar rhamnose. These may be useful when expression has to be reduced because of toxicity or problems with protein folding and solubility.

Normally, the cytoplasmic environment is relatively reducing, preventing the formation of disulphide bonds between cysteine residues in the target protein. This can compromise solubility and hence yields of active protein. Host strains that maintain a more oxidizing cytoplasm, such as Origami and Rosetta-gami or SHuffle (Table 7.2), can facilitate disulphide formation, which can improve yields of some proteins.

7.3.3 Transformation and selection

Plasmids are introduced into *E. coli* cells by either electroporation or heat shock of chemically competent cells. Chemically competent cells are suitable for standard cloning procedures such as the introduction of an expression vector. Typically, calcium chloride or rubidium chloride treatment under ice cold conditions is used to make the cells competent to take up plasmid DNA during a brief heat step lasting for ~60 seconds at 42°C.

Table 7.2 *E. coli* host expression strains

Strain	Comments	Manufacturer
BL21	General expression strain deleted for *lon* and *Omp*T proteases	Various
BL21-A1	T7 expression controlled by *ara*BAD promoter Inducible with arabinose	Invitrogen
BL21 Codon-plus	Enhanced levels of rare tRNAs that can be expression-limiting T7 also expresses versions	Agilent
BL21(DE3)	BL21 background with lambda phage DE3 lysogen for T7 expressing versions are also available	Various
BL21(DE3) pLysE/S	Expression of T7 lysosyme from pLysE and pLysS gives stringent control of T7 RNA polymerase activity	Various
BL21 Rosetta*	Enhanced levels of expression-limiting tRNAs that are rare in *E. coli* but common in eukaryotes Rosetta 2 version has additional tRNA	Novagen
Rosetta-gami	Combines properties of Origami and Rosetta strains	Novagen
BL21-SI	T7 expression controlled by NaCl-inducible promoter	Invitrogen
BL21 (DE3) Star	RNase E mutation gives greater mRNA stability and lifetime	Invitrogen
BL21 Tuner	*lac* permease-deleted strain Inducer uptake (IPTG) is strongly positively correlated with concentration across a wide range Expression is stringently concentration-dependent	Novagen
BLR	*rec*A$^-$ version of BL21 Stabilizes plasmids containing repeat sequences	Novagen
B834 (DE3)	Methionine auxotroph that facilitates labelling of proteins with ^{35}S-methionine and selenomethionine	Novagen
Lemo21(DE3)	For control of T7 polymerase expression T7 lysosyme expression is tightly regulated by rhamnose	NEB
LMG194	Deleted for *ara*BADC For use with pBAD arabinose-inducible system	Invitrogen
NiCo21(DE3)	Minimizes metal-binding protein contaminants in IMAC.	NEB
Origami/Origami 2	Deletion of *trx*B and *gor* genes facilitates disulphide bond formation and improved folding DE3 T7-expressing versions are also available	Novagen
Origami B	Combines the properties of Tuner and Origami cells DE3 T7-expressing versions are also available	Novagen
T7 Express	T7 and *lac*I^q expressing T7 expressed from native *lac* operon, not DE3	NEB
SHuffle	Permits cytoplasmic disulphide bond formation and enhanced protein folding T7-expressing versions are also available	NEB

* Novagen strains are generally available as DE3-, pLysS-, and *lac*I^q-expressing variants.

Transformed cells are selected by plating onto agar plates containing the appropriate antibiotic carried on the expression plasmid.

7.3.4 Induction strategies

A typical protocol for an induction experiment involves growing cells at 37°C in shake flasks to a specific density (typically $OD_{600} = 0.6$). This is followed by addition of inducer (IPTG) to a specific concentration and continued growth during which time the target protein is expressed with the culture typically reaching $OD_{600} = 2$. A number of variables influence protein yield, including inducer concentration, induction time, growth temperature, and culture medium. The optimum conditions will need to be determined empirically.

An alternative to IPTG induction is to use auto-induction strategies (Studier 2000; Grabski et al. 2005) which allow *lac*-regulated expression to switch on automatically as catabolite repression is switched off and the *lac* promoter of the vector or DE3 lysogen is de-repressed. To achieve this, the culture medium contains lactose in combination with glucose and glycerol. The host *E. coli* cells grow to high density, initially using glucose as a carbon source. When glucose is depleted, with cell densities reaching $OD_{600} = 20–50$, the cells switch to using lactose as a carbon source which induces expression of the T7 RNA polymerase from the DE3 *lacUV5* promoter. The advantage of this system is that high levels of protein expression can occur in unsupervised cultures, ready for harvest usually two or more days later. A variety of vector–host strain combinations will work with this system, but pLysE/pLysS strains are not recommended as they will suppress T7 activity in the earlier stages of the auto-induction process and can cause cell lysis. After induction the cells are harvested and lysed, and the recombinant protein is purified usually using an associated affinity tag. See Chapter 8

In many instances, eukaryotic proteins expressed in *E. coli* are produced in an aggregated, insoluble form called inclusion bodies. Inclusion bodies can be purified from cell lysates and solubilized, and the protein can be recovered with its natural fold. The success of this process is easiest to monitor if the protein has a biological activity, but can also be followed using biophysical techniques such as circular dichroism or FTIR spectroscopy that will show the formation of secondary structure as the polypeptide folds. Deacon and McPherson (2011) provide an example of combining a range of the approaches described above, together with the manipulation of RNA secondary structure, to enhance the translation initiation efficiency of a fungal gene to substantially increase its recombinant expression levels in *E. coli*. See Chapter 12

7.4 Yeast expression systems

Yeast systems have the same basic advantages as expression in *E. coli*: well-described genetics, ease of transformation and maintenance, and fast growth to high cell densities on cheap and simple media. Most commonly, strains of the baker's yeast *Saccharomyces cerevisiae* and the methylotrophic yeast *Pichia pastoris* are used. Although *Saccharomyces* has been widely used for many years for protein expression, *Pichia* is fast becoming the 'industry standard' yeast expression system because of the spectacularly high protein

CASE STUDY 7.1

High-yield production of a bacterial enzyme

As part of a study into sphingolipid metabolism, you are tasked with expressing the bacterial enzyme serine palmitoyltranferase that catalyses the formation of the sphingolipid precursor 3-ketodihydrosphingosine from l-serine and palmitoyl-CoA. It is a soluble homodimer of a 45kDa polypeptide, and requires the cofactor pyridoxal 5′-phosphate for enzymatic activity. You are provided with cDNAs for the polypeptide. Your experiments will require >5mg of pure enzyme.

QUESTION

Devise an expression strategy for this enzyme. Explain the rationale behind your choice of expression system. What modifications would make purification of the protein easier? Are any special considerations needed to ensure that the expressed protein is active? What would you do if your system of choice fails to produce protein?

Expression of the bacterial enzyme should be straightforward. Soluble bacterial proteins generally express to high levels in *E. coli*. If enzymically active, the protein may have detrimental effects on cell growth as the high levels could take up much of the cell's serine. For this reason, use a tightly controlled T7 vector in BL21(DE3)pLysS host cells. You may need to modify the 5′ and 3′ ends of the cDNA to facilitate insertion into the T7 vector. In-frame fusion with a polyhistidine, or alternative tags such as GST (glutathione S-transferase) or MBP (maltose binding protein), will facilitate expression of the expressed enzyme and can improve solubility. Initially, test different growth times, concentrations of inducer (IPTG), and induction times. Analyse levels of soluble protein by SDS-PAGE, and further optimize induction conditions. As pyridoxal phosphate is required for activity, you will need to add this to the growth medium since it may be limiting. If you are unable to obtain soluble protein, consider testing different culture media or decreasing growth temperature. If inclusion bodies are formed, you may be able to solubilize these and then refold in the presence of pyridoxal phosphate to obtain active enzyme. You will need to scale up and purify once expression conditions have been optimized.

yields that can be achieved (Romanos 1995; Cereghino and Cregg 2000; Macauley-Patrick et al. 2005; Cregg et al. 2009). Both species are safe to handle, requiring only the lowest category of laboratory containment. A key advantage of yeast systems over *E. coli* is their ability to fold and post-translationally modify expressed proteins from other eukaryotes. A variety of commercially available plasmid vectors can be selected which allow for either intracellular accumulation of the expressed protein or its secretion into the culture media, and significant successes have been achieved for both soluble and integral membrane proteins (Long et al. 2005; André et al. 2006; Shimamura et al. 2011; Hino et al. 2012).

Expression in these yeasts uses plasmid vectors with the cDNA for the 'protein of interest' cloned downstream from a strong transcriptional promoter from a yeast gene (Figure 7.4). Sub-cloning of the cDNA and amplification of plasmid DNA are performed in *E. coli*; therefore an antibiotic resistance gene and an *E. coli* origin of replication are also built into these plasmids. As it is replicated in *E. coli* before transfer and maintenance in yeast, this type of plasmid is often referred to as a **shuttle vector**.

Although some yeast vectors contain antibiotic selection markers, **auxotrophic selection** is more commonly used to identify yeast cells transformed with the expression vector. Strains of yeast are used that have disrupted or deleted genes for key enzymes are used in pathways for the biosynthesis of amino acids (e.g. *TRP1*, *LEU2*, *HIS3* for tryptophan, leucine, and histidine, respectively) or the bases of nucleic acids (*ADE2* and *URA3* for adenine and uracil, respectively). As a result of inactivation of these genes, strains are described as being auxotrophic because they cannot make that specific nutrient compound. The compound must be included in the culture medium for growth to occur. Expression vectors for yeast cells typically contain a gene matched to the auxotrophy of the host strain. Hence successful transformation results in loss of auxotrophy—the strain is now referred to as prototrophic and is able to grow without addition of the amino acid or base compound to the medium.

The steps involved in a yeast expression experiment are similar to those for an *E. coli* experiment.

- A plasmid shuttle vector carrying the cDNA for the 'protein of interest' is constructed, replicated in an *E. coli* cloning strain, and purified.
- The recombinant plasmid is used to transform competent yeast cells, and transformants are selected by an auxotrophic or antibiotic marker.
- The cells are grown to a specific cell density and expression of the recombinant protein is induced.
- The cells are harvested and lysed, and the recombinant protein is purified usually using an associated affinity tag.

7.4.1 *Pichia pastoris*

The methylotrophic yeast *Pichia pastoris* can equal *E. coli* for high-yield and low-cost protein production (Macauley-Patrick et al. 2005; Cregg et al. 2009). Expression levels of up to 10 mg protein per litre of liquid culture, and in some cases exceeding 50 mg/L, have been reported. Similarly high concentrations of biologically active membrane proteins can also be achieved. As *Pichia* is only weakly fermentative, it does not convert its carbon sources into toxic end-products and therefore can grow to very high cell densities provided that the culture medium is thoroughly oxygenated. It has the ability to metabolize methanol to formaldehyde and hydrogen peroxide in the presence of oxygen using two enzyme activities—alcohol oxidases 1 and 2 encoded by the genes *AOX1* and *AOX2*. The promoter for the *AOX1* gene is tightly regulated, being strongly induced by methanol and tightly repressed by glucose. *AOX1* will remain inactive when the yeast is cultured on non-fermentable carbon sources such as glycerol. However, when methanol is present, *AOX1* gene expression will be switched on and can account for up to 5% of the total mRNA and 30% of the total protein in the cell. The activity of *AOX1* is much greater than that of *AOX2*. The Michaelis constant (K_M) of *AOX1* for oxygen is high, indicating it has low affinity for this substrate, but *Pichia* compensates for this by producing very large amounts of the enzyme.

Construction of Pichia *expression vectors*

The commercial license for *Pichia* expression technology is currently held by Invitrogen, (www.invitrogen.com), and a variety of plasmid expression vectors and *Pichia* host strains are available from this supplier. These vectors exploit the inducible *AOX1* promoter to drive expression of recombinant protein (Cereghino and Cregg 2000; Lin Cereghino and Lin Cereghino 2008; Cregg et al. 2009; Life Technologies Inc. 2010d).

Features of the *Pichia* plasmids include the following.

- A bacterial origin of replication and an antibiotic resistance gene to facilitate plasmid production in *E. coli*.
- An *AOX1* promoter included upstream of a multiple cloning site for cDNA insertion and terminator sequence immediately downstream of this. A region of sequence corresponding to the 3′ end of the *AOX1* gene is also included.
- Marker genes that enable selection of transformed cells based on loss of auxotrophy/gain of prototrophy. Vectors typically carry the *HIS4* gene for integration into strains with histidine auxotrophy. Vectors may carry the *ADE2* gene for complementation of an *ade2* mutation (Invitrogen's PichiaPink system). The marker gene for resistance to the antibiotic zeocin carried by the pPICZ plasmids can also be used as a convenient screen for transformants. (Life Technologies Inc. 2010b, c).

The first step is to sub-clone the cDNA for the 'protein of interest' into the shuttle vector of choice, fused if necessary to a sequence encoding an affinity tag to facilitate purification. If secretion from *Pichia* into the culture medium is required, this can be engineered by N-terminal fusion of the signal sequence to the 'protein of interest'. This is discussed further in the section on production of recombinant protein in *Pichia*.

Transformation and selection of Pichia

Pichia does not replicate **episomal DNA** and so expression vectors must be stably integrated into the host cell chromosomal DNA. *Pichia* allows recombination of exogenous linear or circular DNA molecules to insert at defined sites in the host cell genome. To drive integration into the yeast chromosome competent *Pichia* are transformed with plasmid DNA, (Life Technologies Inc. 2010d). Electroporation will generally yield the greatest number of transformants. Incubating cells treated with lithium salts with plasmid DNA–polyethylene glycol mixtures followed by heat shock or using **sphaeroplasts** can also be successful, but produces lower transformation efficiencies.

Commercially available *Pichia* host strains such as GS115 and X33 have a fully active *AOX1* gene, and can grow rapidly with methanol as sole carbon source. These are described as having a Mut$^+$ (*methanol utilization positive*) phenotype. Strains in which *AOX1* is disrupted (such as KM71) grow much more slowly on methanol because of their dependence on the weak *AOX2* activity. These strains have a Mut^s (*methanol utilization slow)* phenotype. Most transformation procedures for *Pichia* target integration to part of the chromosomal *AOX1* gene, which can occur in one of two ways (Figure 7.4). Random single cross-over events between any of the *Pichia*-derived sequences in circular vector DNA and the corresponding sites in the *Pichia* chromosomal DNA will result in complete insertion of the expression vector at that site. The expression vector can also be directed to a particular site in the *Pichia* chromosomal DNA by linearizing it with restriction enzyme prior to transformation. In Figure 7.4(a), linearizing at a site within the *AOX1* promoter region favours integration at the 5′ end of the chromosomal *AOX1* gene. Similarly, linearizing the vector by cutting at a site within the 3′ *AOX1* sequence will favour integration at the 3′ end of the chromosomal *AOX1* gene. Integration at an auxotrophic marker gene distant from the chromosomal *AOX1* gene can also be achieved by cutting at a site within the equivalent gene in the vector. Integration does not affect the

Figure 7.4 Expression vector integration in *Pichia pastoris*. (a) A *Pichia* expression vector contains the cDNA for the 'protein of interest' (POI) under control of the methanol-inducible *AOX1* promoter and upstream of a terminator (TT) sequence, a 3′*AOX1* sequence, and a marker gene for auxotrophic selection or antibiotic resistance. Linear vector DNA is integrated into the host cell genome by single cross-over recombination. The site of integration can be controlled by cutting the vector DNA at different sites with restriction endonucleases. In the example shown, cutting at site 1 in the plasmid *AOX1* promoter sequence facilitates integration at the 5′ end of the host chromosomal *AOX1* gene. Cutting the vector DNA at sites 2 and 3 would favour integration at the 3′ end of the *AOX1* gene or at an auxotrophic marker gene, respectively, indicated by the matching numbered arrows. (b) If the vector DNA is linearized by cutting at site 4, it can integrate by double cross-over between the 5′ and 3′ ends of the *AOX1* gene. The host *AOX1* gene is completely replaced by vector DNA, resulting in a Mut^S phenotype.

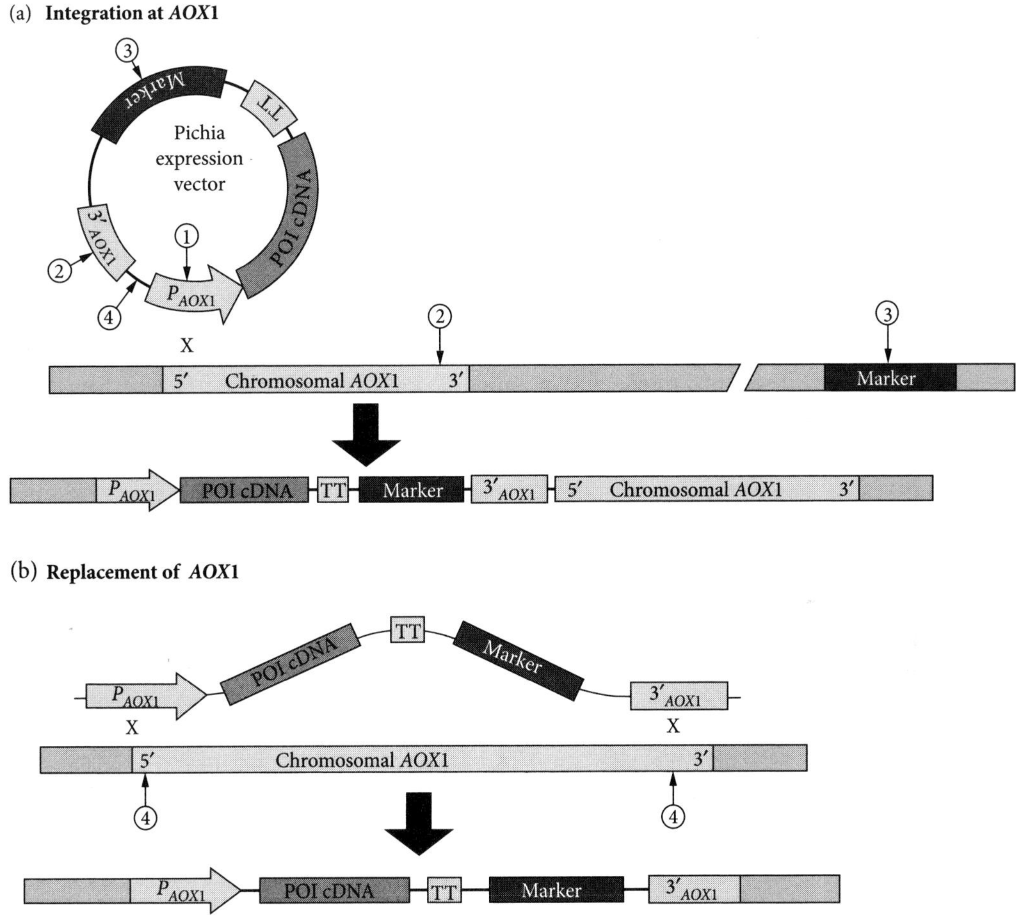

chromosomal *AOX1* gene, and hence the original Mut$^+$ or Mut^s phenotype is retained. An alternative strategy is to completely replace the chromosomal *AOX1* gene with the expression vector (Figure 7.4(b)). By cutting the vector DNA at a site between the *AOX1* promoter and the 3′ sequences it becomes a linear molecule with *AOX1* complementary sequences at both ends. Low-frequency double cross-over events can integrate this linear molecule into the *Pichia* chromosome *AOX1* locus. As the gene is effectively deleted from the host cell, a Mut^s phenotype will result.

Transformants are easily selected because a marker such as an *HIS4* gene is integrated and complements a defective gene in the host. Growth on minimal media without histidine is used to screen for host cells that have integrated the vector. If the 'marker' confers resistance to zeocin, integration is directed towards the host *AOX1* locus and transformed cells are selected on rich media (combinations of yeast extract with peptone supplemented with 1–2% glucose (YEPD)) supplemented with the antibiotic.

Multiple integration events can also occur, with the potential for increased expression levels. As these only occur in 1–10% of transformants, a large number must be screened. With luck, so-called 'jackpot clones' with 10–20 integrations can be found. If zeocin resistance is used, sequential plating onto media with increasing antibiotic concentrations can isolate cells with multiple insertions. If a conventional auxotrophic marker is used, it may be more convenient to screen directly for increased levels of expressed protein by Western or dot blot. An alternative approach is to 'pre-assemble' multiple tandem promoter–cDNA–terminator cassettes into a single expression vector (such as Invitrogen's pAO815) for simultaneous integration into the host. This strategy also has the potential to force the host cell to express a number of different proteins simultaneously, since each expression cassette can contain a different cDNA (Life Technologies Inc. 2010b).

See Chapter 9

Growth and induction of Pichia

Once highly expressing host cells have been found, the optimum growth and induction conditions must be determined (Life Technologies Inc. 2002). Many parameters can be varied, such as the rate of growth and cell density at induction, the timing and length of induction, and the concentration of inducer.

Most laboratories without access to specialized fermentation equipment will use batch culture in baffled conical flasks, partially filled and shaken vigorously to achieve the necessary oxygenation. For most purposes, a minimal medium will give sufficiently high cell densities. For Mut$^+$ strains, 0.5–1% methanol will act as both the carbon source and the inducer, whereas Mut^s strains will require glycerol as the carbon source, with methanol added solely as the inducer. The methanol is both volatile and rapidly metabolized, and hence may need to be topped up to keep the culture going. This is particularly important if the cells are cultured over a number of days, as might occur when the aim is to accumulate secreted protein in the medium. The pH of *Pichia* batch cultures can drop from pH5.5–6 to as low as pH3, which will slow growth and can have an adverse effect on secreted proteins. This can be overcome by monitoring and subsequent adjustment of the culture pH. In some cases, it may be advantageous to grow a large biomass of cells on glycerol medium prior to adding methanol to initiate expression.

Much higher cell densities and protein yields are achieved in bioreactors that permit tight control of growth conditions (Figure 7.5). Instruments set up for *Pichia* growth are temperature-controlled and have direct feed-in of filtered air–nitrogen–oxygen mixtures and methanol. A dissolved oxygen sensor maintains optimal oxygen concentration by regulating the gas input and rate of agitation, and the feedback signal from a pH sensor controls input of alkali solution to maintain a constant pH. Cells and media can be sampled at defined time points and the expressed protein monitored by, for example, SDS-polyacrylamide electrophoresis (SDS-PAGE) and immunoblotting.

Production of recombinant protein in Pichia

Expressed soluble proteins will accumulate in the *Pichia* cytoplasm unless a targeting signal sequence has been engineered into the cDNA of the protein of interest. To recover the expressed protein, yeast cells can be mechanically disrupted by agitation

Figure 7.5 Bioreactor for *Pichia* growth. To provide adequate aeration, the culture is agitated by a motorized stirrer with an impeller (1), with air–nitrogen–oxygen mixtures pumped into the vessel (2) and distributed by a sparge (3). Continuous monitoring of dissolved oxygen concentration and pH (and, in some instruments, the methanol concentration) by probes within the bioreactor (4) feed back information to a control unit (5). This continually adjusts gas input and rate of agitation to achieve a stable optimal dissolved oxygen concentration, and adjusts the feed-in of base (KOH) to maintain a stable pH. Methanol is fed in as the carbon source for Mut$^+$ strains or as a gratuitous inducer for Mut^s cells. Solid lines, information from sensor probes to controller; dotted lines, feedback signals from the controller to the feed-in pumps and stirrer. Lower panels: induction time courses for intracellular (left) and secreted (right) expression of target proteins (arrowed). Cell lysates were analysed by SDS-PAGE. Pictures courtesy of Dr Richard Jones and Dr Jared Cartwright, Department of Biology and Technology Facility, University of York, UK.

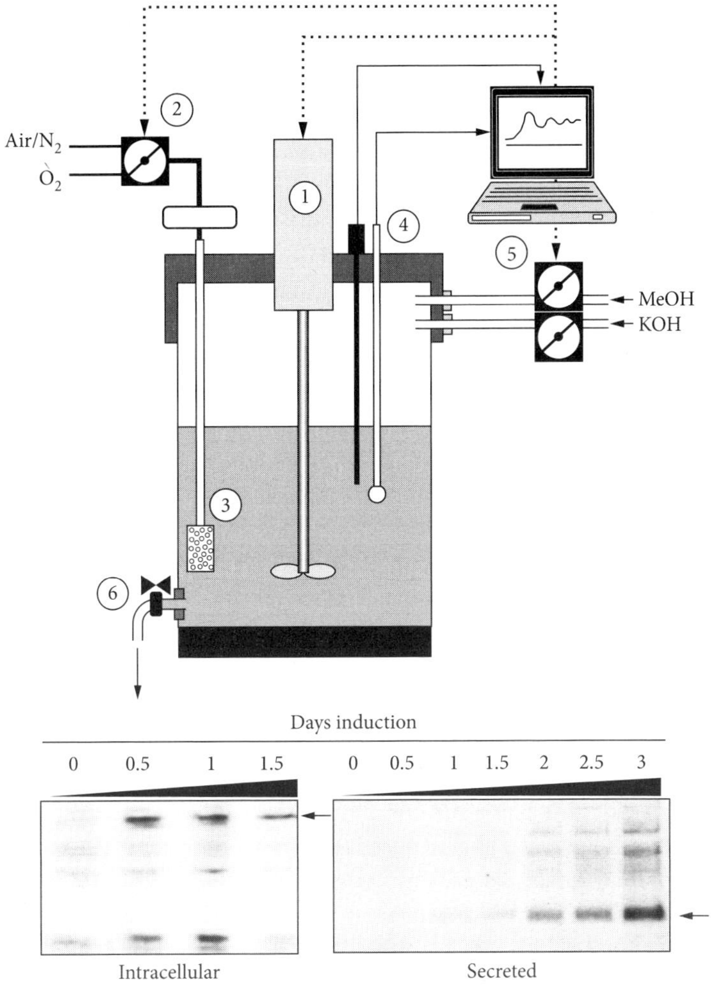

with glass beads or extrusion through a pressure cell. *Pichia* is more difficult to disrupt than *Saccharomyces* and requires, for example, two passages through a pressure cell at ~30 kPa to achieve complete disruption. A gentler approach which releases fewer proteolytic enzymes and contaminating DNA fragments is to make the cells sensitive to osmolysis by removing the cell wall with 'zymolase', a combination of lytic enzymes. Recombinant membrane proteins may reach the plasma membrane, although they can also accumulate in the endoplasmic reticulum and the vacuolar membrane. A membrane fraction can easily be recovered from disrupted yeast cells by centrifugation. *Pichia* cells are particularly efficient at secreting the expressed protein into the culture medium, and this is normally achieved by fusing the 5′ end of the target protein cDNA to the pre-pro-sequence of the *Saccharomyces* α-mating factor peptide. The yeast cell will remove the pre-sequence during the secretion process, usually restoring the natural N-terminus of the expressed protein. A number of other secretion signals can be tested and a commercial set, the PichiaPink™ secretion signal set (Invitrogen) comprises eight different secretion signals that can be cloned in front of the test coding sequence. Targeting the expressed protein for secretion makes cell disruption unnecessary, and the medium can be continuously harvested to give a high yield of protein that is already semi-pure. An added advantage is that it can allow disulphide bonds to form, in many instances an essential step in the maturation of the protein to its fully active state. Yeast cells also recognize the consensus N-glycosylation site Asn-X-Ser/Thr (where X is any amino acid) but will add carbohydrate that is atypical of higher eukaryotes, having a high mannose content and a much larger number of sugar residues. They can also O-glycosylate Ser and Thr sites that would not be glycosylated in the higher eukaryote. This can have an impact on activity of the expressed protein. There have been important developments in this regard with the development of the Pichia GlycoSwitch® Protein Expression system (Research Corporation Technologies). This uses a series of engineered strains that result in human-like glycosylation patterns, thus improving the potential for use of *Pichia* in the production of therapeutic proteins. Proteins can be purified by making use of a suitable purification tag added to the coding sequence during the gene cloning stages.

7.4.2 *Saccharomyces cerevisiae*

Saccharomyces is a good alternative to *E. coli* as an expression host. Typically, *Saccharomyces* strains will grow to high cell densities on rich media (YEPD) at an optimal pH of ~5, giving yields in excess of 10mg protein per litre of culture. There are three types of vectors for introducing the cDNA for a 'protein of interest' into *Saccharomyces*:

- the yeast has its own 2-micron (2μm) 'plasmid', and sequences from this are incorporated into expression vectors to enable replication in yeast (Figure 7.6);
- the vector can contain a sequence from a centromere and the transformed cell will replicate the plasmid as if it were a 'mini-chromosome', i.e. the vector is *episomal*;
- a promoter–cDNA construct can be incorporated directly into the host cell genome via homologous recombination.

Plasmids for expression in Saccharomyces

The majority of commercially available yeast expression vectors (such as Invitrogen's pYES and Stratagene's pESC) are 2μm-based. These have the advantage of high copy number (>40 per cell), which can be important in maximizing protein yield. Centromere-based plasmids differ from the 2μm vectors essentially only in their yeast replication sequences, but are maintained at only one or two copies per cell. This has the advantage of providing more 'physiological' levels of expression; over-expression from high-copy

Figure 7.6 'Shuttle' vectors for *Saccharomyces* expression. (a) Centromere (CEN) and 2μm-based vectors are respectively low- and high-copy plasmids that replicate autonomously in *Saccharomyces*. An *E. coli* origin of replication (ORI) and antibiotic resistance gene (typically β-lactamase for ampicillin resistance (Amp^R)) facilitates replication and selection in *E. coli*. Selection in transformed yeast is based on loss of auxotrophy by the host strain as the vector carries a marker gene that complements the host gene deficiency. Expression of the 'protein of interest' (POI) is driven by a yeast promoter, in this example the galactose-inducible promoter from the *GAL1* gene. (b) Integrating plasmids (YIps) are incapable of autonomous replication in yeast, but drive expression of the POI by integrating an expression cassette into the host genome by homologous recombination between equivalent chromosomal and vector sequences (Marker A). Auxotrophic selection of transformants is possible as the recombination process can either repair a defective version of the chromosomal gene or introduce a new marker (Marker B) into the host chromosome. YIps also contain elements for replication and selection in *E. coli*, deleted here for clarity.

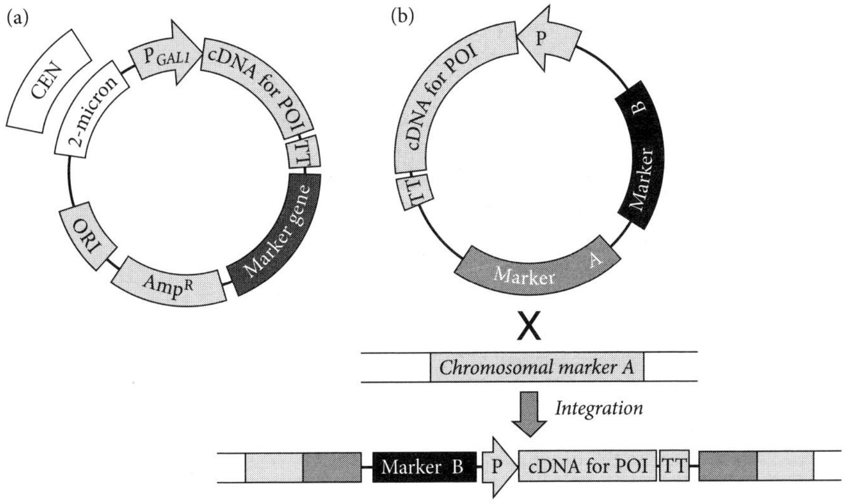

plasmids can sometimes give undesirable changes in phenotype. A second major advantage is that centromere-based plasmids allow stable mitotic segregation, and as a result will continue to be maintained within the dividing yeast cell population without specifically selecting for the marker on the plasmid. In contrast, 2μm-based vectors are much less stable; their presence in the population will decline rapidly if selection for the marker gene on the vector is not maintained, for example by growth in selective minimal medium. A variety of auxotrophic markers for transformant selection are available for both plasmid types.

To drive expression, the cDNA for the 'protein of interest' is ligated into a multiple cloning site downstream of a strong yeast promoter and upstream of a terminator sequence (Figure 7.6(a)). Commercially available yeast expression vectors most commonly include inducible promoters from yeast *GAL* genes (*GAL1* or *GAL10*) which show ~1000-fold repression by glucose and induction by galactose. With these promoters, initiation of expression can be tightly controlled, which is particularly useful if the expressed protein is toxic to the cell. Vectors may instead include a constitutive promoter to drive expression (e.g. derived from *CYC1*, *ADH1*, or genes for highly expressed glycolytic enzymes), allowing constant expression regardless of carbon source.

Greatest genetic stability is achieved by integration of the promoter–target cDNA cassette into the host chromosome (Figure 7.6(b)). To achieve this, the cDNA of interest is inserted into a yeast integrating plasmid (YIp) that lacks the autonomously replicating sequences of other yeast shuttle vectors. This YIp can insert into the host cell genome via the same recombination mechanism as that outlined above for *Pichia*.

Transformation of Saccharomyces

Competent yeast cells can be transformed with recombinant plasmid either by electroporation or by incubating cells treated with lithium salts with plasmid DNA–polyethylene glycol mixtures followed by heat shock. After transformation, cells are plated out onto minimal medium for auxotrophic selection.

Growth and induction

The key factors to consider when selecting conditions for expression are the stability of the vector used to deliver the cDNA and the properties of the promoter. For highly stable centromere-based or integrative vectors, cells can be grown rapidly to high biomass density on non-selective rich media such as YEP with glucose. The less stable 2μm vectors may necessitate growth on selective minimal media to prevent progressive loss of the plasmid from the yeast culture. If the cDNA of interest is downstream of a constitutive promoter, protein expression will be constant and continuous regardless of carbon source (glucose is inexpensive and gives high cell densities). However, for inducible *GAL* promoters, induction with galactose must occur without any repressing glucose present. Two approaches are possible: (1) grow cells to high density on a glucose (or other non-galactose carbon source) medium first, before moving to a galactose medium to induce expression; (2) grow continuously in the presence of galactose, with an additional carbon source that does not influence *GAL* promoter activity (such as raffinose or glycerol) to support growth. Either approach can prove successful for different proteins and different host strains.

Producing recombinant protein in Saccharomyces

The production of proteins from *Saccharomyces* is achieved in essentially the same manner as previously described for *Pichia*. *Saccharomyces* cells are slightly easier to disrupt and recombinant proteins can be recovered from the cytoplasm or membrane fraction. The cells can also be forced to *secrete* the expressed protein into the culture medium using the pre-pro-sequence of the *Saccharomyces* α-mating factor peptide, allowing recovery of the protein from the culture medium. Again, this facilitates disulphide bond formation and glycosylation but of high mannose content atypical of higher eukaryotes.

Determining protein function by complementation assay in Saccharomyces

One of the significant advantages of the *Saccharomyces* system is that in many cases the expressed protein will have a functional homologue in the yeast cell. By expressing the foreign protein in a yeast strain in which the gene for its homologue has been disrupted, it is possible to monitor the function of the expressed protein. In this type of **complementation assay**, recovery of the protein's function may be reflected in the phenotypic properties of the transformed cell, for example growth dependency on a particular nutrient or under particular conditions of pH or ion concentration. Gene disruption in *Saccharomyces* is relatively straightforward, and is achieved by integrating a YIp containing a copy of the relevant yeast gene made non-functional by mutation or disruption with a marker gene.

7.5 Baculovirus–insect larval cell expression system

This system exploits the natural ability of 'baculovirus' (*Autographa californica* multicapsid nucleopolyhedrovirus) to infect cells of larval stage lepidopteran insects, hijacking the transcription and translation apparatus of the cell to produce almost exclusively viral proteins (Murhammer 2007; Jarvis 2009). In the latter stages of the pathogenic cycle, multicapsid baculoviruses accumulate within the host cell, ultimately causing cell lysis and death of the insect. At this stage, the viral particle is encased in a coat containing the protein polyhedrin.

For recombinant protein expression, the strong viral promoter that facilitates high levels of polyhedrin expression is used to drive expression of the 'protein of interest'. The major advantage is the ability to produce high yields of expressed protein (in the range of 20–30% of total cell protein) which is fully functional with a complete repertoire of post-translation modifications. The main disadvantage is that very high cell densities and large culture volumes are not easy to achieve because of the high cost and the need for specialized culture facilities. If scale-up is achievable, this expression system can easily provide the milligram quantities of protein needed for structural biology applications such as X-ray crystallography. Major recent advances in our understanding of the structures of G-protein-coupled receptors were made possible by the high levels of expression achieved in insect cells (Akermoun et al. 2005; Rasmussen et al. 2007; Jaakola et al. 2008). Baculovirus–insect cell expression is compatible with a broad range of commonly used affinity tags and protein fusions that facilitate purification of expressed protein, and can also be set up to allow secretion of protein into the culture medium for easy recovery. Overall, the process is straightforward.

- A small plasmid vector carrying the cDNA for the 'protein of interest' is constructed, replicated in *E. coli*, and purified.
- The recombinant plasmid is recombined with a larger DNA molecule which contains all the essential baculovirus genes. The new recombined DNA molecule is introduced into cultured larval insect cells by the process of transfection, forcing them to produce virus that also contains the cDNA for the target protein.
- Recombinant virus is harvested from the culture medium and can be stored in a refrigerator before being introduced back into an insect cell culture at a later date.
- When infected, the host cells start producing recombinant target protein as the polyhedrin promoter becomes active.

A number of systems for setting up baculovirus expression are commercially available, each with its own advantages. Most laboratories routinely doing insect cell expression will have a preferred 'tried and tested' method that works for their proteins.

The most widely used host cell lines are derived from the larvae of the moths *Spodoptera frugiperda* (Sf9 and Sf21 cells) or *Trichoplusia ni* ('High Five' cells). They grow in specialized but widely available media (such as Grace's Insect Cell Medium), optimally at 27°C, and do not need to be gassed with CO_2 mixtures. For normal growth, fetal bovine serum (5–10%) is added to the medium, but should be excluded if cultures are intended for transfection. Cells can be grown as monolayers, but are easily adapted to suspension culture in spinner flasks, allowing a larger biomass of cells to be grown for high-yield expression.

7.5.1 Constructing a recombinant baculovirus

The first step in constructing a recombinant baculovirus is to sub-clone the cDNA for your 'protein of interest' into a plasmid vector. This 'transfer vector' is then used to shuttle the cDNA into the baculovirus genome. Integration of the cDNA into the baculovirus DNA occurs via one of two mechanisms: recombination or transposition (Figure 7.7).

For recombination, both transfer vector and the baculovirus DNA carry common viral or *lacZ* (β-galactosidase) sequences. When larval insect cells are co-transfected with transfer vector and baculovirus DNA, recombination moves the polyhedrin promoter–cDNA cassette from the transfer vector to the baculovirus DNA. This is shown in Figure 7.7(a) and is typical of systems such as Invitrogen's BacNBlue system, BD Bioscience's BaculoGold, and Clontech's BacPAK (Clontech 2009). These systems incorporate features that eliminate unwanted non-recombinant viruses that do not carry the target protein cDNA or enable recombination-dependent recovery of a gene essential for formation of viable viral particles. The product of this first-stage transfection is recombinant baculovirus that is recovered from the culture medium and stored for later expression experiments.

Figure 7.7 Constructing recombinant baculovirus DNA. (a) Recombination between common sequences in transfer vector and baculovirus DNA. Viral orf603 and orf1629 sequences are common to the transfer vector and baculovirus DNA, allowing recombination *in situ* in the host insect cell. Replacement of a partial orf1629 sequence in the baculovirus DNA by the full sequence in the transfer vector recovers the essential function of this gene, leading to assembly of viable virus only if successful recombination has occurred. Recombination can also occur at the *lacZ* gene, resulting in β-galacosidase activity. The transfer vector also contains an ampicillin resistance gene and origin of replication (ORI) for selection and replication in *E. coli.* (b) Bac-to-Bac™ system. Transposable elements in the transfer vector (Tn7L, Tn7R) drive insertion of an expression cassette comprising a gentamycin resistance gene, the polyhedrin promoter (P_{PolH}) upstream of the cDNA for the 'protein of interest', and a polyadenylation sequence (polyA). Insertion occurs at a mini-attTn7 transposon region in the hybrid plasmid–baculovirus ('bacmid') DNA. *E. coli* carrying recombinant bacmid are *lacZ*⁻ because insertion disrupts this gene.

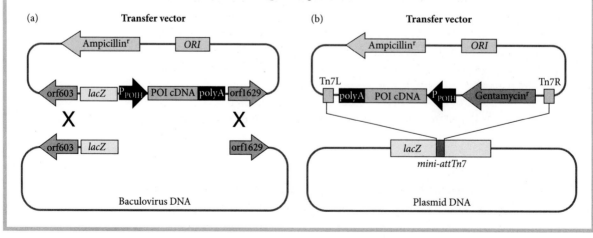

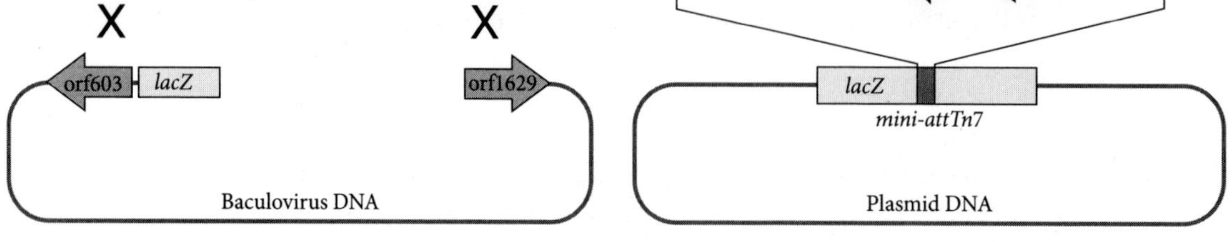

Invitrogen's Bac-to-Bac system (Life Technologies Inc. 2008) uses the reportedly more efficient transposition mechanism to combine transfer vector and baculovirus DNA in specialized *E. coli* cells (DH10Bac cells) to produce 'bacmid' hybrid DNA (Figure 7.7(b)). Sequences in the transfer vector derived from the left and right arms of the Tn7 transposable element drive insertion into a 'mini-transposon' within the bacmid DNA. Insertion disrupts a *lacZ* gene in the bacmid and transfers gentamycin resistance, making the process of selecting true recombinants easy. The new bacmid incorporating the cDNA of interest is used to transfect insect cells, forcing them to produce a stock of recombinant virus, as above.

Transfection of insect cells is performed using cationic lipid-type transfection reagents and requires high densities of insect cells (~2×10^6/mL) in log phase growth with high viability (>95%) for maximum efficiency. Successfully transfected cells appear larger, with a more prominent nucleus. After 24 hours, they take on a granular appearance, as virus buds from the membrane, and may detach if grown as monolayers. Recombinant virus is collected from the culture medium after 3–4 days when cell lysis may have started to occur. Viral stocks can be stored at 4°C for several weeks or at –80°C for longer periods. However, you should avoid repeated freeze–thaw cycles as this will damage the virus. With some practice, any of the baculovirus expression systems can be set up to produce protein in 8–10 weeks after establishing the host insect cell cultures.

7.5.2 Producing recombinant protein

Once the recombinant baculovirus is constructed using either of the approaches outlined above, subsequent steps of virus amplification, titre assay, and infection for protein expression, shown in Figure 7.8, are common to each process.

The first of these common steps is to determine how much recombinant virus has been produced. The titre (concentration) of virus in the medium is determined by plaque assay. Serial dilutions of virus are used to infect monolayers of insect cells, and the infection is allowed to proceed to the point at which the infected cells have lysed, leaving 'holes' (plaques) in the cell monolayer. Virus titre is expressed as 'plaque-forming units' (pfu) per millilitre of culture medium. At this stage, you would be aiming for 10^6–10^7 pfu/mL. Subsequent rounds of infection and virus harvesting can be used to amplify and concentrate the working stock that is later used to drive protein expression.

Expression is initiated by infection of cells at a defined 'multiplicity of infection' (MOI), the number of viruses added to the culture medium per cell. This varies and should be determined for each viral construct with a particular cell line, although an MOI of 10 is often used. Too much virus will cause rapid cell death and lower protein yields. When testing a new viral construct, remember to include a non-recombinant viral control. Protein will be expressed during the later stages of infection, and cells should initially be sampled at intervals of 8–12 hours for up to 4 days for intracellular expression. Protein secretion can be driven by N-terminal fusion of the 'protein of interest' with the presequence of honey bee mellitin (Invitrogen) or viral glycoprotein 67 (BD Biosciences), allowing protein to be recovered from the culture medium. Appearance of secreted protein should be monitored from 1 to 4 days post-infection. 'High Five' cells are reported to be better hosts for secretion.

Typically, cell lysates are analysed by SDS-PAGE to detect expressed protein. If protein levels are low, immunoblotting with a specific antibody may be needed to detect it. Often, low levels of expression occur because of problems with the transfer vector

Figure 7.8 Summary of steps in baculovirus–insect cell expression. Steps are outlined for construction of recombinant baculovirus using either recombination between transfer vector and baculovirus DNA *in situ* in the host insect cell (*recombination*), or via *transposition* between transfer vector and bacmid (Bac-to-Bac™ system). Subsequent steps of virus amplification, titre assay, and infection for protein expression are common to each process.

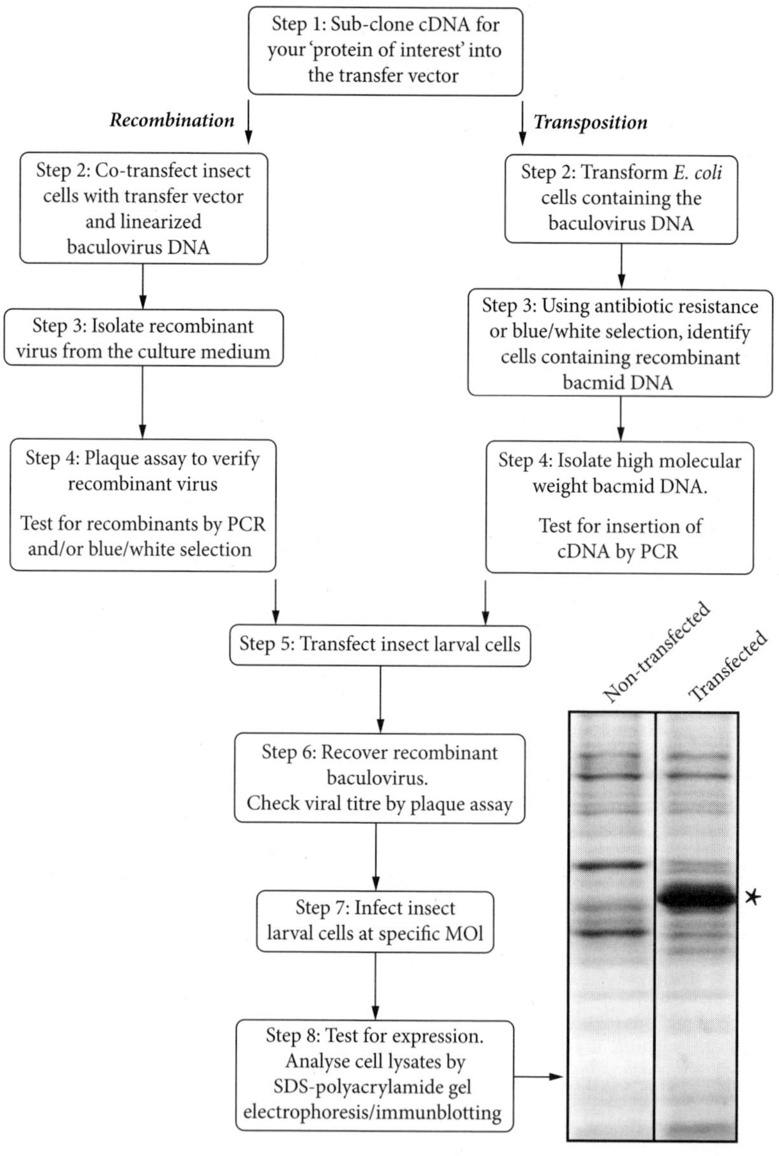

construction. Poor expression can also be caused by low efficiency of infection because incorrectly stored or old virus has been used, or because the host cells have low viability or are at the wrong stage in their growth cycle. These problems are easily remedied. However, poor expression may also be a natural characteristic of the target protein, in which case a different expression system may be required. Table 7.1 summarizes the advantages and disadvantages of each system and should assist you in your selection.

Production of a human membrane protein

CASE STUDY 7.2

The histamine H_2 receptor is involved in gastric acid secretion and is the target for drugs used to treat peptic ulcers and gastric reflux disease. This G-protein-coupled receptor (GPCR) is a glycosylated integral membrane protein comprised of a single polypeptide chain. To aid in the design of new drugs, your research team intends to determine the structure of the human form of this integral membrane protein, and you have been tasked with purifying sufficient protein to start X-ray crystallography studies.

QUESTION

Devise an expression strategy for this protein. Which system(s) would you choose, and why? How would you facilitate purification? What approaches can you use to maximize the yield of expressed protein?

The chances of *E. coli* successfully expressing a human membrane protein are extremely small, and therefore should probably be avoided. Both *Pichia* and insect cells will perform glycosylation and transport the expressed protein to the plasma membrane, and both systems are capable of giving the high yield of protein needed for crystallization trials. Before choosing a system, it is also worth checking the literature for precedents. The structure of the human H_1 receptor has been solved by X-ray crystallography after expression in *Pichia* (Shimamura et al. 2011), so this may be the best place to start. Other GPCRs have been crystallized after expression in *Pichia*, but insect cells have also been successful for this task. Overall, the potential for high levels of expression, ease of set-up, and relatively low cost of scale-up point to initially testing *Pichia*.

Facilitate purification by adding a polyhistidine affinity tag by in-frame fusion of the receptor cDNA with the tag coding sequence. However, this tag can interfere with crystallization, so a proteolytic cleavage site should be incorporated between the receptor and tag to allow its removal after purification (check that the chosen cleavage site does not also occur naturally in the receptor amino acid sequence!). After sub-cloning the cDNA into your *Pichia* vector, transform host cells and screen for cells that have integrated the expression cassette into the *Pichia* genome (by loss of auxotrophy or gained antibiotic resistance). To maximize expression, look for transformed cells carrying multiple integrations ('jackpot clones'). These are easiest to find with vectors that transfer zeocin resistance to the host, since growth on increasing antibiotic concentrations correlates with insert number. Pre-assembly of multiple expression cassettes in the vector prior to transformation can also be used to achieve this effect. Do an initial small-scale screen of a number of transformed host cells under identical growth conditions, and test for expression in cell lysates using SDS-PAGE and immunoblotting (you will need a specific anti-H_2 receptor or anti-polyhistidine tag antibody). After identifying highly expressing cells, optimize expression by performing induction time courses with cells at different stages in the growth cycle. Purify the protein by metal affinity chromatography using the polyhistidine tag after solubilizing the host cell membrane fraction in a non-denaturing detergent.

See Chapter 8

7.6 Chapter summary

- Your choice of expression system will depend on the type and source of the protein you are trying to produce, and what you want to do with it. Common expression systems are *E. coli*-, yeast-, and insect-based systems.

- **Most researchers will attempt protein expression in *E. coli* initially because of the speed and ease with which this system can be set up and tested. You may find it more convenient to go down the route of maximizing yield regardless of the state of the protein before proceeding to refold it from inclusion bodies.**

- **If this strategy does not work, expression in a yeast (*Pichia* or *Saccharomyces*) may be the next choice. Although the molecular biology can be difficult with *Pichia*, high levels of expression, and particularly secretion, of eukaryotic proteins with near-native glycosylation patterns can be achieved with equipment and facilities no more expensive or specialized than are used for *E. coli*.**

- **Insect cell expression offers perhaps the highest likelihood of generating functional eukaryotic protein, although the lead time and need for specialized culture equipment may be a deterrent.**

- **Virtually all proteins will express to some extent in at least one host, and a careful and systematic approach will be needed to find out which host that is. In all systems the incorporation of purification tags will simplify recovery of your recombinant protein.**

References

Akermoun, M., Koglin, M., Zvalova-Iooss, D., Folschweiller, N., Dowell, S.J., and Gearing, K.L. (2005) Characterization of 16 human G protein-coupled receptors expressed in baculovirus-infected insect cells. *Protein Expr Purif* **44**: 65–74.

André, N., Cherouati, N., Prual, C., et al. (2006) Enhancing functional production of G protein-coupled receptors in *Pichia pastoris* to levels required for structural studies via a single expression screen. *Protein Sci* **15**: 1115–26.

Balbás, P. and Lorence, A. (eds) (2004) *Recombinant Gene Expression: Reviews and Protocols*. New York: Humana Press.

Bernaudat, F., Frelet-Barrand, A., Pochon, N., et al. (2011) Heterologous expression of membrane proteins: choosing the appropriate host. *PLoS ONE* **6**: e29191.

Brondyk, W.H. (2009) Selecting an appropriate method for expressing a recombinant protein. In: Burgess, R.R. and Deutscher, M.P. (eds), *Guide to Protein Purification* (2nd edn). San Diego, CA: Academic Press, pp. 131–47.

Cereghino, J.L. and Cregg, J.M. (2000) Heterologous expression in the methylotrophic yeast *Pichia pastoris*. *FEMS Microbiol Rev* **24**: 45–66.

Clontech (2009) BacPAK baculovirus expression system. http://www.clontech.com/US/Products/Protein_Expression_and_Purification/Baculovirus_Expression/BacPAK_Expression_System (accessed 25 June 2012).

Cregg, J.M., Tolstorukov, I., Kusari, A., Sunga, J., Madden, K., and Chappell, T. (2009) Expression in the yeast *Pichia pastoris*. In: Burgess, R.R. and Deutscher, M.P. (eds), *Guide to Protein Purification* (2nd edn). San Diego, CA: Academic Press, pp. 169–89.

de Boer, H.A., Comstock, L.J., and Vasser, M. (1983) The *tac* promoter: a functional hybrid derived from *trp* and *lac* promoters. *Proc Natl Acad Sci USA* **80**: 21–5.

Deacon, S.E. and McPherson, M.J. (2011) Enhanced expression and purification of fungal galactose oxidase in *Escherichia coli* and use for analysis of a saturation mutagenesis library. *Chembiochem* **12**: 593–601.

Grabski, A., Mehler, M., and Drott, D. (2005) The Overnight Express autoinduction system: high-density cell growth and protein expression while you sleep. *Nat Methods* **2**: 233–5.

Hino, T., Arakawa, T., Iwanari, H., et al. (2012) G-protein-coupled receptor inactivation by an allosteric inverse-agonist antibody. *Nature* **482**: 237–40.

IBA (2011) StarGate: the new dimension in combatorial cloning: instruction manual. http://www.iba-lifesciences.com/StarGate_Cloning.html (accessed 25 June 2012).

Jaakola, V.P., Griffith, M.T., Hanson, M.A., et al. (2008) The 2.6 ångstrom crystal structure of a human A_{2A} adenosine receptor bound to an antagonist. *Science* **322**: 1211-17.

Jarvis, D.L. (2009) Baculovirus–insect cell expression systems. In: Burgess, R.R. and Deutscher, M.P. (eds), *Guide to Protein Purification* (2nd edn). San Diego, CA: Academic Press, pp. 191–222.

Katzen, F., Chang, G., and Kudlicki, W. (2005) The past, present and future of cell-free protein synthesis. *Trends Biotechnol* **23**: 150–6.

Life Technologies Inc. (2002) Invitrogen *Pichia* fermentation process guidelines. http://tools.invitrogen.com/content/sfs/manuals/pichiaferm_prot.pdf (accessed 2 July 2012).

Life Technologies Inc. (2008) Invitrogen Bac-to-Bac expression system manual. http://tools.invitrogen.com/content/sfs/manuals/bactobac_topo_exp_system_man.pdf (accessed 2 July 2012).

Life Technologies Inc. (2010a) Invitrogen Gateway® Technology. http://tools.invitrogen.com/content/sfs/manuals/gatewayman.pdf (accessed 28 June 2012).

Life Technologies Inc. (2010b) Invitrogen *Pichia* multi-copy expression kit manual. http://tools.invitrogen.com/content/sfs/manuals/pichmulti_man.pdf (accessed 2 July 2012).

Life Technologies Inc. (2010c) Invitrogen EasySelect *Pichia* expression kit manual. http://www.invitrogen.com/content/sfs/manuals/easy_select_man.pdf (accessed 4 July 2012).

Life Technologies Inc. (2010d) Invitrogen *Pichia* expression kit manual. http://www.invitrogen.com/content/sfs/manuals/pich_man.pdf) (accessed 4 July 2012).

Lin-Cereghino, J. and Lin Cereghino, G.P. (2008) Vectors and strains for expression. In: Cregg, J.M. (ed), *Methods in Molecular Biology: Pichia Protocols*. Totowa, NJ: Humana Press, pp. 111–26.

Long, S.B., Campbell, E.B., and Mackinnon, R. (2005) Crystal structure of a mammalian voltage-dependent shaker family K^+ channel. *Science* **309**: 897–903.

Macauley-Patrick, S., Fazenda, M.L., McNeil, B., and Harvey, L.M. (2005) Heterologous protein production using the *Pichia pastoris* expression system. *Yeast* **22**: 249–70.

Murhammer, D.W. (ed) (2007) *Baculovirus and Insect Cell Expression Protocols*. New York: Humana Press.

Novagen (2011) Novagen pET system manual. http://www.merckmillipore.co.uk/life-science-research/pet-expression-systems. (accessed 24 June 2012).

Rasmussen, S.G., Choi, H.J., Rosenbaum, D.M., et al. (2007) Crystal structure of the human β_2-adrenergic G-protein-coupled receptor. *Nature* **450**: 383–7.

Romanos, M. (1995) Advances in the use of *Pichia pastoris* for high-level gene expression. *Curr Opin Biotechnol* **6**: 527–33.

Shimamura, T., Shiroishi, M., Weyand, S., et al. (2011) Structure of the human histamine H_1 receptor complex with doxepin. *Nature* **475**: 65–70.

Studier, F.W. (2005) Protein production by auto-induction in high density shake cultures. *Protein Expr Purif* **41**: 207–34.

Studier, F.W. and Cregg, J.M. (1986) Use of bacteriophage T7 RNA polymerase to direct selective high-level expression of cloned genes. *J Mol Biol* **189**: 113–30.

Studier, F.W., Rosenburg, A.H., Dunn, J.J., and Dubendorff, J.W. (1990) Use of T7 RNA polymerase to direct expression of cloned genes. In: Goeddel, D.V. (ed), *Gene Expression Technology*. San Diego, CA: Academic Press, pp. 60–89.

Vaillancourt, P.E. (ed) (2003) *E. coli Gene Expression Protocols*. New York: Humana Press.

Wacker, M., Linton, D., Hitchen, P.G., et al. (2002) N-linked glycosylation in *Campylobacter jejuni* and its functional transfer into *E. coli*. *Science* **298**: 1790–3.

Zerbs, S., Frank, A.M., and Collart, F.R. (2009) Bacterial systems for production of heterologous proteins. In: Burgess, R.R. and Deutscher, M.P. (eds), *Guide to Protein Purification* (2nd edn). San Diego, CA: Academic Press, pp. 149–68.

8 Protein purification

Iain Manfield

Chapter overview

A number of different purification methods can be used to isolate a protein of interest from a complex protein mixture. These can be categorized into non-chromatographic and chromatographic methods. The former include techniques such as centrifugation and ultrafiltration, whilst the latter refers to techniques in which proteins are passed through a column packed with a stationary material.

In this chapter we will describe the common chromatography methods used in protein purification: affinity chromatography, ion exchange chromatography, hydrophobic interaction chromatography, and size exclusion chromatography. In addition, we will outline the key questions to consider before embarking on a protein purification strategy.

LEARNING OBJECTIVES

This chapter will enable the reader to:

- **describe the principles underlying different protein purification methods;**
- **consider the key questions that should be addressed when designing a protein purification strategy—the intended application of the protein, the source of the protein, and what is known about the characteristics of the target protein;**
- **design a multi-step purification strategy for a protein of interest including making sample conditions compatible with the subsequent step.**

8.1 Introduction

Studying the structural and functional properties of a protein is an essential part of understanding how cells and organisms work. As cells contain thousands of different proteins, performing these analyses requires that the protein of interest is purified from a complex mixture. Proteins can be purified on the basis of their different properties, commonly size, charge, ligand-binding specificities, or hydrophobicity. The method(s) selected will depend on the target protein characteristics (if known), the source of the

See Chapter 7

protein, which may be a natural biological sample or heterologously expressed in either a prokaryotic or eukaryotic expression system, and why you are purifying the protein, i.e. the intended applications.

See Chapter 9

See Chapters 11, 12, and 13

There are a number of reasons why you may want to purify a protein. For example; you may wish to determine the structure of the protein or its amino acid sequence. You may be interested in studying a particular biological property, such as enzymatic activity, or generating an antibody against the protein. The intended application will in turn define the level of purity and the amount of protein required. For example, structural biology projects using X-ray crystallography and NMR spectroscopy and binding assays to determine stoichiometry, affinity, kinetic, and thermodynamic information require a high level of protein purity, as contamination with misfolded protein could reduce the quality of the crystals produced (for crystallography studies), or interfere with affinity measurements between interacting molecules.

In terms of the amount of protein required, some applications such as Western blotting may require only micrograms of protein, whilst others, such as raising antisera, crystallography, enzyme assays, and analysis of interactions, will require milligrams of protein. Milligrams of protein can be made using equipment available in many research laboratories, whilst the kilograms of protein required for medical treatments require bioprocess equipment used in pharmaceutical production plants (Noble 2006). This chapter will focus on laboratory-based purification schemes.

Any protein purification strategy will include some common steps:

(i) consideration of the amount of protein and the level of purity required;

(ii) identification of key properties of your target protein (if known) that are different from the non-target proteins and hence can be used to identify the most appropriate purification method(s);

(iii) a method for assessing the amount of target protein recovered during each purification stage.

These steps are discussed further in the following sections.

8.2 Devising a purification strategy

Purification of proteins from natural sources typically starts with a non-chromatographic method such as fractional precipitation (using ammonium sulphate or polyethylene glycol), heat precipitation, or solvent extraction of the crude cell extract followed by chromatography for further purification (Harris 2006; Burgess 2009a; GE Healthcare Life Sciences 2009). However, over-expression of the target protein in *E. coli* or another heterologous host (Burgess and Deutscher 2009) with an **affinity tag** is the most common starting point for a protein purification strategy. This involves cell lysis and extraction of soluble proteins, followed by purification of target protein using affinity chromatography.

See Chapter 7

This and other chromatography purification methods—the focus of this chapter—are summarized below and described further in section 8.5.

- **Affinity chromatography** uses the ability of a target protein to bind to a specific ligand that is immobilized on a chromatography matrix. The bound protein is then eluted under conditions that displace the ligand–target protein interaction. This is depicted schematically in Figure 8.1.

- **Ion-exchange chromatography** (IEX) uses charged groups attached to a matrix to capture proteins with regions of complementary charge. Increasing salt concentrations are then used to displace the bound protein and elute it from the column.
- **Hydrophobic interaction chromatography** (HIC) utilizes hydrophobic interactions between protein and matrix to bind protein under high salt concentration conditions. The bound protein is then eluted by reducing the salt concentration which weakens the interactions between the protein and the matrix.

An alternative purification approach to capturing proteins on a solid support (matrix) is to separate the molecules by size and shape using a sieve or filtering effect. This approach is used in:

- **Gel filtration chromatography** (or **size exclusion chromatography** (SEC)). The sample is passed through a porous matrix which excludes molecules of a particular size from entering the pores whilst smaller molecules penetrate the pores. Hence molecules are separated so that the larger proteins are eluted from the column first followed by the smaller proteins.

To purify a protein of interest to the required level of purity and to remove aggregated and potentially inactive molecules a combination of different chromatography steps are commonly used. The flow-chart in Figure 8.2 outlines the steps of a typical purification strategy starting with cell lysis and protein solubilization followed by an initial **capture step**, an **intermediate purification step**, and a final **polishing purification step**. The aim

Figure 8.1 Schematic diagram showing the separation of a tagged protein by affinity chromatography. Protein sample is applied to the chromatography matrix under conditions which favour target protein binding to the ligand immobilized on the matrix. The many unbound proteins lacking complementarity to the ligand are then washed out and the bound proteins are eluted from the column under conditions which disrupt the ligand–protein interaction.

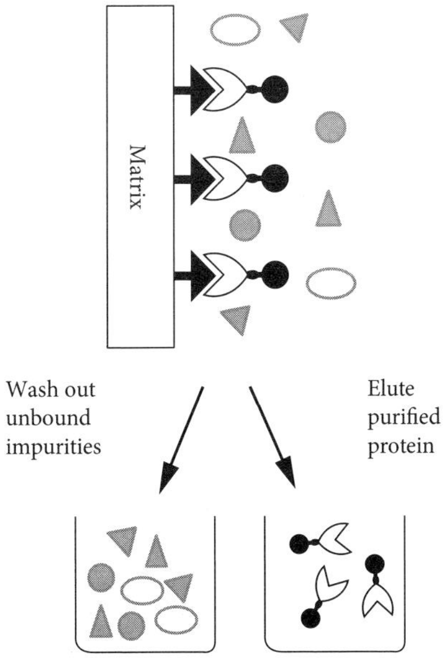

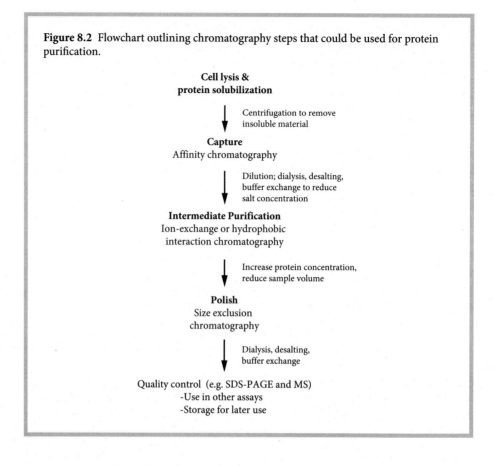

Figure 8.2 Flowchart outlining chromatography steps that could be used for protein purification.

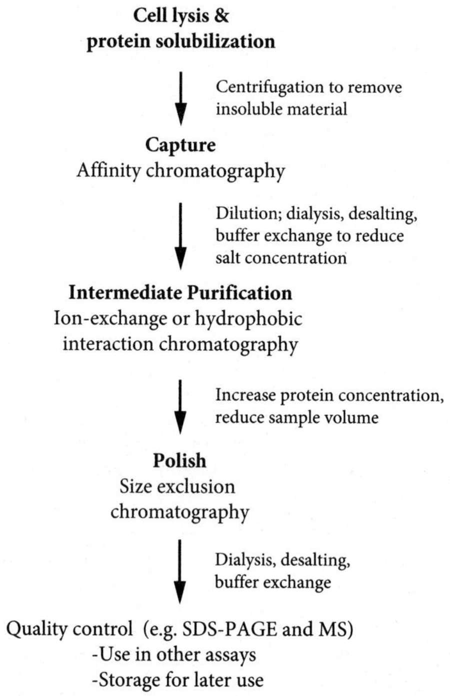

of the capture stage is to remove the target molecule from contaminating impurities as quickly as possible, especially those likely to be the most damaging to the protein of interest. Affinity chromatography can be a very good option at this stage. At the intermediate purification stage, different protein properties are utilized to remove impurities carried through from the capture step. A final polishing step is then used to obtain fully active pure target protein. At this stage more expensive chromatography matrices may be used that are not appropriate for the capture step and which may take more time to run.

A good purification strategy will utilize a minimum number of steps and recover as much of the target protein as possible from the starting material. However, it is important to remember that there is a trade-off between purity and yield, with higher levels of purity often accompanied by lower protein yields. A compromise may be made such that a higher yield is obtained but at reduced levels of purity if, for example, it proves difficult to remove an impurity which interferes minimally with the activity in the intended application.

8.2.1 Linking steps together in a purification strategy

When designing a purification strategy it is necessary to ensure that the sample conditions (e.g. buffer composition, salt concentration, and pH) are compatible with the subsequent step in the strategy. Typical salt and protein concentrations for binding and **elution** of protein during chromatography are outlined in Table 8.1. Ideally, the eluate

Table 8.1 Comparison of different chromatographic purification methods

Chromatographic separation method	When you would use this technique	Advantages	Disadvantages	Binding and elution conditions
Affinity chromatography	Early capture stage to remove many impurities from a crude extract	Specificity allows removal of many impurities	Affinity matrices are often more expensive and less stable than IEX or HIC matrices	Dilute protein loaded in medium to high salt Elution of more concentrated protein in medium to high salt
Ion exchange (IEX)	Commonly-used intermediate purification step	Useful for any protein when appropriate conditions are selected IEX matrices have high capacity and good chemical stability Low cost	Lacks the specificity of affinity chromatography	Dilute protein loaded in low salt Elution of more concentrated protein in medium to high salt
Hydrophobic interaction (HIC)	Alternative intermediate purification step	Useful for any protein when appropriate conditions are selected HIC matrices have high capacity and good chemical stability Low cost	Lacks the specificity of affinity chromatography	Dilute protein loaded in high salt Elution of more concentrated protein in medium to low salt
Size exclusion/gel filtration (SEC)	Final polishing step for the most demanding applications	Removes aggregated protein	Only small sample volumes can be loaded (<< column volume) Large well-packed columns required	Concentrated protein loaded Elution of more dilute protein Flexibility in salt concentration for elution possible

from one step will be an appropriate starting material for the next step. If this is not the case, methods such as dialysis, ultrafiltration, and desalting are available for changing buffer composition and protein concentration.

Dialysis is often used to change solution composition by placing the sample in a 'sausage-skin'-like tube with a membrane which is permeable to small molecules (such as NaCl and buffers) but retains macromolecules above a certain molecular weight. The tube is placed in a beaker with a large volume of the dialysis buffer—in this instance the buffer to be used at the next purification stage. The small molecules from both the sample and the dialysis buffer diffuse across the membrane, slowly reaching equilibrium (Figure 8.3(a)).

Ammonium sulphate precipitation can be used to reduce the sample volume and hence increase protein concentration. After centrifugation, the protein precipitate is re-dissolved in a smaller volume of buffer. Alternatively, **ultrafiltration** uses a molecular weight cut-off membrane (like dialysis) but with centrifugation or pressure from a gas cylinder to drive liquid through the membrane. The membrane retains the macromolecules so that their concentration increases as the volume decreases (Figure 8.3(b)). However, the concentration of the small molecules does not change during ultrafiltration. Multiple cycles of concentration and dilution with a different buffer, in a process called **buffer exchange**, can be used to change the concentration of small molecules. Gel filtration columns can also be used to change buffers. Small columns filled with a

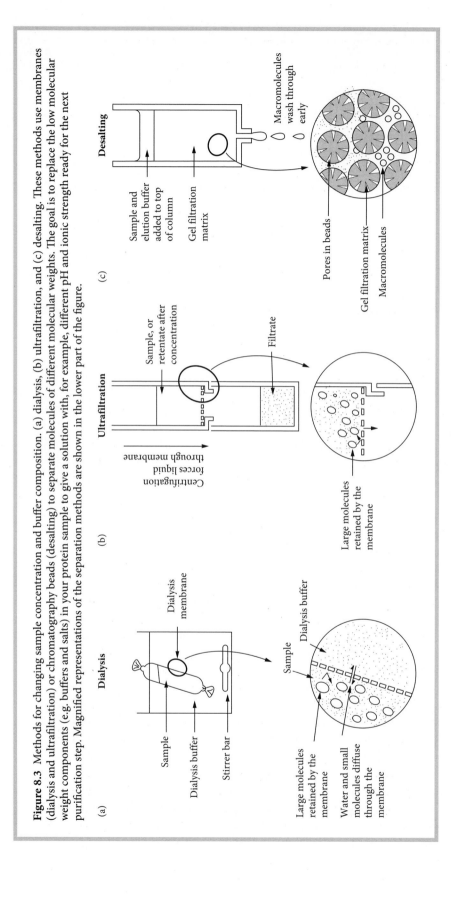

Figure 8.3 Methods for changing sample concentration and buffer composition. (a) dialysis, (b) ultrafiltration, and (c) desalting. These methods use membranes (dialysis and ultrafiltration) or chromatography beads (desalting) to separate molecules of different molecular weights. The goal is to replace the low molecular weight components (e.g. buffers and salts) in your protein sample to give a solution with, for example, different pH and ionic strength ready for the next purification step. Magnified representations of the separation methods are shown in the lower part of the figure.

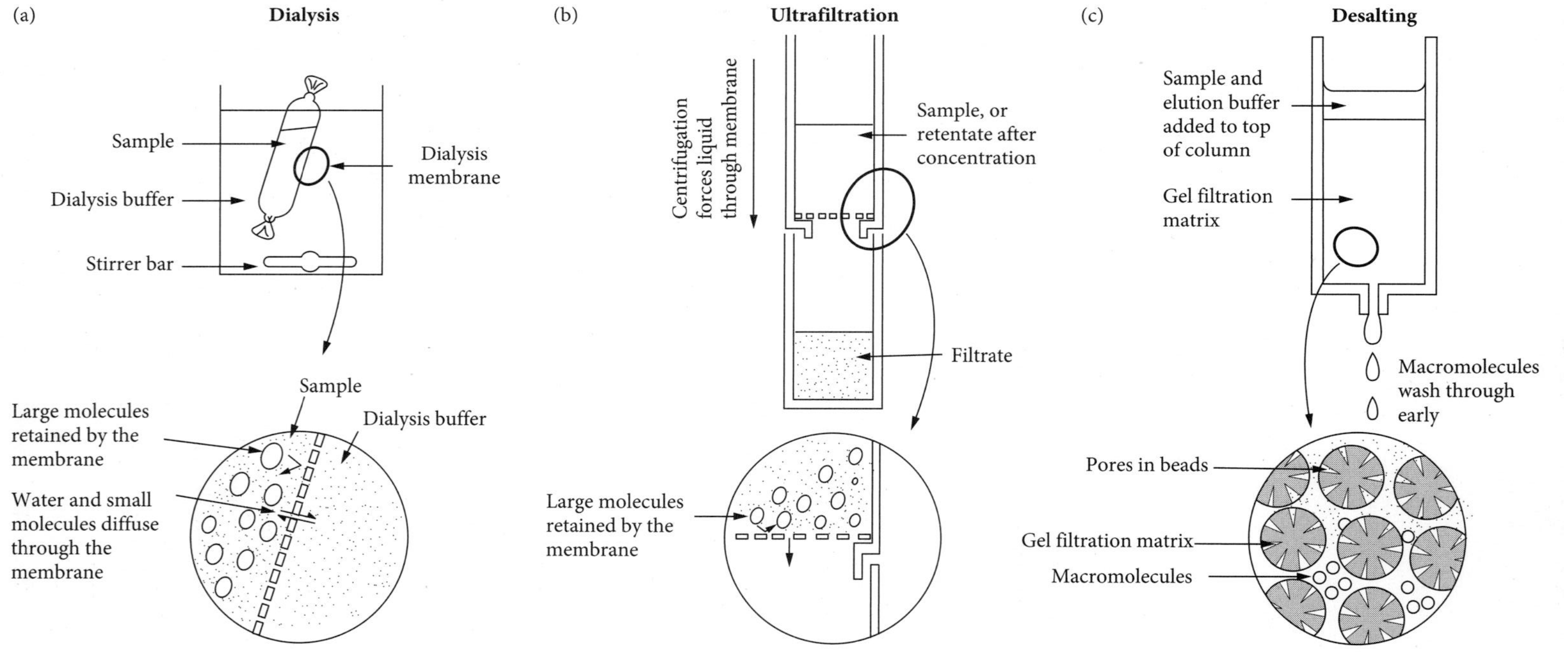

matrix are used to separate macromolecules from all other molecules below ~2 kDa. This approach is called **desalting** as it removes the salts from the sample (Figure 8.3(c)).

8.3 Cell lysis and disruption

The first step in a protein purification strategy is to prepare a cell extract, i.e. a solution of cellular proteins from your source material which can immediately be passed through a column containing a chromatography matrix (Bonner 2007; Grabski 2009; GE Healthcare Life Sciences 2010a). When preparing a cell extract, you should:

- select a method that will lyse the cells efficiently with minimum damage to the proteins;
- use conditions that retain the integrity, folding, and solubility of the protein of interest;
- use conditions which allow protein capture on chromatography columns without the need to change solution conditions.

If the starting material is a liquid sample, such as plasma, it can be used directly in chromatography. However, when proteins are to be purified from cells, the cells must be lysed (broken open) to release the contents into solution so that chromatography can be performed. Mammalian cells grown in tissue culture can be lysed gently using detergents to dissolve the cell membrane, producing a solution of the cell contents.

See Chapter 15

Breaking open the cell wall of *E. coli*, yeast, or plant cells often requires more aggressive treatment. Sonication using high-frequency ultrasound is popular as it is easy to use and the equipment is cheap. However, it can heat the sample, increasing the activity of proteases, and can unfold native protein structure. An alternative approach more commonly used is to digest bacterial cell walls with the enzyme lysozyme in combination with a mild detergent to disrupt the cell membrane. The advantages and disadvantages of this and other methods used to lyse cells are outlined in Table 8.2. Table 8.3 summarizes additional reagents that are commonly used during cell lysis and disruption procedures and are described further below.

A **buffer** is included in the extraction solution to maintain the solution at the required pH value (Dawson et al. 1989; Stoll and Blanchard 2009). Typically a pH of 7–8 is used for cell lysis with Tris or phosphate as the buffering ions. However, proteins are characterized by their **isoelectric point** (pI), reflecting the relative composition of differently charged amino acids. The pI of a protein is the pH at which it has no overall charge. This is also the pH at which the protein has lowest solubility and therefore different pHs (with appropriate buffer) can be used for different proteins to enhance solubility.

The charged groups on proteins can reduce protein solubility by interacting with oppositely charged groups on a different protein. To prevent this, salts are used in the extraction buffer, providing counter-ions—this is called **salting in**. Sodium chloride (NaCl) is the most commonly used salt. Importantly, the Na^+ and Cl^- ions are in the middle of the Hofmeister or lyotropic series of ions, with NH_4^+ and SO_4^{2-} at one end tending to precipitate proteins, while guanidinium and thiocyanate (SCN^-) at the other end of the series tend to solubilize proteins, potentially with a denaturing effect (GE Healthcare Life Sciences 2010a). Exploiting the effects of different buffers and

Table 8.2 Comparison of different approaches used to lyse cells

Cell lysis method	Advantages	Disadvantages
Sonication	Low-cost equipment Easy to use Fragments DNA, reducing sample viscosity	Can heat sample, causing protein denaturation Can generate bubbles, denaturing proteins
Digestion of cell wall and genomic DNA with enzymes	Cheap Easy to process multiple samples in parallel Easily scalable for different sample volumes	Incubation steps required to allow enzyme activity may also allow proteases to work if incompletely blocked by inhibitors
French pressure cell	Fragments DNA, reducing sample viscosity	More difficult to process multiple samples or samples with a small volume Expensive equipment
Freeze–thaw	Cheap	Time-consuming May require nuclease digestion of DNA
Grinding with liquid nitrogen using a pestle and mortar	Good method for disrupting structure of multicellular tissues (e.g. plant leaves)	Cell pastes freeze into hard lumps which are difficult to grind Liquid nitrogen hazard

Table 8.3 Common reagents included in cell lysis and extraction procedures

Component	Purpose	
Buffer	Maintain pH of solution Typically pH 7–8 is used for cell lysis with Tris or phosphate as buffering ions, but different pHs may be used for different proteins	
NaCl	Provides counter-ions to enhance protein solubility ('salting in')	
Protease inhibitors	To protect target protein from degradation by proteases Different classes of protease inhibitors are available:	
	PMSF, AEBSF, benzamidine	Serine protease inhibitors
	E-64, antipain, chymostatin	Cysteine protease inhibitors
	Pepstatin	Aspartyl protease inhibitors
Detergents	Disrupt cell membranes to release cell contents Help to solubilize proteins Mild detergents, such as Tween 20 and Triton X-100 are commonly used	
EDTA, EGTA	Chelates metal ions Inhibits metal dependent protease activity	
Dithiothreitol (DTT), 2-mercaptoethanol, TCEP	Reducing agents: maintain cysteine residues in reduced form	
Glycerol	Can stabilize protein folding	

counter-ions on solubility is an essential part of the toolbox of protein purification techniques.

To protect the protein against degradation by proteases during purification, it is often critically important to include protease inhibitors (Table 8.3) in extraction buffers and chromatography buffers (Grabski 2009) to prevent protease activity.

A **reducing agent** is often included in cell lysis and chromatography buffers to prevent oxidation damage—for example, conversion of cysteine amino acid side chains to disulphide groups which can lead to protein inactivation and inappropriate multimerization. Commonly used reducing agents are dithiothreitol (DTT), 2-mercaptoethanol, and TCEP. Other chemicals can be added on a case-by-case basis depending on the individual properties of a protein. For example, glycerol is used to stabilize protein folding, and Ca^{2+} or Zn^{2+} may be required as structural cofactors.

A number of proteins show poor solubility and this is especially common when proteins are over-expressed in *E. coli*. This can be due to aggregation and misfolding of the protein into **inclusion bodies** which require aggressive treatments to break them up, resulting in the unfolding of the proteins. Extensive (and time-consuming) screening is then performed attempting to refold the proteins effectively (Willis et al. 2005; Burgess 2009b). Strong detergents such as SDS and sarkosyl can denature (unfold) proteins, whilst milder detergents can solubilize proteins without unfolding them. Empirical testing of a range of different detergents is often performed to find those giving soluble active protein and which are compatible with the planned chromatography method (Link, 2009; GE Healthcare Life Sciences 2010a).

8.4 Performing chromatography

8.4.1 Chromatography supports

Protein purification is usually performed using a solid support, also called a **matrix** or **resin**, and involves the differential binding of molecules to functional groups attached to the solid support, or differential filtering of molecules by size and shape. The properties of the support are an important component of the quality of any purification achieved. They include **binding capacity** (how many milligrams of protein will bind per millitre of matrix), the stability of the matrix during chromatography, cleaning (can it be re-used?), and the cost of the matrix.

In many protein purification methods, the matrix is based on beads made from a range of polymers. The beads present a large surface area contributing to the binding capacity, whilst the liquid channels between the beads contribute to the **flow rate** (the speed at which solutions can be washed through a column).

The first chromatography matrices were based on irregular materials such as cellulose fibres and hydroxyapatite which have poor flow properties. Now polymers such as dextran, agarose, and polyacrylamide are used, with brand names such as Sepharose, Sephadex, Sephacryl, Biogel, and Ultrogel. Other materials such as silica, monoliths, and polystyrene divinylbenzene are used typically in HPLC (high-performance liquid chromatography) columns. These materials are more rigid and offer better separation properties and faster flow rates, but are more expensive.

8.4.2 Using chromatography matrices: manual and automated approaches

Proteins can be purified manually using the batch method or in column format, or by using an automated approach.

In the batch method, beads equilibrated with appropriate buffer are added to the protein sample in a tube and incubated for long enough to capture the protein. Unbound proteins are then removed by washing the beads with wash buffer and the bound protein is eluted by applying an elution buffer (or buffers). Batch purification is simple to use, cheap, scalable for different sample volumes, and allows parallel processing of multiple samples. However, it is an inefficient way of washing the matrix and elutes the protein in larger volumes than other approaches.

A more efficient purification approach is to run the sample, wash buffer, and elution buffers in turn through a column under gravity or manually using a syringe. This is shown diagrammatically in Figure 8.4(a).

However, the best approach is to use a chromatography machine (Figure 8.4(b)) which is comprised of the following components:

- a pump to run solutions through a column and mix gradients of changing eluent composition;
- a detector to record UV and/or visible absorbance, perhaps at multiple wavelengths;
- pH or conductivity meters to monitor the progress of column equilibration and elution;
- a fraction collector which allows many small-volume samples to be collected during elution;
- a computer to control the process, which allows the same method to be run reproducibly with analysis and storage of the data recorded.

8.5 Chromatographic separation methods

The range of chromatography based methods that are commonly used in protein purification strategies are discussed in this section. A comparison of the different chromatography methods is given in Table 8.1

8.5.1 Affinity chromatography

Affinity chromatography exploits the ability of the target protein to bind to a specific ligand which is immobilized on a chromatography matrix (Figure 8.1). In this method, the protein sample is applied to the matrix under conditions that favour the binding of the target protein. Unbound sample is washed out (wash phase), and then the bound target protein is eluted (elution phase) under conditions that displace the ligand–target protein interaction. This separation method removes more impurities in one step than will typically be achieved using other chromatography methods such as ion-exchange or hydrophobic interaction.

Figure 8.4 Approaches to using chromatography matrices. (a) Chromatography under gravity or using a syringe are often used for step elution using affinity matrices (along with the batch approach described in the text). (b) Automated computer-controlled system which is useful for gradient elution (e.g. IEX purification of target protein from complex mixtures).

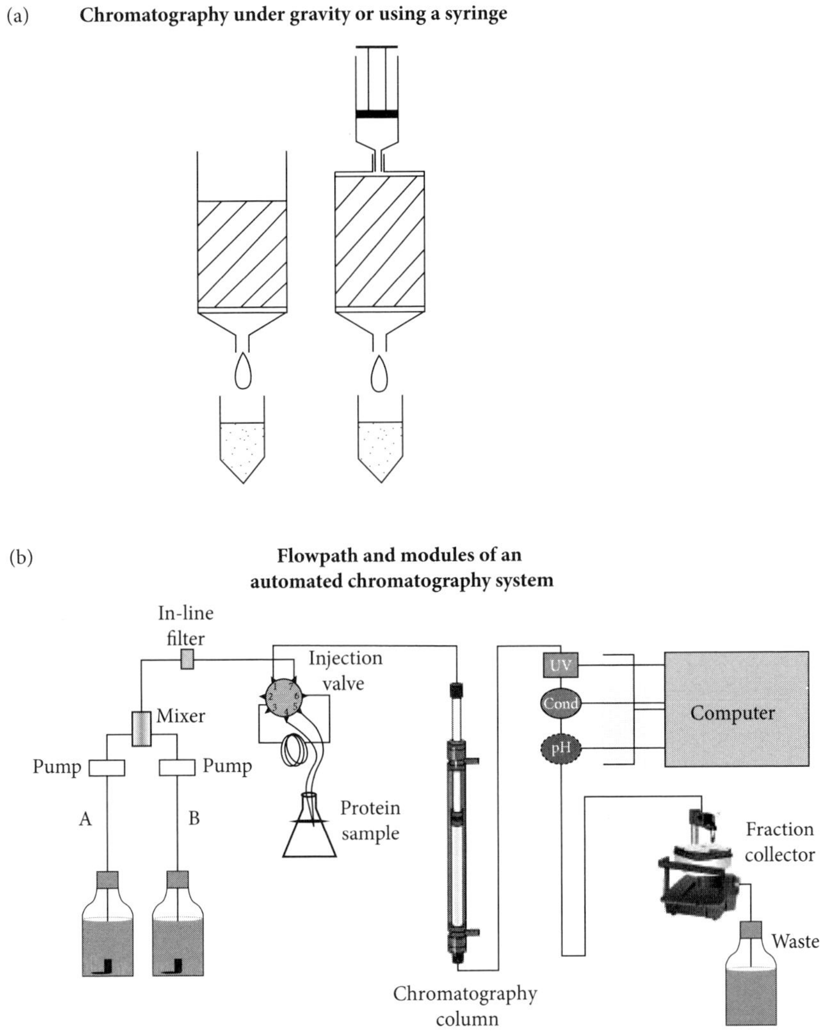

The ligand-binding part of the target protein is usually a fusion tag such as the small hexahistidine tag (His-tag) or larger tags such as the maltose binding protein (MBP) or thioredoxin tags.

See Chapter 7

Proteins carrying a His-tag are purified on a chromatography matrix carrying metal ions (e.g. Ni^{2+} or Co^{2+}) chelated by, for example, iminodiacetic acid or **nitrilotriacetic acid** (NTA) (Block et al. 2009; GE Healthcare Life Sciences 2009). The His-tagged protein in the sample binds to the metal ion, whilst most of the remaining impurities wash straight through the column—the **flow-through**. The bound protein is then eluted by washing the matrix with a solution of imidazole which competes to bind to the immobilized metal ion, displacing the target protein. Alternatively, the bound protein can be eluted by reducing the pH to protonate the histidine side chain, weakening its interaction with the immobilized metal ion and hence displacing the protein.

Elution can be performed using step elution or an automated chromatography system which allows you to program a gradient of increasing imidazole concentration, eluting different proteins in turn. Passing the eluted material through a UV detector generates a chromatogram—a trace of the change in protein concentration over time with changing imidazole concentration. An example of such a chromatogram is shown in Figure 8.5(a). By collecting the eluted material in small volumes, or **fractions**, the purity and activity of each individual fraction can be checked. Other tags used for recombinant proteins, including MBP and GST (glutathione-*S*-transferase), can be purified in a similar way to proteins containing the His-tag but have their own affinity matrices (sugar-based dextrin and the tri-peptide glutathione, respectively) to which the target protein can bind and subsequently can be eluted. An example of a chromatogram of the step elution of an MBP-tagged protein is shown in Figure 8.5(b). These chromatograms are discussed further in Box 8.1.

Affinity chromatography can also be used to purify proteins that do not carry a fusion tag. For example, antibodies are often purified using a matrix to which Protein A or Protein G is attached. These proteins bind the common, Fc, part of the antibody. Bound antibodies are eluted from these columns using a low-pH buffer which partially disrupts the structure of the proteins and weakens the interaction. Similarly, transcription factors can be purified on DNA columns and eluted using high-concentration salt solutions. These approaches contrast with the use of a specific ligand to compete for binding (Angal et al. 2001).

i BOX 8.1

Comparison of chromatograms obtained from an automated chromatography system: gradient and step elution

Figure 8.5(a) shows a chromatogram obtained using a gradient of increasing imidazole concentration to elute a His-tagged protein bound to Ni^{2+} on a Ni-NTA column. Black and grey lines indicate the UV absorbance at 280nm and 260nm, respectively, measuring protein concentration as the protein is eluted. Multiple peaks are observed. The early fractions contain the weakly bound impurities and appear as a small absorbance peak (low protein concentration) on the chromatogram. The later fractions will contain the target protein and should be the largest absorbance peak (higher protein concentration) on the chromatogram. The broken line represents the changing concentration gradient of imidazole used for selective elution of proteins from the matrix.

Figure 8.5(b) shows a chromatogram obtained using step elution of an MBP-tagged protein from a dextrin affinity matrix. Again, black and grey lines indicate the UV absorbance at 280nm and 260nm, respectively, measuring protein concentration as it is eluted. In this example, a single absorbance peak is observed corresponding to the eluted protein. The broken line represents the switch from wash to elution buffer. Step elution works well if few impurities bind to the matrix; gradient elution is preferable if impurities bind to the matrix but can then be removed selectively by increasing the concentration of the elution solution using a gradient.

Figure 8.5 Comparison of chromatograms obtained from an automated chromatography system. (a) gradient elution of a His-tagged protein on an Ni-NTA column. (b) Step elution of an MBP-tagged protein from a dextrin affinity matrix. Black and grey lines indicate the UV absorbance at 280nm and 260nm, respectively, with values in milli-absorbance units on the left hand *y*-axis. The broken lines indicate the shape of the theoretical gradient—the changing concentration of the elution solution of imidazole and maltose in (a) and (b), respectively.

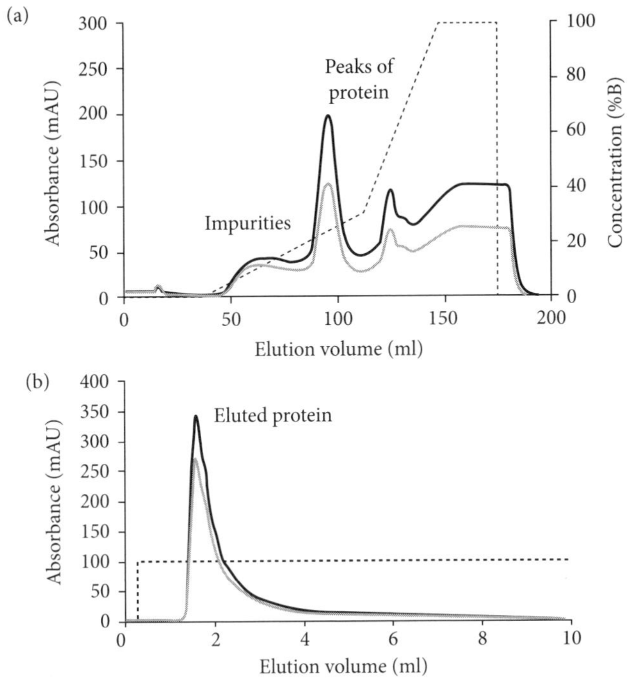

8.5.2 Ion-exchange chromatography

Ion-exchange (IEX) chromatography separates proteins on the basis of charge. IEX matrices carry small simple chemical groups giving the beads either positive or negative charges. IEX matrices can be positively charged (anion exchange) or negatively charged (cation exchange) (Table 8.4). If cation exchange is used, positively charged molecules bind to the negatively charged chemical groups on the solid support. In contrast, in anion exchange, negatively charged molecules bind to the positively charged chemical groups on the solid support. Anion and cation exchange matrices can be further categorized as 'strong' or 'weak', depending on whether the matrix retains the charge over a wide pH range or over a smaller pH range respectively (Jungbauer and Hahn 2009). In IEX chromatography proteins are normally bound to the IEX matrix at low salt concentrations and the bound protein is eluted using solutions of increasing salt concentration (or changing pH). This may be done as a series of step elutions or as a gradient if an automated system is being used.

Table 8.4 Different types of ion exchange chromatography matrices

	Anion exchange (positively charged)		Cation exchange (negatively charged)	
Property	'Weak'	'Strong'	'Weak'	'Strong'
Matrix	DEAE (diethyl aminoethyl)	Q (quaternary amine)	CM (carboxymethyl)	S, SP (sulpho, sulphopropyl)
Recommended buffers	Tris		HEPES, phosphate	

Selecting the IEX matrix

Two key factors influence protein binding to IEX matrices: one is the pH of the solution in which you have your protein (and therefore the charge on the protein *and the matrix*) and the second is the ionic strength of this solution (as higher concentrations of salts compete with the protein for binding to the matrix).

Individual polar groups in amino acids are characterized by their pKa, the pH at which the chemical groups are half charged and half uncharged. The combined properties of these many charged groups give rise to the isoelectric point (pI) of a protein, the pH at which a protein has no overall net charge.

Initial selection of which IEX matrix to use is based on the pI value of the protein. If the pI is not known, it can either be predicted from the sequence or you can resort to using empirical binding screens.

1. A predicted pI can be obtained by pasting the amino acid sequence (and any tag) in a web tool such as Expasy. A pH value is then selected which is likely to give a charge sufficient to bind the protein to the IEX matrix of the opposite charge. This is usually one pH unit above or below the pI.

2. Alternatively, optimum binding conditions can be determined empirically by testing protein binding to the matrix at a range of salt concentrations and pH values. This is shown diagrammatically in Figures 8.6(a) and 8.6(b) respectively. The chromatography matrix is represented as circles and the protein as grey colour in solution or captured on the beads where conditions are appropriate. A range of salt concentrations should be included because, as you will see from Figure 8.6(a), low salt concentrations favour protein binding to the matrix, whilst higher salt concentrations tend to elute the protein.

As a guideline, for proteins with pI 6–8, choose a pH one pH unit above or below the pI. For strongly acidic or strongly basic proteins, a pH of 7–8 would be used to avoid damaging effects of extreme pH.

Ion exchange chromatography is widely used as it is likely that conditions can be found where any protein will bind to an IEX matrix of one charge or another. The matrices are low cost and have a high binding capacity. In addition, the simplicity of the functional groups in IEX matrices provides greater chemical stability than many affinity chromatography matrices (although less specificity), so they are less likely to be damaged by components in the sample and can be cleaned more aggressively when necessary.

8.5.3 Hydrophobic interaction chromatography

Hydrophobic interaction chromatography (HIC) separates proteins based on their differences in hydrophobicity. In HIC, matrices are functionalized with hydrophobic

Figure 8.6 Small-scale screen to determine optimum binding conditions. (a) salt concentration screen for IEX; (b) pH screen for IEX; (c) salt concentration screen fo HIC. Suitable conditions of salt concentration and pH for chromatography can be determined by equilibrating the chromatography matrix in solutions at a range of conditions and adding the protein extract (represented by the grey colour in each tube). At appropriate conditions, the protein binds to the beads (filled circles, protein bound to beads; open circles, chromatography matrix without protein bound). The indicated concentrations and pH values are for illustration only. Different conditions will be selected based on an analysis of the properties of the protein of interest and the chromatography matrices relevant to your experiment. After binding, the remaining solution or beads can be assayed for protein concentration or activity, or by SDS-PAGE.

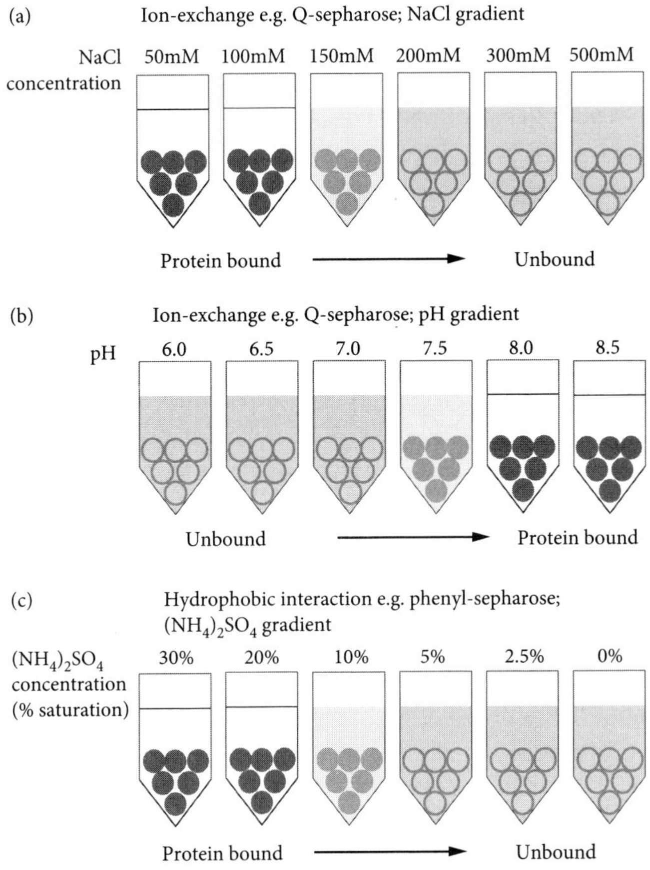

groups (non-polar) to which hydrophobic parts of the proteins bind. Two types of HIC matrix are commonly used: straight chain (alkyl) groups, either butyl (4-carbon) or octyl (8-carbon), or a phenyl group which is a 6-carbon aromatic ring. These two group types provide different selectivities, with the phenyl group selecting for proteins with accessible aromatic amino acid side chains.

In HIC the sample is applied to the column at a high salt concentration (e.g. 1–2M ammonium sulphate) which favours binding of hydrophobic regions of the protein to the hydrophobic ligands attached to the matrix. Proteins are then eluted by reducing the salt concentration. The principle underpinning protein binding in HIC is that in an aqueous environment the hydrophobic groups on proteins are normally buried in hydrophobic environments away from the aqueous solvent. By increasing the salt concentration, the salts interact with the water molecules which in turn strengthen the interaction of the hydrophobic regions on the protein surface with the hydrophobic ligand attached to the matrix.

As the hydrophobic amino acids are normally buried away from the solvent, it is difficult to predict how the target protein may interact with a HIC matrix. Therefore binding conditions for a protein are identified empirically by testing protein binding to beads over a range of salt concentrations, although pH could also be tested. This is shown schematically in Figure 8.6(c).

HIC is a valuable linking step in a purification strategy. For example, HIC can be used following ammonium sulphate precipitation as the high ionic strength of the resulting supernatant is likely to be suitable for binding protein to a HIC matrix. As the sample is eluted at low ionic strength in HIC, it can subsequently be loaded onto an IEX matrix without the need to change sample conditions.

8.5.4 Gel filtration/size exclusion chromatography

In contrast to chromatographic techniques, which utilize differential *binding* of molecules to the chromatography matrix, in SEC molecules do not bind to the matrix but are differentially *filtered* based on their individual size and shape (Cutler 2006; GE Healthcare Life Sciences 2010b).

In SEC, porous matrices with a defined molecular weight fractionation range are packed into a column, equilibrated with buffer, and the sample is passed through the column. Molecules above the fractionation size of the matrix are excluded from entering the pores and are washed out before the smaller molecules which penetrate the pores and are therefore eluted later. In SEC, a single buffer is used during the purification; this **isocratic elution** contrasts with the **gradient elution** of IEX and HIC. In addition, relatively long narrow columns are used to provide better resolution of the different proteins.

Molecules that do not enter the beads elute in the **void volume** (V_0 or 'vee nought'). This is the minimum volume needed to wash any unbound molecules through a column. For any matrix, each molecule will have its own **elution volume** (V_e), the amount of eluent required to wash out the particular molecule from the column. The smallest molecules in the sample, such as buffers, salts, reducing agents, and protease inhibitors, are washed out is called the total volume (V_t).

In SEC, the void volume is usually determined by passing a solution of a very high molecular weight polysaccharide called blue dextran through the column. Aggregated protein in your sample may also elute in the void volume. If your target protein elutes in a lower V_e than expected, it could be because it naturally forms an ordered multimer. This information is important for understanding the activity of the protein.

SEC can be used with pure protein samples to separate aggregates of misfolded proteins from correctly folded proteins by selecting an appropriate SEC matrix. It can also be used to separate proteins from a complex protein mixture. However, the resolution of proteins by SEC is relatively poor, which means that proteins must have quite different sizes to be separated effectively.

Box 8.2 shows example chromatograms obtained using IEX and SEC purification steps.

Analytical use of SEC

As well as being an important purification tool, SEC is often used analytically to determine the (apparent) molecular weight (MW) of a protein in solution.

Example chromatograms obtained from IEX and SEC purification steps

BOX 8.2 i

Figure 8.7(a) shows an example of a chromatogram obtained using IEX purification. This image shows the second purification step of the protein purified by affinity chromatography in Figure 8.5(a) (and therefore is an intermediate purification step). The broken line represents the gradient in salt concentration programmed into the computer and the dotted line shows the actual increase in conductivity as the salt concentration rises. The bold traces show the change in absorbance at two wavelengths (280nm and 260nm), detecting molecules as they are eluted from the column. Two peaks are observed in the chromatogram. The first larger peak is the target protein, and the second smaller peak is some DNA which this DNA-binding protein picked up from *E. coli* during expression and which carried through the previous affinity step. Importantly, this impurity would not be seen on SDS-PAGE gels, but its removal provides a purer and more stable protein.

Figure 8.7(b) shows a chromatogram of a protein sample separated using SEC. The dotted line shows the conductivity measuring the salt concentration whilst the solid line indicates the concentration of protein. In the example shown, the protein is eluted at two elution volumes. The first peak represents misfolded protein eluting around the 10mL

Figure 8.7 (a) absorbance trace during gradient elution of molecules from an anion exchange matrix; (b) isocratic elution of a protein using a size exclusion column. Black and grey lines indicate the UV absorbance at 280nm and 260nm with values in milli-absorbance units on the left hand y-axis; on the right hand y-axis, dotted lines indicate the shape of the theoretical gradient while the dashed line shows the conductivity, measuring changes in salt concentration reflecting the actual gradient.

(a)

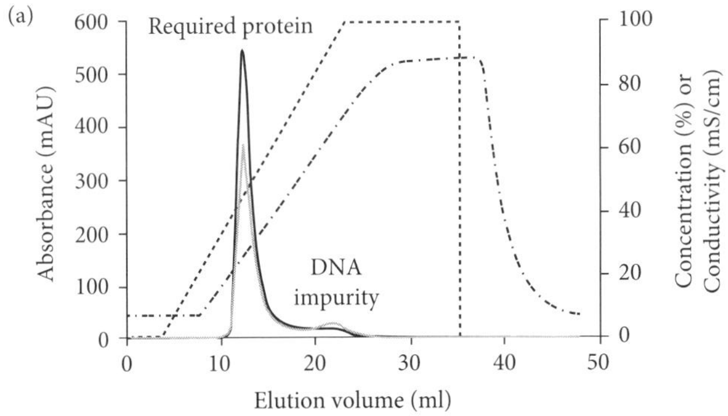

(b)

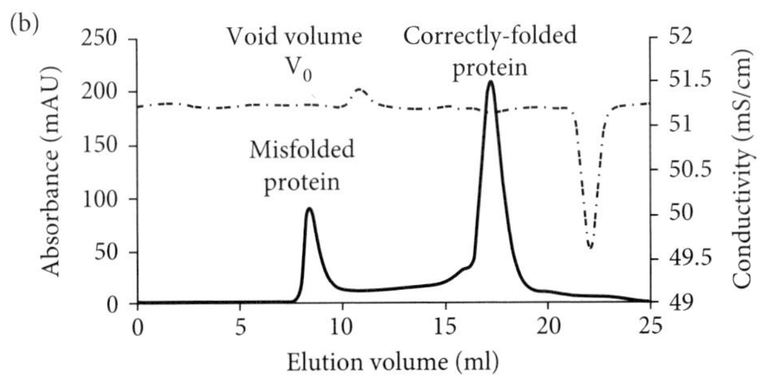

elution volume representing V_0. The second peak, between 15 and 20mL, is the correctly folded target protein eluted at its expected V_e. The lowest molecular weight species are eluted just before V_t, represented by the trough in the conductivity signal in this example. Removing the aggregated misfolded protein ensures that the purified sample is now likely to be fully active.

Commercially prepared columns pre-packed with small homogenously sized beads are best used for analytical experiments. First, a calibration curve is plotted of $\log_{10}$ (MW) against K_{av}, a function describing how much time each protein is retained in the pores of the column. To construct the calibration curve, sets of commercially available standard proteins of known molecular weight, are passed through the column and the elution volume (V_e) is noted for each protein. The void volume (V_o) is determined using the high molecular weight compound blue dextran. The total column volume is denoted by V_t. K_{av} is then calculated using the following formula:

$$K_{av} = (V_e - V_0)/(V_t - V_0)$$

From the elution volume for your protein, $\log_{10}$ (MW) can then be read from the calibration graph or calculated from the equation for the line of best fit for the standard curve. This will give you the apparent molecular weight which may match your expected molecular weight obtained from SDS-PAGE, the amino acid sequence, or mass spectrometry (MS) data, suggesting that the protein is monomeric. If it does not, and the ratio of apparent and expected molecular weights is close to an integer, this may suggest that your protein is multimeric (e.g. a dimer, trimer, or tetramer in solution). However, non-integer numbers (e.g. 2.5) are possible and could have a number of explanations including subunits in exchange between multimers and monomers or asymmetric shape. Therefore any evidence of multimerization from SEC is best checked using additional approaches such as analytical ultra-centrifugation (AUC), native MS, or chemical cross-linking.

See Chapter 13

Peak shape on the chromatogram is an important diagnostic tool for the quality of chromatography, especially in SEC. The sample is loaded in a small volume, but diffusion, heterogeneity in bead size, and differences in the path length taken by proteins should give a broader, but symmetrical, bell-shaped peak. Any 'tail' is indicative of an unusual event, for example protein interacting with the matrix, a crack in the packing of the column, or protein multimers associating and dissociating while the sample is washing through the column.

8.6 Measuring protein recovery

It is important to measure the amount of target protein recovered with each purification step. You may use an assay of target protein activity if one is available, although it is possible that impurities may mask protein activity. However, SDS-PAGE with Coomassie blue staining is the most routinely used method of tracking protein purification. Total protein concentration can also be measured at each purification stage using colorimetric assays such as Bradford or BCA. Measuring the UV absorbance at 280nm is the most useful with pure protein (Noble and Bailey 2009), and recording the absorbance spectrum (e.g. 220–350nm) provides additional protein quality information. From your quantitation of target protein and of total protein at each step, you will be able to calculate percentage protein recovery and degree of purification compared with the starting extract.

When you have a protein of a suitable purity, biophysical analyses can be undertaken to further characterize it or it can be used for quality control of new batches. Mass spectrometry is most commonly used to check that the molecular weight matches the expected weight. Gel filtration (and AUC) provides information about molecular weight/shape and aggregation in solution. Circular dichroism (CD) spectroscopy is often used to demonstrate protein folding, which can also be determined using differential scanning calorimetry (DSC).

See Chapter 12

Planning a protein purification strategy

CASE STUDY 8.1

You have identified a protein that you think is important in regulating correct cell division and differentiation in stem cells. Knocking down expression of the transcript coding for the protein of interest using RNAi shows that cell growth is affected. Transfecting cells with the gene coding sequence fused to the reporter gene *GFP* and subsequent visualization by fluorescence microscopy shows that the expressed protein localizes to the nucleus. It is evident from the published literature that the gene can carry a number of disease-causing mutations. You wish to undertake a range of structural studies (e.g. X-ray crystallography and NMR spectroscopy) to determine how the mutations may affect protein structure and hence protein function.

See Chapter 6

QUESTION

Outline some of the key factors you would need to consider in purifying your protein of interest.

These structural experiments will require milligrams of protein. You will not be able to purify sufficient protein from the natural source as you cannot grow enough of the cells in tissue culture. Therefore, expression of the recombinant protein in *E. coli* would be your first choice. It is possible that the protein is subject to post-translational modifications and hence *E. coli* will not produce these correct modifications. However, the power of heterologous recombinant protein expression in *E. coli* is likely to allow you to make protein with which you can answer many questions about its activity.

See Chapter 7

It will be necessary to clone the gene coding sequence (of the wild-type gene and mutant forms) fused to an affinity tag sequence in tpa plasmid vector. Affinity to chromatography can then be used as an initial (capture) purification step to separate the tagged proteins from the non-tagged proteins. As crystallization requires very pure protein and removal of any misfolded protein, an additional intermediate purification step, such as ion-exchange or hydrophobic interaction chromatography, may be required to remove the remaining impurities. Looking at the sequence of the normal and mutant proteins may help you decide which types of IEX or HIC matrix to use, which may well be different for each protein. It may also be sufficient to omit the intermediate purification step and go directly to a polishing gel filtration step. This pure active protein can then be concentrated and used in screens to find conditions under which the protein solution forms crystals.

See Chapter 12

Once you have a strategy giving you pure protein for crystallization, the same strategy can be applied to make and purify protein for other biophysical analyses. As well as preparative SEC, you could use an analytical SEC column with appropriate molecular weight standards to obtain information on protein multimerization. Knowing if your protein is a monomer, a dimer, or a larger multimer can provide valuable information about protein function. For example, you might find that the mutant proteins do not readily form multimers. The mutant proteins may also fold into an active conformation less effectively. You could obtain this information from crystallography or NMR spectroscopy studies. Other approaches to assessing protein folding include circular dichroism spectroscopy and fluorescence spectroscopy. These can be used over a range of temperatures, which may suggest differences in thermal stability and secondary structure.

See Chapters 11 and 12

8.7 Chapter summary

- A number of different chromatography-based purification methods can be used to purify a protein of interest. These include affinity chromatography, ion exchange chromatography, hydrophobic interaction chromatography, and size exclusion chromatography.
- Before embarking on a purification strategy, a number of key questions should be considered: the intended application of the protein, the source of the protein, what is known about the characteristics of the target protein, and the amount of protein required.
- When designing a purification strategy it is necessary to ensure that the sample conditions (e.g. buffer composition, salt concentration, and pH) are compatible with the subsequent step in the purification strategy. Three approaches are commonly used to achieve this compatibility: dialysis, ultra-centrifugation, and desalting.
- Approaches to keeping protein soluble and active throughout purification must be considered. Similarly, measuring the activity and purity of the protein at each stage is important for assessing the power of your purification strategy.

References

Angal, S., Dean, P.D.G., and Roe, S.D. (2001) Purification by exploitation of activity. In: Roe, S. (ed.) *Protein Purification Techniques* (2nd edn) Oxford: Oxford University Press, pp. 213–40.

Block, H., Maertens, B., Spriestersbach, A., et al. (2009) Immobilized-metal affinity chromatography (IMAC): a review. In: Burgess, R.R. and Deutscher, M.P. (eds), *Guide to Protein Purification* (2nd edn). San Diego, CA: Academic Press, pp. 439–74.

Bonner, P.L.R. (2007) *Protein Purification: The Basics*. London: Taylor & Francis.

Burgess, R.R. and Deutscher, M.P. (eds) (2009) *Guide to Protein Purification* (2nd edn). San Diego, CA: Academic Press.

Burgess, R.R. (2009a) Protein precipitation techniques. In: Burgess, R.R. and Deutscher, M.P. (eds), *Guide to Protein Purification* (2nd edn). San Diego, CA: Academic Press, pp. 331–42.

Burgess, R.R. (2009b) Refolding solubilized inclusion body proteins. In: Burgess, R.R. and Deutscher, M.P. (eds), *Guide to Protein Purification* (2nd edn). San Diego, CA: Academic Press, pp. 259–82.

Cutler, P. (2001) Chromatography on the basis of size. In: Roe, S. (ed.) *Protein Purification Techniques* (2nd edn) Oxford: Oxford University Press, pp. 187–201.

Dawson, R.M.C., Elliott, D.C., Elliott, W.H., and Jones, K.M. (1989) *Data for Biochemical Research* (3rd edn). Oxford: Oxford University Press.

GE Healthcare Life Sciences (2009) *Recombinant Protein Purification Handbook 2009*. GE Healthcare, product code 18-1142-75. http://www.gelifesciences.com/protein-purification

GE Healthcare Life Sciences (2010a) *Protein Sample Preparation Handbook 2010*. GE Healthcare, product code 28-9887-41. http://www.gelifesciences.com/protein-purification

GE Healthcare Life Sciences (2010b) *Gel Filtration: Principles and Methods 2010*. GE Healthcare, product code 18-1022-18. http://www.gelifesciences.com/protein-purification

Grabski, A.C. (2009) Advances in preparation of biological extracts for protein purification. In: Burgess, R.R. and Deutscher, M.P. (eds), *Guide to Protein Purification* (2nd edn). San Diego, CA: Academic Press, pp. 285–304.

Harris, E.L.V. (2006) Concentration of the extract. In: Roe, S. (ed.) *Protein Purification Techniques* (2nd edn) Oxford: Oxford University Press, pp. 111–54.

Jungbauer, A. and Hahn, R. (2009) Ion-exchange chromatography. In: Burgess, R.R. and Deutscher, M.P. (eds), *Guide to Protein Purification* (2nd edn). San Diego, CA: Academic Press, pp. 349–72.

Linke, D. (2009) Detergents: an overview. In: Burgess, R.R. and Deutscher, M.P. (eds), *Guide to Protein Purification* (2nd edn). San Diego, CA: Academic Press, pp. 603–18.

Noble, B. (2006) Scale-up considerations. In: Roe, S. (ed.) *Protein Purification Techniques* (2nd edn) Oxford: Oxford University Press, pp. 241–54.

Noble, J.E. and Bailey, M.J.A. (2009) Quantitation of protein. In: Burgess, R.R. and Deutscher, M.P. (eds), *Guide to Protein Purification* (2nd edn). San Diego, CA: Academic Press, pp. 73–96.

Stoll, V.S. and Blanchard, J.S. (2009) Buffers: principles and practice. In: Burgess, R.R. and Deutscher, M.P. (eds), *Guide to Protein Purification* (2nd edn). San Diego, CA: Academic Press, pp. 43–56.

Willis, M.S., Hogan, J.K., Prabhakar, P., et al. (2005) Investigation of protein refolding using a fractional factorial screen: a study of reagent effects and interactions. *Protein Sci* **14**: 1818–26.

9 Antibodies as research tools

Gavin Allsop and John Colyer

Chapter overview

Direct measurement of individual proteins in a complex mixture is required in many areas of biological research, but in most cases is not possible based on the intrinsic properties of the individual protein. Antibodies provide a solution to this conundrum: they can bind very specifically to individual components of a system and permit measurement of the behaviour of that component alone. In this chapter we will look at the way antibodies can be used as research tools to describe the behaviour of specific proteins within complex biological systems (cell, tissue, or whole organism). We will provide practical insight into the manufacture of a new antibody, or antibody alternative, describe three of the most common experimental applications of antibodies (Western blotting, immunofluorescence microscopy, and co-immunoprecipitation), and discuss how to confirm the specificity of measurements made in each format of work.

LEARNING OBJECTIVES

This chapter will enable the reader to:

- refresh their understanding of what an antibody is;
- understand how antibodies are used in molecular biology and why they are an integral component of biomolecular research;
- gain an insight into the work and care required for the production of a good-quality antibody;
- understand the limitations of antibody use as dictated by their intrinsic properties;
- describe how to address these limitations to obtain reliable datasets.

9.1 Introduction

See Chapter 14

As biological systems contain a large numbers of different proteins (>5000), we require a mechanism by which to identify a particular protein in this mixture. The way to do this is to take advantage of a unique measurable characteristic of the target protein, and measure it. Unfortunately, very few proteins contain unique intrinsic properties that are directly measurable, and although the spectroscopic properties of each protein are not equivalent, differences between them are usually insufficient to facilitate their identification. There are exceptions to this, an example being the green fluorescent protein (GFP). This protein contains a chromophore made up of cyclization of three residues within the primary sequence. The chromophore is able to absorb light at two wavelengths (395nm and 475nm for wild-type GFP) and fluoresce at a longer wavelength (509nm). Since no other protein in the system shares this property, GFP can be visualized in a living specimen (cell, tissue, or whole animal) from its fluorescent emission.

An indirect identification strategy is required for other proteins. Antibodies are a group of proteins that can be used as research tools for this purpose. Each antibody binds to a precise 3D structure on the target protein, and the fit with the antibody binding site and target molecule is very exact. Moreover, antibodies are capable of detecting targets with high specificity. For example, a change in the stereochemistry of a target is sufficient to prevent recognition and binding (Dutta et al. 2003). This specific antibody–target interaction can be measured in a number of different ways, and we will explore some of these in this chapter.

Since antibodies are used to detect individual target proteins indirectly, we assume that the signals generated from them faithfully represent the characteristics of their target, i.e. their concentration and location. This assumption is key to all antibody-based experimentation, and must be verified strictly and frequently if we are to obtain critical and useful information. Before we look at the production and characterization of a new antibody, it is worth refreshing our understanding of antibodies in general.

9.2 Antibody structure and function

Antibodies form an important part of the defence mechanism of an animal (the host) to pathogenic organisms and other foreign agents (Elgert 1998). They recognize a specific structural feature on a target molecule on the foreign agent or pathogen, and this is the same property that is exploited in the use of an antibody as a research agent. *In vivo*, the host expresses a number of different classes of antibodies (IgG, IgA, IgD, IgE, and IgM) which are members of the immunoglobulin protein family. These proteins share a common structure, but differ in the number of core (target-binding) subunits and involvement of accessory proteins. Each of these immunoglobulin classes contributes to different aspects of the host defence process or provides defence at different locations in the body.

A single antibody molecule can recognize just one target structure or **antigen**. However, the antibody contains multiple binding sites, or **paratopes**, for the target molecule (IgG, IgD, and IgE have two paratopes, IgA has four paratopes, and IgM has ten paratopes), but these are identical and thus will bind to different copies of the same antigen.

An organism is likely to encounter many different foreign agents and pathogens in its lifetime, all of which will have different molecular structures. Therefore, to mount an effective defence response, the host requires antibodies capable of recognizing any foreign agent likely to be encountered. This is achieved by the expression of a vast collection of closely related antibody molecules (10^9 different antibody sequences) by the host. Each antibody in the collection (or library) differs in the structure of the antigen binding site (or paratope). Each member of the library recognizes a different antigen, increasing the probability that the host contains at least one antibody that will recognize each new pathogen that it encounters.

To describe the structure of an antibody, we will focus on an IgG molecule, as these are most commonly used as research reagents. Although a variety of IgG subtypes exist (IgG1, IgG2a, IgG2b, and IgG3—classified according to their heavy-chain sequences), and minor differences are known between these subtypes, they share a common structure and function. Thus we will examine just one subtype, IgG1.

A series of representations of an IgG1 molecule are shown in Figure 9.1. An IgG molecule comprises two heavy and two light chains (Figure 9.1(a)). Each heavy chain is 446 residues in length and comprises four distinct protein domains—a variable domain at the N-terminus and three constant region domains in series. The variable domain differs amongst IgG1 molecules of the host, and these differences underpin the molecular recognition of different antigen targets. The sequences of the constant region domains are preserved across all IgG1 molecules. These domains are not involved in antigen binding, but are involved in common downstream processes that catalyse destruction of the antigen (on occasion, a pathogenic organism). The light chain displays a similar modular structure, with a variable region at the N-terminus and a constant region at the C-terminus. Both heavy and light chains contain three short protein sequence segments that are particularly variable in sequence between different members of the IgG library. The regions called the **hypervariable regions** or **complementary determinant regions (CDRs)**, comprise sequences of just 5–16 residues each, and form the paratope or binding site for the antigen.

The modular domain structure is displayed schematically in Figure 9.1(b), and an example of the 3D structure of an IgG molecules solved by X-ray crystallography is shown in Figure 9.1(c). Here, two copies of the heavy chain and two copies of the light chain come together to form a stable hetero-tetramer that is shaped like the letter Y. The N-termini of both subunits are at the far end of each arm of the immunoglobulin molecule. This arrangement positions the six CDR sequences (three from heavy chain and three from the light chain) close together in space to form the paratope (or antigen binding site) at the far end of each arm of the immunoglobulin. This tertiary structure is stabilized by extensive non-covalent interactions between the subunits and by covalent bonds (disulphide bonds) between the subunits and within each subunit. The orderly structure of the heavy chain is interrupted near the centre of the molecule (by residue number) to form the hinge region, a somewhat disordered or flexible section of protein structure. This allows considerable movement of the two arms relative to each other and to the trunk.

In some experimental applications, you will also come across the use of fragments of an immunoglobulin molecule: Fab, (Fab′)2, and scFvs.

- Fab fragments arise from the digestion of an IgG with papain (Figure 9.1(b)). This enzyme hydrolyses the heavy chain at a site N-terminal to the disulphide bonds that link the heavy chains and thus cleave each arm from the Fc region of the immunoglobulin. The Fab fragment contains one paratope.
- $F(ab')_2$ is similar to Fab but contains two paratopes. This fragment is achieved by the digestion of IgG with pepsin, an enzyme that cleaves the heavy chain on the C-terminal side of the inter-heavy-chain disulphide bonds (Figure 9.1(b)). An Fc molecule is not generated as this part of the IgG molecule contains many pepsin cleavage sites and therefore is digested into small inactive pieces.

Figure 9.1 Representations of an IgG1 molecule. (a) Location of CDRs in the light and heavy immunoglobulin chains. Two types of polypeptides make up an IgG molecule, a light chain (L) and a heavy chain (H). Complementary determinant regions (CDRs) in the variable regions of the light chain V_L and the heavy chain V_H are shown. (b) Schematic diagram of an IgG molecule. Two heavy chains (black outline) and two light chains (grey outline) come together to form a stable Y-shaped structure through non-covalent interactions (disulphide bonds). The constant (C) and variable (V) regions of both heavy and light chains are labelled (C_H1–3, V_H, V_H, and V_L respectively). The resulting structure can be physically divided into two functional regions through chemical digestion at the hinge region. Papain liberates Fab and Fc fragments, whereas pepsin liberates F(ab′)$_2$ fragments only (see text for details). (c) Crystal structure of an IgG molecule. The structure of IgG1 (PDB:1IGY) (McPherson et al. 1998) shows the heavy chains (black) and the light chains (grey). The CDR regions that fold to form the unique antibody binding sites are represented by spheres. Protein A binding site is also shown.

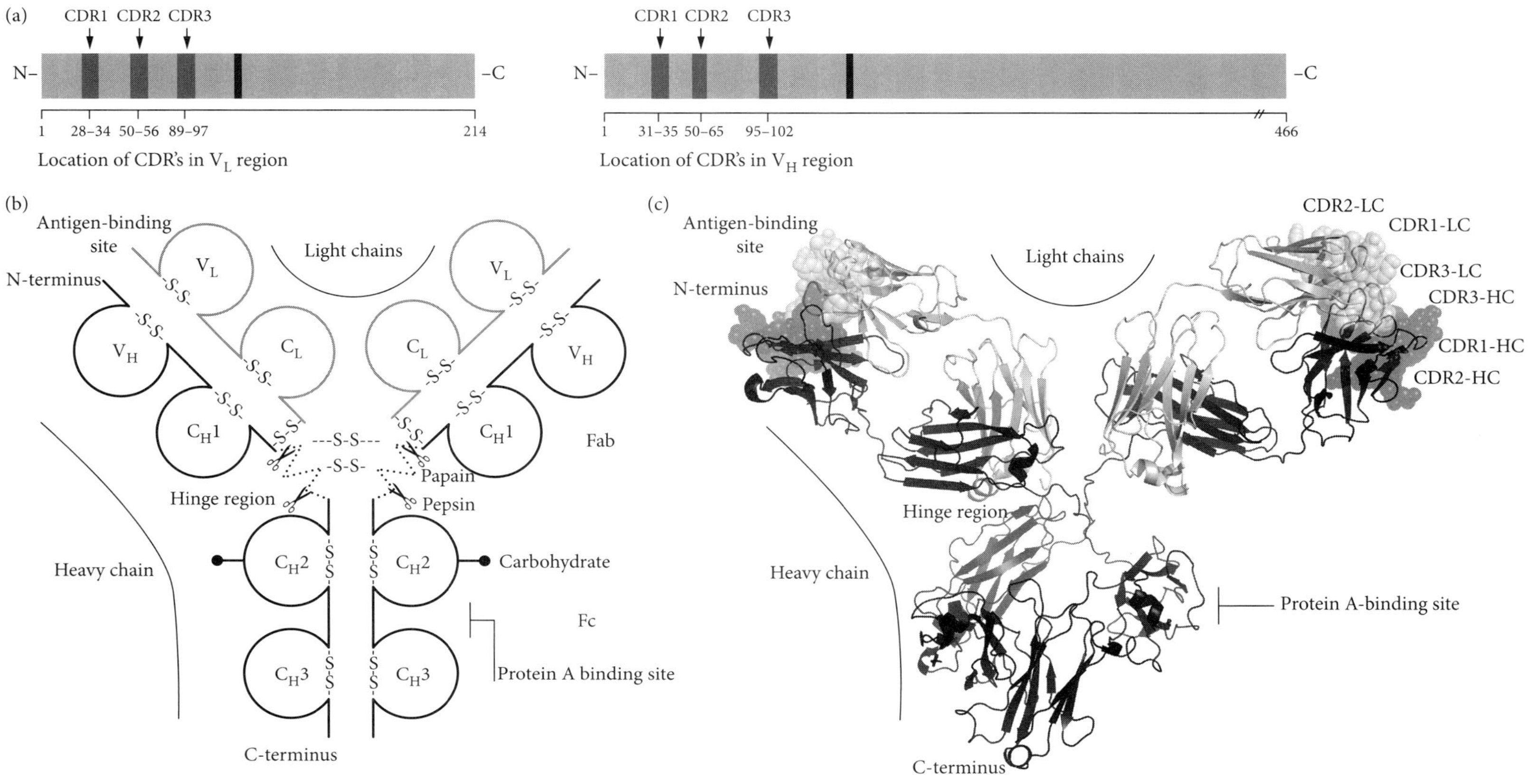

- **The single-chain antibody fragment** (scFvs) is a derivative of the IgG molecule achieved by the expression of a gene encoding an engineered antibody fragment (Maynard and Georgiou 2000). As the paratope of an antibody is contained within the three CDR loops of one domain of the heavy chain and within the three CDR loops of one domain of the light chain, it is possible to create a gene encoding just the variable domain of the heavy chain linked covalently to the variable domain of the light chain. This is the basis of scFvs molecules—a minimum scaffold containing the six CDR loop features in appropriate positions. These can be expressed from a synthetic gene in a variety of host cells and the protein purified by standard means. They contain a single paratope for each protein monomer.

See Chapters 7 and 8

- Further single-binding-site antibody derivatives can be formed by chemical reduction of native IgG molecules by treatment with a reducing agent (β-mercaptoethylamine). This reduces the disulphide bonds, preferentially the inter-subunit bonds (Hermanson 1996). The reduced IgG molecule retains its shape and function because of the extensive non-covalent contacts throughout the structure and the preservation of intramolecular disulphide bonds (under gentle reduction conditions). Chain separation is possible and the accessibility of reduced thiols (the reduced cysteine side chain) provides a means of selective chemical immobilization of half-IgG molecules to a surface for sensor or other applications.

In summary, we can view an IgG molecule as a large protein (150kDa) with a stable structure that brings together six loops at the tip of each of two arms of the molecule to form the binding site (paratope) for a particular antigen (**epitope**). The binding site forms a series of non-covalent interactions with different parts of the antigen, the number and strength of which determine the affinity of the interaction. Other sites on the immunoglobulin protein, such as the constant regions, can also participate in binding to partner molecules. This feature is exploited in experimental formats to detect antibodies through the use of labelled binding partners (protein A, protein G, secondary antibodies). However, you should remember that the IgG molecule displays a large surface area with patches of charged, polar, and hydrophobic chemical features. These can interact with partner proteins non-selectively. This has significant implications for many research applications with antibodies, as the discrimination between specific (paratope-binding partners) and non-specific (other site-binding partners) interactions is key. We will outline the approaches enabling specific and non-specific interactions to be distinguished in this chapter.

9.3 Target detection and visualization

All valid research performed with antibody-based technologies requires that *all signals generated in an experiment arise from the specific interaction of the primary antibody with its target epitope.*

This cannot be assumed to be true, and has to be validated for each experiment. A schematic diagram of a typical antibody-based experiment (a Western blot) is shown in Figure 9.2. This usually involves immobilizing the target protein onto a solid surface such as a membrane (in a Western blot), a microtitre plate (in an ELISA), or the surface of a bead (in immunoprecipitation). A primary antibody specific to the target protein is applied which binds to the available epitope within the structure of the target protein. In the particular experiment shown in Figure 9.2, a complex protein sample (cardiac sarcoplasmic reticulum) containing phosphorylated phospholamban (the target protein) is immobilized onto a PVDF membrane. For simplicity, just one antibody molecule binding to one target protein molecule is displayed.

Some immunoassay formats detect the binding of an antibody to an antigen physically associated with a surface (microbalance, cantilever, or chip surface). These are called label-free assays and measure the change in mass on the surface (Su et al. 2000) or the interference with a physical phenomenon caused by the increased thickness of the surface layer (antigen coating going to antigen–antibody complex) (Cush et al. 1993).

The more common techniques (ELISA, dot blot, Western blot, immunohistochemistry, and immunofluorescence microscopy) do not visualize the binding of primary antibody directly, but employ a secondary (labelled) reagent that binds to the primary antibody (Figure 9.2). Such reagents can be further antibodies, called **secondary antibodies**, which have been raised to the appropriate immunoglobulin class and host (animal) species of the primary antibody (e.g. goat anti-rabbit IgG). Alternatively, the secondary reagent could be specific to the primary antibody but of bacterial origin, such as protein A or protein G. Whichever secondary reagent is used, a detectable label is *covalently attached* to it to enable the complex to be detected. Labels can be an enzyme (peroxidase, alkaline phosphatase), a dye (fluorescein, Alexafluor, Cy3, Cy5, Li-Cor dyes), or a radionuclide (^{125}I). If the label is an enzyme, an appropriate substrate (chromogenic or chemiluminescence) is needed to visualize and quantify the complex. These substrates are invisible, but produce a detectable product upon catalysis by the enzyme. You can read more about the visualization of labels in Chapter 18.

The advantage of using secondary reagents is their convenience, as they can be used against any primary antibody from the same animal host and immunoglobulin class (e.g. rabbit IgG). In addition, as multiple secondary antibody molecules can bind to a primary antibody molecule, considerable amplification of the signal is achievable.

9.3.1 Testing the specificity of an antibody

A control experiment is necessary to verify that the label is faithfully recording the presence of the target antigen (protein), i.e. the interaction between antibody and antigen is specific. A pair of identical experiments can be performed in parallel: in one experiment the primary antibody is exposed to the target antigen, and in the other the primary antibody is exposed to excess quantities of target antigen in solution using a soluble recombinant antigen or epitope peptide (as shown in Figure 9.2). The experiment is then completed for both samples and the signals recorded. In the presence of excess soluble epitope (target antigen), the antibody will be completely inhibited and hence any signals observed will be a consequence of non-specific interactions. These can arise from either the primary antibody or the secondary antibody.

A comparable experiment to verify the specificity of observations can be performed with ELISA, dot blot, immunoprecipitation, and immunomicroscopy experiments and similarly, excess quantities of target antigen are used in solution. These should always be performed to validate a new assay, and it should be repeated whenever a component of the assay is changed (biological specimen, reagents of the assay, even assay conditions (such as changes to buffers and timings)).

Figure 9.2 Schematic diagram of a Western blot. (a) In this example, proteins from 4μg of cardiac sarcoplasmic reticulum (cSR) extract ± CaMKII treatment were separated by SDS-PAGE and transferred to a PVDF membrane. The membrane was incubated with 5% milk protein in TBS (a blocking step) and then with a primary antibody raised in rabbit against phospholamban phosphorylation (PLB^P) at threonine-17 (PLB-PT-17). The membrane was then incubated with a goat anti-rabbit secondary antibody HRP conjugate. This complex is visualized by a signal generated through the oxidation of ECL reagent by HRP using H_2O_2 and captured on an X-ray film. A single band corresponding to PLB^P (at threonine 17) at around 26kDa is observed. (b) Specificity of the antibody to PLB^P (at threonine-17) was tested using two controls: a protein sample containing unphosphorylated PLB, and a competitive inhibition experiment where the primary antibody incubation step included 1μM of the epitope peptide in solution. There is a complete inhibition of the signal in the inhibition experiment, suggesting that the antibody is specific for the target protein.

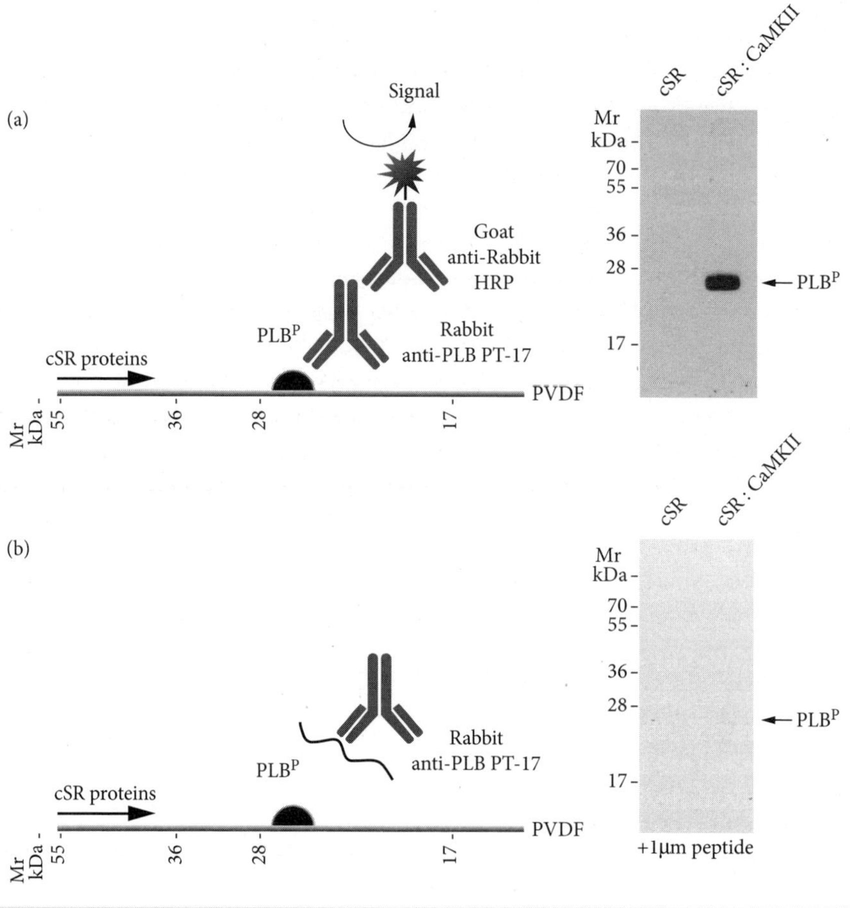

9.4 Do I need to make a new antibody?

The manufacture and characterization of a new antibody is feasible, but it is expensive and time-consuming. It should be undertaken if essential, but first you should look for alternative routes forward.

Epitope tags: an antibody to recombinant forms of a target protein is not essential. Recombinant proteins can be expressed fused to an epitope tag at the N- or C-terminus. These tags are short sequences (6–20 residues) against which antibodies are routinely available. A list of different epitope tags and a review of their use can be found in Brizzard (2008).

See Chapter 7

There are some research questions that cannot be addressed using the epitope tag antibody approach. For example, when the structure or function of the target protein is affected by the expression of the tagged version of the protein, or when comparing naturally occurring target protein in different sample groups (e.g. patients and controls), stages of development, or organisms exposed to different stresses. Here an antibody to an epitope in the intact target molecule is required.

Pre-existing antibodies: antibodies to the target protein may already exist. They could be in the possession of individual academic groups or in the catalogues of commercial antibody suppliers. In either case, once identified it is very likely that you will be able to access these reagents for your research. Academics are duty bound to make their reagents freely available. This is often an explicit requirement of journals that publish the original research. Most academics are also pleased to collaborate with groups wishing to build upon their former discoveries or advances. It is becoming increasingly common for the sharing of reagents to be made formal with a legal agreement between the parties—the **Material Transfer Agreement** (MTA). Similarly, commercial organizations will supply the reagents for a fee. Where reagent supply is not the commercial focus of a company, it may provide the materials free under an MTA.

When identifying antibodies from commercial reagent companies you should search a variety of commercial databases (Antibodypedia, Labome, 1DegreeBio, and CiteAb, to name a few) as the coverage of each database is incomplete and not identical. Having found antibody reagents of interest, you need to assess the quality and suitability of these before purchasing. This assessment should be thorough and critical, and should rely on the data provided by the company, the peer-reviewed publications that have utilized the particular reagent, and the specificity test that we have outlined in this chapter. This assessment needs to be performed for an antibody from whatever source (commercial supplier, generous academic collaborator, or colleague) as the cost of omitting it is high. This cost is measured in time lost to assay optimization for a reagent that may finally be rejected as inappropriate for your research programme.

If the data to convince you of the specificity of an antibody are not available from the supplier, you should find another supplier. It is not uncommon for researchers to report the loss of many months of time and effort as they evaluate new antibodies which turn out to be of poor quality or specificity.

9.4.1 Making a new antibody yourself

If an antibody is essential for the research programme you plan to execute and a suitable antibody does not already exist, you will need to make and characterize a new one. There are a number of different antibodies that you could make: polyclonal antibody, monoclonal antibody, and recombinant antibody alternative. These are described below.

Polyclonal antibodies represent a preparation of immunoglobulin molecules that are derived from many different B-lymphocyte cell lines. The preparation is mainly IgG molecules and will contain a number of different CDR loop sequences allowing different

IgG molecules to recognize different epitopes on the antigen. The polyclonal antibody preparation is the crudest antibody preparation used in research, as it will contain antibodies to non-target proteins (to which the host animal was previously exposed) as well as antibodies to the target protein. This can be a real problem for data interpretation in a new system of study, and is one of the reasons we need to apply the binding specificity test (see section 9.3.1) at all times in an antibody-based research plan.

In some instances polyclonal antibodies are affinity purified to improve the quality of the reagent. Affinity purification on immobilized target protein permits enrichment of relevant target-specific antibodies from the polyclonal antibody pool. However, it should be noted that the term affinity purified is not used only in this context—purified on immobilized antigen. It is also used to describe antibody purified by protein A or protein G affinity chromatography. These techniques purify all immunoglobulin molecules of a particular class or subclass, irrespective of the epitope they recognize.

The property of a polyclonal antibody preparation is the sum of the properties of the individual antibodies within that polyclonal response. As such, it will not have a single well-defined affinity for the target, nor will all components bind to the same features on the target. In some applications this is a distinct advantage. The range of affinities in the antibody preparation can extend the dynamic range of measurements possible, as some antibodies are high affinity and some are low. It can also increase the probability of success in assays involving fixation of the specimen, as some epitopes to some of the polyclonal antibody components will not be corrupted by the processing or fixation of the specimen. The polyclonal antibody preparation is also the quickest and cheapest to perform, and hence it is often the first approach tried on a new project.

A **monoclonal antibody** is the immunoglobulin produced by one particular B-lymphocyte cell line. As such, all copies of the monoclonal antibody have the same primary sequence and react with the same affinity to the same epitope structure. The advantages offered by a monoclonal antibody are specificity and consistency between batches. Moreover, the process of selection of monoclonal antibody cell lines allows you to identify antibody lines with properties that meet exacting specifications, for example antibodies with an affinity for a particular epitope or antibodies that can discriminate between two very similar epitope structures. Thus the properties of the resulting antibody can be better controlled than those of a polyclonal antibody preparation.

It should be noted that monoclonal antibodies are still capable of participation in specific and non-specific interactions. Therefore it cannot be assumed that all proteins recognized by a monoclonal antibody contain the target epitope. That needs to be verified, as described already (Figure 9.2). It should also be noted that some methods of production of monoclonal antibody protein can result in contamination with other immunoglobulin molecules, thereby adding to the possibility of non-specific results. For example, monoclonal antibody purified from ascites fluid can be contaminated with serum antibodies from the host, and monoclonal antibodies produced in cell culture with serum supplements may contain antibody protein from the host of the serum supplement (often bovine). If the monoclonal antibody is required in pure form, it can be produced in cell culture in serum-free conditions, either by expression by the original hybridoma cell line, or following cloning of the appropriate gene expressed in a suitable mammalian host cell (such as Chinese hamster ovary (CHO) cells).

Recombinant antibody alternatives: as antibodies are effectively a protein apparatus that creates a binding site for a particular partner epitope from a series of loop structures that are close in 3D space, antibody-like properties can be built into a variety of alternative protein and non-protein (DNA, RNA) structures. Several excellent reviews of these antibody alternative technologies are available: protein (Nygren 2008); DNA (Bock et al. 1992); RNA (Bunker and Stockley 2006). In all cases the antibody alternatives contain regions of variable sequence (amongst members of a library) which form the paratope structure. Libraries of high diversity, equivalent to the immune system of a mammal, are created and screening strategies employed to identify polyclonal pools or monoclonal binding partners to a target antigen.

These methods have some clear advantages. They eliminate the need for animal experimentation, which is a desirable outcome. In addition, synthetic libraries may contain binding-partner molecules for antigen that are not immunogenic in a host animal species. The screening of synthetic libraries is well controlled; it can be configured to alter the stringency of selection and introduce counter-selection strategies to enhance the power of discrimination of the antibody-like product molecule.

Making a polyclonal antibody

A polyclonal antibody to any material recognized by the immune system of the host animal can be made. This can include small molecules, or **haptens**, such as dyes, modified amino acids, metabolites, signalling molecules, peptides, and post-translationally modified peptides, or large molecules such as proteins, protein complexes, and organelles. The polyclonal antibody is created by the repeated exposure of an animal to the immunogen at intervals and in preparations (of adjuvant) that have been shown to maximize the production of antibodies. A typical schedule in rabbits or other large mammals (sheep, goat) involves an initial immunization challenge (of 150μg of the immunogen) followed by booster challenges (of a similar dose of immunogen) at intervals of approximately 4 weeks over a period of 3–4 months. The timing is set to coincide with the antibody response solicited; repeat immunizations occur as the antibody concentration in the bloodstream (to your immunogen) declines. Blood is taken from the host 7–14 days after a booster immunization, and serum is prepared by allowing the blood to clot and then separating the serum from clotted components by centrifugation (polyclonal antiserum). An IgG class polyclonal antibody response to the target immunogen will be evident from boost 1 onwards.

The success of an antibody production schedule is dictated by good planning and involves the following steps:

- identification of an appropriate immunogen;
- preparation of pure immunogen;
- immunization of the host animal;
- characterization of the antibody.

Identification of an appropriate immunogen

To identify an appropriate immunogen, one must establish what properties the antibody should possess, decide whether to challenge an animal with the full-length protein of interest as an immunogen (Colyer et al. 1989), or the DNA encoding this protein sequence (Lin et al. 1998), or segments of the protein (peptides or domains) (Drago and Colyer 1994), and consider the immunogenic potential of the approach selected. These factors are explained in Case Study 9.1.

CASE STUDY 9.1

Identification of an appropriate immunogen

Let us assume that we want to make an antibody to a particular protein, say SERCA2, the Ca^{2+} transport protein of the sarcoplasmic reticulum of certain muscle cells. A number of questions arise.

QUESTION 1

Do we want the antibody to recognize all isoforms of SERCA2? Do we want the antibody to recognize SERCA2 but not the closely related SERCA enzymes SERCA1 and SERCA3 (that are expressed in other cell types)?

The answer to these questions will depend upon your research goals. Let us assume that your goal is to define where SERCA2 is expressed in the body, and how this changes throughout development of the organism. What then are the properties of the antibody you require? The antibody should recognize all isoforms of SERCA2 (SERCA2a, SERCA2b, SERCA2c) in any organism or organisms that you might want to study, but it should not recognize SERCA1, SERCA3, or any other closely related protein. Once this initial answer is clear, we will have to consider whether it is feasible to produce an antibody that fits the criterion set.

QUESTION 2

In the case of SERCA2, should we challenge an animal with either the full-length protein of interest as an immunogen or the DNA encoding this protein sequence or segments of the protein?

To define how likely each approach would be to the goals of this project (to define where SERCA2 is expressed in the body), an analysis of the protein sequence of all SERCA proteins should be conducted using bioinformatic tools. The sequence of all isoforms of the study subject (SERCA1, SERCA2, and SERCA3) can be compared, and the comparison extended to all species of interest in your immediate and longer-term research. We will be studying the sequences with a view to identifying whether immunization with the whole protein is likely to be successful or whether we have to immunize with parts of the protein (domains) or small parts of the protein (peptide sequences). If a high degree of identity exists between SERCA1, SERCA2, and SERCA3, the polyclonal antiserum that we produce by immunization with SERCA2 will probably contain antibody molecules that recognize all three versions of the enzyme (SERCA1, SERCA2, SERCA3). Such a polyclonal antibody would not permit examination of our research goal and therefore we would reject the strategy of immunizing an animal with the whole-protein (SERCA2) immunogen. To satisfy our research goal, we need to immunize with domains or peptide sequences that are present in all isoforms of SERCA2, but which are not present in other SERCA proteins or other proteins in general. Some examples of sequences are shown in Figure 9.3. The greater the difference in primary sequence between multiple segments of related proteins, the more likely it is that a polyclonal antibody will react with only one form. Having selected a set of peptides or domains by this approach, one should use additional bioinformatic analysis to discover whether similar sequences occur in other less closely related proteins (e.g. BLAST search).

QUESTION 3

If a peptide sequence is chosen as the most attractive immunogen, is it likely to be immunogenic and accessible to an antibody?

A variety of algorithms have been developed to consider the immunogenic potential of a given sequence, for example Epitopia (Rubinstein et al. 2009) and PEPOP (Moreau et al. 2008). Access of the antibody to its epitope is needed for a successful immunoassay,

Figure 9.3 Protein sequence alignments for the design of immunizing peptides for SERCA antibodies. Examples highlight peptides suitable for the production of antibodies against all SERCA isoforms (top left), SERCA1-specific (top right), SERCA 2a-specific (bottom left), and all three SERCA 2 isoforms (bottom right). Sequence alignments were carried out using the Muscle Multiple Protein Sequence Alignment tool in Jalview. For a colour reproduction of this figure, please see Plate 5.

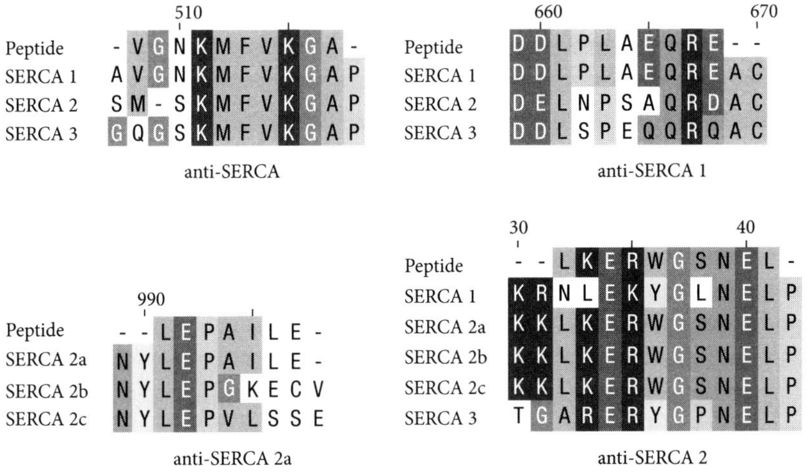

which is best satisfied by a feature on the external surface of the target protein facing the solvent. There is good alignment of the chemical properties of surface exposure and immunogenicity, which augurs well for this anti-peptide approach.

Short peptides are not immunogenic if introduced to the host alone. This is because they contain too few epitopes to stimulate the full cascade of events required to produce an antibody. This cascade represents a set of interactions between the immunogen and macrophages or antigen-presenting cells, T cells, and B cells. This cascade stimulates the appropriate B cells (which recognize an epitope in the immunogen) to proliferate and switch to the production of IgG class antibodies specific to this epitope. The smallest protein shown to elicit an immune response is the 52-residue long protein phospholamban (Suzuki and Wang 1986). It was purified from SDS-PAGE gels and used to immunize mice for the production of monoclonal antibodies. Often a research plan requires the use of shorter peptide sequences than 52 residues. In these cases, short peptides need to be coupled to large immunogenic proteins and presented as a covalent complex. In this way the peptide immunogen is able to activate the B-cell receptor, and the carrier protein, or a fragment therefrom, will activate the additional receptors in the antigen presentation cascade. A variety of carrier proteins are suitable, with a popular choice being keyhole limpet haemocyanin (KLH). A carrier protein that is irrelevant to your research system should be selected so that antibodies to this protein component will not create problems in your experiments (for example, KLH is a good choice provided that you never plan to analyse samples from the keyhole limpet or similar animals).

Preparing pure immunogen

Once the immunogen has been identified, the next stage is to prepare pure immunogen. If it is a whole protein, immunization of the host animal can either be performed with pure protein immunogen (the higher the purity the better: >98% pure by SDS-PAGE (see section 9.5.2) and mass spectrometry), or with naked DNA encoding the gene for

See Chapter 13

the immunogen protein. Injection of DNA into the muscle of an animal results in the expression of the target protein by the host and stimulates an antibody-based immune response to this foreign protein (Lin et al. 1998).

If the immunogen is a peptide, it will be purified by HPLC to high levels (>99% checked by mass spectrometry) and covalently attached to a carrier protein using commercially available cross-linkage reagents (details of the chemistry are given by Hermanson (1995)). The N- and C-termini of the peptide are blocked during chemical synthesis to eliminate the positive and negative charges at the N- and C-termini, respectively, to mirror the intact protein. Covalent attachment to the carrier protein usually exploits a unique chemical feature of the peptide. For instance, an additional cysteine residue can be included at either the N- or C-terminus of the peptide (where no other cysteine exists in the sequence) to facilitate conjugation to the carrier protein through the reactive side chain of the cysteine. Alternatively, in a peptide that lack lysine residues, the N-terminus can be left unmodified, and amine-directed conjugation chemistry between the N-terminus of the peptide and the carrier protein can be used to form the immunogen complex.

Immunization of the host animal

We are now ready to immunize our host animals. Healthy rabbits with good genetic variability (many bucks in the breeding programme) kept in a conventional animal unit have a strong immune system and respond well to immunization. A single animal is all that is required for an immunization protocol, although there will be minor differences in response between different animals and so it is not uncommon for several (two or three) animals to be used in parallel. A further check worth performing is a screen of the pre-immune status of possible animals. Many animals already contain antibodies to components in your tissue or cell system that will interfere with the specificity and usefulness of your antibody-based research. Screening for such antibodies in the pre-immune serum of rabbits will allow you to select animals with no existing antibodies reactive to components of your research system (Figure 9.4).

Once your animal(s) for immunization has been selected, high-quality antibodies appear in the serum within 5 weeks (first immune bleed). We advocate two more boosts at 4-week intervals with termination of the programme at that point. We have seen no significant improvement in immune response with longer immunization routines. A total of 70mL of immune serum per rabbit can be attained with this protocol. These are highly stable proteins and retain their activity for at least 20 years at −80°C.

Characterization of the polyclonal antibody

It is necessary to test any new antibody against a peptide–protein conjugate first, where the protein is different from the immunization protein (e.g. BSA (bovine serum albumin) in this test and KLH for the immunogen). Once antibodies to the target peptide have been confirmed, analysis of the target protein in cell and tissue extracts can follow. In each case a competitive experiment should be conducted in parallel to differentiate between specific and non-specific protein components in the immunoassay (see section 9.3.1). The ideal situation is one where the raw antiserum can be used in your experiments. This is possible if no other antibodies to components in your experimental system are present, other than those directed to the immunizing epitope. The steps and issues discussed above are geared to making this a likely outcome. However, if the antiserum contains antibodies to many different components, a programme of affinity purification of the antibodies to your particular antigen can be performed to obtain highly specific polyclonal antibodies to your target.

See Chapter 8

Figure 9.4 Pre-immune rabbit screen. Pre-immune serum from two rabbits was used in a Western blot against 50µg of rat brain extract (sample A), 50µg of rat heart extract (sample B), AD-293 cell extract (sample C), and 5µL of *E. coli* cell (sample D). There are significant differences in staining pattern between the animals and this highlights the importance of performing a pre-immunization screen to allow animal selection. If rabbit 2 were chosen to raise an antibody against a neural or cardiac protein, the resulting product would include antibodies non-specific to the target protein. In order to produce an antibody of exceptionally high quality, there should be no visual bands on a Western blot of a pre-immune screen.

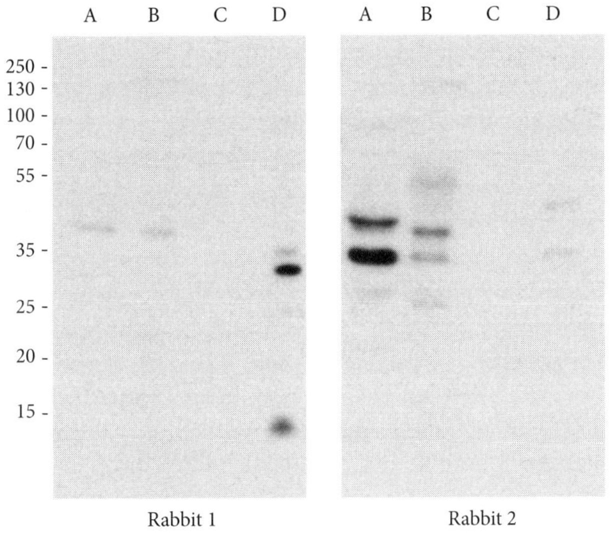

Monoclonal antibodies

The criteria outlined above for selecting an appropriate immunogen apply equally to the generation of monoclonal antibodies. Screening and counter-screening strategies can be devised to isolate monoclonal antibodies that differentiate between two similar components.

Monoclonal antibodies can be produced in mice (Colyer et al. 1989), rats (Knox 1997), and rabbits (Huang et al. 2005). Once a good immune response has been achieved upon challenge with an immunogen, the B-lymphocytes from the donor (spleen) are harvested and fused with a myeloid cell line (such as NS/1) to produce a hybrid cell with properties of antibody secretion and constant proliferation. These hybrid cells (hybridoma) are the monoclonal cell lines, and a variety of cloning strategies are available to isolate individual cell lines (monoclonal) expressing the antibodies of interest (the screening results). Once cloned, the cells can be stored long term in liquid nitrogen and provide a constant supply of a single-antibody molecule.

9.5 Common experimental platforms for immunoassay

In an immunoassay an antibody is used to identify a biomarker (a molecule of biological origin, for example a target protein) in a specimen of interest. Immunoassays tend to be extremely sensitive, capable of detecting a few (70) attomoles (7×10^{-17} moles) of target protein (Bromage et al. 2009) depending on the affinity of the antibody and the amplification achieved by the detection system. However, the assays are not always as specific as required. This can be due to incomplete knowledge of the properties of the antibody or its use in a new system or setting. We have already discussed how specificity can be interrogated using a competitive immunoassay—the specificity test.

A second way of improving the information content of a result is to create an assay that explores two independent features of the target protein (e.g. the simultaneous binding of two antibodies), or combining measurements of a physical property of the target protein and its recognition by an antibody (molecular weight (Mr) and immunoreactivity). Assessment of two parameters simultaneously improves the confidence with which the data can be interpreted but is not a particularly stringent test. The specificity test, on the other hand, provides data of greater quality.

9.5.1 Dot blot/reverse-phase lysate array (RPA)

The dot blot (or **r**everse-**p**hase **l**ysate array (RPA)) is a simple, rapid, and highly sensitive technique. It involves the deposition of samples onto a surface in an array. This can be as simple as spotting microlitre quantities of sample onto nitrocellulose membrane in a dot blot, through to the robotic spotting of 250pL of cell lysate per spot in a high-density array (Sevecka and MacBeath 2006). The remaining sites on the surface are then blocked to minimize non-specific binding and the target protein in the samples is detected using an antibody.

This technique is extremely useful for the rapid review of a large number of samples, where positive results will be investigated further using more discriminating techniques. This is an important approach in systems biology research, as many of the more discerning techniques, such as Western blotting, have low sample throughput (10–25 samples per gel) in the configuration used by most researchers. Dot blotting is also useful when the quantity of sample is small, as it has the highest sensitivity of all immunoassays (lysate from 10^3 cells is sufficient (Ciaccio et al. 2010)).

Dot blotting also serves as a useful initial screen of large numbers of samples against large numbers of antibodies/target proteins. The antibodies appropriate for the screen need to be validated in a separate and more discerning assay, and positive results need to be confirmed by an independent approach. It is likely that a large proportion of significant changes will be overlooked at this stage because of false-negative results. An example of antibody validation is given in Box 9.1.

9.5.2 Western blot

The Western blot or immunoblot is probably the most commonly used immunoassay in biological research—84% of antibody users perform Western blots (http://www.biocompare.com). It provides information on two independent parameters: the molecular weight (Mr) of the protein, based on the separation of sample components by SDS-PAGE, and the relative abundance of the protein recognized by the antibody, based on the intensity of signal produced.

In SDS-PAGE, proteins are denatured using **SDS** (sodium dodecyl sulphate), a strong negatively charged detergent that binds uniformly to the protein chain and denatures it.

Validation of antibodies for use in high-throughput RPA

BOX 9.1

Sevecka and MacBeath (2006) used the high-throughput RPA technique to analyse cell signalling states in cancer cells in an effort to define new therapeutic strategies. Focusing on the epidermal growth factor signalling pathway, they first validated, by Western blotting, 61 antibodies to proteins and phosphorylation sites within this pathway. Thirty-four of these antibodies detected a single protein in cancer cell lysates of the appropriate molecular weight, and were considered to be suitable for use in the RPA assay. To validate the RPA assay, a comparison of the immunoassay signal was made between the RPA assay and a Western blot for a collection of samples containing different quantities of biomarker in each sample. Four antibodies gave equivalent results in both assays, i.e. RPA and Western blot assays reported the same alteration in concentration of target protein between samples. Eight further antibodies displayed the same pattern of response in both assays although the dynamic range of signals in RPA assays was compressed, and 22 antibodies reported no change in signal in RPA whereas they defined a clear change in signal (concentration of biomarker) in Western blot assays. Thus in this study approximately half the antibodies available could not be used in an RPA format assay. Twelve out of the 34 that could be used in RPA generated robust results, i.e. qualitatively similar to alternative immunoassays with the same reagents—a success rate of 36%. The remaining 22 antibodies gave false-negative readings with RPA.

Hence the charge-to-mass ratio of all proteins becomes similar, as does their shape. As the movement of a molecule in an electric field (electrophoresis) is a function of the size, shape, and charge of that molecule, this reduces the factors affecting rates of movement of a molecule in an electric field to mass only. Hence the protein samples are separated by mass in 1D SDS-PAGE. If 2D electrophoresis is used, separation is first by isolectric focusing (IEF) followed by SDS-PAGE. Once proteins have been separated by electrophoresis, they are transferred to the surface of a membrane (nitrocellulose, nylon, or polyvinylidene fluoride) by electrophoresis. This application of an electric field perpendicular to the plane of the SDS-PAGE gel elutes the negatively charged proteins from the SDS-PAGE onto the membrane which is positioned between the SDS-PAGE gel and the anode. Once on the surface of the membrane, the immunoassay is performed as shown in Figure 9.2.

See Chapter 14

Apparent Mr is calculated from the distance moved by the target protein during the SDS-PAGE electrophoresis phase of the experiment. The distance moved is compared with the distance moved by several standard proteins that are also included in the experiment, and a plot of distance moved against $\log_{10}$ Mr is used to calculate the Mr of the protein. Apparent Mr means the molecular weight of a protein estimated from an SDS-PAGE analysis.

How accurate are these measurements of molecular weight? SDS-PAGE reports apparent Mr to approximately 1–2kDa (cf mass spectrometry which reports to <1Da). It is reasonably accurate for many proteins, and much of the research in cell biology relies upon this. However, for some proteins apparent Mr and theoretical Mr are in very poor agreement. Glycosylated proteins often display an apparent Mr very different from their true Mr, and this can appear at an unexpected position in a Western blot. They appear much heavier than expected, which is probably due to the bulkiness of the carbohydrate chains of the glycoprotein altering the size and shape of the denatured protein. They often also appear as a diffuse ('fuzzy') band in the Western blot, which suggests a high

See Chapter 13

degree of heterogeneity in their Mr and thus their structure. Removal of the carbohydrate from the protein (by enzymatic cleavage) corrects many of these anomalies.

The apparent Mr of membrane proteins is often inaccurate (Ragan 1986), perhaps due to atypical binding of detergent by the highly hydrophobic sections of these proteins. Finally, some proteins appear resistant to the denaturing power of SDS detergent. Figure 9.5 shows a Western blot of the cardiac protein phospholamban, which is a protein of 52 residues. The theoretical Mr of phospholamban is 6080Da. However, the protein migrates in two forms, one migrating at an Mr of 25–30kDa (the pentamer) and the other at 6kDa (the monomer). The latter is not seen in Figure 9.2 as the cardiac sarcoplasm reticulum samples were not heat treated. Both of these bands are specific, and the ratio of the two sizes can be altered by sample pre-treatment (boiling, increasing SDS-to-protein ratio, addition of non-ionic detergents). Additional bands of Mr between these two extremes can be captured using these treatments (Figure 9.5(a)), leading to the description of a series of oligomer states of phospholamban of dimer, trimer, tetramer, and the native pentamer structure. Therefore this protein retains significant structure in the presence of SDS such that protein–protein interactions can still occur, resulting in high-order oligomer species. Phospholamban shows one further structural change that is resistant to SDS. The protein is a substrate for a number of kinases (cAMP-dependent protein kinase, Ca^{2+}/calmodulin-dependent kinase II) and these phosphorylate Ser-16 and Thr-17, respectively. Figure 9.5(b) displays a Western blot of cardiac muscle

Figure 9.5 Western blot of the cardiac protein phospholamban multimers and phosphorylation states. (a) Phospholamban (PLB) forms various homo-multimers and these forms can be identified by varying sample treatment conditions. PLB monomer and pentamer can be seen following a 30°C incubation, whereas the monomer, dimer, trimer, tetramer, and pentamer are all seen following a 60°C incubation. (b) PLB is a substrate for a number of kinases (cAMP-dependent protein kinase, Ca^{2+}/calmodulin-dependent kinase II) and these phosphorylate Ser-16 and Thr-17, respectively. The two phosphorylation sites can be seen in the monomer (labelled 1 and 2), and these sites are subsequently apparent in the pentamer (0–10).

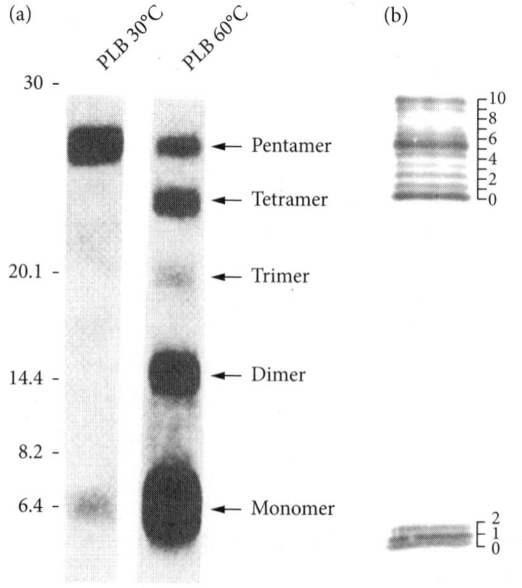

sarcoplasmic reticulum, stained for phospholamban. The monomer and pentamer of phospholamban are visible. However, the pentamer is now separated into a number of well-resolved species.

What distinguishes one species from another? Clearly, each pentameric species displays a different Mr. The difference in Mr is a unit of approximately 1kDa; however, the chemical basis that distinguishes one from the other is the addition of a phosphate group to either residue Ser-16 or residue Thr-17. Phosphate addition adds 80Da of mass, which is too small a change to be seen by SDS-PAGE. However, phosphorylation also changes the structure of the protein, and the increase in Mr in steps of 1–2kDa represents pentameric species with different numbers of subunits in the phosphorylated and unphosphorylated structure.

9.5.3 Immunoprecipitation

Immunoprecipitation is a technique which allows separation of a target protein from a complex protein mixture. In contrast with a Western blot, the antibody is initially immobilized rather than the target protein. First, specific antibodies are coupled non-covalently to beads coated with protein A or protein G. These are proteins of bacterial origin that interact with the **Fc region** of an antibody. The interaction of protein A with the Fc portion of an antibody leaves the antigen-binding portion of the antibody available to bind antigen present in a sample. In practice, a biological sample containing the target protein is exposed to the antibody–protein A beads. Target protein molecules interact with the paratope of the immobilized antibody and are captured on the bead. These can be harvested by low-speed centrifugation (facilitated by the heavy agarose beads) (Figure 9.6) and the identity of proteins captured analysed by SDS-PAGE or mass spectrometry.

A variety of similar techniques are in common use (e.g. GST (glutathione S-transferase) pull-down, co-immunoprecipitation). With a GST pull-down experiment, no antibodies are used. Instead a recombinant protein (**bait**) fused to a GST tag is immobilized on a bead via an interaction of GST with its ligand, glutathione. Next, a complex biological specimen is mixed with the GST–bait beads and proteins able to interact with the bait (or GST, or the bead itself) will be captured and harvested as above.

See Chapter 10

Co-immunoprecipitation is a similar technique, but here the purpose is to purify a target by immunoprecipitation and analyse what *else* has been purified along with the target. A standard immunoprecipitation experiment is performed, although in this experimental mode the binding of target protein to its bead-bound antibody creates a new 'bait' (analogous to the bait in GST pull-down). Protein associated with the bait will also be captured on the antibody–bead, and these proteins are detected by SDS-PAGE, Western blotting, or mass spectrometry. These co-immunoprecipitating proteins may be physiologically relevant partners of the target protein. In this way a description of macromolecular complexes can be made.

As previously outlined, the signals generated from any experiment must be thoroughly validated using a series of controls. To ensure that neither bait nor binding partner interacts with the beads or antibodies non-specifically, three controls are set up:

- a no-antibody control whereby the proteins from a biological specimen are incubated with beads only;

Figure 9.6 The co-immunoprecipitation assay. (a) An antibody specific for a target protein is non-covalently linked to a bead and a complex protein mixture is applied (jigsaw pieces). The target protein is captured by the specific antibody, but remains bound to its binding partners. Bound beads are separated from the complex biological system by centrifugation of the heavy beads. (b) The separated bead–antibody–protein complexes are then separated by SDS-PAGE and the Mr is identified by gel staining or Western blotting using antibodies directed against the binding partners.

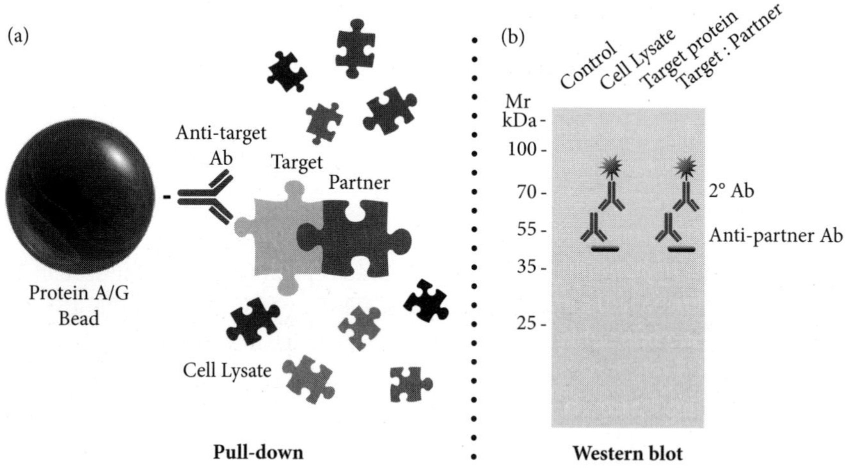

- a test pull-down using an irrelevant antibody from the same species and of the same form (e.g. affinity purified IgG);
- a pull-down using a protein sample that does not contain the target protein.

For a good overview of these techniques see Bonifacino *et al.* (2001).

9.5.4 Immunomicroscopy

Immunofluorescence (IF) microscopy is the second most common immunoassay (76.8% of antibody users perform IF microscopy (http://www.biocompare.com)), and follows a similar principle to a Western blot. A specific primary antibody is used to detect a target protein (antigen) and a secondary antibody conjugate specific for the Fc region of the primary antibody is used to produce a signal. The conjugate is a fluorescent label such as FITC, and hence the sample is visualized using fluorescence microscopy. In this technique, the sample is an intact cell or tissue preparation, and the target is the native protein (Figure 9.7).

See Chapter 17

This assay relies on a single parameter of the target protein only—its ability to bind to an antibody. As with dot blot experiments (see section 9.5.1), such assays have low stringency and are prone to errors. The specificity of an antibody is paramount and, as in Western blotting, is examined by co-incubation of the primary antibody with excess antigen in solution and the specimen simultaneously. As before, specific staining will be blocked but non-specific staining will be retained. Additional controls that improve the critical interpretation of IF data include the manipulation of the target protein

Figure 9.7 Immunofluorescence microscopy. (a) In order to visualize the cellular localization pattern of a target protein, cells grown in tissue culture are seeded onto a glass coverslip and fixed in paraformaldehyde. Following a permeabilization step (with a mild detergent), the sample is incubated with primary antibody against the target protein and counter-stained using fluorescently conjugated secondary antibodies. (b) The localization pattern of the target protein is then visualized using fluorescence microscopy. In this example, a myoblast has been fixed and stained with an antibody against vinculin (a protein associated with focal adhesion complexes). The secondary antibody carries the fluorescent label Alexa fluor 488. The fluorescence micrograph shows vinculin staining along focal adhesion complexes in the protruding lamellipodia and trailing edge of the myoblast. For a colour reproduction of this figure, please see Plate 6.

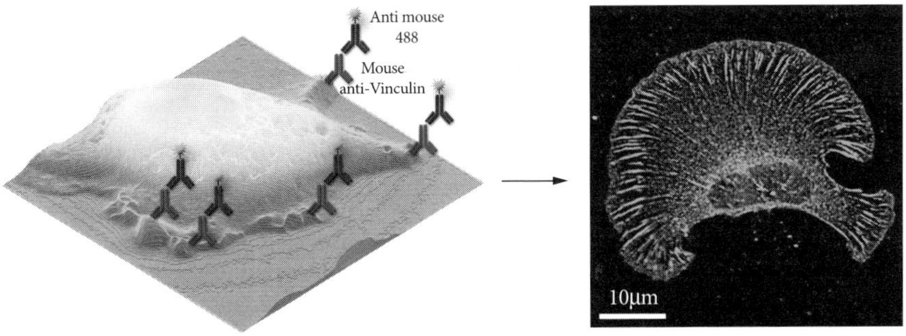

Fixed cell stained with mouse vinculin primary Ab and counter stained with anti-mouse Alexa fluor 488 secondary Ab

Cellular localisation of Vinculin visualized using fluorescence microscopy

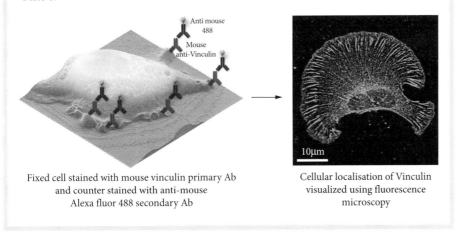

expression level. This can be by upregulation (by exogenous expression of recombinant target protein) or downregulation (e.g. by siRNA interruption in expression).

See Chapter 6

9.6 Chapter summary

- Antibody-based research permits the examination of the role of individual components in complex biological systems. It does this by creating a unique and measurable feature on the target protein (the protein being studied) which permits its measurement.
- The measurement of the target is always indirect (we are actually measuring the antibody partner), and thus the equivalence of antibody signal and target behaviour must always be verified.
- Many simple tests are available to confirm the specificity of an antibody in your experimental system and thereby critically examine the fundamental assumption that antibody signals faithfully report target activity.
- Finally, a variety of experimental platforms which can be deployed to address the role of particular target proteins in biological or medical phenomena are available for antibody-based research.

References

Bock, L.C., Griffin, L.C., Latham, J.A., Vermaas, E.H., and Toole, J.J. (1992) Selection of single-stranded DNA molecules that bind and inhibit human thrombin. *Nature* **355**: 356–66.

Bonifacino, J.S., Dell'Angelica, E.C., Springer, T.A. (2001) Immunoprecipitation. *Current Protoc Mol Biol* **48**: 10.16.1–29.

Brizzard, B. (2008) Epitope tagging. *BioTechniques* **44**: 693–5.

Bromage, E., Carpenter, L., Kaattari, S., and Patterson, M. (2009) Quantification of coral heat shock proteins from individual coral polyps. *Mar Ecol Prog Ser* **376**: 123–32.

Bunker, D.H. and Stockley, P.G. (2006) Aptamers come of age–at last. *Nat Rev Microbiol* **4**: 588–96.

Ciaccio, M.F., Wagner, J.P., Chuu, C-P., Lauffenberger, D.A., and Jones, R.B. (2010) Systems analysis of EGF receptor signaling dynamics with micro-Western arrays. *Nat Methods* **7**: 148–55.

Colyer, J., Mata, A.M., Lee, A.G., and East, J.M. (1989) Effects on ATPase activity of monoclonal antibodies raised agains (Ca^{2+}-Mg2^{+})-ATPase from rabbit skeletal muscle sarcoplasmic reticulum and their correlation with epitope location. *Biochem J* **262**: 437–47.

Cush, R., Cronin, J.M., Stewart, W.J., Maule, C.H., Molloy, J., and Goddard, N.J. (1993) The resonant mirror: a novel optical biosensor for direct sensing of biomolecular interactions. Part I: Principle of operation and associated instrumentation. *Biosens Bioelectron* **8**: 347–53.

Drago, G.A. and Colyer, J. (1994) Discrimination between two sites of phosphorylation on adjacent amino acids by phosphorylation site-specific antibodies to phospholamban. *J Biol Chem* **269**: 25 073–7.

Dutta, P., Tipple, C.A., Lavrik, N.V., et al. (2003) Enantioselective sensors based on antibody-mediated nanomechanics. *Anal Chem* **75**: 2342–8.

Elgert, K.D. (1998) Antibody structure and function. *Immunology: Understanding the Immune System*, Chapter 4. Chichester: Wiley-Blackwell.

Harris, L.J., Skaletsky, E., and McPherson, A. (1998) Crystallographic structure of an intact IgG1 monoclonal antibody. *J Mol Biol* **275**: 861–72.

Hermanson, G.T. (1996) *Bioconjugate Techniques*. San Diego, CA: Academic Press.

Huang, Z., Zhu, W., Szekeres, G., and Xia, H. (2005) Development of new rabbit monoclonal antibody to estrogen receptor: immunohistochemical assessment on formalin-fixed, paraffin-embedded tissue sections. *Appl Immunohistochem Mol. Morphol* **13**: 91–5.

Knox, J.P. (1997) The use of antibodies to study the architecture and developmental regulation of plant cell walls. *Int Rev Cytol* **171**: 79–120.

Lin, Y.L., Chen, L.K., Liao, C.L., et al. (1998) DNA immunization with Japanese encephalitis virus nonstructural protein NS1 elicits protective immunity in mice. *J Virol* **72**: 191–200.

Maynard, J. and Georgiou, G. (2000) Antibody engineering. *Annu Rev Biomed Eng* **2**: 339–76.

Moreau, V., Fleury, C., Piquer, D., et al. (2008) PEPOP: computational design of immunogenic peptides. *BMC Bioinformatics* **9**: 71.

Nygren, P.A. (2008) Alternative binding proteins: affibody binding proteins from a small three-helix bundle scaffold. *FEBS J* **275**: 2668–76.

Ragan, C.I. (1986). In: Ragan, C.I. and Cherry, R.J. (eds). *Techniques for the Analysis of Membrane Proteins*. London: Chapman & Hall, pp. 1–25.

Rubinstein, N.D., Mayrose, I., Martz, E., and Pupko, T. (2009) Epitopia: a web-server for predicting B-cell epitopes. *BMC Bioinformatics* **10**: 287.

Sevecka, M. and MacBeath, G. (2006) State-based discovery: a multidimensional screen for small-molecule modulators of EGF signaling. *Nat Methods* **3**: 825–81.

Su, X., Chew, F.T., and Li, S.F.Y. (2000) Design and application of piezoelectric quartz crystal-based immunoassay. *Anal Sci* **16**: 107–14.

Suzuki, T. and Wang, J.H. (1986) Stimulation of bovine cardiac sarcoplasmic reticulum Ca^{2+} pump and blocking of phospholamban phosphorylation and de-phosphorylation by a phospholamban monoclonal antibody. *J Biol Chem* **261**: 7018–23.

10 Measuring protein–protein interactions: qualitative approaches

Johanna M. Avis and Sanjay Nilapwar

Chapter overview

A protein is often implicated in a biological pathway, or is identified in a cell function assay, but no clear role can be ascertained from its sequence analysis. One starting point for understanding a protein's function is to determine its interactions with other molecules. Once interaction partners are identified, further investigation of the relative **affinity** and **specificity** of a protein–protein interaction can be undertaken to increase our understanding of the role of the interaction in the overall functioning of the protein. This chapter outlines the main broadly qualitative methods used to detect and measure protein–protein interactions.

LEARNING OBJECTIVES

This chapter will enable the reader to:

- summarize a variety of protocols used for qualitative protein–protein interaction analysis;
- describe the information retrievable using the different methods (for example, can a method identify new binding partners, or can it explore factors determining the specificity of a protein interaction?);
- make an informed choice of method selection and consequently plan a work scheme to probe a particular protein interaction;
- interpret the data obtained after application of a method and thus extract information about the protein interaction.

10.1 Introduction

You have an interesting protein perhaps from previously published data, from database information, or from proteomic analysis—and you wish to explore its interactions. The role of the protein is not defined and sequence analysis may, or may not, reveal that it contains one or more protein interaction modules, i.e. a tertiary fold known to mediate such interactions in other proteins, for example the **SH3 WW domain** (Bhattacharyya et al. 2006). There are two main scenarios for the stage of functional characterization you may be at: either you have no knowledge of the molecules your protein can directly interact with, or you may have identified potential interaction partners. Whichever scenario applies to you, the techniques described in this chapter can be optimized to obtain useful information regarding your protein's interactions. Your choice will depend on the questions you are addressing and the level of detail to which a molecular interaction requires probing. The techniques described in this chapter—pull-down methods, phage display, and yeast two-hybrid systems—are largely used for 'qualitative' assessment of protein interaction. Quantitative approaches that seek to put numerical values on protein interaction affinities and potentially probe the intricate molecular recognition mechanism at an interface are described in Chapter 11. As you will discover, however, some methods overlap the qualitative–qualitative boundary.

10.2 Pull-down methods

The **pull-down assay** is a qualitative method for detecting a protein–protein interaction *in vitro* and is similar to **immunoprecipitation**. As with immunoprecipitation, it relies on a high-affinity interaction, but uses a **bait** protein instead of an antibody. It can be used to confirm an interaction between two proteins, or as a screen to identify possible unknown interaction partners, for example within a cell lysate. This section provides a brief overview of the pull-down system and its applications.

See Chapter 9

In a pull-down assay, the protein of interest is known as the 'bait' protein. This bait is tagged and then captured by immobilization on a resin bearing an affinity molecule for that tag. **Affinity tags** are covalently joined to either the N- or C-terminus of the bait protein, often by fusion of the coding sequence in recombinant expression constructs (plasmids), giving rise to **fusion proteins**. Expression of **recombinant proteins** is described in Chapter 7 of this book. The most common fusion tags are glutathione S-transferase (GST) (Kaelin et al. 1991; Harper and Speicher 1997; Zhang et al. 2009), the hexahistidine $(His)_6$-tag (Ahmed et al. 2008; Block et al. 2009; Dennis et al. 2009), and the Strep-tag (Zhou et al. 2004; Sermwittayawong and Tan 2006). Alternatively, a protein can be **biotinylated** at the N-terminus by reaction with available commercial biotin esters (Ducoux et al. 2001). Table 10.1 summarizes these common tags, the relevant affinity resin, and the elution method.

See Chapter 7

Table 10.1 Fusion tags exploited for pull-down assays

Tag	Affinity resin	Elution method
GST	Glutathione Sepharose™	Reduced glutathione
His_6	Immobilized metal ion (Ni^{2+}, Cu^{2+}, Zn^{2+}, Co^{2+})	EDTA or reduced pH (<6.0)
Strep	StrepTactin (engineered streptavidin)	Desthiobiotin (biotin analogue)
Biotinylated	Streptavidin-agarose	Reduced pH (<3.0) or high salt (to elute only the 'prey' protein(s))

Often a specific protease cleavage site between the 'tag' and the protein product can be exploited as an alternative means of releasing a complex from the affinity resin.

Once a tagged protein is bound to the affinity resin, the resin is washed with a buffer to remove non-specifically bound proteins and incubated with a solution containing the potential interacting protein ('prey'), which could be a component of a mixture (such as a cell lysate) or a purified protein, followed by a further wash step to remove unbound protein. If a protein binds to the bait, then this protein is retained on the resin by virtue of the interaction. The bound protein complexes can then be analysed by direct loading onto an SDS-polyacrylamide gel. For visualization of proteins in SDS-polyacrylamide gels, protein staining is usually possible if the quantities you are working with are high enough (see below). Alternatively, if a candidate interacting protein is known, **Western blotting** or radiolabel detection can be used to ascertain its retention on the resin. Figure 10.1 summarizes the experimental scheme. If characterization of the protein–protein complex further downstream is required (e.g. in functional assays), instead of using

See Chapter 9

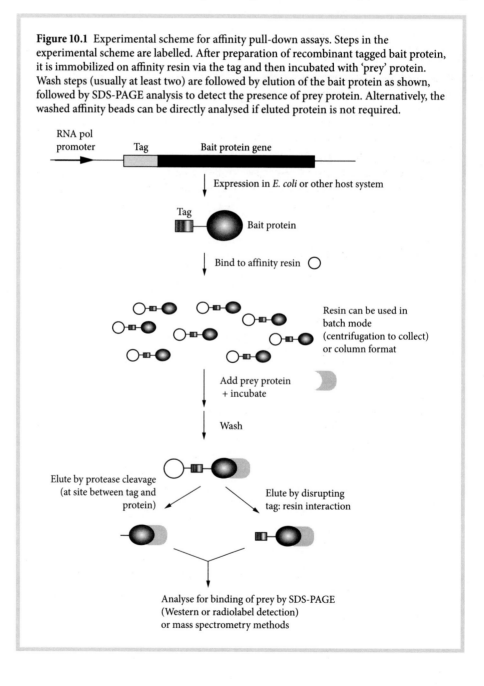

Figure 10.1 Experimental scheme for affinity pull-down assays. Steps in the experimental scheme are labelled. After preparation of recombinant tagged bait protein, it is immobilized on affinity resin via the tag and then incubated with 'prey' protein. Wash steps (usually at least two) are followed by elution of the bait protein as shown, followed by SDS-PAGE analysis to detect the presence of prey protein. Alternatively, the washed affinity beads can be directly analysed if eluted protein is not required.

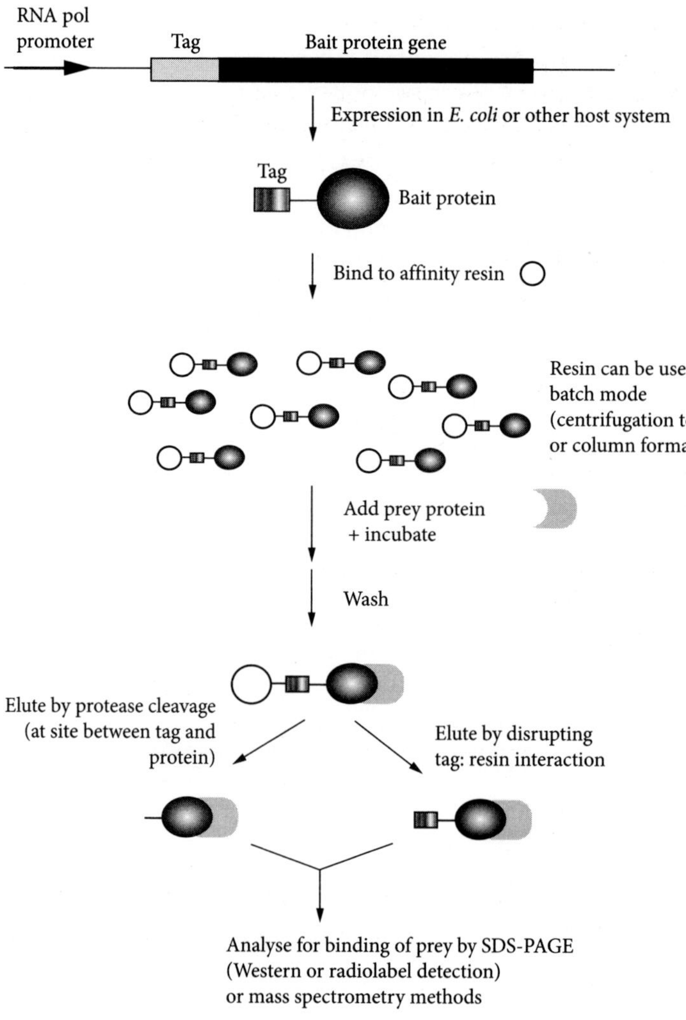

SDS-PAGE sample buffer to elute the complex and analysing on a gel (thereby denaturing the proteins and effectively destroying them), each affinity method offers specific elution methods to displace the tagged bait protein with its prey (Table 10.1).

Pull-downs can be attempted using proteins expressed in mammalian, yeast, or bacterial cells, although in practice the assay usually works best for proteins expressed in *E. coli* as higher quantities are obtainable. Lower protein quantities obtained upon expression in yeast, insect, or mammalian cells, or use of *in vitro* translated proteins, would require sensitive detection methods in the final gel analysis (as mentioned earlier). Case Study 10.1 demonstrates the type of data a pull-down assay typically generates, how the data can be interpreted, and the experimental controls that are necessary.

CASE STUDY 10.1

GST pull-down assay data for subunit interactions in a macromolecular complex

Figure 10.2 shows the results obtained upon SDS-PAGE analysis after screening proteins known to normally reside within a macromolecular complex (RNase MRP) for interactions with other subunits of this complex (Aspinall et al. 2007) using the GST pull-down approach. In this case, each of the other subunits (GST-tagged) are used as bait to pull down an untagged radiolabelled subunit (prey).

All RNase MRP subunits are GST-tagged and immobilized on glutathione Sepharose 4B (GE Healthcare) for use as bait. Figure 10.2(a) shows the data obtained after pull-downs using the bait against the radiolabelled Snm1 prey protein. If the radiolabelled Snm1 interacts with an immobilized GST-fusion protein, then a band should be observed

Figure 10.2 GST pull-down assay data. (a) Data set obtained from pull-downs using radiolabelled Snml subunit as prey with each of the other subunits as bait (GST tagged). (b) Pull-down data set using radiolabelled Popl as prey.

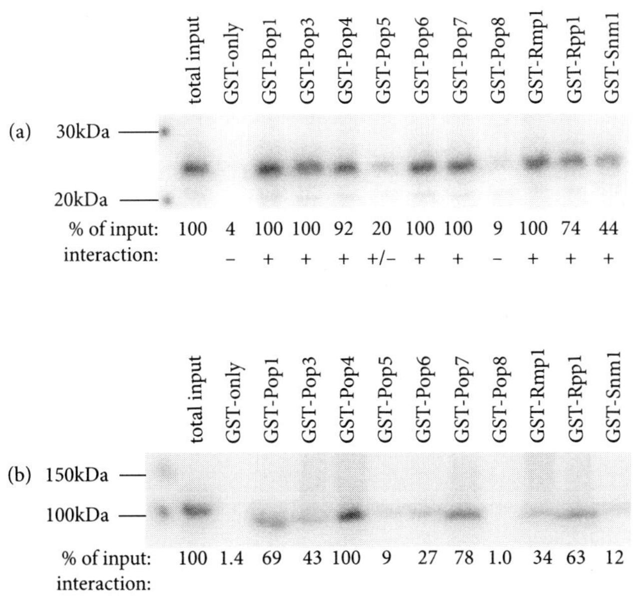

(at the MW for Snm1: ~24kDa). A Typhoon (GE Healthcare) scanner was used to observe and measure the intensity of bands on the gel via phosphorimaging.

QUESTION 1
Which lane(s) represents a control and why?

When considering the control, you need to think about what else the prey protein might be able to interact with. You are looking for an interaction with another RNase MRP subunit (bait). However, Snm1 could be retained on the glutathione Sepharose beads and pulled down due to association with the GST component of the RNase MRP GST-fusion proteins, or non-specific association with the glutathione beads (a false-positive result). The 'GST-only' lane represents a control which shows that no interaction occurs between the Snm1 prey and GST (or the beads) and that Snm1 is removed during the wash steps (Figure 10.1). Thus the bands observed on incubation of Snm1 with RNase MRP subunit GST-fusions are likely to be due to a genuine interaction.

QUESTION 2
The intensity of the bands observed is related to an input amount. Interactions were assigned as strong (+, at least 40% of input protein co-precipitated), weak (+/–, 20–40% of input protein co-precipitated), or none (–, <20% of input protein co-precipitated). What do you think was used as the input sample in this case?

The input lane is loaded with a quantity of radiolabelled Snm1 equivalent to that incubated with glutathione Sepharose beads in the pull-down experiments. Thus it gives a band of an equivalent intensity to that obtained if 100% of Snm1 protein is retained on the beads after incubation and washes. Band intensities for incubation of Snm1 with GST-fusions can be compared with this level to estimate % Snm1 retained.

QUESTION 3
The data in Figure 10.2(b) represent interactions observed for Pop1 (MW ~102kDa), another RNase MRP subunit. Pop1 interacts with other subunits a little less promiscuously than Snm1. Identify the subunits with which Pop1 interacts and use the symbols +, –, and +/– to assign an interaction strength.

Pop1 makes strong (+) interactions with the RNase MRP subunits Pop3, Pop4, Pop7, Rpp1, and itself. Weak (+/–) interactions are made with subunits Pop6 and Rmp1. Pop1 does not interact with the remaining subunits (Pop5, Pop8, Snm1).

10.2.1 Factors to consider in pull-down experiments

When designing a pull-down experiment there are a number of factors you should consider. These include the experimental controls that are necessary, as described in Case Study 10.1, the type of interaction between the bait and prey proteins, and the possible states the proteins could adopt.

The type of interaction between bait and prey proteins could be either **obligate** or **transient**, which in turn will determine the buffer conditions that should be selected when eluting the protein. For example, if the interaction is very stable (expected of an obligate interaction), a complex is likely to be detected in a range of buffer conditions and perhaps at quite low levels of protein. However, proteins that naturally undergo transient interaction, such as signalling molecules, may require more defined salt concentrations, pH, and other buffer components. As the lower-affinity constants are usually associated with transient interactions, you may also need to increase the quantity of protein used if results are negative at first attempt. Protein cross-linking prior to the pull-down may also stabilize transient interactions and enable their detection. Whilst the pull-down method is not quantitative, it can be used to estimate the strength of an interaction by experimenting with different solutions to elute the prey protein from the bait. In changing elution buffer conditions, you should also assess whether anything is known about the likely protein interaction interface. For example, is it **hydrophobic** in nature or dominated by **polar** residues? Increasing the salt concentration would have opposing effects on hydrophobic versus polar interactions.

See Chapter 8

An additional factor to consider in pull-down experiments is the possible different states a protein can adopt, such as its phosphorylation state, or the need for cofactors or an energy source. For some types of protein, if it is possible to trap the protein in one state or another, pull-downs could be a very useful method by which to screen for interaction partners that depend on its state. For example, a GTPase in its GTP- or GDP-bound form could be fused to GST and used in pull-downs to identify candidate GEF (guanine nucleotide exchange factor) or GAP (GTPase activating) proteins (Ortiz et al. 2002; Gu et al. 2006; Kanno et al. 2010).

There are potential drawbacks to the pull-down method, the main one being how to interpret a negative result. The fusion of your bait protein to a tag, particularly a large one such as GST, could influence its folding and structure, or block the binding surface by steric hindrance and hence prevent the prey protein from binding. Furthermore, whilst in principle the scheme (Figure 10.1) makes the experiment look straightforward, tagged proteins can bind too strongly to the affinity resin and may prove difficult to elute. Suppliers of affinity resins (e.g. GE Healthcare glutathione Sepharose 4B beads) offer troubleshooting guidelines, but problems can persist.

False positives also occur in pull-down assays. Protein interaction studies can be particularly prone to a 'stickiness' problem. A protein that can mediate a number of interactions with target proteins is more likely to make non-specific interactions because it has more clusters of surface residues which can make non-covalent interactions. The difference between a non-specific and a specific interaction would probably only become apparent at a downstream characterization phase. Furthermore, nucleic acid can give rise to false positives as it frequently co-purifies with protein; this is a particular problem with basic proteins, especially during affinity purification methods under mild elution conditions. In Case Study 10.1 steps were taken to remove nucleic acid, and other researchers have documented methods to avoid this problem (Nguyen and Goodrich 2006). However, despite the potential difficulties, affinity pull-downs are an effective way of assessing protein–protein interactions.

10.3 Phage display for analysis of protein interactions

Phage display is the longest-established of the **protein display** methods. Protein display refers to a technique that can generate very large numbers (millions) of proteins of varied sequence and present these for screening against a particular binding target (bait). In phage display, a library is constructed in which the proteins or peptides are expressed on the surface of a bacteriophage (via fusion to a phage coat protein) and are amplified through culture growth. The phage-displayed proteins are then tested for binding

Table 10.2 Protein display methods

Display method	No. of protein sequences in one library	Site of protein folding*	Advantages
Phage	$>10^{11}$	Extracellular (protein requires secretion)	Large library size Useful when nothing is known regarding a protein interaction as a larger number of residue positions and side chains can be screened Large library size can also enhance selection of a 'rare' sequence, (e.g. for identification of specific inhibitors/ ligands of exceptionally high affinity)
Yeast	$\leq 10^{6}$	Intracellular (cell is lysed)	Post-translation modifications are possible (Does your protein require them, or not, for interactions? Is this an interesting aspect to explore now or in future?) Amenable to cell-sorting screening methods (e.g. FACS) (high throughput is therefore possible)
Bacteria	$\leq 10^{10}$	Intracellular	Large library size Amenable to cell-sorting screening methods (e.g. FACS)
Ribosome	$>10^{12}$	*In vitro* (cell-free system). Protein must fold during translation but does not require secretion/ extraction from a cell	Cloning/transformation step not required—therefore: • simpler method • large library size (see advantages of large library size given earlier)

*Consideration of whether the polypeptide undergoing 'display' may fold correctly or not when produced through a particular pathway (see Box 10.1).

to a bait protein, with those that bind undergoing further rounds of amplification and selection. Importantly, proteins in the library that possess the desired activity can be readily identified, since the display system provides an easy link back to the encoding nucleic acid, which is sequenced. In addition to phage display, which was first developed in the mid-1980s (Smith 1985; Barbas et al. 1991), display methods now also include yeast, bacteria, and ribosome (Yan and Xu 2006; Gai and Wittrup 2007; Daugherty 2007; Dreier and Plückthun 2011). The key aspects of these display methods are compared in Table 10.2.

Phage display might be the method of choice when your research goals include one or more of the following:

- identification of candidate binding partners for a protein of interest;
- identification of key specificity/affinity determinants between two known binding partners;
- engineering proteins with novel binding properties (including enzyme **inhibitors**).

10.3.1 The stages of a phage display experiment

A typical phage display experiment progresses through three stages:

- a **cDNA library** expressing protein or peptide sequences is constructed;
- the library is tested for binding to your bait protein in rounds of selection (Figure 10.3);
- the selected clones are analysed further (e.g. by sequencing).

Each of these stages is described in turn below.

Figure 10.3 *In vitro* selection using phage display. (a) Simplified structure of an M13 bacteriophage showing the key coat proteins, most of which have been tested for fusion with protein libraries. The pVIII and pIII coat proteins are currently the most frequently used, depending on the size of the protein fusion and the stringency of binding selection required (see text). (b) Selection cycle employed during a phage display experiment. A phage library (in this case with the protein library displayed as fusion to pIII) is screened for binding the target protein (immobilized in polystyrene dishes) by incubation followed by washes (panning). Phages that remain bound are cultured (amplified) and taken through a round of panning again. Generally, two to eight rounds of selection are performed.

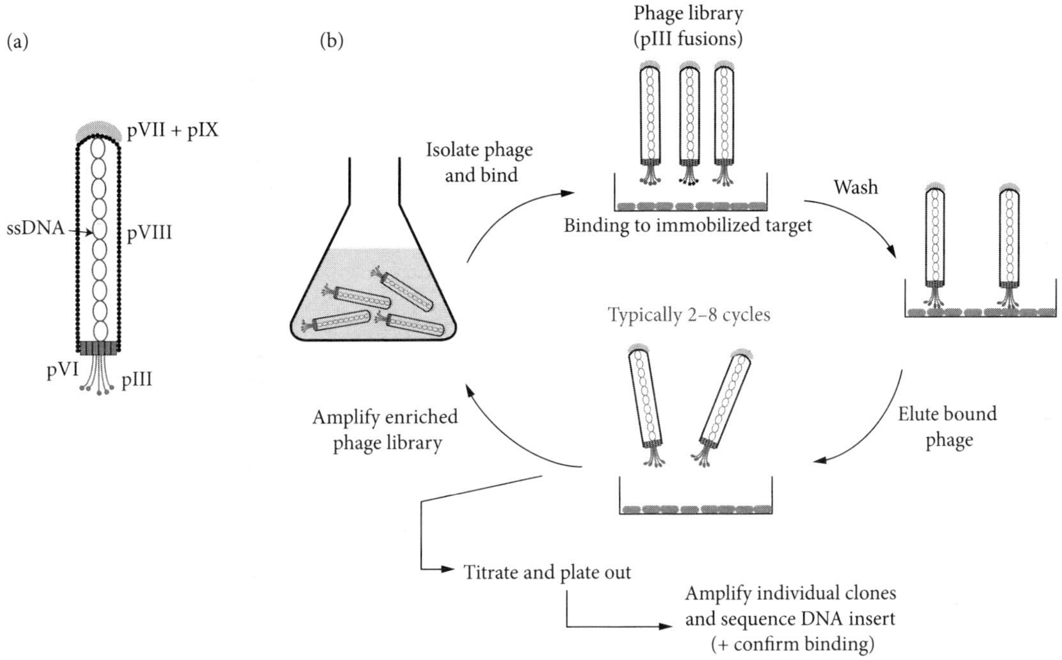

Constructing phage display libraries

The library of protein sequences is expressed on the surface of a **bacteriophage**. Bacteriophages are viruses used to infect *E. coli*. The genomes of the most commonly employed non-lytic filamentous phages (e.g. M13) encode five structural capsid (or coat) proteins, to which the sequence library can be fused, leading to expression of a fusion **coat protein**. Thus the protein library is displayed on the surface of the phage. Although all coat proteins have been tested in display systems, the favoured choices are pVIII and pIII (Figure 10.3(a)). The results obtained with the two choices are quite different: pVIII is the major coat protein and will present multiple copies of the library protein on the phage surface, whilst pIII is generally considered monovalent, expressing only one copy of the protein on the phage surface. Thus a library generated using pIII fusion will generally be a more stringent test for binding to your bait, although it can be too stringent. For example, weaker binding proteins that may be identified when the bait is fused to the pVIII protein may be missed using pIII protein fusion. However, if an interaction is detected using the pIII system, it is more likely to be a specific relatively high-affinity interaction. Furthermore, the size of the fusion may influence your choice; pVIII is a large structured protein essential for phage assembly and is less tolerant of large fusions.

See Chapter 1

Libraries can be cDNA libraries coding for an initial library of proteins or a library comprising variants of a single protein. The former is used when a target protein for your bait is unknown and needs to be identified. The latter is used when the binding partner is known but you wish to characterize, for example, binding epitopes or functional 'hot-spots'. The library could also be a random peptide library (i.e. a short stretch of random amino acid sequence) rather than a whole protein. This would be used when nothing is known about the binding preferences of your protein and/or you are interested in finding a ligand, perhaps as a research tool or a lead to therapeutic inhibitors.

Libraries are generally constructed in small plasmid vectors called 'phagemids' (Bass et al. 1990; Qi et al. 2012), which are capable of replication as plasmids or as phage DNA when *E. coli* is infected with 'helper phage'. As with plasmids, phagemids have an origin of replication, an f1 origin for single-strand replication, and packaging into phage particles and an antibiotic selection marker. Also present are restriction sites which enable cloning of the DNA insert coding for your library in frame with the gene for the chosen coat protein, downstream of a promoter, thereby encoding a fusion protein. Therefore the library may be amplified and stored in double-strand 'plasmid' form, or taken up in single-strand form by helper phages which replicate and exit *E. coli* cells with the library protein displayed on their surface.

If you are generating a library of sequence variants for a known binding partner, a further consideration is the randomization of the binding **epitope**. That is, how many positions within the epitope should you randomize, and should you create a library where any amino acid could occur at those positions, or a reduced selection? If full degeneracy is used at each changed position, cloning technology generally restricts you to randomizing six positions in your epitope which gives a theoretical library size of 1×10^9 DNA clones, representing ~6×10^7 protein variants (Scholle et al. 2005). More positions can be varied if selective degeneracy is used, or if complete amino acid coverage at each position is not required (see section 10.3.2). Whilst construction of the library is a slow technical step, once it is constructed useful data can be obtained in a matter of weeks, assuming no further bottlenecks (see Box 10.1). Table 10.2 offers some additional guidance on whether to opt for phage display or another protein display method, but you will need to investigate the choices yourself.

Selection against a bait and analysis of positive clones

Once the library is constructed, it is screened for binding the target protein. The target (bait) protein is immobilized in polystyrene dishes and incubated with phage library, followed by washes (**panning**) to remove unbound phage. Phage that remains bound is then cultured (amplified) via infection of a host *E. coli* strain and taken through a round of further panning (Figure 10.3(b)). Generally, two to eight rounds of selection are performed in this way, after which positive clones are identified by DNA sequencing. On a high-throughput format, microplates are coated with bait protein and incubated with aliquots of phage library. Non-binding phage is washed out of the microplate wells and bound phage is amplified for another round of panning. As with GST pull-downs (see section 10.2), the stringency of the binding selection can be altered by using different buffers (e.g. altered salt content, pH).

To confirm and identify positive binding interactions, an ELISA step can be included after washing the non-bound phage. The ELISA utilizes an antibody against the phage coat protein, together with a secondary antibody that is conjugated to an enzyme that facilitates a spectroscopic assay (in a **plate reader**) for the presence of phage. After the final round of selection, DNA is extracted from bound phage for sequencing, enabling identification and analysis of the selected binding epitope.

BOX 10.1

Secretion bottlenecks and using scaffolds in a phage display system

A secretion bottleneck – The filamentous phage used by most researchers fuses the protein library to a phage coat protein which is secreted into the **periplasm** via the SecB pathway and then presented on the phage surface. In this secretion pathway, unfolded protein is threaded through the cell membrane which then folds extracellularly. Therefore if a library protein is a rapidly folding intracellular protein, it might not be successfully secreted and displayed. To broaden the use of phage display, alternative secretion pathways have been developed which use either the Tat (twin-arginine translocation) (Speck et al. 2011) or SRP (signal recognition particle) dependent pathways (Steiner et al. 2006). The SRP-dependent co-translational secretion system appears to work well for different types of protein (intracellular and extracellular) and thus could be a good starting point for a new protein.

An alternative to changing the secretion pathway utilized by the filamentous phage coat proteins is to use another type of phage that **lyses** the host cell to release the displayed protein and does not require a through-membrane secretion mechanism at all. A system based on the T7 phage is popular for this purpose (Rosenberg et al. 1996; Herman et al. 2007). Note that library construction using T7 rather than the phagemid system for filamentous phage is technically a little harder.

Scaffolds for generation of novel binding proteins – If generation of novel binding proteins (perhaps as therapeutic inhibitors) is a goal, different **scaffolds** for presenting the binding epitope library can be considered. The most obvious scaffold is the antigen-binding fragment (Fab) of immunoglobulins (Griffiths et al. 1994; Lee et al. 2004; Fellouse et al. 2007), which is already tailored to accept multiple variations in its **complementarity determining regions** (CDRs). This scaffold can be restricted further to the variable regions of the light chain or heavy chain only (Hoet et al. 2005). However, a number of non-antibody-based scaffolds have also been tested successfully. Choice of scaffold may require consideration of the shape of the putative interaction surface of your bait protein, since engineered protein scaffold libraries will have different topologies which impose different restraints on the shape of the binding region (Sidhu and Koide 2007; Nygren and Skerra 2004).

See Chapter 9

10.3.2 Combinatorial mutagenesis and high-throughput sequencing in phage display

As discussed, phage display is a library-based method for finding new binding proteins, either native or engineered. It can also provide information on important amino acid residues required for the binding interaction and define binding interface 'hot spots'. Combinatorial mutagenesis strategies (which change amino acid identity at several sites simultaneously) can be particularly useful in characterizing the protein interaction interface. Such strategies need consideration of library size and number of mutated positions (see section 10.3.1), but phage display offers the advantage of a rapid binding assay prior to protein purification.

A starting point to mapping the protein–protein interaction interface is to use an alanine scanning approach. In this approach amino acids across a chosen sequence window are sequentially replaced with alanine, a substitution which effectively removes any original functional group. The effects of this replacement on binding screens are then observed. This alanine scanning approach may identify 'hot spots' on a protein surface, which can then be explored further through different amino acid substitutions. In phage display, alanine scanning has been modified to 'shotgun alanine scanning' (Weiss et al. 2000; Sidhu et al. 2003), where codon degeneracy is applied such that certain positions in a protein library will encode for either alanine or the wild-type residue. After library screening, the ratio of wild-type to mutated protein in the pool of selected binders is measured. The ratio is taken as a measure of the contribution of the wild-type residue to the **free energy** of the binding interaction. The greater the ratio, the more the wild-type residue contributes to the interaction in terms of favourable free-energy change. The approach can identify a hot spot, but generally works better using fewer rounds of selection (with consequently a greater number of remaining 'binders') than is the case where the goal is to select for the tightest and most selective binders, i.e. when the objective is to survey the binding interface and yield mutant frequency data that can be statistically analysed at a number of residue positions. Fewer selection rounds will provide the necessary larger dataset.

See Chapter 4

A larger dataset, or larger number of selected 'binders', presents its own technical challenges, such as the high volume of DNA sequencing required. Recently, use of phage display to generate combinatorial libraries, combined with new high-throughput sequencing (HTS) methods (Schuster 2008; Metzker 2010) has led to 'deep mutational scanning' of protein interaction interfaces (Araya and Fowler 2011). The new HTS techniques enable even larger libraries of variant proteins than those generated by shotgun scanning to be sequenced both prior to and following enrichment. As hundreds of thousands, or even millions, of variants can be sequenced, analysis of mutational frequency across an interaction interface that can be larger and/or employ full degeneracy at each position (i.e. not just alanine substitution) can be undertaken. The PDZ and WW domains are examples of protein interaction modules to which such 'deep mutational scanning' methods have been applied (Ernst et al. 2010; Fowler et al. 2010). These are exciting new developments in the protein interaction field, but they rely on access to HTS technology and place greater emphasis on the computational analysis of extremely large datasets. Such developments increase the depth to which phage display can probe the molecular recognition process at an interaction interface and take the method more towards one of a quantitative nature. Certainly, the phage display method will have utility in protein interaction studies for years to come.

CASE STUDY 10.2 Characterizing protein–protein interactions in intracellular signalling pathways

The aim of the experiment is to explore the ligand specificity of a protein interaction domain within a newly identified regulator of a signalling pathway. The domain is of a family that occurs frequently in signalling molecules—the SH3 domain. The SH3 domain binds a linear (with respect to primary structure) epitope.

QUESTION

Design a workflow from library design to screening that will identify potential interaction targets for this SH3 domain. In your workflow, think about the choice of fusion system and secretion pathway, and whether to use a 'scaffold' for the epitope.

What is the consensus target sequence for SH3 domains? You will find that it is a short linear peptide sequence. Hence you could consider fusion of the peptide sequence, with certain positions randomized (mutated) to pVIII protein in the first instance (low stringency), followed by construction of pIII fusion libraries for higher stringency selection.

Selection against the bait SH3 domain containing protein and sequencing of 'hits' should identify preferred amino acids in the X (randomized) positions of the consensus SH3 target sequence. Mining of sequence data banks (e.g. UniProt or NCBI Protein database) for the selected sequence would be the next step towards identifying potential interaction partners. You could also consider the shape of the epitope. Would it make more sense to use a system that just expresses linear peptide, or to tether the peptide into a more restricted cyclic conformation (see peptide phage libraries available from New England Biolabs)? It is probably not necessary to consider tethering the peptide into a folded protein scaffold in this case since SH3 binding epitopes are usually well exposed (within looped regions) in protein structures.

Alternatively, you could choose to screen the SH3 domain against a cDNA library of proteins (Malini et al. 2011). This would pull out potential interacting proteins, which could be analysed for possession of a consensus SH3 domain target sequence.

10.4 Yeast two-hybrid assay

The yeast two-hybrid system can evaluate protein–protein interactions within the context of a yeast cell. It can be used to assay for an interaction between two defined proteins or to screen for interacting partners (prey) for one known protein (bait). The assay relies on transcriptional activation of a **reporter gene** only when a protein–protein interaction occurs. A number of different transcription activation systems are utilized for the yeast two-hybrid assay, but the most common are the *GAL4* or *LexA* systems (Chien et al. 1991; Toby and Golemis 2001; Daines et al. 2002). Here, we focus on the *GAL4* system to illustrate the concept underpinning the method.

The Gal4 protein is a transcription activator that has two domains—a DNA binding domain (DBD) that mediates binding to the *GAL4* promotor region of a gene, and an activation domain (AD) that coordinates assembly of transcription complexes, including RNA polymerase II. These two domains must be in physical contact for transcription to occur. In the yeast two-hybrid system, which is available from a number of commercial sources, the two domains of the Gal4 protein are separated by expression of their coding sequences on different plasmids. These plasmids carry restriction sites that enable cloning of genes for the protein(s) of interest (X and Y) as a translational fusion with either the DBD or the AD. Instead of a gene for a single protein, X or Y may be a library. On transformation of the plasmids into yeast, the fusion proteins are expressed. If protein X interacts with protein Y, the two domains of Gal4 can also interact and transcription of a reporter gene, driven by the *GAL4* promotor, is activated. This is summarized in Figure 10.4. Note that the Gal4 DBD can also be used in combination with other compatible activation domains such as VP16, derived from the herpes simplex virus. If a library is used, the experiment can lead to identification of novel partners for a known bait protein. If X and Y are two suspected binding partners, the yeast two-hybrid assay will confirm the interaction. By combining the assay with mutagenesis of one or other partner, the yeast two-hybrid assay can also be used to map the sites of protein interaction in X and Y.

See Chapter 7

See Chapter 3

Figure 10.4 The principle of the yeast two-hybrid experiment. (a) Positive control: expression of the Gal4 protein containing both its DNA binding domain (DBD) and activation domain (AD) induces transcription of a reporter gene by RNA polymerase II, as normal. (b) Plasmids bearing *one* of either the GAL4 DBD or the GAL4 AD coding regions (fused to the proteins of interest, X and Y respectively) will produce only the corresponding part of the Gal4 protein which, by itself, cannot activate reporter gene transcription. However, if both plasmids are transformed into yeast cells, and X and Y interact, DBD and AD are brought into close enough proximity to facilitate recruitment of RNA polymerase II and activate reporter gene transcription.

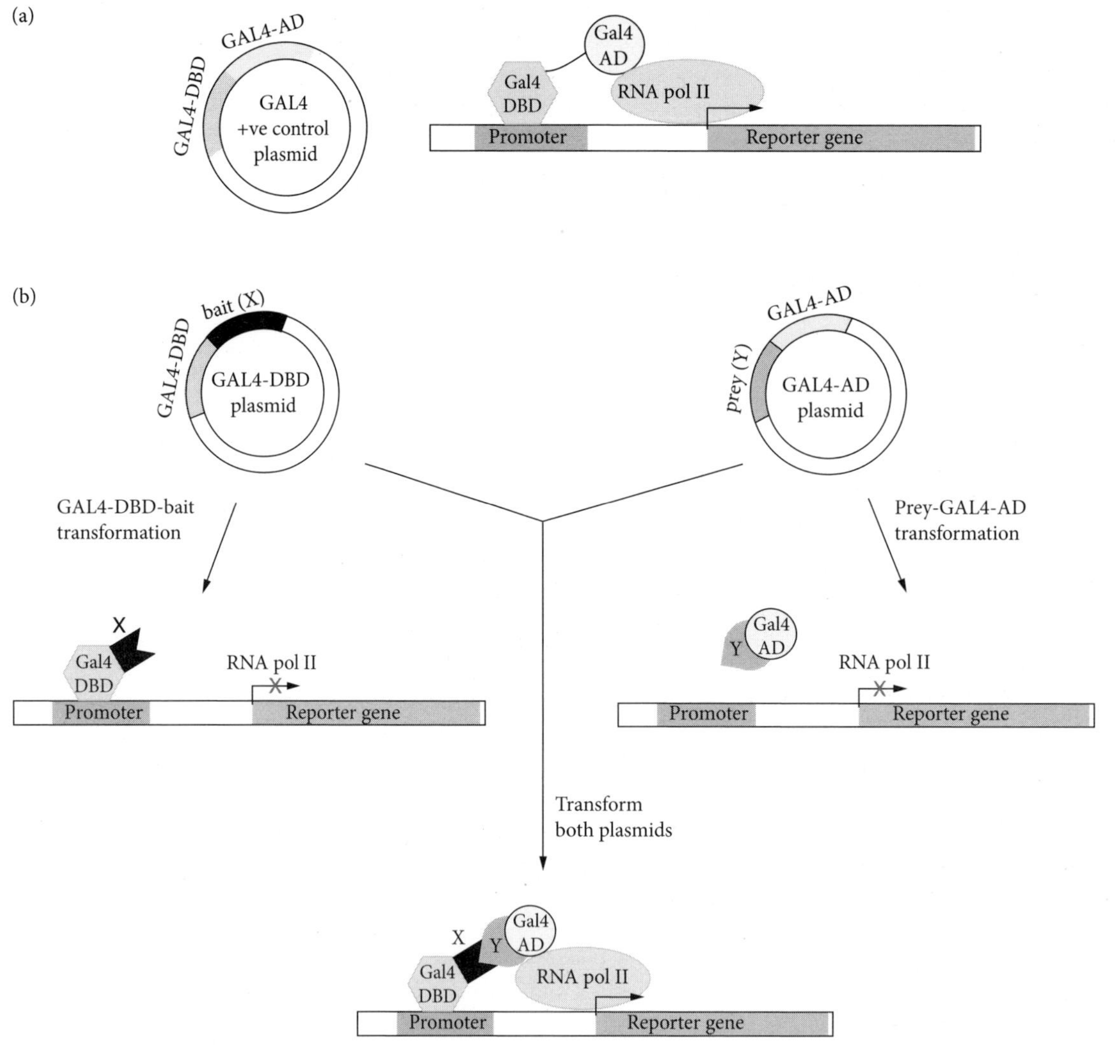

10.4.1 Successful implementation and interpretation of the assay

Although yeast two-hybrid systems have been used extensively to detect protein–protein interactions, a limitation of the systems is the potential generation of false positives. False positives (Bartel et al. 1993; Gietz et al. 1997) can arise, for example, via an ability of the test protein (X or Y) to trans-activate transcription via non-specific interaction either with DNA sequences within the reporter gene promoter or with the AD. This is evidently a problem when you want to test for an interaction between two potential protein partners. Hence controls are needed in which both possible combinations of fusion constructs (i.e. DBD-X–DBD-Y and AD-X–AD-Y) are made and tested for such transactivation of the reporter gene. The potential for non-specific associations is enhanced as the bait and prey proteins are over-represented in the cell because of over-expression. Indeed, over-expression could also cause non-specific interaction with the partner protein that the assay falsely reports as a 'hit'. Such false positives can be a particular problem when the prey is a library of proteins arising from a cDNA library cloned into plasmids. However, in an investigation of a 1:1 protein–protein interaction, further experiments probing its specificity can be undertaken to validate an interaction. Proof of specificity can be obtained by generating variants of your proteins of interest in which residue changes have been introduced, or even predicted interaction domains have been deleted, and performing the yeast-two hybrid analysis on these variants (Litchfield et al. 1996).

The classic reporter gene used in the *GAL4* system is *LacZ* encoding β-galactosidase, which can utilize X-Gal as a substrate to generate a blue colour **colorimetric assay**) (see Case Study 10.3). Variations on the use of *LacZ* exist, such as the fluorimetric luciferase assay or fluorimetric/colorimetric assays specific to a commercial kit. For example, the Clontech Matchmaker™ kit uses *MEL-1*, encoding α-galactosidase (Aho et al. 1997), which also generates a blue colour when the appropriate substrate is supplied in the growth medium. In recent years, yeast strains which incorporate additional reporter genes have been developed specifically for the two-hybrid assay, primarily to reduce the detection of false positives. The first of these was the *HIS3* gene, used in a strain that is normally unable to synthesize histidine. Hence, if you screen for growth of yeast colonies on media lacking histidine, the only yeast to grow will be that in which *HIS3* expression is activated, again via interaction of the DBD and AD fusion constructs. Additional reporter genes, such as *HIS3*, are usually placed downstream of different promoters to that used for *LacZ*, but are modified to contain recognition sequences for the Gal4 DBD. The likelihood of two proteins, X and Y, undergoing a non-specific interaction with each other and/or different promoters to induce expression of two reporter genes is reduced and therefore false positives are reduced. Most modern yeast two-hybrid kits now have up to four reporter genes to improve stringency further. Variations of the assay have been developed to detect protein–DNA interactions (one-hybrid) (Li and Herskowitz 1993; Inouye et al. 1994) and protein–RNA interactions (three-hybrid, or tribrid) (SenGupta et al. 1996). Tribrid systems have also been developed to analyse the effects of post-translational modifications, such as phosphorylation, on protein–protein interactions (Osborne et al. 1995).

Screening for protein–protein interaction by yeast two-hybrid analysis

CASE STUDY 10.3

Figure 10.5 illustrates the data that can be obtained from screening *LacZ* reporter gene activity after conducting a yeast two-hybrid experiment. In this particular experiment, proteins with binding domains for the translation factor eIF4E were investigated. One of these proteins was a human protein called 4E-BP1 (4E-binding protein 1) and the other a protein from yeast called p20. More details are given in the legend to Figure 10.5.

Qualitative screens for β-galactosidase activity are generally performed via a colony filter lift in which, after transformation with plasmid constructs, yeast colonies are transferred

Figure 10.5 Yeast 2-hybrid analysis: interaction of 4E-binding proteins with eIF4Es. Colony lift filter assay for β-galactosidase activity (activity gives blue colonies). 4E binding proteins (yeast p20 (*S. cerevisiae (s.c)*) human 4E-BP1) are fused to the AD and screened with yeast and human DBD–eIF4E constructs for interaction (shown schematically beneath the assay results). See Case Study 10.3 for further description of the experiment. For a colour reproduction of this figure please see Plate 7.

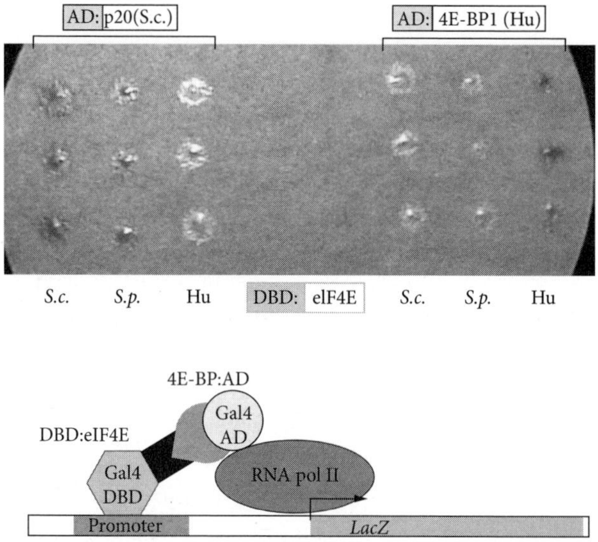

to filter paper simply by resting the paper on the agar plate (but think about how you will mark and orientate the 'lift' and trace back a result from the filter paper to the position of a colony on the original plate). After brief immersion of the filter paper in liquid nitrogen to permeabilize the cells, the paper is incubated with a buffer containing X-gal, a substrate for β-galactosidase that generates a blue product. Notice the development of blue colour in some yeast colonies on the paper in Figure 10.5; these are the colonies in which a positive interaction is occurring due to activation of β-galactosidase.

QUESTION

Some of the results are expected, but others are less so. Inspect Figure 10.5. What conclusions can you draw?

Both *S. cerevisiae* p20 protein and human 4E-BP1 give a positive interaction (blue colonies) with eIF4E proteins. Hence, as expected, the presence of eIF4E binding domains in these proteins is reflective of a real ability to interact with eIF4E.

The additional information the data provide concerns species specificity. It appears that human 4E-BP1 only gives blue colonies (interacts) with human eIF4E protein, and not with the homologue from either *S. cerevisiae* or *S. pombe*. *S. cerevisiae* p20 protein interacts with *S. cerevisiae* eIF4E and gives less intense blue colonies (weaker interaction?) with eIF4E of *S. pombe*. The p20 protein does not interact with human eIF4E. Hence there are obvious species differences between the interaction interfaces of these 4E binding proteins with eIF4E.

More details on research related to these data can be found in Hughes et al (1999) and Ptushkina et al. (1999). The former paper includes a HIS3 screen. Note that better comparison of relative β-galactosidase activity, and therefore of protein–protein interaction affinities, is achievable via more quantitative photometric methods of monitoring product formation (Mockli and Auerbach 2004; Oender et al. 2006).

In summary, the yeast two-hybrid assay has been used extensively to detect protein–protein interactions. It is useful for investigation of 1:1 interactions, particularly in cases where you might find it hard to obtain a protein in recombinant form for GST pull-downs (see section 10.2) or more quantitative assessment of its interactions. The assay is also useful where you have a protein implicated in a cellular function or pathway and may want to identify interacting partners from an expressed cDNA library to explore its function further (Legrain and Selig 2000).

10.5 Chapter summary

- A range of methods, which can be broadly divided into qualitative and quantitative methods, are commonly used to detect and analyse protein–protein interactions. Qualitative methods include pull-down assays, phage display, and yeast two-hybrid systems.
- The qualitative methods have a variety of requirements with respect to production of translational fusions of proteins and the quantity and quality of protein required.
- The methods also differ with respect to whether interactions are detected *in vitro* or *in vivo*, their capacity for generating data, and the time-scale in which this can be achieved. With all methods, care is needed in experimental design to ensure detection of a genuine positive interaction.
- As the network of protein–protein interactions in a cell is critical to understanding whole-cell and organism function, qualitative methods such as those described are essential current tools for probing biological systems.

References

Ahmed, N.T., Gao, C., Lucker, B.F., Cole, D.G., and Mitchell, D.R. (2008) ODA16 aids axonemal outer row dynein assembly through an interaction with the intraflagellar transport machinery. *J Cell Biol* **183**: 313–22.

Aho, S., Arffman, A., Pummi, T., and Uitto, J. (1997) A novel reporter gene MEL1 for the yeast two-hybrid system. *Anal Biochem* **253**: 270–2.

Araya, C.L. and Fowler, D.M. (2011) Deep mutational scanning: assessing protein function on a massive scale. *Trends Biotechnol* **29**: 435–42.

Aspinall, T.V., Gordon, J.M.B., Bennett, H.J., et al. (2007) Interactions between subunits of *Saccharomyces cerevisiae* RNase MRP support a conserved eukaryotic RNase P/MRP architecture. *Nucleic Acids Res.* **35**: 6439–50.

Barbas, C.F., III, Kang, A.S., Lerner, R.A., and Benkovic, S.J. (1991) Assembly of combinatorial antibody libraries on phage surfaces: the gene III site. *Proc Natl Acad Sci USA* **88**: 7978–82.

Bartel, P.L., Chien, C.-T., Sternglanz, R., and Fields, S. (1993) Elimination of false positives that arise in using the two-hybrid system. *BioTechniques* **14**: 920–4.

Bass, S., Greene, R., and Wells, J.A. (1990) Hormone phage: an enrichment method for variant proteins with altered binding-properties. *Proteins* **8**: 309–14.

Bhattacharyya, R.P., Reményi, A., Yeh, B.J., and Lim, W.A. (2006) Domains, motifs, and scaffolds: the role of modular interactions in the evolution and wiring of cell signaling circuits. *Annu Rev Biochem* **75**: 655–80.

Block, H., Maertens, B., Spriestersbach, A., et al. (2009) Immobilized-metal affinity chromatography (IMAC): a review. *Methods Enzymol* **463**: 439–73.

Chien, C.T., Bartel, P.L., Sternglanz, R., and Fields, S. (1991) The two-hybrid system: a method to identify and clone genes for proteins that interact with a protein of interest. *Proc Natl Acad Sci USA* **88**: 9578–2.

Daines, D.A., Granger-Schnarr, M., Dimitrova, M., and Silver, R.P. (2002) Use of LexA-based system to identify protein–protein interactions in vivo. *Methods Enzymol* **358**: 153–61.

Daugherty, P.S. (2007) Protein engineering with bacterial display. *Curr Opin Struct Biol* **17**: 474–80.

Dennis, M.D., Person, M.D., and Browning, K.S. (2009) Phosphorylation of plant translation initiation factors by CK2 enhances the in vitro interaction of multifactor complex components. *J Biol Chem* **284**: 20 615–28.

Dreier, B. and Plückthun, A. (2011) Ribosome display: a technology for selecting and evolving proteins from large libraries. *Methods Mol Biol* **687**: 283–306.

Ducoux, M., Urbach, S., Baldacci, G., et al. (2001) Mediation of proliferating cell nuclear antigen (PCNA)-dependent DNA replication through a conserved $p21^{Cip1}$-like PCNA-binding motif present in the third subunit of human DNA polymerase δ. *J Biol Chem* **276**: 49 258–66.

Ernst, A., Gfeller, D., Kan, Z., et al. (2010) Coevolution of PDZ domain-ligand interactions analyzed by high-throughput phage display and deep sequencing. Mol Biosyst **6**: 1782–90.

Fellouse, F.A., Esaki, K., Birtalan, S., et al. (2007) High-throughput generation of synthetic antibodies from highly functional minimalist phage-displayed libraries. *J Mol Biol* **373**: 924–40.

Fowler, D.M., Araya, C.L., Fleishman, S.J., et al. (2010) High resolution mapping of protein sequence–function relationships. *Nat Methods* **7**: 741–6.

Gai, S.A. and Wittrup, K.D. (2007) Yeast surface display for protein engineering and characterization. *Curr Opin Struct Biol* **17**: 467–73.

Gietz, R.D., Triggs-Raine, B., Robbins, A., Graham, K.C., and Woods, R.A. (1997) Identification of proteins that interact with a protein of interest: applications of the yeast two-hybrid system. *Mol Cell Biochem* **172**: 67–79.

Griffiths, A.D., Williams, S.C., Hartley, O., et al. (1994) Isolation of high affinity human antibodies directly from large synthetic repertoires. *EMBO J* **13**: 3245–60.

Gu, Y., Li, S., Lord, E.M., and Yang, Z. (2006) Members of a novel class of *Arabidopsis* rho guanine nucleotide exchange factors control rho GTPase-dependent polar growth. *Plant Cell* **18**: 366–81.

Harper, S. and Speicher, D.W. (1997) Expression and purification of GST fusion proteins. In: Coligan, J.E. Dunn, B.M. Speicher, D.W., and Wingfield, P.T. (eds), *Current Protocols in Protein Science*. New York: Wiley, pp. 6.6.1–21.

Herman, R.E., Badders, D., Fuller, M., et al. (2007) The Trp cage motif as a scaffold for the display of a randomized peptide library on bacteriophage T7. *J Biol Chem* **282**: 9813–24.

Hoet, R.M., Cohen, E.H., Kent, R.B., et al. (2005) Generation of high affinity human antibodies by combining donor-derived and synthetic complementarity-determining-region diversity. *Nat Biotechnol* **23**: 344–8.

Hughes, J.M., Ptushkina, M., Karim, M.M., Koloteva, N., von der Haar, T., and McCarthy, J.E. (1999) Translational repression by human 4E-BP1 in yeast specifically requires human eIF4E as target. *J Biol Chem* **274**: 3261–4.

Inouye, C., Remondelli, P., Karim, M., and Elledge, S. (1994) Isolation of a cDNA encoding a metal response element binding protein using a novel expression cloning procedure: the one-hybrid system. *DNA Cell Biol* **13**: 731–42.

Kaelin, W.G., Jr, Pallas, D.C., DeCaprio, J.A., Kaye, F.J., and Livingston, D.M. (1991) Identification of cellular proteins that can interact specifically with the T/E1A-binding region of the retinoblastoma gene product. *Cell* **64**: 521–32.

Kanno, E., Ishibashi, K., Kobayashi, H., Matsui, T., Ohbayashi, N., and Fukuda, M. (2010) Comprehensive screening for novel rab-binding proteins by GST pull-down assay using 60 different mammalian Rabs. *Traffic* **11**: 491–507.

Lee, C.V., Sidhu, S.S., and Fuh, G. (2004) Bivalent antibody phage display mimics natural immunoglobulin. *J Immunol Methods* **284**: 119–32.

Legrain, P. and Selig, L. (2000) Genome-wide protein interaction maps using two-hybrid systems. *FEBS Lett* **480**: 32–6.

Li, J.J. and Herskowitz, I. (1993) Isolation of ORC6, a component of the yeast origin of recognition complex by a one-hybrid system. *Science* **262**: 1870–3.

Litchfield, D.W., Slominski, E., Lewenza, S., Narvey, M., Bosc, D.G., and Gietz, R.D. (1996) Analysis of interactions between the subunits of protein kinase CK2. *Biochem Cell Biol* **74**: 541–7.

Malini, E., Maurizio, E., Bembich, S., Sgarra, R., Edomi, P., and Manfioletti, G. (2011) HMGA Interactome: new insights from phage display technology. *Biochemistry* **50**: 3462–8.

Metzker, M.L. (2010) Sequencing technologies—the next generation. *Nat Rev Genet* **11**: 31–46.

Mockli, N. and Auerbach, D. (2004) Quantitative β-galactosidase assay suitable for high-throughput applications in the yeast two-hybrid system. *BioTechniques* **36**: 872–6.

Nguyen, T.N. and Goodrich, J.A. (2006) Protein–protein interaction assays: eliminating false positive interactions. *Nat Methods* **3**: 135–9.

Nygren, P.A. and Skerra, A. (2004) Binding proteins from alternative scaffolds. *J Immunol Methods* **290**: 3–28.

Oender, K., Niedermayr, P., Hintner, H., et al. (2006) Relative quantitation of protein–protein interaction strength within the yeast two-hybrid system via fluorescence beta-galactosidase activity detection in a high-throughput and low-cost manner. *Assay Drug Dev Technol* **4**: 709–19.

Ortiz, D., Medkova, M., Walch-Solimena, C., and Novick, P. (2002) Ypt32 recruits the Sec4p guanine nucleotide exchange factor, Sec2p, to secretory vesicles: evidence for a Rab cascade in yeast. *J Cell Biol* **157**: 1005–15.

Osborne, M.A., Dalton, S., and Kochan, J.P. (1995) The yeast tribrid system: genetic detection of trans-phosphorylated ITAM-SH2-interactions. *Biotechnology* **13**: 1474–8.

Ptushkina, M., von der Haar, T., Karim, M.M., Hughes, J.M., and McCarthy, J.E. (1999) Repressor binding to a dorsal regulatory site traps human eIF4E in a high cap-affinity state. *EMBO J* **18**: 4068–75.

Qi, H., Haiquin, L., Qiu, H.-J., Petrenko, V., and Liu, A. (2012) Phagemid vectors for phage display: properties, characteristics and construction. *J Mol Biol* **417**: 129–43.

Rosenberg, A., Griffin, K., Studier, W., et al. (1996) T7Select® phage display system: a powerful new protein display system based on bacteriophage T7. *Innovations* **6**: 1–6.

Scholle, M.D., Kehoe, J.W., and Kay, B.K. (2005) Efficient construction of a large collection of phage-displayed combinatorial peptide libraries. *Comb Chem High Throughput Screen* **8**: 545–51.

Schuster, S.C. (2008) Next-generation sequencing transforms today's biology. *Nat Methods* **5**: 16–18.

SenGupta, D.J., Zhang, B., Kraemer, B., Pochart, P., Fields, S., and Wickens, M. (1996) A three-hybrid system to detect RNA-protein interactions in vivo. *Proc. Natl Acad Sci USA* **93**: 8496–501.

Sermwittayawong, D. and Tan, S. (2006) SAGA binds TBP via its Spt8 subunit in competition with DNA: implications for TBP recruitment. *EMBO J* **25**: 3791–800.

Sidhu, S.S. and Koide, S. (2007) Phage display for engineering and analyzing protein interaction interfaces. *Curr Opin Struct Biol* **17**: 481–7.

Sidhu, S.S., Fairbrother, W.J., and Deshayes, K. (2003) Exploring protein–protein interactions with phage display. *ChemBioChem* **4**: 14–25.

Smith, G.P. (1985) Filamentous fusion phage: novel expression vectors that display cloned antigens on the virion surface. *Science* **228**: 1315–17.

Speck, J., Arndt, K.M., and Muller, K.M. (2011) Efficient phage display of intracellularly folded proteins mediated by the TAT pathway. *Protein Eng Des Sel* **24**: 473–84.

Steiner, D., Forrer, P., Stumpp, M.T., and Plückthun, A. (2006) Signal sequences directing cotranslational translocation expand the range of proteins amenable to phage display. *Nat Biotechnol* **24**: 823–31.

Toby, G.G. and Golemis, E.A. (2001) Using the yeast interaction trap and other two-hybrid-based approaches to study protein–protein interactions. *Methods* **24**: 201–17.

Weiss, G.A., Watanabe, C.K., Zhong, A., Goddard, A., and Sidhu, S.S. (2000) Rapid mapping of protein functional epitopes by combinatorial alanine scanning. *Proc Natl Acad Sci USA* **97**: 8950–4.

Yan, X. and Xu, Z. (2006) Ribosome-display technology: applications for directed evolution of functional proteins. *Drug Discov Today* **11**: 911–16.

Zhang, F., Fan, Q., Ren, K., and Andreassen, P.R. (2009) PALB2 functionally connects the breast cancer susceptibility proteins BRCA1 and BRCA2. *Mol Cancer Res* **7**: 1110–18.

Zhou, D., Noviello, C., D'Ambrosio, C., Scaloni, A., and D'Adamio, L. (2004) Growth factor receptor-bound protein 2 interaction with the tyrosine-phosphorylated tail of amyloid β precursor protein is mediated by its Src homology 2 domain *J Biol Chem* **279**: 25 374–80.

Measuring protein–protein interactions: quantitative approaches

11

Johanna M. Avis and Sanjay Nilapwar

Chapter overview

Proteins impart function to biological systems by forming 'gateways' into cells, mediating cell–cell communication, transmitting signals into cells, directing gene expression, and performing and regulating metabolism. This list is not complete, but from it we can begin to appreciate the network of protein interactions that comprise a biological system. Whilst qualitative analysis of protein interactions helps us start to locate a protein within an interaction network, a full understanding of what guides one protein to interact with another (**specificity**), how persistent the resulting interaction is, and how it could be regulated requires a deeper and more precise quantitative analysis of the binding events.

See Chapter 10

LEARNING OBJECTIVES

This chapter will enable the reader to:

- summarize a variety of protocols used for quantitative measurement of protein–protein interactions, namely fluorescence titrations, isothermal titration microcalorimetry, and surface plasmon resonance;
- make an informed choice of method selection to probe a particular protein interaction and plan a work scheme that investigates the protein interaction;
- interpret the data obtained after application of a method and thus extract information about the protein interaction—for example, determine binding constants, kinetic constants, and thermodynamic parameters.

11.1 Introduction

Once you have identified the biological pathway or process that your protein is involved in, or at least have some degree of information regarding its potential interactions, you are in a position to start using quantitative tools for investigation of protein–protein interactions. For example, you may know that it possesses one or more known **protein**

interaction modules and have access to genetic interaction data that suggest potential interacting partners that contain binding motifs for these modules. Alternatively, a protein interaction may be well established in the literature, perhaps via yeast two-hybrid analysis, but greater functional insight could be gained by analysing the affinity and specificity of the interaction. The affinity of a protein–protein interaction can indicate whether it is a tight interaction, and therefore the proteins most likely exist as a stable complex in the cell, or a weaker **transient** interaction, which implies regular association–dissociation of the protein partners. The latter would be required for regulatory proteins involved in a cellular signalling pathway, for example. By combining measurement of binding parameters with site-directed mutagenesis of recombinant proteins, an interaction study can probe the molecular recognition determinants at an interaction interface, providing insight into specificity. All methods in this chapter can measure binding constants (i.e. they can *quantitate* an interaction). Some can also provide data on thermodynamic and kinetic parameters. For proteins that could be a potential therapeutic target, the more information we have on their protein interactions, the better equipped we are to effectively block a key interaction and hence prevent their action. This chapter outlines some of the key methods to consider for more quantitative analysis of a protein–protein interaction, namely fluorescence spectroscopy, isothermal titration microcalorimetry, and surface plasmon resonance.

See Chapter 10

See Chapter 3

11.2 Fluorescence spectroscopy

11.2.1 Using intrinsic protein fluorescence to monitor binding

Intrinsic protein fluorescence can be used to monitor a protein–protein or protein–ligand binding event relatively quickly and easily (Eftink 1997; Groemping and Hellmann 2005).

Fluorescence is a particular type of spectroscopic property that involves absorption of electromagenetic radiation to excite a **fluorophore**, which then emits energy as fluorescence at a lower-energy longer wavelength. The **fluorescence** of a compound or molecule is multidimensional, in that it can be described in terms of quantum yield (the ratio of emitted to absorbed quanta), anisotropy, decay time, excitation and emission intensity, and wavelength (λ_{max}). All these features can be measured using fluorescence spectrophotometers which are available in most research laboratories. Many proteins are intrinsically fluorescent because of the presence of the aromatic residue tryptophan (Trp). Tyrosine (Tyr) can also impart weakly fluorescent properties to a protein. An important property of Trp fluorescence is that it is exceptionally sensitive to its micro-environment, particularly the **dielectric constant**. Excitation of a protein at a near-UV wavelength of 280nm (near the excitation maximum of Trp) will lead to **fluorescence emission maxima** (λ_{max}) which can vary from 308 to 350nm (blue to red region of the electromagnetic spectrum), depending on the local Trp environment, its polarity, and/or the presence of any nearby chemical groups that can quench Trp fluorescence. Examples of potential effective quenchers are apolar aromatics, histidine, carboxylate groups, and disulphide bonds. Generally, a Trp buried in an apolar environment will emit at shorter wavelengths, whereas a solvent-exposed Trp will have a higher λ_{max}. However, when considering the use of intrinsic protein fluorescence as a means of monitoring a binding event, it is less important to understand the theoretical basis of fluorescence than to understand its utility and the analysis of data that it provides. In this respect, we should note that the **intrinsic fluorescence** (usually monitored via the emission λ_{max} and/or fluorescence intensity) of a protein can change if a Trp near an interaction surface becomes more exposed to or buried from solvent when a partner protein binds. Even if a Trp is not directly affected by binding of the partner protein, a binding event could induce a conformational change in the protein that affects the environment of a Trp, such that a change in fluorescence properties accompanies binding. Indeed, protein–protein interactions usually follow **induced-fit** mechanisms. It is not necessary to know which Trp residue is responsible for the change in fluorescence reading on binding. Therefore, we do not describe the physiochemical basis for Trp fluorescence any further in this section, but instead address ways in which intrinsic protein fluorescence can facilitate monitoring of interactions and determination of binding constants. A fluorescence study of protein–protein interactions is more feasible if only one interaction partner is fluorescent. However, even if both proteins contain Trp residues, the spectral properties may not overlap and experiments can be designed to focus on one protein or the other.

11.2.2 Monitoring binding events

Figure 11.1 shows possible modes of binding between protein interaction modules and putative peptide sequences. The protein possesses two **WW domains** which show preference for binding a conserved linear PPxY **sequence motif** (where x is any amino acid). However, the exact target specificity is unknown in this example. Different peptides representing PPxY motifs from various putative target proteins can be used in fluorimetry assays to monitor binding to the WW protein. The experimental set-up for this protein–protein interaction investigation can be equated to that of a protein–ligand interaction, since short peptides can be introduced as the binding partners. Different interaction possibilities are demonstrated pictorially in Figure 11.1, which shows the following models, where n is the number of binding sites on the WW protein.

Model A: single-site binding ($n = 1$).

Model B: two identical non-interacting binding sites ($n = 2$).

Model C: two non-identical non-interacting binding sites ($n = 2$).

We will explore how to study this example interaction using fluorimetry and how far we can distinguish between the possible models.

Starting a fluorescence study

When starting a fluorescence study with a protein, the key aspects to check are:

(a) the wavelengths of excitation and emission which give 'maximal' peaks (λ_{max});

(b) the 'resting state' (no binding) fluorescence signal of the protein is proportional to its concentration, which ensures that corrections can be made to the recorded fluorescence signal that allow for the change caused by sample dilution as increasing volumes of the binding partner (ligand) are added during a titration;

(c) whether the binding partner has fluorescent properties (contains a fluorophore).

Figure 11.1 The different binding models for an example protein–peptide interaction. The example is a protein containing two WW domains. Each WW domain has the potential to provide a binding site for a given peptide sequence. The diagrams show three different models for the addition of peptides: (a) Single binding site for WW protein-peptide interaction, n = 1; (b) Two identical binding sites for peptide (on each WW domain), n = 2; (c) Two non-identical peptide binding sites, n = 2. Peptide is binding to each WW domain but clearly at different sites.

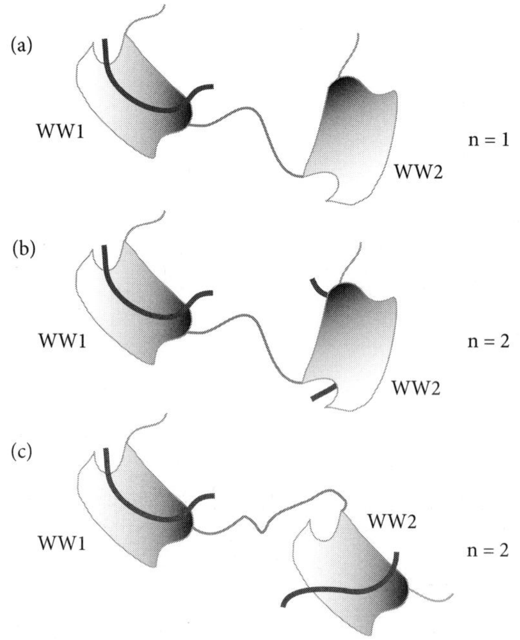

In our example, maximal protein fluorescence intensity is achieved at an excitation wavelength of 280nm. However, the peptide ligands all contain a Tyr, which also excites near 280nm and hence could interfere with the fluorescence signal upon titration. How this potential problem could be avoided is described in Case Study 11.1.

CASE STUDY 11.1

Fluorophores in both interacting partners

The data in Figure 11.2 are fluorescence emission scans of different WW domain proteins and of a number of peptides.

QUESTION 1

Can you identify the curves representative of protein fluorescence due to Trp and those caused by Tyr fluorescence in peptides?

The excitation λ_{max} is near 280nm for both Tyr and Trp, although Tyr peaks around 274nm and Trp can sometimes peak slightly above 280nm. One Tyr is also less intense in fluorescence than one Trp. Trp emission wavelengths are in the range 308–350nm, depending on the chemical environment (see main text). In WW domains, a Trp resides on the binding surface. A solvent-exposed Tyr has an emission peak nearer 303nm. From

Figure 11.2 Fluorescence emission scans for binding partners: WW domains and peptide ligands. Raw fluorescence emission scans, superposed for a range of tested peptides (containing the PPxY motif) and different WW domain-containing proteins. Emission was recorded (300–360nm) using an excitation wavelength of 280 nm. The concentration of WW protein was 10μM, and the concentration of peptides was at least double this. (Given names for peptides and WW domain proteins are not of importance here.) For a colour reproduction of this figure, please see Plate 8.

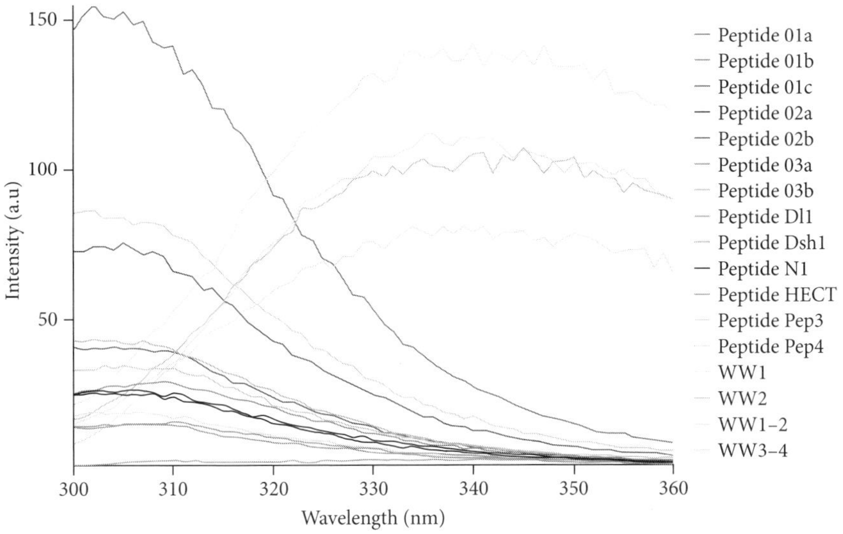

this information you should be able to deduce which scans are associated with the WW protein and which with the peptides. The answer to this is quite self-evident since the traces are labelled, but you can also see that the set of peptides give an emission peak just after 300nm, corresponding to the fluorescence of the Tyr residue they contain. The WW domain protein has a fluorescence emission peak dominated by its Trp residues which emit at 340nm.

QUESTION 2

What problems could the spectral overlap cause for interpretation of spectroscopy readings during peptide titrations?

You want to see a change in fluorescence as you add peptide, with this change increasing as more peptide is added, up to saturation of the binding sites. However, an increase in total recorded fluorescence at a certain wavelength may simply be a consequence of adding another fluorescent species, rather than the result of a specific binding event. Therefore in all fluorescence experiments you need to be constantly aware of other components in the system that might have overlapping fluorescence properties or might absorb some of the exciting light, causing what is known as the 'inner filter effect'.

QUESTION 3

Can you propose an alteration to the spectroscopy experiment that could minimize these problems?

In this example you could consider changing the excitation wavelength. When an excitation wavelength of 290nm is used, the Trp fluorescence signal is diminished but remains significant, whilst that of the single Tyr in the ligand becomes negligible.

Binding experiments

To obtain a quantitative measure of the binding affinity between the WW protein and the peptides tested, each peptide is simply added to the WW protein and the change in fluorescence signal recorded using single-wavelength emission readings (at 340nm). The WW domains are so called because of the presence of two conserved Trp residues, one of which is on the binding surface and makes contact with the interacting partner(s). Hence the local environment of this surface Trp will undoubtedly change if a binding event occurs and a corresponding change in fluorescence will result. Titrations are carried out to record fluorescence changes at different ligand (peptide) concentrations, which enable binding curves of normalized fluorescence against ligand concentration to be plotted. You can then make a decision regarding the likely binding model (Figure 11.1), and subsequently test the fitting of binding equations to your data for each possible model.

Model A: single binding site (n = 1)

In this model it is assumed that only one of the WW domains in the protein binds the peptides tested. The model can be described by the following simple equation:

$$\mathrm{P} + \mathrm{L} \rightleftharpoons \mathrm{PL} \quad (11.1)$$

where P is the WW protein and L is the interacting molecule (which could be another protein, and in this case is a peptide ligand). An equilibrium for this process is described by:

$$K_a = [\mathrm{PL}]/[\mathrm{P}][\mathrm{L}] \quad \text{or} \quad K_d = [\mathrm{P}][\mathrm{L}]/[\mathrm{PL}] \quad (11.2)$$

where K_a and K_d are *association* and *dissociation* constants, respectively.

The relationship between fluorescence intensity (F) and complex formation is given by:

$$F = X_P F_P + \Sigma X_{PL} F_{PL} \quad (11.3)$$

where X is the fraction of a species present. That is, equation (11.3) represents the fluorescence intensity that results from the fraction of free protein and the sum of the fluorescence arising from the fraction of protein bound to ligand, as a complex is formed.

The fraction terms are derived from:

$$P_{Tot} = [\mathrm{P}] + [\mathrm{PL}]$$
$$L_{Tot} = [\mathrm{L}] + [\mathrm{PL}]$$

Given that the fraction of P bound to L (X_{PL}) can be expressed as a function of total concentrations of P (P_{Tot}) and L (L_{Tot}), and K_a, the following single-site binding relationship can be derived:

$$F = (F_P + F_{PL} K_a[\mathrm{L}]/(1 + K_a[\mathrm{L}]) \quad (11.4)$$

where F_P and F_{PL} are the fluorescence intensities of the free and complexed protein, respectively, and [L] is free ligand concentration. For convenience, the fluorescence is normalized by dividing the observed fluorescence upon adding ligand by the reading without the ligand to give:

$$F' = (1 + F_{PL}K_a[L])/(1 + K_a[L]) \quad (11.5)$$

where F′ represents 'normalized' fluorescence. Plotting F' versus [L] enables the parameter K_a to be obtained.

Model B: two identical non-interacting binding sites (n = 2)

In this case, the binding model assumed is:

$$P + L \underset{}{\overset{K_a}{\rightleftharpoons}} PL \underset{}{\overset{K_a}{\rightleftharpoons}} PL_2 \quad (11.6)$$

Since the two possible ligand-bound protein species are identical, with identical association constants, the model is fairly simple. The fitting equation becomes:

$$F' = (1 + \Delta F'_{PL} \times 2K_a[L])/(1 + K_a[L]) \quad (11.7a)$$

or

$$\Delta F = \Delta F_{PL} \times 2K_a[L])/(1 + K_a[L]) \quad (11.7b)$$

Again, the known (measured) variables can be graphically plotted and fitted to either equation to obtain K_a.

Model C: two non-identical non-interacting sites

This model is far more complex and there are different ways of deriving the binding equations that describe it. Essentially, you need to consider that, since the binding sites are non-identical, different complexed species are formed, and this affects the distribution of species and partition functions. Equation (11.8) represents such a model (Eftink 1997):

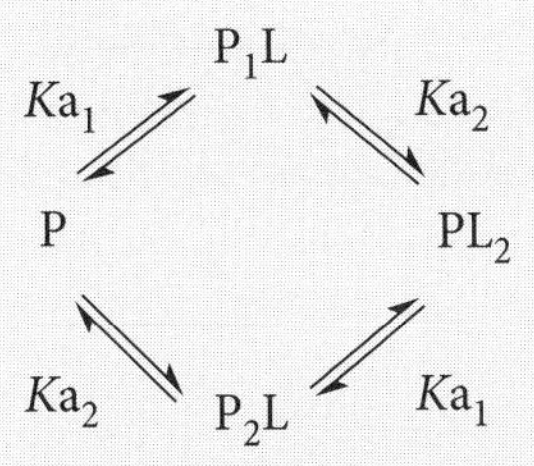

(11.8)

$$F'\varpi = (1 + F_{PL1}K_{a1}[L] + F_{PL2}K_{a1}K_{a2}[L]^2)/(1 + K_{a1}[L] + K_{a1}K_{a2}[L]^2) \quad (11.9)$$

The fit of binding curves to equations for the different models will help establish the number and type of interaction sites in a particular case and provide the binding constants. Figure 11.3 (Case Study 11.2) shows example binding curves obtained for binding peptides to WW-domain-containing proteins, each fitted using the equation for the single-binding-site model (Djiane et al. 2011).

Concentration of binding partner (ligand)

An aspect that is frequently neglected in all the preceding equations is that [L] represents free ligand. In many instances, fluorescence intensities are great enough and binding is strong enough for ligand to be used in excess of the protein, such that [L] can be equated to $[L]_{Tot}$ (the known amount of ligand added in the experiment). However, free-ligand concentration may also require calculation because it cannot be equated to $[L]_{Tot}$. The latter applies to the WW domain example presented here. The exposed Trp has a low fluorescence intensity that increases on binding the peptide. The low 'resting state' fluorescence means that relatively large concentrations of protein have to be used (micromolar). Furthermore in this specific case, the binding affinities for peptide ligands are relatively weak, also in the micromolar range. Hence it is incorrect to equate $[L]_{Tot}$ to [L]. These weak affinities are functionally relevant since WW domains are signalling modules and hence the binding partner can dissociate from the protein more easily. Free-ligand concentrations for use in the binding equations for the single-site model are calculated using quadratic equations (Eftink 1997):

$$[L] = 0.5 \times \{-b + (b^2 - 4[L]_{Tot}/K_a)^{1/2}\} \tag{11.10}$$

where $b = ([P]_{Tot} - [L]_{Tot} + 1/K_a)$, and $[P]_{Tot}$ and $[L]_{Tot}$ are total protein and ligand concentrations respectively.

In summary, as with any quantitative method, mathematical fitting of binding models and selection of the most appropriate model can be difficult. However, once the binding mode is established and potentially proved via additional experimental routes, fluorescence recordings are relatively simple. The sensitivity of Trp fluorescence means that the quantities of protein and binding partner required are often small (micromolar upper concentrations range) and therefore the method is relatively reagent inexpensive. You are less likely to face the problems of protein aggregation that occur at the high concentrations often required for less sensitive quantitative methods, and the protein does not require immobilization or any tags, both of which could potentially interfere with a macromolecular interaction.

CASE STUDY 11.2 Determining the number of binding sites from binding curves

Figure 11.1 shows different model binding scenarios. The protein has two WW domains that are likely to have similar specificity for peptide ligands. The data give a good fit to a single-site binding model (Figure 11.3) when recombinant proteins containing single domains from the protein are tested (WW1 and WW2). Recombinant protein constructs containing two (tandem) WW domains (WW1-2 and WW3-4) also fit well to a single-site binding model.

QUESTION

Does a good fit to the single-binding-site model for the tandem WW domain proteins mean that only one WW domain is binding the peptides tested, i.e. only one WW domain is folded and active? Could there be another explanation and how could you test this experimentally?

In this experiment, there are two potential binding sites for the peptide ligand on the WW1-2 (and WW3-4) protein(s). However, the data (Figure 11.3) imply a single-site binding model (as in model A, Figure 11.1). It could be the case that, because of a different specificity, the given peptide is only binding to one of the WW domains. However, if binding of peptide to the two sites is identical, such that the K_a values are also similar or identical, it is not always possible to distinguish which binding model applies—model A or model B (Figure 11.1). To distinguish, you would need to carry out further experiments. Possible experiments could include truncating or mutating the protein

Figure 11.3 Binding curve obtained in fluorescence experiments. The data shown represent binding of a peptide (containing a PPxY motif) by recombinant WW-domain-containing proteins, possessing either one (WW1 and WW2) or two (WW1-2 and WW3-4) WW domains. Normalized fluorescence intensity is plotted against concentration of peptide, and equation (11.5) (single-binding-site model) is used to fit a curve through the data points and obtain K_a. Note that free-ligand (peptide) concentrations were calculated (equation (11.9)) for the plot.

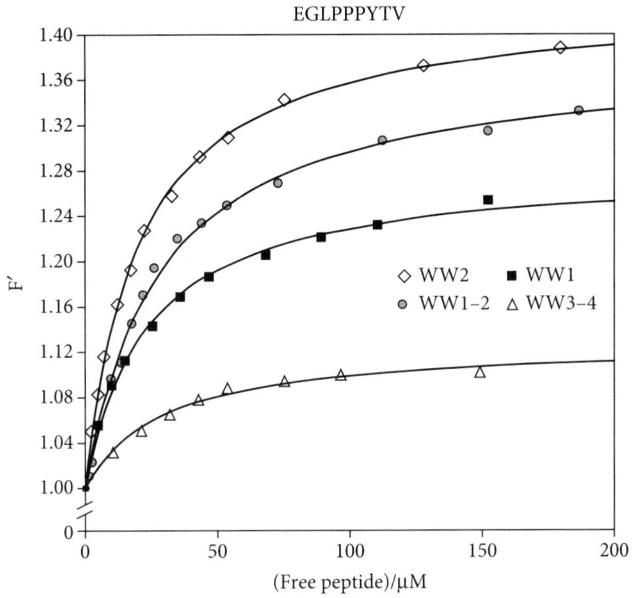

construct and observing consequent effects on peptide binding. An alternative would be to titrate the peptide and record chemical shift changes on one or more WW domain binding surfaces using protein NMR (nuclear magnetic resonance). This case study is based on Jennings et al. (2007) and Djiane et al. (2011).

See Chapter 12

11.2.3 Other fluorescence-based methods

The example of WW domains binding peptide ligands illustrates how changes in the intensity of intrinsic protein fluorescence can be exploited to monitor the formation of complexes. The approach was feasible in this particular example, since small peptides with a domain known to bear an exposed Trp on its binding surface could be used as interacting partners. In the absence of intrinsic fluorescence changes, it is possible to attach small fluorophores to peptides and proteins and use extrinsic fluorescence change to monitor interactions (see Loving et al. (2010) for a good recent review). Further fluorescence techniques (Yan and Marriot 2003) can exploit other fundamental photophysical principles, including Förster resonance energy transfer (FRET) (Jares-Erijman and Jovin 2003) and fluorescence polarization (FP) (Goulko et al. 2008; Rossi and Taylor 2011), to monitor interactions and protein dynamics. The most widely used approach to measuring the association of two relatively large molecules of similar size (such as proteins) is fluorescence polarization, a measure related to the extent to which an excited fluorophore retains the orientation of the exciting light in its emission. This quantity will be small for a rapidly rotating molecule, and large for a less mobile one. Obviously, when two proteins associate and one of the binding partners is fluorescent, the polarization value will increase upon titration of the other partner. The method is highly sensitive (down to low nanomolar range) and can be used to derive both equilibrium and kinetic binding data.

FRET is non-radiative energy transfer from an excited fluorophore to a **chromophore** (which may also be fluorescent) in a lower energy state. The efficiency of the energy transfer depends on the distance between the chromophores, and FRET is now very commonly used to monitor molecular interactions and conformational changes. Of course, it does depend on incorporating suitable acceptor–donor partner chromophores into a protein, or proteins. In the context of quantitating protein–protein interactions, perhaps the most popular method currently is the fusion of potential interacting proteins, X and Y, to yellow fluorescent protein (YFP) and cyan fluorescent protein (CFP), both variants of the well-known green fluorescent protein (GFP). FRET will occur between YFP and CFP when distances between the proteins reduce to less than ~10nm (for an example of use, see Martin et al. (2008)). If fusion of protein partners is undesirable, it may be possible to label them with appropriate acceptor–donor fluorescent dyes (small molecules) (Hieb et al. 2012). As with all fluorescence-based methods, FRET offers the advantage of high sensitivity and hence can be used with small sample quantities, yet can deliver precise binding data.

11.3 Isothermal titration microcalorimetry

Isothermal titration calorimetry (ITC) is a technique in which changes in heat, or **enthalpy**, are measured as a way of monitoring a binding event. Any binding event between chemical species (such as protein molecules, with their array of functional groups) will involve changes in enthalpy (ΔH) due to the changes in non-covalent interactions (hydrogen bonds, salt bridges, London dispersion forces) that occur when two molecules associate in aqueous solution. The amount of heat liberated to or taken up from the surroundings upon molecular interaction is equal to the extent of reaction (or interaction) that has occurred. In recent years, development of sensitive instrumentation to detect tiny heat changes associated with biomolecular interactions has led to the wide use of ITC to investigate protein–protein, protein–DNA/RNA, protein–small molecule, and small molecule–small molecule interactions, as well as enzyme kinetics (Ladbury and Chowdhry 1996; Torres et al. 2010; Ghai et al. 2012). A typical ITC set-up for measuring a protein–protein interaction is described in this section and the types of information that can be obtained from such a technique are highlighted.

11.3.1 Instrumentation and method

A schematic diagram of a typical modern calorimetric instrument is shown in Figure 11.4. This type of **calorimeter** is based on a cell feedback principle that measures the differential heat effects between a reference and a sample cell. A constant power is applied to the reference cell, which activates the power feedback circuit for maintenance of the temperature in the sample cell. The resting power applied to the sample cell is the baseline signal.

When a binding or chemical reaction occurs in the sample cell, heat will be generated or absorbed. The cell feedback system will respond by correcting this heat change (increasing or decreasing the temperature of the sample cell) to keep the temperature differential between the two cells constant. This heat compensation is detected by the instrumentation, and thus heat associated with a reaction can be measured. Highly sensitive modern instruments have the ability to measure heat changes as low as 0.1μcal (0.4μJ) and thus can directly enable determination of binding (association) constants K_a of up to $10^9 M^{-1}$ (nanomolar range dissociation constants: $K_d = 1/K_a$) (Freyer and Lewis 2008). Competitive binding experiments can facilitate measurement of even stronger binding events.

In an actual experiment of ligand (L) binding to macromolecule (P), the heat absorbed or evolved (Q) during a calorimetric titration is proportional to the fraction of bound ligand and can be represented by (Pierce et al. 1999):

$$Q = \frac{V_0 \Delta H_b [M]_t K_a [L]}{(1 + K_a [L])} \tag{11.11}$$

Figure 11.4 Diagram of ITC cells and syringe. A schematic diagram of an isothermal microcalorimeter, based on the system supplied by Microcal (www.microcalorimetry.com), showing the two cells (reference and sample) across which a constant temperature differential is maintained by the system. A syringe rotates in place during the ITC experiment to provide continuous mixing in the ITC sample cell. The plunger is computer-controlled and injects precise volumes of ligand.

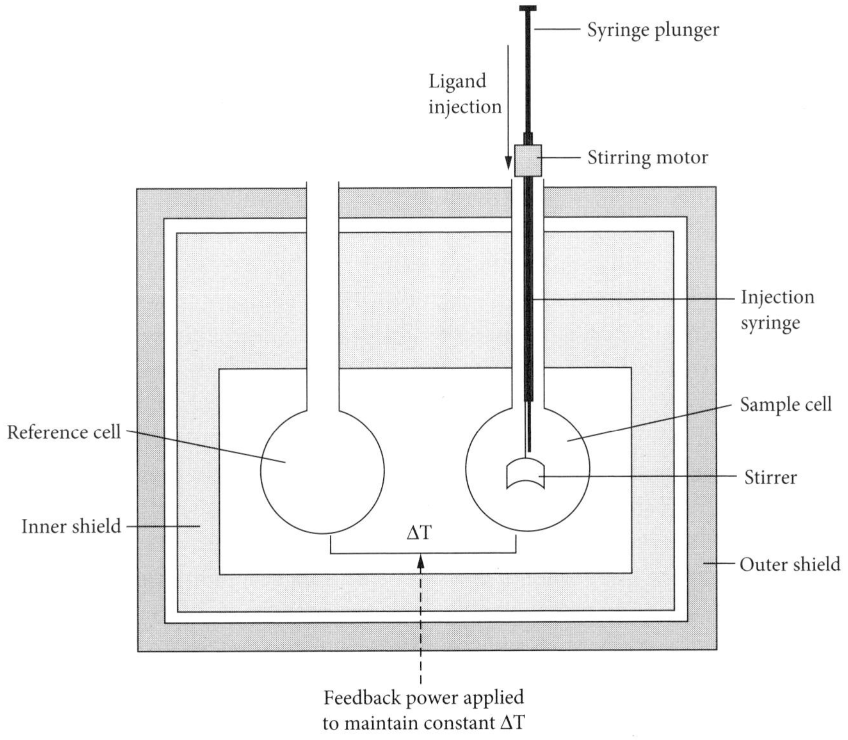

where V_0 is the cell volume, ΔH_a is the binding enthalpy per mole of ligand, $[M]_t$ is the total macromolecule concentration including bound and free fractions, K_a is the association constant, and [L] is the free-ligand concentration. (Note the similarity of this equation to equation (11.5) which relates fluorescence changes to ligand binding to a single site.) As the ligand concentration increases, the macromolecule becomes saturated (no further binding sites are available) and subsequently less heat is evolved or absorbed on further addition of titrant. Consequently, the heat evolved for all cumulative titrations can be described by:

$$Q = \frac{V_0[M]_t \sum n_i \Delta H_i K_{ai}[L]}{(1 + K_a[L])} \tag{11.12}$$

where n is the number of binding sites and i is the injection number.

Fitting the data enables K_a for an interaction to be obtained. The following equations can be used to obtain $T\Delta S$:

$$\Delta G^\circ = -RT \ln K_a \tag{11.13}$$

$$\Delta G^\circ = \Delta H - T\Delta S^\circ \tag{11.14}$$

where ΔG is the free energy of the reaction, ΔS is the **entropy**, and T is the temperature in kelvins.

How is an ITC experiment performed in practice? Typically, the sample cell of an ITC instrument is filled with the macromolecule of interest in the chosen buffer system, leaving the reference cell untouched (usually containing only water). The ligand, which can be another macromolecule or small molecule, is injected into the sample cell using a syringe with a long needle which has a stirring paddle attached to the bottom end and is continuously rotated during an experiment (~30 rotations per minute), leading to complete mixing in the cell within a few seconds after an injection. The mechanical heat of stirring is constant and becomes part of the resting baseline. The ITC instrument can be programmed to perform injections of defined volumes of ligand at constant time intervals. With optimal performance (short equilibration time), a complete binding isotherm can be determined within 1–2 hours (Lewis and Murphy 2005; Ghai et al. 2012).

Calorimetric binding experiments need precision and care since non-covalent binding heats are intrinsically small, typically in the range 21–42kJ·mol^{-1}. Stepwise ligand addition also produces additional heat effects arising from dilution and mixing, which contribute to the heat change values. These can be subtracted by conducting control experiments to measure heats of dilution separately. Some of these effects can also be minimized by preparing the macromolecule and ligand in exactly the same buffer (pH, salt concentration), perhaps by using the **dialysate** of one to dissolve the other, or dialysing both into the same solution (Ghai et al. 2012).

For protein–protein interactions, as well as obtaining the binding affinity, ITC provides the thermodynamic parameters to indicate whether the interaction is enthalpy or entropy driven. Entropy tends to have a major influence where larger protein interfaces are involved, relative to protein–small molecule interactions, but generalizations are dangerous and you are referred to some original literature on using ITC to investigate protein–protein interaction (Cooper 1999; Velazquez-Campoy et al. 2004). Furthermore, each year the *Journal of Molecular Recognition* publishes a review of the literature regarding ITC applications (e.g. Ghai et al. 2012), which is an extremely useful resource for those embarking on an ITC study.

ITC is a quick and usually semi-automated quantitative method for detecting and measuring interaction between molecules in solution, that is not dependent on any fusion protein expression to 'tag' a protein, and does not immobilize one of the binding partners. Indeed, no modification of either binding partner is needed to detect an interaction. It is becoming less reagent-expensive and throughput is becoming higher as instrumentation develops (Hansen et al. 2007; Torres et al. 2010), although the quantities and concentrations required may still be prohibitive in some cases. The method not only provides a K_d for an interaction, but can provide insight into the mechanism of an interaction by providing information on thermodynamic parameters (see Case Study 11.3). If you have binding partners that are soluble in the millimolar to nanomolar range, then you are certainly encouraged to try this method.

CASE STUDY 11.3

Deciphering the thermodynamics of geldanamycin and its analogues binding to N-Hsp90

This case study illustrates the method for an antibiotic–protein interaction. Whilst this is not a protein–protein interaction, the principle of the binding experiment and data analysis is the same. The protein is the N-terminal domain (recombinant) of the heat shock protein Hsp90 (N-Hsp90). The antibiotic ligands used are geldanamycin and its analogues 17-(allylamino)-17-demethoxy-geldanamycin (17-AAG) and 17-*N,N*-dimethylaminoethyl-amino-17-demethoxy-geldanamycin (17-DMAG) (Figure 11.5). These ansamycin antibiotics have been shown to have potent anti-tumour properties via disruption of the cell's protein-folding machinery, particularly for Hsp90.

The isothermal titration calorimeter (MicroCal, GE Healthcare) was set to automatically inject antibiotic from the syringe into the sample chamber housing the N-Hsp90 (both the protein and ligand were at known concentrations in identical buffers). The raw data (Figure 11.6) illustrate the decreasing enthalpy change as binding sites on N-Hsp90 become saturated. Data fitting is automated using the MicroCal Origin™ software that comes with the instrumentation. The derivation and explanation of the equations and data fitting are available on the MicroCal product website, as well as in Freyer and Lewis (2008).

Table 11.1 gives details of the binding constants and associated thermodynamic changes on interaction of the different antibiotics with N-Hsp90. At 25°C, geldanamycin binds to N-Hsp90 with a K_d of 1.05 μM, with ΔH being highly favourable at -34.35 ± 0.53kJ·mol^{-1} and a negligible $T\Delta S$ of -0.26kJ·mol^{-1}. A major source contributing to the negative ΔH values is usually considered to be van der Waals and hydrogen-bonding interactions (Sturtevant 1977).

Binding of 17-AAG and 17-DMAG to N-Hsp90 at 25°C leads to an increase in K_d to 2.48μM and 2.2μM, respectively.

? QUESTION

Analyse the data in Table 11.1 and try to rationalize the differences observed for binding of the ligands (structures given in Figure 11.5) to the protein N-Hsp90. For example, consider the question: Does binding relate to the nature of the differing substituent group (hydrophobicity, steric bulk) and if it does, why might this be? You will need to take thermodynamic parameters into account.

Both analogues bind approximately twofold less tightly to N-Hsp90. For most purposes, this information about an interaction may be sufficient. However, further analyses may

Figure 11.5 Chemical structures of ligands bound to N-Hsp90 in ITC studies. The ligands used are the antibiotics geldanamycin and its analogues 17-(allylamino)-17-demethoxy-geldanamycin (17-AAG) and 17-*N,N*-dimethylaminoethyl-amino-17-demethoxy-geldanamycin (17-DMAG). The molecules differ by the R group attached at position 17, as shown.

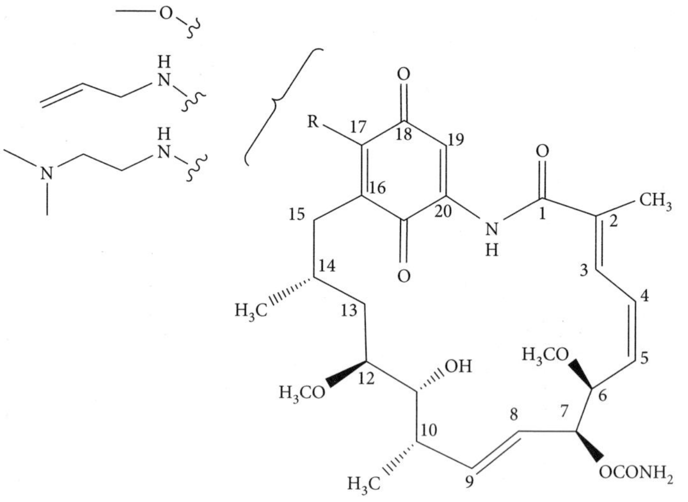

R = OCH_3, Geldanamycin
R = $NHCH_2CH{=}CH_2$, 17-(allylamino)-17-demethoxy-geldanamycin (17-AAG)
R = $NHCH_2CH_2N(CH_3)_2$, 17-*N,N*-dimethylaminoethyl-amino-17-demethoxy-geldanamycin (17-DMAG)

be worthwhile, as we attempted here. First, and interestingly, the enthalpy (ΔH) of the reaction becomes slightly less favourable with the analogues than with geldanamycin, whereas there is an increase in $T\Delta S$. Geldanamycin, which is less hydrophobic than the other two analogues, shows negligible $T\Delta S$, whereas 17-DMAG, which is marginally the most hydrophobic of the antibiotics, binds to N-Hsp90 with the most favourable $T\Delta S$ of 7.55kJ·mol^{-1} at 25°C. Non-covalent interaction has undoubtedly changed slightly with the analogues (less negative ΔH), highlighting the role of the group at the 17 position of the molecule in interactions with N-Hsp90. Why are the $T\Delta S$ values different?

A further observation you might make is that K_d increases when the group at position 17 contains N instead of O and is of increased steric bulk: $K_{CH3\text{-}O\text{-}}$ (geldanamycin) < $K_{(CH3)2\text{-}N\text{-}CH2\text{-}CH2\text{-}NH\text{-}}$ (17–DMAG) ≈ $K_{CH2=CH\text{-}CH2\text{-}NH\text{-}}$ (17-AAG). Perhaps increasing the steric bulk of the substituents leads to progressive distortion of the N-Hsp90 binding pocket and its surroundings, which is reflected in the increasing $T\Delta S$ of these ligands and is not entirely compensated by the ΔH contribution, ultimately affecting ΔG (Table 11.1). However, this explanation is not evident on close examination of the X-ray crystallographic structure of geldanamycin and 17-DMAG bound to N-Hsp90 (Stebbins et al. 1997; Jez et al. 2003), where the backbone of the N-Hsp90 remains relatively unaffected. Perhaps the entropy contribution is linked solely to the 'removal' of the more hydrophobic analogues from solvent exposure? Other studies have similarly shown large differences in thermodynamic signature upon binding of ligands that vary only relatively conservatively (with small modifications), where X-ray structures could not reveal a structural basis for varying ligand selectivity (Renatus et al. 1998; Bohm et al. 1999). Thus ITC can provide additional insight into molecular recognition mechanisms.

Figure 11.6 Raw microcalorimetry data and fitting. (a) Raw data for the titration of 17-DMAG into N-Hsp90 solution at 25°C in Tris-HCl buffer (pH 8.00). Note how the recorded heat accommodation required after each injection of ligand (represented by the 'spike' in μcal/sec) decreases as binding sites are filled (reaching saturation). (b) Binding isotherm derived from the raw data, corrected for the heats of dilution. The line represents the least-squares fit to the single-site binding model obtained using the Microcal Origin™ software supplied with the instrumentation (see equations (11.10) and (11.11)). ΔH and K_a are readily obtained from the fit.

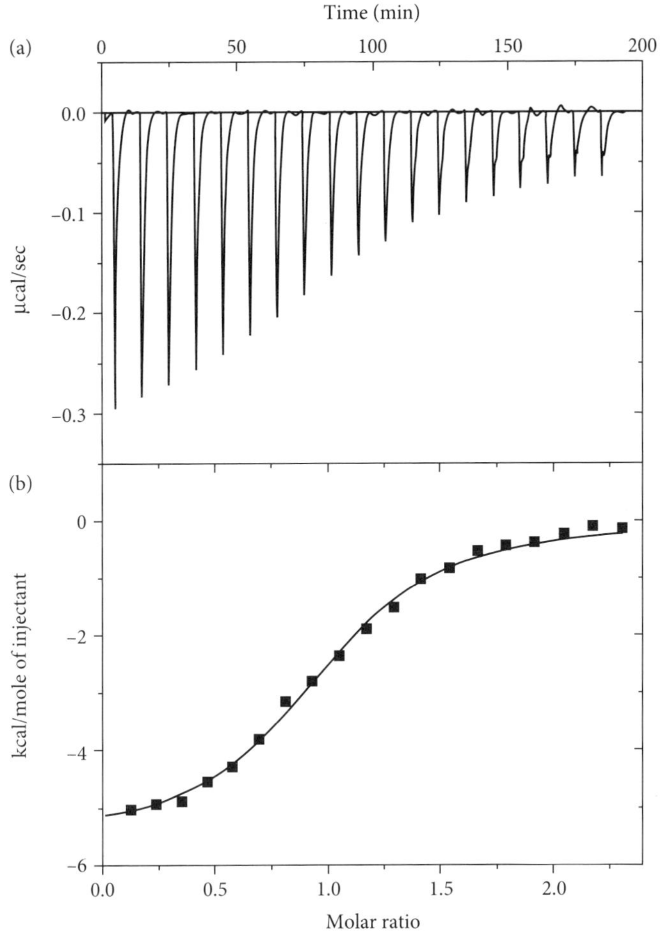

Table 11.1 Summary of the thermodynamic parameters for the binding interaction of N-Hsp90 with geldanamycin analogues

Ligand	$K_a \times 10^4$ (M^{-1})	K_d (μM)	ΔH_{obs} (kJ·mol^{-1})	$T\Delta S_{obs}$ (kJ·mol^{-1})	ΔG_{obs} (kJ·mol^{-1})
Geldanamycin	95.17 ± 7.44	1.05	-34.35 ± 0.53	-0.26	-34.09
17-AAG	40.32 ± 3.45	2.48	-27.97 ± 0.42	4.02	-31.99
17-DMAG	43.73 ± 3.04	2.2	-24.60 ± 0.37	7.55	-2.15

Parameters are for experiments performed in 20mM Tris buffer (pH 8.00). Error values are standard deviation from the mean for triplicate experiments. Stoichiometry (n) values (not listed) are close to 1.0 for all ligands.

11.4 Surface plasmon resonance

Surface plasmon resonance (SPR) is a non-invasive optical measuring technique which measures the biomolecular mass concentration in close contact or proximity with specially prepared metal surfaces. As with microcalorimetry, SPR does not require any labelling of the interacting molecules. In an SPR instrument, one molecule (the ligand) is immobilized over a sensor surface and its corresponding binding partner (the analyte) is injected in buffer or aqueous solution under continuous flow through a **microfluidic** delivery system. As the mobile analyte binds to the surface-immobilized ligand, it results in accumulation of analyte over the binding surface and a subsequent increase in the **refractive index**. This change in refractive index is measured in real time. Real-time SPR can measure surface protein concentrations from as little as few picograms per square millimetre and can monitor changes in surface concentrations on a time scale down to 0.1s (Millot et al. 1999). SPR technology allows determination of binding specificity, kinetics, and concentration using analytes ranging in size from about 150 to 10^6g/mole (daltons) or more (Rich et al. 2009). The technique does not require absolutely pure preparations of protein and can be used to measure binding constants between known binding partners, or to establish the presence of a binding partner in a crude mixture such as a cell lysate.

11.4.1 Instrumentation and measurements

SPR instrumentation can be used to investigate interactions involving small molecules, nucleic acids, proteins, protein conjugates, lipid micelles, and even larger particles such as viruses and whole cells. Several real-time SPR instruments are commercially available, including those manufactured by Biacore (GE Healthcare), Biorad Inc., Reichert Technologies, and Bionavis Limited. These SPR devices range from simple hands-on systems to much more expensive high-throughput multichannel devices with integrated robotics. The main components of SPR-based analytical systems are:

- an optical system for detection of the SPR signal;
- a liquid-handling system;
- a data collection and analysis system with appropriate software for data capture and analysis;
- the sensor chip with a specific surface where the protein–protein interaction takes place.

A typical optical system used in a Biacore SPR instrument is shown in Figure 11.7.

The SPR analytical system uses high-efficiency near-infrared light-emitting **diodes**. A wedge-shaped beam of light is focused on to a sensor chip, giving a fixed range of incident angles. The reflected light or SPR response is monitored continuously by a fixed array of light-sensitive diodes covering the whole wedge of reflected light. Sophisticated computer algorithms are then employed to automatically interpolate and calculate the angle at which minimum light reflection occurs (the SPR angle or resonance angle) to a high degree of accuracy. By using a combination of a wedge of incident light and a fixed array of detectors, the response angle is monitored in real time without physically

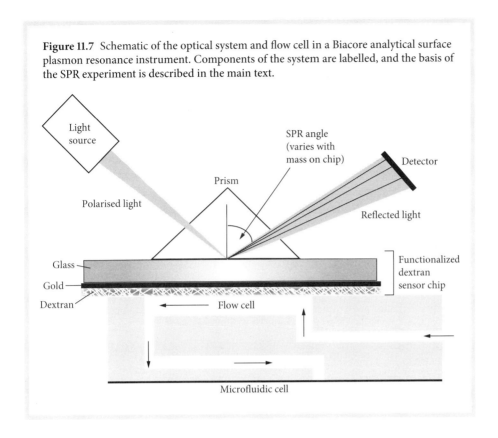

Figure 11.7 Schematic of the optical system and flow cell in a Biacore analytical surface plasmon resonance instrument. Components of the system are labelled, and the basis of the SPR experiment is described in the main text.

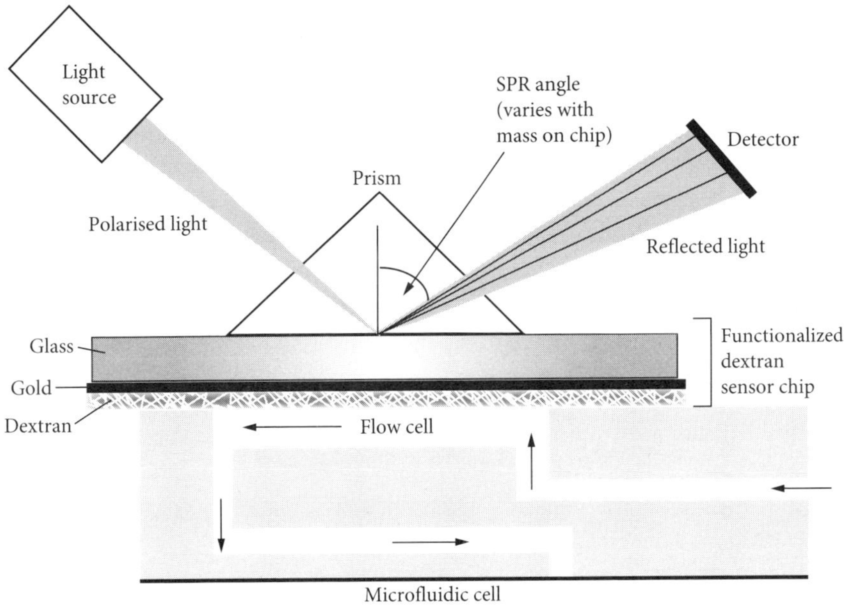

moving any light source, sensor chip, or detector. The data generated from the SPR instrument is the response difference (response units (RU)) versus time. The time resolution data are then decoded and analysed with specialized mathematical algorithms embedded in the instrument handling software.

How does the resonance angle, and therefore RU, enable protein interactions to be monitored? Put simply, the resonance angle is related to the refractive index of the chip surface, which in turn is related to the mass of biomolecule(s) present on the chip. After immobilization of one binding partner ('ligand') on a chip, a baseline RU measurement is made. The resonance angle of the sample will change if the biomolecular mass on the chip changes. Hence, if binding occurs on injecting analyte (the binding partner, or a protein mixture containing potential binding partners), it will result in a change in refractive index and a change in the resonance angle with time (as more molecules bind). Hence RU can be plotted against time to obtain the association rate; this is a sensorgram, which is recorded in real time by the SPR instrumentation (see Figure 11.8 for a theoretical example).

One response unit represents an approximate change in surface concentration of about 1pg/mm^2 for most biomolecules (Albrecht et al. 2002). After binding, i.e. once association has reached the steady-state condition, the flow of analyte is stopped and a constant buffer flow is maintained to establish whether complex dissociation can then be detected, again via a change in RU (due to loss of mass at sensor surface).

The liquid-handling system in SPR instruments comprises an integrated microfluidic cartridge (IFC) and computer-controlled precision pumps for delivery of samples to the sensor chip with varying instrument-dependent levels of automation. Samples are generally guided over the sensor surface with the IFC and transported using syringes

Figure 11.8 A typical SPR sensorgram. After ligand has been immobilized using the relevant chemistry, the baseline RU is taken. If injection of analyte results in binding to ligand on the sensor chip surface, the mass transfer to the sensor chip causes an increase in RU. This increase can be monitored in real time and is represented by the 'association' phase on the schematic diagram (note that the clarity of this phase and whether or not kinetic analysis proves straightforward will depend on the particular ligand–analyte system). The steady-state RU value represents the amount of analyte bound at equilibrium, and this value is used to determine K_d using a 1:1 binding model (see main text). Further buffer flow may enable observation of a dissociation phase (and associated kinetics), as shown.

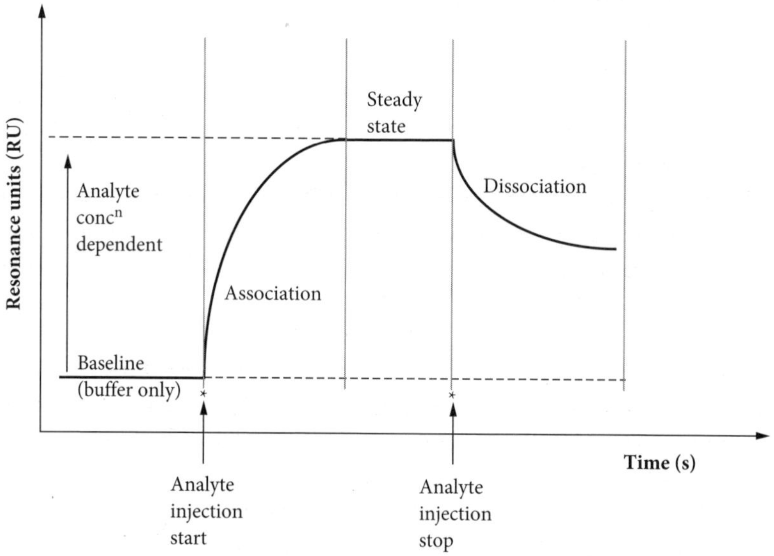

or peristaltic pumps, either with or without pneumatic valves and sample loops. In many of the newer and more sophisticated instruments, inline degassers are included to remove dissolved gases, which might have artefactual effects on the recorded data, from the samples/buffers. Analyte samples are injected through selected individual **flow cells** (of which there are usually four—described below) on the sensor chip. The IFC system ensures that accurate and reproducible analyte is delivered over the sensor surface, which advantageously also results in low sample consumption.

The sensor chip and methods of ligand immobilization

The sensor chip is the specific surface where the actual binding event takes place. It is also the main signal transducer in the real-time SPR instrument. The sensor chip is effectively a tiny glass slide coated with gold on one face. This gold face is further functionalized to facilitate covalent (direct) or non-covalent (indirect) immobilization of a ligand (usually protein).

The protein is usually directly immobilized via linkage to a layer of carboxymethyl dextran (CM) on the sensor chip (Clow et al. 2008). The CM-coated surface also provides a hydrophilic environment for the interaction and increases the binding capacity of the surface. Other available sensor chips, such as SA (streptavidin) and NTA (nitrilo-triacetic acid), enable indirect capture of the protein via an affinity interaction with biotinylated or His-tagged proteins, respectively. The use of NTA chips is a popular choice because of the frequency with which His-tags are used on recombinant proteins and the ease with which NTA sensor chips can be completely regenerated and used several times, even with different His_6-tagged proteins (described below). A disadvantage of these chips is leaching of ligand from the NTA surface, although newer Tri-NTA chips from Biorad (Biorad Inc. 2012) have been used to overcome this. The usual method of capturing GST-tagged proteins is to derivatize a CM surface with anti-GST antibody first, which enables subsequent indirect capture of the tagged protein via the antibody–epitope interaction. The same antibody-based approach could be used with other fusion tags. Sensor chips which can incorporate proteins into lipid mono- or bilayers are now available.

Although usually straightforward, successful protein (ligand) immobilization requires some planning. You will need to appreciate the chemistry involved in the immobilization and observe buffer characteristics that could interfere, such as use of Tris-(amine) buffers where amine coupling to CM chips is being performed, or the use of DTT in metal affinity approaches. Furthermore, does the chemistry require extremes of pH that might affect the structure or solubility of your protein? Glycerol- or sucrose-containing buffers should normally be avoided, as they can increase the viscosity of the running buffer and result in higher baseline noise.

Although pure protein is not needed for SPR, the microfluidic (IFC) system present in the Biacore instrument, and most others, makes the SPR instrument susceptible to blockages by insoluble proteins, protein aggregates, and impurities. A sample preparation step of either filtration (0.2μm) or centrifugation (14 000g for 4–8min) and degassing of buffers is strongly advised to ensure even flow in the Biacore systems. Each Biacore sensor chip has four separate flow cells, three of which can be used for varying the concentration of interacting analyte or immobilizing three different ligands. The fourth flow cell can be left blank or used for immobilizing dummy ligand (i.e. a non-interacting ligand) as a negative control flow cell.

Regeneration of an SPR sensor chip for repeated experiments

Sensor chips are a fairly costly consumable, and ligand immobilization can be the slowest step of the experimental process. Thus it is sensible to try to regenerate the chip, removing the interacting partner (analyte) but retaining the immobilized ligand, to enable a series of experiments to be conducted. Alternatively, the ligand could also be stripped away from the chip, if possible, and the chip used again with either the same or a different ligand. Of course, conditions used to remove bound analyte must not affect the architecture and stability of the immobilized ligand. Hence, regenerating conditions, which can completely dissociate the binding complex yet maintain the integrity of the chip matrix and its attached ligand, are usually mild. As the physical forces responsible for a binding event are often unknown, regeneration conditions that disrupt binding generally need to be evaluated empirically. A common starting point (which often works) is 10mM glycine pH 2.5, and a drop in the pH of the glycine buffers (pH 2.0, pH 1.5) can often disrupt tighter interactions. The bound protein becomes partly unfolded and positively charged in these buffers, and can easily be removed. The ligand, whilst also likely to be affected, will usually regain its native charge/architecture on resumption of neutral buffer conditions. Other regeneration conditions include high salt concentration, high (basic) pH, surfactants, and ionic solutions. A chelating agent such as EDTA is effective at regenerating NTA sensor chips. EDTA will chelate the coordinating metal, thus also dislodging the immobilized ligand. The NTA chips can be reloaded with metal ions for indirect immobilization of the same or different ligands for repeated experimentation.

Data analysis: do they interact?

The analysis performed will depend on the information required from the experiment. If you are attempting to establish whether two molecules interact, or if there is a binding partner for your ligand in a cell lysate, all you require is an observed change in RU. However, remember that you need controls to exclude false positives (no ligand, no analyte, inclusion of an inhibitor). Conversely, if no binding is detected, a positive control (e.g. using the same ligand with a known analyte and vice versa) is recommended, if such a control is available.

If you require quantitative data, as in Case Study 11.4, the easiest method is to derive an affinity constant from a binding model. The binding model (equation (11.15)), which is referred to in the case study, applies to most SPR experiments for measuring protein–protein interactions, which generally involve a 1:1 stoichiometric ratio:

$$\text{Bound} = \frac{[\text{A}]\,\text{Max}}{[\text{A}] + K_\text{d}} \qquad (11.15)$$

where 'bound' is the RU measured, Max is the maximum response (RU), and [A] is the concentration of the injected analyte. The values of parameters Max and K_d are obtained upon fitting of the data to equation (11.15) using non-linear regression software. A kinetic analysis, requiring accurate k_{on} and k_{off} values is also possible, although it presents more of a challenge and success depends on the nature of the interaction and how tightly the molecules associate. There are plenty of resources in the literature and online to help with this, if it is your objective (Lipschultz et al. 1997; Schuck 1997; McDonnell 2001; Jason-Moller et al. 2006; Venkatraman Girija et al. 2010). In particular, Biacore maintains a useful web site (http://www.biacore.com/) which includes an extensive (and regularly updated) list of publications that have utilized SPR.

SPR is a powerful tool for the analysis of protein–protein interactions. It is particularly useful where small amounts of ligand/analyte are available (which would make ITC non-viable) and for measurement of low-affinity interactions. It is also a quick method for establishing structural integrity of a recombinant protein and for a quick 'yes' or 'no' result with respect to ascertaining whether two molecules can interact. Measurement of binding affinities is possible relatively easily, and the method can yield further information (kinetics, thermodynamics, activation energy) if care and effort are taken with a suitable system. The method is not suitable where analytes are small (<800Da) (ITC would be the better method in such a case), and cannot really be regarded as high-throughput, in which case yeast two-hybrid or phage display screens would be more appropriate.

CASE STUDY 11.4

Analysis of the interaction between the GST-PcrV fusion protein antigen and an antibody fragment (Fab)

Figure 11.9 illustrates a sensorgram for the binding response of a GST-PcrV fusion protein to immobilized antibody fragment (Fab) using a Biacore instrument (A100).

Figure 11.9 Binding sensorgrams for a Fab fragment–antigen interaction. Changes in RU recorded over time (sensorgram) upon flow of HBS-EP buffer (Biacore) containing analyte (GST-PcrV fusion protein, 30μL/min) over a sensor surface with immobilized fragment antibody (Fab). The different association and dissociation curves are for experiments using increasing concentrations (6.25, 12.5, 25, 50, and 200nM) of GST-PcrV (utilizing different flow cells on the chip). Buffer flow for the binding phase was continued for 2min and dissociation for 3min. The chip surface was regenerated with 1M guanidine–HCl. Reprinted from Rich, R.L., Papalia, G.A., Flynn, P.J., et al (2009) A global benchmark study using affinity-based biosensors. *Anal Biochem* **386**: 194–216. Copyright (2009), with permission from Elsevier.

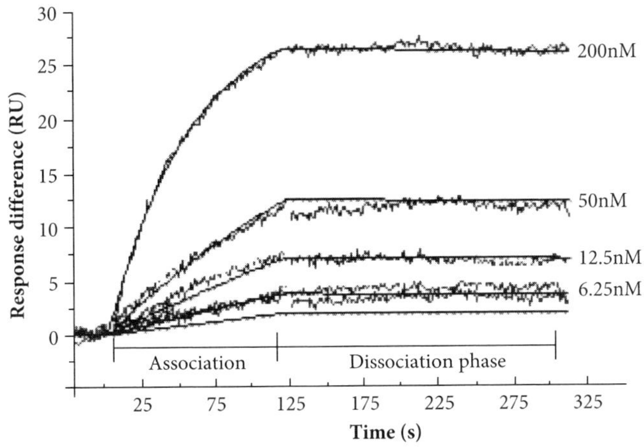

Recombinant PcrV possesses an N-terminal glutathione S-transferase (GST) purification tag (Papalia et al. 2006). The Fab fragment is human Fab 1A8, with the specificity of mouse MAb 166, isolated from an epitope-focused library of human Fab fragments by antibody 'humaneering' (Baer et al. 2009). The Fab 1A8 fragments were immobilized on a Biacore CM5 dextran-matrix-coated sensor chip. A 1:1 mixture of EDC/NHS (ethyl(dimethylaminopropyl)carboiimide–*N*-hydroxysuccinamide) was passed over the sensor chip at a rate of 10μL/min for 10 minutes (to activate the chip surface), followed by injection of Fab at a rate of 10μL/min for 5 minutes in 10mM sodium acetate pH 5.0 buffer. This resulted in immobilization of around 552RU on the Biacore CM5 chip surface.

See Chapter 9

A concentration series of GST-PcrV fusion protein ranging from 200nM to 6.25nM was then passed over the chip for 2 minutes at 30μl/min. This was followed by injection of buffer for 3 minutes and is denoted as a dissociation phase in Figure 11.9. The chip was then regenerated using a 1 M guanidine hydrochloride solution, which resulted in complete dissociation of the bound GST-PcrV fusion protein.

The PcrV-Fab binding/dissociation event was evaluated using a 1:1 Langmuir fitting algorithm (see main text) with mass transfer parameters using BIA evaluation software (Biacore). A K_d of 1.02nM was obtained.

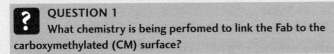

QUESTION 1
What chemistry is being perfomed to link the Fab to the carboxymethylated (CM) surface?

The NHS/EDC reagents are commonly used for covalent coupling of proteins to surfaces. Together, these reagents activate the carboxylate group present in the carboxymethyl dextran that covers the CM5 chip surface for ready formation of an amide linkage to proteins via reaction with primary amine groups (i.e. lysine residues, amino terminus).

QUESTION 2
Roughly how much Fab protein (in pg/mm^2) is immobilized?

You have been previously informed that one RU corresponds to ~1pg/mm^2 of protein immobilized on a chip surface.Therefore the amount of protein immobilized here is ~552pg/mm^2. You can appreciate how small a quantity of protein this is, and therefore how sensitive the method is.

QUESTION 3
How could the four flow cells on the sensor chip have been utilized to conduct the experiment?

The experimental plan should consider how to utilize four parallel flow cells. For example, one flow cell could lack immobilized Fab (negative control), and another could contain immobilized Fab but the binding experiment could be performed with GST as a test for non-specific interaction of ligand with the tag component of the recombinant PcrV (another negative control). The remaining two flow cells could be used for passing over different concentrations of analyte (GST-PcrV).

11.5 Chapter summary

- Quantitative methods commonly used to analyse protein–protein interactions include fluorescence spectroscopy, isothermal titration calorimetry, and surface plasmon resonance.
- The methods have different requirements with respect to protein quantity and quality (purity), and at least one should suit the biological system you are investigating. On the whole, the most thorough quantitative studies will use a combination of methods to measure binding parameters.
- As protein interactions are central to the functional genomics puzzle, new technologies, such as biolayer interferometry, microscale thermophoresis (Lin et al. 2012), and single-molecule atomic force microscopy (combined with fluorescence), are continually emerging and could be considered in addition to those outlined here.

See Chapter 12

References

Albrecht, M.T., Wang, W., Shamova, O., Lehrer, R.I., and Schiller, N.L. (2002) Binding of protegrin-1 to *Pseudomonas aeruginosa* and *Burkholderia cepacia*. *Respir Res* **3**: 18.

Baer, M., Sawa, T., Flynn, P., et al. (2009) An engineered human antibody fab fragment specific for *Pseudomonas aeruginosa* PcrV antigen has potent antibacterial activity. *Infect Immun* **77**: 1083–90.

Biorad Inc. (2012) ProteOn Sensor Chips. Bulletin 5404B. (http://www.bio-rad.com).

Bohm, M., Sturzebecher, J., and Klebe, G. (1999) Three-dimensional quantitative structure-activity relationship analyses using comparative molecular field analysis and comparative molecular similarity indices analysis to elucidate selectivity differences of inhibitors binding to trypsin, thrombin, and factor Xa. *J Med Chem* **42**: 458–77.

Clow, F., Fraser, J.D., and Proft, T. (2008) Immobilization of proteins to Biacore sensor chips using *Staphylococcus aureus* sortase A. *Biotechnol Lett* **30**: 1603–7.

Cooper, A. (1998) Microcalorimetry of protein–protein interactions. *Methods Mol Biol* **88**: 11–22.

Djiane, A., Shimizu, H., Wilkin, M., et al. (2011) Su(dx) E3 ubiquitin ligase-dependent and -independent functions of polychaetoid, the Drosophila ZO-1 homologue. *J Cell Biol* **192**: 189–200.

Eftink, M.R. (1997) Fluorescence methods for studying equilibrium macromolecule–ligand interactions. *Methods Enzymol* **278**: 221–57.

Freyer, M.W. and Lewis, E.A. (2008) Isothermal titration calorimetry: experimental design, data analysis, and probing macromolecule/ligand binding and kinetic interactions. *Methods Cell Biol* **84**: 79–113.

Ghai, R., Falconer, R.J., and Collins, B.M. (2012) Applications of isothermal titration calorimetry in pure and applied research: survey of the literature from 2010. *J Mol Recognit* **25**: 32–52.

Goulko, A.A., Zhao, Q., Guthrie, J.W., Zou, H., and Le, X.C. (2008) Fluorescence polarization: recent bioanalytical applications, pitfalls, and future trends. In: Wolfbeis, O.S. (ed), *Springer Series on Fluorescence*. New York: Springer, pp. 303–22.

Groemping, Y. and Hellmann, N. (2005) Spectroscopic methods for the determination of protein interactions. *Curr Protoc Protein Sci* Chapter 20: Unit 20.8.

Hansen, L.D., Russell, D.J., and Choma, C.T. (2007) From biochemistry to physiology: the calorimetry connection. *Cell Biochem Biophys* **49**: 125–40.

Hieb, A.R., D'Arcy, S., Kramer, M.A., White, A.E., and Luger, K. (2012) Fluorescence strategies for high-throughput quantification of protein interactions. *Nucleic Acids Res* **40**: e33.

Jares-Erijman, E.A. and Jovin, T.M. (2003) FRET Imaging. *Nat Biotechnol* **21**: 1387–95.

Jason-Moller, L., Murphy, M., and Bruno, J. (2006) Overview of Biacore systems and their applications. *Curr Protoc Protein Sci* Chapter 19: Unit 19.13.

Jennings, M.D., Blankley, R.T., Baron, M., Golovanov, A.P., and Avis, J.M. (2007) Specificity and autoregulation of Notch binding by tandem WW domains in suppressor of Deltex. *J Biol Chem* **282**: 29 032–42.

Jez, J.M., Chen, J.C., Rastelli, G., Stroud, R.M., and Santi, D.V. (2003) Crystal structure and molecular modeling of 17-DMAG in complex with human Hsp90. *Chem Biol* **10**: 361–8.

Ladbury, J.E. and Chowdhry, B.Z. (1996) Sensing the heat: the application of isothermal titration calorimetry to thermodynamic studies of biomolecular interactions. *Chem Biol* **3**: 791–801.

Lewis, E.A. and Murphy, K.P. (2005) Isothermal titration calorimetry. *Methods Mol Biol* **305**: 1–16.

Lin, C., Melo, F.A., Ghosh, R., et al. (2012) Inhibition of basal FGF receptor signaling by dimeric Grb2. *Cell* **149**: 1514–24.

Lipschultz, C.A., Li, Y., and Smith-Gill, S. (1997) Experimental design for analysis of complex kinetics using surface plasmon resonance. *Methods* **20**: 310–18.

Loving, G.S., Sainlos, M., and Imperiali, B. (2010) Monitoring protein interactions and dynamics with solvatochromic fluorophores. *Trends Biotechnol* **28**: 73–83.

Martin, S.F., Tatham, M.H., Hay, R.T. and Samuel, I.D.W. (2008) Quantitative analysis of multi-protein interactions using FRET: application to the SUMO pathway. *Protein Sci* **17**: 777–84.

McDonnell, J.M. (2001) Surface plasmon resonance: towards an understanding of the mechanisms of biological molecular recognition. *Curr Opin Chem Biol.* **5**: 572–7.

Millot, M.C., Martin, F., Mangin, C., Levy, Y., and Sebille, B. (1999) Use of polymethacryloyl chloride to immobilize proteins onto gold surfaces: detection by surface plasmon resonance. *Material Sci Eng C* **7**: 3–10.

Papalia, G.A., Baer, M., Luehrsen, K., Nordin, H., Flynn, P., and Myszka, D.G. (2006) High-resolution characterization of antibody fragment/antigen interactions using Biacore T100. *Anal Biochem* **359**: 112–19.

Pierce, M.M., Raman, C.S. and Nall, B.T. (1999) Isothermal titration calorimetry of protein–protein interactions. *Methods* **19**: 213–21.

Renatus, M., Bode, W., Huber, R., Sturzebecher, J., and Stubbs, M.T. (1998) Structural and functional analyses of benzamidine-based inhibitors in complex with trypsin: implications for the inhibition of factor Xa, tPA, and urokinase. *J Med Chem* **41**: 5445–6.

Rich, R.L., Papalia, G.A., Flynn, P.J., et al. (2009) A global benchmark study using affinity-based biosensors. *Anal Biochem* **386**: 194–216.

Rossi, A.M. and Taylor, C.W. (2011) Analysis of protein–ligand interactions by fluorescence polarization. *Nat Protoc* **6**: 365–87.

Schuck, P. (1997) Reliable determination of binding affinity and kinetics using surface plasmon resonance biosensors. *Curr Opin Biotechnol* **8**: 498–502.

Stebbins, C.E., Russo, A.A., Schneider, C., Rosen, N., Hartl, F.U., and Pavletich, N.P. (1997) Crystal structure of an Hsp90–geldanamycin complex: targeting of a protein chaperone by an antitumor agent. *Cell* **89**: 239–50.

Sturtevant, J.M. (1977) Heat capacity and entropy changes in processes involving proteins. *Proc Natl Acad Sci USA* **74**: 2236–40.

Torres, F.E., Recht, M.I., Coyle, J.E., Bruce, R.H., and Williams, G. (2010) Higher throughput calorimetry: opportunities, approaches and challenges. *Curr Opin Struct Biol* **20**: 598–605.

Velazquez-Campoy, A., Leavitt, S.A., and Freire, E. (2004) Characterization of protein–protein interactions by isothermal titration calorimetry. *Methods Mol Biol* **261**: 35–54.

Venkatraman Girija, U., Furze, C., Toth, J., et al. (2010) Engineering novel complement activity into a pulmonary surfactant protein. *J Biol Chem* **285**: 10 546–52.

Yan, Y and Marriott, G. (2003) Analysis of protein interactions using fluorescence technologies. *Curr Opin Chem Biol* **7**: 1–6.

Structural analysis of proteins: X-ray crystallography[1], NMR[2], AFM[3], and CD spectroscopy[4]

12

Thomas Edwards[1], Arwen Pearson[1], Gary Thompson[2], Arnout Kalverda[2], Oliver Farrance[3], David Brockwell[3], and Gareth Morgan[4]

Chapter overview

Structural knowledge of biological macromolecules such as proteins is fundamental to understanding their function and the mechanisms underpinning their activities. Studying the structure of a protein can provide us with information such as size, conformation, protein-folding, interactions with other molecules, and changes in structure in response to changing environmental conditions. Structural information can be obtained at different levels of organization ranging from atomic structures of individual molecules using techniques such as X-ray crystallography and nuclear magnetic resonance (NMR), through structures of macromolecular assemblies using atomic force microscopy (AFM) and electron microscopy, to organelle and cellular structures using electron and fluorescence microscopy approaches.

This chapter will provide a brief overview of the X-ray crystallography, NMR, and AFM. Electron and fluorescence microscopy are covered in Chapter 17. In addition, this chapter will describe circular dichroism (CD) spectroscopy, a technique used to study the secondary structure content of proteins and how this changes when a protein unfolds or binds to another molecule.

LEARNING OBJECTIVES

This chapter will enable the reader to:

- describe the basic principles that underpin how X-ray crystallography, NMR, AFM, and CD spectroscopy work;
- summarize the types of information that each of these techniques can provide and how the information may complement or validate that obtained from the different techniques;
- select the most appropriate technique(s) to answer a defined research question and outline a workflow for the experiment.

12.1 X-ray crystallography

Ever since father and son W.H. Bragg and W.L. Bragg solved the atomic structure of sodium chloride in 1913, leading to the 1915 Nobel prize, X-ray crystallography has been used to determine the relative position of atoms in molecules. This work on small molecules has now been extended to macromolecules, with the first peptide structures (glucagon and insulin) and the first protein structure (myoglobin) solved in the 1950s, and the first enzyme structure (lysozyme) solved in 1965.

Of all the techniques available to investigate the structures of individual proteins, nucleic acids, and their complexes—X-ray crystallography, NMR, electron microscopy, small-angle X-ray scattering (SAXS), and CD—X-ray crystallography provides the highest **resolution**. Therefore it is the method of choice for the study of previously undetermined protein structures. These experiments generate precise high-resolution data allowing a **model** of the sample to be built with high confidence in the positioning of each atom. This is important when attempting to understand protein–ligand complexes and enzyme mechanisms, and high resolution is critical for the expanding field of structure-based drug design. Here we provide a brief summary of macromolecular X-ray crystallography. Those interested in a more detailed description, may wish to consult Rhodes (2006), Wlodawer et al. (2008), and Rupp (2009).

12.1.1 Atomic structure determination: an overview

A structural determination experiment seeks to provide a detailed 3D image of the sample under study. Cellular structure can be probed at relatively long length scales (microns to millimetres) using microscopy, where visible light is refracted through a series of lenses to form a magnified image. However, visible light cannot be used to resolve the structure of a molecule at the atomic level as the distance between atoms in a molecule is on a much shorter length scale (the length of a carbon–carbon bond is ~0.15nm). Hence, we must use X-rays to resolve such finely spaced features. These are **photons** (light) with wavelengths between 0.01 and 0.5nm (0.1–5Å). Structural biologists usually use the angstrom (Å) to describe size, distances, and resolution in molecular structures (1Å = 0.1nm). Due to their short wavelength, X-rays cannot be focused with lenses to directly form a high-resolution image of the molecule that is being studied. Instead, the scattered, or diffracted, rays are mathematically treated to yield a 3D map of the electron density within the molecule. As the inter-atomic bonds are formed by electrons, we are able to determine the relative positions of the atoms within the molecule by tracing the electron density. This is described further later in this section (step 3).

Like all photons, X-rays interact with the electron orbitals around each atom. However, photons are tiny compared with an atom and so the probability that an X-ray will encounter an atom when passing through a molecule is extremely small—most X-rays go straight through the sample. In fact, less than 1% of incident X-rays contribute to the measured **diffraction pattern**. Even with modern high-brilliance X-ray sources, the diffraction from a single molecule is too weak to be measured. In order to amplify the diffracted signal, we use **crystals**. A crystal is an ordered array of the molecule of interest. Each molecule in the crystal lattice diffracts the X-rays, resulting in a summed signal that is strong enough to be detected.

There are four steps to an X-ray crystallography experiment: crystallization, collecting diffraction data, atomic structure determination, and refinement, validation, and deposition of atomic structure.

Step 1. Crystallization

The first step in an X-ray crystallography experiment is to grow a crystal of the macromolecule under investigation. Growing crystals of macromolecules remains one of the major bottlenecks in protein structure determination. This is because, even with knowledge of the structure, we are currently unable to predict the conditions under which a macromolecule will crystallize. Therefore the quickest way to identify crystallization conditions is to screen a wide range of solution parameters, with the aim of identifying the optimum conditions for crystal growth.

A crystal can be thought of as an ordered precipitate rather than the amorphous precipitate that is observed when a protein aggregates or denatures. The basis of all crystallization experiments is controlled dehydration, where solvent is progressively removed from around the macromolecule, concentrating it until crystallization occurs. Therefore a primary component of all crystallization solutions is the precipitant (which can be a salt or a polymer such as polyetheylene glycol). This sequesters solvent, driving the macromolecules closer together. Due to their surface charge, macromolecules will electrostatically repel each other; therefore a second important component of the crystallization solution is the buffer. The net charge of a protein is zero (i.e. interparticle repulsion is minimal) at its pI. By modulation of the pH of the crystallization solution the surface electrostatics of the target macromolecule can be altered in order to promote crystallization. Finally, crystallization solutions often contain a variety of small molecule additives, such as salts, small carbohydrates, and organic solvents. These affect the behaviour of both the macromolecule and the solution, for example by masking surface charges, and can be vital for crystal growth to occur. Other factors that affect crystallization, by affecting the solubility of the macromolecule, include temperature, volume, the type of crystallization experiment used, the redox environment, and even how the macromolecule and crystallization solutions are mixed. For successful crystallization, the difference between shaken and stirred is an important one!

Most macromolecular crystallization experiments use a vapour diffusion approach (Figure 12.1). Here the crystallization solution is placed into the well of a crystallization plate, this is the reservoir. Then a small volume (usually 1–2μL) of the crystallization solution is mixed with a similar volume of the macromolecule-containing solution, the resulting droplet is either suspended above or placed on a little step over the reservoir, and the well is sealed. As the concentration of precipitant is lower in the droplet than in the reservoir, a vapour equilibrium is set up where water slowly leaves the droplet, concentrating the macromolecule. If the conditions are just right, crystals will appear.

One of the most important factors for crystals to form is the purity of the macromolecule. Proteins should be at least 95% pure (even a trace impurity can prevent crystal formation), well-folded, and monodisperse in solution. Before embarking on a crystallization project you should characterize your protein carefully. For example, if you are trying to crystallize an enzyme, in addition to assessing its purity by SDS-PAGE you should also check its specific activity, oxidation state (if it contains a metal or redox cofactor),

Figure 12.1 Illustration of a hanging-drop vapour diffusion experiment. The droplet containing protein and precipitant solution is suspended over a reservoir of precipitant solution in a sealed chamber. The arrows within the chamber indicate the direction of vapour diffusion as the protein and precipitant concentrations in the droplet increase.

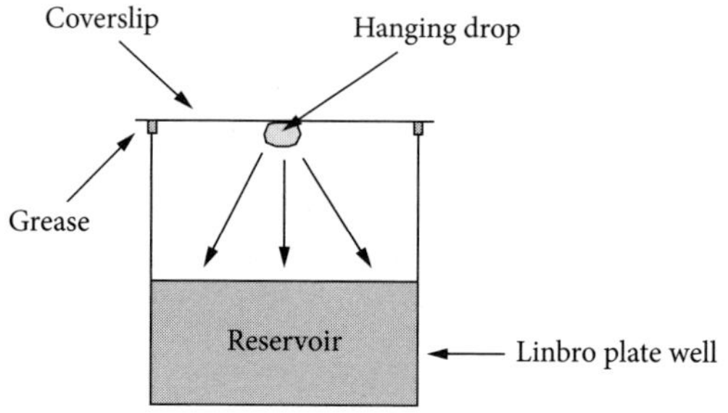

whether any post-translational modifications (i.e. glycosylation or phosphorylation) are present, and whether its kinetic parameters are correct. Samples often show variability from preparation to preparation, which can have a significant detrimental effect on crystal formation projects, and hence you should track everything that you do and observe your sample carefully. For example: Was it frozen? How many days since it was purified?

If you have been unable to obtain crystals, even with a pure sample, all is not lost. With modern molecular biology tools you can generate mutant, or truncated, proteins that contain (hopefully) well-folded domains and hence improve the chances of crystal formation. Bioinformatics and biochemical analyses, such as limited proteolysis, can help guide the design of plasmid constructs from which the proteins are generated. An alternative approach is to try and crystallize a homologue of the protein from a closely related organism where sequence homology suggests that the protein will have the same fold.

See Chapter 7

Step 2. Collecting diffraction data

In order to collect diffraction data the crystal is placed in an X-ray beam which is usually monochromatic, i.e. of a known fixed wavelength (Figure 12.2). The majority of the photons pass straight through the crystal, but a small fraction are scattered (diffracted) by the atoms within. Photons are small enough to have both a wave and a particle nature. When thinking about diffraction, it is easiest to think of them as waves. A characteristic property of any set of waves is that they will interfere with each other. This interference is additive, giving a greater amplitude where their displacements are in **phase** (constructive interference) and a smaller amplitude where their displacements are out of phase (destructive interference). As the intensity of an electromagnetic wave is proportional to the square of the amplitudes of its component waves, the regions of constructive and destructive interference appear as regions of enhanced and diminished intensities, respectively. Diffraction occurs when a set of waves that are in phase (all interfering in a constructive manner) encounter an object in their path which scatters the waves and causes some to become out of phase with others. If a detector is then placed beyond the object the varying intensities that result from the diffraction can be recorded as a varying pattern of dark and light. The resulting pattern of spots is called a diffraction pattern

Figure 12.2 (a) Schematic diagram of a diffraction experiment. The X-rays hit the crystal and the majority pass straight through and are absorbed by the beamstop. A small fraction of the X-rays are scattered by the crystal and are collected on the detector. (b) Example of a diffraction pattern. The white shadow is the beamstop which is blocking the direct beam. The dark spots are reflections resulting from constructive interference of the diffracted X-rays.

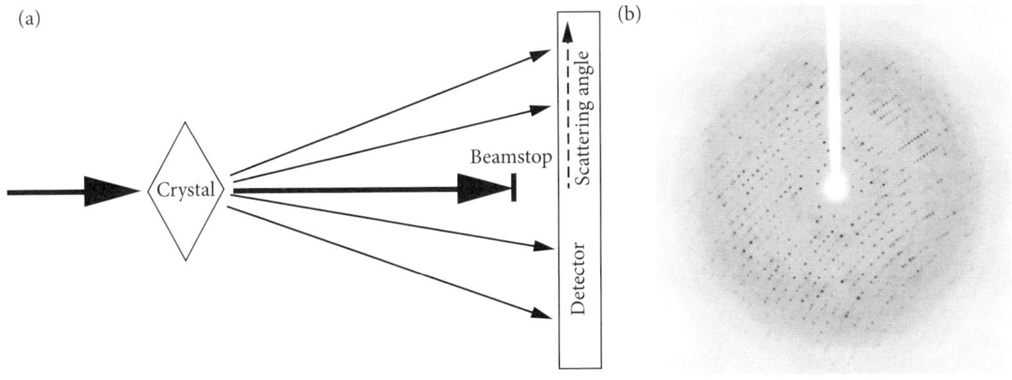

(Figure 12.2). Every atom that scatters an incoming X-ray contributes to each spot on the diffraction pattern.

The resolution of the data, i.e. the finest detail that can be resolved in the resulting **electron density map**, is determined by the maximum scattering angle at which data are recorded. This relationship between scattering angle and resolution was defined by Bragg in 1912 and is given by:

$$n\lambda = 2d \sin\theta \qquad 12.1$$

where λ is the wavelength of the incident X-rays, n is an integer, θ is the scattering angle, and d is the minimal spacing that can be resolved.

A single diffraction pattern only tells us about the contents of the lattice viewed from one direction. As macromolecules are three-dimensional, the crystal is rotated in order to record diffraction from all orientations of the lattice. In general, the crystal is rotated over successive small angles while each diffraction pattern is recorded. The square root of the intensity of each spot is then taken to determine the amplitude of the resultant wave at that point. All these data are then combined (i.e. each spot position and amplitude is recorded) to produce the final diffraction dataset.

Data collection is usually carried out at a 100K (degrees Kelvin). The crystal is normally mounted in a nylon loop and is held in place by the surface tension of a thin film of crystallization solution (Figure 12.3). The crystal is then flash cooled in liquid nitrogen before being mounted into a dry nitrogen gas stream, maintained at 100K by a cryostat.

Step 3. Atomic structure determination

To determine the structure of a macromolecule from X-ray diffraction data an electron density map must be calculated. An atomic model can then be built into the electron

Figure 12.3 A crystal mounted in a nylon loop and stabilized at 100K in a dry nitrogen gas stream.

density using the experimenter's knowledge of the sequence of the protein or nucleic acid and its standard chemistry. An example of an electron density map is shown in Figure 12.4. It is obtained by recombining the scattered waves that have been measured in the diffraction experiment into an image of the scattering molecule. Unlike light microscopy, this cannot be done using lenses; instead, we must use a series of mathematical operations as a 'lens'.

The amount of electron density at any point within the crystal lattice is related to the resultant scattered wave measured in the diffraction pattern by the function:

Figure 12.4 The final result of an X-ray crystallography experiment. The left panel shows the electron density; the connectivity and position of atoms are clear. The right panel shows the molecular model that has been built into the electron density by the crystallographer. For a colour reproduction of this figure, please see Plate 9.

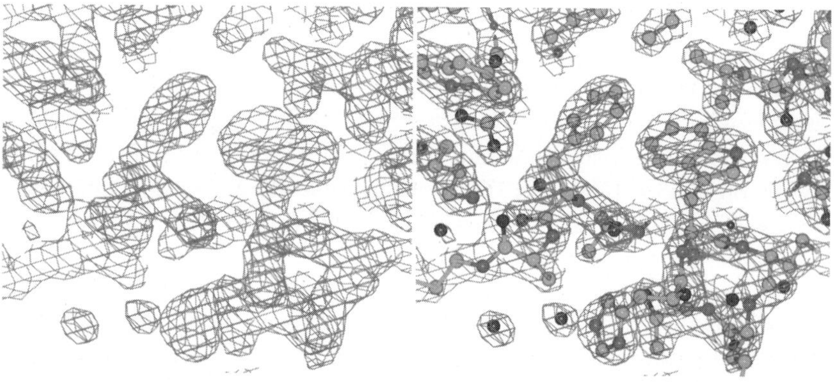

$$\rho(x,y,z) = \frac{1}{V}\sum_{hkl} F_{hkl} e^{-2\pi i(hx+ky+lz)} \qquad 12.2$$

where $\rho(x,y,z)$ is the electron density at a point (x,y,z), V is the volume of the repeating unit of the crystal lattice, F_{hkl} is the structure factor of the scattered wave summed over all reflections, and the final term is an operator that relates real space (electron density) to reciprocal space (the diffraction pattern). F_{hkl} is a complex term which contains both the amplitude of the resultant wave and its phase. We can measure the amplitude directly from the diffraction pattern; however, the phase, although implicit in the measured intensity of each reflection, cannot be directly measured by X-ray detectors and hence is lost from the experiment. This is known as the phase problem and means that the electron density of the protein under study cannot be calculated directly from the initial diffraction data. First, we need to conduct other experiments that allow us to estimate the phases. Only then can an electron density map be generated in order to build a model of our structure.

Determining the correct phase is hugely important for obtaining the correct structure. Calculating an electron density map using the amplitudes from one molecule (A) and the phases from a second molecule (B) results in electron density approximating molecule B. This is because phases 'dominate' the appearance of the electron density map produced by the Fourier transform and therefore a 'reasonable' estimate of the phases is required to produce an interpretable map.

Several methods can be used to resolve the phase problem so that an atomic structure can be determined, and these will be only briefly described here. They fall into three classes: direct methods, experimental phasing, and molecular replacement. Sometimes a single method alone is not enough to solve a structure unambiguously, and a combination of approaches is required. Each of these techniques yields an estimate of the phases, so that an interpretable electron density map can be obtained. The phases are then improved by iterative rounds of model rebuilding followed by new electron density map calculations using improved phases.

Direct methods are routinely used to solve small molecule structures in chemical crystallography. However, as they require very high resolution data and are rarely applied to macromolecular crystallography, we will not discuss them further here.

Experimental phasing takes advantage of the effect of electron-dense (high atomic number) atoms on the measured diffraction pattern. The differences in amplitudes measured with and without the electron-dense atom (or at two wavelengths at and far from the elemental absorption edge) can be used to locate the positions of electron-dense atoms within the lattice. Once these have been identified, they provide an initial estimate of the phases that can be used to calculate a first electron density map.

Molecular replacement (MR) estimates the phases of the protein structure under study using the phases from a known structure. The known structure must be a molecule that closely resembles the one under study, for instance the same protein from a different source, or a protein with high sequence homology. The availability of suitable 'model' molecules is continually increasing as more and more structures are deposited in the Protein Data Bank (PDB) (http://www.pdb.org). The observed amplitudes of the unknown molecule are combined with the phases from the known structure to generate a first electron density map. Care must be taken here to avoid model bias, and molecular replacement solutions are often combined with experimental phasing data to provide additional confirmation that the correct solution has been obtained.

Step 4. Refinement, validation, and deposition of atomic structure

The phases obtained from experimental phasing or molecular replacement often have large errors, and so it is necessary to improve these in order to obtain an accurate final model of the structure by **refinement**. If the initial phases used to calculate the electron density map are close enough to the true phases, it is possible to begin building a structural model. This can be done manually using model-building programs or in an automated fashion with expert software systems.

Once as much as possible of the model has been built into the density, refinement of the model is carried out. The aim of refinement is to improve the agreement of the observed and calculated structure factor amplitudes (F_{obs} and F_{calc}). Restraints based on known protein chemistry parameters, such as bond angles, lengths, and torsions, are put on the movement of atoms in the molecule during refinement. The amount of weight given to these restraints can be varied in the refinement process, depending on the resolution of the data collected. For instance, at high resolution the clarity of the electron density maps allows a higher confidence in the position of the atoms within the density. Therefore less weight can be given to the geometric restraints. Conversely, at low resolution, the position of individual atoms in the electron density is not as well resolved. Therefore, higher weighting must be given to the geometric restraints.

Following initial refinement, the phases used to generate the next map are calculated from the refined model itself. Therefore these phases should be more accurate and so electron density maps should become clearer. This allows more model building to take place, which can then be refined again in an iterative manner. The quality of the model is assessed after each round of refinement using a pair of statistics known as R_{work} and R_{free}:

$$R = \frac{\sum |F_{obs}| - |F_{calc}|}{\sum |F_{obs}|} \qquad 12.3$$

These measure how well the two terms F_{obs} and F_{calc} agree. As the accuracy of the model increases, the difference between these two terms decreases and therefore the ***R* factors** improve. Although model building and refinement could continue indefinitely, a generally recognized endpoint is when the alteration of a model no longer results in decreased *R* factors. The final value for the *R* factors for any given model depends on the resolution of the data. For instance, *R* factors are generally lower for high-resolution structures because of the quality of the observed data.

Structure validation is the final stage of structure determination. In a fully refined structure, deviations from 'normal' chemistry (bond angles, lengths, torsions) should be minimal. For instance, the vast majority of the amino acid residues should reside in the preferred and allowed regions of a **Ramachandran plot**; there should be no steric clashes between any residues/solvent molecules in the structure. There are now a number of servers online which allow structure validation.

The final completed and validated model is deposited in the Protein Databank which is the single public repository for all X-ray crystal structures. From the PDB you can download 3D coordinates of all deposited models, and more recent depositions will also include **structure factor** data, i.e. reduced experimental data that will allow recalculation of electron density maps.

12.2 Nuclear magnetic resonance

Solution-state nuclear magnetic resonance (NMR) is currently the only method by which high resolution biomolecular structures, including those of proteins, RNA, DNA, and sugars, can be determined at atomic resolution in the solution state. Along with X-ray crystallography, NMR provides most of the atomic resolution biomolecular structures in the PDB, with the current split being approximately 10:1 in favour of crystallography. In addition, the NMR technique allows a wide range of complementary information to be studied. This includes the size of molecular motions at very short times ranging from picoseconds to milliseconds (which are important for ligand binding and catalysis), the presence of remaining structure in disordered proteins, fast mapping of the location of ligand-binding sites, and assessment of protein-folding. These additional insights have been of huge importance in studying the molecular processes of a number of diseases, including SARS, prion-based diseases such as BSE/Creutzfeldt–Jakob disease, which derive from intrinsically disordered or partially unfolded proteins, and a variety of amyloid-based pathologies including Parkinson's disease. When compared with other solution-state methods such as SAXs, FRET, IR, CD, or UV, most of this information can be measured at a large number of atomic sites within a molecule. The following references describe protein NMR techniques in greater detail than is covered here: Pochapsky and Pochapsky (2006), Downing (2010), and Roberts and Lian (2011).

12.2.1 What NMR measures

At its simplest, the NMR signal can be attributed to the presence of an intrinsic angular momentum or **spin** within the nuclei of atoms, which causes them to behave like small bar magnets.

In the absence of an external magnetic field the nuclei are free to take up all orientations and no signals are observed. However, when placed in a magnetic field, nuclei are restricted in the orientations they may take up because of **quantization**. For most biologically important nuclei (hydrogen (^{1}H), nitrogen (^{15}N), and carbon (^{13}C)) only two orientations, α and β, are possible. These two states have different energies as they orientate in the magnetic field. The α-state aligns with the external magnetic field and has a lower energy, whilst the second β-state aligns against the external magnetic field and has a higher energy. The difference in energy between the two states depends on the strength of the external magnetic field in which the atom is placed as well as the nucleus under study. The energy difference is measured as a frequency in hertz (Hz) and magnetic fields are measured in tesla (T).

Transition (resonance) between the α- and β-states—from the lower to the higher energy state as well as from the higher to the lower energy state—can be induced by applying electromagnetic radiation at the radiofrequency that is equal to the energy difference between the two states for a particular nucleus type. These range from 100–1000MHz for ^{1}H nuclei to 50–100MHz for ^{15}N nuclei and 125–250MHz for ^{13}C nuclei when using magnetic field strengths of 2.3–23T. For each different transition where radio-energy is absorbed, an NMR signal or peak is observed in the NMR spectrum and measured using the spectrometer. It is common to use the ^{1}H frequency of a spectrometer as shorthand for the magnet's field strength, so a spectrometer with a field strength of 11.5T will often be called a 500MHz spectrometer.

12.2.2 Chemical shifts

Although the frequencies of transitions predominantly depend on the external magnetic field strength, there is still a variation in the frequencies measured for the various atoms, or more strictly their nuclei, within a molecule even for the same type of nucleus at the same magnetic field strength. This is due to the small magnetic fields produced by electrons from within the atom and electrons from neighbouring local double bonds, including carbonyl groups and aromatic rings, which can add to or subtract from the applied magnetic field. This secondary effect is called the **chemical shift** (designated by the symbol δ) and is measured in changes of parts per million (ppm) of the main field strength or spectrometer frequency. The reason for measuring the chemical shift in this way is to make the value independent of the magnetic field strength used.

As atoms within different chemical groups have distinct chemical shifts, this enables an atom to be identified from its signal. For example, ^{1}H signals from amides (–CO–NH–CA–) in proteins appear over a range of shifts from 5 to 8ppm and can easily be distinguished from the ^{1}H signals from methyls (CH_3) which appear at around 1ppm. Therefore each peak in a simple 1D NMR spectrum (Figure 12.5(a) corresponds to one atom and its environment (i.e. the neighbouring atom and/or chemical group) within a molecule. Thus from NMR spectra we can identify the type of molecule present in a sample (an RNA has a very different spectrum to a protein) and distinguish between biomolecules with different primary sequences (including single-site mutations), and from changes in the chemical shifts identify where ligands bind to a molecule.

12.2.3 Couplings

The 1D NMR spectrum of even a small protein (Figure 12.5(a)) contains a large number of peaks within a small region, which leads to overlap, making the signals hard to distinguish. As an example, the peaks labelled 'A' and 'B' in Figure 12.5(a) are from separate amides in a protein and overlap or are **degenerate** with a number of other peaks originating from different amides. To overcome this problem, another energy change called **scalar coupling** or **J coupling**, which is even smaller than the chemical shift, is used to provide a multi-dimensional or *n*D **correlation spectrum** ($n = 2 - 4$). In these spectra peaks only appear if there is a chemical bond or a small number of chemical bonds between neighbouring groups. The scalar coupling is caused by neighbouring nuclei influencing each other through the electrons in chemical bonds. Small changes in the measured energy differences occur, depending on whether the connected neighbouring spin is α or β. Using NMR experiments based on these couplings such as the **h**etronuclear single **q**uantum **c**orrelation **s**pectrum (**HSQC**) it is possible to transmit information about connections between specific types of nuclei. NMR experiments have been designed to use specific sets of couplings, thus allowing signals to be connected through the bonding network in a molecule in a specific way.

To demonstrate how preliminary spectra are acquired, assignments are made, and the binding of a ligand (Case Study 12.1) can be examined at a molecular level using NMR, we will focus on the published study of the protein K7 (Kalverda 2009) and the peptide it

Figure 12.5 Typical 1D and 2D NMR data and their interpretation. (a) A typical 1D ^{1}H NMR spectrum of a protein. Peaks arising from amide, aromatic, Hα, and methyl groups are labelled. The positions of the signals arising from 'A' and 'B' shown in (b) are labelled. (b) Schematic diagram of the bonds in a protein backbone leading to the correlations shown in the bottom panel. The active amide bonds are labelled with their $^{1}J_{HN}$ coupling (92 Hz). (c) ^{1}H–^{15}N HSQC spectrum of the protein K7 showing two sets of correlated shifts: 'A–D' and 'C–B'.

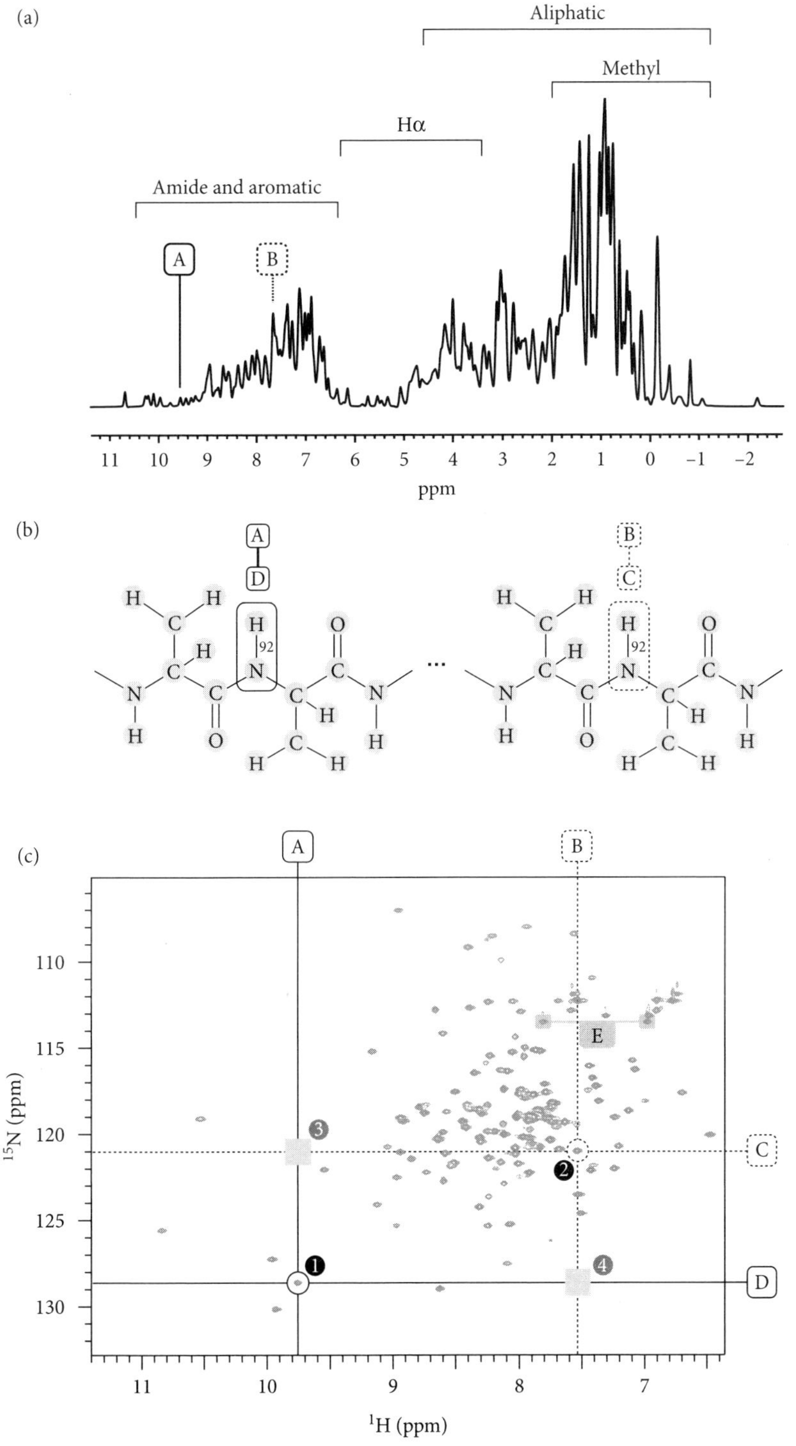

binds to derived from the N-terminus of DDX3 helicase. Additional information on this protein and other examples of where NMR has been used to study molecular dynamics, unfolded states, and transiently populated states of proteins can be found on the Online Resource Centre.

Measuring a $^1H-^{15}N$ HSQC fingerprint spectrum using $^1H-^{15}N$ couplings

To provide a concrete example of the use of scalar couplings, Figure 12.5(b) shows the $^1H-^{15}N$ HSQC spectrum of the protein K7. The experiment only measures signals from ^1H and ^{15}N nuclei connected by a $^1H-^{15}N$ scalar coupling of ~90Hz. This limits the signals measured to those of the nuclei in backbone amides and the side chains of the amino acids asparagine, glutamine, and tryptophan, which all have a coupling of ~90Hz across the N–H bond. The spectrum is a 2D spectrum because it correlates signals from two types of nuclei (^1H and ^{15}N). Each distinct spot or 2D peak in the spectrum arises from an N–H-bonded pair of atoms in a distinct environment within the protein. The chemical shift ranges measured for ^1H and ^{15}N are quite different, and are quite specific for these nuclei and the environment created by the amide bond and the structure of the protein.

To follow a specific connection, the vertical line labelled 'A' will cross all peaks within the spectrum with a ^1H chemical shift of 9.75ppm. In this case, there is a single peak (labelled 1). This peak is also on the horizontal line labelled 'D', which has a ^{15}N chemical shift of 128.65ppm. The shifts of the ^1H and ^{15}N nuclei at 9.75ppm and 128.65ppm, respectively, are described as being correlated and must be from the same residue and from an N and H connected via a single bond. One such NH pair is highlighted and labelled on the structure above the spectrum. A second correlation (labelled 2) is highlighted between the ^1H and ^{15}N shifts of a second amide from a different residue (chemical shifts labelled 'B' and 'C' at ^1H 7.54ppm and ^{15}N 121.0ppm respectively). The spectrum clearly shows there are no peaks where the shift lines 'A–C' and 'B–D' cross (labels 3 and 4) indicating that nucleus 'A' is not connected to nucleus 'C' and, equivalently, nucleus 'B' is not connected to nucleus 'D'. Using this technique we have started to make connections between chemical shifts using the chemical bonds in the molecule. This technique is used extensively during **chemical shift assignment** (see section 12.2.4).

The $^1H-^{15}N$ HSQC spectrum is often used as a convenient **fingerprint** of a protein as it has one signal for each residue except proline, which has no amide proton, and the N-terminus. Side-chain $^1H-^{15}N$ peaks of asparagine and glutamine appear as pairs, such as the pair highlighted and labelled 'E'. Side-chain peaks are also detected for tryptophan, and sometimes from the side-chain nitrogens of histidine, lysine, and argine, but these all have rather distinctive chemical shifts. Hence the number of 2D peaks in the spectrum which are not from side chains should be given by:

$$\text{number of peaks} = (\text{number of residues}) - (\text{number of prolines}) - 1$$

where −1 is included to account for the N-terminal amine group, which is not observed because of the fast exchange of protons with the solvent.

A correct residue count in this spectrum is generally used as a good indication that the sample is suitable for NMR analysis. If the peaks in this spectrum are broader than expected, missing, or heavily overlapped, this can indicate that the protein is unfolded, partially folded, or undergoing significant motions on millisecond time-scales. Examples of ^1H–^{15}N spectra from a folded mainly α-sheet protein, a mainly β-sheet protein, an unfolded protein, and a molten globule protein can be found on the Online Resource Centre.

Labelling samples and acquiring trial ^1H–^{15}N HSQC spectra

The use of ^1H–^{15}N HSQC spectra depends on the presence of a ^1H–^{15}N single-bond scalar coupling (often abbreviated to $^1J_{NH}$ = 92Hz, where the superscript 1 indicates a coupling over a single bond and NH indicates the connected atoms). Many common biomolecular NMR experiments used with protein and RNA samples utilize single-bond couplings between different types of nuclei. There couplings are not present unless the protein is labelled with stable isotopes. Therefore many NMR experiments require **stable isotope labelling** with ^{15}N and/or ^{13}C nuclei. Stable isotope labelling completely replaces the naturally occurring, but NMR inactive, ^{12}C and ^{14}N nuclei with ^{15}N and/or ^{13}C nuclei. Labelling is typically achieved by expressing a protein using a plasmid-based expression system in *E. coli* in a minimal medium such as M9. This medium, unlike the Luria Betani (LB) medium commonly used to grow *E. coli*, contains no amino acids and uses ^{15}N-labelled ammonium chloride and/or ^{13}C-labelled glucose as the sole nitrogen or carbon source. Preparation of labelled protein usually takes approximately the same amount of time as for unlabelled protein (typically 1–2 weeks), but is more expensive because of the cost of the labelled chemicals. Experimental measurement times for ^1H–^{15}N HSQC spectra are typically between 20 minutes and 16 hours depending on molecular weight and sample concentration.

See Chapter 7

Labelling with ^{15}N and ^{13}C gives good spectra for proteins with masses up to 20kDa (up to 200 residues), and, with the correct selection of experiments, signals from almost all ^1H, ^{15}N, and ^{13}C nuclei in a protein can be individually detected. Larger proteins with masses above ~25kDa require the use of specialized experiments based on a technique called transverse relaxation optimized spectra (TROSY), and deuteration is necessary to obtain suitable spectra (Downing, 2010). This is because slow **molecular tumbling** due to the increase in molecular mass broadens the NMR peaks and makes them harder to detect. Deuteration replaces non-amide ^1H nuclei with deuterium nuclei (^2H); it is typically achieved by expressing proteins in D_2O instead of H_2O and may require use of deuterium-labelled sugars and amino acids. Typically, proteins with masses above 30–40kDa take considerably longer to study and hence are much more time-consuming to work with.

NMR sample volumes are typically 300μL and the pH of the sample is usually between 3 and 8 depending on whether the species being studied is stable. Samples with a pH above 8 are not generally used as the fast exchange rate of the amide protons with water broadens the peaks enough to make many NMR signals undetectable. For proteins that are not sufficiently soluble or show poor spectra, a number of approaches are available to improve solubility and spectral quality. Typical approaches include changing the pH, temperature, and salt concentration, adding additives such as glycerol, and expressing truncated proteins, mutants, or related proteins such as the same protein from a thermostable species.

12.2.4 Assignments

Obtaining a high-quality ^1H–^{15}N HSQC fingerprint spectrum of a protein is usually only the first step in an NMR analysis. To use the information, for example to monitor if and where a ligand binds or to calculate a structure, assignments must be made. Chemical shift assignment (often shortened to just 'assignment') is the process used to connect a chemical shift in a spectrum with the specific atom in the protein structure that it arises from. An example of an assigned spectrum is shown in Figure 12.6(a). This is an expansion of the HSQC in Figure 12.5(b) around the peak labelled 2 (^{15}N = 120.96ppm, ^{1}H = 7.52ppm). Without assignment, all that is known is that peak 2 is from an amide bond in the protein which is in a distinct environment. After assignment, we can identify it as being from amino acid residue aspartate at position 136 in the protein sequence (ASP 136).

Methods of assignment

A number of methods of assignment are available. Published chemical shifts for proteins which have already been assigned usually will be available from the BioMagResBank (BMRB). Otherwise, assignments have to be made by analysing spectral data.

Most biomolecules are assigned by walking along the backbone of the molecule connecting chemical shifts from the current residue and neighbouring residues. These connections are made using an extension of the HSQC spectrum that correlates three atoms. These are usually the amide ^{1}H and ^{15}N shifts from one residue and one shift from $^{13}C\alpha$, $^{13}C\beta$, or ^{13}CO.

The names of the assignment spectra describe which chemical shifts are correlated (Cavanagh et al. 2006). HNCA spectra measure ^{1}H and ^{15}N amide shifts and a $^{13}C\alpha$ shift, all from the same residue (a second weak peak with the $^{13}C\alpha$ shift from the previous residue is also measured, but we shall ignore this here). The HNcoCA spectrum measures ^{1}H and ^{15}N amide shifts from the same residue and the $C\alpha$ shift of the previous residue (the chemical shift of the intervening co is not measured and hence is given in lower case). As the same $C\alpha$ shift is common between the residues, we can use it to make a connection. These are 3D spectra and are displayed as strips of data with $^{13}C\alpha$ shifts on the y axis and ^{1}H chemical shifts on the x axis. The ^{15}N chemical shift is not displayed but just used as an index, as shown in Figure 12.6(d).

To give a worked example, we will follow the assignment of a few residues. It should be noted that in Figure 12.6(c) the HNCA and HNcoCA strips are shown separated to allow the correlations to be seen clearly. However, in Figure 12.6(d) they are shown overlaid as they would normally be seen during assignment. The amide ^{1}H and ^{15}N shifts designated 'A' and 'B' belong to residue 137. If this ^{1}H shift is followed vertically down, there is a peak labelled 1 in the HNcoCA spectrum at the chemical shift corresponding to $C\alpha$ (55.31ppm $^{13}C\alpha$). This is the chemical shift of the $C\alpha$ in the previous residue (residue 136). Following the vertical line 'D–C' which is from the ^{1}H and ^{15}N shifts from residue 136, there is another $^{13}C\alpha$ peak in the HNCA which is also on the chemical shift line 1. This is the $C\alpha$ of residue 136 again, and since it aligns with the peak on the HNcoCA of residue 136 we can make a connection between the two residues.

By repeating the steps described above, in some cases it is possible to assign a whole protein using an HNCA and an HNcoCA spectrum. However, for most proteins there are enough accidental overlaps in $C\alpha$ chemical shifts to ensure that the results will not be unique. Therefore further pairs of spectra, such HNCO–HNcaCO and HNCaCB–HNcocaCB must be measured. These correlate amide ^{1}H and ^{15}N chemical shifts with $C\beta$ and CO shifts which are common between two residues. However, prolines which do not have an amide ^{1}H proton and missing peaks can also lead to breaks in the assignment. To

Figure 12.6 Assignment of chemical shifts from 3D NMR spectra. (a) Region from the ^{1}H–^{15}N HSQC spectrum of the protein K7 (Figure 12.5(b) for the whole spectrum). Assigned peaks are labelled and the amide peak of ASP 136 and its shifts are highlighted. (b) The network of strong heteronuclear scalar couplings used to assign the backbones of a protein (couplings are shown in Hz with their bonds). The couplings used in the HNCA are highlighted in grey while the couplings used in the HNcoCA are outlined in black. (c) Overview of the backbone assignment process using HNCA and HNcoCA strip spectra of K7. The bonds whose couplings are used in an HNCA are shown in grey, whereas those for the HNcoCA are outlined in black. Connections are labelled and matched to those in (d). Note that the schematic HNCA (grey) and HNcoCA (white) spectra are separated for clarity, but in (d) are shown overlaid. (d) The same assignment as in (c) shown on actual HNCA and HNcoCA spectra. Chemical shifts and correlations are labelled. The HNCA spectrum is shown as contours and the HNCA spectrum is shown with its peaks filled in. The chemical shifts of residue 136 are highlighted, as are the connections shown in (c).

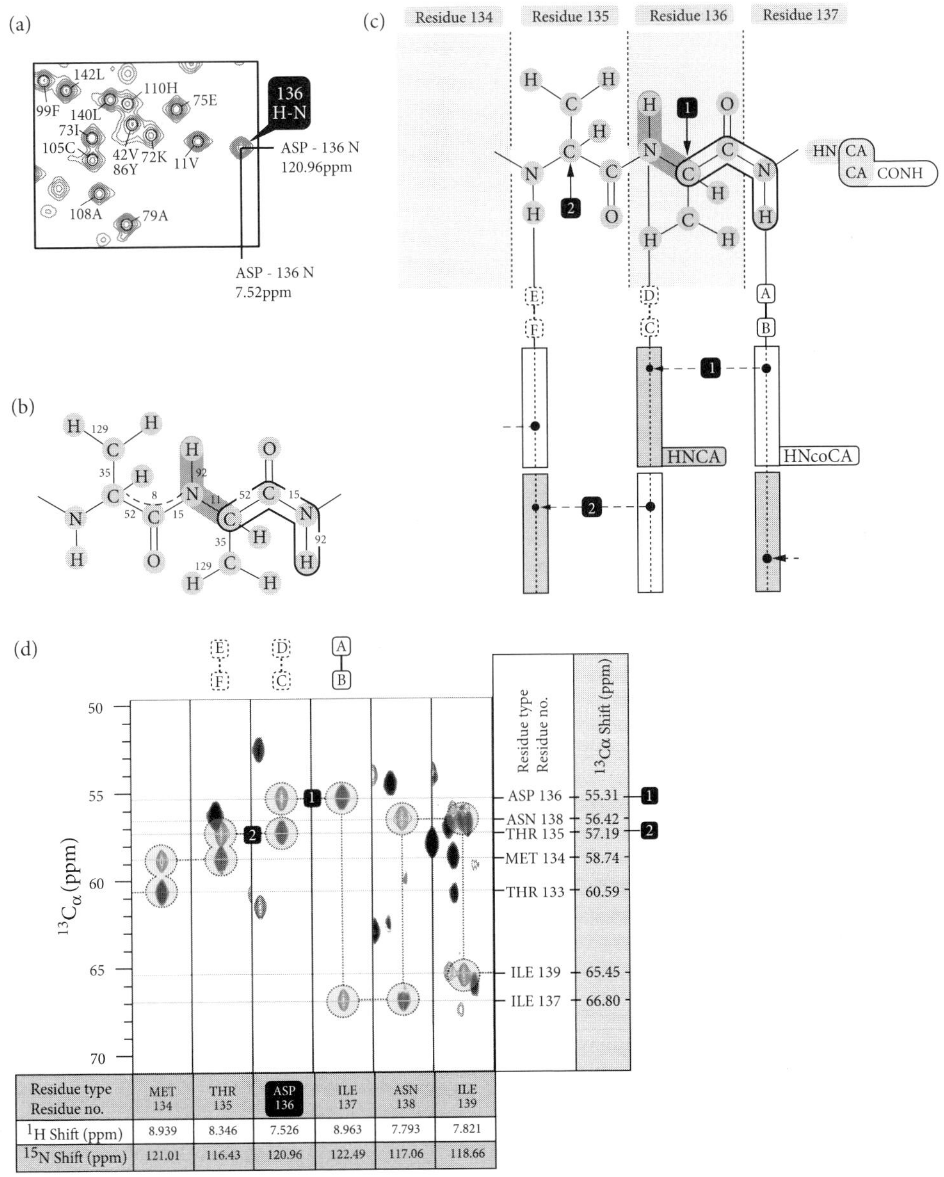

Residue type Residue no.	MET 134	THR 135	ASP 136	ILE 137	ASN 138	ILE 139
^{1}H Shift (ppm)	8.939	8.346	7.526	8.963	7.793	7.821
^{15}N Shift (ppm)	121.01	116.43	120.96	122.49	117.06	118.66

connect the resulting fragmentary assignments to the sequence of the protein, the distinctive Cα chemical shifts of glycine, threonine, and serine and the Cβ shifts of alanines are used to connect amide shifts to residue types which can be matched to the protein's amino acid sequence.

Once assigned the Cα, Cβ, CO, and Hα chemical shifts of residues can be used to determine the secondary structure of a protein without structure calculation. Typically, secondary structure content is calculated by database matching using a program such as TALOS. Having understood how to analyse NMR spectra you should now look at Case Study 12.1 which shows how to use chemical shift mapping to isolate where a protein and a ligand interact. You should also use the online resources to find out about further uses of NMR.

CASE STUDY 12.1

Measuring ligand binding

NMR is an extremely useful tool for measuring ligand binding. The simplest method used, which only requires the unbound or **apo state** of the protein to be assigned is called conservative **chemical shift mapping** (Williamson et al. 1997) (Figure 12.7(a)). As chemical shifts are strongly dependent on the environment of the nucleus being measured, even small changes in backbone conformation, non-bonded interactions, the orientation of side chains, and (especially for amides) hydrogen bonding that occur on ligand binding are clearly discernible by movements of peaks in the HSQC spectrum. These changes in the spectrum are usually local to the ligand binding site, unless major conformational changes occur upon ligand binding. Chemical shift mapping is usually carried out by comparing ^{1}H–^{15}N HSQC spectra (Figure 12.7(a)) measured before and after adding enough ligand to saturate the binding site. For ^{1}H–^{15}N HSQC spectra a 5:1 weighted sum of the change in the ^{1}H and ^{15}N shifts for each peak is used to allow for the larger range of the ^{15}N chemical shifts. Usually the bound state is not assigned, and hence the smallest possible shift change which moves an assigned peak measured without the ligand to a position in the bound spectrum where there is a peak is used to look for the ligand binding site.

As an example, the results of shift mapping for the protein K7 bound to the small peptide ligand DDX3 are shown in Figure 12.7. The chemical shift changes measured between the unbound (black) and bound-state spectra (red) are shown in Figure 12.7(a) and are plotted against the primary sequence of the protein in Figure 12.7(b). Chemical shift changes which are significantly above random noise and the protein's secondary structure elements are highlighted. The same significant shift changes are mapped onto the NMR structure of the protein in Figure 12.7(c) which shows a clear cluster of shift changes on α-helices 1 and 5 towards the bottom of the protein. A subsequent crystal structure (Oda et al. 2009) of the K7-DDX3 protein ligand complex (Figure 12.7(d)) shows that the residues from K7 which are within 6Å of the ligand DDX3 are consistent with the chemical shift mapping.

12.3 Atomic force microscopy

Atomic force microscopy (AFM) is an imaging technique that can provide structural information at intermediate resolution (1–100nm) and bridges the information gained from high-resolution techniques, such as X-ray crystallography (12.1) and NMR (12.2), and lower-resolution techniques, such as light microscopy. AFM can be used to provide a 3D profile of the surface features (**topography**) of the sample under study. It can also be used to study the mechanical properties of **single molecules**. Both of these applications are described in this section.

Figure 12.7 Determining the binding site of the ligand DDX3 on K7 by chemical shift mapping. (a) ^{1}H–^{15}N HSQC spectrum of K7 with (red) and without (black) the ligand DDX3, showing chemical shift changes. The cut-out highlights the changes for two peaks representing the amino acid tryptophan (W) at positions 36 and 112 in the protein sequence. For clarity, the rest of the assignments are not shown. (b) Significant chemical shift differences plotted on the sequence of K7. The secondary structure elements of K7 are shown schematically in yellow and labelled at the top; they are also shown with a grey vertical panel shading. Significant changes in chemical shift are shown as purple bars and all others as black bars. (c) NMR structure of K7 with the significant chemical shift changes mapped onto the structure in pink and the α-helices 1, 5, and 6 labelled. (d) Subsequent structure (by X-ray crystallography) of K7 with DDX3 bound. Residues in K7 within 6Å of the ligand site are highlighted (blue) and the bound DDX3 ligand is shown (yellow, red, and blue sticks). For a colour reproduction of this figure, please see Plate 10. Figures 12.7(b) and (c) reprinted from Kalverda, A.P., et al. (2009) *Journal of Molecular Biology* **385**: 843–53 with permission from Elsevier. Figure 12.7(d) reprinted from Oda, S., et al. (2009) *Structure* **17**: 1528–37 with permission.

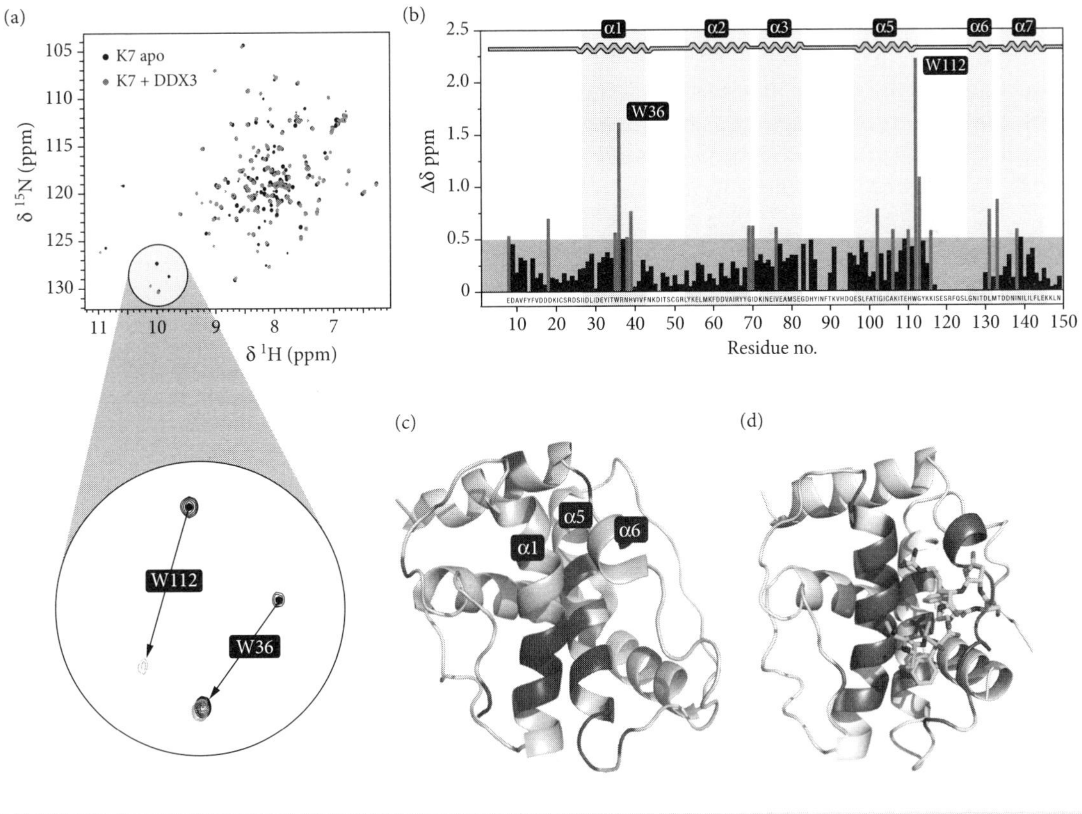

12.3.1 What is AFM?

At its simplest, AFM (Atomic force microscope/microscopy) can be described as dragging a very sharp needle (known as a stylus or tip) across the surface of a sample immobilized onto an inert substrate. By monitoring the height of the AFM tip as it moves over the sample it is possible to build a line-by-line picture of the surface features (topography) of the sample. These 'top-down' images can, for example, reveal arrangement of one or more proteins in a lipid membrane (Scheuring and Dufrene 2010) or the architecture and aggregation mechanism of fibrin clot formation (Abou-Saleh et al. 2009).The level of detail in the sample that can be revealed by an AFM (its resolution) is primarily determined by the sharpness of the tip and the hardness of the sample, and AFM can now be used routinely to obtain images of hard crystalline structures, typically with nanometre resolution. Since its invention in the early 1980s, AFM has been used extensively in the materials and electronics industries. However, it can only achieve a similar resolution for biological samples which form close-packed 2D crystals, such as membrane proteins (Engel and Gaub 2008; Buzhynskyy et al. 2011). The resolution is much lower for most biological samples, which are generally softer; a resolution below 10nm is typically difficult to achieve on a standard instrument, but this is strongly dependent on the properties of the sample.

The advantage of AFM is that it can be used to gain structural information on samples under biologically relevant conditions (i.e. in aqueous solution) or in air. A second major strength of AFM is that it can be used to gain information about the mechanical properties of a sample, sometimes at the single-molecule level. In addition, sample preparation is relatively easy (by immobilization onto varied substrates such as mica and silicon nitride) and a wide range of samples (of various thickness) can be visualized, ranging from single antibodies and protein polymers such as collagen, to bacterial cells undergoing division (Touhami et al. 2004; Kailas et al. 2009) and ex vivo tissues. Finally, AFM can be used in conjunction with complimentary techniques such as fluorescence microscopy and other optical techniques (Gaiduk et al. 2005; Madl et al. 2006). This allows, for example, a specific cellular compartment to be probed by AFM after being identified using fluorescently labelled antibodies specific for that compartment.

See Chapter 17

12.3.2 How an AFM works

An AFM can be regarded as a mechanical **scanning probe** that reveals the topography of an object by monitoring the vertical movement of the AFM tip as it follows the contours of the object. As the differences in height of surface features are typically small (often below 1×10^{-9}m or 1nm), the vertical movement of the tip is amplified using an **optical lever**. This amplification is achieved by reflecting a laser beam from the top of the AFM tip, which is located at the end of a stiff cantilever, onto a photodetector located at a fixed distance from the tip (Figure 12.8). Thus measurement of the change in position of the laser beam on the detector allows calculation of the change in tip **deflection** and therefore changes in height between two points on the surface. To generate an image, the tip is moved a small distance relative to the sample in the xy plane and the process is repeated. A computer is used to calculate the height and position of the tip for each scan, building the image line by line by a process called raster scanning.

This simplified description implies that the AFM tip is passively dragged along the surface during imaging. In reality this is not the case, and the AFM can be operated in different ways to increase its data acquisition rate, decrease sample damage, and gain data in addition to surface topography. Here we focus on the two most common imaging methods, contact mode and tapping mode AFM.

12.3.3 Using AFM for imaging

Contact mode

Contact mode is the simplest method of imaging. In this technique the change in height of the tip is monitored as it is scanned over the sample surface. One approach to contact mode operation is simply to bring the tip into contact with the sample surface and then

Figure 12.8 Schematic diagram showing the components of an AFM. The position of the tip is calculated by measuring the displacement of the laser reflected from the back of the tip onto the photodetector. This feedback signal is used (via a computer) to change the distance between the AFM cantilever and the scanning stage to maintain a constant degree of cantilever deflection as it moves across the sample surface. The height change required to maintain this constant deflection is used to produce the AFM height image as the tip is scanned across the sample surface.

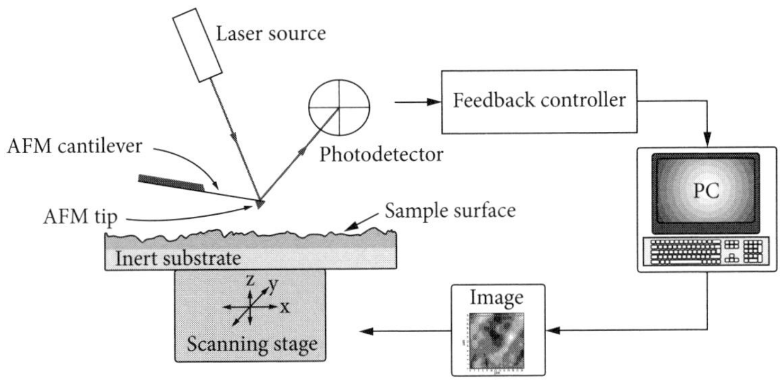

raster scan the sample while keeping the scanning stage at a fixed distance below the tip (Figure 12.9(a), top). This method, termed constant-height contact mode, is useful for rapidly obtaining high-resolution images of the relatively stiff and hard samples that are of interest to the materials and electronics industries, but is of little use for imaging soft biological samples (proteins, lipid bilayers, cells, and tissues). This is because the tip can exert a significant force onto the sample, leading to damage of soft samples. This is especially true for samples with large variations in surface height, as the force exerted on to the sample is directly proportional to the deflection of the tip and therefore the change in height of the sample. A simple way to minimize the **tip-sample interaction** force is to maintain a constant deflection of the AFM cantilever during the scanning process irrespective of the topography of the sample. To do this, feedback electronics (known as a **feedback loop**) is used to move the scanning stage up or down to maintain a constant tip–sample interaction force (Figure 12.9(a), bottom). This is known as either constant-deflection or constant-force AFM and is usually the mode of operation implied when contact mode AFM is referred to. The advantage of this technique is that it leads to less sample degradation and improves the imaging of soft biological samples. The disadvantage is that a longer time is needed to acquire an image (i.e. reduced time resolution) compared with constant-height methods because of the time required for the feedback loop to operate. An image obtained using contact mode AFM is shown in Figure 12.10(a).

Tapping mode

The tip–sample interaction can also be minimized during AFM imaging by rapidly oscillating the AFM cantilever, with an amplitude of typically ~100nm in air and much less in liquid, as it scans across the sample (Figure 12.9(b), top). This mode of AFM operation is known as the 'tapping mode' or intermittent contact mode AFM. How does this prevent sample damage? As we have seen for constant-force mode methods, sample damage is decreased if the interaction force is minimized, which in turn is minimized by

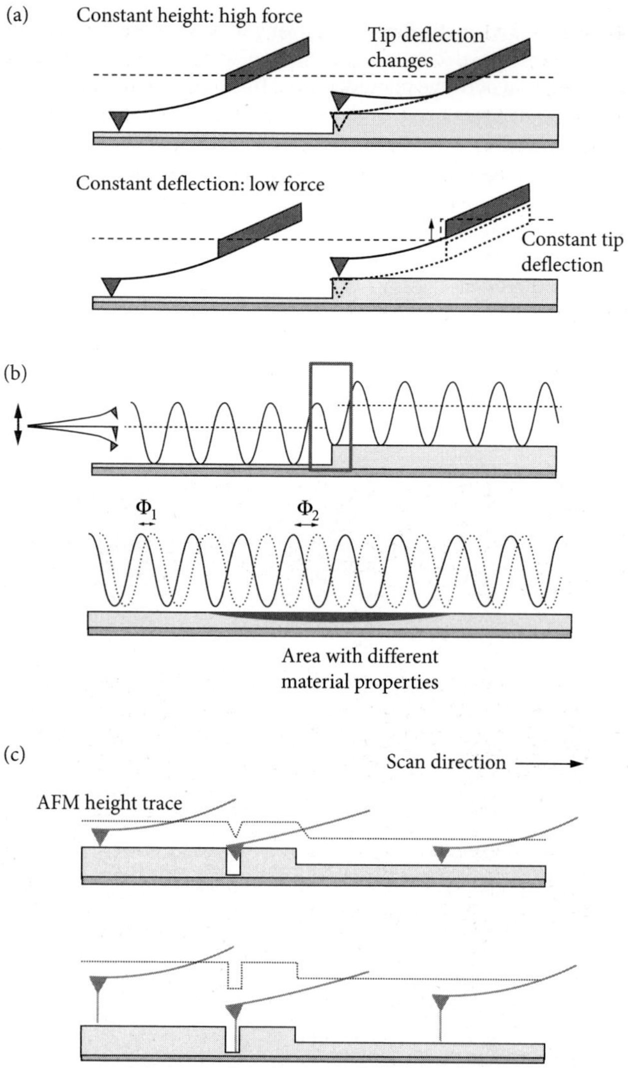

Figure 12.9 Outline of contact and tapping mode AFM imaging methods and the effect of aspect ratio on image resolution. (a) In constant-height contact mode (top) the interaction force between the tip and the surface increases on encountering a step on the sample surface. In constant-deflection or constant-force mode (bottom) the interaction force is maintained at a constant value on encountering the same step. In this case the change in distance between the scanning stage and the cantilever substrate is used to produce the AFM height image. In each diagram the position of the cantilever substrate and AFM tip before encountering the surface feature is shown by a broken line. (b) Top: in intermittent contact mode AFM, the AFM tip is oscillated at high frequency (the mean position of the tip is shown as a broken line). Upon hitting a step the decrease in cantilever oscillation amplitude (enclosed in a rectangle) is detected and the feedback loop maintains constant amplitude by increasing the distance between the scanning stage and the cantilever substrate. This change in distance is used to produce the AFM height image. Bottom: measuring the phase signal. The difference in the position of the maxima and minima of the drive (solid wave) and response signals (broken wave), known as the phase (Φ) increases upon contact with a change in the material properties of the sample surface (dark shaded area). (c) Representation of the effects of tip shape on image quality (tip convolution effects).

reducing cantilever deflection. In tapping mode AFM, the high-frequency oscillation allows the interaction force to be monitored rapidly as measurement of the amplitude of the oscillation reveals how close the tip is to the surface: when far from the sample's surface, the amplitude of the oscillation will remain constant; when close to the surface, the full amplitude of the cantilever's oscillation is prevented as the tip makes intermittent contact with the sample. The closer the tip is to the sample, the greater the reduction in amplitude. By making use of a feedback loop it is possible to set the cantilever oscillation amplitude to a slightly lower value than the full amplitude of the oscillation when away from the surface. Thus the AFM tip 'feels' its way along the surface, rapidly responding to increases or decreases in its amplitude by moving the scanning stage closer to, or further from, the surface. In this mode of AFM operation smaller forces are applied to the sample and the AFM tip is only in contact with the sample for a short part of its oscillation cycle. This has the effect of reducing sample degradation. Therefore tapping mode

Figure 12.10 AFM can be used to image a wide variety of biomolecules and biological materials. (a) High-resolution contact mode AFM image of a membrane protein complex involved in bacterial photosynthesis, light harvesting, and charge separation. A central RC-LH1 complex is surrounded by seven LH2 complexes. (b) Tapping mode AFM image of an Alzheimer plaque directly excised from a human brain. (c), (d) height and phase image of a model cell membrane (under buffer) consisting of a mixture of DOPC (1,2-dioleoyl-sn-glycero-3-phosphocholine), sphingomyelin, and cholesterol. This mixture phase separates into two co-existing liquid phases—isolated liquid-ordered domains (L_o phase) surrounded by a continuous liquid-disordered (L_d) phase at room temperature. The phase contrast image indicates a difference in mechanical properties between the two phases. The height cross section of this model membrane (bottom) shows that the L_o domains are 0.8nm greater in depth than the L_d domains. The position of cross-section is shown on the height image. Image (a) reprinted from Scheuring, S., et al. (2007) *J Struct Biol* **159**: 268–76, with permission from Elsevier. Images (b), (c), and (d) courtesy of Dr Simon Connell, University of Leeds.

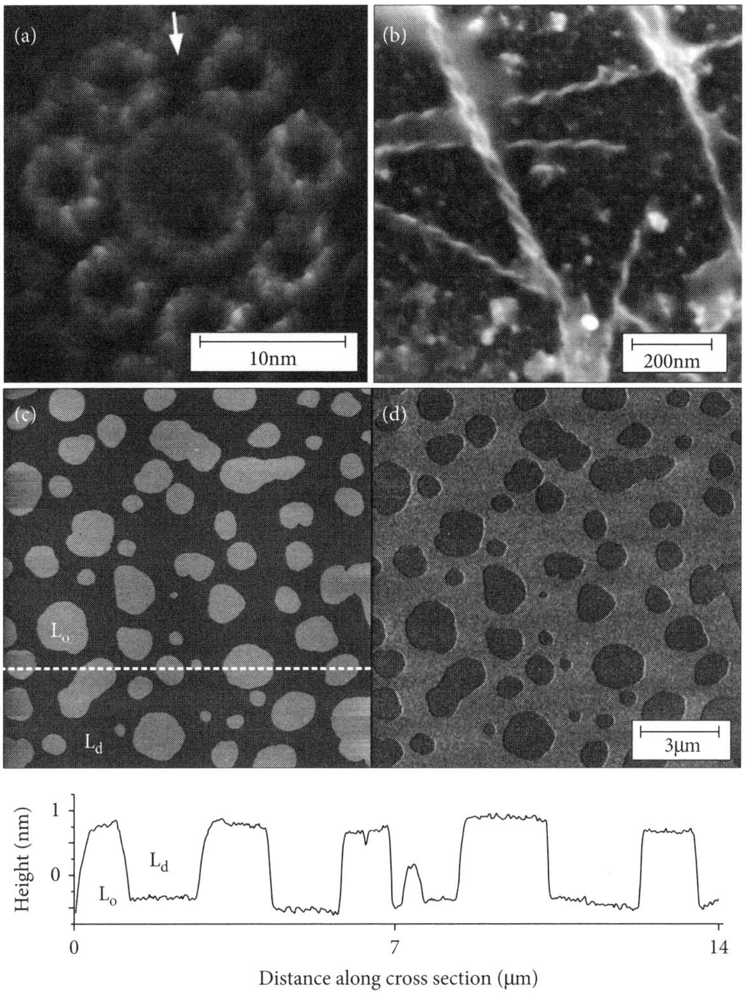

AFM can be used to image more weakly immobilized samples and, with careful selection of imaging parameters, is gentle enough to observe single protein molecules bound to DNA and single polymer chains and ordered protein aggregates (Figure 12.10(b)) as well as lipid bilayers (Figures 12.10(c) and 12.10(d)).

Tapping mode AFM can also be used to acquire additional information on the mechanical properties of a sample by monitoring the phase difference between the signal used to drive the oscillating cantilever and its actual response. The softer or stickier a particular region of a sample, the further behind (or out of phase) the response of the cantilever will be relative to the driving signal for the cantilever's oscillations (Figure 12.9(b), bottom). In the absence of solvent (for biological samples, this is water) the phase signal is directly related to the energy dissipated away from the AFM cantilever into the sample. By monitoring this phase signal it is possible to map the surface mechanical properties of a sample at the same time as obtaining topographic information, and this map is known as the 'phase image' (Figure 12.10(d)). However, the necessity of a liquid environment for biological samples presents several problems. First, the oscillation of an AFM cantilever under water is **damped** and a larger driving signal is required to obtain adequate tip oscillation amplitude. This is problematic because it exacerbates the excitation of the liquid environment that is also induced by the driving signal. Reflections of these induced acoustic waves throughout the liquid interfere with the cantilever's oscillation. Consequently, phase measurements under liquid yield only qualitative data.

However, advances in cantilever excitation methods, the measurement of higher vibrational cantilever modes, and the manufacture of cantilevers of different shapes and sizes with ultrasharp tips have enabled improvements in temporal or image resolution and reliable cantilever excitation under liquids. In addition, real-time data can be gathered. For example, the Ando group have recently used 'high-speed AFM' to acquire 'movies' of myosin walking along surface-immobilized actin in real time (Kodera et al. 2010).

12.3.4 Resolution of AFM images

For an ideal sample, the resolution of an AFM image is ultimately limited by the tip's sharpness, which is quantified by its **radius of curvature** (R_C). The role that tip sharpness (also known as the **aspect ratio**) plays in determining the resolution is shown schematically in Figure 12.9(c). Imagine a sample that has deep channels or valleys with steep sides; the ability of a tip to map the valley floor is determined by steepness of the valley relative to the sharpness of the AFM tip (compare Figure 12.9(c), top and bottom). The broader the tip, the shallower the feature becomes until the tip is so large that it does not enter the valley at all. In addition to errors in estimating the depth (and, following a similar argument, height) of topographical features, broader tips also distort the profile of surface topography as the sides of the tip prevent the apex from closely mirroring the surface features.

Tips with R_C = 10nm are routinely used, and sharper tips (R_C = 2nm) are now commercially available. The sharpness can also be increased by sculpting standard tips using etching methods (e.g. reactive ion etching) or by derivatizing with thin but strong **carbon nanotubes**. These dimensions are smaller than the wavelengths of light used for almost all spectroscopic methods, hence AFM has the potential to achieve very high resolution—even atomic resolution under appropriate conditions (Sokolov et al. 1999; Fukuma 2010). However, the highest-resolution AFM images are obtained under vacuum as atmospheric moisture and even the interaction of the cantilever with the surrounding gas can reduce resolution. Unfortunately, biological samples are far from ideal as most are soft (deform and become damaged during imaging) and require an aqueous environment to remain functional. These constraints limit the resolution of biological samples when observed with AFM to ~1nm at best to typically a few nanometres.

12.3.5 Using the AFM as a force sensor

As the tip directly contacts the sample, AFM can be used to measure the mechanical properties of, and the effects of force on, single biomolecules (DNA and proteins), tissues such as collagen, and whole cells. At the cellular level, the magnitude of the applied force is usually extremely small (~5–100 pN). AFM is able to quantify these small forces because the cantilever behaves like a small spring whose displacement (d) is proportional to the magnitude of the applied force (F):

$$F = kd \qquad 12.4$$

where k is a proportionality constant. Therefore if the proportionality constant (the spring constant in this case) and displacement are known, the applied force can be calculated.

One parameter, **Young's modulus** (E), can be measured by looking at the force exerted on the AFM tip as it is pushed into a sample (Figure 12.11(a)). Young's modulus describes the **elastic** properties of materials and is related to its **stiffness**. For comparison, Young's moduli for steel, collagen, and living cells are 200GPa, 1GPa and 1–100kPa, respectively. By looking at force–extension profiles as a function of the rate at which the AFM tip is pushed into and pulled away from the sample, it is also possible to extract information regarding the sample's **viscoelastic** response. Viscoelastic materials display more liquid-like (viscous) or more solid-like (elastic) behaviour depending on the timescale they are observed over. Such methods can discriminate between cancerous and non-cancerous cell types as each cell type displays a distinct force–extension profile (Lekka et al. 1999a, b).

The ability to manipulate and measure the mechanical properties of single biomolecules by AFM or with **optical and magnetic tweezers** has led to the development of single-molecule force spectroscopy. In this technique, AFM is used to measure the forces required to unfold single-protein molecules that are immobilized between the AFM tip and substrate (Figure 12.11(b), left). Such data have revealed that some proteins are activated by force and some resist mechanical **deformation**, while other proteins unfold under low force but refold rapidly when the perturbation is removed. These latter proteins can act as mechanical safety valves that prevent cellular damage. More information on this topic can be found elsewhere (Crampton and Brockwell 2010).

Instead of examining the unfolding force of the non-covalent interactions within a folded protein, the force required to break apart specific non-covalent interactions between a chemical group immobilized on an AFM tip and a sample surface can also be measured as the tip is scanned across the surface of a biological sample such as a cell or tissue. This technique is known as recognition microscopy. Functional groups can include oligonucelotides, peptides, and proteins, and these can be used to identify and localize cell surface receptors specific for these ligands. Measurement of the force required to break the receptor–ligand complex apart (the rupture force) can be used to obtain insight into how mechanical signals are detected and relayed at the cellular level (Hinterdorfer and Dufrene 2006). A related technique, dynamic force microscopy, measures how the rupture force of a protein and its ligand, which are immobilized on the AFM substrate and AFM tip, respectively, varies as a function of the speed at which the complex is pulled apart. The effect of a mechanical force on the complex can be identified by comparing the dissociation rate of a complex in the absence and presence of the force. Surprisingly, applying a mechanical force to some complexes involved in cell

Figure 12.11 Using AFM as a force sensor. (a) Left: comparison of force–distance profiles obtained when the AFM tip is pressed onto the surface on an *E. coli* cell immobilized on a silicon substrate (green line) and directly onto the silicon surface (red line). Distance is plotted as tip–sample separation; hence a value of zero indicates that the tip is resting on the surface with zero deflection (and hence no force), and a negative value indicates that the cantilever is deflecting upwards (and therefore applying a force that increases as the cantilever moves closer to the surface). Measuring the gradient of the force versus tip–sample separation profiles at negative values of tip–sample separation allows the stiffness to be calculated and demonstrates that *E. coli* cells are softer than the silicon substrate. Right: repeating this experiment over a grid generates low-resolution images that can reveal the mechanical properties of a sample (if measuring indentation) or the presence of a specific interaction (if measuring an adhesion force). In this case, the distance between the tip and the silicon substrate upon reaching a pre-set deflection force is plotted (the larger the distance, the lighter the colour). The image shows a cluster of ellipsoid *E. coli* cells immobilized on the substrate. Red and green circles show where the force–distance profiles shown on the plot were measured. (b) Unfolding a single polyprotein chain using an AFM. Left: cartoon of the sequential unfolding of individual domains (shown in green) of the polyprotein as the AFM tip is retracted from the substrate surface. The tip is pushed onto a surface that is covered with a polyprotein that is tethered to the surface at one end (1). If the tip binds to the other end of the polyprotein, retraction of the tip subjects each domain to an incrementing force (2) until one domain suddenly unfolds (asterisk), increasing the end-to-end length of the polyprotein which reduces the force applied to the other domains (3). This process is repeated until all domains are unfolded. The final high-force rupture event occurs when any of the interactions that link the substrate to the tip fails. Right: a force–extension profile for the unfolding of a single polyprotein comprising five protein L domains (Brockwell et al. 2005). Each unfolding event is identified by an asterisk. The distance between each unfolding event is proportional to the number of amino acids released from the native state upon unfolding. The numbers refer to the unfolding sequence described above. For a colour reproduction of this figure, please see Plate 11.

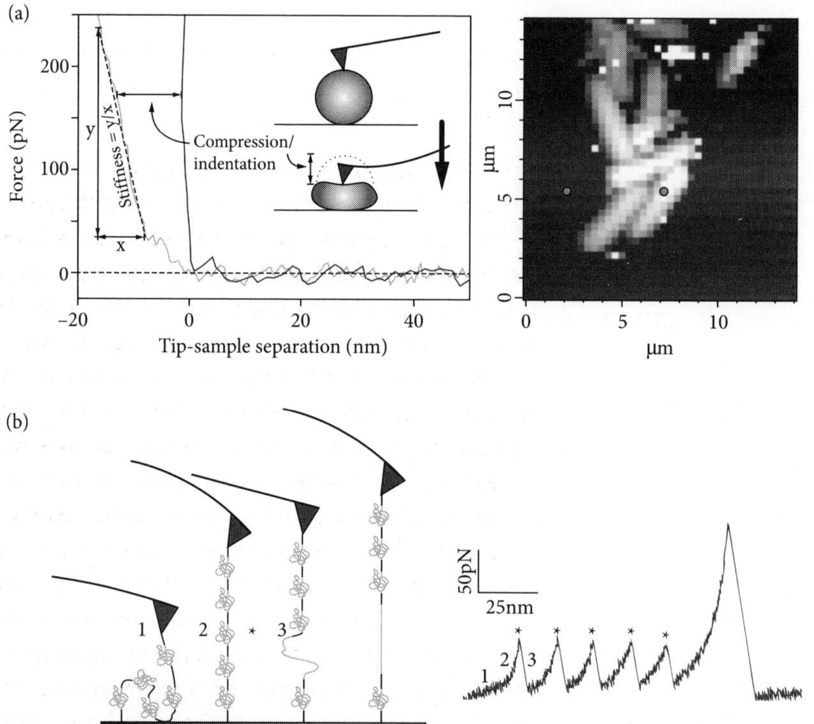

adhesion strengthens the interactions between the proteins involved. This allows a white blood cell or a bacterium to resist **hydrodynamic flow** in a blood capillary (Thomas 2009) or the urinary tract (Thomas 2009), respectively. See Zlatanova et al. (2000) for a more detailed account of this area of research.

CASE STUDY 12.2

Determination of the subunit architecture and stoichiometry within a protein complex

Many proteins associate with one or more other proteins to form functional complexes. These complexes can comprise a series of different proteins, identical proteins, or mixtures of both. For example, the F_1 component of ATP synthase comprises three pairs of alternating α- and β-subunits arranged around the end of an elongated γ-subunit. In order to understand their function it is often useful to map the structural arrangement of subunits within a complex. While this information can be obtained using X-ray crystallography, the inability to form suitable crystals precludes this method.

QUESTION 1
How can AFM be used to determine the stoichiometry and arrangement of different subunits within a membrane protein complex?

Many membrane proteins can be incorporated into liposomes in a functional state which can then be rolled out onto an inert flat substrate. Alternatively, membranes can be solubilized in detergent and immobilized on an inert substrate. If the protein complexes are packed into the membrane at a density sufficient to form a close-packed array, direct imaging by contact mode AFM may yield a top-down image from which the arrangement of subunits can be visualized.

If each subunit type is similar in topology or sufficient resolution cannot be obtained, an indirect method can be used that utilizes both the specificity of antibodies for their epitopes and their relatively large size. Each subunit should be genetically modified to express an epitope for a distinct monoclonal antibody at a solvent-accessible region of the protein. After immobilization, different samples of the membrane are incubated with each antibody before extensive wash steps are performed.

Antibodies are resolvable under AFM imaging and so counting the number of a particular antibody attached to each complex will reveal the number of that particular subunit. Repeating this experiment with different labels will yield the stoichiometry directly.

If subunits have more than one copy in the complex, measuring the angles between identical subunits for each label allows the arrangement of subunits within the complex to be deduced. Note that if only one subunit is present in the complex separate labels that can be distinguished under AFM are required and multiple subunit types must be labelled under the same experiment in order to calculate subunit arrangement.

QUESTION 2
What caveats should be considered during data interpretation?

Any labels lying directly below or above the complex may be difficult to resolve, in contrast with those lying to the side of the complex. Solubilization, immobilization, and introduction of labels may influence subunit arrangement.

An approach similar to this is described by Barrera et al. (2008).

12.4 Circular dichroism spectroscopy

Circular dichroism (CD) spectroscopy is a simple but powerful technique for observing the structures of biomolecules, and how they change in response to stimuli. Changes in structure are important for many aspects of molecular function, including catalysis, ligand binding, and regulation, and CD spectroscopy can provide a rapid way of measuring conformational changes associated with these activities. A major use for CD spectroscopy in biology is to measure protein folding and stability. Proteins are often only marginally stable and their folded structures can be altered, leading to misfolding and aggregation which in turn can lead to diseases such as Alzheimer's and Parkinson's. Thus CD spectroscopy can provide useful insights into how disease-causing mutations, for example, can affect the folding and stability of particular proteins. However, CD spectroscopy is a low-resolution method and therefore cannot provide the residue-specific resolution of techniques such as X-ray crystallography (12.1) and NMR (12.2). The basic principles underpinning CD measurements and how the data can be interpreted are described in the following sections.

12.4.1 Interaction of circularly polarized light with chiral molecules

Circular dichroism is the interaction of circularly polarized light with **chiral molecules**. Light is made up of electric and magnetic fields, which oscillate around the light path. In 'normal' unpolarized light, the electric and magnetic components are randomly oriented, perpendicular to the direction of the light beam, and average out. Polarization means that the waves are aligned. Plane polarization is where the waves have a fixed orientation, whereas you can imagine circular polarization as the alignment moving around the light beam. Light can be plane polarized in any direction, but circular polarization must be rotating either left or right (Figure 12.12). Unpolarized light is made up of an equal balance of left and right circular polarized components, and the process of polarization effectively removes one part of the light, leaving a net rotation. (Fortunately, it is not necessary to understand the physics to record or interpret the data.) This 'handedness' means that left or right circular polarized light interacts differently with chiral molecules that have asymmetric structures.

Many biomolecules are chiral. For example, all known proteins are made from L-amino acids. Bonds around chiral centres absorb left and right circular polarized light differently, and this is what a CD spectometer (also known as a spectrapolarimeter) measures. A spectrapolarimeter's light source switches between left and right polarized light at 50kHz and the machine compares the sample's absorbance of each type of light. The difference between the two signals is called the **ellipticity** of the sample. It is designated by the symbol θ (theta) and is measured in degrees (or millidegrees as the signal is generally very small). The ellipticity is corrected for the concentration of protein, the number of chromophores (light-absorbing groups such as peptide bonds) in the protein, and the pathlength of the cuvette to give the **mean residue ellipticity** (MRE, [θ]) of the sample:

$$[\theta] = 100 \times \theta / c \times l \times (n - 1) \qquad (12.5)$$

Figure 12.12 How CD works. Top: polarization of light describes the orientation of the electric and magnetic fields (lines) that make up a beam of light, shown by the arrows. Bottom: CD measures the difference in absorbance of left and right circularly polarized light by biomolecules in the sample to generate an ellipticity spectrum. CD also measures the sum absorbance of left and right circularly polarized light by biomolecules in the sample to generate a sum (total) absorbance spectrum.

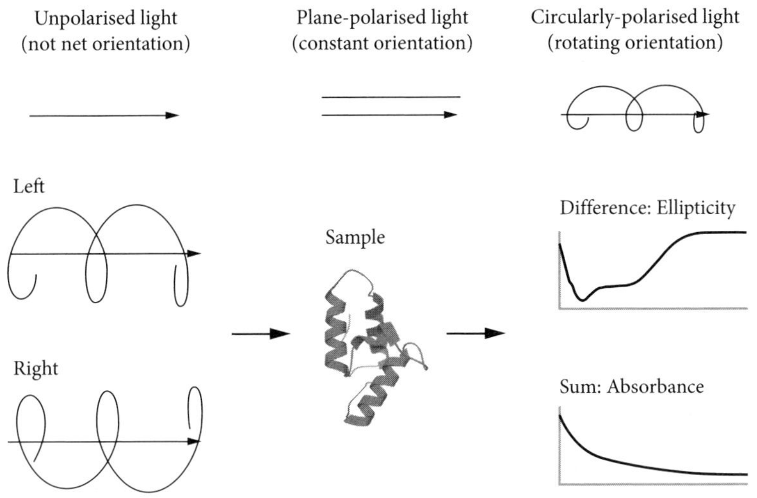

where θ is the measured ellipticity (in degrees), c is the concentration of protein (in moles per litre), l is the pathlength of the cuvette (in centimetres), and n is the number of residues in the protein. The factor of 100 corrects for the pathlength being measured in centimetres. The units of MRE are deg cm^2 $dmol^{-1}$.

As well as ellipticity, the spectrometer also measures the sum of the absorbances of left- and right-polarized light, which is the total absorbance of the sample. This is reported as a voltage. If the absorbance is too high, not enough light reaches the detector and hence the ellipticity measurement cannot be made accurately. This is described further below.

12.4.2 Assessing protein secondary structure by CD

The main application of CD spectroscopy in biology is measuring the secondary structure content of proteins (α helices and β sheets), and how this changes when a protein unfolds or binds to another molecule. It is also possible to observe other processes, such as packing of aromatic groups and the binding of cofactors (Kelly et al. 2005).

Any protein in solution can be studied by CD. There is no upper size limit, and it is possible to observe membrane proteins if they are solubilized in detergents or purified in liposomes. However, since it is not possible to differentiate between proteins by CD, it is best used on purified proteins. Generally, CD is used to characterize pure proteins which have been identified previously. By combining CD detection with **stopped-flow fluidics**, it is possible to observe rapid changes in conformation. CD is complementary to other optical spectroscopy methods such as fluorescence or infra-red in that it can rapidly measure structural properties in unmodified biomolecules (Jackson and Mantsch 1995). A related technique, linear dichroism, uses linearly polarized light to study fibrous macromolecules, but it is not commonly used for biomolecules (Dafforn and Rodger 2004). Here, we will focus on using CD to measure protein backbone conformations as one illustration of the use of CD spectroscopy.

12.4.3 The CD spectrum

Secondary protein structure is made up of repeating patterns of protein backbone phi/psi angles (Fersht 1999). Peptide bonds absorb UV light at wavelengths between 190 and 230nm. Unstructured bonds rotate freely and have no CD signal. However, when bonds are constrained in protein structures, they have a fixed orientation that can be detected by CD. The CD spectra of the secondary structures (the α-helix and the β-sheet) are shown in Figure 12.13. In a CD spectrum, wavelength (*x*-axis) is plotted against **mean residue ellipticity** (MRE) (*y*-axis). The important features are (1) the curves are negative for folded proteins in the region between 200 and 250nm, and (2) α-helices and β-sheets have distinctive shapes. In particular, look at how the α-helical spectrum (solid line in Figure 12.13) has a double-dip shape with MRE minima at 208 and 222nm, whereas β-sheets (broken line in Figure 12.13) have a single negative MRE minimum at 215nm. In contrast, unstructured or unfolded proteins (dotted line in Figure 12.13) have very little signal at wavelengths greater than 210nm. Remember that CD spectra show the difference between two absorbance values. The direction of the peaks is arbitrary but consistent. By convention, folded proteins have a negative signal between 200 and 240nm. The maxima and minima represent wavelengths at which differential absorbance of circularly polarized light is the strongest, which depends on the electron density distribution of the peptide bond. The *y* axis in Figure 12.13 shows the MRE, which is the average

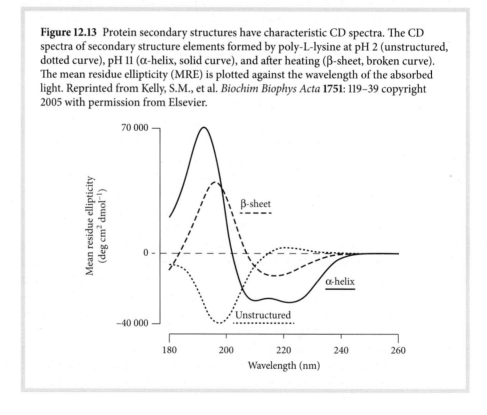

Figure 12.13 Protein secondary structures have characteristic CD spectra. The CD spectra of secondary structure elements formed by poly-L-lysine at pH 2 (unstructured, dotted curve), pH 11 (α-helix, solid curve), and after heating (β-sheet, broken curve). The mean residue ellipticity (MRE) is plotted against the wavelength of the absorbed light. Reprinted from Kelly, S.M., et al. *Biochim Biophys Acta* **1751**: 119–39 copyright 2005 with permission from Elsevier.

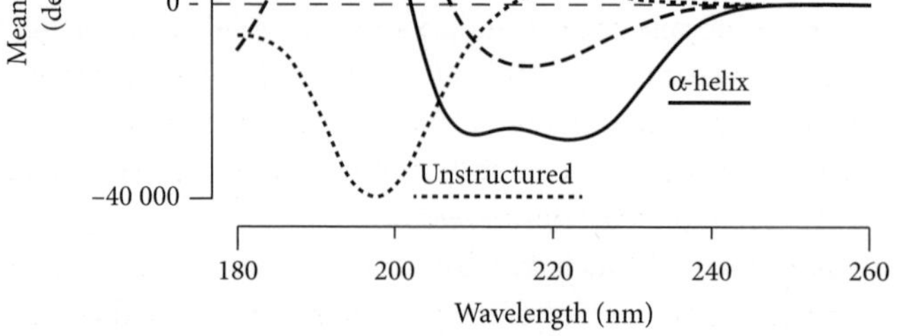

ellipticity of each peptide bond in the protein. Calculating MRE allows spectra from different proteins to be compared, as well as correcting for concentration. MRE is a measure of structure. A purely α-helical protein would have an MRE of about −30 000deg cm^2 $dmol^{-1}$ at 222nm, so an MRE value of −10 000 suggests that a protein is 33% helical. MRE does not depend directly on the stability of a protein, but *changes* in MRE can be used to calculate stability.

Since most proteins are composed of regular secondary structure elements, we can estimate the secondary structure content of a protein by fitting its CD spectrum to a combination of the spectra of isolated units of secondary structure. There are several web-based calculators for this purpose, such as Birkbeck College's DichroWeb (http://dichroweb.cryst.bbk.ac.uk/html/home.shtml). However, in order to obtain an accurate estimate, it is important to acquire high-quality data and to measure the concentration of the protein accurately.

In parallel with the CD signal (MRE spectrum), the CD spectrometer also measures total absorbance and displays it as a voltage (called HT, for 'high tension', or 'Dynode' on different spectrometers). This is an indicator of the reliability of the CD signal and discussed further in the next section.

12.4.4 Using CD to measure protein folding and stability

A CD spectrum showing α-helix or β-sheet character is a good indication that the protein is at least partially folded, but does not necessarily tell you if it is folded to its native state. In most cases, the magnitude of the CD signal does not correlate with stability. Therefore, to measure stability, you must measure how the ellipticity changes when the protein is destabilized, for instance by heating, changing the pH, or adding **denaturants** such as urea. This is illustrated in Figure 12.14 using the protein Im7 as an example.

Figure 12.14 shows the characteristic MRE minima at 208 and 222nm for α-helices (refer back to Figure 12.13), as you would expect from its helical crystal structure at 0M urea. Adding the denaturant urea to the protein causes it to unfold, so that its helical CD signal (MRE) is reduced, while its absorbance (HT) increases. By titrating urea into the protein (taking care to keep the protein concentration constant by diluting from a common stock solution), it is possible to work out the urea concentration at which the protein is 50% unfolded or determine the equilibrium constant for the folding–unfolding reaction, which are both measures of the protein's stability. The workflow and data from this experiment are shown in Figure 12.15; look at how the ellipticity has a distinct change at around 4M urea. The sigmoidal shape of the curve shows that the protein unfolds in a cooperative all-or-nothing way, characteristic of folded globular proteins. Altering the protein's stability by mutation, changing the temperature, or adding molecules that bind to it will cause the protein to unfold at higher or lower concentrations of urea, showing that it is more stable or less stable, respectively. An example of this kind of study is comparison of the stabilities of wild-type and mutant proteins, where the mutation is associated with disease. If the mutant protein is destabilized, or even fails to fold entirely, this could be the cause of the disease.

Many proteins do not fold to a specific structure in isolation. These 'intrinsically unstructured' or 'natively disordered' proteins often only fold into their active conformations when they bind to their biological partners. This recognition process is particularly

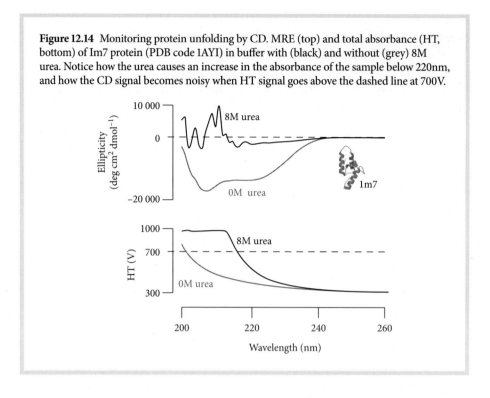

Figure 12.14 Monitoring protein unfolding by CD. MRE (top) and total absorbance (HT, bottom) of Im7 protein (PDB code 1AYI) in buffer with (black) and without (grey) 8M urea. Notice how the urea causes an increase in the absorbance of the sample below 220nm, and how the CD signal becomes noisy when HT signal goes above the dashed line at 700V.

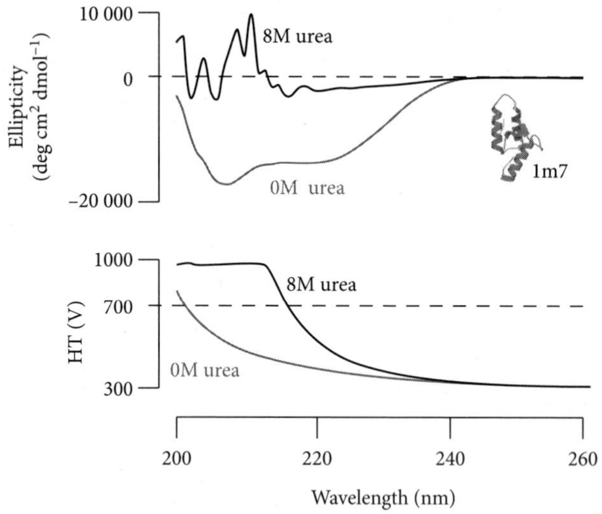

important for eukaryotic signalling proteins, and may help in recognition and regulation of targets (Dyson and Wright 2005). CD can show that one protein folds when it binds to another by measuring the ellipticity of the individual proteins and their complex. If the complex has more secondary structure than the sum of the isolated proteins, one protein must be folding as it binds. Again, it is crucial to keep the concentrations of the two proteins identical between samples.

12.4.5 Acquiring good CD data

To obtain good CD data, we need a sample of pure protein solution that is free from other molecules that also absorb in the peptide bond region of the UV spectrum. A major problem is chloride ions, which are present in many buffers used for biological samples and absorb in this region. Phosphate buffers do not absorb UV light at these wavelengths, which makes them useful for CD. If a high ionic strength is required to keep the protein soluble, sodium chlorate, which is transparent above 190nm, is a useful alternative to sodium chloride. Generally this problem is worse at shorter wavelengths. If the sample absorbs too much UV light, the subtle difference between the absorbance of left and right polarized light is swamped. Therefore it is good practice to show the HT data to demonstrate that the CD signal is reliable. At too high an absorbance, the data become unreliable and noisy; this can be seen in Figure 12.14 as HT passes 700V. Very accurate determination of secondary structure needs CD data extending to 190nm, which can be difficult to acquire. Narrow cuvettes (less than 1mm pathlength) can help, as long as the protein sample is concentrated enough to measure the CD effect.

Measuring protein concentration accurately is important if different protein samples are to be compared using CD. The magnitude of the CD signal is directly proportional to protein concentration, which is difficult to measure accurately. It is often more useful

Figure 12.15 Workflow for a protein stability CD experiment. Samples of protein at different concentrations of urea have their ellipticity measured at 222nm (where the helix signal is large but the absorbance from urea is still low), and the data are plotted to give the sigmoidal curve at the bottom of the figure. The data can be fitted to an equation describing the unfolding transition; see Fersht (1999) for more details.

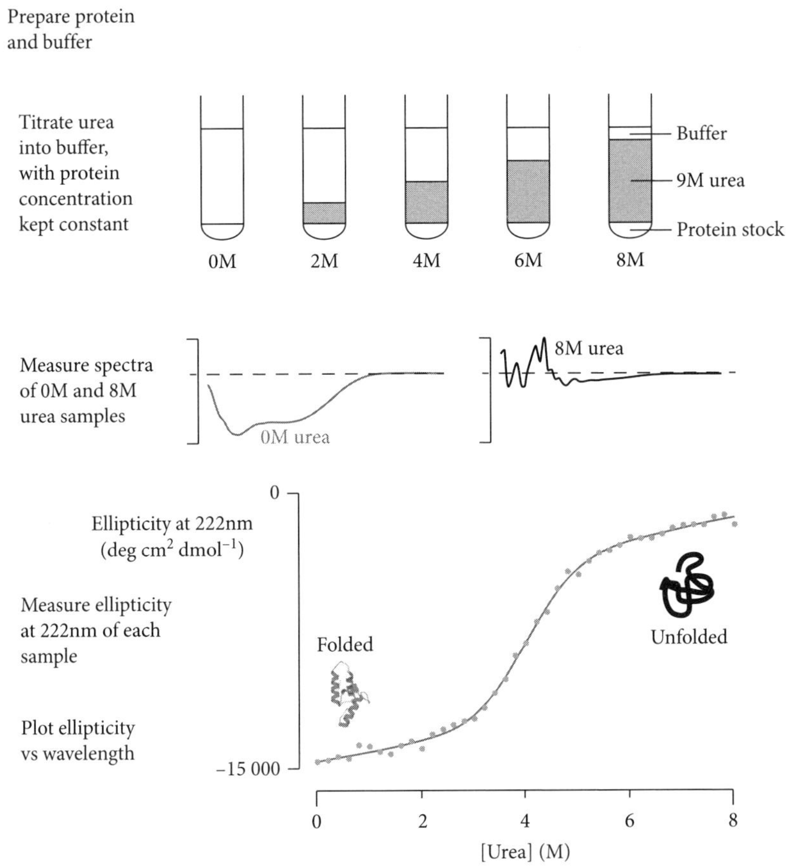

to look at relative changes in ellipticity, for instance between samples of protein from the same stock in different buffers, or with and without ligand. For a typical experiment, a concentration of 1–2mM peptide bonds (which works out to a protein concentration of around 0.1–0.2mg/mL) in a 1mm pathlength cuvette will give a reasonable signal. This information is generally known for the protein of interest before studying it with CD as it can easily be calculated from the protein's sequence.

12.5 Chapter summary

- **X-ray crystallography experiments generate precise high-resolution data allowing a model of the sample at the atomic level to be built with high confidence. Therefore it is the method of choice for the study of previously undetermined protein structures.**

- **In addition to structure determination at atomic resolution, solution-state NMR also allows a wide range of complementary information to be studied. This includes the size of molecular motions, the presence of remaining structure in disordered proteins, fast mapping of the location of ligand-binding sites, and assessment of protein-folding.**

- **AFM bridges the structural information gained by high-resolution techniques such as X-ray crystallography and NMR and lower-resolution techniques such as light microscopy. AFM can be used to provide a 3D profile of the topography of the sample under study and to investigate the mechanical properties and the effects of force on single proteins and complexes, cells, and tissues.**

- **CD spectroscopy is a simple way to measure the conformation and stability of proteins in solution. However, it is a low-resolution method and therefore cannot provide the residue-specific resolution of techniques such as NMR and X-ray crystallography.**

✓ References

Abou-Saleh, R.H., Connell, S.D., Harrand, R., et al. (2009) Nanoscale probing reveals that reduced stiffness of clots from fibrinogen lacking 42 N-terminal B beta-chain residues is due to the formation of abnormal oligomers. *Biophys J* **96**: 2415–27.

Barrera, P.N., Betts, J., You, H., et al. (2008) Atomic force microscopy reveals the stoichiometry and subunit arrangement of the $\alpha_4\beta_3\delta$ $GABA_A$ receptor. *Mol Pharmacol* **73**: 960–7.

Brockwell, D.J., Beddard, G.S., Paci, E., et al. (2005) Mechanically unfolding the small, topologically simple protein L. *Biophys J* **89**: 506–19.

Buzhynskyy, N., Liu, L.-N., Casuso, I., and Scheuring, S. (2011) High-resolution atomic force microscopy of native membranes. In: Dufrene, Y. (ed.), *Life at the Nanoscale*. Singapore: Pan Stanford, pp. 21–44.

Cavanagh, J., Fairbrother, W.J., Palmer, A.G., III, Skelton, N.J., and Rance, M. (2006) *Protein NMR Spectroscopy: Principles and Practice* (2nd edn). San Diego, CA: Academic Press.

Crampton, N. and Brockwell, D.J. (2010) Unravelling the design principles for single protein mechanical strength. *Curr Opin Struct Biol* **20**: 508–17.

Dafforn, T.R. and Rodger, A. (2004) Linear dichroism of biomolecules: which way is up? *Curr Opin Struct Biol* **14**: 541–6.

Downing, A.K. (ed) (2010) *Protein NMR Techniques* (2nd edn). New York: Humana Press.

Dyson, H.J. and Wright, P.E. (2005) Intrinsically unstructured proteins and their functions. *Nat Rev Mol Cell Biol* **6**: 197–208.

Engel, A. and Gaub, H.E. (2008) Structure and mechanics of membrane proteins. *Annu Rev Biochem* **77**: 127–48.

Fersht, A. (1999) *Structure and Mechanism in Protein Science: A Guide to Enzyme Catalysis and Protein Folding*. New York: W.H. Freeman.

Fukuma, T. (2010) Water distribution at solid/liquid interfaces visualized by frequency modulation atomic force microscopy. *Sci Technol Adv Mater* **11**(3).

Gaiduk, A., Kühnemuth, R., Antonik, M., and Seidel, C.A. (2005) Optical characteristics of atomic force microscopy tips for single-molecule fluorescence applications. *Chemphyschem* **6**: 976–83.

Hinterdorfer, P. and Dufrene, Y.F. (2006) Detection and localization of single molecular recognition events using atomic force microscopy. *Nat Methods* **3**: 347–55.

Jackson, M. and Mantsch, H.H. (1995) The use and misuse of FTIR spectroscopy in the determination of protein structure. *Crit Rev Biochem Mol Biol* **30**: 95–120.

Kailas, L., Ratcliffe, E.C., Hayhurst, E.J., Walker, M.G., Foster, S.J., and Hobbs, J.K. (2009) Immobilizing live bacteria for AFM imaging of cellular processes. *Ultramicroscopy* **109**: 775–80.

Kalverda, A.P., Thompson, G.S., Vogel, A., et al. (2009) Poxvirus K7 protein adopts a Bcl-2 fold: biochemical mapping of its interactions with human DEAD box RNA helicase DDX3. *J Mol Biol* **385**: 843–53.

Kelly, S.M., Jess, T.J., and Price, N.C. (2005) How to study proteins by circular dichroism. *Biochim Biophys Acta* **1751**: 119–39.

Kodera, N., Yamamoto, D., Ishikawa, R., and Ando, T. (2010) Video imaging of walking myosin V by high-speed atomic force microscopy. *Nature* **468**: 72–6.

Lekka, M., Laidler, P., Gil, D., Lekki, J., Stachura, Z., and Hrynkiewicz, A.Z. (1999a) Elasticity of normal and cancerous human bladder cells studied by scanning force microscopy. *Eur Biophys J* **28**: 312–16.

Lekka, M., Lekki, J., Marszalek, M., et al. (1999b) Local elastic properties of cells studied by SFM. *Appl Surf Sci* **141**: 345–9.

Madl, J., Rhode, S., Stangl, H., et al. (2006) A combined optical and atomic force microscope for live cell investigations. *Ultramicroscopy* **106**: 645–51.

Oda, S., Schroder, M., and Khan, A.R. (2009) Structural basis for targeting of human RNA helicase DDX3 by poxvirus protein K7. *Structure* **17**: 1528–37.

Pochapsky, T. and Sondej Pochapsky, S. (2006) *NMR for Physical and Biological Scientists*. New York: Garland Science.

Rhodes, G. (2006) *Crystallography Made Crystal Clear* (3rd edn). San Diego, CA: Academic Press.

Roberts, G. and Lian, L.-Y. (eds) (2011) *Protein NMR Spectroscopy: Practical Techniques and Applications*. Basingstoke: Wiley–Blackwell.

Rupp, B. (2009) *Biomolecular Crystallography: Principles, Practice, and Application to Structural Biology*. New York: Garland Science.

Scheuring, S., Boudier, T., and Sturgis, J.N. (2007) From high-resolution AFM topographs to atomic models of supramolecular assemblies. *J Struct Biol* **159**: 268–76.

Scheuring, S. and Dufrene, Y.F. (2010) Atomic force microscopy: probing the spatial organization, interactions and elasticity of microbial cell envelopes at molecular resolution. *Mol Microbiol* **75**: 1327–36.

Sokolov, I.Y. and Henderson, G.S. (1999) Theoretical and experimental evidence for true atomic resolution under non-vacuum conditions. *J Applied Phys* **86**: 5537–40.

Thomas, W.E. (2009) Mechanochemistry of receptor-ligand bonds. *Curr Opin Struct Biol* **19**: 50–5.

Touhami, A., Jericho, M.H., and Beveridge, T.J. (2004) High-resolution atomic force microscopy of native membranes atomic force microscopy of cell growth and division in *Staphylococcus aureus*. *J Bacteriol* **186**: 3286–95.

Williamson, R., Carr, M.D., Frenkiel, T., Feeney, J., and Freedman, R. (1997) Mapping the binding site for matrix metalloproteinase on the N-terminal domain of the tissue inhibitor of metalloproteinase-2 by NMR chemical shift perturbation. *Biochemistry*, **36**: 13 882–9.

Wlodawer, A., Minor, W., Dauter, Z., and Jaskolski, M. (2008) Protein crystallography for non-crystallographers, or how to get the best (but not more) from published macromolecular structures. *FEBS J* **275**: 1–21.

Zlatanova, J., Lindsay, S.M., and Leuba, S.H. (2000) Single molecule force spectroscopy in biology using the atomic force microscope. *Prog Biophys Mol Biol* **74**: 37–61.

Mass spectrometry 13

James R. Ault and Alison E. Ashcroft

Chapter overview

This chapter focuses on the use of **mass spectrometry (MS)** for the analysis of biomolecules. The basic components of a mass spectrometer are the ionization source, into which the **analyte** is introduced, the mass-to-charge (m/z) analyser, and a detector, all of which are under data system control. We will discuss the importance of sample preparation and describe the two ionization techniques recommended for the analysis of biomolecules—electrospray ionization and matrix-assisted laser desorption ionization. Details of the range of m/z analysers which are in common use in this field will also be given.

We will then move on to describe some of the applications of mass spectrometry in biomolecular research. The technique is most commonly used to determine the molecular mass of a biomolecule with high sensitivity and excellent mass accuracy, but can also be used to generate a plethora of other information for many applications. Here, we will describe peptide sequencing, the study of non-covalent interactions in biomolecular complexes, and the detection of pharmaceutical products in river water.

LEARNING OBJECTIVES

This chapter will enable the reader to:

- understand the basic principles of how mass spectrometers work;
- describe how samples are prepared for analysis;
- describe the types of bioanalyses which can be carried out by mass spectrometry including
  - how a molecular mass is obtained from a mass spectrum
  - how mass spectrometers can be used to fragment, or dissociate, analytes and generate structural information (e.g. identification of metabolites, or peptide sequencing)
  - how MS can be used to investigate non-covalent interactions of biomolecules.

13.1 Introduction

13.1.1 What is mass spectrometry?

Mass spectrometry (MS) is an analytical technique that can be used to determine the molecular mass of chemical compounds, such as small organic molecules (e.g. drugs, hormones, and metabolites) or much larger biomolecules (e.g. lipids, carbohydrates, oligonucleotides, and proteins), with good mass accuracy and high sensitivity. Biochemists, molecular biologists, chemists, and clinicians are amongst those who use MS for a wide variety of reasons to gain information about the identity and structure of biomolecular compounds. The analyses undertaken can include, for example, checking the molecular mass and purity of a recombinant protein or observing protein–ligand binding. More in-depth analyses may involve identification of a number of proteins within a mixture, characterizing the structure of metabolites, or determining the sequence of amino acids or nucleotides in peptides and oligonucleotides, respectively.

Mass spectrometers measure the **mass-to-charge ratio** (or mass per unit charge, denoted ***m/z***) of analyte ions in the gas phase. From this measurement, the molecular mass of the analyte can be determined. Any analyte under investigation must be introduced into the mass spectrometer and transferred from the solid, liquid, or solution phase into the gas phase as charged ions before a mass measurement can be performed. The following steps all take place in a full analysis:

- introduction of the analyte into the mass spectrometer;
- transfer of the analyte from the solid, liquid, or solution phase into the gas phase;
- conversion of analyte molecules into charged ions (the ions produced may be positively or negatively charged depending on the voltages applied during this ionization step);
- transfer of the analyte ions from the **ionization source** into the analyser region of the mass spectrometer;
- separation of the analyte ions according to their mass-to-charge ratios;
- detection of the analyte ions and production of the *m/z* spectrum.

The basics of how each of these steps is carried out, with particular emphasis on the analysis of biomolecules, are described in the following sections.

13.1.2 Uses of mass spectrometry

Mass spectrometers are used in many research, industrial, regulatory, and clinical laboratories for a range of scientific applications including (Watson and Sparkman 2007):

- biological and biochemical analyses;
- security and terrorism prevention;
- drug testing in sports;
- pharmaceuticals (drug discovery, metabolism, pharmacokinetics);
- forensic science;

- environmental monitoring;
- geology and space exploration;
- engineering.

Mass spectrometers have excellent sensitivity and a fast speed of analysis. These instruments have a number of key characteristics that make them particularly suited to a wide variety of applications in routine high-throughput screening as well as in research situations. An important example of a routine mass spectrometric analysis is that of human screening for in-born errors of metabolism. This involves the analysis of blood samples taken from newborn babies within a few days of birth to screen for specific metabolic disorders such as medium-chain acyl-coenzyme A dehydrogenase deficiency (MCADD) (Wilcken et al. 2003).

Mass spectrometers are amenable to a wide variety of compound classes and are able to analyse not only single analytes but also mixtures of analytes. Molecular mass determination is probably the most common use of a mass spectrometer, enabling the user to determine the molecular mass and relative abundance of each of the components in the sample under investigation. However, mass spectrometers are also used to determine other information about biomolecular samples, such as the following.

- Sample purity: a mass analysis will also confirm that no unexpected species/ impurities are present in the sample.
- Detection of post-translational modifications (PTMs) in proteins. The presence and location of PTMs cannot be determined from genomic information. By direct analysis of proteins and peptides, MS can be used to detect mass shifts that indicate the presence, number, and type of PTMs. By MS fragmentation of the modified analyte, the exact location of the modification within the protein's sequence can be established (see section 13.4).
- Reaction monitoring. MS can be used to monitor biological reactions (e.g. enzyme catalysis) and assembly processes (e.g. construction of biomolecular complexes) over time. By removing aliquots of the sample for mass analysis at intervals during a reaction or an assembly pathway, the presence, absence, or change in abundance of signals arising from the co-populated heterogeneous ensemble of starting material, intermediate species, and reaction/assembly products can be monitored.

13.1.3 How do mass spectrometers work?

Many different types of mass spectrometer are available, but all have the same central components that are required for their operation. These components are illustrated in Figure 13.1 and comprise a method of introducing the sample into the instrument, an ionization source to ionize neutral molecules, a mass analyser to separate the ions produced according to their m/z ratio, a detector and data system to record a signal and intensity for each of the separated ions and generate m/z spectra, and a vacuum system to ensure that ions can travel through the instrument unhindered by collisions with gas molecules.

Sample introduction into the mass spectrometer is the first step required for a mass spectrometric analysis. Samples in solution can be infused into the instrument using a pumping system, such as a syringe driver, whilst solid samples are generally placed on a

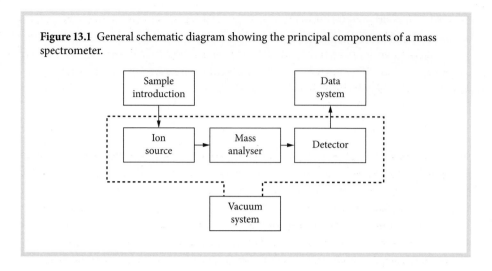

Figure 13.1 General schematic diagram showing the principal components of a mass spectrometer.

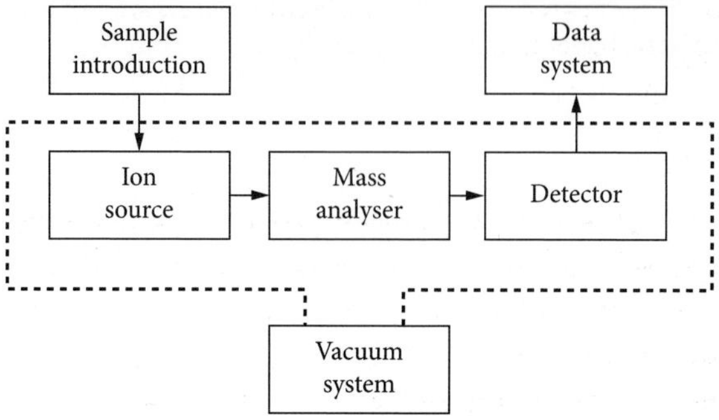

target or probe, such as a stainless steel plate, which is then inserted into the instrument. This introduction step is commonly combined with the ionization step in which the neutral analyte molecules are converted into ions. The ionization step can produce singly or multiply charged ions. Details of the most common ionization techniques used for biological applications are given in section 13.2. The **mass analyser** is the region of the mass spectrometer that distinguishes ions with different m/z ratios. The analyser works on the principle that gas phase ions present in a vacuum can be manipulated with electric and magnetic fields to help determine information about their molecular mass and structure. The **detector** measures electronically the ion separation achieved during mass analysis, either by impact of the ions with the surface of the detector or by induction of a current in a detector electrode as the ions oscillate nearby. The **data system** processes the amplified signals received from the detector to produce the **mass spectrum**, which represents m/z versus ion intensity. The mass analyser and detector (and frequently the ionization source as well) are enclosed within a **vacuum system** to ensure ions are able to travel through the entire length of the instrument and be detected without experiencing collisions with background gas molecules which would impede their progress and prevent them from reaching the detector.

13.2 Sample preparation and ionization

13.2.1 Sample ionization

The first stage in a mass spectrometric analysis is sample ionization where analytes are transferred to the gas phase as ions. Many large biomolecules and biomolecular complexes are non-volatile and therefore are not easily transferred to the gas phase. Any energy introduced to such analytes (e.g. through heating) may cause degradation of these large ensembles. However, a number of ionization techniques that can achieve transfer to the gas phase without degradation have been developed over the past three decades. These techniques are collectively known as 'soft ionization' techniques as they introduce very little energy into the system, and they enable most types of biomolecules, not only small drug metabolites but also mega-Dalton protein complexes such as virus capsids (Uetrecht et al. 2008), to be transferred intact to the gas phase as ions for *m/z* analysis. The two most widely used techniques for biomolecular analyses are **electrospray ionization** (ESI) (Fenn et al. 1989) and **matrix-assisted laser desorption ionization (MALDI)** (Karas and Hillenkamp 1988).

Electrospray ionization

Electrospray ionization (ESI) is an atmospheric pressure ionization technique which is well suited to the analysis of polar molecules of mass ranging from <100Da to >10 000 000Da, giving a high degree of mass accuracy (error ≤0.01%).

In standard ESI (Yamashita and Fenn 1984), the analyte is dissolved in a polar volatile solvent and pumped through a narrow stainless steel capillary at a flow rate between 1μL/min and 1mL/min (Figure 13.2). A high voltage of 3–4kV is applied to the tip of the capillary, and as a consequence of this strong electric field the sample emerging from the tip is dispersed into an aerosol of tiny highly charged droplets, a process that is aided by a coaxially introduced nebulizing gas flowing around the outside of the capillary. This gas, usually nitrogen, helps to direct the spray emerging from the capillary tip towards the mass spectrometer. The charged droplets diminish in size by solvent evaporation, assisted by a warm flow of nitrogen, known as the drying gas, which passes across the front of the ionization source. Eventually, charged sample ions, free from solvent, are released from the droplets and pass through an orifice into the analyser of the mass spectrometer, which is held under high vacuum. Nanospray ionization (Wilm and Mann 1996) is a low-flow-rate version of electrospray ionization, which can also be used. Further information concerning the mechanism of production of ions from droplets containing solvent and analyte can be found in Dole et al. (1968) and Iribarne and Thomson (1976).

The high voltage that is applied to the capillary can be either negative or positive, and this results in the production of positively and negatively charged ions, respectively. The

Figure 13.2 Principles of electrospray ionization. Figure courtesy of Lydia Young, University of Leeds.

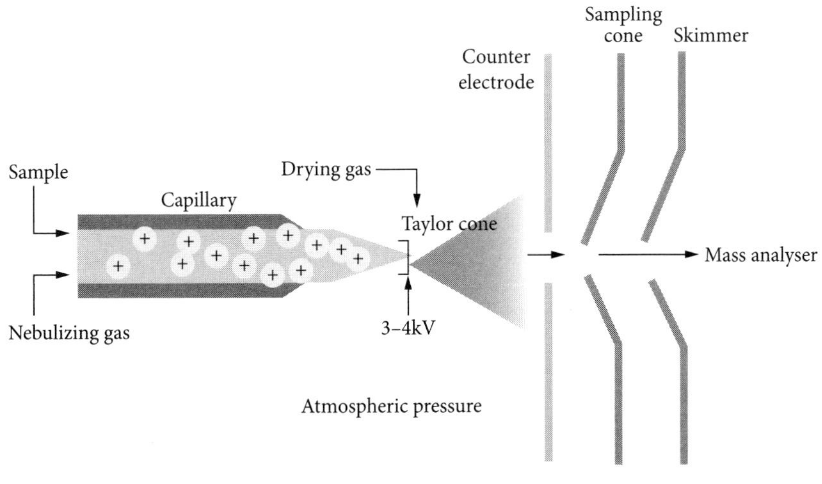

mode that is selected for a particular analysis will depend on the chemical nature of the analyte. For example, proteins are analysed primarily in the positive mode because of their multiple basic sites that are amenable to protonation. Some small molecules (e.g. drug metabolites) which contain carboxylic acid groups are best analysed in the negative mode as they can easily be ionized by removal of a proton from the –COOH group to give a negatively charged carboxylate anion ($-COO^-$).

When ESI is used, analytes with a molecular mass ≤1200Da generally give rise to singly charged molecular-related ions, usually protonated molecular ions with the formula $(M + H)^+$ in the positive ionization mode and deprotonated molecular ions with the formula $(M - H)^-$ in the negative ionization mode. Samples with a molecular mass >1200Da tend to give rise to multiply charged molecular-related ions such as $(M + nH)^{n+}$ in the positive ionization mode and $(M - nH)^{n-}$ in the negative ionization mode.

Figure 13.3 shows an example of an ESI mass spectrum of ubiquitin, a protein with a molecular mass of 8.6kDa (8564.0Da). The *x*-axis shows the *m/z* of the analytes present in the sample. The *y*-axis is a measure of the intensity of the signals generated at the detector for each *m/z* analysed in the experiment. This is usually labelled 'relative abundance', 'percentage', or simply '%' as the most intense signal in the spectrum is normalized to 100% abundance and is called the **base peak** (in this example *m/z* 779.58). All other signal intensities are shown relative to the base peak. An important concept to understand when analysing a mass spectrum is that a sample containing a single analyte may produce multiple signals in the mass spectrum. Protein analysis using ESI (in positive ionization mode) leads to ions with multiple charges through addition of protons to the basic sites in the protein, such as arginine and lysine side chains, as well as the nitrogen atoms in the amide bonds. However, not all basic sites will undergo protonation in every single molecule in the sample, which leads to the generation of a series of charge-state signals. As a mass spectrometer measures *m/z*, the spectrum will show signals for each of the protein's charge states generated during ionization. In this example, the *m/z* spectrum for ubiquitin shows a series of charge states ranging from +5 to +13 (Figure

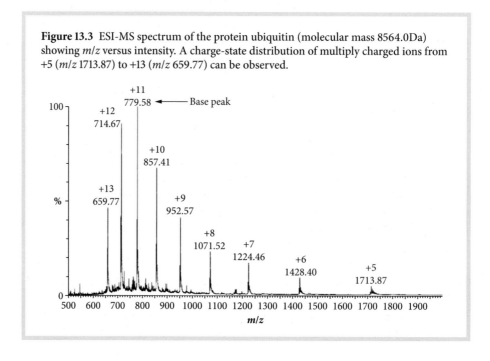

Figure 13.3 ESI-MS spectrum of the protein ubiquitin (molecular mass 8564.0Da) showing *m/z* versus intensity. A charge-state distribution of multiply charged ions from +5 (*m/z* 1713.87) to +13 (*m/z* 659.77) can be observed.

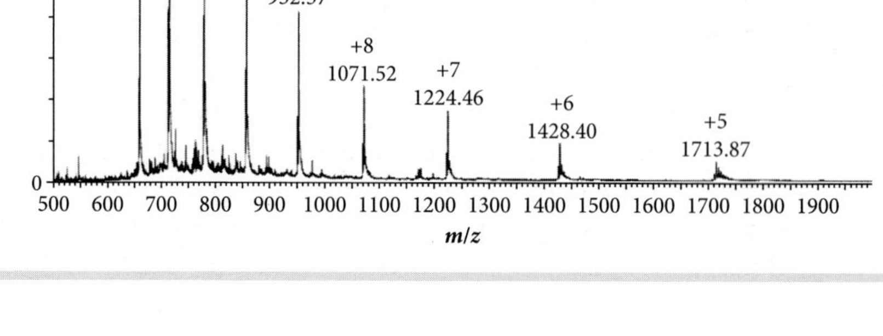

13.3). Each charge state is defined by $(M + nH)^{n+}$, where M is the mass of the protein, H is the mass of hydrogen, and n is the number of protons (and hence charges) attached. The following equation is used to determine the molecular mass of the protein from the m/z spectrum containing multiply charged ions:

$$M = \left(\frac{m}{z}\right)_1 n - nH^+ \qquad 13.1$$

Taking the signal at m/z 857.41 as an example, this value should be multiplied by the number of charges given (+10) and, assuming that the charges have arisen from attachment of protons, subtracting the mass of these 10 protons as follows:

$$M = (857.41 \times 10) - (10 \times 1.0079) = 8564.0\text{Da}$$

where M is the mass of the protein, n is the number of charges, and H^+ is the average mass of a proton (1.0079Da).

If there were a number of proteins present in the sample, the m/z spectrum would exhibit a distribution of charge-state ions representative of each protein present. An example of calculation of the charge state of the ions is shown in Box 13.1.

Matrix-assisted laser desorption ionization (MALDI)

Matrix-assisted laser desorption ionization (MALDI) (Hillenkamp et al. 1991) is well suited to the analysis of thermolabile non-volatile organic compounds, especially those with a high molecular mass, and is used successfully for the analysis of proteins, peptides, glycoproteins, oligosaccharides, and oligonucleotides. The mass accuracy depends on the type and performance of the analyser of the mass spectrometer, but most modern instruments are capable of measuring masses to within 0.01% of the molecular weight of the sample, at least up to ~40 000Da.

For MALDI analyses, a solution of the analyte is premixed with a solution of a small light-absorbing organic compound, usually containing an aromatic group and a carboxylic acid moiety, known as a matrix. A small aliquot of the mixed solution (~1μL) is placed on a metal strip known as the MALDI target and the solvent is allowed to evaporate. The dried mixture on the target is then inserted into the high vacuum of the ionization source of the mass spectrometer. Here, the analyte and matrix mixture is bombarded with laser light to bring about sample ionization. The matrix transforms the laser energy into excitation energy for the sample, which leads to 'sputtering' where analyte and matrix ions are ejected from the surface of the solid mixture (Figure 13.4). In this way, energy transfer is efficient and the analyte molecules are spared excessive direct energy that may otherwise cause decomposition. Most commercially available MALDI mass spectrometers use a pulsed nitrogen laser of wavelength 337nm.

Like ESI, MALDI is a 'soft ionization' method. However, MALDI results predominantly in the generation of singly charged molecular-related ions (protonated $(M + H)^+$ or deprotonated $(M - H)^-$) in the positive or negative ionization mode, respectively, regardless of the molecular mass of the analyte. Hence the spectra are relatively straightforward to interpret. Figure 13.5 shows the MALDI-MS m/z spectrum obtained from the positive ionization mode analysis of a mixture of six peptides: Arg8-vasopressin, somatostatin, adrenocorticohormone (ACTH) (fragment 1-24), bovine insulin β-chain,

BOX 13.1

Calculating the charge state of the ions in the *m/z* spectrum

When analysing proteins such as ubiquitin, the question arises: 'How do we know what the charge states of the ions in the *m/z* spectrum are?' After analysis of a protein the raw spectrum provides signals of certain *m/z* ratios with their intensities, but does not indicate the number of charges on each peak. In order to calculate the molecular mass of the protein from the *m/z* mass spectrum, we assume that adjacent signals in the spectrum differ by a single charge. Therefore, if we say that the number of charges on the signal at $(m/z)_1$ 1071.52 is *n* and then assume that the number of charges on the adjacent signal at $(m/z)_2$ 952.57 is $n + 1$, we can determine the molecular mass *M*. We can use equation (13.1) for each of these two charge states:

$$M = \left(\frac{m}{z}\right)_1 n - nH^+$$

$$M = \left(\frac{m}{z}\right)_2 (n+1) - (n+1)H^+$$

Then, combining the two equations and eliminating *M*, we obtain

$$\left(\frac{m}{z}\right)_1 n - nH^+ = \left(\frac{m}{z}\right)_2 (n+1) - (n+1)H^+$$

and inserting our *m/z* ratios, we can calculate the number of charges on $(m/z)_1$:

$$(1071.52)\ n - n\ (1.0079) = (952.57)\ (n+1) - (n+1)\ (1.0079)$$

$$1071.52n - 1.0079n = 952.57n + 952.57 - 1.0079 - 1.0079n$$

$$1070.50n = 951.56n + 951.56$$

$$118.94n = 951.56$$

$$n = 951.56/118.94 = 8$$

Now we know the value of *n*, we can substitute this back into equation (13.1) to calculate the molecular mass of ubiquitin:

$$M = \left(\frac{m}{z}\right)_1 n - nH^+$$

$$M = (1071.52 \times 8) - (8 \times 1.0079)$$

$$M = 8564.0\text{Da}$$

insulin, and hirudin. The MALDI matrix used for the analysis was α-cyano-4-hydroxy-cinnamic acid. Each peptide gives rise to a single peak corresponding to its protonated molecular ion. To determine the molecular mass you need to subtract the mass of a proton from each *m/z* value.

Figure 13.4 Principles of matrix-assisted laser desorption ionization. Figure courtesy of Lydia Young, University of Leeds.

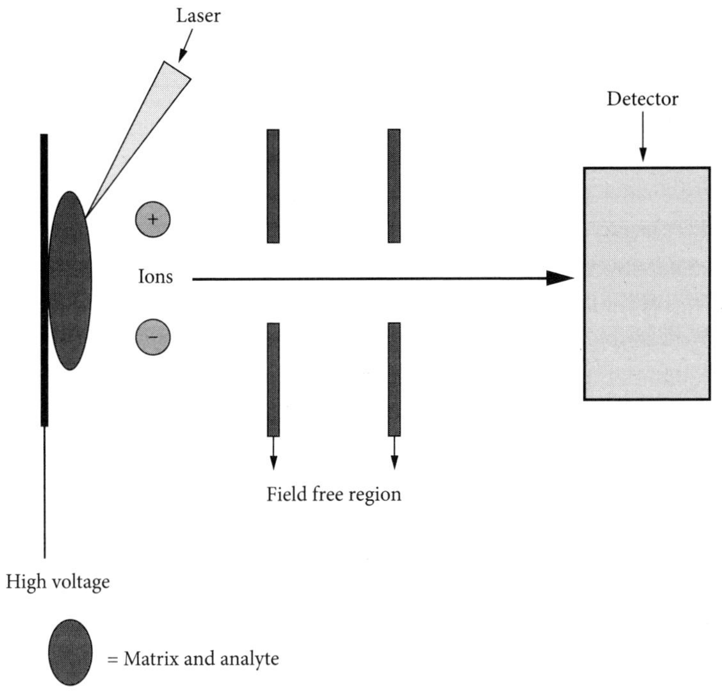

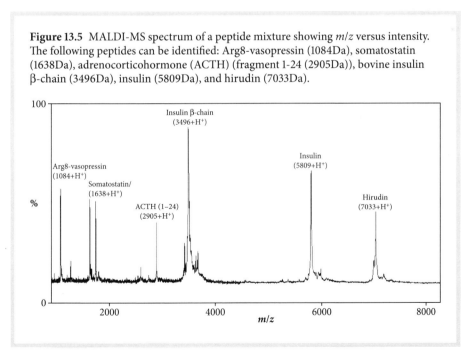

Figure 13.5 MALDI-MS spectrum of a peptide mixture showing *m*/*z* versus intensity. The following peptides can be identified: Arg8-vasopressin (1084Da), somatostatin (1638Da), adrenocorticohormone (ACTH) (fragment 1-24 (2905Da)), bovine insulin β-chain (3496Da), insulin (5809Da), and hirudin (7033Da).

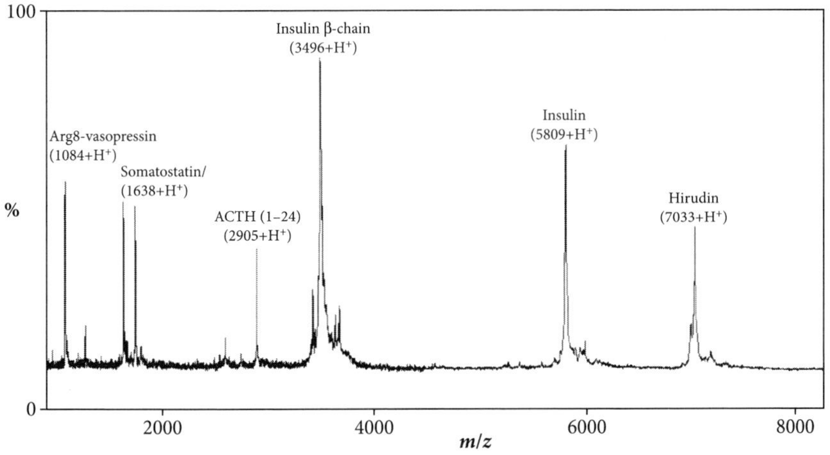

13.2.2 Sample preparation

A number of factors need to be considered when preparing a sample for mass spectrometry.

- What type of compounds will be analysed?
- Which method of sample ionization will be used?
- Should positive or negative ionization be used?
- What is the concentration of the analyte?

The answers to these questions will depend on the analyte under study and the type of information required. Some guidelines are presented in the remainder of this section.

Proteins have many suitable sites for ionization as all the backbone amide nitrogen atoms are susceptible to protonation, as are the basic amino acid residues, in particular lysine and arginine which contain basic amine functionalities, and the N-terminus. In contrast, oligonucleotides and oligosaccharides have many functional groups amenable to deprotonation, including phosphate and hydroxyl groups. Hence proteins and peptides are usually analysed under positive ionization conditions and oligosaccharides and oligonucleotides under negative ionization conditions. In all cases, the *m/z* scale must be calibrated by analysing a standard sample of known mass, ideally of a similar type to the sample being analysed (e.g. a protein calibrant for a protein sample), and then applying a mass correction to the analyte data.

ESI-MS and MALDI-MS are highly sensitive analytical techniques, but the sensitivity deteriorates when the biomolecular analyte is accompanied by non-volatile buffers and other additives, especially in the case of ESI. Hence these contaminants should be removed as much as possible before the MS analysis using dialysis, chromatographic techniques, or buffer exchange into volatile buffers such as ammonium formate, ammonium acetate, or ammonium hydrogen carbonate, which do not interfere with the mass spectrometric analyses.

See Chapter 8

For positive ionization ESI-MS, the analyte should be dissolved at a concentration of ~ 1–10pmol/μL in a volatile solvent (water, acetonitrile, methanol, chloroform, or mixtures thereof), and a trace (0.1–5%) of formic or acetic acid can be added to aid protonation of the sample. In negative ionization mode the same range of solvents can be used but a trace (1–5%) of ammonia solution or a volatile amine (e.g. triethylamine) can be added to aid deprotonation of the analyte. Ideally, at least 1μL of solution is required for analysis.

Similarly, for MALDI-MS, the analyte is dissolved in an appropriate volatile solvent, usually adding a trace (~1%) of trifluoroacetic acid if positive ionization is being used, at a concentration of ~1–10pmol/μL. An aliquot (1–2μL) of this is removed and mixed with an equal volume of a solution containing a large excess of a matrix. A range of compounds can be used as matrices: sinapinic acid is a common one for protein analyses, while α-cyano-4-hydroxycinnamic acid is often used for peptide analyses. An aliquot (1–2μL) of the final solution is applied to the sample target which is allowed to dry prior to insertion into the high vacuum of the mass spectrometer.

13.2.3 Non-covalent mass spectrometry

The sample analyses described so far have involved subjecting the analyte to denaturing solvent conditions in which, if the sample is a protein or a biomolecular complex, any tertiary or quaternary structure is lost. Increasingly, ESI-MS is being used to probe the 3D architecture of biomolecules and macromolecular complexes. To achieve this, the analyte is maintained in solution in conditions as near to physiological as possible so that any non-covalent interactions are maintained intact throughout the analysis process (e.g. 10–500mM ammonium acetate or ammonium hydrogen carbonate at pH 7). This has proved a successful technique for a wide range of analyses including the study of protein-folding mechanisms, protein–ligand binding, and the analysis of intact biomolecular complexes involving proteins, nucleotides, metal ions, and small ligands (Loo 1997; Ashcroft 2005).

Proteins in their native state, or at least containing a significant amount of tertiary structure, tend to produce a charge-state distribution of multiply charged ions that cover a smaller range of charge states (say two or three) compared with denatured proteins. These charge states tend to have fewer charges than the same protein analysed under denaturing conditions because of the inaccessibility of many of the potential protonation sites. This effect is illustrated in Figure 13.6 which shows m/z spectra of the protein horse heart myoglobin acquired under different solvent conditions. Analysis of the protein in 1:1 v/v acetonitrile–0.1% aqueous formic acid at pH 3 (i.e. conditions under which non-covalent interactions are lost) gave a Gaussian-type distribution of charge states with charges ranging from $n = +8$ at m/z 2119.87 to $n = +23$ at m/z 737.96, centring on n values between +18 and +20 (Figure 13.6(b)). The molecular mass measured for this protein is 16 950Da. Analysis of the protein under aqueous conditions at pH 7 (i.e. where non-covalent interactions are maintained) led to the generation of fewer charge states, from $n = +7$ at m/z 2510.42 to $n = +9$ at m/z 1952.76, centring at $n = +8$ (Figure 13.6(a)). Not only has the charge state distribution changed, the molecular mass is now 17 566Da which represents an increase of 616Da and indicates that the native haem ligand remains bound to the protein. On closer inspection of the m/z spectrum of the denatured protein, an intense ion signal at m/z 616 can be observed, which corresponds

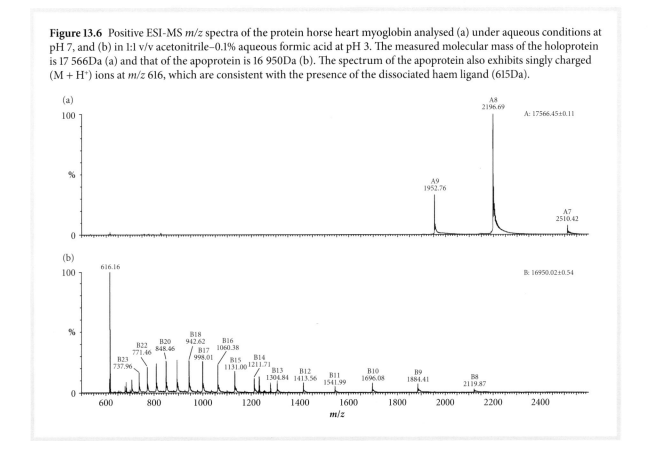

Figure 13.6 Positive ESI-MS m/z spectra of the protein horse heart myoglobin analysed (a) under aqueous conditions at pH 7, and (b) in 1:1 v/v acetonitrile–0.1% aqueous formic acid at pH 3. The measured molecular mass of the holoprotein is 17 566Da (a) and that of the apoprotein is 16 950Da (b). The spectrum of the apoprotein also exhibits singly charged ($M + H^+$) ions at m/z 616, which are consistent with the presence of the dissociated haem ligand (615Da).

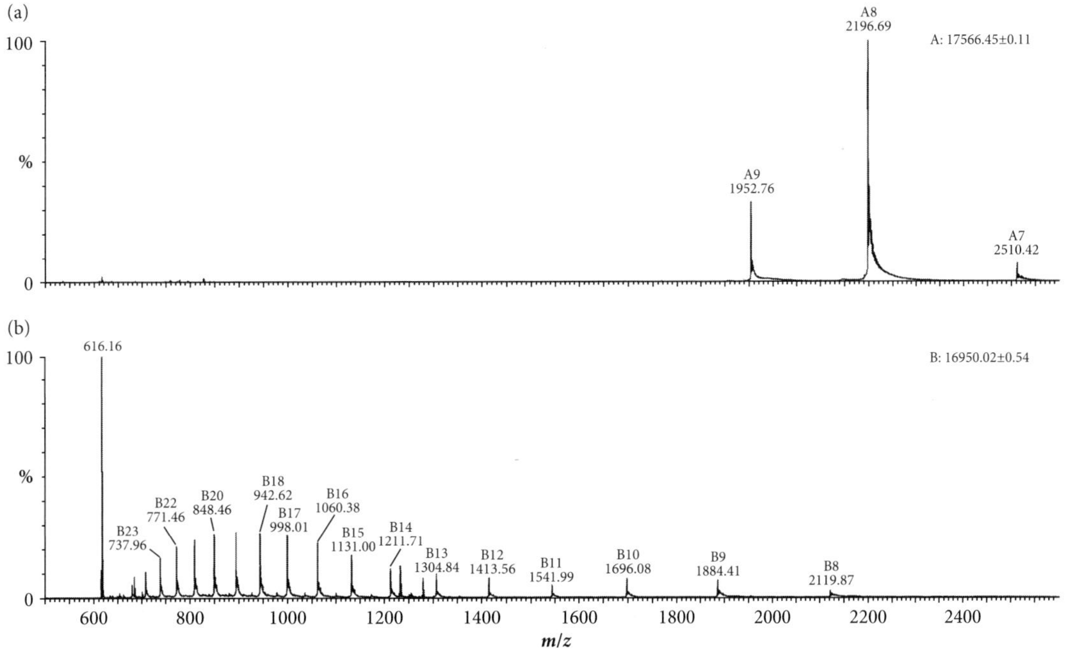

to the (M + H^+) ions of the unbound haem residue (molecular mass 615Da). Many types of protein complexes can be observed in this way, including protein–ligand, protein–peptide, protein–metal, and protein–RNA macromolecules.

13.3 Mass analysis

13.3.1 General features of mass analysers

As we have seen, to determine the molecular mass of an analyte, the m/z of the analyte's ions in the gas phase must be measured. This process is called mass analysis and it exploits the well-defined behaviour of ions when they experience electric or magnetic fields, or a combination of the two. For example, depending on the mass analyser used, the ions may travel at a velocity depending on their m/z ratio, or alternatively the frequency with which the ions oscillate about a particular point may be used to measure the m/z ratio (see section 13.3.2).

Many mass analysers are available commercially. Each manipulates the ions in a unique way to measure the m/z ratio and each has particular applications for which they are well suited. A number of features of a mass analyser that can be considered when choosing the most appropriate one for an analysis (Yates et al. 2009).

- **m/z range limit:** the maximum m/z range over which an analyser can measure. Most mass analysers have a fixed m/z range of a few thousand, but in some cases there is a theoretically unlimited mass range.
- **Transmission:** the ratio of the number of ions reaching the detector to the number entering the mass spectrometer. If only a small fraction of the ions reach the detector, the overall sensitivity of the instrument will be low and a higher analyte concentration is required.
- **Mass accuracy:** the difference between the measured mass and the theoretical mass.
- **Resolution:** a measure of the ability of an instrument to distinguish two analytes of different m/z.
- **Dynamic range:** the concentration range over which the increase in signal is proportional to the increase in concentration of the analyte.

While mass spectrometry is well suited to measuring the m/z values of multiple analytes within a mixture, the resolution of the instrument determines the smallest difference in m/z ratio that two analytes can have before they can no longer be distinguished from each other. Most modern mass spectrometers have sufficient resolution to resolve the individual carbon isotopes of an analyte of mass about 9000Da or less. Figure 13.7(a) shows a common method used to calculate the resolution between two signals of very similar m/z. The resolution of two peaks of equal intensity at m/z and $m/z + \Delta m/z$ with a 10% valley between the two signals is given by $M/\Delta M$. Figure 13.7(b) illustrates the resolution achieved using two different instruments. The m/z mass spectrum was obtained from the protein cytochrome c and shows the +17 charge-state ions. The grey trace was acquired on an instrument with sufficient resolution to separate and distinguish each of the carbon isotope signals; whereas the instrument used to acquire the black trace could not achieve baseline separation on the same sample.

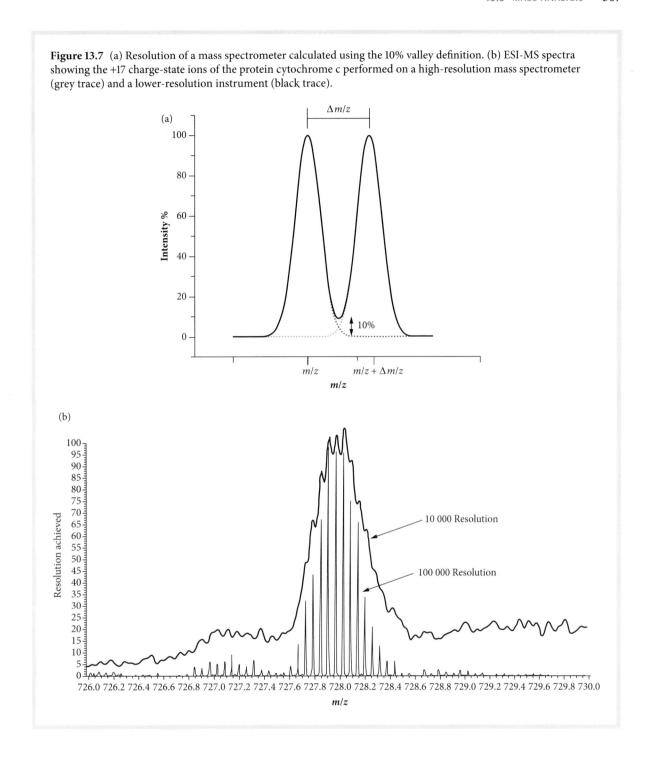

Figure 13.7 (a) Resolution of a mass spectrometer calculated using the 10% valley definition. (b) ESI-MS spectra showing the +17 charge-state ions of the protein cytochrome c performed on a high-resolution mass spectrometer (grey trace) and a lower-resolution instrument (black trace).

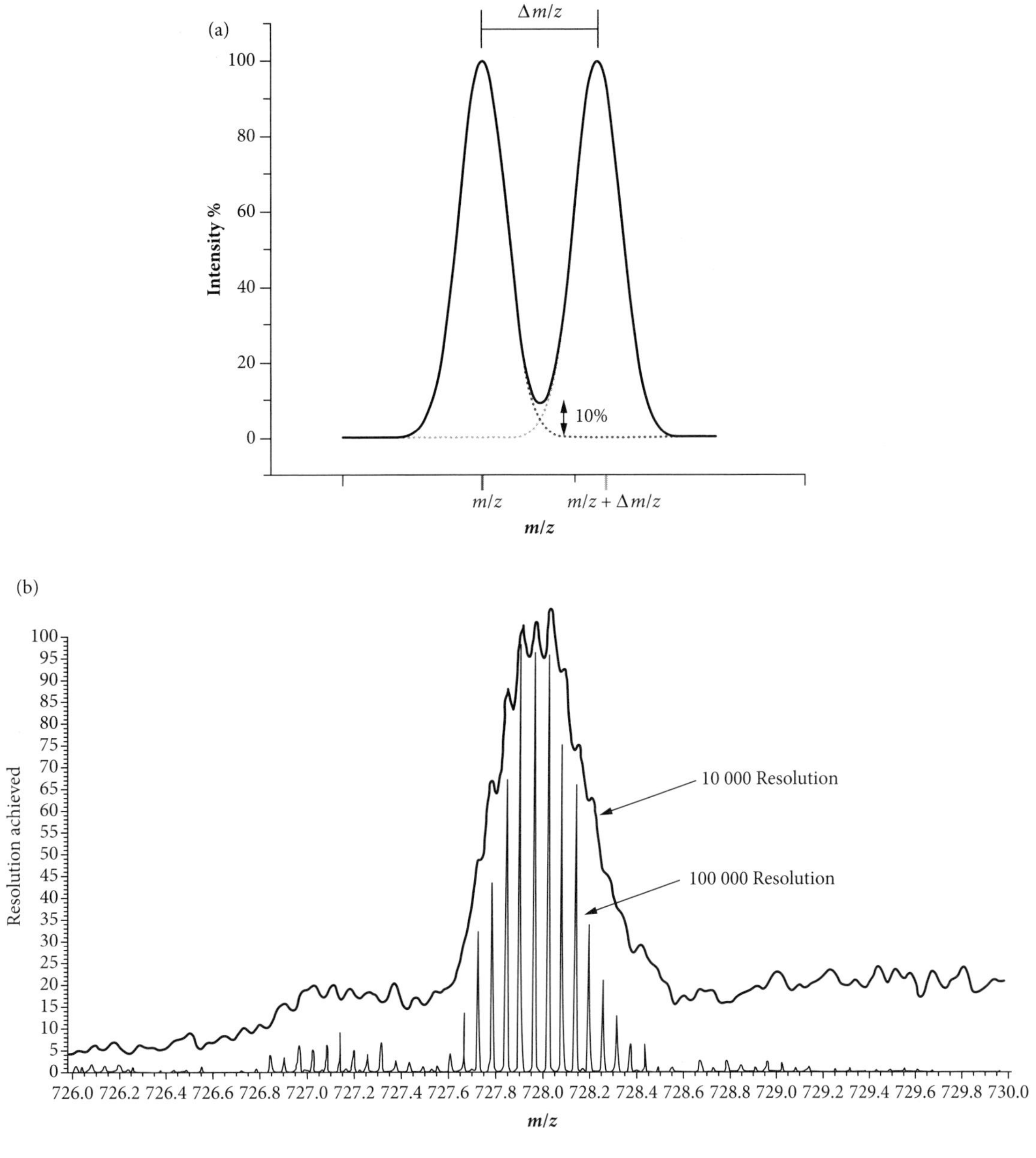

13.3.2 Types of mass analyser

The various mass analysers can be classed broadly as scanning and beam analysers or trapping mass analysers. Scanning and beam analysers separate ions of different *m/z* based on different flight trajectories or velocities. In trapping analysers, the ions oscillate around a central point or axis in a trap, with the oscillation frequency of an ion being

Table 13.1 A comparison of the most commonly used analysers in biomolecular mass spectrometry

Analyser	Mass range (m/z)	Resolution (at m/z 1000)	Mass accuracy (<1000Da)	Applications	Advantages	Limitations
TOF	≤32 000	30 000	1–5ppm	Intact proteins and non-covalent complexes, peptide and oligonucleotide sequencing, small molecule identification	High sensitivity, large mass range, very versatile	Limited dynamic range
Quadrupole	50–4000	5000	100ppm	Intact protein and small molecule mass measurement, quantification	Good dynamic range	Limited m/z range, limited sensitivity
FT-ICR	100–10 000	>500 000	<1ppm	Intact protein sequencing, elemental composition determination (petrochemicals)	High mass accuracy, excellent resolution	Slow measurements, expensive to buy and maintain
Orbitrap	≤4000	≤100 000	<1ppm	High-throughput proteomics, quantification	High sensitivity, high mass accuracy, high throughput	Limited m/z range

characteristic of that particular ion's m/z. The mass analysers discussed here (Table 13.1) are some of the most common instruments used for the analysis of biomolecules. Other analysers are available but are uncommon for biological analysis.

Time-of-flight mass analyser

One of the most common analysers used for the analysis of biological molecules is the **time-of-flight (TOF) analyser** (Wolff and Stephens 1953). The concept of TOF is simple. The ions are all given the same amount of energy, and the time it takes for the ions to traverse a drift tube and reach the detector at the other end depends on their m/z ratio. This is because ions of the same energy but different m/z have different velocities and hence will take different periods of time, depending on their m/z, to traverse the drift tube. The time from when the ions are given the energy to when the ions reach the detector is measured and the signals converted to a m/z spectrum by the data system.

TOF analysers have high transmission and theoretically unlimited mass range, which is why they are well suited to the analysis of large biomolecules. Protein complexes >10MDa have been analysed using TOF instruments (Uetrecht et al. 2010). Historically, their major drawback was poor resolution, especially at higher m/z, because of small differences in the amount of energy received by each ion. However, instrumental developments have much improved TOF resolution sufficiently for most applications (Wiley and McLaren 1955; Mamyrin et al. 1973; Coles and Guilhaus 1993; Brown and Lennon 1995; Krutchinsky et al. 1998a,b).

Quadrupole analysers

Quadrupole mass analysers (Paul and Steinwedel 1953, Finnigan 1994) comprise four electrically conductive rods of circular cross-section arranged symmetrically about the ion beam axis. The pairs of rods that are diametrically opposite to one another are connected electronically, and each pair generates an oscillating electric field which is slightly different from that generated by the other pair of electrodes. The combination of the two electric fields generated by the two pairs of electrodes causes the charged ions to oscillate about the beam axis as they travel through the quadrupole. Only ions of a particular *m/z* will have a stable trajectory for a given combination of the electric field, and hence fully traverse the quadrupole. The other ions entering the quadrupole will have unstable trajectories and will be neutralized by collision with one of the rods, never reaching the detector. The quadrupole can be thought of as a filter, only allowing one specific *m/z* through the analyser at a time, depending on the combination of electric fields applied. To create a mass spectrum of all *m/z* values in the required mass range the electric fields are changed incrementally (or 'scanned'), thereby transmitting each *m/z* sequentially. Quadrupole analysers are useful for the mass measurement of small molecules and proteins, having good mass accuracy but limited sensitivity as only ions of one *m/z* value are transmitted at any one time. Quadrupole analysers are often used as the first analyser in a tandem mass spectrometer to select particular ions within a mixture. The mass spectrometrist can then carry out further analyses on these selected ions, for example using a second mass analyser in the same instrument (see section 13.4).

Fourier transform–ion cyclotron resonance (FT-ICR)

The **Fourier transform-ion cyclotron resonance** (FT-ICR), or simply ICR, mass analyser (Sommer et al. 1951) and the **orbitrap** (Makarov 2000) are the most common of the trapping instruments. These analysers operate by measuring the frequency of oscillation of the ions within a trap or cell. ICR mass analysers have the highest resolution and mass accuracy of all mass analysers. Their basic mode of operation exploits the fact that charged particles in a uniform magnetic field attain a circular motion perpendicular to the magnetic field. The detected current is subject to a mathematical transformation called the Fourier transform (Comisarow and Marshall 1974a,b) to produce the final mass spectrum. Due to the excellent mass accuracy of FT-ICR instruments, they have the ability to determine the elemental composition of analytes to aid in the identification of molecules and are particularly useful in the field of **proteomics** where mixtures of many peptides have to be identified unambiguously. The high resolution of these instruments allows analysis of exceptionally complex mixtures, such as oil samples containing complex mixtures of polymers of different lengths and complex mixtures of peptides. However, scanning in an FT-ICR instrument is much slower than with other mass analysers and thus they are not suitable when speed of analysis is a major consideration—for example, when **liquid chromatography** (LC) is coupled with mass spectrometry analysis.

Orbitrap

Like the FT-ICR analyser, the orbitrap (Makarov 2000) uses ion oscillations inside a trap to perform mass analysis. Instead of a magnetic field, the orbitrap uses purely electrostatic fields to cause the ion to oscillate both around and along the central 'spindle' electrode of the trap. The frequency of oscillation along the spindle is characteristic of the *m/z* of the ion. Again, Fourier transform is used to convert the image current into an *m/z* spectrum. The mass accuracy and resolution are high and are more than sufficient for most applications. The main limitation of the orbitrap is the mass range of *m/z* 4000; hence these instruments are not particularly suitable for the analysis of intact proteins or large biomolecular complexes. Orbitrap instruments are primarily used in proteomics analyses, where high-throughput and high-confidence identification of peptides are of paramount importance.

See Chapter 14

13.4 Tandem mass spectrometry

13.4.1 What is tandem mass spectrometry?

Tandem mass spectrometry (MS/MS) is used to generate structural information about an analyte by fragmenting specific sample ions inside the mass spectrometer and identifying the resulting fragment ions. The information can then be pieced together to generate structural details regarding the intact molecule. MS/MS can also be used to detect a specific compound within a complex mixture because of the compound's unique and characteristic fragmentation patterns. Often, MS/MS is carried out on a mass spectrometer which has more than one analyser, in practice usually two. The two analysers are separated by a collision cell into which an inert gas (e.g. argon, xenon) is admitted to collide with the sample ions selected for scrutiny and bring about their fragmentation. The analysers can be the same (e.g. quadrupole–quadrupole) or different (e.g. quadrupole–TOF). Fragmentation experiments can also be performed on certain single-analyser mass spectrometers such as ion trap instruments.

13.4.2 Inducing dissociation

A wide range of techniques have been developed for inducing the dissociation of ions within a mass spectrometer (Jones and Cooper 2011): collision induced dissociation (CID), electron capture dissociation (ECD), electron transfer dissociation (ETD), infra-red multiphoton dissociation (IRMPD), blackbody infra-red radiative dissociation (BIRD), and surface-induced dissociation (SID). The availability of any particular technique will depend on the individual mass spectrometer being used.

13.4.3 Scan modes

A range of MS/MS experiments can be performed depending on the analytical question to be answered and the type of instrument being used. These experiments include the following:

- **Product ion scanning** can be used to help determine the overall structure of an analyte. The analyte ions, termed 'precursor ions', are isolated and fragmented in the collision cell of the mass spectrometer. Therefore all the fragment ions detected arise directly from the isolated precursor ions and generate a fingerprint pattern specific to the compound under investigation. This experiment is useful for sequencing peptides, oligonucleotides, and oligosaccharides, in addition to providing structural information for small organic molecules.

- **Precursor ion scanning** is particularly useful for detecting, or monitoring, a specific class of compounds within a mixture. In such cases, each member of the class dissociates to produce a common fragment ion. Examples of this are the specific detection of phosphorylated peptides originating from a tryptic digest of a protein (phosphotyrosine residues fragment to produce an immonium ion at *m/z* 216) or of glucuronide conjugates in biological samples (which fragment to lose a glucuronic acid moiety with *m/z* 178). All precursor ions are fragmented, but a signal in the mass spectrum is only registered at the precursor's *m/z* if this precursor has dissociated to produce the characteristic fragment that is being monitored.
- **Constant neutral loss scanning** is similar to precursor ion scanning. However, instead of looking for a particular fragment ion, a specific loss of mass from the precursor is monitored. An example of this type of experiment would be to monitor all the carboxylic-acid-containing species within a mixture of small molecules. Carboxylic acids fragment by losing a (neutral) molecule of carbon dioxide (CO_2), which is equivalent to a loss of 44Da, and the experiment would record all the precursor ions which fragment to lose 44Da.
- **Selected/multiple reaction monitoring** (SRM/MRM) is used to detect one particular analyte specifically within a complex mixture. It involves monitoring a known precursor ion of specified *m/z* undergoing dissociation to produce a known fragment ion, also of specified *m/z*. Hence the analyte must be well characterized prior to this experiment, which is used for unambiguous confirmation of the presence of a compound in a matrix (e.g. testing for a specific drug in complex matrices such as blood or urine samples).

13.4.4 Peptide sequencing by tandem mass spectrometry

One of the most common applications of MS/MS in biochemical areas is the product ion scanning experiment which is particularly useful for peptide and oligonucleotide sequencing.

Peptides fragment in a well-documented manner (Roepstorff and Fohlman 1984; Johnson and Biemann 1989). The protonated molecules fragment along the peptide backbone and also undergo some side-chain fragmentation (Ashcroft and Derrick 1990). There are three different types of chemical bonds along the amino acid backbone, all of which can undergo fragmentation by MS/MS: the NH–CH, CH–CO, and CO–NH bonds. Each bond breakage gives rise to two species, and only the charged species are detected. A charge can stay on either of the two fragments, depending on the chemistry and relative proton affinity of the two species. Hence, in Figure 13.8, there are six possible fragment ions for each backbone cleavage site. The **a**, **b**, and **c** ions have the charge retained on the N-terminal fragment, and the **x**, **y**, and **z** ions have the charge retained on the C-terminal fragment. The fragmentation pattern can also vary from one instrument type to another. For example, on a quadrupole–quadrupole or quadrupole–TOF tandem mass spectrometer, collision-induced dissociation is usually used and, as the fragmentation is of low energy, the most common cleavage sites are at the CO–NH bonds which give rise to the **b** and/or **y** ions. High-energy dissociation techniques, such as ECD and ETD, tend to generate **c** and **z** ions predominantly, but not

Figure 13.8 Peptide sequencing by tandem mass spectrometry (MS/MS). The **a**, **b**, and **c** ions have the charge retained on the N-terminal fragment, and the **x**, **y**, and **z** ions have the charge retained on the C-terminal fragment. Both the **y** and the **c** ions result from cleavage of the peptide backbone accompanied by the gain of two hydrogen atoms.

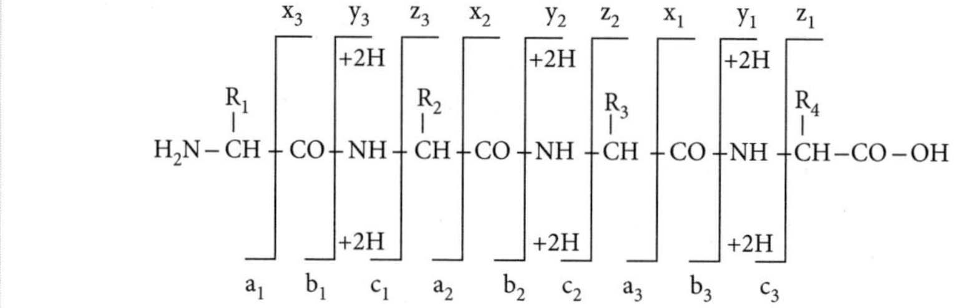

exclusively. Note that both the **y** and the **c** ions result from cleavage of the peptide backbone accompanied by the gain of two hydrogen atoms.

An example of an MS/MS product ion spectrum is illustrated in Figure 13.9 for the peptide [Glu[1]]-fibrinopeptide B. The molecular mass of the peptide was confirmed as 1569.7Da from the protonated molecular ions ($M + H^+$) at m/z 1570.7 using ESI-MS. These ions were selected for transmission through the first analyser of a tandem quadrupole–TOF mass spectrometer and then fragmented in the argon-filled collision cell. Their fragments were then analysed by the second analyser to produce the MS/MS m/z spectrum. The sequence (i.e. amino acid backbone) ions have been identified and labelled. In this example the peptide fragmented predominantly at the CO–NH bonds and gave both **b** and **y** ions (often either the **b** series or the **y** series predominates, sometimes to the exclusion of the other). The mass difference between adjacent members of a series can be calculated. For example, the difference between the singly charged ions corresponding to **b5** and **b4** is 497.21 – 382.19 = 115.02Da, which is indicative of an aspartic acid (D) amino acid residue, and similarly **y8** – **y7** = 942.45 – 813.40 = 129.05 Da which is indicative of a glutamic acid (E) residue. In this way, using either the **b** series or the **y** series, the amino acid sequence of the peptide can be determined and was found to be EGVNDNEEGFFSAR (note that the **y** series reads from right to left!).

See Chapter 14

Peptide sequencing is often used in the field of proteomics to identify proteins. Typically, the protein under scrutiny is enzymatically digested to generate a number of peptides. Trypsin is the enzyme of choice because each proteolytic fragment generated will contain a basic arginine or lysine amino acid residue at its C-terminus and thus is suitable for positive ionization mass spectrometric analysis. The mixture of peptides resulting from the protein digest is analysed by ESI-MS or MALDI-MS, and the molecular masses of all the proteolytic fragments are measured to produce collectively a peptide mass fingerprint. Following this, the protonated molecular ions of each of the peptides in the mixture can be subjected independently to MS/MS sequencing. Together, the peptide mass fingerprint and the sequence information can be used to search databases and identify the original protein.

Figure 13.9 Peptide sequencing by ESI-CID MS/MS using a tandem quadrupole–TOF mass spectrometer. The product ion spectrum of [Glu[1]]-fibrinopeptide B (1569.7Da) confirms the amino acid sequence: EGVNDNEEGFFSAR.

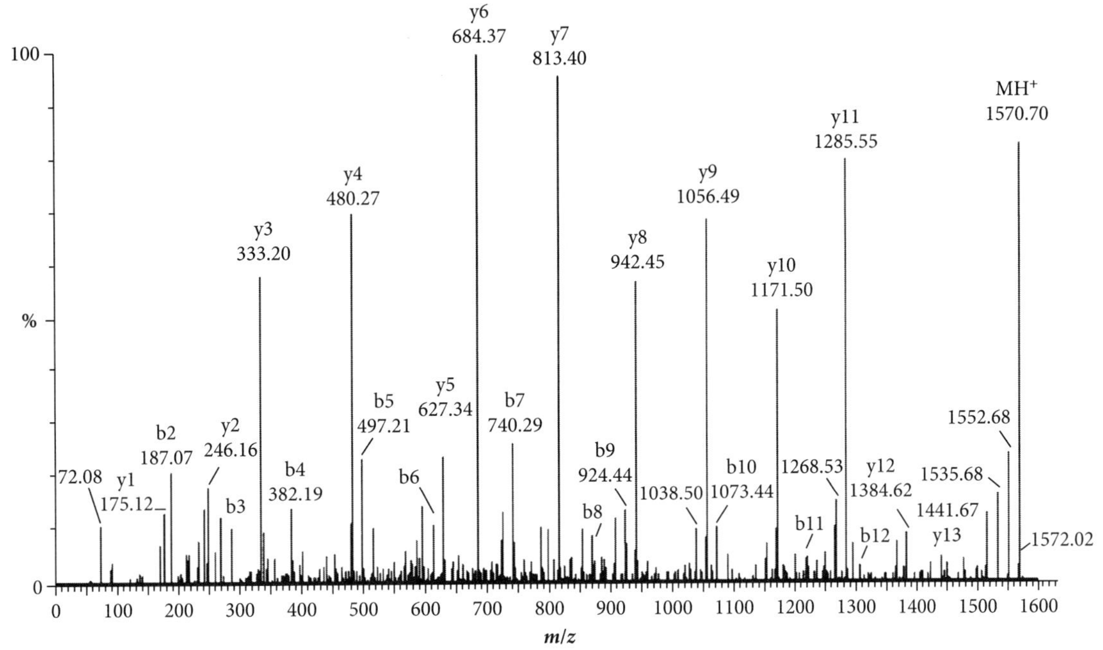

13.5 Chapter summary

- Mass spectrometry is used to measure the mass-to-charge ratio (*m/z*) of analyte ions in the gas phase, from which the molecular mass can be determined.
- Sample preparation is an important aspect of a mass spectrometry analysis.
- Soft-ionization techniques such as ESI and MALDI are used to transfer analyte molecules into the gas phase as ions.
- A number of mass analysers are available to measure *m/z* ratios. The most appropriate one to use depends on the application.
- Some analytes produce multiple signals in the mass spectrum because different molecules of the same analyte can pick up different numbers of charges during ionization.
- Non-covalent interactions of biological complexes can be maintained during ionization and so information can be obtained about the tertiary and quaternary structures of biomolecular complexes.
- Tandem mass spectrometry (MS/MS) experiments can be used to determine information about analyte structure by causing the precursor ion to fragment. This is particularly useful for determining the amino acid sequences of proteins and peptides.

CASE STUDY 13.1 Quantification of propanolol and mefenamic acid in river water

The aim of the analysis is to identify and quantify the levels of certain pharmaceuticals in river water samples that were removed from a specific sampling site at hourly intervals over a 24-hour period. In this case we are looking at the levels of the beta-blocker propanolol and the anti-inflammatory mefenamic acid to see how their levels vary over the course of the sampling period.

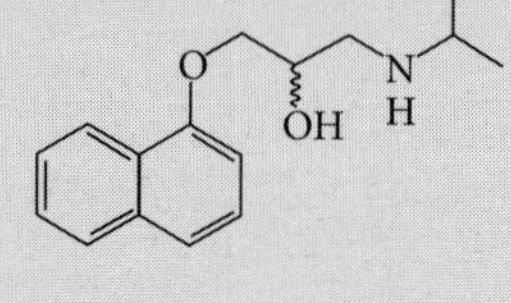

Propranolol

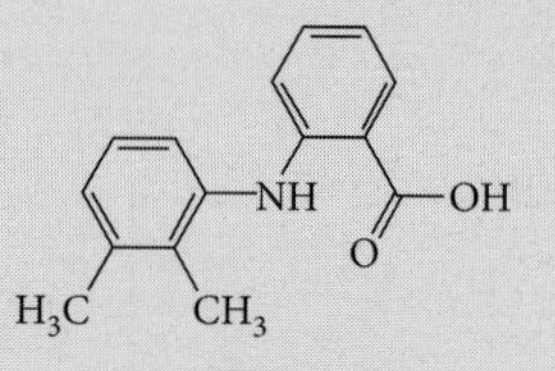

Mefenamic acid

QUESTION 1
How can LC-MS help to confirm the identity of the pharmaceuticals present?

In this case we want to detect known compounds that have been well characterized previously. By analysing standard samples of propanolol and mefenamic acid, we will already know the LC retention time and the elemental composition (from the exact molecular mass measurements) of the two compounds under scrutiny. We will also know the m/z values of any fragment ions produced during MS/MS experiments. Thus we can confirm the presence or absence of the target analytes by analysing the river samples and looking for signals that have the same characteristics as the standards.

QUESTION 2
What would be the most appropriate type of mass spectrometry and MS/MS experiment to use for this study?

As we are looking for two specific compounds within river water samples that are likely to contain many other compounds, we want to ensure that our instrument can detect our two target analytes specifically. This can be achieved by choosing the most appropriate MS/MS experiment. Precursor ion scanning is useful for unknown analytes, but if the molecule's fragmentation behaviour is well characterized, you do not necessarily want to spend time detecting the whole range of fragments produced. Precursor ion scanning and neutral loss experiments are useful for detecting specific classes of compounds. For example, mefenemic acid fragments to lose 44Da, indicative of the presence of a carboxylate group on the molecule. However, there may be other compounds present with a similar m/z which also contain this functional group, leading to ambiguities in assignment. In this respect, selected (or multiple) reaction monitoring using unique fragmentation transitions would be the method of choice for highly selective detection. For even greater confidence, a combination of two or more unique transitions can be monitored at the expected retention time of the analyte.

The most appropriate instrument to use would be a tandem mass spectrometer with two quadrupole mass analysers (called a tandem or triple quadrupole mass spectrometer, as the collision cell is often a quadrupole design although it is not used as an m/z analyser to separate ions). Quadrupole instruments have a very good dynamic range (suitable for quantification) and can be set to transmit a single m/z making this ideal for SRM or MRM analyses where each quadrupole is set to transmit a different m/z corresponding to a particular analyte's transition from precursor ion to product ion observed during fragmentation.

CASE STUDY 13.2

How to determine whether biomolecular complexes are non-covalently or covalently bound

The aim of the experiment is to differentiate between covalent and non-covalent bonds which can exist between different subunits in biomolecular complexes. In this example, the biomolecular system under scrutiny was the coat protein building block for the bacteriophage Q-beta. The protein monomer has a molecular mass of 14 123.0Da and 180 copies of the protein monomer assemble to produce a single phage capsid with spherical icosahedral geometry.

QUESTION 1
Are the protein monomers which assemble to form the phage capsids held together by covalent or non-covalent bonds?

In this case we know that 180 copies of the protein monomer comprise a single phage capsid and, from electron microscopy, that the capsid is of icosahedral geometry. We want to determine how the proteins are held together: possibilities of bonding include intermolecular disulphide bridges linking two protein monomers through cysteine residues, or non-covalent bonds (i.e. electrostatic interactions, hydrogen bonding, and/or hydrophobic interactions).

To determine if disulphide bonds play a role in the assembly of this phage capsid, a sample of the capsid was treated with the reducing agent dithiothreitol. The resulting mass spectrum showed evidence of the protein dimer alone, indicating that the protein dimer is stable to reduction and hence is not held together by disulphide bonds. Further treatment of this sample by dilution into a denaturing solution (organic solvent plus dilute formic acid) led to a mass spectrum containing ions originating from the protein monomer alone, suggesting that the dimer is held together by non-covalent bonds which are susceptible to denaturants.

QUESTION 2
Can we use mass spectrometry to prove which bonds are present?

In a separate experiment, the capsid was first denatured by dilution into organic solvent and aqueous formic acid. The mass spectrum obtained on this sample contained sample ions consistent with the presence of protein monomer and pentamer, indicating that the pentamer is stable to denaturants. Further treatment of this sample with dithiothreitol led to a mass spectrum containing only the protein monomer, and hence we can conclude that the pentamer is held together by covalent disulphide bonds.

As the experiments involved analysing biomolecules under both denaturing and non-denaturing conditions, electrospray ionization was the method of choice as the sample can be analysed from a range of volatile solutions over a wide pH range. The mass spectrometer used had a TOF analyser to ensure that the high-*m*/*z* ions (i.e. with few charges) produced under non-denaturing conditions could be detected (Ashcroft et al. 2005).

References

Ashcroft, A.E. and Derrick, P.J. (1990) Four-sector tandem mass spectrometry of peptides. In: Desiderio, D.M. (ed), *Mass Spectrometry of Peptides*. Boca Raton, FL: CRC Press.

Ashcroft, A.E. (2005) Recent developments in electrospray ionization mass spectrometry: noncovalently bound protein complexes. *Nat Prod Rep* **22**: 452.

Ashcroft, A.E., Lago, H., Macedo, J.M., Horn, W.T., Stonehouse, N.J., and Stockley, P.G. (2005) Engineering thermal stability in RNA phage capsids via disulphide bonds. *J Nanosci Nanotechnol* **5**: 2034–41.

Brown, R.S. and Lennon, J.J. (1995) Mass resolution improvement by incorporation of pulsed ion extraction in a matrix-assisted laser desorption/ionization linear time-of-flight mass spectrometer. *Anal Chem* **67**: 1998–2003.

Coles, J. and Guilhaus, M. (1993) Orthogonal acceleration—a new direction for time-of-flight mass spectrometry. Fast, sensitive mass analysis for continuous ion sources. *Trends Anal Chem* **12**: 203–13.

Comisarow, M.B. and Marshall, A.G. (1974a) Fourier transform ion cyclotron resonance spectroscopy. *Chem Phys Lett* **25**: 282–3.

Comisarow, M.B. and Marshall, A.G. (1974b) Frequency-sweep Fourier transform ion cyclotron resonance spectroscopy. *Chem Phys Lett* **26**: 489–90.

Dole, M., Mack, L.L., Hines, R.L., Mobley, R.C., Ferguson, L.D., and Alice, M.B. (1968) Molecular beams of macroions. *J Chem Phys* **49**: 2240–9.

Fenn, J. B., Mann, M., Meng, C.K., Wong, S.F., and Whitehouse, C.M. (1989) Electrospray ionization for mass spectrometry of large biomolecules. *Science* **246**: 64–71.

Finnigan, R.E. (1994) Quadrupole mass spectrometers. *Anal Chem* **66**: 969A–75A.

Hillenkamp, F., Karas, M., Beavis, R.C., and Chait, B.T. (1991) Matrix-assisted laser desorption/ionization mass spectrometry of biopolymers. *Anal Chem* **63**:1193A–1203A.

Iribarne, J.V. and Thomson, B.A. (1976) On the evaporation of small ions from charged droplets. *J Chem Phys* **64**: 2287–94.

Johnson, R.S. and Biemann, K. (1989) Computer program (SEQPEP) to aid in the interpretation of high-energy collision tandem mass spectra of peptides. *Biol Mass Spectrom* **18**: 945–57.

Jones, A.W. and Cooper, H.J. (2011) Dissociation techniques in mass spectrometry-based proteomics. *Analyst* **136**: 3419–29.

Karas, M. and Hillenkamp, F. (1988) Laser desorption ionization of proteins with molecular masses exceeding 10,000 daltons. *Anal Chem* **60**: 2299–301.

Krutchinsky, A.N., Chernushevich, I.V., Spicer, V.L., Ens, W., and Standing, K.G. (1998a) Collisional damping interface for an electrospray ionization time-of-flight mass spectrometer. *J Am Soc Mass Spectrom* **9**: 569–79.

Krutchinsky, A.N., Loboda, A.V., Spicer, V.L., Dworschak, R., Ens, W., and Standing, K.G. (1998b) Orthogonal injection of matrix-assisted laser desorption/ionization ions into a time-of-flight spectrometer through a collisional damping interface. *Rapid Commun Mass Spectrom* **12**: 508–18.

Loo, J.A. (1997) Studying noncovalent protein complexes by electrospray ionization mass spectrometry. *Mass Spectrom Rev* **16**: 1–23.

Makarov, A. (2000) Electrostatic axially harmonic orbital trapping: a high-performance technique of mass analysis. *Anal Chem* **72**: 1156–62.

Mamyrin, B.A., Karataev, V.I., Schmikk, D.V., and Zagulin, V.A. (1973) The mass-reflectron, a new nonmagnetic time-of-flight mass spectrometer with high resolution. *Sov. Phys. JETP* **37**: 45–8.

Paul, W. and Steinwedel, H. (1953) Ein neues massenspektrometer ohne magnetfeld. *Z Naturforsch A* **8**: 448–50.

Roepstorff, P. and Fohlman, J. (1984) Proposal for a common nomenclature for sequence ions in mass spectra of peptides. *Biol Mass Spectrom* 11: 601.

Sommer, H., Thomas, H.A., and Hipple, J.A. (1951) The measurement of eM by cyclotron resonance. *Phys Rev* **82**: 697–702.

Uetrecht, C., Versluis, C., Watts, N.R., Wingfield, P.T., Steven, A.C., and Heck, A.J. (2008) Stability and shape of hepatitis B virus capsids in vacuo. *Angew Chem Int Ed* **47**: 6247–51.

Uetrecht, C., Watts, N.R., Stahl, S.J., Wingfield, P.T., Steven, A.C., and Heck, A.J.R. (2010) Subunit exchange rates in Hepatitis B virus capsids are geometry- and temperature-dependent. *Phys Chem Chem Phys* **12**: 13 368–72.

Watson, J.T. and Sparkman, O.D. (2007) *Introduction to Mass Spectrometry: Instrumentation, Applications, and Strategies for Data Interpretation* (4th edn). Chichester: Wiley–Blackwell.

Wilcken, B., Wiley, V., Hammond, J., and Carpenter, K. (2003) Screening newborns for inborn errors of metabolism by tandem mass spectrometry. *N Eng J Med* **348**: 2304–12.

Wiley, W.C. and McLaren, I.H. (1955) Time-of-flight mass spectrometer with improved resolution. *Rev Sci Instrum* **26**: 1150–7.

Wilm, M. and Mann, M. (1996) Analytical properties of the nanoelectrospray ion source. *Anal Chem*, **68**:1–8.

Wolff, M.M. and Stephens, W.E. (1953) A pulsed mass spectrometer with time dispersion. *Review Sci Instrum* **24**:616–17.

Yamashita, M. and Fenn, J.B. (1984) Electrospray ion source: another variation on the free-jet theme. *J Phys Chem* **88**: 4451–9.

Yates, J.R., Ruse, C.I., and Nakorchevsky, A. (2009) Proteomics by mass spectrometry: approaches, advances, and applications. *Annu Rev Biomed Eng* **11**: 49–79.

14 Proteomic analysis

Richard Unwin

Chapter overview

In this chapter, we will look at methods for analysing the total protein content of cells or tissues, the **proteome**. Study of the proteome, or **proteomics**, is important. While genes may provide the blueprint for what a cell may do, it is the proteins that are produced which actually determine all aspects of cell behaviour—size, shape, activity, and response to external and internal factors. In short, it is the proteins we express which make us what we are. By comparing the levels and activity of those proteins in response to experimental conditions, for example diseased tissue versus normal, or cells with and without a particular gene, we can begin to piece together how proteins work in concert to ensure that cells and organisms function properly.

In this chapter, we will describe some of the most common methods for analysing the proteome, noting their strengths and weaknesses such that we can choose and use the most appropriate method for providing an answer to a particular biological question.

LEARNING OBJECTIVES

This chapter will enable the reader to:

- describe what proteomics research is, and how it is applied;
- summarize the major steps of a variety of different proteomics workflows;
- select an appropriate workflow to answer a defined biological problem, and provide reasons for making that selection;
- extract information about how a biological system is working from proteomics data.

14.1 Introduction

Just as the genome is defined as the genetic information within the cell, so the protein content of a cell is called its proteome. Likewise, while genomics is the study and analysis of the genome, the analysis of the proteome is called proteomics. However, the similarity between the two ends there. The genome of an organism is static, with almost every type of cell in a single organism containing an identical copy of the same genes. Nevertheless, from this genome a dazzling array of processes and regulatory events can take place, such that only a selected and tightly regulated portion of the genes in the genome are switched 'on' any cell at any time, generating a transcriptome. From there a second, equally vast set of regulatory events decide which transcripts are turned into proteins, and how much protein is made. While this ensures that each cell type in a body has its own distinctive set of expressed proteins, the story does not end there. A myriad of further post-translational modifications to the expressed proteins can then occur. These may include cleavages of the protein, or the addition of phosphate, sugars, lipids, other proteins, or any one of up to 200 other chemical reactions which control the fate and activity of each protein molecule, including where in the cell the protein goes, what it will bind to, its enzymatic activity, and its lifespan. It has been estimated that humans, whose genome contains around 20 000–25 000 protein coding genes, can use these additional modifications to make over a million different proteins (Jensen 2004).

Therefore we require a number of different tools to study the proteome depending on whether we are interested in what proteins are present or we are interested in a particular type of protein or modification, although these two questions are not necessarily mutually exclusive. A range of tools is also useful since, while the behaviour of a length of DNA is largely predictable if you know its sequence, the same cannot be said about proteins. Proteins are chains made up of any combination of 20 very different amino acids and can be any length. As a result, they have an enormous array of biochemical properties which makes it virtually impossible to study all proteins at once. Proteins can range in size from tens to tens of thousands of amino acids (the smallest human protein, histone H2A, is 66 amino acids long; the largest is titin, a whopping 34 350 amino acids long) and may be positively or negatively charged, may be soluble or insoluble in water, and can incorporate non-protein components such as lipids and metals.

Proteomics is still a rapidly evolving field with many new and improved technologies being developed all the time. However, almost all proteomics experiments must incorporate some standard steps.

- Extraction of protein from the cell/tissue/biofluids in a buffer which can maximize extraction and keep protein in solution, yet is compatible with the next step.
- Separation, usually by either running through a gel or by chromatography; this is a critical step to 'spread out' the proteins so that you can see many at once.
- Labelling or visualization of the proteins, which is especially important if you want to get an idea of how the *amount* of a protein is changing in response to a particular factor.
- Identification of the proteins, usually by mass spectrometry.

See Chapter 13

- Analysis of large and complex patterns of protein identification and expression data, usually computer-driven.

The order in which these steps are carried out can vary, depending on the technology used, but as a variety of proteomics workflows are described in this chapter we shall see that all steps are equally important and selection of the right methods in the right combination is critical to allow us to learn about how the dynamic proteome responds to environmental or biological factors.

14.2 Gel-based proteomics

14.2.1 Two-dimensional gel electrophoresis (2-DE)

Gel electrophoresis was introduced in Chapter 9 as a means of separating a mixture of proteins on the basis of their size prior to detection of a specific protein by Western blot. However, this separation on the basis of size does not provide enough resolution to allow comparison of the levels of many proteins simultaneously. To do this, proteins can be separated using two distinct properties, namely their mass *and* their charge, using **two-dimensional electrophoresis (2-DE)**. In this method, which was first described by O'Farrell (1975), proteins from a single sample are first separated out on the basis of their charge in a long thin strip of gel, which contains a fixed (immobilized) pH gradient (Westermeier et al. 1983). Proteins migrate to their isoelectric point (pI), i.e. the pH at which their overall charge becomes zero. This is known as **isoelectric focusing** (IEF). This thin strip of gel is then placed on top of a large acrylamide gel slab, and proteins migrate out perpendicular to the IEF and are further separated on the basis of their size. This can be seen in the first part of Figure 14.1, where a mixture of proteins migrate into different positions on the gel depending on their pI and mass, allowing you to separate two (or more) proteins of similar mass which would migrate to the same place in a standard gel. Gels are then stained with a protein-specific stain, most commonly silver or **Coomassie blue**, to reveal that the proteins have formed a pattern of spots. This can be seen in the lower half of Figure 14.1, where each spot represents a different protein form. The same protein can appear as several spots if it has several different modified forms. This often occurs as a spot 'train' with the size of the spot being proportional to the amount of protein. By comparing spot patterns for different samples and looking at which spots differ between the two, it is possible to identify differences in protein expression.

Typically, samples from cultured cells generate around 2000 distinct protein spots on these gels, although of course this does not represent all the proteins in the cell. However, the method can be tailored towards certain (or more) proteins. In the first (IEF) dimension, it is possible to separate proteins over very short pH ranges so that, instead of using a pH range which will display all proteins (e.g 3–10), you can use several 'zoom' gels, which have shorter ranges, i.e. split the sample between gels of range 3.5–4.5, 4.5–5.5, 5.5–6.7, 6.2–7.5, 7–11. You could also alter the strength of the second-dimension gel to preferentially resolve very small proteins (increased acrylamide concentration) or very large proteins (decrease acrylamide concentration).

In addition, the type of stain used can allow different populations of proteins to be seen. While silver is the most sensitive and therefore the most commonly used total visible protein stain for 2D gels, some people use Coomassie blue if sensitivity is not an issue. This is because silver can lead to higher background staining and is less compatible with methods for protein identification, as described later in this chapter. New generations of fluorescent stains, such as Sypro Ruby, are even more sensitive, but expensive scanning equipment is required to see the fluorescent spots (Patton 2002).

Figure 14.1 The upper panel shows how 2-DE can resolve many proteins on the basis of their charge and then their size. Proteins are first separated on the basis of their charge on a thin gel strip. This strip is then placed on top of a slab of gel and the proteins are separated in a perpendicular direction on the basis of the mass. Staining these gels with (in this case) silver gives a unique pattern of spots, as shown in the lower panel, with proteins spread out across the gel and the size/intensity of the spot indicating the amount of that protein in the sample.

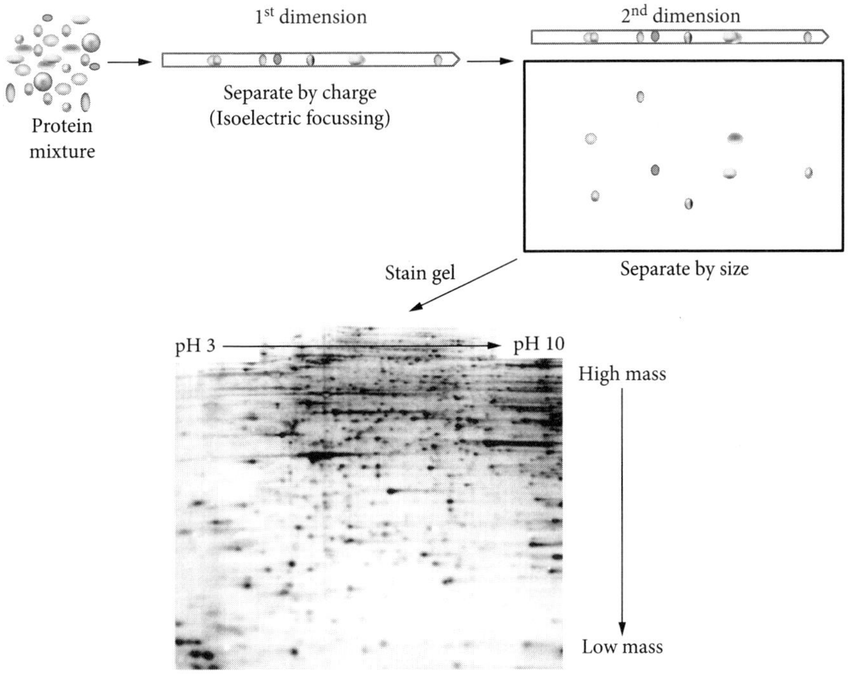

14.2.2 Difference in-gel electrophoresis (DiGE)

Possibly the greatest technical difficulty in using 2-DE is that the spot patterns can be very hard to compare, especially if you have a lot of gels. Different gels may run slightly differently, and it is important to make sure that you are comparing the same protein spot across all gels. Very sophisticated computer software has been written which can take scanned gel images and 'warp' them, effectively pulling and stretching the gel image until the spots overlay. However, there is still some degree of uncertainty, particularly if, in two separate samples, a particular protein is differentially modified which slightly moves its position in the gel. It then becomes very difficult to decide if a small shift is because of a real biological change, or is just a slight difference between the way two gels have run.

One way of overcoming this problem was developed in 1997 and is called difference in-gel electrophoresis (DiGE) (Unlü et al. 1997). Up to three samples are labelled with different fluorescent dyes which emit light at different wavelengths (red, blue, and green) and are mixed and separated on the same gel. As they are on the same gel, all proteins will move to exactly the same place. By scanning the gel under different wavelengths, you can produce three gel images (one 'red', one 'blue', and one 'green') which theoretically should overlay perfectly and enable spot shifts to be seen more easily. If an experiment requires multiple gels, the third DiGE channel may contain a reference sample (often a mixture of all samples in the experiment) so that all gels have the same sample on them, allowing simpler matching and improved quantification (Friedman and Lilley 2008). In practice, the need for running samples and experiments more than once means that DiGE does not eliminate the effect of gel-to-gel variation, but it can certainly help to reduce it. However, the dyes are more expensive than the stains and a special fluorescent scanner is needed to image the gels, so some laboratories still prefer to use visible stains such as silver.

2D gel electrophoresis has many advantages over other large-scale protein screening technologies such as those described later in the chapter. One of the main advantages is that as you are separating and comparing intact proteins, it is possible to detect changes in protein modification without prior knowledge of what that modification might be. As we shall see in the next section, gel-free approaches tend to require that proteins are digested into peptides before they are analysed, immediately losing any information regarding situations where the same protein has two or more patterns of post-translational modification. However, the technical difficulties surrounding analysis of very large or very small proteins or membrane proteins, gel matching, the need for further (albeit relatively straightforward) experiments to identify the protein contained in spots whose intensity changes, and the relative lack of penetration into the proteome (a 2D-gel will display perhaps 600–1000 proteins in its 2000 spots, whereas gel-free approaches can commonly quantify over 2000 proteins) mean that 2-DE has fallen out of fashion somewhat in recent times. That said, it still has a significant role to play in certain experimental situations, especially in the study of the regulation of post-translational modification, where it provides a protein-level view of differentially modified, and thus presumably functionally different, forms of protein.

14.3 Protein identification by tandem mass spectrometry (MS/MS)

While the analysis of 2D gels can tell you which spot is more or less intense or is shifting between experimental conditions, it does not tell you what the actual protein is. To identify proteins, the method of choice is called **mass spectrometry (MS)**. MS is a method in which a mass spectrometer is used to calculate the masses of molecules in a sample. Since mass spectrometers are generally better at measuring smaller molecules (say, with a molecular weight <3000–4000Da) a protein, or mixture of proteins, is digested into smaller fragments or peptides before being analysed. This involves incubating the protein with an enzyme, usually trypsin, which cuts the protein up at specific points—in the case of trypsin, this is after any lysine or arginine residue (except those that are immediately followed by proline). A protein in a gel spot can be simply identified by cutting the spot out of the gel with a scalpel, removing the stain, and then leaving the gel piece in trypsin overnight to cut up the protein, with the peptides diffusing out of the gel. The liquid is then analysed by **tandem mass spectrometry** to identify the peptides it contains, and hence the protein in the gel piece.

Briefly, peptides are **ionized** as they are put into the mass spectrometer and their mass is measured. Individual peptides are then isolated within the instrument using a mass 'filter' called a **quadrupole**, and are fragmented along their peptide backbone, usually by collision with an inert gas such as nitrogen or resonance excitation within an ion trap. The masses of the peptide fragments are then measured and the pattern of fragments can be used to 'read' the amino acid sequence of the peptide, either manually or using sophisticated computer programs (Steen and Mann 2004). The sequence obtained can then be used to identify the peptide from which it came by searching databases of protein sequences. This method is referred to as tandem mass spectrometry (MS/MS) as two mass measurements are made. The first is a measurement of the intact peptide (ion) and ensures that only this specific mass is fragmented (MS^1), and the second is the mass measurement of the peptide fragments (MS^2).

See Chapter 13

14.3.1 Liquid chromatography–tandem mass spectrometry (LC–MS/MS)

Identification of peptides from an individual gel spot by mass spectrometry is probably the 'simplest' situation for proteomics because the spot usually contains only one or a few major proteins. In many cases, however, we wish to identify many proteins simultaneously from a more complex sample, or identify more peptides from the proteins in a 2D gel spot to provide a more confident identification or structural information. In order to achieve this, we need to separate the peptides from the sample so that only a few peptides at a time are presented to the MS. This gives the MS more chance to identify more peptides, and it also separates abundant peptides from less abundant ones, thus improving the sensitivity of the analysis.

The methodology most commonly used to achieve this separation is high-performance **liquid chromatography** (HPLC). This is a key technology in proteomics as good chromatography plays a major role in obtaining high-quality data. The type usually employed is reverse phase chromatography, where peptides are separated based on their hydrophobicity, since the buffers employed for this kind of separation are compatible with peptide ionization in the mass spectrometer. Peptides are loaded onto the column in buffer containing an acid (usually trifluoroacetic acid or formic acid) and a small amount of an organic solvent (usually acetonitrile or, less commonly, methanol). They bind to the column and the salts used during the sample preparation/digestion are washed away. Peptides are then eluted from the column by introducing a gradient of increasing organic solvent; for example the acetonitrile content of the buffer may increase from 5% to 40% over a fixed time. As the concentration begins to increase, hydrophilic peptides, which are weakly bound to the column begin to elute, followed by the more hydrophobic peptides. Peptides eluting off the column are introduced directly into the mass spectrometer by connecting the end of the column to either an electrospray needle (more commonly) or a robot which spots sample onto a solid support for MALDI (matrix-assisted laser desorption ionization) analysis. Performing the separation 'on-line' with the mass spectrometer in a single experiment is referred to as liquid chromatography–tandem mass spectrometry (LC–MS/MS).

See Chapter 8

See Chapter 13

Peptide chromatography on-line with the MS is usually performed at high pressures in very narrow columns (75µm internal diameter) and at very low flow rates (200–300nL of liquid through the column per minute). However, settings such as gradient, flow rate, column size, and chromatographic media all play a role in obtaining a good separation. An added advantage of HPLC is that the peptides elute from the chromatography column in around 100–200nL, whereas in the original sample each peptide is present in a tens to hundreds of microlitres of sample. Therefore each peptide is much more concentrated when it is introduced into the mass spectrometer, providing another increase in sensitivity.

14.3.2 Two-dimensional liquid chromatography

For very complex mixtures, just performing LC–MS/MS still does not spread out the peptides sufficiently. For example, a whole mammalian cell may contain 15 000 proteins which, upon digestion, generates a mixture of 750 000 peptides (assuming a conservative average of 50 peptides per protein). This means that in a standard 1 hour run you would need to identify peptides at a rate of around 250 per second, which is well beyond the limits of any modern instrument. Therefore in order to identify more peptides (and hence proteins), we need to spread out the sample even further. To do this you can use a second, and different, mode of chromatography to fractionate the peptides in the sample, much the same as you use two modes of electrophoresis to separate out proteins in 2-DE. Therefore peptides are fractionated in one direction using one type of chromatography, commonly separation by charge using strong cation exchange (SCX) or alternatively reverse phase columns run with high-pH buffers, as shown in Figure 14.2. Peptide fractions are collected from this column and then analysed individually by LC–MS/MS. This approach is called two-dimensional liquid chromatography (2D-LC or LC-LC–MS/MS). It is also often, incorrectly, described as MudPIT (multidimensional protein identification technology) which differs from other 2D-LC workflows since MudPIT has both LC columns in line so that one elutes directly onto the other (Washburn et al. 2001). The use of two different modes of chromatography is sufficient to take the number of proteins identified in a sample from the hundreds into the thousands, and is critical to obtaining deep penetration into the most complex protein samples.

14.4 Post-translational modifications

Although the methods described in the previous sections are ideally suited to the identification of protein species, they are generally poor at defining important post-translational modifications. In 2-DE, a protein with several modifications may appear as several spots, each corresponding to a different form, but at that stage it is impossible to know what particular modification is responsible for the spot shift. In MS/MS methods, the modified peptides are vastly outnumbered by unmodified peptides, such that a typical 2D-LC–MS/MS analysis may identify 5000 or 10 000 peptides, of which only a handful are phosphorylated, one of the most common types of modification. As such, specific methods need to be used to allow us to look directly at protein modifications.

14.4.1 Protein/peptide enrichment

The simplest way of looking at a particular post-translational modification is to extract proteins carrying that modification from the sample. One of the most commonly used affinity reagents is antibodies. Antibodies have been made which react specifically with

Figure 14.2 Schematic representation of two-dimensional liquid chromatography. A mixture of protein is digested with an enzyme such as trypsin to generate a very complex mixture of peptides. These peptides are first separated by liquid chromatography, usually strong cation exchange (SCX) which separates peptides on the basis of their charge. Fractions are collected, and then each fraction in turn is separated using a different mode of chromatography, usually reverse phase. Peptides elute sequentially according to their hydrophobicity and can be directly analysed by mass spectrometry.

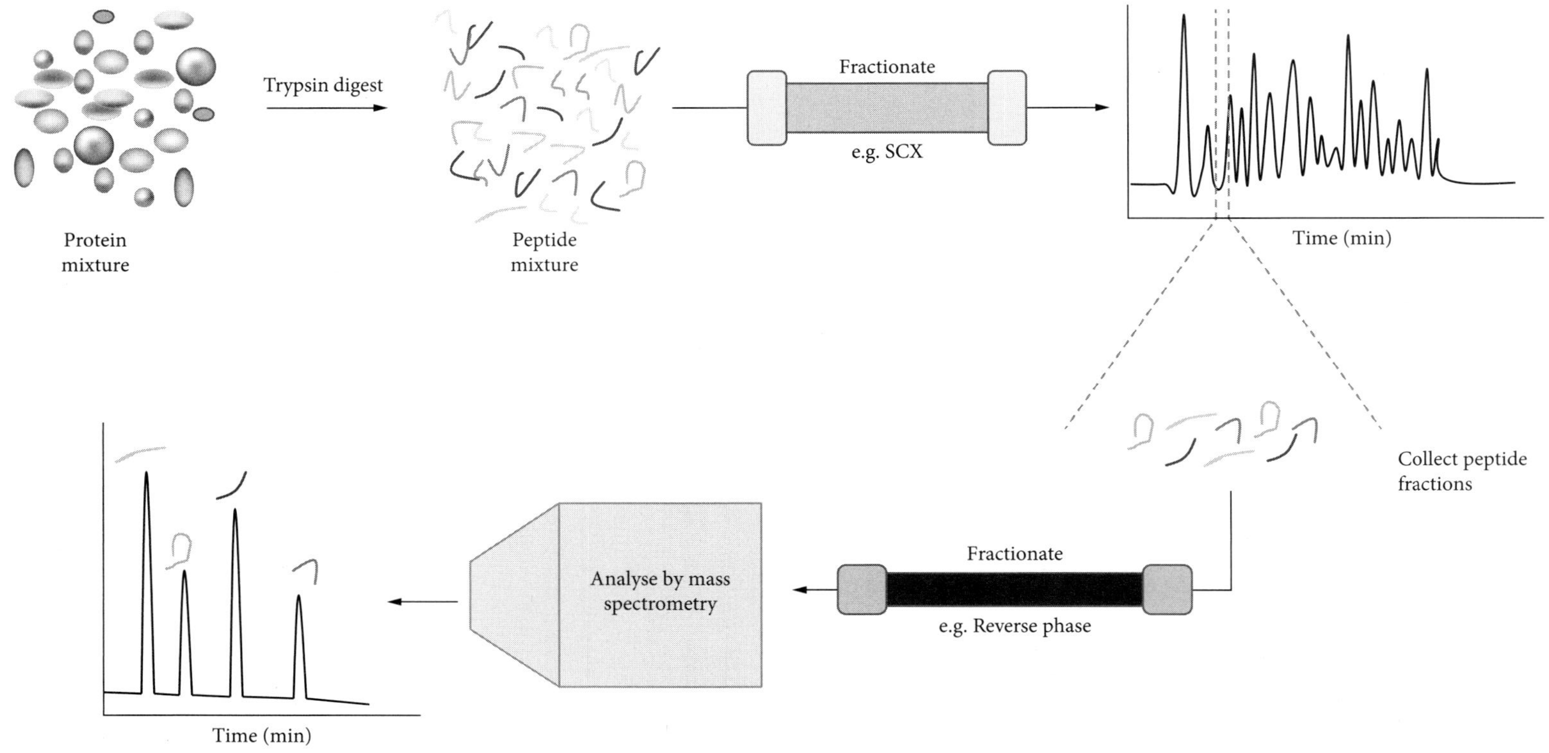

See Chapter 9

a number of post-translational modifications, such as phosphorylated or nitrated tyrosine, acetylated or methylated lysine, ubiquitin, and SUMO. These antibodies can be used to **immunoprecipitate** populations of target proteins, using the antibodies to extract proteins or peptides carrying a certain modification from the sample. Extracted proteins/peptides can then be analysed as an enriched 'sub-proteome'. In cases where no antibodies are available, other affinity reagents can be used in their place. For example, phosphorylation on serine, threonine, and tyrosine can be enriched using titanium dioxide or immobilized metal affinity chromatography (IMAC), where metals ion such as Fe(III), Ni(II), or Ga(III) are immobilized onto an inert support which will bind phosphoproteins/peptides (Dunn et al. 2010). Similarly, glycoproteins can be enriched using a variety of lectins, specific sugar-binding proteins which can be bound to a solid support to enable enrichment of proteins with different classes of sugar modification (Abbott and Pierce 2010). Such enriched populations of peptides can then be analysed by mass spectrometry, increasing the number of peptide identifications containing the modification of interest.

14.4.2 Specific detection: in gels

While gels (1D or 2D) are commonly stained for total protein by silver or Coomassie blue, or by using fluorescently tagged proteins (DiGE), it is also possible to use reagents which specifically and selectively stain proteins carrying a certain modification. Specifically, stains such a ProQ Diamond for phosphorylation and ProQ Emerald for glycoproteins have been used to compare levels of these modifications in 2D gel spots. Both stains are fluorescent and allow subsequent staining for total protein, which is important as it allows modified proteins whose expression is changed to be distinguished from proteins whose expression is unchanged but whose modification status is changed. In addition to specific stains, it is also possible to 'Western blot' proteins from 2D gels onto nitrocellulose or polyvinylidene fluoride (PVDF) membrane and probe spot patterns with antibodies directed at specific modifications, although identification of proteins using such an approach is technically challenging.

14.4.3 Specific detection: mass spectrometry

See Chapter 13

A mass spectrometer can be set up to target peptides which have a specific post-translational modification. The two most common methods for this are precursor ion scanning and neutral loss scanning. In each case, the mass spectrometer is set to detect a fragmentation which is specific to the post-translational modification of interest, either by detection of an ion which is generated by the modification itself (precursor ion scanning) or by looking for the loss of a specific modification mass from the intact peptide (neutral loss scanning). In cases where the protein target is known and the aim is to identify where it is modified, it is also possible to perform **selected reaction monitoring** (SRM) MS, where the MS is set up to detect specific peptides, in this case post-translationally modified versions of peptides from the protein of interest. Positive signals from precursor ion scanning, neutral loss scanning, or SRM can be used to trigger an MS/MS analysis of the potentially modified peptide to allow confirmation of the peptide identity and the site and nature of the modification.

14.5 Quantitative mass spectrometry

Thus far, mass spectrometry has been described for the identification of proteins in a sample, either directly or following 1D-LC for simple mixtures, or using 2D-LC for more complex samples. However, these methods do not allow comparison of protein levels between samples per se. However, there are a range of methods which can be employed to allow protein quantification by mass spectrometry. These methods can be classified as *global*, where the aim is to quantify as many proteins in the sample as possible (hypothesis generation), or *targeted*, where the aim of the experiment is to quantify a specific protein of interest (hypothesis-driven). In all cases such experiments, especially global profiling, generate large and complex datasets with plenty of opportunities for the introduction of experimental errors, and as such it is critical that experiments of this nature are properly designed in terms of knowing how much variation there is if you measure the same protein in the same sample multiple times, and trying to minimize this variation. Repeating profiling experiments, in particular, is key to obtaining reliable data for further more targeted and hypothesis-driven experiments.

14.5.1 Label-free protein quantification

There are two common methods for performing protein relative quantification without chemically labelling the proteins/peptides first. The first, spectral counting, is effectively based on counting the number of peptides (or MS/MS spectra) associated with a given protein in a standard MS/MS run. The more abundant a protein, the more peptides (MS/MS spectra) will be identified from it. While this method is relatively easy to implement, it is not possible to determine by how much the levels of a protein differ, and it tends to be relatively low resolution, in that it is only really good at detecting very large changes in protein expression.

The second label-free approach relies on the analysis and comparison of the signal intensity for each peptide in the MS phase of the experiment (detection of intact peptide prior to MS/MS fragmentation).This is illustrated in Figure 14.3, where several different protein samples can be processed and analysed by LC-MS. The size of each peptide peak is extracted from the spectrum and compared using dedicated computer software to produce a ratio for each peptide (defined by mass and chromatographic retention time) in each sample. These experiments require very careful control as each sample is analysed sequentially, and so minor changes in the performance of the chromatography or mass spectrometry can affect signal intensity. Sophisticated computer algorithms are required to be able to determine over the course of entire samples what is a 'peptide' peak, to characterize that peak with respect of its mass (m/z) and its retention time on column, and then to be able to find the same peptide in all samples in a run, despite the fact that the m/z and retention time will not be identical (although they will be very similar).The software then needs to be able to predict and 'iron out' the inevitable noise and drifts in the data caused by running samples sequentially over an extended period, making it difficult to control this technique and to obtain reliable data, although it is possible with good experimental design and controls.

Figure 14.3 Example workflow for label-free quantification by mass spectrometry. Each sample is analysed individually by mass spectrometry and the signal intensity for each peptide is calculated. Intact peptide intensity from many samples is compared, with sophisticated computer algorithms producing a list of peptides (defined by their mass and chromatographic retention time) along with the relative amount in each sample. In this case the computer will detect an increase in the amount of the blue protein and a decrease in the amount of yellow protein in sample C. For a colour reproduction of this figure, please see Plate 12.

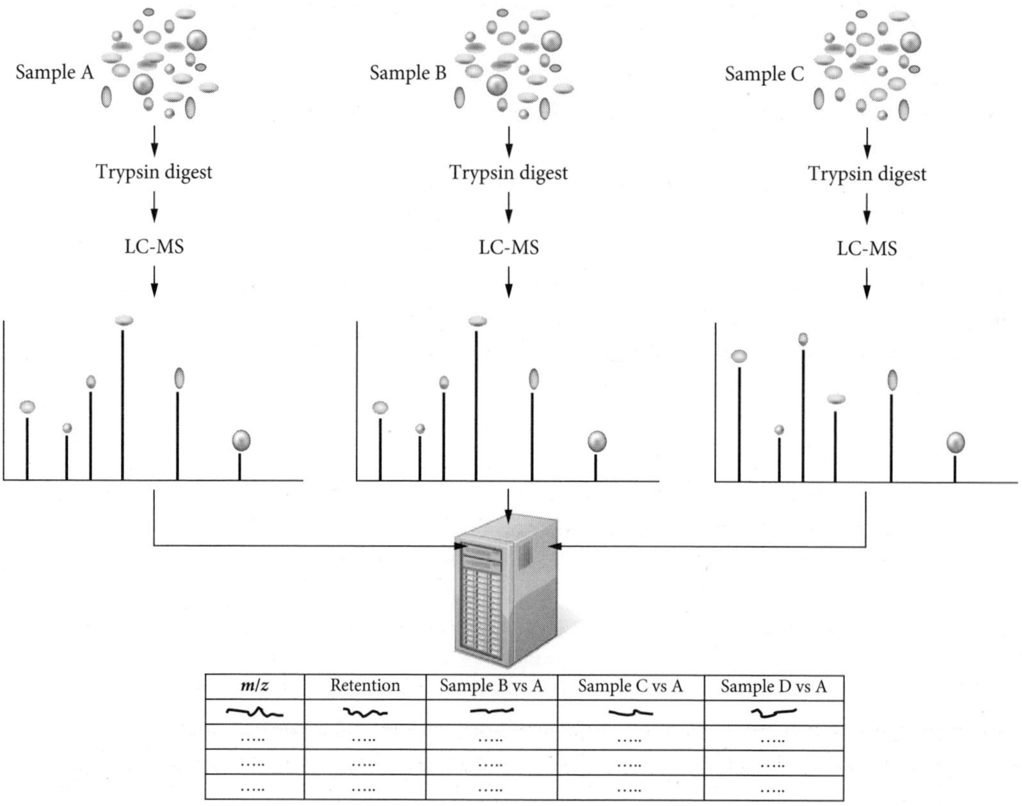

The advantage of this approach is that it is possible to compare many samples in a single experiment, whereas this is more difficult for the labelling experiments described later in this section. Also, it does not rely on obtaining successful MS/MS identification of the peptide sequence in the first analysis; peptides which are identified as being potentially differentially expressed can be identified in a second selective MS/MS experiment. The disadvantages of this approach are that it tends to be less accurate and more 'noisy' than labelling approaches, and so label-free methods are less likely to detect smaller changes in protein expression than labelling approaches, although in this respect comparison of peptide intensity is more sensitive than spectral counting. It is also the case that there is benefit in knowing which peptides do not change. For labelling approaches, where all peptides are identified by MS/MS, it is possible to detect outlier peptides such as protein splice variants of modifications where some peptides from a protein will change but others will not.

14.5.2 Stable isotope labelling by amino acids in culture (SILAC)

In order to negate the effects of minor variations between experimental runs, it is possible to combine samples and analyse them simultaneously. To enable this, each sample must be labelled or tagged so that it is possible to see which sample any peptide came from, and to be able to compare the amount of that peptide in each sample. Two types of labelling are currently used. The first introduces a mass difference between the same peptide from different samples by the addition of heavy isotopes, and the second uses isobaric chemical tags. There are several methods of introducing a heavy label into peptides (see Box 14.1), but the most common is stable isotope labelling by amino acids in culture (SILAC) (Ong et al. 2002).

In SILAC labelling, cells are cultured in either normal media or media containing specific amino acids (usually arginine or lysine) which contain 'heavy' isotopes, i.e. $^{13}C_6$-Lys, in which six of the carbon atoms have mass 13 (rather than the normal 12) and therefore is 6Da heavier than 'normal' lysine, or $^{13}C_6$, $^{15}N_4$-Arg, which is 10Da heavier than normal arginine. These amino acids are incorporated into proteins made as the cells expand, and after about five passages labelling should be complete, i.e. all lysines or arginines are 'heavy'. It is then possible to wash the medium away and mix the 'heavy' and 'light' samples together, extract proteins, digest, and analyse the peptides by LC–MS/MS. As is shown in Figure 14.4, the mass spectrum contains a series of 'mass pairs'. This is because the same peptide sequence from the two samples will behave identically, since they are processed and analysed in tandem and are of the same basic chemical composition, but because of the label they will appear as a 'doublet' in the mass spectra, with the heavy and light forms separated by 6Da or 10Da. Comparison of the intensity of the heavy and light peaks allows the relative amount of that peptide between samples to be calculated.

See Chapter 15

BOX 14.1

Methods for introducing a heavy label into peptides for relative quantification

Stable isotope labelling with amino acids in culture (SILAC) – SILAC uses amino acids containing heavy stable isotopes such as ^{13}C or ^{15}N in cell culture medium (Ong et al. 2002). After a sufficient number of cell divisions, the heavy-isotope-labelled amino acid is completely integrated into the cellular proteome. This method is popular as it allows samples to be pooled prior to any processing, although it is limited to experiments in cell culture systems.

^{18}O water during digestion – Performing proteolysis in water containing ^{18}O results in the transfer of the 'heavy' oxygen to the carboxy terminus of each digestion product (Yao et al. 2001). Although this method can be used for all samples, it is difficult to obtain complete labelling since, once sample are mixed, back exchange of ^{18}O with ^{16}O can occur. In addition, the relatively small mass difference can sometimes hamper quantification of larger peptides.

^{13}C acrylamide labelling – The heavy and light acrylamides differ by 3Da and are relatively cheap reagents for modifying cysteine residues (Faca et al. 2006). The modification reaction is easy and efficient, but again this technology limits you to comparing two samples simultaneously, and the relatively small mass difference can sometimes hamper quantification of larger peptides.

Isotope-coded affinity tags – This was one of the first methods of introducing a heavy label into peptides (Gygi et al. 1999). These reagents consist of three reactive groups, one to modify cysteine, a linker which contains the heavy (or light) isotopes, and a biotin moiety. This allows labelled peptides to be enriched by binding the biotin to streptavidin, 'cleaning up' the sample in the process. A cleavable form also has be used where the biotin is cut off after purification to reduce the size of the modification. Generally, this approach is more expensive than ^{18}O or ^{13}C-acrylamide labelling and has few advantages. Also, some proteins do not contain a Cys residue and some contain only one, so quantification of such proteins is either impossible or unreliable since it is based on only a single measurement.

Figure 14.4 Example of a workflow for SILAC quantification. One sample is labelled with heavy stable isotopes, which induces a mass shift allowing samples to be pooled and processed together. Pooled 'heavy' and 'light' samples are analysed by MS, with the same peptide from each sample appearing as a doublet in the mass spectrum. Analysis software detects these 'mass pairs', compares the relative levels of the 'light' and the 'heavy' version of the peptide, and thus calculates the ratio of peptide amount between the two samples.

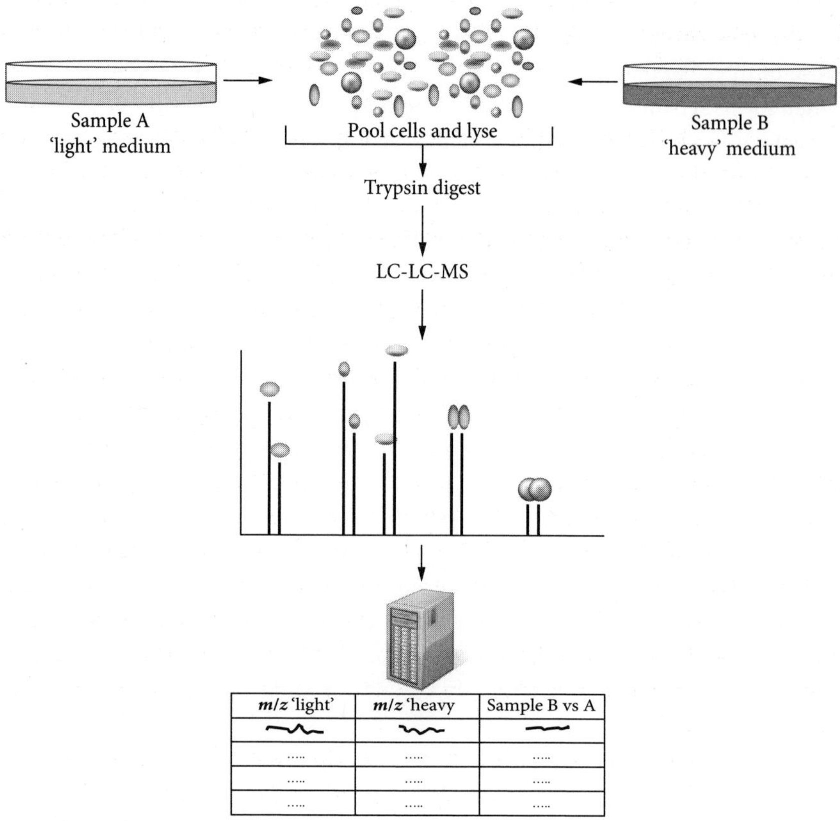

SILAC has become the method of choice for many types of experiment, particularly those where sample handling prior to mass spectrometry is complex, such as protein purification or subcellular fractionation, since samples are combined at the beginning of the process and therefore every step is identical in both samples, minimizing variations which can result from basic sample preparation. However, experiments with more than three samples are difficult as this is currently the limit of how many samples can be pooled. It is possible to compare more conditions by including the same sample in every set of three as a reference, although there is some loss of data in this procedure as the MS will not necessary identify and quantify exactly the same peptides in each set of three pooled samples. Also, since the labelling is based on cell culture, this method is not suitable for the quantification of proteins from, for example, clinical or animal tissues or biofluids. Other methods of introducing a heavy isotope, which usually involve chemical reaction of the peptides with heavy and light versions of the same compound, can be used for these types of sample.

14.5.3 Isobaric tags for quantification in MS/MS

An alternative to using heavy isotopes to introduce a mass difference between samples is to use chemical tags which are isobaric, i.e. with the same overall mass. These tags, the most common of which are called iTRAQ (isobaric tags for relative and absolute quantification), are again chemically identical overall, and have the same mixture of heavy and light atoms, but are located in different places around the tag (Ross et al. 2004). This can be seen at the top of Figure 14.5, which shows the structure of the iTRAQ label. On the right-hand side of the molecule is a chemical group (the protein reactive group) which will react with the N-terminus of every peptide in the sample (and the free amine in the side chain of lysine residues). The centre and left-hand parts of the molecule are the 'reporter' and 'balance' regions, respectively. These have different masses for the different tags. The reporter groups for the eight-sample iTRAQ tag are 113, 114, 115, 116, 117, 118, 119, or 121Da. The mass of the balance group is also different between the tags, but in such a way that the combined mass of the balance plus reporter groups is the same (i.e. if the reporter is 113 the balance is 192, if the reporter is 114 the balance is 191, and so on). As can be seen in the lower part of Figure 14.5, in which only three of eight samples are shown for clarity, all peptides from each sample are tagged with this reagent and are then mixed together and analysed by LC–MS/MS. As the tags are isobaric, the same peptide from each sample appears at the same mass and so samples are indistinguishable in the MS phase of the experiment. When this peptide is selected for MS/MS, the tags fragment at the same time as the peptide and release the reporter fragment. Therefore by comparing the amount of each reporter you get an indication of how much peptide is present versus other tagged samples. The balance group is lost and the peptide fragments are still at the same mass, generating data on the sequence of the peptide at the same time as its relative abundance.

iTRAQ offers several advantages over SILAC and heavy-isotope-labelling approaches. First, it is applicable to all sample types, including primary and clinical material. Second, it enables comparison of up to eight samples, allowing you to study time-courses in more detail, to compare more experimental conditions, or to build in more controls (e.g. replicates of the same sample) to provide greater confidence in the data. Third, because of the isobaric nature of the tags there is a slight improvement in overall sensitivity, since

Figure 14.5 Example of a workflow for iTRAQ quantification. The top panel shows the structure of the iTRAQ tags. The reporter and balance groups have different masses for the different tags but the same overall mass, and the protein reactive group binds the tag to each peptide. The bottom panel shows the iTRAQ workflow, with up to eight different samples iTRAQ labelled, pooled, and analysed by MS/MS. In the MS scan (intact peptides), each peak is the sum of all eight samples. Upon selection and fragmentation of one of the peptides, sequence ions are still summed from all eight samples but the reporters are released, giving relative quantification of that peptide. In this case, the peptide is higher in samples A, C. and F (corresponding to the first, third, and sixth quantification bars).

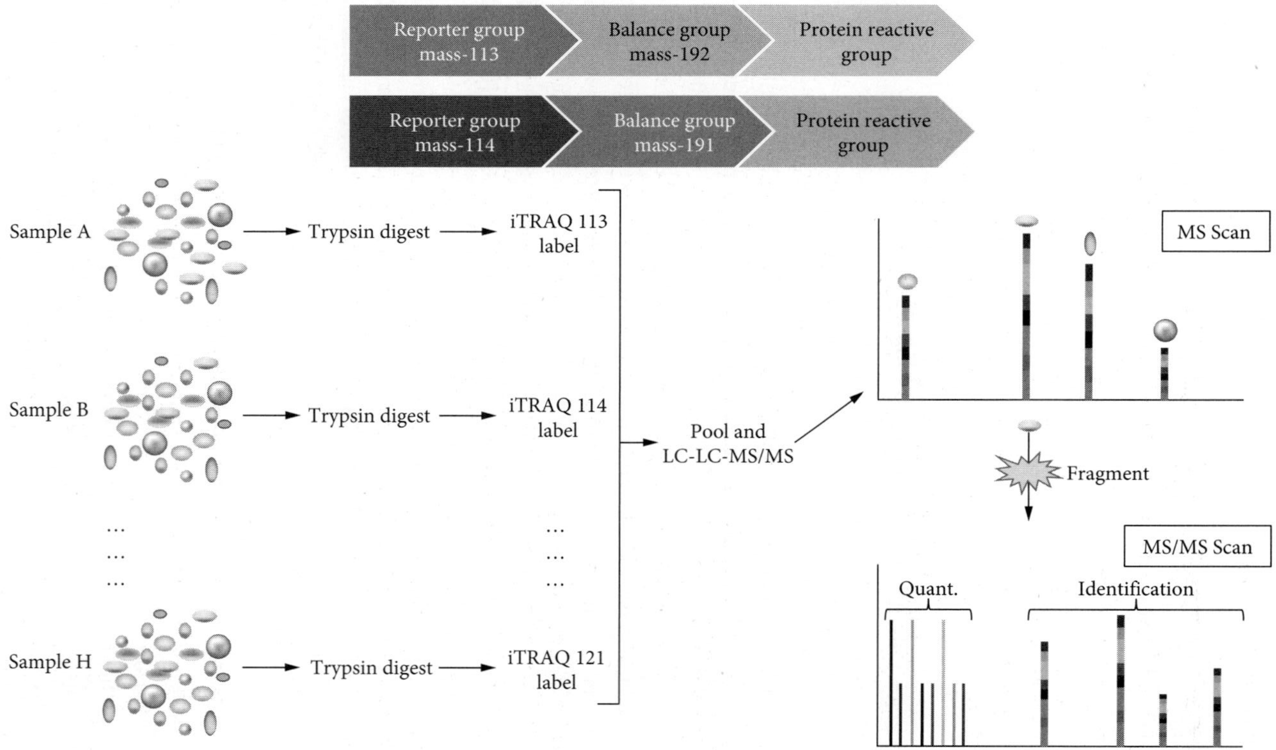

all signals for the intact peptide and peptide fragments are the sum of that peptide from eight individual samples. This is particularly important if the amount of sample is limited (e.g. a rare cell type from primary tissue). The major drawback of iTRAQ is that the labelling generally takes place after proteins have been isolated, enriched (where appropriate), and digested, and so any variation introduced during these steps can mask real biological changes. Therefore it is important to minimize these variations with careful experimental design and control.

CASE STUDY 14.1

Comparison of protein expression in primitive haematopoietic cells from mice

The aim of the experiment is to determine what protein changes occur as stem cells begin to differentiate into blood cells in mouse bone marrow. This process is important in regulating the control of blood cell production, and understanding the signals that trigger stem cell differentiation could be useful in transplantation medicine. It is also important to understand these processes, as many leukaemias have stem cell origins and share similarities. Therefore having an understanding of normal stem cell biology helps our understanding of leukaemia cell biology. Primitive bone marrow cells can be isolated by flow cytometry and provided as a dry cell pellet, but they are rare—it is only possible to obtain a maximum of ~5 million cells of each type, defined as 'early' progenitor cells (more primitive) and 'late' progenitors (beginning to differentiate) for the whole experiment, which will give only around 50–60 μg of protein.

See Chapter 16

QUESTION

Which is the most appropriate proteomics methodology to use for this study?

Consider the options. Since the protein is from a primary source, SILAC–MS is not an option. Likewise, because of the small amount of material available, label-free MS approaches are less appropriate since it would be hard to run the required number of replicates, given the extra noise generated by these methods compared with labelled methods. This leaves heavy isotope labelling by ^{18}O or isobaric tagging MS, or 2D-PAGE as viable options. Given the small amount of sample available, 2D-PAGE or DiGE may be less suitable. Although spots could be visualized on a gel using this approach, there is probably too little sample to obtain good identifications on spots identified as 'changing' due to sample losses going into and back out from the gel. Therefore a gel-free approach may be best. Since isobaric tagging can impart a small sensitivity increase because signals in the MS are summed, this is probably the method of choice for such a study.

CASE STUDY 14.2

Identification of signalling events in the cell nucleus

The aim of the experiment is to identify phosphorylation events which occur in a cell-line model of cancer, with cells transfected with either empty vector (as a control) or an oncogene. Many oncogenes are known to affect cell signalling pathways, and many kinases are known to be dysregulated during tumour development. Understanding which pathways are switched on or off by the oncogenes reveals novel targets for therapy. This cannot be achieved by expression proteomics alone, as the levels of the signalling proteins may not change and measurement of downstream protein changes may not be able to tell us exactly which pathways have been activated and how. As such, specific detection of phosphorylation changes is more likely to be informative in this instance. The cells grow quickly and well in culture, so there is no real upper limit on the amount of protein which can be generated. The oncogene product is known to localize to the nucleus, and so signalling events are of most interest here.

? QUESTION

Which is the most appropriate proteomics methodology to apply for this study? Design a workflow from collecting the cells in culture to analysing them by LC-LS MS/MS or 2D-PAGE, whichever is most appropriate.

Consider the options. Specific phosphorylation events are probably better probed by MS rather than gel-based methods, since MS approaches allow enrichment of phosphopeptides (rather than proteins) and therefore can enable relative quantification of phosphorylation on a specific amino acid, rather than the total amount of phosphate on a whole protein. Again, the poorer quantitative accuracy and single-peptide nature of the quantification makes label-free MS approaches unsuitable for this analysis. Whilst isobaric tagging is generally useful for analyses of time-course data, the binary nature of the experiment above, plus the fact that the scarcity of phosphopeptides requires large amounts of starting material to enrich from, makes SILAC the method of choice here. This allows large amounts of sample to be generated for each cell type, and the sample can be pooled early so that any downstream manipulations do not significantly increase experimental noise. Therefore an appropriate workflow would be as follows.

- Harvest and pool SILAC-labelled cells.
- Isolate nuclei using standard biochemical methods (hypotonic buffer to 'burst' cells, keeping nuclei intact, and then isolating nuclei by centrifugations).
- Lyse nuclei and extract protein. Digest protein into peptides using trypsin.
- Enrich phosphopeptides using IMAC or TiO_2 affinity chromatography.
- Analyse enriched phosphopeptides by LC-LC–MS/MS.

NB: It is critical in this experiment to analyse the total peptide fraction as well. This allows you to distinguish between a change in the phosphorylation status of a static protein population and the increased expression of a phosphoprotein.

14.5.4 Selected reaction monitoring for targeted quantification

The approaches described above all deal with so-called 'global' profiling, i.e. the identification and relative quantification of as many different peptides (hence proteins) as possible from a complex mixture. One drawback of this approach is that the proteins identified and quantified are largely unpredictable. Analysing the same sample twice will identify different (but overlapping) lists of proteins in each run, with all proteins tending to be the most abundant in the sample. There are several reasons for this, the main one being the selection of which peptides to fragment, and if fragmentation occurs when sufficient peptide is being eluted to provide good identification. What this means is that if you wish to quantify *specific* proteins or peptides in many samples, this method is likely to be unsuccessful. However, it is possible to set up a mass spectrometer to specifically target and quantify peptides of interest, using a scan mode called selected reaction monitoring (SRM).

See Chapter 13

SRM on a triple quadrupole-type mass spectrometer is a highly sensitive mode of analysis which involves two mass filters. As shown in Figure 14.6, in this type of scan the first mass filter (quadrupole) is set to allow only the intact peptide through, which is then fragmented, and the second filter is set to allow only a specific fragment through to the detector. The two masses are known as an SRM transition. A signal is only registered by the detector if the specific fragment ion is generated from that specific parent. This makes the MS analysis both more selective and more sensitive, as most of the

Figure 14.6 A triple quadrupole mass spectrometer performing an SRM (selected reaction monitoring) scan set up to specifically detect a peptide with *m/z* 785.9 and which generates a fragment with *m/z* 684.4. The mass spectrometer has two mass filters. The first only allows through ions which match the mass of the intact peptide of interest (*m/z* 785.9). These ions are then fragmented, and the second mass filter lets through only products with a mass matching the product from the peptide of interest (*m/z* 684.4). In this way, the detector only registers the presence of peptides with a given mass-fragment pair, and has high levels of sensitivity due to the filtering out of other 'background' ions. Several 'positive' signals from the same peptide are required to confirm the presence of that peptide in the sample.

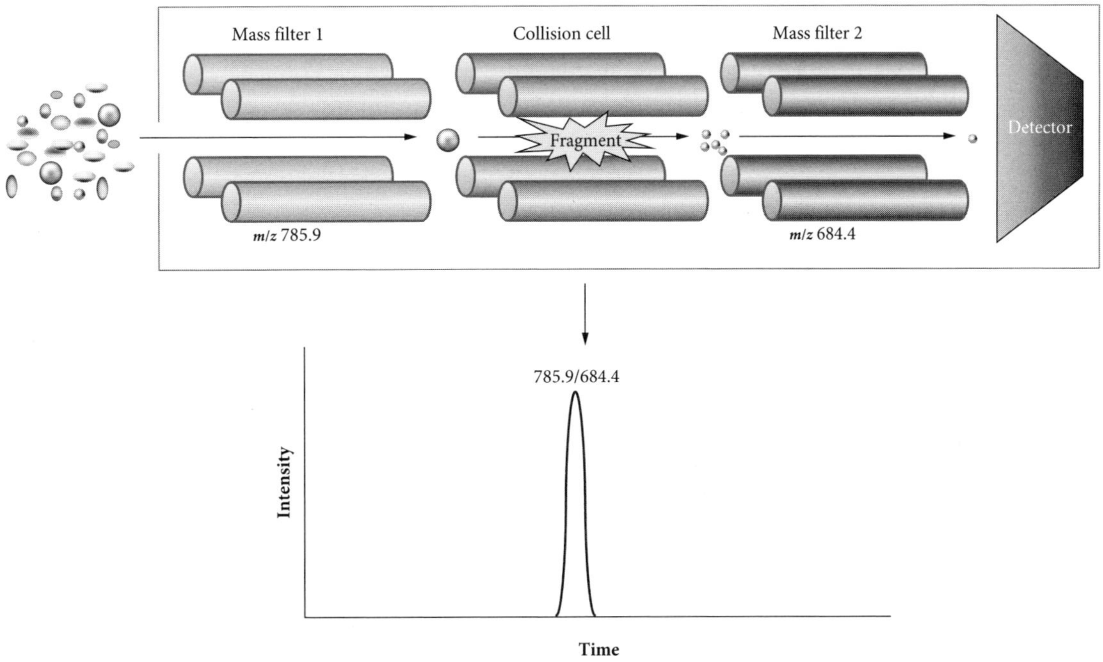

background signal is filtered out and therefore smaller signals are revealed. This mode of analysis is also more accurate in terms of quantification than other methods, and therefore is suitable for label-free comparison of selected proteins from many samples.

SRM methods are valuable in that they allow quantification of several specific peptides in a single run in the absence of antibodies. Common uses are the quantification of a panel of proteins, for example components of a pathway or of potential diagnostic/prognostic markers, or the quantification of a post-translationally modified peptides, for example key phosphorylation events in a signalling cascade. The latter application is particularly important as antibodies specific for a particular modification are frequently unavailable.

Good experimental set-up, based on selection of peptides and peptide transitions, is key for SRM quantification. In a complex mixture many peptides will be of similar mass and some of these may share some fragment ion masses, and some proteins (particularly isoforms) may share peptide sequences. Peptides which are specific to a particular protein are known as **proteotypic peptides**. Some peptides are unsuitable for SRM targets as they can easily (and irreproducibly) be modified during sample preparation, for example by oxidation of methionine residues. Also, a single SRM transition may not be selective or specific for the target peptide, and it is common for at least three transitions to be used to monitor each peptide, the correct peptide being identified when all three transitions are positive at the same point during an LC-MS run. It is also important for quantitative accuracy to be confirmed for each transition, usually by analysing either synthetic versions of each peptide spiked into a standard sample or a dilution series of the same sample.

To build a full SRM method, the list of transitions is combined and the MS performs them sequentially, with each pair of filters set at these fixed defined masses for a specific time (the 'dwell time'). Dwell time is important, and increasing the dwell time increases sensitivity (more ions will hit the detector the longer the transition is being analysed), but it is important for all the transitions to be performed within a time-scale which allows multiple measurements of each transition across the chromatographic elution peak. For example, if each peptide elutes over a 30-second window, each transition must be performed at least once every 3 seconds to ensure 10 points across the peak to give good quantification. One recent way of allowing increasing dwell times (or allowing more

Table 14.1 Summary of mass spectrometric methods for proteome relative quantification

Method	Advantages	Disadvantages
Label-free	Useful for comparison of many samples. Cheap. Applicable to all sample types. Does not require MS/MS identification of peptides for successful quantification. Higher overlap of quantified peptides between runs?	Quantification less accurate, therefore more prone to false positive or false negative results. Data prone to error from instrument drift over time. Not all peaks are peptides. Not all differences can be subsequently identified.
Heavy-isotope labelling (SILAC)	Samples labelled early, so less prone to variation in sample preparation. Cultured cells so large amounts of sample can be labelled (SILAC only). Quantification in MS does not absolutely require high quality MS/MS every time.	Expensive. Applicable to cultured cells only. Some cells hard to completely label, so labelling has to be optimized for each cell type. Limited to two or three samples per experiment.
Heavy-isotope labelling (non-SILAC)	Relatively cheap. Applicable to all sample types. Quantification in MS does not absolutely require high quality MS/MS every time.	Limited to two samples per experiment. Quantitative accuracy relies on complete chemical labelling. Labelling of peptides so can be affected by variation introduced during sample preparation.
Isobaric tagging	Ability to quantify up to eight samples at once. Applicable to all sample types. Allows inclusion of biological and technical replicates and controls allows good data quality control. Pooling isobarically labelled samples offers a small increase in sensitivity if sample is limited.	Relatively expensive? Labelling of peptides so can be affected by variation introduced during sample preparation. Requires good optimal MS/MS to acquire both sequence and quantification ions. Requirement for MS/MS for quant potentially reduces overlap in identified peptides between replicate experiments
Selected Reaction Monitoring	Highly sensitive and specific. Linear over a wider range of concentrations. Targeted towards the same peptide in every sample. Allows targeted quantification of a specific protein/peptide of interest without the need for an antibody Useful for comparison of many samples. Applicable to all sample types.	Need to know the sequence and optimize fragmentation pattern and chromatographic retention time for each peptide. Limited number of peptides can be analysed in any one experiment.

transitions into an experiment) is 'scheduling', so that any particular transition is only set during the period when the peptide is expected to elute. This means that it is not being scanned during the rest of the run, releasing additional time for other peptide transitions. As a result of all these conditions, SRM method design depends strongly on computer programs to organize the transitions into a method, and extract the signal from each transition to allow the relative abundance of peptide levels to be compared between samples.

Mass spectrometric methods for proteome relative quantification are summarized in Table 14.1.

14.5.5 'Absolute' protein quantification

The methods described in the previous sections all provide relative quantification, i.e. they provide a ratio of the amount of protein in two samples, but they do not tell us how much of each protein is there as an absolute amount. In many experiments, knowledge of the exact concentration of a protein is not necessary. However, being able to calculate protein concentration in a sample is important, especially for the measurement of protein in clinical samples such as plasma/serum or urine, which has to be able to produce comparable results in many laboratories over long periods of time. It is possible to achieve this using a technique called **stable isotope dilution,** or AQUA (Absolute QUAntification).

This method is really an amalgamation of stable isotope labelling relative quantification (see section 14.5.2) and targeted SRM analysis (see section 14.5.4). To calculate the quantity of protein present, it is first necessary to obtain a synthetic version of the peptide(s) of interest which are made to a known concentration and contain stable heavy isotopes, such that the standard (heavy) peptide behaves in the same as the endogenous (light) peptide but has different mass. In much the same way that stable isotope labelling allows comparison of two complex mixtures using SILAC, you then spike the heavy-labelled standard peptide(s) into the sample at a known concentration. As you know which specific peptides you need to analyse, SRM transitions for the light analyte and the heavy standard are optimized and used to obtain data on how much signal is generated by these peptides. Comparing the intensity of the light and heavy peptide gives a ratio of how much peptide is present in the sample versus how much was spiked in. Since you know how much standard peptide was present, you can easily calculate how much endogenous peptide there is.

This method was originally known as 'absolute' quantification, to highlight the fact that you can obtain a measure of the concentration of a protein, to distinguish it from 'relative' quantification. However, there is significant debate as to how 'absolute' this value really is, since it depends to a great extent on the purity, stability, and quantification of the internal standard peptide, and on how efficiently the endogenous peptide is released from the cell during lysis and from the intact protein during proteolytic digestion. Since these losses are difficult, if not impossible, to determine, the value obtained is not really an absolute quantity. However, this argument is really a side issue—the key question is whether the method is fit for purpose. Studies have shown that, using the same methods and standardized 'heavy' reference peptides, it is possible for several laboratories to assign the same concentrations to proteins in the same samples (Addona et al. 2009), and so this method remains a valuable tool for proteomics research.

14.6 Data analysis

The performance of successful proteomics experiments depends critically on good sample preparation, experimental design, and optimally set-up data-collection workflows, whether by 2D-PAGE or LC–MS/MS (Cairns 2011). However, it is equally important to be able to understand the outputs of these experiments and to be able to analyse the vast amounts of data which they generate in a proper manner. Regardless of the experimental set-up, it must be possible to answer a few key questions. How do you define a protein 'change'? Are my results reliable and due to biology, rather than technical issues? What do the data tell me about the biology of my system?

Taking the first question of how you define a protein change, this requires you to work out how much 'noise' is in the system. For relative quantification, most experiments are pairwise comparisons (even if you have more conditions, for example a time-course, data have to be analysed as a series of pairwise comparisons (e.g. time 0 vs. time 1; time 1 vs. time 2; time 0 vs. time 2)). Analysis of replicate samples will never provide a ratio of exactly 1.0 for each protein, even though it is the same sample being compared against itself. Instead you obtain a distribution of values around 1 which represents the experimental noise. By determining this noise you can then look at each pairwise analysis of interest and identify which gel spot/protein ratios lie outside this experimental noise, such that these changes are unlikely to have occurred by chance and are likely to be due to the biology of the system under study.

The second key question about whether data are reliable is harder to address. Of course, repeating experiments is critical and changes which appear time and time again are likely to be very reliable. However, proteomics analysis may not identify a protein which was identified and quantified as changing in the first experiment in a subsequent replicate. These 'missing values' present a problem, as there is no evidence that the original result is *not* reproducible. Performing sufficient replicates to ensure that all data points are replicated at least once, and preferably twice, is a time-consuming and costly undertaking. In this case (and if the protein in question is of interest), validation by a targeted approach, such as SRM or Western blotting, would be necessary.

Another question over reliability is the accuracy of peptide identification from MS/MS spectra. All database search engines provide a probability score that a peptide is correctly identified. In addition, it is possible to search data against a protein database which contains all sequences in reverse as well as in the correct orientation. Peptide 'hits' are ranked according to their probability score and the percentage false discovery rate (%FDR) at any given probability score can be calculated by dividing the number of matches to the reverse database by the total number of peptide matches. In this way you can generate a cut-off at which most peptide identifications are likely to be real, usually a confidence of >95% or an FDR of <5% (Nesvizhksii 2010). However, even at scores where there is a 20% FDR, 80% of the data are probably 'real', but these data are often discarded. These data may contain valuable information, especially if the same protein/peptide has been identified at higher confidence in the replicate, and so manual inspection of the data and comparison with other spectra from the same peptide can be used to increase the confidence in identification. This is most pertinent when studying post-translational modifications. Current software tools are good at telling if a peptide is modified, but are less good at telling where. Modification site assignments should almost always by checked manually to ensure correct assignment.

The final question, 'What do the data tell me about my biological system?', is really the most important—the process of turning data into information. When faced with a list of proteins that change under certain conditions, how does one work out what that means for the cell or tissue under study. To answer this question the experimenter primarily needs a good understanding of the biology of the system, to see how the proteins which change can contribute the observed changes in phenotype.

Proteins in the cell do not act alone. They are small parts of much larger pathways and networks. As such, several databases of pathways exist such as Gene Ontology (GO), Kyoto Encyclopaedia of Genes and Genomes (KEGG), and PANTHER, to name just three. Data from proteomics experiments can be cross-referenced with these databases, assigning lists of proteins to the pathways, networks, and processes in which they act. Quantification data can then be mapped onto this network to look for coordinated changes of, for example, several proteins involved in a specific biological process or pathway, which then becomes the target for further experiments. Likewise proteins can be grouped according to which signalling pathways and/or transcription factors are responsible for their regulation/production, and the pattern of proteins changed can act as a surrogate for these factors, even if the factors themselves are not identified or quantified.

Overall, however, the success of a proteomics experiment is determined not only by the choice of sample type, processing workflows, or analysis methods. A major factor is what you do with the data afterwards—analysing the data correctly, knowing what the data can (and cannot) tell you, and using this information to generate good testable hypotheses to further enhance our knowledge of biology.

14.7 Chapter summary

- Proteomics is the study of protein expression.
- Experiments can be *global*, i.e. they aim to quantify as many proteins as possible from a set of samples, or *targeted*, i.e. specific proteins are analysed and quantified.
- Proteomics methods can be gel-based, where intact proteins are separated and the patterns of protein expression are compared, or gel-free, where proteins are digested into peptides and levels of peptides used as surrogates for protein level.
- Several mass spectrometry methods can be used for peptide relative quantification. The choice of method depends on the type and number of samples to be analysed.
- Analysis of protein post-translational modifications is important, and requires specific tools such as affinity chromatography or modification-specific antibodies to enrich or to target modified proteins/peptides.
- Sample preparation, good experimental design, and good set-up of analysis tools are all critical in obtaining high-quality proteomics datasets.
- Data analysis is key, especially given that global proteomics experiments are generally used for the generation of new hypotheses about biological processes.

References

Abbott, K.L. and Pierce, J.M. (2010) Lectin-based glycoproteomic techniques for the enrichment and identification of potential biomarkers. *Methods Enzymol.* **480**: 461–76.

Addona, T.A., Abbatiello, S.E., Schilling, B., et al. (2009) Multi-site assessment of the precision and reproducibility of multiple reaction monitoring-based measurements of proteins in plasma. *Nat Biotechnol* **27**: 633–41.

Cairns, D.A. (2011) Statistical issues in quality control of proteomic analyses: good experimental design and planning. *Proteomics* **11**: 1037–48.

Dunn, J.D., Reid, G.E., and Bruening, M.L. (2010) Techniques for phosphopeptide enrichment prior to analysis by mass spectrometry. *Mass Spectrom Rev* **29**: 29–54.

Faca, V., Coram, M., Phanstiel, D., et al. (2006) Quantitative analysis of acrylamide labeled serum proteins by LC–MS/MS. *J Proteome Res* **5**: 2009–18.

Friedman, D.B. and Lilley, K.S. (2008) Optimizing the difference gel electrophoresis (DiGE) technology. *Methods Mol Biol* **428**: 93–124.

Gygi, S.P., Rist, B., Gerber, S.A., Turecek, F., Gelb, M.H., and Aebersold, R. (1999). Quantitative analysis of complex protein mixtures using isotope-coded affinity tags. *Nat Biotechnol* **17**: 994–9.

Jensen, O.N. (2004) Modification-specific proteomics: characterization of post-translational modifications by mass spectrometry. *Curr Opin Chem Biol*, **8**: 33–41.

Nesvizhskii, A.I. (2010) A survey of computational methods and error rate estimation procedures for peptide and protein identification in shotgun proteomics. *J Proteomics.* **73**: 2092–123

O'Farrell, P.H. (1975) High resolution two-dimensional electrophoresis of proteins. *J Biol Chem* 250: 4007–21.

Ong, S.E., Blagoev, B., Kratchmarova, I., et al. (2002) Stable isotope labeling by amino acids in cell culture, SILAC, as a simple and accurate approach to expression proteomics. *Mol Cell Proteomics* **1**: 376–86.

Patton, W.F. (2002) Detection technologies in proteome analysis. *J Chromatogr B* **771**: 3–31.

Ross, P.L., Huang, Y.N., Marchese, J.N., et al. (2004). Multiplexed protein quantitation in *Saccharomyces cerevisiae* using amine-reactive isobaric tagging reagents. *Mol Cell* Proteomics **3**: 1154–69.

Steen, H. and Mann, M. (2004) The ABC's (and XYZ's) of peptide sequencing. *Nat Rev Mol Cell Biol* **5**: 699–711.

Unlü, M., Morgan, M.E., and Minden, J.S. (1997) Difference gel electrophoresis: a single gel method for detecting changes in protein extracts. *Electrophoresis* **18**: 2071–7.

Washburn, M.P., Wolters, D., Yates, J.R., 3rd. (2001) Large-scale analysis of the yeast proteome by multidimensional protein identification technology. *Nat Biotechnol* **19**: 242–7.

Westermeier, R., Postel, W., Weser, J., and Görg, A. (1983) High-resolution two-dimensional electrophoresis with isoelectric focusing in immobilized pH gradients. *J Biochem Biophys Methods.* **8**: 321–30.

Yao, X., Freas, A., Ramirez, J., Demirev, P.A., and Fenselau, C. (2001) Proteolytic 18O labeling for comparative proteomics: model studies with two serotypes of adenovirus. *Anal Chem* **73**: 2836–42.

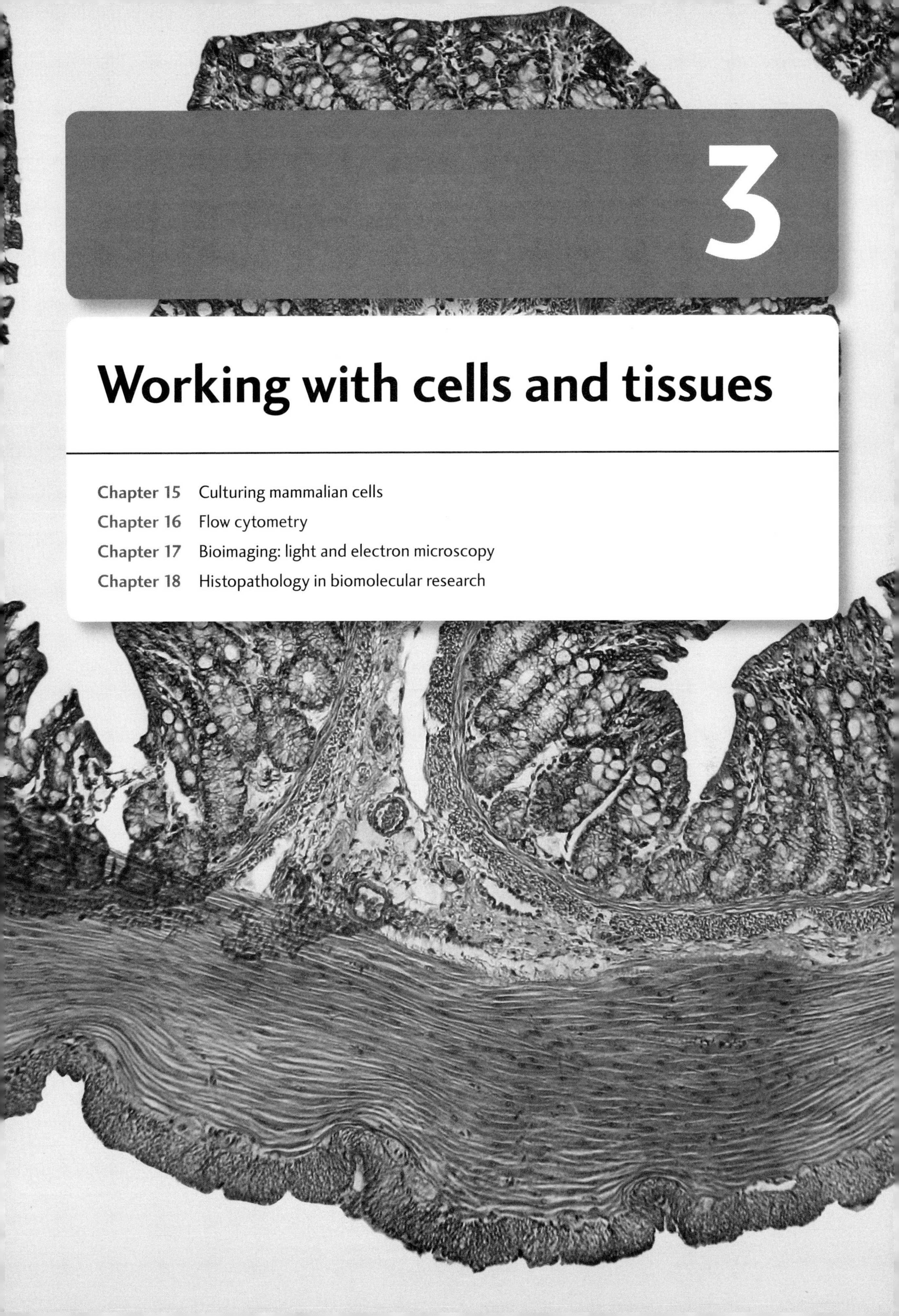

3

Working with cells and tissues

Culturing mammalian cells

15

Geoffrey J. Pilkington, Suzanne M. Birks, and Laura K. Donovan

Chapter overview

Cell culture is one of the major tools used in cellular and molecular laboratories, and has paved the way for many ground-breaking discoveries and scientific advances since it was first established in 1907. In a little over a hundred years the method of cell culture has undergone many advances and is now routinely used in a variety of biological disciplines such as large-scale manufacturing of vaccines, antibodies, and therapeutic proteins, drug screening and development, and model systems. Culturing cells requires a constant temperature in addition to a supply of gases and nutrients. The latter are provided in the growth medium, which can vary considerably in constituents and resulting pH depending on the specific cell types required (e.g. stem cells). This will be discussed in more depth later in this chapter. However, the technique of culturing cells is not without its problems and cultures can often be compromised through contamination. Such problems and how they can be overcome will also be discussed.

LEARNING OBJECTIVES

This chapter will enable the reader to:

- describe how cell cultures are obtained from tissue and how **cell lines** can be established using defined media and environmental conditions;
- state the general basic components of growth media;
- communicate the procedure for **subculturing** cell lines and explain how the phases of growth dictate the most effective time for subculture;
- list the applications of cell cultures and discuss how cell culture has contributed to breakthroughs in modern medicine;
- discuss how the specific conditions for stem cell growth can be replicated *in vitro* using various culturing techniques, defined media, and growth environments;
- state the ways in which **stem cells** are isolated from cell cultures and expanded;
- identify the ways in which cell cultures may be compromised and how these limitations can be overcome.

15.1 Introduction

In 1907, Ross Harrison, a Professor at Johns Hopkins University, published the first paper detailing a procedure for maintaining cells outside the body. At the time, the paper was viewed as a curiosity, but was later to be named as one of 'medicine's 10 greatest discoveries'. It built on the principles initiated by Wilhelm Roux who managed to remove the neural plate from a chick embryo and maintain it in warm saline solution for ten days. This paved the way for some of science's most ground-breaking technologies, such as the production of monoclonal antibodies, insulin-producing pancreatic cells, and the latest developments in using human **embryonic stem cells** for therapeutic purposes (e.g. to treat spinal injury patients).

Harrison used the **hanging drop** method which involved placing tissue fragments onto a cover slip and allowing them to coagulate in the presence of lymph before turning the cover slip over into the well of a glass depression slide (Harrison 1907). This technique was quickly developed by Carrel, who started using plasma instead of lymph and was able to grow chicken, rat, dog, and human tumours. He was able to establish the first cell lines by cutting the tumours into small pieces which were placed in fresh plasma, thus producing the first protocol for subculturing (Carrel and Burrows 1911). The advent of D-flasks (Carrel 1923) meant that cells could be grown in larger quantities of growth medium and so could be maintained more easily. This technique was adopted by most cell culture laboratories around the world and remained the standard until the 1950s.

Cells isolated from donor tissue today are established as cultures in one of two ways. Tissue is either mechanically or enzymatically disaggregated, and the cell suspension is allowed to adhere to the culture vessel to form a **monolayer** and then subcultured to form a cell line. Alternatively, tissue fragments are allowed to adhere to the vessel surface, known as the primary explant; the cells that migrate out are referred to as the 'outgrowth' and can be subcultured to form a cell line. Figure 15.1 illustrates how a primary cell culture is obtained from tissue in our laboratory. This process involves several steps: washing the tissue with a balanced salt solution, dissecting the tissue and resuspending it as a cell suspension in fresh serum-rich medium. It is essential to remember that cell cultures obtained in this way must be screened for **Mycoplasma** infection at the earliest opportunity (see section 15.6.1). In addition to this basic procedure, it is possible to separate out the cell types of interest, using automated magnetic cell sorting (MACs™) or fluorescence-activated cell sorting (FACS), and enrich the culture using specialist growth media and growth factors. This is described in section 15.5.1.

See Chapter 16

Eagle first discussed the use of defined media for the growth of specific cell types in a paper published in *Science* (Eagle 1955). He demonstrated that whilst cells could be maintained for short periods in a balanced salt solution containing serum, additional specific requirements were required for long-term culture. Indeed, the majority of cell cultures today are grown in a variation of basal Eagle's minimum essential medium (EMEM) known as Dulbecco's modified Eagle's medium (DMEM). However, a huge variety of other media are also available, such as RPMI, F-10, F-12, Opti-MEM, neurobasal, and optimizer for the expansion of specific subsets of cells.

Figure 15.1 Flow diagram describing how a primary culture is obtained from a tissue sample.

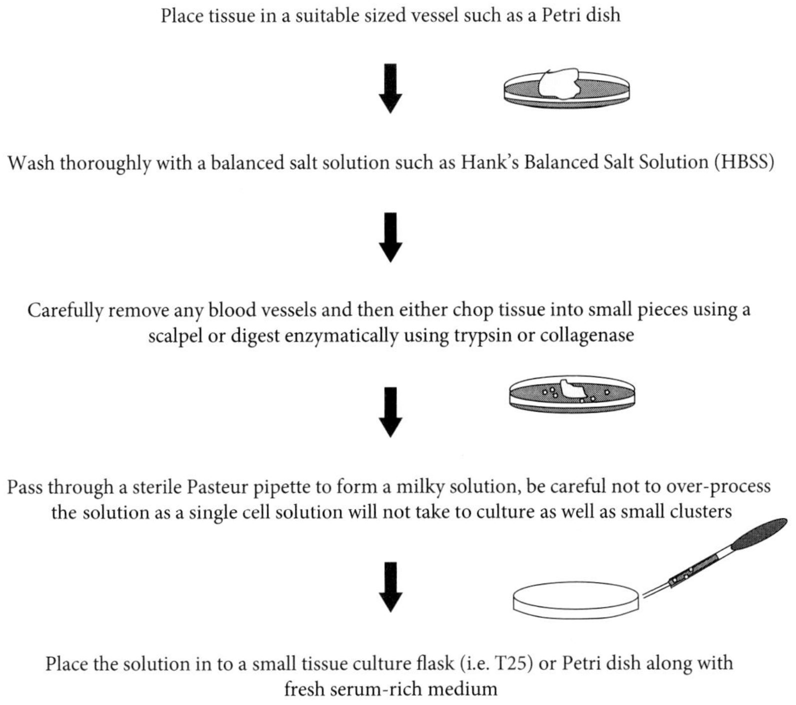
Place tissue in a suitable sized vessel such as a Petri dish
Wash thoroughly with a balanced salt solution such as Hank's Balanced Salt Solution (HBSS)
Carefully remove any blood vessels and then either chop tissue into small pieces using a scalpel or digest enzymatically using trypsin or collagenase
Pass through a sterile Pasteur pipette to form a milky solution, be careful not to over-process the solution as a single cell solution will not take to culture as well as small clusters
Place the solution in to a small tissue culture flask (i.e. T25) or Petri dish along with fresh serum-rich medium

15.2 Culturing cells *in vitro*: essential principles

15.2.1 The growth curve

The growth of cells in culture follows a characteristic **growth curve** as shown in Figure 15.2. This is a semi-logarithmic plot that shows how the **cell density** increases over time. The first part of the curve, which is fairly flat, represents the **lag phase**. During this period growth is slow as the cells are adapting to their new environment—recovering from subculture, restructuring their cytoskeleton, and secreting new extracellular matrix (ECM) proteins for anchorage to the bottom of the vessel in preparation for a period of exponential growth. The **log phase** is represented by the steep gradient on the curve. This is the period in which the cells **proliferate** exponentially; cells are actively dividing and quickly deplete the nutrients in the growth medium as well as producing toxic metabolites. This continues until either the medium is spent (i.e. one or more essential nutrients have been completely consumed) or the surface of the culture vessel is completely covered (the cells have reached **confluency**). Figure 15.3 shows adherent CC-2565—a normal human astrocyte cell line growing in a monolayer at around 80% confluency.

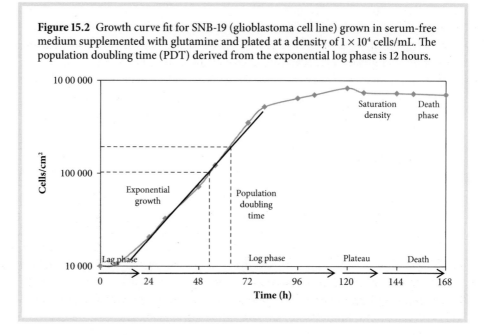

Figure 15.2 Growth curve fit for SNB-19 (glioblastoma cell line) grown in serum-free medium supplemented with glutamine and plated at a density of 1×10^4 cells/mL. The population doubling time (PDT) derived from the exponential log phase is 12 hours.

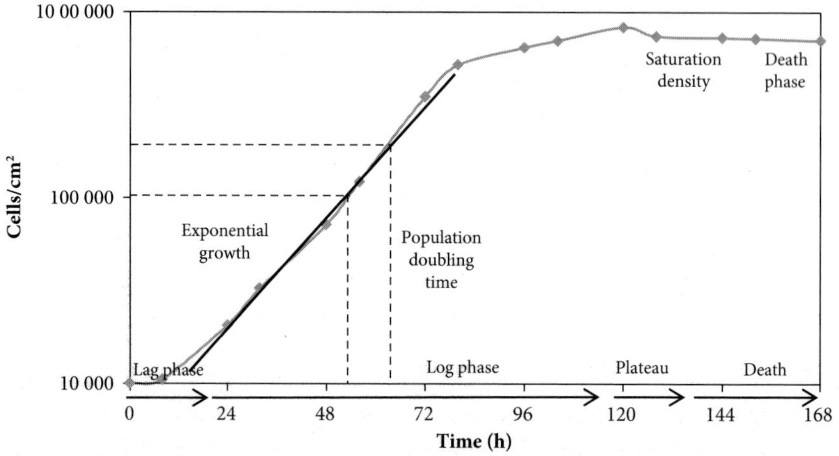

At this point the cells will enter the stationary phase where they cease to proliferate. This plateau phase is shown on the graph by a levelling off of the curve. Eventually cells will enter the decline phase where cell death occurs and thus the cell density decreases. The population will double over a period of time that can be defined by collecting growth data and preparing a growth curve. This doubling time is characteristic for each cell line and, since environmental conditions such as oxygen content, temperature, media type, and serum content also greatly affect the cell growth rate, it is advisable to collect growth curve data for each cell line and condition that you use. This can then be used to plate cells at the correct **seeding** density for experiments.

Figure 15.3 Phase contrast microscopy image of CC-2565 (normal human cerebral astrocytes) growing in culture in DMEM containing 10% FBS. This culture is around 80% confluent.

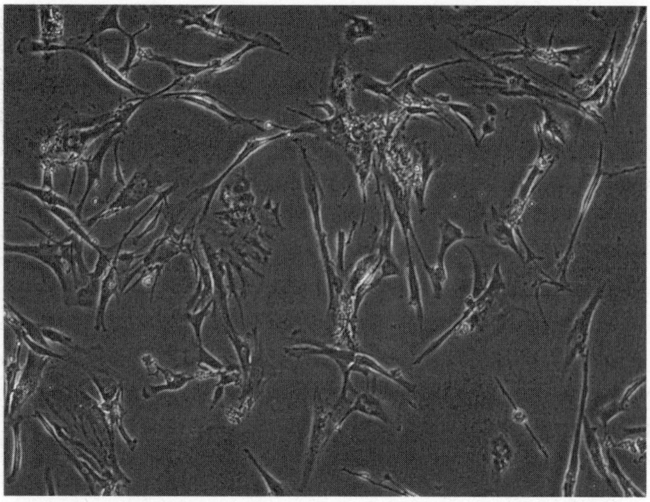

15.2.2 Cell culture requirements

As previously mentioned, the type of growth media used, as well as oxygen content and temperature, can greatly affect the growth rate of the cell culture as well as selectively promote subsets of cells or cell behaviour. The growth media you use for cell culture will normally have the same basic components. Essential conditions, such as pH, temperature, and O_2/CO_2 conditions, must also be met in order to ensure that cell cultures grow effectively.

Growth media

The carbohydrate requirement essential for respiring cells is achieved by the addition of around 1–4.5g/L glucose. Amino acids needed for optimal cell growth are supplied by glutamine supplementation to a final concentration of 2mM. Growth factors such as FGF-2 (fibroblast growth factor) and EGF (epidermal growth factor) are normally supplied by the addition of serum or added individually. For example, fetal bovine serum (FBS) is frequently added to growth medium because of its high growth factor content and buffering ability. However, in protein-sensitive or binding assays the presence of serum is a hindrance as it can not only promote the expression of some molecules unnaturally but also bind to molecules added to the culture (e.g. in drug studies, **transfection**, and **transduction**). Furthermore, it can complicate protein purification procedures. There are many alternatives available today (such B-27, B-18, G5, or N2) which abolish the need for serum in culture, and in some cases are actually superior in promoting cell growth and viability. However, as with most things, there is a cost implication to using these.

pH

pH should be maintained between 7.2 and 7.4, and a buffering system is normally supplied by the bicarbonate content in media when grown in 5% CO_2 conditions. The presence of inorganic salts (e.g. K, Ca, Na) also helps to regulate osmotic pressure and membrane potential. Phenol red can be added to the media as a pH indicator, but this is sometimes omitted in absorbance-sensitive and oestrogen-sensitive experiments.

Temperature

Mammalian cells are cultured at 37°C to mimic the conditions of the donor's environment. This temperature ensures that the cells continue to grow and function in a manner similar to their *in vivo* status (e.g. enzymatic reactions and protein synthesis). Temperature will be controlled by the incubator settings, which should be regularly calibrated and serviced.

O_2/CO_2 conditions

Mammalian cell cultures normally require ~20% O_2 content and ~5% CO_2—a condition referred to as normoxic. Hypoxic (≤5% O_2) and anoxic (≤1% O_2) conditions are occasionally used to mimic certain *in vivo* conditions. **Hypoxia** switches cell respiration from aerobic to anaerobic, which will affect the transcription of specific genes involved in adaptation and growth, whilst anoxic conditions, which usually result in swelling, membrane blebbing, changes in mitochondrial membrane potential, and the induction of apoptosis, provide the opportunity to study trauma injuries and pathogenesis such as myocardial infarction and brain tumours. Oxygen tensions will be discussed in more depth in section 15.5.2.

15.2.3 Subculturing

Subculturing or **passaging** a cell culture refers to the removal of cells from their current growth media and seeding into multiple new culture vessels containing fresh media. Since the growth cycle is repeated for each subculture, this enables further propagation and expansion of the culture. Generally subculture should take place under the following conditions

- Before the growth medium has been exhausted: this is indicated by a drop in pH due to lactic acid build-up which is identified by a colour change in the media due to the presence of phenol red (if phenol red is used).
- Before the cells reach confluency, i.e. when they are in the log phase. If cells are allowed to reach confluency, **contact inhibition** will occur, causing the cells to stop growing, and the culture can take a while to recover even if it is re-seeded.

Cell cultures are grown either as adherent monolayers or in suspension. Subculturing cells in suspension requires that you first spin down the culture before splitting the cells between new culture vessels. The subculture of adherent cells first requires **dissociation** of the cells from the bottom of the culture vessel. This is achieved either enzymatically using trypsin, EDTA, or a trypsin substitute, or non-enzymatically using detergent or a cell scraper. The cell suspension is then gently spun down to 'harvest' the cells and the resulting pellet is reserved and slowly resuspended in fresh media in a new culture vessel. Each time subculturing is carried out the passage number increases. For example, a primary culture established from disassociated tissue is referred to a P-0; however, after subculture this becomes P-1. Caution should be taken during the subculturing process as some cells are very sensitive to centrifugal and shear force and can be damaged—a centrifugal force of 80–150 $\times g$ is usually satisfactory. If the cells are required for use in experiments, they should be harvested in the same way and then seeded at the optimal cell density for the experiment in appropriate culture vessels. The optimal cell density is usually stated by the manufacturer for commercially available kits, and otherwise should be determined experimentally. You will need to take into account the population doubling time (as discussed in section 15.2.1) so as to ensure that the correct cell density is reached at the required time. In order to seed cells at a specific density, it is necessary to count the cells in the population and then dilute as appropriate. This can be done using the trypan blue exclusion test either by eye using a haemocytometer under a phase contrast microscope or using an automated cell counter. Figure 15.4 shows you how to obtain information on the total number of cells and the number of viable cells in your culture using a haemocytometer. Case study 15.1 shows you how to use cell viability data to calculate dilutions for different seeding densities.

15.2.4 Cryopreservation of cells

When continuously culturing any cell line, it is prudent to preserve a portion of the cells at their current passage number in case any problems arise with the culture such

Figure 15.4 Flow diagram with worked examples to show how to obtain a total cell count and viable cell count of a cell culture using a haemocytometer and how to transform these figures to calculate percentage cell viability.

Prepare cells for subculture by dissociating cells from culture vessel, centrifuging and resuspending in 1ml medium

Remove 100µl from the suspension and mix with 100µl Trypan blue for live/dead cell count

Prepare the haemocytometer and fill each chamber using a small sterile glass pipette and allow the suspension to be drawn out by capillary action

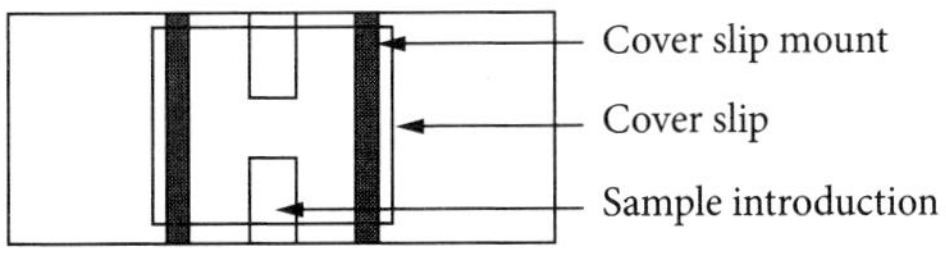

Under the microscope focus the 10X objective on the grid on the haemocytometer and count the number of cells in each corner chamber. Make a note of the live and dead cells (live cells do not take up the Trypan blue as they have an intact membrane)

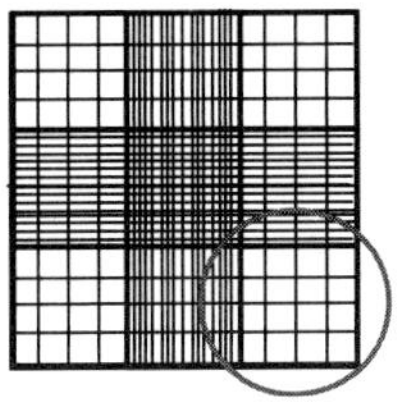

In order to keep things consistent, when counting in any square, count the cells in the following manner: include (√)cells which touch the middle line on the top and left, discount (×) cells which touch the middle line on the bottom and right

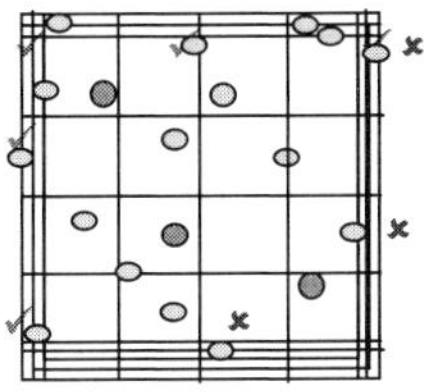

Calculate the number of cells/ml: each chamber represents [number of cells] × 10^4/ml

as contamination, senescence (see section 15.6.2), or equipment failure. This can be achieved by harvesting the culture as described above and then resuspending the cells at 1×10^6 – 1×10^7/mL in a cryoprotectant freezing medium. This special medium is available commercially, or it can be made in-house by the adding of 5–10% dimethyl sulphoxide (DMSO) or glycerol to your normal culture medium, which will prevent the formation of intracellular crystalline structures and thus protect cells from crystal-induced damage. Vials should then be slowly cooled at a rate of around 1–3°C/min to –80°C (this is easily achieved with a passive freezer which uses an isopropanol bath to gradually cool vials in a –80°C freezer) before transferring to liquid nitrogen for long-term storage.

CASE STUDY 15.1 Calculating cell viabilities and using dilutions for different seeding densities

1. Calculate the total number of cells, number of viable cells, and viability for the following cell counts.

Cell line 1		
Square	Number cells	Viable cells
1	152	149
2	169	155
3	171	163
4	144	140

Cell line 2		
Square	Number cells	Viable cells
1	60	52
2	57	51
3	75	70
4	81	76

Cell line 1: mean number of cells = 159, mean number of viable cells = 151, viability = 95%

Cell line 2: mean number of cells = 68, mean number of viable cells = 62, viability = 91%

2. Assuming that you would like to seed both cell lines at a density of 5×10^5 and 1×10^4 with an overall volume of 2mL and 100μL, respectively, think how you would dilute the suspension appropriately and what volume of the suspension you would need to add to the cell culture vessel.

Cell line 1

- Mean number of viable cells = 151.
- Viable cells/mL: $151 \times 2 = 302 \times 10^4$ (151×2 to account for trypan blue dilution).
- $302 \times 10^4 = 3.02 \times 10^6$/mL.
- First dilute suspension 1:3 to obtain 1×10^6/mL (dilution 1) (i.e. add 2 mL media to suspension).
- For 5×10^5, remove 1mL of dilution 1 and dilute 1:4 to obtain 2.5×10^5, and then add 2mL of this suspension to the culture vessel.
- For 1×10^4, remove 1mL of dilution 1 and dilute 1:10 to obtain 1×10^5, and then add 100μL of suspension to the culture vessel.

Cell line 2

- Mean number of viable cells = 62.
- Viable cells/mL: $62 \times 2 = 124 \times 10^4$ (62×2 to account for trypan blue dilution).
- $124 \times 10^4 = 1.24 \times 10^6$/mL.
- First dilute suspension 1:1.24 to obtain 1×10^6/mL (dilution 1) (i.e. add 240μL media to suspension), and then repeat steps for 5×10^5 and 1×10^4 dilutions as for cell line 1.

Archived cells can then be recovered when required and a new stock of cells generated. The vials of cells to be resurrected from liquid nitrogen storage should be thawed quickly and then slowly diluted 1:20 in fresh culture medium. This will dilute the DMSO or glycerol to a concentration that can be tolerated by the cells, and diluting slowly prevents osmotic shock. However, you will still need to carry out a media change after 24 hours, or as soon as the cells have become adherent, to prevent toxicity from the cryoprotectant.

15.3 General applications

In a little over a hundred years the principles of cell culture have been developed and expanded to the point where it is now routinely used in a whole host of biological fields. Not only are cells propagated from donor tissue for use in basic research on all kinds of diseases, but they are now frequently used in the research laboratory for gene expression studies and drug screening. For example, gene sequences of interest can be expressed by cells growing in culture, or knocked out using gene silencers via transfection and transduction, and the effects on various aspects of biological behaviour (such as viability and apoptosis, migration, angiogenesis, and differentiation) examined using various assays, some of which are explored in this book. Transfection refers to the introduction of a gene sequence into a eukaryotic cell by non-viral means, i.e. using a plasmid, whereas transduction uses viral particles to carry the gene sequence of interest into the host cell nucleus. Serum-free medium should be used to culture cells prior to, during, and after transfection or transduction in order to reduce the binding of the plasmid or viral particles to the high level of proteins in serum-containing medium; Opti-MEM is frequently used to enhance the uptake into the cells and improve transfection/transduction efficacy. These procedures will result in a transient transfection of the cell line since the gene sequence is not usually incorporated into the host genome, and as such the DNA sequence and its products will degrade over time. However, if the sequence is required to be incorporated into the genome it is possible to generate stable cell lines expressing the gene of interest by selecting out those cells that are transfected with a marker. This is usually done by incorporating an antibiotic resistance gene into the plasmid and then treating the culture with the antibiotic (e.g. geneticin). This selective pressure means that only cells with a stable transfection will survive and be propagated further.

See Chapter 6

See Chapter 1

Other biological practices, such as recombinant protein production for therapeutic uses (e.g. insulin), vaccine production, antibody production, and tissue engineering also use cell culture technology, albeit on a far larger basis than has already been discussed. Scaling up cell culture to the level demanded by these processes can be done in the laboratory by changing the culture vessel to multilayer flasks or even roller bottles which greatly increase the surface area available to adherent cells; cells grown in suspension can be cultivated in large quantities in spinner flasks. It should be noted that these options are offered as solutions to non-specialist laboratories. For cell culture on a truly large scale it is possible to use a bioreactor, which has a culture volume of 100–10 000L, for both adherent (with the addition of micro-carrier beads) and suspension cells.

See Chapters 7 and 9

15.4 Human embryonic and adult stem cells: culture, isolation, and expansion

Stem cells are distinguished from other cell types by two key characteristics. First, they are unspecialized cells (i.e. they cannot perform specialized functions, such as in heart muscles) with the innate ability to perpetuate themselves through self-renewal (through mitosis), often following extended periods of inactivity. Second, under precise physiological or experimental environments they can be stimulated to generate mature cells of a specific tissue or organ through a process known as **differentiation**.

Stem cells are found in multicellular organisms such as mammals. Two main types of stem cell exist in mammals: embryonic stem cells and non-embryonic **'adult' stem cells**. The fundamental difference between embryonic and adult stem cells is their differentiation capacity. Embryonic stem cells have the capacity to generate all the cells and tissues which make up the entire embryo, and hence they are defined as being **totipotent** (or **pluripotent** if cultured *in vitro*). In contrast, adult stem cells have limited differentiation capacity in that they can only differentiate into cell types of their tissue or organ of origin. Table 15.1 lists the different terms used to describe the **potency**, i.e. the differentiation potential, of a cell.

As stem cells can proliferate an infinite number of times and cells such as embryonic cells have the potential to form any cell type of the body, these cells potentially present access to unprecedented levels of tissue from the human body. As a result, they can help support:

- research into the functionality of human tissues;
- material for drug testing, preventing reliance on animal models;
- transplantation therapies to treat degenerative diseases such as Parkinson's disease.

15.4.1 Embryonic stem cells

Embryonic stem cells are derived from the inner cell mass of pre-implantation embryos. Every cell of this embryo (*blastomere*) is undifferentiated and has yet to commit to cellular differentiation. Maintenance of these cells under appropriate *in vitro* culture conditions will sustain the proliferative capacity of these cells indefinitely.

In 1998, James Thomson and colleagues published the primary paper detailing the derivation of human embryonic stem cells (hESCs), 7 years after the derivation of mouse embryonic stem cells. They showed that human ES cell lines can be cultured *in vitro* whilst maintaining the potential to form all three of the embryonic germ layers—the

Table 15.1 Terms used to describe the differentiation capacity of stem cells

Differentiation potential	Number of cell types produced	Example of cell type
Totipotent	All	Zygote
Pluripotent	All except cells of the embryo	Cultured embryonic stem cells
Multipotent	Many	Haematopoietic stem cells
Oligopotent	Few	Neural stem cells
Quadripotent	4	Epithelial cells
Tripotent	3	Cultured neural stem cells
Bipotent	2	Mesenchymal stem cells
Unipotent	1	Skin cells
Nullipotent	0	Red blood cells

ectoderm, the mesoderm, and the endoderm (Thomson et al. 1998). Thomson's method of embryonic stem cell expansion has been routinely adopted by clinics to treat certain types of infertility (***in vitro* fertilization**).

Culture of embryonic stem cells

Human ESCs are produced by growing a pre-implantation stage embryo in a tissue culture flask or Petri dish containing fresh nutrient-rich culture medium. To prevent hESCs from spontaneously differentiating into specialized cell types and to maintain pluripotency, the lower surface of the tissue culture vessel is usually layered with non-proliferative cells such as embryonic fibroblasts (Amit et al. 2004), bone marrow stromal cells (Zhang et al. 2001), foreskin fibroblasts (Amit et al. 2003), or human cell lines D551/CCL-10 and CCL-2552 (Wang et al. 2005) pre-treated to prevent cell division. These cells act as a feeder layer, providing the hESCs with an adhesive surface to which they can attach as well as providing a necessary source of growth factors and cytokines for their growth (Xu et al 2001).

An alternative to feeder layers are 'feeder-free' culture systems in which hESCs are grown on commercially available basement membranes, such as Laminin, Matrigel, Fibronectin, or Vitronectin, in serum-free medium supplemented with basic fibroblast growth factor (bFGF), transforming growth factor β (TGF-β), and leukaemia inhibitory factor (LIF). This has the advantage of avoiding the introduction of unwanted immunogens and viruses to the hESCs from the feeder layer.

Human ESCs will begin to differentiate, forming embryoid bodies, if they are removed from feeder layers and cultured in suspension on non-adherent substrate. These embryoid bodies will then undergo further spontaneous differentiation to produce specific cell types, but in an uncontrolled way (Eiselleova 2009). Therefore, to maintain stem cells in an undifferentiated state it is necessary to culture them in conditions which prevent spontaneous differentiation. If differentiated cells are required, hESC cultures can be supplemented with specific growth factors to induce differentiation into specific cell types (Schuldiner et al. 2000; Trounson 2006). Figure 15.5 provides an overview of how hESCs are derived from a blastocyte, expanded *in vitro*, and differentiated into organ/tissue-specific cell types. The use of growth factor treatment to direct cell-specific differentiation is described in section 15.5.2.

15.4.2 Adult stem cells

Adult stem cells are undifferentiated cells found amongst differentiated cells in a tissue or an organ. As described earlier, the significant difference between adult and embryonic stem cells is that adult stem cells can only differentiate into cell types of their tissue or organ of origin, whereas embryonic stem cells have the ability to become all cell types. The fundamental functions of adult stem cells are to sustain and repair the tissue of origin throughout the lifetime of the organism (Reya et al. 2001). Adult stem cells are diffused throughout the tissue and organs of the mammalian body, and behave very differently depending upon their local microenvironments. For example, haematopoietic (blood-forming) stem cells (HSCs) are continuously produced within the bone marrow where they undergo differentiation into mature blood cells. In contrast, neural stem cells, which are constantly being generated in the brain, differentiate into the three major cell types of the brain: nerve cells (neurons), astrocytes, and oligodendrocytes.

Figure 15.5 Flow diagram describing how human embryonic stem cells are derived, expanded, and differentiated *in vitro* from a blastomere.

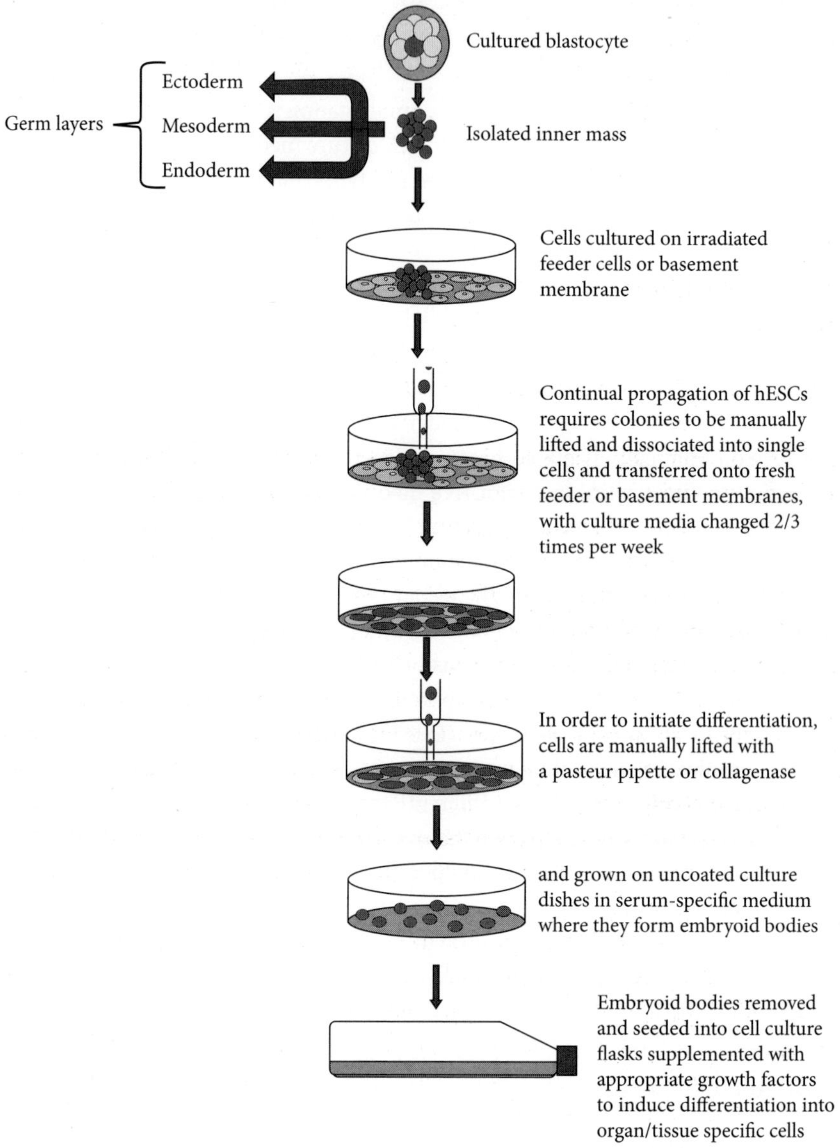

Adult stem cells are extremely rare; for example, only 1 in 15 000 cells in the adult bone marrow are estimated to be HSCs and are able to generate terminally differentiated cells with mature phenotypes that have the capacity to undergo specific functions appropriate to the tissue or organ of origin. A cell's phenotype refers to the physical characteristics of the cell: its morphology, its interactions with other cells and the extracellular matrix, the proteins (or antigenic markers) present on the cell surface (extracellular) or within the cell (intracellular), which are often used for characterization, and the behaviour of the cell (e.g. its migration, proliferation, and synaptic interactions).

Specific adult stem cells can be isolated, cultured *in vitro*, and manipulated by the addition of growth factors or genes, or by decreasing the oxygen tension (hypoxia). To achieve successful growth and expansion of stem cells, the biochemical and mechanical niche (that is the microenvironment) which controls adult stem cell survival, self-renewal, and differentiation *in vivo* must be re-created *in vitro* (Allaire 2011).

In the next section we describe how stem cells—embryonic and adult—can be isolated from mixed populations of cells and the culturing techniques that can be used for the expansion and differentiation of stem cells.

15.5 Stem cell culture protocols

15.5.1 Embryonic and adult stem cell isolation

As discussed earlier, adult SCs and hESCs cultured *in vitro* without the use of specific growth factors and environmental manipulations (such as hypoxia) or feeder layers, respectively, have the potential to spontaneously differentiate, a trait which may not always be required. Therefore it may be necessary to isolate stem cells from the mixed population of undifferentiated, partially differentiated, and terminally differentiated cells. Two suggested methods of isolating stem cells are magnetic cell sorting (AutoMACS™) and fluorescent activated cell sorting (FACS™).

See Chapter 16

In AutoMACS™ separation, as illustrated in Figure 15.6, stem cells (i.e. the target cells) are selectively 'tagged' with fluorochrome-conjugated stem cell marker and anti-fluorochrome magnetic-labelling microbeads. The whole cell population is then passed through a MACS column matrix which delivers a magnetic field strong enough to retain the labelled cells (positive cells), while the unlabelled cells (negative cells) pass through without the compromise of viability. The labelled stem cells are then eluted separately to the rest of the cell population.

In FACS™ the stem cells are again 'tagged' with a fluorochrome-conjugated antibody directed at a protein uniquely expressed within the stem cells. A laser light excites the fluorochrome which emits light at a particular wavelength, so the cells of interest can be identified. For isolation, the individual cells are applied with a charge; positively charged cells are attracted to the negatively charged plate and negatively charged cells are attracted to the positively charged plate, each deflected into waiting sample tubes. This is shown diagrammatically in Figure 15.7.

15.5.2 Expansion and differentiation of embryonic and adult stem cells

There are four key factors to take into account when considering the expansion and differentiation of stem cells: the type of growth medium used, the specific growth factors used, the culture technique, and the oxygen tensions under which the stem cells are cultivated. The conditions selected will be determined broadly by whether your aim is to grow large numbers of unspecialized stem cells or whether you are aiming to produce differentiated cells of a specific cell type.

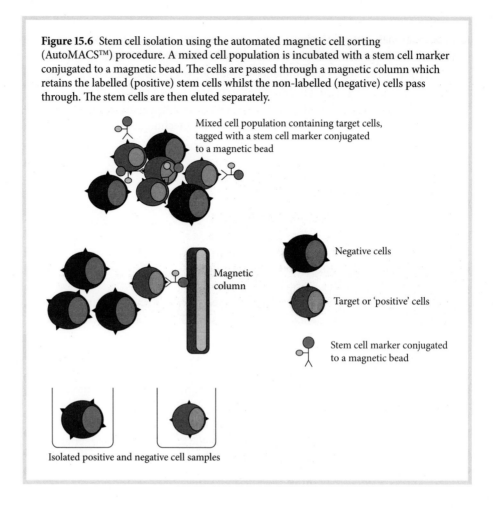

Figure 15.6 Stem cell isolation using the automated magnetic cell sorting (AutoMACS™) procedure. A mixed cell population is incubated with a stem cell marker conjugated to a magnetic bead. The cells are passed through a magnetic column which retains the labelled (positive) stem cells whilst the non-labelled (negative) cells pass through. The stem cells are then eluted separately.

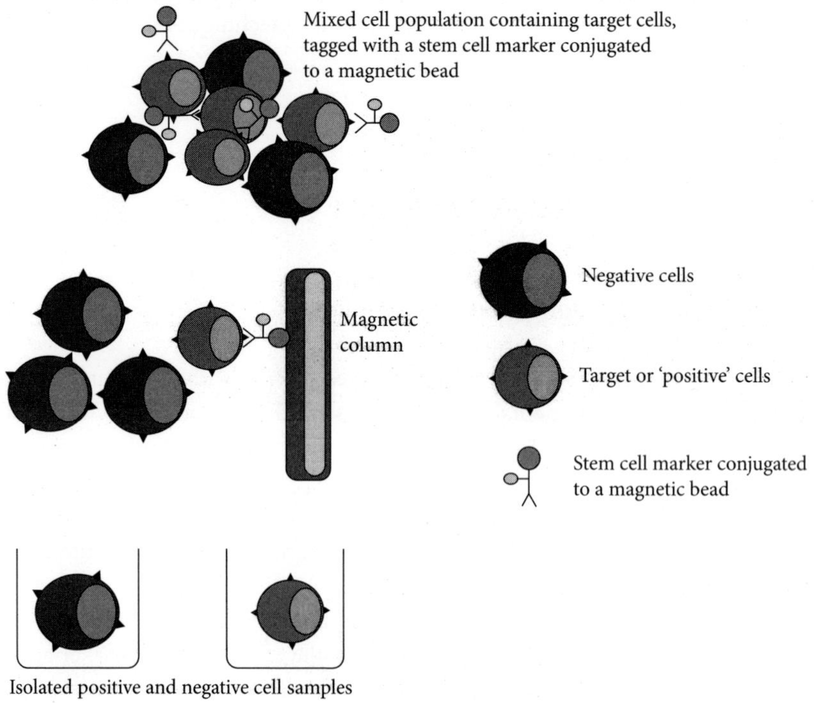

Growth media and growth factors

As previously mentioned, the type of growth medium used, as well as the growth factors or cytokines, can significantly affect the growth rate of the cell culture as well as selectively promote or expand subsets of cells. Although the majority of cell cultures are grown in a variation of EMEM, a large variety of other media (such as RPMI, F-10, F-12, Opti-MEM, neurobasal), designed and optimized to support the expansion of specific stem cell populations, are also available. Growth factors, usually steroids or hormones, are critical for a variety of cellular processes and can promote proliferation, differentiation, or maturation depending upon the growth factor and target cell type. For example, bone morphogenic proteins induce bone cell differentiation, epidermal growth factor maintains a highly proliferative unspecialized state, and vascular endothelial growth factors stimulate blood vessel formation (angiogenesis). Growth factors such as FGF-2 and EGF are frequently used to stimulate stem cell proliferation and expansion. They are usually added in conjunction with a serum-free supplement (such as B-27, B-18, G-5, or N-2), abolishing the need for serum in the culture, the presence of which facilitates stem cell outgrowth and differentiation (Brewer 1995). Table 15.2 gives a brief overview of variations in stem cell media, growth factors, and cytokines required for mesenchymal,

Figure 15.7 Stem cell isolation using the fluorescence-activated cell sorting procedure (FACS™). A mixed cell population is incubated with a stem-cell-associated antibody conjugated to a fluorophore. A laser light excites the fluorochrome, which emits light at a particular wavelength so that target cell is identified. For isolation, the individual cells are applied with a charge; positively charged cells are attracted to the negatively charged plate and negatively charged cells are attracted to the positively charged plate, and each is deflected into a sample tube.

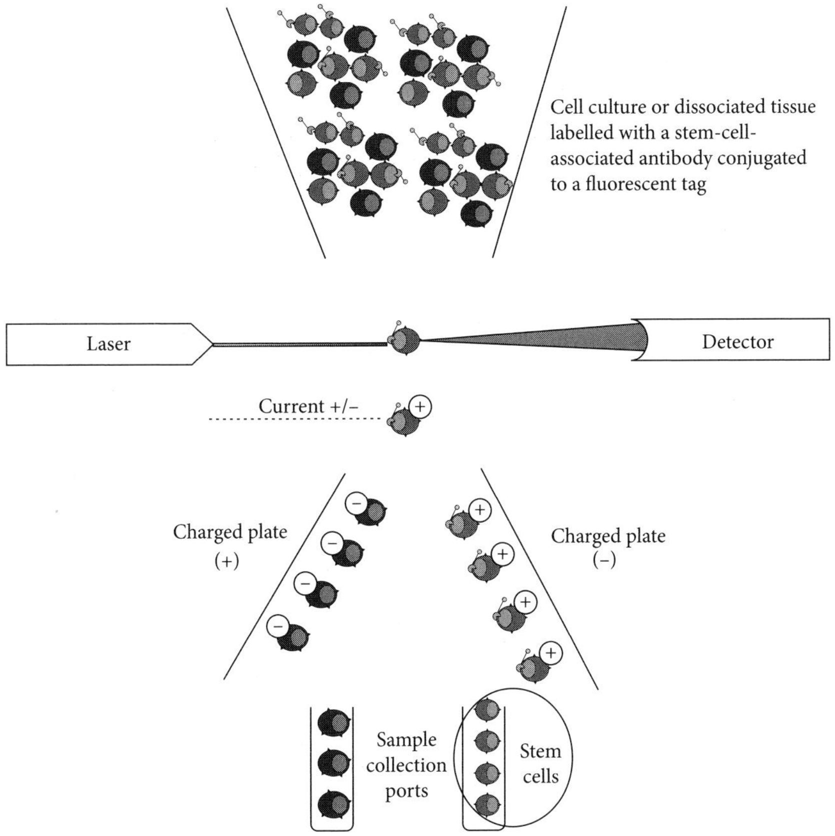

embryonic, neural, and haematopoietic stem cell expansion and maintenance in an undifferentiated state. Table 15.3 lists the growth factor treatments that can be used to direct cell-specific differentiation.

Culturing techniques

The majority of stem cells are grown as adherent monolayers, although a small number of cultures require alternative culture conditions for optimal growth and expansion. For example, embryonic stem cells are produced and expanded using 'feeder-layers' or basement membrane substrates providing the cells with an adhesive surface to which they can attach. In contrast, a non-adherent microenvironment created using spinner flasks promotes the formation of embryoid bodies and subsequent embryonic stem cell differentiation, albeit in an uncontrolled manner.

Table 15.2 Embryonic and adult stem cell media plus supplement variations for embryonic mesenchymal, neural, and haematopoietic stem cell expansion (undifferentiated state) *in vitro*

Stem cell type	Basal media	Media supplements
Embryonic stem cell	Iscove's modified Dulbecco's media (IMDM)	Fetal bovine serum (embryonic stem cell qualified) Transforming growth factor-β Basic fibroblast growth factor Epidermal growth factor Leukaemia inhibitory factor
Mesenchymal stem cell	MesenPro RS™ media	MesenPro RS™ growth supplement Epidermal growth factor Interleukin-1 Interleukin-4 Interleukin-7
Neural stem cell	Neurobasal™ media	Serum-free supplement Epidermal growth factor Basic fibroblast growth factor Leukaemia inhibitory factor
Haematopoietic stem cell	AIM V® media MarrowMAX™ bone marrow media	Interleukin-2 Interleukin-4 Platelet-derived growth factor Tumour necrosis factor α Transforming growth factor

Another culture technique, known as the hanging drop technique is used to generate non-adherent spherical clusters of cells. This technique is an excellent method for the continued division and expansion of neural stem cells. A spherical cluster of neural stem cells, commonly referred to as a neurosphere, is generated when a small drop of cell suspension is inverted, resulting in cell settling due to gravity and cells adhering to each other (cell-to-cell adhesion). Once generated, the neurospheres exert high proliferation rates, maintaining the stem cell in an undifferentiated and unspecialized state. Figure 15.8 shows an example of a neurosphere generated by the hanging drop technique.

Oxygen tensions

Oxygen homeostasis is not only crucial to all organisms which rely upon aerobic respiration, but is a critical environmental factor that regulates the fate of stem cells. Oxygen is the terminal electron acceptor for the electron transport chain in the mitochondria during aerobic respiration, and therefore is essential for cellular metabolism.

Hypoxia is defined as a reduction in normal oxygen partial pressures (or a decrease in oxygen tension) for specific tissues or cells and may occur under specific stresses such as high altitude, anaemia, infection, tumour development, or tissue injury. Very few tissues or cells are exposed to atmospheric oxygen levels (~159mmHg or 21% O_2); the exceptions are the outer layers of skin, the respiratory tract, and the cornea of the eye (Watt et al. 2009). Instead, the physiological oxygen tensions for the majority of mammalian

Table 15.3 Suggested primary inducer for the differentiation of human embryonic stem cells into specific organ/tissue cell types

Human embryonic stem cells			References
Embryonic germ layer	**Associated tissue or organ**	**Suggested primary inducers**	
Ectoderm (external layer)	Neuroectoderm Midbrain neural cells Forebrain and midbrainTH$^+$ neurons TH$^+$ neurons Neural crest Oligodendrocytes Motor neurons	Noggin Stromal-derived inducing activity Fibroblast growth factor 2, fibroblast growth factor 8, sonic hedgehog Fibroblast growth factor Stromal-derived inducing activity, bone morphogenetic protein 4, sonic hedgehog Basic fibroblast growth factor, epidermal growth factor, retinoic acid. Retinoic acid, fibroblast growth factor 2, sonic hedgehog, brain-derived neurotrophic factor, glial cell-derived neurotrophic factor, insulin-like growth factor-1	Pera et al., 2004 Perrier et al. 2004 Perrier et al. 2004 Yan et al. 2005 Schulz et al. 2003 Schulz et al. 2004 Mizuseki et al. 2003 Nistor et al. 2005 Li et al. 2005
Mesoderm (middle layer)	Keratinocytes Cardiomyocytes Blood Blood Blood	P68 expression Co-culture with endoderm-like cell line Bone morphogenetic protein 4, stem cell factor, interleukin 3, interleukin 6, granulocyte colony-stimulating factor Vascular endothelial growth factor Bone morphogenetic protein 4 or Activin A	Green et al. 2003 Mummery et al. 2002 Mummery et al. 2003 Chadwick et al. 2003 Cerdan et al. 2004 Ng et al. 2005a,b
Endoderm (internal layer)	Hepatocytes Pancreas and β-cells	Sodium butyrate, DMSO, or hepatocyte growth factor Basic fibroblast growth factor, nicotinamide	Lavon et al. 2004 Segev et al. 2004

cells and tissues is between 1% and 13%, varying with cell type, location within the body, proliferation state, distance from the blood supply, amount of mitochondria within the cell or tissue, and stage of development (i.e. embryogenesis and organogenesis) (Okazaki and Maltepe 2006).

Hypoxia is a physiologically essential element of the microenvironment. It is hypothesized that low oxygen concentrations protect stem cells from the potential hazardous effects of oxygen radicals. For example, in the female reproductive organs oocytes and the developing embryo are exposed to oxygen levels between 1% and 5.5% (Webster and Abela 2007). Furthermore, the bone marrow of adult mammals is also subjected to reduced oxygen concentrations in a bid to protect the harbour of haematopoietic stem cells (Keith and Simon 2007). In the presence of hypoxia, cells must either build resistance or respond quickly to varying levels of oxygen; hence cells have adapted effective mechanisms such as vasodilation, increased blood vessel formation, increased glycolysis, and the oxygen-dependent hydroxylation of hypoxia-inducible factors (HIFs). HIFs are proteins stabilized with decreasing oxygen tensions which affect the transcription of specific genes involved in adaptation and growth mechanisms such as proliferation, self-renewal, energy metabolism, motility, and invasion. Therefore, as well as protecting stem cells from the potential damaging effects of radicals, hypoxia and hypoxic-inducible

Figure 15.8 A sphere of stem cells, generated by the hanging drop technique, maintained in normal growth medium.

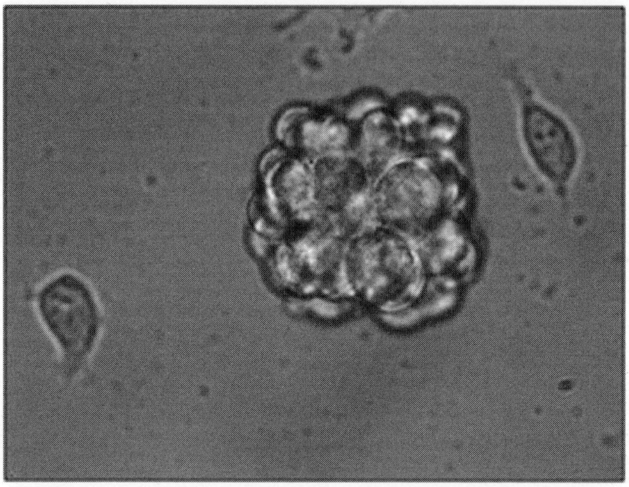

factors maintain stem cells in a highly proliferative unspecialized state by altering specific gene expression associated with an immature cell state such as the Notch signalling pathway.

A hypoxic incubator is frequently used to mimic the hypoxic *in vivo* microenvironment of stem cells. To calculate the appropriate oxygen tension for *in vitro* stem cell culture conditions the partial pressure of oxygen (PPO_2) *in vivo* should be determined initially (Sullivan et al. 2006; Csete 2005). Table 15.4 shows the variation of the partial pressure of oxygen and the associated oxygen tension in different mammalian tissues and environments. The equation below shows you how to use these data to calculate desired partial pressure of oxygen and oxygen tension for different *in vitro* culture conditions:

$$\text{Oxygen tension} = \frac{PPO_2(\text{desired})}{PPO_2(\text{dry air})} \times \text{percentage of } O_2 \text{ (dry air)}$$

15.6 Limitations of cell culturing methods

15.6.1 Aseptic technique and contamination

Microbial contamination and cross-contamination are commonly encountered problems in cell culturing and will compromise the culture and the quality of the data generated.

Table 15.4 Comparison of gaseous percentages and partial pressures in various tissues with those in dry air

	Nitrogen	Oxygen	Water	Total
Dry air				
Percentage (%)	79	21	0	100
Partial pressure (mmHg)	600.4	159.4	0	759.8
Blood				
Percentage (%)		9.9–13.2		
Partial pressure (mmHg)		75–100		
Adult brain				
Percentage (%)		1.5–7		
Partial pressure (mmHg)		11.4–53.2		
Placenta at first 10 weeks of pregnancy				
Percentage (%)		3.2		
Partial pressure (mmHg)		~25		
Placenta at 12–40 weeks of pregnancy				
Percentage (%)		7.9–8.5		
Partial pressure (mmHg)		60–65		

Microbial contamination

Cells in culture are highly susceptible to contamination with fungi, bacteria, and viruses which are often transferred to the cultures from, for example, dirty incubators and work surfaces, non-sterile media, reagents, and pipettes. It is easily the most commonly encountered problem in laboratories and thus it is essential to develop and maintain good **aseptic** working practice from the outset. This includes maintaining a sterile work area, using sterile handling techniques, and using sterile reagents, which all help to reduce the probability of infection.

- Sterile working area—the most effective way to maintain a sterile working area is to use a laminar flow cabinet. This is a specially designed hood that draws air from the outside and passes it through a high-efficiency particulate air (HEPA) filter before blowing very gently in a laminar flow towards the user. It should be located in an area designated for cell culture only (i.e. no through traffic) and be wiped down with 70% ethanol before and after each use. Additionally, the cabinet is fitted with a UV lamp which provides extra protection by sterilizing the air and surfaces between uses.
- Sterile handling techniques and sterile reagents—there are several pieces of protective clothing that can be worn when culturing cells. such as a Howie lab coat, disposable nitrile gloves, and a disposable face-mask. Reagents and plastics used in the culturing process also need to be sterile; many plastic products are supplied in sterile conditions (such as individually wrapped plates, flasks, and serological pipettes). Additionally, solutions to be used in sterile conditions should either be autoclaved or, if the volumes are small, passed through a sterile filter using a syringe.

The addition of antibiotics (such as penicillin, streptomycin, and amphotericin) to the growth medium reduces the possibility of microbial contamination. However, their routine use is not recommended for several reasons: they may interfere with the biological processes under investigation, their constant use allows low-level infections to persist and go undetected, and finally antibiotic-resistant strains can develop. Therefore their use should be short term only.

Microbial contamination of a cell culture is usually easily detected by eye (e.g. the medium may appear cloudy or have changed colour) or visualized under a microscope. However, contamination with *Mycoplasma* is far more difficult to detect in culture as it rarely produces visible signs of infection. First detected in cell culture in 1956, slow-growing *Mycoplasma* cells, which are <1μm in size, can exist in the cell culture without causing cell deterioration or death but can significantly alter the metabolic behaviour of the cells. For example, they can slow the rate of proliferation considerably as well as inducing chromosome mutations (such as deletion, duplication, or translocation of part of a chromosome). Figure 15.9 illustrates a cell culture infected with *Mycoplasma*. Whilst this culture appears to grow normally under the phase contrast microscope, staining the cells with Hoechst 33258 produces a characteristic blue spotting of the entire cell and a halo effect around the cell nuclei which is indicative of *Mycoplasma* contamination. It is advisable to screen all new material received in the laboratory for *Mycoplasma* as well as routinely checking for contamination, which can be done quickly and easily by fluorescence staining with Hoechst 33258, PCR, ELISA, or a number of commercially available kits such as MycoAlert® (Lonza).

Cross-contamination

An additional problem that may be encountered is cross-contamination with other cell lines and sometimes even between species. This has been the subject of many published papers and calls for action due to the large-scale problem and its spurious effects on

Figure 15.9 Example of SNB-19 cells (a glioblastoma) infected with *Mycoplasma*. The cell nuclei are stained with Hoechst blue which has also stained the *Mycoplasma*, resulting in the typical dotted effect throughout the cells. Scale bar, 16μM. For a colour reproduction of this figure, please see Plate 13.

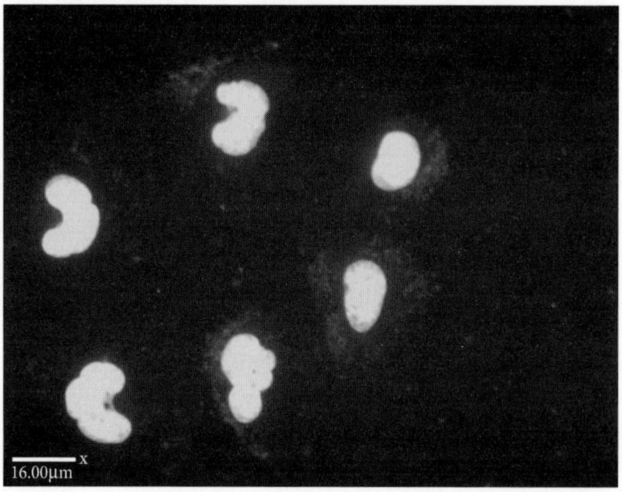

scientific data. The commonly used **HeLa** cell line (Skloot 2010), established in 1952, is one such example of cross-contamination with other cell lines. This highlights the scale and effect of the problem, since many 'HeLa contaminants' (cells labelled as HEK, HBT-3, and MA160 have all been found to contain chromosomes characteristic of HeLa) are in use and researchers not always aware of the mis-identity.

Furthermore, inter-species contamination can also occur and has been the topic of considerable attention in recent years, more so since it was revealed that the human **glioblastoma** cell lines U251 and SNB-19 deposited with the European Collection of Cell Cultures had in fact characteristically become the same cell line because of cross-contamination at some stage, as they showed very similar karyotypes when analysed by SKY, CGH, and FISH (Kubota et al. 2001; Garraway et al. 2005).

Cross-contamination of cultures can be avoided by, for example, handling only one cell line at a time and reserving different bottles of culture reagents for each cell line. Researchers should be aware of these issues and take extra precaution to prevent cross-contamination and test and authenticate established cell lines received into their laboratory.

Short tandem repeat profiling can be used to authenticate a cell line, as well as commercially available kits which can authenticate human origin by PCR amplification of specific DNA fragments unique to human, mouse, or rat. Figure 15.10 shows four human glioblastoma brain tumour cell lines and two human normal astrocyte cell lines that have been authenticated by PCR amplification. PCR products have been run on an agarose gel along with mouse and rat DNA as controls. The cells of human origin are not contaminated, which is evident by the bands running at 500, 300, and 200bp as expected, while the mouse and rat bands have significantly different profiles.

See Chapter 2

15.6.2 Loss of cell characteristics: genetic drift and senescence

There are several disadvantages in maintaining a cell line in continuous culture. First, it increases the risk of microbial and/or cross-contamination, and secondly, and significantly more difficult to treat, is risk of **genetic drift** and senescence. In genetic drift a number of genetic mutations accumulate, which leads to the loss of characteristics that are of specific interest in the study (e.g. surface antigens and secreted molecules).

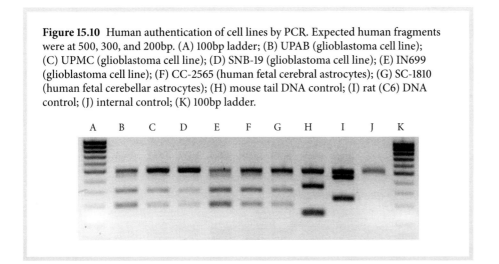

Figure 15.10 Human authentication of cell lines by PCR. Expected human fragments were at 500, 300, and 200bp. (A) 100bp ladder; (B) UPAB (glioblastoma cell line); (C) UPMC (glioblastoma cell line); (D) SNB-19 (glioblastoma cell line); (E) IN699 (glioblastoma cell line); (F) CC-2565 (human fetal cerebral astrocytes); (G) SC-1810 (human fetal cerebellar astrocytes); (H) mouse tail DNA control; (I) rat (C6) DNA control; (J) internal control; (K) 100bp ladder.

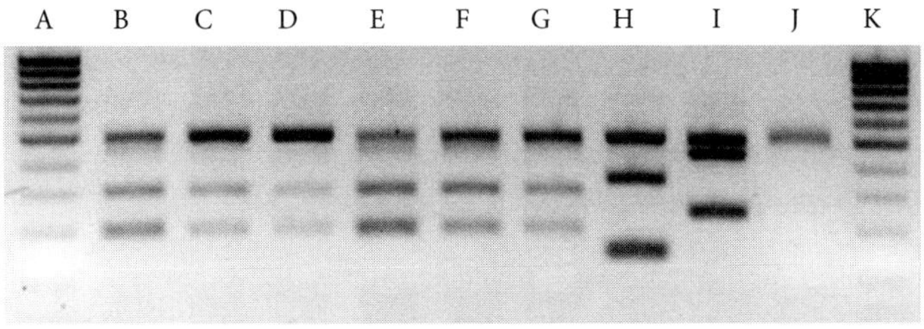

Senescence occurs when the cells have undergone their maximum number of population doublings because the telomeres have reached their critical minimum length, leaving the cells no longer able to divide—the **Hayflick limit**—and thus proliferation ceases. These cell lines, which have a finite population-doubling number are known as **non-immortal cells**. **Immortal cell lines** are those that can circumvent the Hayflick limit and thus are able to proliferate continuously. Continuous proliferation can be achieved by stable transfection of cells with the telomerase gene hTRT. Telomerase maintains the telomere length, thus allowing population doubling to continue indefinitely and so the cells avoid senescence. In the culture of normal non-immortal cell lines the only way to prevent senescence is to maintain a frozen bank of the cell line at different passage numbers so that you are able to maintain some consistency in your experiments, or go back to the stock if contamination occurs or you reach the finite population doubling number.

15.7 Ethical and storage issues

Cells that are obtained from human donors and cultured in the laboratory to be used in experimental procedures are governed by the Human Tissue Act 2004. This covers England, Wales, and Northern Ireland and regulates all activities involved with the removal, storage, use, and disposal of human tissues from both living and deceased persons. The patient is required to sign a form, which outlines in lay terms the objectives of the research, to authorize the use of their tissue in research and the relevant hospital committee must pass approval. Permission to store tissues for research is also required by the laboratory in the form of either a licence or NHS-REC (National Health Service Research Ethics Committee) approval. Samples that are received in the laboratory should be recorded anonymously, i.e. by assigning a number and recording the tissue site rather than the patient name.

15.8 Chapter summary

- The technique of culturing cells has undergone several advances since its advent in the early twentieth century. Cell cultures are now routinely established from tissue samples and expanded to form cell lines which can be cryopreserved and banked.
- Growth media have several basic components which aid cell growth by supplying essential growth factors and amino acids. Media can be defined for expanding the culture of specific cell types.
- Cell growth occurs in lag, log, and plateau phases. Subculturing, in which cells are dissociated from the culture vessel and split to form new cultures, should be carried out in the log phase of exponential growth.
- Cell culturing is used in many fields of biological research: gene expression, recombinant protein production, and tissue engineering are a few examples. Cell cultures can be 'scaled up' to aid this.

- Stem cell research is a large and active discipline in itself. It requires specific culturing techniques, media, and growth environments in order to replicate the *in vivo* situation.
- Stem cell cultures can be established by sorting the cells using specific cell markers and purifying the culture using FACS™ or MACs™.
- Cross-contamination and microbial contamination are common features in cell culturing and will compromise the culture. Methods of avoiding these include use of an aseptic technique, routine screening, and genetic fingerprinting.

✓ References

Allaire, A., Luong, M.X., and Smith, K.P. (eds) (2011) *Concepts in Stem Cell Culture*. Hoboken, NJ: John Wiley.

Amit, M., Margulets, V., Segev, H., et al. (2003) Human feeder layers for human embryonic stem cells. *Biol Reprod* **68**: 2150–6.

Amit, M., Shariki, C., Margulets, V., and Itskovitz-Eldor, J. (2004) Feeder layer- and serum-free culture of human embryonic stem cells. *Biol Reprod* **70**: 837–45.

Brewer, G.J. (1995) Serum-free B27/neurobasal medium supports differentiated growth of neurons from the striatum, substantia nigra, septum, cerebral cortex, cerebellum, and dentate gyrus. *J Neurosci Res* **42**: 674–83.

Carrel, A. (1923) A method for the physiological study of tissues in vitro. *J Exp Med* **38**: 407–18.

Carrel, A and Burrows, M.T. (1911) An addition to the technique of the cultivation of tissues in vitro. *J Exp Med* **14**: 244–7.

Cerdan, C., Rouleau, A., and Bhatia, M. (2004) VEGF-A165 augments erythropoietic development from human embryonic stem cells. *Blood* **103**: 2504–12.

Chadwick, K., Wang, L., Li, L., et al. (2003) Cytokines and BMP-4 promote hematopoietic differentiation of human embryonic stem cells. *Blood* **102**: 906–15.

Csete, M. (2005) Oxygen in the cultivation of stem cells. *Ann NY Acad Sci* **1049**: 1–8.

Eiselleova, L., Matulka, K., Kriz, V., et al. (2009) A complex role for FGF-2 in self-renewal, survival, and adhesion of human embryonic stem cells. *Stem Cells* **27**: 1847–57.

Eagle, H. (1955) Nutrition needs of mammalian cells in tissue culture. *Science* **122**: 501–4.

Freshney, R.I. (ed) (2007) *Subculture and Cell Lines*. Hoboken, NJ: John Wiley.

Garraway, L.A., Weir, B.A., Zhao, X., et al. (2005) Lineage addiction in human cancer: lessons from integrated genomics. *Cold Spring Harb Symp Quant Biol* **70**: 25–34.

Green, H., Easley, K., and Iuchi, S. (2003) Marker succession during the development of keratinocytes from cultured human embryonic stem cells. *Proc Natl Acad Sci USA* **100**: 15 625–30.

Harrison, R. (1907) Observations on the lung developing nerve fiber. *Anat Rec* **1**: 116–28.

Keith, B. and Simon, M.C. (2007) Hypoxia-inducible factors, stem cells, and cancer. *Cell* **129**: 465–72.

Kubota, H., Nishizaki, T., Harada, K., et al. (2001) Identification of recurrent chromosomal rearrangements and the unique relationship between low-level amplification and translocation in glioblastoma. *Genes Chromosomes Cancer* **2**: 125–33.

Lavon, N., Yanuka, O., and Benvenisty, N. (2004) Differentiation and isolation of hepatic-like cells from human stem cells. *Differentiation* **72**: 230–8.

Li, X.J., Du, Z.W., Zarnowska, E.D., et al. (2005) Specification of motor neurons from human embryonic stem cells. *Nat Biotechnol* **23**: 215–21.

Mizuseki, K., Sakamoto, T., Watanabe, K., et al. (2003) Generation of neural crest-derived peripheral neurons and floor plate cells from mouse and primate embryonic stem cells. *Proc Natl Acad Sci USA* **100**: 5828–33.

Mummery, C., Ward, D., van den Brink, C.E., et al. (2002) Cardimyocyte differentiation of mouse and human embryonic stem cells. *J Anat* **200**: 233–42.

Mummery, C., Ward-van Oostwaard, D., Doevendans, P., et al. (2003) Differentiation of human embryonic stem cells to cardiomyocytes: role of coculture with visceral endoderm-like cells. *Circulation* **107**: 2733–40.

Ng, E.S., Azzola, L., Sourris, K., Robb, L., Stanley, E.G., and Elefanty, A.G. (2005a) The primitive streak gene Mixl1 is required for efficient haematopoiesis and BMP4-induced ventral mesoderm patterning in differentiating ES cells. *Development* **132**: 873–84.

Ng, E.S., Davis, R.P., Azzola, L., Stanley, E.G., and Elefanty, A.G. (2005b) Forced aggregation of defined numbers of human embryonic stem cells into embryoid bodies fosters robust, reproducible hematopoietic differentiation. *Blood* **106**: 1601–3.

Nistor, G.I., Totoiu, M.O., Haque, N., Carpenter, M.K., and Keirstead, H.S (2005) Human embryonic stem cells differentiate into oligodendrocytes in high purity and myelinate after spinal cord transplantation. *Glia* **49**: 385–96.

Okazaki, K. and Maltepe, E. (2006) Oxygen, epigenetics and stem cell fate. *Regen Med* **1**: 71–83.

Pera, M.F., Andrade, J., Houssami, S., et al. (2004) Regulation of human embryonic stem cell differentiation by BMP-2 and its antagonist noggin. *J Cell Sci* **117**: 1269–80.

Perrier, A.L., Tabar, V., Barberi, T., et al. (2004) Derivation of midbrain dopamine neurons from human embryonic stem cells. *Proc Natl Acad Sci USA* **101**: 12 543–8.

Reya, T., Morrison, S.J., Clarke, M.F., and Weissman, I.L. (2001) Stem cells, cancer, and cancer stem cells. *Nature* **414**: 105–11.

Schuldiner, M., Yanuka, O., Itskovitz-Eldor, J., Melton, D.A., and Benvenisty, N. (2000) Effects of eight growth factors on the differentiation of cells derived from human embryonic stem cells. *Proc Natl Acad Sci USA* **97**: 11 307–12.

Schulz, T.C., Palmarini, G.M., Noggle, S.A., Weiler, D.A., Mitalipova, M.M., and Condie, B.G. (2003) Directed neuronal differentiation of human embryonic stem cells. *BMC Neurosci* **4**: 27.

Schulz, T.C., Noggle, S.A., Palmarini, G.M., et al. (2004) Differentiation of human embryonic stem cells to dopaminergic neurons in serum-free suspension culture. *Stem Cells* **22**: 1218–38.

Segev, H., Fishman, B., Ziskind, A., Shulman, M., and Itskovitz-Eldor, J. (2004) Differentiation of human embryonic stem cells into insulin-producing clusters. *Stem Cells* **22**: 265–74.

Skloot, R. (2010) *The Immortal Life of Henrietta Lacks*. New York: Broadway.

Sullivan, M., Galea, P., and Latif, S. (2006) What is the appropriate oxygen tension for in vitro culture? *Mol Hum Reprod* **12**: 653.

Thomson, J.A., Itskovitz-Eldor, J., Shapiro, S.S., et al. (1998) Embryonic stem cell lines derived from human blastocysts. *Science* **282**: 1145–7.

Trounson, A. (2006) The production and directed differentiation of human embryonic stem cells. *Endocr Rev* **27**: 208–19.

Wang, Q., Fang, Z.F., Jin, F., Lu, Y., Gai, H., and Sheng HZ (2005) Derivation and growing human embryonic stem cells on feeders derived from themselves. *Stem Cells* **23**: 1221–7.

Watt, S.M., Tsaknakis, E., Forde, S.P., Corpentes, L. (2009) Regulation by Stem Cell Nichis: Stem Cells Hypoxia, and Hypoxia-Inducible Factors. In Rajasekhar, V.K., Venuri, M.C. (Eds), Regulatory Network, in Stem Cells (pp. 211-31). New York: Humana Press.

Webster, W.S. and Abela, D. (2007) The effect of hypoxia in development. *Birth Defects Res C Embryo Today* **81**: 215–28.

Xu, C., Inokuma, M.S., Denham, J., et al. (2001) Feeder-free growth of undifferentiated human embryonic stem cells. *Nat Biotechnol* **19**: 971–4.

Yan, Y., Yang, D., Zarnowska, E. D., et al. (2005) Directed differentiation of dopaminergic neuronal subtypes from human embryonic stem cells. *Stem Cells* **23**: 781–90.

Zhang, Y., Mao, N., Li, X.S., et al. (2001) Maintaining growth of long-term culture initiating cells from human cord blood on feeder layers of bone marrow stromal cells transfected with FL and/or TPO genes. *Zhongguo Shi Yan Xue Ye Xue Za Zhi* **9**: 97–100.

16 Flow cytometry

John Lawry*

Chapter overview

Flow cytometry is a powerful technique used for counting and sorting cells based on their size or protein expression. It employs laser technology to facilitate rapid and simultaneous multiple analyses from cells suspended in a fluid stream. Flow cytometry can detect a range of cell sizes from very small cells, such as bacteria (1.0μM), to larger cells, such as white blood cells (8μM). Cell populations are distinguished from one another by the use of fluorescent dyes, which may stain cellular components directly (e.g. DNA or mitochondria) or indirectly using, for example, fluorescently conjugated **monoclonal antibodies**. Possible outcomes include cell sorting, cell cycle analyses, **cell viability** analysis, and **immunophenotyping**.

In this chapter, we will first look at how a flow cytometer works, how data are presented and what they mean, and then describe three key applications of these instruments: immunophenotyping, cell cycle analysis, **apoptosis**, and viability. We will then focus on the unique feature of these instruments, namely single-cell sorting.

LEARNING OBJECTIVE

This chapter will enable the reader to:

- understand how flow cytometers work and how they could be applied to your research;
- recognize the importance of sample preparation when using a flow cytometer;
- know that fluorescence selection depends upon the laser and detector choice within the cytometer, and that this guides the dyes used in any application;
- design experiments on immunophenotyping, DNA cell cycle analysis, apoptosis, and cell viability, and interpret the data generated;
- describe the basic processes leading to cell sorting.

* The opinions expressed in this chapter are those of the author and not necessarily those of BD Biosciences.

16.1 Flow cytometers: how they work

The reason we have flow cytometers today is due to three key developments in the 1950s. The first was work by Friedman (1950) who identified new fluorescent dyes based on acid fuchsin and acridine yellow to detect uterine cancer cells in cervical smears by fluorescence microscopy. The second was the development by Coons and Kaplan (1950) of isocyanate forms of fluorescein as a fluorescent antibody conjugate, and the third was the production by Coulter (1956) of a flow system in which cells passed through a narrow orifice and the resulting change in electrical impedance enabled them to be counted. When computers became more sophisticated during the 1960s instruments resembling modern flow cytometers were produced, first by Kamentsky et al (1965), who combined the Coulter counter with image analysis to classify cervical cancer cell suspensions, and then by Fulwyler (1965), who produced the first **cell sorter** using the same process of **electrostatic deflection** that is used today.

A flow cytometer consists of four components: fluidics, lasers, optics, and electronics. In a flow cytometer, the sample—a single-cell suspension—is carried by sheath fluid (usually saline solution) through a flow cell. Lasers (used as a source of illumination) positioned within the cytometer cause light scattering and fluorescence as the cells pass through the laser beam. This light is gathered and directed through photomultiplier tubes (PMTs)—the optical component—and finally expressed graphically, as digital data, on a computer. The components of the flow cytometer and how they work are described in sections 16.1.1–16.1.3.

16.1.1 Fluidics

Almost all flow cytometers require a sheath fluid which runs through the whole system. This is usually pressurized between 4 and 70**psi**, depending on the instrument, and is normally an **isotonic saline solution** or phosphate-buffered saline (PBS), but in some cases it can be filtered water. However, in cell sorters the sheath fluid must be saline as it has to carry a small electric charge as part of the droplet charging process. This process will be explained in more detail in section 16.6.

The fluidic system is illustrated in Figure 16.1. At its most basic, it is simply the flow of sheath fluid from the storage tank to the waste tank via the flow cell. Most cytometers use a pressurized system to 'push' saline through. However the BD FACS Verse™ utilizes suction, where the fluid is drawn through the cytometer with the sheath system under a vacuum.

The sample is 'injected' into the sheath stream at the flow cell, and pressurized so that it is just above that of the sheath fluid. However, the sample pressure can be regulated by the user over a range of flow rates (typically 12–120μL/min) so as to optimize the event rate, i.e. the number of cells passing through the flow cell. Samples are often limited to being taken up from 12 × 75mm (5mL) polystyrene or polypropylene tubes, but some cytometers are able to take sample from microtitre plates or larger sample containers (e.g. 15mL conical-based centrifuge tubes).

The sample undergoes **hydrodynamic focusing** when it arrives in the centre of the conical channel, to form the central core of the stream of sheath fluid. In the conical component, the flow rate of the sheath fluid running against the outer edge slows down

Figure 16.1 Schematic diagram of the fluidics of the flow cell. The sample is first 'injected' into the centre of a stream of saline, and then undergoes hydrodynamic focusing (A) before passing along the central channel of the flow cell (B), through the path of the laser beams (C), and then to waste.

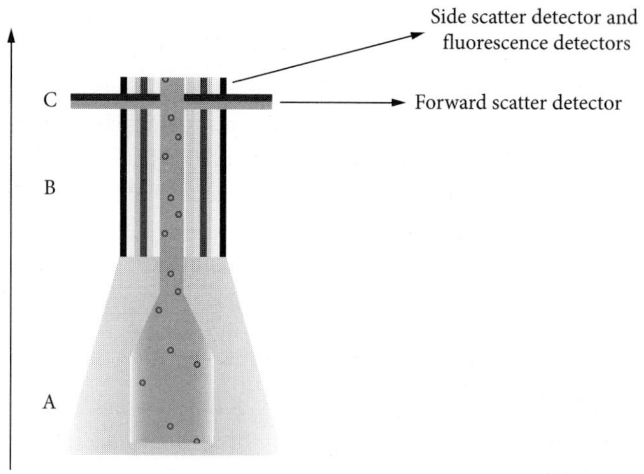

slightly due to drag. Therefore the sheath fluid in the centre flows with a higher velocity, and as this is where the sample is injected, the result is that the random mass of cells entering the conical channel is 'teased' apart to form a column of mostly single cells. This results in a stream of sheath fluid with a central core of sample which then progresses into a channel in the cuvette of the flow cell ready for the next stage of laser illumination, and fluorescence and light scatter detection.

All flow cytometers require one key feature of the sample—it has to be a single-cell suspension. This is easily achieved with samples such as blood, sperm ejaculates, synovial fluid, bacterial cell suspensions, and marine samples. However, cells which clump together in large masses, or cells which have been cultured as monolayers, or cells derived from a tissue sample will need to be processed to achieve a single-cell suspension. This could involve enzymatic treatment, mechanical dissociation, or using a complex buffer in which cells are resuspended with additives such as serum (reduces cell clumping), EDTA (reduces cell–cell adhesion), and DNase (to break down DNA released by dead cells which would otherwise causes cells to clump together). Finally, before being run through a flow cytometer all samples should be filtered to remove large cell aggregates and sample debris to reduce the chance of blocking sample injection tubes, sample lines, and the flow cell.

16.1.2 Lasers and optics

The standard illumination source for flow cytometers is the **laser**, an acronym for **l**ight **a**mplification by the **s**timulated **e**mission of **r**adiation. Lasers provide pure single-wavelength light. Although a few are tunable, most only provide one single fixed wavelength which is in a coherent (synchronized) and polarized (aligned) waveform.

Table 16.1 lists the wavelengths of the most frequently used lasers in a flow cytometer. Some instruments may only incorporate one laser, but many are now capable of

Table 16.1 Laser options

Laser wavelength (nm)	Colour
350	Ultraviolet
405	Violet
457	Navy blue
488	Blue
561	Yellow/green
635	Red

supporting multiple lasers, typically three although five is not unusual in large research instruments, and some custom flow cytometers can support seven lasers.

When a particle or cell passes through the laser beam three effects can be measured on a flow cytometer.

- Forward or low-angle light scatter is light which passes through the sheath fluid and then through the cell, in which it undergoes refraction. The larger the cell, the greater the refraction, and therefore **forward light scatter** (FSC) is indicative of the particle size. However, it cannot be used to measure absolute dimensions as different particles have different refractive indices, as well as being of different sizes.
- Wide-angle or **side scatter** (SSC) is produced by photons of light being 'bounced' off robust structures such as cell walls, nuclear membranes, and cytoplasmic organelles, and is captured by detectors perpendicular (at right angles) to the path of the laser beam. Hence the term 'side scatter' or sometimes 'right-angled light scatter'. Side scatter is a measure of the optical granularity or density of a particle and is frequently used to distinguish cells such as lymphocytes, monocytes, and granulocytes in blood samples, as they demonstrate increasing cytoplasmic and nuclear complexity and therefore greater side scatter properties.
- **Fluorescence** is the release (emission) of photons of light from a dye molecule which has been excited to an unstable high energy state and which releases energy (in the form of photons and heat) when returning to a more stable ground state. This emitted light has a specific wavelength and can be used in flow cytometry to investigate particular cell parameters. A large numbers of fluorescent dyes are available for use in flow cytometry. Each has a specific **excitation maximum**, i.e. the optimal laser wavelength required to excite the dye, and a specific **emission maximum**, i.e. wavelength of light at which maximum release of photons is observed. In flow cytometry, fluorescent dyes can be conjugated to **monoclonal antibodies** (MoAbs) which have been generated to identify and bind to specific antigens expressed by the cell (**fluorescent conjugate**). This is discussed further in section 16.3. Other dyes used in flow cytometry may be functional in that they bind to, and therefore label, particle characteristics such as DNA, RNA, intracellular calcium levels, membrane potential, and cytoplasmic pH. Table 16.2 lists the commonly used fluorescent dyes conjugated to MoAbs and the commonly used DNA dyes. It is possible to combine several dyes which are excited by the same laser (the same excitation maximum), but which emit at different wavelengths, to 'tag' or label a cell and generate a multicolour flow cytometry sample. Using multiple colours is now the norm, with cells being labelled with six, eight, ten, and more dyes.

Table 16.2 Excitation and emission maxima of fluorescent dyes commonly conjugated to MoAbs and commonly used DNA dyes

Dye*	Excitation maximum (nm)	Emission maximum (nm)
Fluorescent dyes commonly conjugated to MoAbs		
Cascade Blue	377	420
Alexa Fluor 405	401	421
Pacific Orange	400	551
BD Horizon™ V450	404	448
BD Horizon™ V500	415	500
Pacific Blue	401	452
AmCyan	458	491
FITC (fluorescein isothiocyanate)	493	525
Alexa Fluor 488	495	519
Phycoerythrin (PE)	496	578
Alexa Fluor 488	488	519
PE-Texas Red	496	615
BD Horizon™ PE-CF594	496, 564	612
PE Cy5	496	667
PerCP	482	678
PerCp Cy5.5	482	695
PE Cy7	496	785
APC	650	660
BD APC H7	650	785
Alexa Fluor 647	650	668
Alexa Fluor 700	696	719
Commonly used DNA dyes		
Propidium iodide	540	625
Ethidium bromide	518	610
Mithramycin	421	575
Hoechst 33342	352	460
DAPI (4′,6-diamidino-2-phenylindole)	365	450
7-Actinomycin D	550	655
DRAQ5	488 and 633	670

* Pacific Blue, Pacific Orange, Alexa Fluor, and Texas Red are trademarks of Molecular Probes Inc., Cy is the trademark of Amersham Biosciences Corporation, CF is the trademark of Biotium Inc., and BD is the trademark and property of Becton, Dickinson and Company.

When fluorescent dyes are used, the optical detectors (see section 16.1.3) cannot distinguish the different emission spectra on their own. Therefore each detector will have a specific optical filter placed in front of it to limit the wavelength range it receives, and several will effectively 'divide up' the full visible range of wavelengths so that the whole spectrum can be analysed. Three types of optical filters are typically used in flow cytometers: shortpass filters, longpass filters, and bandpass filters. Shortpass filters have specific 'metal' coatings applied to the face of the glass to allow the passage of light below a specified wavelength, whilst longpass filters allow wavelength of light above a specified wavelength to pass through. Bandpass filters are a combination these and allows light of a defined wavelength range to pass through.

A diagram of longpass and bandpass filters is shown in Chapter 17 (Figure 17.2).

16.1.3 Electronics

Flow cytometers utilize photomultiplier tubes (PMTs) and photodiodes to detect light scatter and fluorescence. Each PMT has a photon-sensitive face, in which an electron cascade is initiated when photons hit it. This generates a small pulse which is amplified and processed through electronic data acquisition cards (DACs). Older flow cytometers measure data relating to the intensity of the fluorescent signal as an analogue pulse, whilst modern systems 'stream' data as digital signals which allows much higher event rates (number of cells to be measured).

When a cell passes through the laser beam, light scattering occurs and any fluorescent dyes present will be excited and subsequently emit fluorescence. A PMT will detect a signal in real time, which will commence as the cell or particle enters the beam, increase as the particle reaches the focal point of the laser beam, and then fall as the particle exits the laser beam. Therefore a signal pulse will last a certain time (pulse width) and reach a maximum intensity (pulse height), and hence it can be integrated (analogue systems) or summated (digital streaming systems) to produce a pulse area. These pulses are subsequently processed so that the data can be expressed graphically. The various ways in which data can be displayed and analysed are discussed in section 16.2.

As intensity of labelling can differ widely between cell types, and also between fluorescent dyes; the signal from each detector can be amplified by the operator. Therefore signal strength can be set to a level appropriate for the assay, relative to either the biological expression of the molecule or the level of the background **autofluorescence** of the particle or cell.

16.2 Data display and analysis

Let us assume that our cytometer has six fluorescent detectors plus forward and side scatter detectors. As a cell passes through the laser, each detector will receive a signal. It may be a low background level signal if there is no fluorescence of the particular wavelength that the PMT is designed to measure, or it may be very bright if a specific dye matches the wavelength that the PMT is designed to measure. Therefore as each cell passes through the flow cytometer, a single data point will be measured by every detector in the instrument. These are stored in chronological order and the file is called a **list mode data file**. Data from the whole sample can be expressed from just *one detector* as a histogram plot as shown in Figure 16.2. In this example cells are stained with a fluorescently labelled antibody directed against the CD4 **cell surface marker** (antigen).

Figure 16.2 Graphical representation of flow cytometry data. Example of a single-parameter fluorescence histogram or frequency distribution on a logarithmic scale (x-axis). Isolated lymphocytes are stained with a CD4 antibody conjugated to PE. Peak 1 represents a population of cells with low-intensity CD4 staining and peak 2 represents a population of cells with high-intensity CD4 staining. Note: the vertical grid lines are a useful aid to positioning the populations but are not used for population statistics.

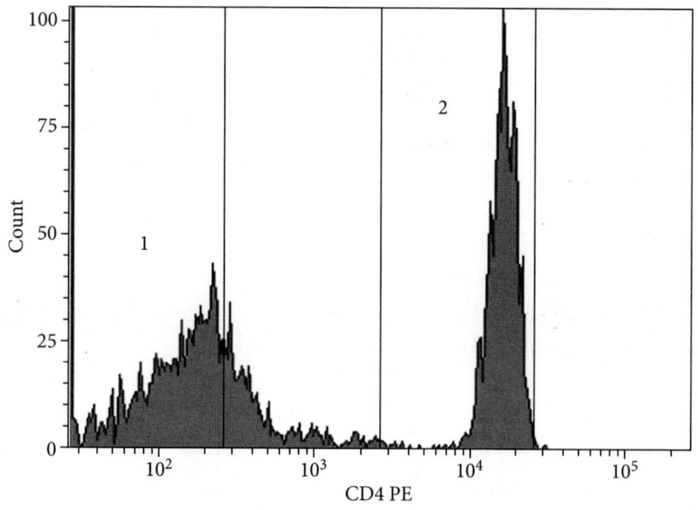

Each peak represents a distinct population of cells; CD4-positive cells (high fluorescence intensity) and CD4-negative cells (low fluorescence intensity). In contrast, a sample from *two detectors* is expressed as dual-parameter correlated data (Figure 16.3). This may be a simple dot plot, a density plot, or a contour plot as you can see from the examples shown. These were all taken from a blood sample in which the red blood cell population had been removed, as otherwise the red cells would outnumber all other cell types and make it hard to resolve them.

Data can be expressed on a logarithmic scale, as in Figure 16.2, or on a linear scale, as in Figure 16.3. In addition to log or linear scales, most recent software packages enable a 'hybrid' display to be produced, with the lower end of the scale being shown as linear data, and the rest as a logarithmic display. These are termed bi-exponential or logical scales.

The ability to modify the scales of plots is important, as some biological parameters can have enormous value ranges, whilst others are relatively restricted. The measurement of cellular DNA through the cell cycle is an example of a restricted range, and there is also a linear relationship between populations within the sample (i.e. as the DNA content doubles the intensity of staining doubles). These factors mean that data should ideally be expressed on a linear scale so that we can clearly measure DNA doubling between G1 phase cells and G2 phase cells. In contrast, a logarithmic scale would normally be chosen for a cell sample stained with antibody as there is a large dynamic difference between unlabelled cells and labelled cells, with a several thousand-fold increase in the measured fluorescence. This would be difficult to visualize on a linear scale. The visual

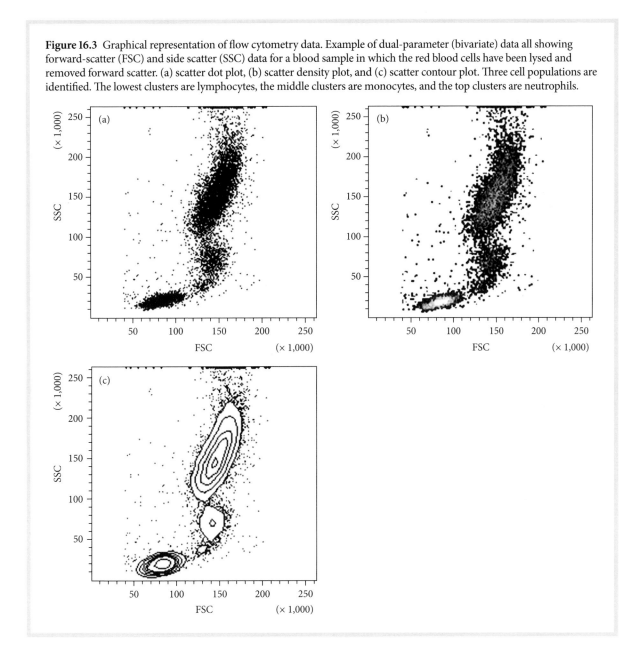

Figure 16.3 Graphical representation of flow cytometry data. Example of dual-parameter (bivariate) data all showing forward-scatter (FSC) and side scatter (SSC) data for a blood sample in which the red blood cells have been lysed and removed forward scatter. (a) scatter dot plot, (b) scatter density plot, and (c) scatter contour plot. Three cell populations are identified. The lowest clusters are lymphocytes, the middle clusters are monocytes, and the top clusters are neutrophils.

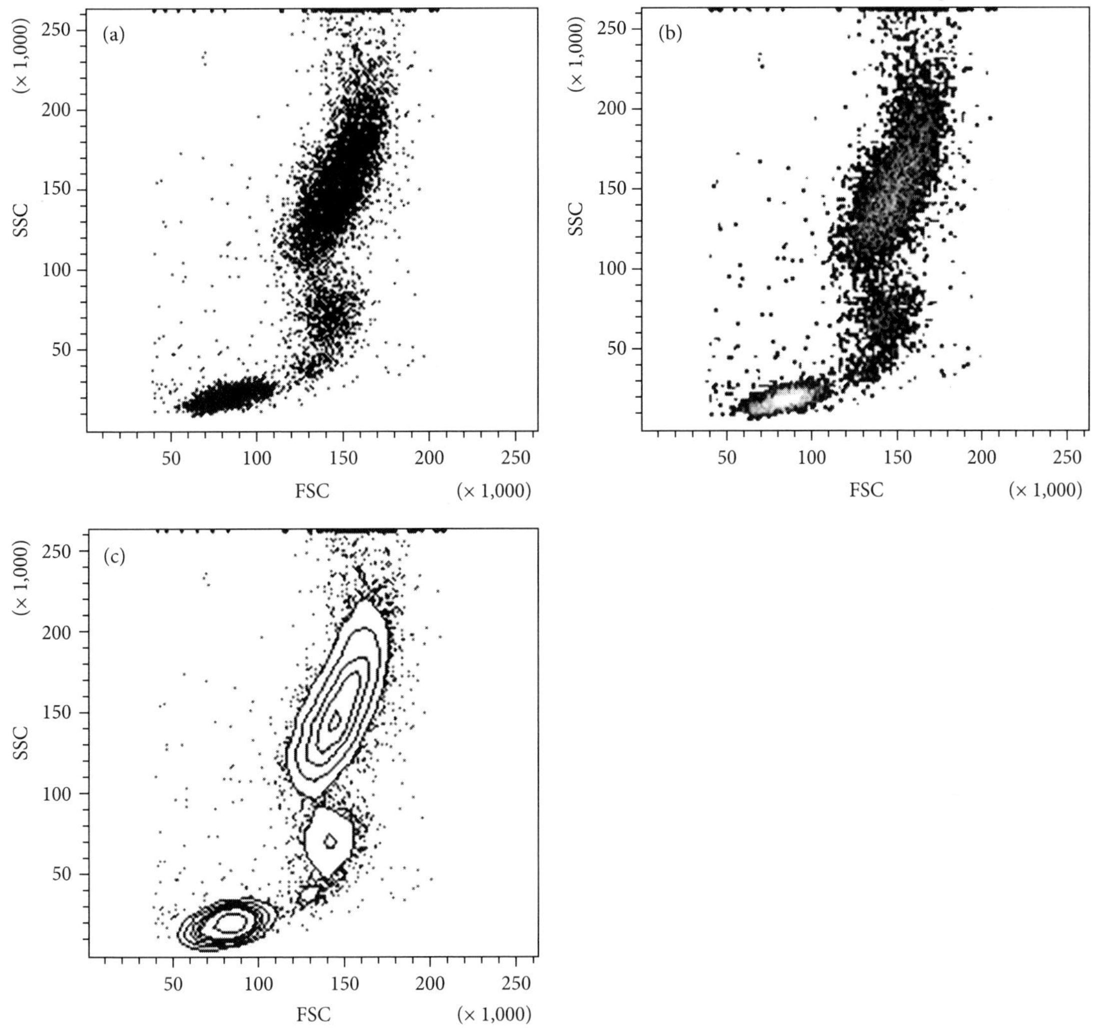

differences between linear and logarithmic displays of the same data are compared in Figure 16.4.

16.2.1 Gating cell populations

By drawing regions (gates) or interval gates on dual-parameter and single-parameter plots, respectively, statistical data can be obtained. These data may include percentage population data or calculations including the mean, geometric mean, median, mode, standard deviation, and coefficient of variation of the dataset. Gates are usually applied

Figure 16.4 Comparison of linear and logarithmic displays (*y*-axis). Example data plots from BrdU (cell proliferation) stained cells. (a) DNA dye propidium iodide (PI) measured along the *x*-axis, and FITC-labelled BrdU measured as a linear scale on the *y*-axis; (b) the same data with logarithmic scaling on the *y*-axis. Populations can be resolved on the logarithimic scale showing (1) G0/G1 phase cells, (2) S phase cells, and (3) G2/M phase cells.

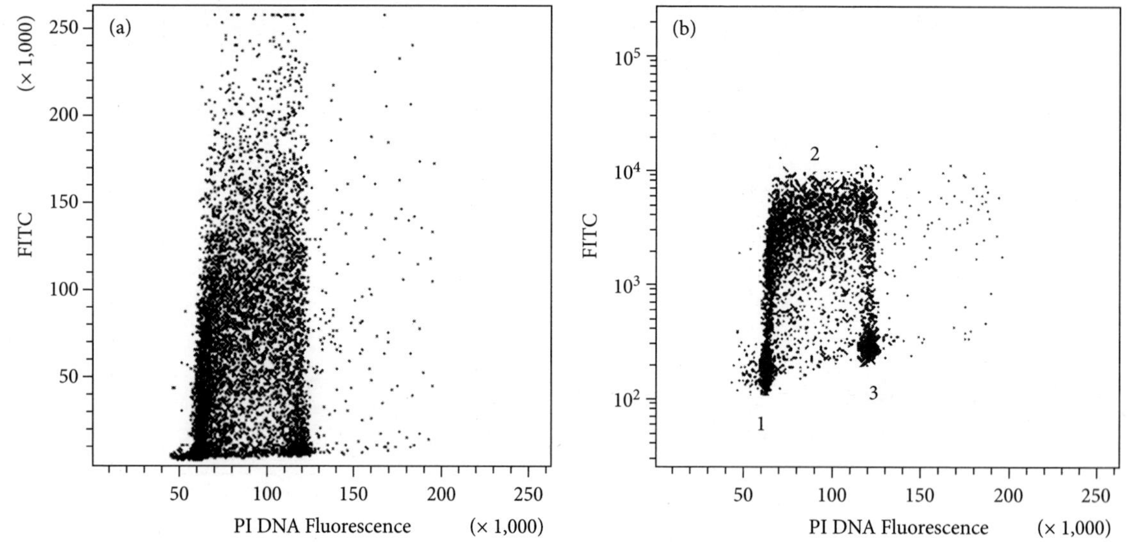

sequentially to data. For example, the scatter display may be gated to analyse only the lymphocytes and then fluorescent data can be visualized and further gates drawn around MoAb-positive and MoAb-negative lymphocyte populations. This step can be followed by further gate steps to restrict the data displayed in subsequent plots or for statistical analysis. This form of analysis is termed hierarchical gating. Figure 16.5 shows such a gating strategy for a blood sample. In this experiment the aim is to identify subpopulations of lymphocytes (e.g. B cells and T cells) from the blood sample. To achieve this the sample is labelled with a cocktail of antibodies specific to the different lymphocytes and the data are analysed by gating. The first 'selection' or gate is drawn to identify lymphocytes labelled with CD45 (leucocyte common antigen) (Figure 16.5(a)) and the gate colour for the data points (dots) is visualized on a light scatter plot (Figure 16.5(b)). This is termed 'back gating' and helps us draw an accurate scatter plot gate for the cells we want to focus on. The next step is to draw plots to show the fluorescent (antibody binding) data, and to use the previously drawn gates as 'filters' (Figures 16.5(c) and 16.5(d)). In this way, we can restrict the data shown on the fluorescent plots to only that which originates from within the scatter gates. The result is that we only measure antibody binding data from, for example, the lymphocytes and not from the other cells present in our whole-blood sample. Finally, we can draw new gate regions on the fluorescent plots in order to measure specific antibody fluorescence, and hence visualize and quantify the lymphocyte subsets to give a total view of T cells, B cells, and NK cells in the sample. This is one example of 'hierarchial gating'.

Gates can also be the defining parameter for a cell sort. If the computer identifies a cell which has a data point within a gate, it will be tracked, and as it becomes surrounded by a droplet, it can then be actively sorted into a specific container (see section 16.6).

Figure 16.5 Example of the workflow required for hierarchical gating in a blood sample labelled with a cocktail of six antibodies so that each lymphocyte subset can be identified. #### in the statistics table denotes an invalid population calculation. For a colour reproduction of this figure, please see Plate 14.

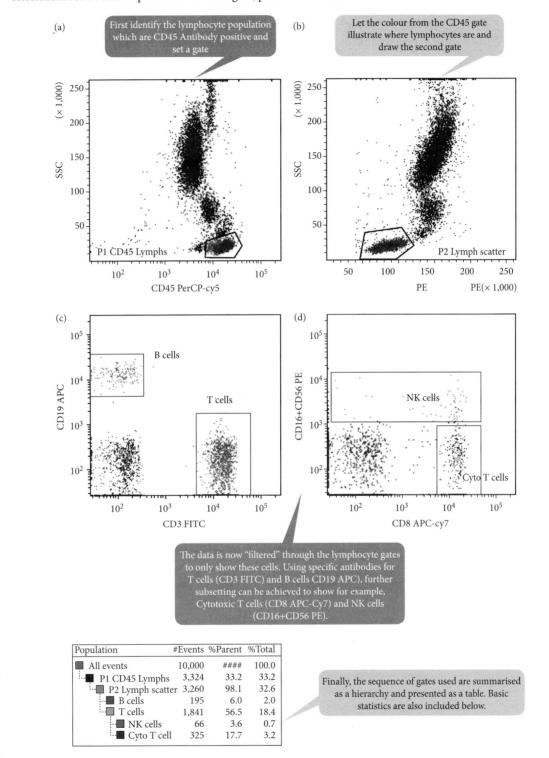

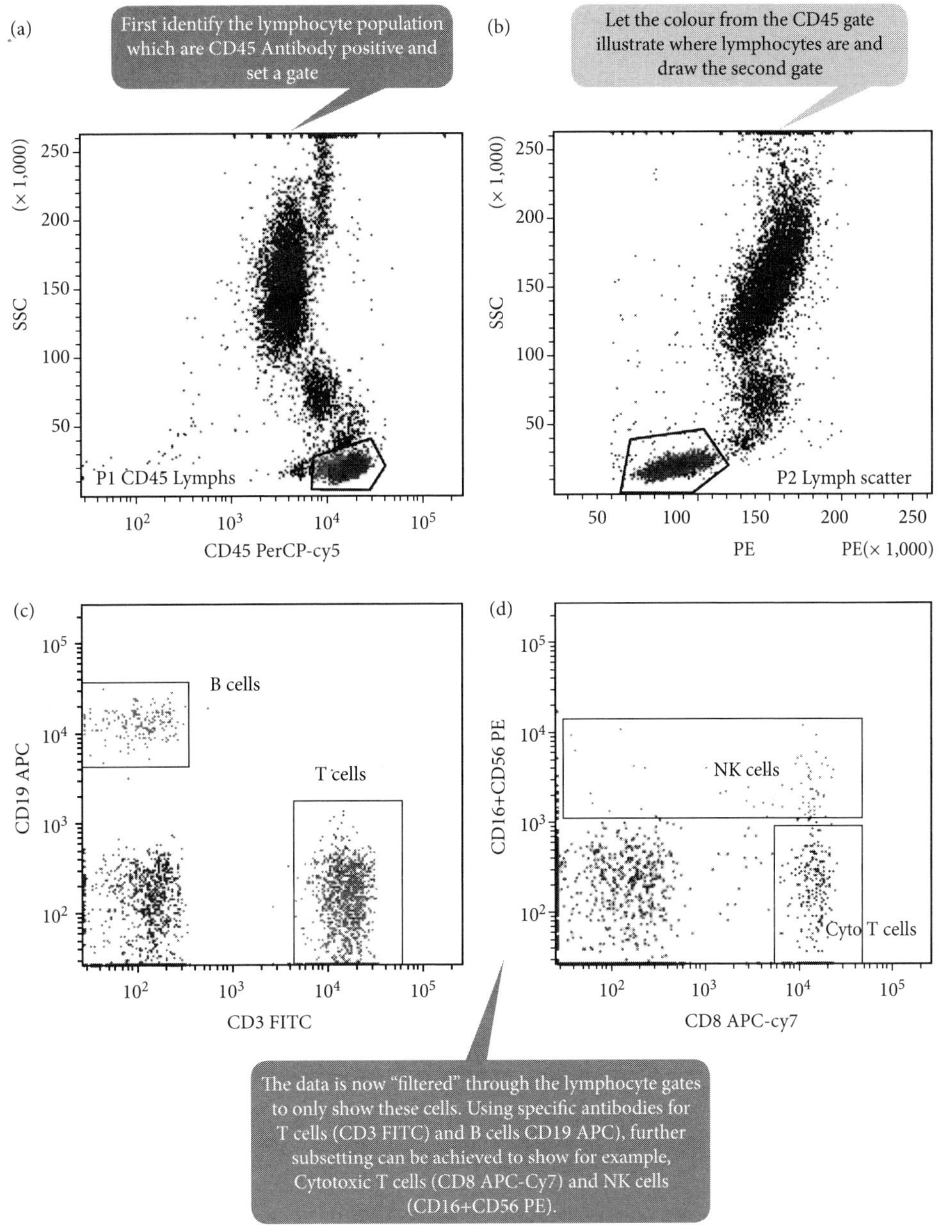

Population	#Events	%Parent	%Total
All events	10,000	####	100.0
P1 CD45 Lymphs	3,324	33.2	33.2
P2 Lymph scatter	3,260	98.1	32.6
B cells	195	6.0	2.0
T cells	1,841	56.5	18.4
NK cells	66	3.6	0.7
Cyto T cell	325	17.7	3.2

Finally, the sequence of gates used are summarised as a hierarchy and presented as a table. Basic statistics are also included below.

BOX 16.1

Where can I find out more?

- **Institute core facilities** Purdue University is probably the most famous flow cytometry resource, but search any university site for 'core facilities' for some excellent information on applications and protocols, and helpful advice.
- **Company websites** Helpful tools such as fluorescence spectral viewers and tools to help devise antibody panels.
- **Reagent manufacturers** Detailed fluorochrome and antibody information.
- **International Society for the Advancement of Cytometry (ISAC)** is the worldwide society for flow cytometerists. It was specifically created for information exchange, running meetings and conferences, and publishing its own scientific journal *Cytometry*.
- **Royal Microscopical Society (RMS)** Based in Oxford, UK, the RMS is the host for a UK flow cytometry section responsible for running many meetings and taught courses.
- **Local societies in your area** Many local groups have been established to network flow cytometry laboratories and cytometrists in cities and larger geographical regions. They often have their own websites which will contain details of meetings and other useful information.

16.3 Flow cytometry application: immunophenotyping

Every cell expresses a variety of membrane-bound cell-surface proteins—markers. These are usually antigen sites, but also include membrane-bound receptors or ligands. Each cell carries a unique combination of markers—'fingerprints'—specific to a particular cell type, and this property can be exploited in flow cytometry to identify subsets of cells in heterogenous samples such as blood. Identification based on the particular antigens or markers carried by cells is called **immunophenotyping** and involves labelling cell samples with reagents, such as antibodies, cytokines, or growth factors, conjugated to a fluorescent dye which binds to specific antigens on the cell surface, or sometimes within the cytoplasm. The samples are then analysed using a fluorescent microscope or by flow cytometry (Maecker et al. 2012) and the cells are classified by determining the antigen(s) expressed on the cell surface or intracellularly. Flow cytometry is considerably faster than fluorescent microscopy as thousands, rather than tens, of cells can be rapidly detected and analysed. In addition, all the cells in what might be a very complex sample can be identified and quantified. Furthermore, the intensity of staining can be measured—this is proportional to the amount of reagent bound to the cell and therefore is indicative of the level of marker expression.

See Chapter 17

Immunophenotyping is commonly used in the investigation of disease, particularly the diagnosis of haematological malignancies such as leukaemias and lymphomas (Sullivan and Wiggers 2000; de Tute 2011). Cancer cells frequently express different markers to normal cells, and may also demonstrate different antigen density and therefore staining intensity. This difference is used to distinguish normal from abnormal cells. It is also possible to follow disease progression and patient relapse by tracking the difference in antigen expression over time. Immunophenotyping is not limited to the diagnosis and monitoring of haematological cancers but can also be used to monitor other diseases such as HIV (Pattanapanyasat 2012).

16.3.1 Classification of markers: the cluster differentiation system

Cluster differentiation (CD) is a classification system for identifying the markers (antigens) found on cell surfaces based on the antibodies which recognize them. This system was initiated in 1982 at the first International Human Leucocyte Differentiation Antigens (HLDA) Workshop (http://www.HCDM.org) in response to the expanding number of cell surface antigens being discovered through antibody binding. These antibodies were being generated by different laboratories around the world with personalized names, and it was difficult to assess if several antibodies were recognizing the same cell surface marker or not. To avoid confusion and to provide a standard language in which scientists could communicate data, the CD system was established. In this system, groups of antibodies with a similar reaction pattern are assigned a CD number. CD antigens are used to define cells based on the molecules present on their cell surface. For example, helper T cells are defined as CD4+ and cytotoxic T cells as CD8+. Since the first HLDA meeting, over 363 antibodies (2010 data) have been classified to identify cell subtypes (such as T cells and B cells), activation markers, chemokine receptors, cytokine receptors, and other functional or descriptive molecules expressed by cells.

The CD classification of MoAbs has had a tremendous impact on diagnostic applications in haematology, immunology, and oncology. First, it has enabled commercial companies to produce a large range of reagents, such as antibodies, cytokines and growth factors, conjugated to a range of fluorescent dyes. Second, these reagents have enabled the complex analysis of cells using several MoAbs per sample, each with a conjugate with different excitation and emission spectral properties. Third, clinical research has generated detailed lists of CD markers for the classification of specific disease states including leukaemia and lymphoma. At the latest European consortium 'Euroflow' (http://www.euroflow.org) (McLaughlin et al. 2008a, b) panels of reagents, divided into primary screening panels and then secondary panels, have been defined for a range of disease states. A primary panel may be four reagents to identify the cell type, namely B cell (CD19), T cell (CD3), NK cell (CD16), and monocyte (CD14), whilst the secondary panel will include phenotype-specific markers, lineage-specific markers, and maturation stage markers. These can be used to diagnose disease; for example, chronic lymphocytic leukaemia is characterized by a typical phenotype of CD5+, CD19+, CD22−, and CD23+, and is distinguished from mantle-cell lymphoma cells which are typically CD5+, CD19+, CD22+, and CD23− (Dillman 2008). Panels of reagents are also important in monitoring clinical progression and minimal residual disease, where a small but potentially deadly clone of cancer cells persists after treatment and may eventually expand causing patient relapse (Al-Mawali et al. 2009). Table 16.3 lists the common CD reagents used in determining the phenotype of leukaemias and lymphomas. Although a single marker can be chosen, it is more common to use a panel to enable identification of increased or decreased expression of particular markers as the disease progresses.

Multicolour flow cytometry is the largest growth area in flow cytometry, with research institutes competing for the greatest number of MoAb combinations in one sample tube. Diagnostic panels have evolved from 'simple' two or three colours, to five (Alamo and Melnick 2000) and above. Table 16.4 lists a typical range of eight fluorochromes capable of being mixed in one sample tube, as their fluorescence is spread over the fluorescent spectrum, and their excitation is divided between three different lasers. One of the most impressive is a 17-colour panel to analyse the immune system (Perfetto et al. 2004). However, it is important in flow cytometry to optimize the sensitivity of instruments to

Table 16.3 Lineage markers commonly used to determine the phenotype of leukaemias and lymphomas

Cell type	Marker
B cells	CD19, CD20, CD22, CD79a
T cells	CD3, CD5, CD7
Neutrophils	CD15, CD11b, MPO
Monocytes	CD13, CD14, CD15, CD36, CD64, CD163
NK cells	CD16, CD56, CD3 negative
Erythroid cells	CD36
Platelets	CD41, CD42, CD61
Plasma cells	CD38, CD45, CD56, CD138
B-ALL	CD10, CD19, CD34, TdT
T-ALL	CD5, CD7, cytoplasmic CD3
Mantle cell NHL	κ, λ, CD5, CD19, CD20
Follicular NHL	CD10, CD19, CD20, CD43

MPO, myloperoxidase; TdT, terminal transferase.

analyse very dim fluorescence and distinguish it from that emitted by unlabelled autofluorescent cells (Chase and Hoffman 1998). Furthermore, as all fluorescent dyes fluoresce over a large part of the spectrum, adding multiple dyes results in overlap of their fluorescence emission. This leads to difficulties, with detectors measuring the fluorescence from the dye of interest from the overlapping fluorescence of other dyes, This phenomenon is termed spillover, and can be corrected by the process of compensation. Compensation is the electronic subtraction of the 'false' fluorescence signal a detector receives, so that it can then reliably measure the 'true' fluorescence of the dye in the assay (Bagwell and Adams 1993; Mahnke and Roederer 2007).

In addition to surface staining of cells with MoAb panels, cell membranes can be fixed and permeabilized to enable intracellular staining of cytokines, oncoproteins, and nuclear proteins alone, in combination, or in parallel to surface staining antibodies. Figure 16.6

Table 16.4 Possible fluorochrome selection for a single-tube eight-colour immunophenotyping panel based on the most commonly used laser combinations in current flow cytometers: violet (405nm), blue (488nm), and red (635nm)

Dyes used	Laser line
Pacific Blue or V450	Violet
Pacific Orange or V500	Violet
FITC or AlexaFluor 488	Blue
PE	Blue
PerCP or PerCP Cy5.5	Blue
PE Cy7	Blue
APC or AlexaFluor 647	Red
APC H7	Red

Figure 16.6 Example of cytokine staining with two anti-cytokine antibodies (IL-2 PE and IFNγ APC). The staining distribution shows four cell populations. (a) IFNγ low and IL-2 low; (b) IFNγ high and IL-2 low; (c) IFNγ low and IL-2 high; (d) IFNγ high and IL-2 high.

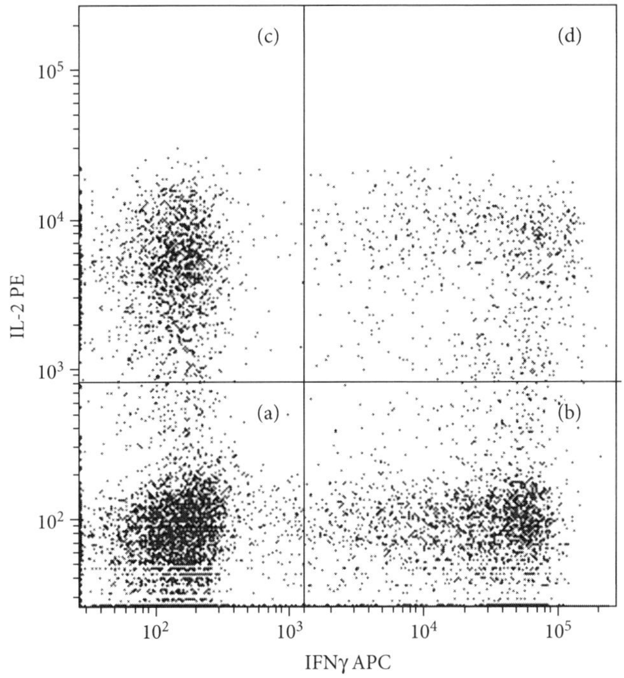

shows a common application with separated lymphocytes labelled with anti-cytokine (IL-2 and IFNγ) fluorescently conjugated antibodies. Commercial reagent suppliers now have excellent kits and detailed protocols for these assays. Many of these assays involve a single fix and permeabilizing incubation period, followed by a wash and centrifugation stage, and then an antibody-labelling step before analysis on the cytometer.

Immunophenotyping

CASE STUDY 16.1

The aim of the experiment is to identify the subtypes of lymphocytes in a sample of blood, and to identify their relative percentages.

QUESTION 1
Which reagents do I select?

You will need to outline which cell types may be present, and then purchase a fluorescently labelled monoclonal antibody matched for this population. For example, CD3 for T cells, CD 19 for B Cells, CD14 for monocytes, and CD16 for NK cells.

Each antibody will have to be conjugated to a different fluorochrome, so you will need to find out which lasers and which fluorochrome detectors your instrument has. Typically, you may decide that one reagent will be conjugated to FITC, one to PE, and one to

PerCP (all measured by a blue laser), and the fourth to APC which is measured by a red laser.

QUESTION 2
How do I stain my sample?

- Pipette 50μL blood into a 5mL polystyrene tube and add 20μL of fluorescent reagent. Control tubes will have just one reagent added; the test sample will have all the reagents added.
- Allow antibodies to bind to their antigen by incubating for 15min.
- Lyse the red blood cells using a lysing buffer.
- Wash the samples by centrifugation and resuspend in 1mL buffer.
- Analyse on a flow cytometer.

You will be able to distinguish lymphocytes, monocytes, and neutrophils on a dual-parameter forward and side scatter plot. Analysis regions can be drawn around each population and used as 'data filters' to demonstrate the fluorescent distributions of the T, B, and NK cells. Draw statistical regions around each of these to quantify each cell type in the blood sample.

16.4 Flow cytometry application: cell cycle analysis

The DNA cell cycle is a process whereby cells undergo cell replication by synthesizing chromosome copies before cell separation (mitosis). This is an essential feature that can be measured by flow cytometry. Cells at the G0/G1 phase of their cell cycle have the normal number of chromosomes for that cell. In normal (diploid) human cells the number of chromosomes is 26 and the cell will be in the diploid phase (referred to as $2n$). When cells undergo replication, they double their chromosome number ($4n$) during the synthesis phase (S phase) and reach the G2/mitosis phase. If stoichiometric DNA-specific dyes, such as those listed in Table 16.2, are added to the cell sample, there will be a measurable doubling of fluorescence intensity from the G1 to the G2 phase linked to the doubling of the DNA content of the cell. This fluorescence intensity, measured on a linear scale, enables quantification of the cells in each phase of the cell cycle.

16.4.1 DNA cell cycle data

There are several dedicated DNA cell cycle software packages capable of de-convoluting the DNA histogram which can provide precise quantification of the percentage of cells in each phase of the cell cycle (G1, S, G2/M), as well as being able to quantify cellular debris, apoptotic cells, and complex multiploid samples. Figure 16.7(a) shows a flow cytometric view of DNA stained (with propidium iodide (PI)) cells. Cells in the G1 phase are represented by the first peak ($2n$), cells in the G2/M phase are represented by the second peak ($4n$), and the central 'spread' of data between the two peaks denotes the S phase cell population. In contrast, Figure 16.7(b) shows a software-generated DNA profile of a complex tumour sample containing a combination of diploid and aneuploid cell populations. This type of multiploidy is a common phenomenon in tumours where normal

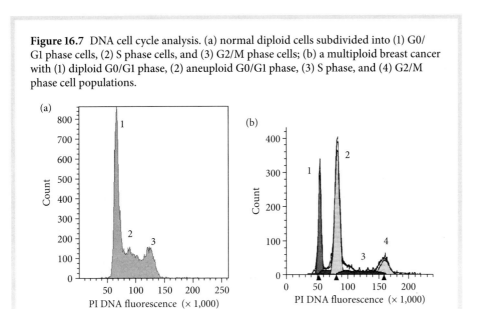

Figure 16.7 DNA cell cycle analysis. (a) normal diploid cells subdivided into (1) G0/G1 phase cells, (2) S phase cells, and (3) G2/M phase cells; (b) a multiploid breast cancer with (1) diploid G0/G1 phase, (2) aneuploid G0/G1 phase, (3) S phase, and (4) G2/M phase cell populations.

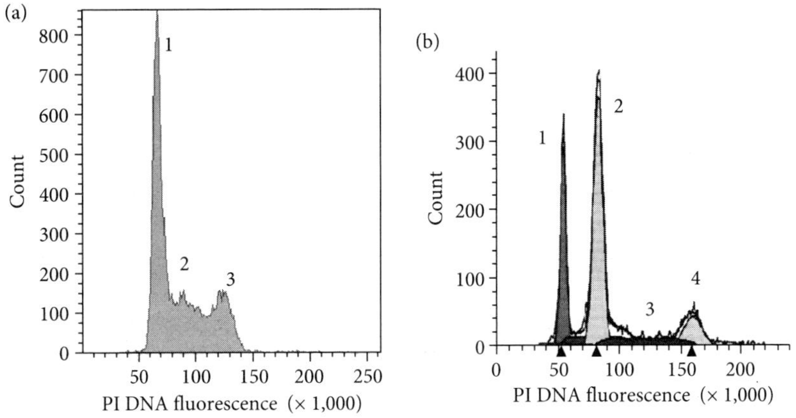

stromal or infiltrative cells are present with the tumour cells. The normal cells will have a non-dividing diploid (normal) chromosome number, whilst the tumour populations exist as clones of cells with both normal and abnormal karyotypes.

The convention on the nomenclature of tumour ploidy defines several karyotypes which include the following.

- Haploid: these are germinal cells with 50% of the normal chromosome number.
- Diploid (2*n*): these are normal cells referred to as having a DNA index of 1.0.
- Aneuploid: these are cells which have a higher DNA index than normal diploid cells (1.1–1.9).
- Tetraploid (4*n*) these cells have a DNA index of 2.0 and therefore have the same total DNA content as a diploid cell reaching the G2/M phase, but comprise cells starting the cell cycle with four copies of each chromosome.
- Multiploid is a heterogeneous mixture of cells with multiple karotypes.

16.5 Flow cytometry application: apoptosis and viability assessment

Apoptosis, also termed programmed cell death, is normally a well-controlled and highly regulated process leading to the elimination of an 'unwanted' cell. Without this process, many stages of embryogenesis would not be complete and immune responses would not be terminated. For example, it is apoptosis that changes the early embryonic hand and foot from being webbed to having distinctive fingers and toes, and it is apoptosis which reduces lymphocyte numbers to a normal level after lymphoproliferation in response to inflammation and infection. Failure to initiate apoptosis can induce many disease states, of which the best documented are arthritis and cancer, where the normal constraints on the lifespan of a differentiated cell are lost and hence the cell is no longer eliminated by programmed cell death.

Cells undergoing apoptosis can be characterized by a series of biochemical and morphological changes. The morphological changes that define apoptosis include cell shrinkage and chromatin condensation, followed by DNA fragmentation. The cell then breaks up to form apoptotic bodies which are eliminated by phagocytosis (Kerr et al. 1972; Wyllie et al. 1980). Apoptosis can occur via two biochemical pathways—extrinsic and intrinsic (Fulda and Debatin 2006). The extrinsic pathway involves the activation of death receptors such as TNF, whilst the intrinsic pathway involves the release of cytochrome C from the mitochondria. Both pathways converge at the caspases (cysteinyl **asp**artic acid prote**ases**), which are proteases that cleave cellular proteins and are responsible for some of the morphological changes associated with apoptosis. Several of these changes that can be analysed on a flow cytometer as monoclonal antibodies are now commercially available for most of the identified pathway proteins. For intrinsic pathways, it may be relevant to measure cytoplasmic and nuclear proteins such as p53, BCL-2, and pRb, in which case cells will require fixing and permeabilizing, whilst extrinsic pathways can be studied by measuring external cell membrane proteins such as TNF, Fas, and TRAIL receptors on live cells. In addition, flow cytometry can be used to measure pro-caspase proteins, phosphorylated proteins, active caspase enzymes, and their target substrates (poly-ADP-ribose polymerase (PARP) which is a DNA repair enzyme that is inactivated during apoptosis) (Darzynkiewicz et al. 1992). Common apoptotic parameters which can be measured by flow cytometry are summarized in Table 16.5, and some of these are described in further detail below.

Table 16.5 The apoptotic pathway and targets for flow-cytometric analysis

Parameter measured	Method	Comments
Plasma membrane asymmetry	Annexin V staining; DNA dye exclusion (e.g. PI, 7-AAD)	• Quick and inexpensive method for assessing early stage apoptosis • Can only be used on live unfixed cells; difficult to use with cultured adherent cells • Can be combined with surface markers to phenotype the apoptotic and non-apoptopic cells
Mitochondrial membrane integrity	YO-PRO and DiOC6 staining	• Quick and inexpensive method for assessing early-stage apoptosis • Can only be used on live unfixed cells; difficult to use with cultured adherent cells
DNA fragmentation	Sub-G1 peak TUNEL assay	• Both assays assess late-stage apoptosis • Measuring sub-G1 peak is inexpensive but TUNEL is not • Both methods use fixed cells • TUNEL enables analysis of cell cycle phases • Sub-G1 peak is not always good on some cell types • Cell cycle data will also be available with sub-G1 analysis.
Activation of caspases	Stain cells with antibodies against phosphorylated proteins, or stain cell lysates using CBAs which use fluorescent beads to immobilize and then quantify the analyte	• Sensitive assays; good for early apoptosis detection. • Hard to combine surface staining with phosphorylation antibodies • Bead assays destroy cell morphology • Only used on fixed cells

CBA, cytometric bead array.

16.5.1 Early apoptosis

Measuring externalization of phosphatidyl serine

An early event in apoptosis is the externalization of phosphatidyl serine molecules from the inner to the outer cell membrane. The protein Annexin V binds specifically to phospatidyl serine molecules, and when conjugated to a fluorescent dye this early apoptotic marker can be identified on cells and quantified by flow cytometry. Annexin V is usually used in combination with a viability dye. This is a DNA dye used on unfixed un-permeabilized cells so that viable cells exclude the dye, but dead and dying cells take up the dye which binds to their DNA. Viable cells are therefore dye-negative, whilst dead or dying cells (including apoptotic cells) are dye-positive. Thus this assay can resolve viable cells (Annexin V-negative, DNA dye-negative), early apoptotic cells (Annexin V-positive, DNA dye-negative), and mid to late apoptotic cells which are positive for both Annexin V and the DNA dye (Figure 16.8(a)). This is a very fast assay (total of 15 minutes incubation with both reagents, and then the sample is ready to analyse on the cytometer), but can only be used with fresh (non-fixed) samples. It is also hard to use with cultured cell suspensions which are grown as adherent monolayers. This is because the process of stripping cells off the tissue culture plastic can damage cell membrane integrity, which leads to uncontrolled entry of the DNA viability dye into the cell, resulting in fluorescence and hence mimicking a dead or dying cell.

Measuring plasma membrane permeability

Early-stage apoptosis can also be detected using the dye YO-PRO-1 Iodide (registered product of Molecular Probes Inc.) in combination with the DNA dye propidium iodide (PI). YO-PRO-1 Iodide is one of a family of monomeric cyanine nucleic acid stains which are able to enter apoptotic cells before propidium iodide and therefore it identifies early apoptotic cells (YO-PRO-1-positive, PI-negative) from viable cells (YO-PRO-1-negative, PI-negative) and dead cells (YO-PRO-1-negative, PI-positive). Example data are shown in Figure 16.8(b).

Measuring mitochondrial membrane permeability

Apoptotic cells rapidly depolarize their mitochondrial membrane with a corresponding increase in membrane permeability, and consequently an increase in the release of cytochrome c and pro-caspases into the cytoplasm. This depolarization event can be measured by flow cytometry using specific cationic markers such as JC-1 and DiOC6. JC-1 is a lipophilic dye which selectively enters the mitochondria and changes its fluorescence from red to green as the membrane potential decreases during apoptosis progression. DiOC6 is also a lipophilic fluorescent stain which labels membranes and other hydrophobic structures. It is a green fluorescent dye which easily diffuses into cells and throughout the cytoplasm, where it becomes incorporated into membranes or binds to protein molecules, at which point its natural fluorescence increases. Dye binding, and therefore fluorescence, is enhanced when mitochondrial membranes are intact, and fluorescence intensity falls when the mitochondrial membrane depolarizes such as during apoptosis.

16.5.2 Late apoptosis: DNA fragmentation

During apoptosis, nucleases are activated which degrade the higher-order **chromatin** structures of DNA to produce fragments initially between 50 and 300kb in size and then further cleaved to fragments of ~200bp in length. These fragments can be visualized as 'DNA ladders' by agarose gel electrophoresis using ethidium bromide or other appropriate staining, although this is not a standard technology for all cell types as some exhibit poor results. DNA fragmentation can also be monitored by flow cytometry using DNA intercalating dyes such as propidium iodide. Apoptotic cells appear as a 'Sub-G1 peak'. This is a small DNA peak visualized to the left of the normal DNA G0/G1 peak for the cell sample when fluorescence intensity is plotted against DNA content (Figure 16.8(c)). Again, not all cell types reveal this population of apoptotic cells. In addition, necrotic cells can also appear as a sub-diploid peak, and hence this may limit the accuracy of the technique.

The formation of large numbers of DNA fragments leads to the presence of large numbers of exposed DNA 3′-OH (hydroxyl) ends. It is this feature which has enabled the development of DNA end-labelling assays, or ***in situ*** **e**nd **l**abelling (**ISEL**), which can be adapted for measurement by fluorescence microscopy or, more frequently, by flow cytometry (Gorczyca et al. 1992, 1993).

The **TUNEL** assay (**T**dT-mediated **dU**TP **N**ick **E**nd **L**abelling) (Li et al. 1995) is one form of an end-labelling assay which is used to measure DNA fragmentation and hence apoptosis. This method utilizes the enzyme terminal deoxynucleotidyl transferase (TdT) to catalyse the template-independent addition of the nucleotide deoxyruridine triphosphate (dUTP) to the 3′-OH ends of double- or single-stranded DNA, irrespective of whether the ends are recessed, overhanging, or blunt. Flow-cytometric assays use fluorescently labelled dUTP and, together with a DNA dye, are able to distinguish non-apoptotic cells (dUTP-negative) from apoptotic cells (dUTP-positive) which have a high fluorescence intensity. The two populations can easily be distinguished, and a statistical region or gate can be drawn to quantify them (Figure 16.8(d)).

16.5.3 What is viability and how is it commonly measured?

A viable cell will show cell membrane integrity, and only permit normal transportation of molecules across the lipid bilayer. In contrast, non-viable cells loose this integrity, and hence dyes normally associated with DNA binding used for cell cycle analysis can freely penetrate the cell and nuclear membranes and accumulate. Therefore DNA dyes can be used to quantify the number of viable cells (dye-negative) and non-viable cells (dye-positive). Cells which are just starting to die will have an intermediate level of dye bound to their DNA, and consequently an intermediate level of fluorescence staining, as they still have some metabolic activity and are able to eliminate some dye using drug-resistant mechanisms located in the cell membrane. The most commonly used DNA dyes for measuring cell viability are those excited by a blue laser (488nm wavelength) such as propidium iodide, and 7-Actinomycin D (7-AAD). Cell viability data are plotted as a histogram: fluorescence intensity on a log scale versus cell numbers (see Figure 16.9). The viable cells can be positioned in the first log decade, with dying cells seen as a population to the right as they increase their DNA dye content and become slightly

Figure 16.8 Examples of data plots obtained using four different methods for the analysis of apoptosis by flow cytometry. All samples are Jurkat cells (a T cell leukaemia cell line).

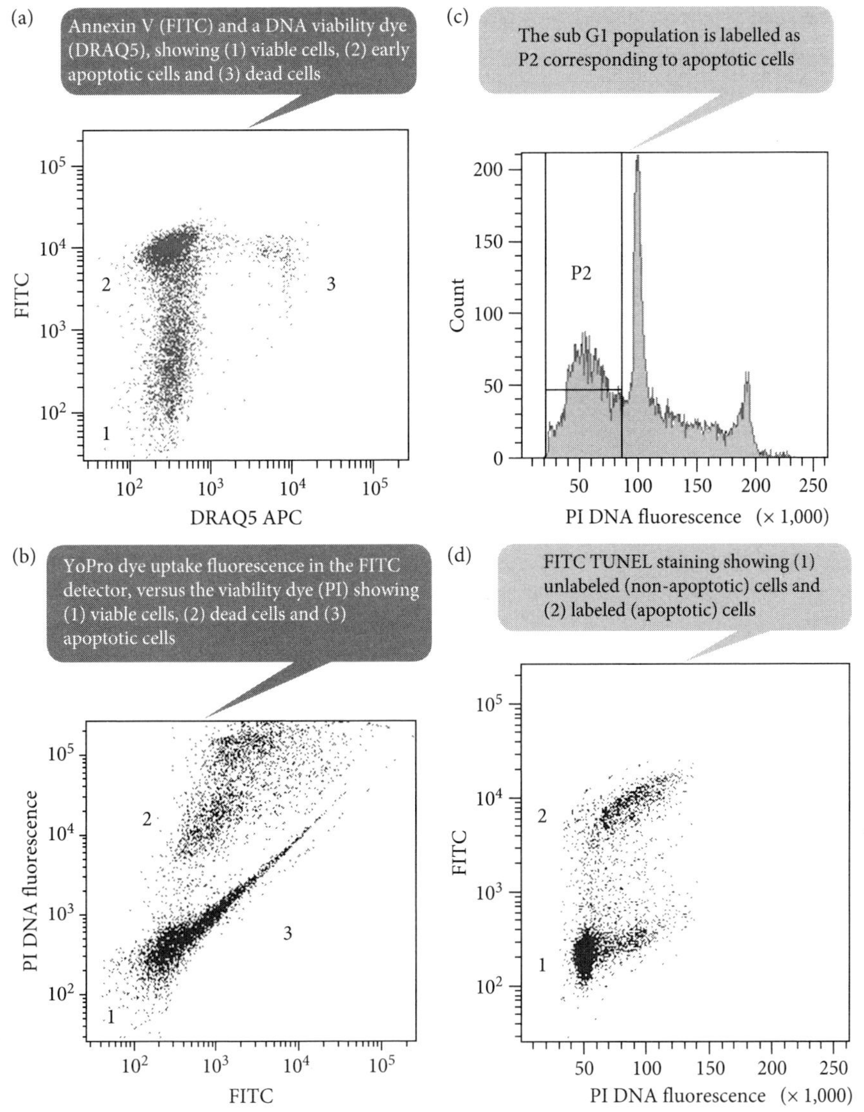

(a)
Annexin V (FITC) and a DNA viability dye (DRAQ5), showing (1) viable cells, (2) early apoptotic cells and (3) dead cells
FITC
DRAQ5 APC
(b)
YoPro dye uptake fluorescence in the FITC detector, versus the viability dye (PI) showing (1) viable cells, (2) dead cells and (3) apoptotic cells
PI DNA fluorescence
FITC
(c)
The sub G1 population is labelled as P2 corresponding to apoptotic cells
P2
Count
PI DNA fluorescence (× 1,000)
(d)
FITC TUNEL staining showing (1) unlabeled (non-apoptotic) cells and (2) labeled (apoptotic) cells
FITC
PI DNA fluorescence (× 1,000)

fluorescent, whilst dead cells are fully labelled with the DNA dye and are therefore even further along the intensity scale.

16.6 Cell sorting

Flow cytometry can be used to separate and collect subpopulations of cells based on particular biochemical and/or physical characteristics of the cells. The first commercial

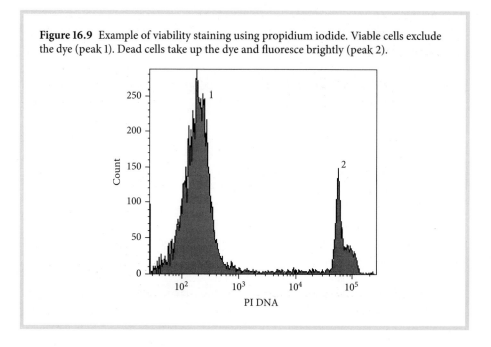

Figure 16.9 Example of viability staining using propidium iodide. Viable cells exclude the dye (peak 1). Dead cells take up the dye and fluoresce brightly (peak 2).

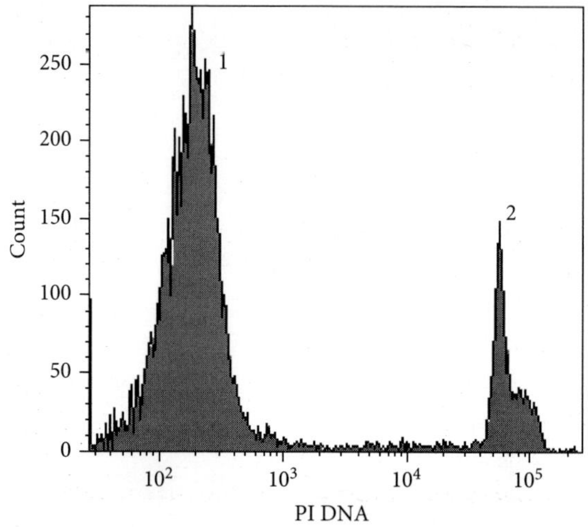

flow cytometers (c. 1970) were all cell sorters, and the flow-cytometric analyser (with no sorting capacity) did not emerge until later. Previously, single-cell selection was only possible using extremely time-consuming methods, including manually 'picking' cells with a pipette. Clearly, the numbers of sorted cells were very small. With the advent of flow cytometry and the cell sorter, samples could be screened at rates of several hundred per second, and highly purified populations (~99%) could be isolated and collected. Given time, and a patient operator, many millions of cells could be accumulated for further investigation and research.

16.6.1 A simple overview of the cell-sorting process

The cell sample is first prepared as a single-cell suspension in a buffer suitable for the cell type (usually saline or cell culture media), filtered through a sieve to remove cellular aggregates and debris, and the tube is then placed on the sample injection port of the flow cytometer. The cells travel through the sample line and are injected into the centre of the stream of sheath fluid (saline) in the nozzle assembly. This saline stream, with its central sample core, passes through a fine orifice contained within the nozzle assembly. As the saline stream emerges from the nozzle, a jet-in-air is produced which is directed downwards to a waste collection point. The whole nozzle assembly is integrated with a transducer. This comprises a piezoelectric quartz crystal which can be induced to vibrate with a frequency and amplitude controlled by software or by the operator. This forces the stream to break up into droplets typically at a distance of 1–2cm from the nozzle. The cell is contained within the droplet, and it is the droplet–cell unit that is sorted.

Nozzle orifices are available in a range of sizes (e.g. 50, 70, 85, 90 100, and 130μm diameter) and size selection is linked to the size of the cell. Small cells and particles can be run through any nozzle size, but with larger cells, a larger nozzle is selected so that the flow of the stream through the orifice is not obstructed. Typically, cells and particles up to the size of a lymphocyte (approximately 6μm) run very well through nozzles of up to 70μm. Monocytes and dendritic cells require a larger nozzle (e.>100μm), whilst large cells such as plant protoplasts, megakaryocytes, and marine algae require even larger nozzle sizes.

Saline, in which the cells are suspended, is an electrolyte and therefore will conduct electricity. If a charge is momentarily introduced into the stream, any droplets breaking off during the charge period will retain the charge. If such droplets free fall between charged deflection plates (one plate positive and the other negative) drops of opposite polarity will be attracted towards the appropriate plate, thus producing side streams on either side of the uncharged central stream of drops. The central stream is directed towards a waste collection tube, whilst the side streams are directed towards a collection device. The process is shown schematically in Figure 16.10.

When planning a cell-sorting experiment, it is first necessary to identify a distinguishing feature of the cell that will enable it to be selected and sorted from the cell mixture. Features may include, for example, being antibody positive, apoptotic, or in phase G0/G1 of the cell cycle. Flow-cytometric cell sorting is invaluable in stem cell research as complex panels of antibodies can be used together to stain peripheral blood, cord blood, or bone marrow samples and identify these early progenitor cells. They are usually very rare events (<0.1% frequency) and therefore essentially undetectable by microscopy. The

Figure 16.10 Schematic diagram of the components of a cell sorter showing the formation of a stream of droplets gives rise to left and right sort streams which fall into collection tubes, and the drop delay calculation which is taken from the distance between the interrogation laser and the stream/droplet break-off point.

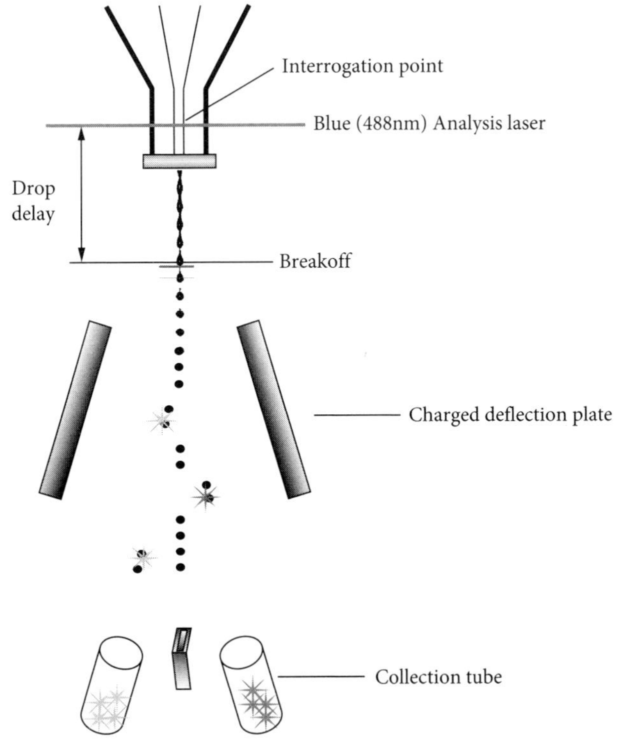

selected cellular property is identified and marked on a dot plot or histogram as a region or gate by the operator (e.g. antibody positive), and any cell falling in this region is then 'nominated' to enter into the sort process. Second, we need to identify in which direction the droplet enveloping our chosen cell is to be sorted (left-hand or right-hand side). Third, we make a calculation to define how long it will take for the cell to travel from the laser interrogation point to the end of the intact stream. At this moment the cell is enveloped by a saline droplet, and is just about to break from the stream. This is termed the break-off point and break-off distance. Fourthly, the sorting electronics is synchronized to create a charge which passes down the saline stream at the exact time that the droplet is about to break off. The droplet now carries a charge.

We now have our 'captured cell', contained within a charged droplet, free-falling down the sort chamber between the charged deflection plates where it is deflected towards the plate of opposite polarity and collected in the collection device.

Cell sorters have grown in complexity, ability, and speed over the last 40 years, and most are now able to run four-way sorting (two side streams to the left, and two to the right) or even six-way sorting. Cell sorters can now sort into 5, 15 or 50mL tubes, as well as cryovials, Eppendorf tubes, and serum tubes. Special devices enable cells to be sorted onto glass slides, Petri dishes, and a range of microtitre plates (6-, 12-, 24-, 48-, 96-, and 384-well plates). Cell sorters can be totally sterilized for aseptic sorting and tissue culturing, and some can be contained within a category 2 safety cabinet to allow high-risk cell sorting. Samples on the cell sorter can be automatically agitated and cooled, and collection devices can be cooled and even linked to 'hotel' systems for high-throughput sorting into plates which are inserted and removed by robot.

In addition, software is available to monitor the sort set-up and make corrections for drift during the sort without operator intervention. Drift occurs because environmental changes in temperature or pressure regulation of the sheath tank, or the level of saline in the tank, can all induce slight changes in the break-off distance. The same software will also terminate the sort and render both the original sample and the sorted samples safe in the event of a total blockage of the nozzle.

16.6.2 Tips for good cell-sorting results

No matter how sophisticated the cell sorter or its sort-monitoring software, the key factor in generating good sorts is sample preparation. Two key points in cell sample preparation are as follows.

- Cell samples need to be single-cell suspensions. Therefore most samples require filtration before being run on a flow cytometer. Sample material must be passed through cell strainers, filter devices, or nylon or stainless steel mesh to remove large cell aggregates, connective tissue, and debris.
- In addition, the buffer in which the cell sample is finally resuspended can be modified to maintain a single-cell suspension and also maintain viability. Additives such as serum, EDTA, and even DNase can be included to stop cells sticking to each other, and also to the sample tube, and to reduce the influence of dead-cell DNA on generating cell clumps. Furthermore, a buffer based on tissue culture HEPES can help maintain a constant pH during the sort process (which may last several hours) and thereby reduce cell death for those cells susceptible to pH change.

16.6.3 Alternative sorting devices

In 1995 BD produced the FACSort™ analyser with sort capabilities, which evolved into the sort module in the FACS Calibur™ and introduced a novel sorting device called the 'catcher tube'. Cell samples pass from the test tube through the flow cell and the interrogation lasers as normal, but rather than simply pass to the waste tank, the sample stream is interrupted by a small catcher, rather like a funnel attached to a sample line, so that cells of interest are 'scooped' out of the stream. The device is based on electromagnetic vibration, and so it is very limited in terms of movement speed. A maximum of 300 'catches' per second are possible, with typical cell sort rates nearer to 150 cells per second. However, only one cell (or particle) population can be isolated per sort, as the primary aim of this device is to isolate a small sample of a cell type for further study. Although not designed to handle millions of cells like the jet-in-air systems, these systems have been used with success in many laboratories.

CASE STUDY 16.2

Cell sorting

You may have a suspension of cells with several cell types present, but you are only interested in analysing one population. You cannot do this with a heterogeneous mixture, so you will need to run the sample through a flow cytometer cell sorter to extract the population you need and analyse it as a pure sample.

QUESTION 1
How do I identify the population of interest?

You may be able to run the cell suspension through the flow cytometer and visualize scatter differences which set apart the cells you are interested in. More likely though, you will need to identify a suitable surface marker (e.g. antigen, surface receptor complex) that can be targeted as a label for the cell population you require.

You then need to find a commercial source for this reagent, preferably in a fluorescently conjugated format to simplify staining procedures, and identify a staining procedure. You will also need a negative (unstained) control sample to allow you to distinguish true positive cells from autofluorescent cells.

QUESTION 2
What is the sorting work flow?

Count the cells in the sample to identify how much reagent will be required as well as the scale and duration of the sort. Do you have a million cells or 50 million cells? Do you require one cell sorted (e.g. for PCR or single-cell cloning) or a million cells for functional studies?

- Harvest your cells, sieve them to remove cell aggregates, and resuspend them to give a convenient cell concentration (e.g. 1 million cells per millilitre, or 10 million cells per millilitre, depending on the scale of the sort).
- Take a small aliquot of cells and pipette them into a 5mL tube for your control sample.
- Centrifuge the remainder of the cells, resuspend the pellet, and add the fluorescent reagent(s) required to identify the population of interest.
- Allow the reagents to stain by incubating the sample for 15–30min. Wash off excess reagent by centrifugation and resuspend the sample to give a good sort cell concentration (as above).
- Run the samples on the flow cytometer, use regions and gates to identify the population(s) of interest, and assign them to a sort stream (such as left or right).
- Calibrate the cell sorter ready for sorting (calculate the drop delay) and start the sort process into suitable collection devices.

16.7 Chapter summary

- Flow cytometry is a powerful technology for analysing multiple cell parameters such as light scattering properties and fluorescent dye binding in single cells in suspension.
- Flow cytometry can be used to immunophenotype cells. This involves labelling cells with antibodies directed against particular antigens conjugated to a fluorescent dye followed by flow-cytometric analysis. The cells are then classified by determining the protein expressed on the cell surface or intracellularly.
- Immunophenotyping is commonly used in the diagnosis of haematological cancers and to monitor progression and relapse.
- Flow cytometry can be used to quantify the percentage of cells in each phase of the cell cycle (G1, S, G2/M) and apoptotic cells in complex tumour samples which have both normal and abnormal karyotypes.
- Measuring cell viability and apoptosis is a common application of flow cytometry. A number of different cell viability and apoptosis parameters can be assessed, including plasma membrane asymmetry, mitochondrial membrane integrity, DNA fragmentation, and the expression of proteins, associated with apoptotic pathways, such as the caspases.
- Cell sorting is a unique feature of flow cytometers that enables cells with particular characteristics to be selected and sorted from a mixture of cells. Distinguishing characteristics include, for example, expression of proteins specific to a particular cell type.

References

Alamo, A.L. and Melnick, S.J. (2000) Clinical application of four and five-color flow cytometry lymphocyte subset immunophenotyping. *Cytometry* **42**: 363–70.

Al-Mawali, A.A., Gillis, D., and Lewis, I. (2009) The role of multiparameter flow cytometry for detection of minimal residual disease in acute myeloid leukaemia. *Am J Clin Pathol*, **131**: 16–26.

Sullivan, J.G. and Wiggers, T.B. (2000) Immunophenotyping leukemias: the new force in hematology. *Clin Lab Sci* **13**: 117–22.

Bagwell, C.B. and Adams, E.G. (1993) Fluorescence spectral overlap compensation for any number of flow cytometer parameters. *Ann NY Acad Sci* **667**: 167–84.

Chase, E.S. and Hoffman, R.A. (1998) Resolution of dimly fluorescent particles: A practical measure of fluorescence sensitivity. *Cytometry* **33**: 267–79.

Coons, A.H. and Kaplan, M.H. (1950) Localisation of antigen in tissue cells (II). Improvements in a method for the detection of antigen by means of fluorescent antibody. *J Exp Med* **91**: 1–13.

Coulter, W.A. (1956) High speed automated blood cell counter and cell size analyzer. *Proc Natl Electronics Conf* **12**: 1034–42.

Darzynkiewicz, Z., Bruno, S., Del Bino, G., et al. (1992) Features of apoptotic cells measured by flow cytometry. *Cytometry* **13**: 795–808.

de Tute, R.M. (2011) Flow cytometry and its use in the diagnosis and management of mature lymphoid malignancies. *Histopathology* **58**: 90–105.

Dillman, R.O. (2008) Immunophenotyping of chronic lymphoid leukaemias. *J Clin Oncol* **26**: 1193–4.

Friedman, H.P. (1950) The use of ultraviolet light and fluorescent dyes in the detection of uterine cancer by vaginal smear. *Am J Obstet Gynecol* **59**: 852–9.

Fulda, S and Debatin, K.M. (2006) Extrinsic versus intrinsic apoptosis pathways in anticancer chemotherapy. *Oncogene* **25**: 4798–811.

Fulwyler, M.J. (1965) Electronic separation of biological cells by volume. *Science* **150**: 910–11.

Kamentsky, L.A., Melamed, M.R., and Derman, H. (1965) Spectrophotometer: new instrument for ultra-rapid cell analysis. *Science* **150**: 630–1.

Gorczyca, W., Bruno, S., Darzynkiewicz, R.J., Gong, J., and Darzynkiewicz, Z. (1992) DNA strand breaks occurring during apoptosis: their early in-situ detection by the terminal deoxynucleotidyl transferase and nick translation assays and prevention by serine protease inhibitors. *Int. J. Oncol* **1**: 639–48.

Gorczyca, W., Gong, J., and Darzynkiewicz, Z. (1993) Detection of DNA strand breaks in individual apoptotic cells, by the in-situ terminal deoxynucleotidyl transferase and nicktranslation assay. *Cancer Res* **52**: 1945–1951

Kerr, J.F.R., Wyllie, A.H., and Currie, A.R. (1972) Apoptosis: a basic biological phenomenon with wide ranging implications in tissue kinetics. *Br J Cancer* **26**: 239–57.

Li, X., Traganos, F., Melamed, M.R., and Darzynkiewicz, Z. (1995) Single-step procedure for labelling DNA strand breaks with fluorescein or BODIPY conjugated deoxynucleotides: detection of apoptosis and bromodeoxyuridine incorporation. *Cytometry* **20**: 172–80.

Maecker, H.T., McCoy, J.P., and Nussenblatt, N. (2012) Standardising immunophenotyping for the Human Immunology Project. *Nat Rev Immunol* **12**: 199–200.

Mahnke,Y.D. and Roederer, M. (2007) Optimizing a multicolor immunophenotyping assay. *Clin Lab Med*: **27**: 469–85.

McLaughlin, BE., Baumgarth, N., Bigos, M., et al. (2008a)A nine-color flow cytometry for accurate measurement of T cell subsets and cytokine responses. Part I: Panel design by an empiric approach. *Cytometry A*, **73**: 400–10.

McLaughlin, B.E., Baumgarth, N., Bigos, M., et al. (2008b) A nine-color flow cytometry for accurate measurement of T cell subsets and cytokine responses. Part II: Panel performance across different instrument platforms. *Cytometry A*, **73**: 411–20.

Pattanapanyasat, K. (2012) Immune status monitoring of HIV/AIDS patients in resource-limited settings: a review with an emphasis on CD4+ T-lymphocyte determination. *Asian Pac J Allergy Immunol* **30**: 11–25.

Perfetto, S.P., Chattopadhyay, P.K., and Roederer, M. (2004) Seventeen colour flow cytometry: unraveling the immune system. *Nat Rev Immunol*, **4**: 648–55.

Wyllie, A.H., Kerr, J.F.R., and Currie, A.R. (1980) Cell death: the significance of apoptosis. *Int Rev Cytol* **68**: 251–306.

17 Bioimaging: light and electron microscopy

Gareth Howell and Kyle Dent

Chapter overview

The term 'bioimaging' in this chapter refers to the use of light and electron microscopy techniques to image biological structures, from whole organisms through to single cells and even single molecules within cells. These techniques enable us to visualize a number of biological processes such as protein transport, development or the effect of disease, and mutations on a cellular and subcellular scale. In addition, it can provide us with structural information on cells, organelles, and individual macromolecular complexes. This chapter will focus on high-resolution fluorescent and electron microscopy covering some of the important technologies commonly utilized in the study of cells and protein structures such as confocal and deconvolution microscopy, as well as transmission and scanning electron microscopy. We will describe the basic structure of these microscopes and how images are generated and visualized. We will also discuss the common applications of these technologies in modern biological research.

LEARNING OBJECTIVES

This chapter will enable the reader to:

- understand the process of **fluorescence** and describe how this property can be utilized in imaging structures within cells;
- describe the basic structure of a fluorescence microscope and explain how an image is generated;
- describe the limitations of fluorescence microscopy, the improved technologies that have been developed, and their applications;
- describe how transmission electron microscopes and scanning electron microscopes work and the types of information that can be obtained using these instruments;
- describe the potential of new techniques such as **correlative light electron microscopy** (CLEM) in imaging experiments.

17.1 Fluorescence microscopy

17.1.1 Introduction to fluorescent molecules

Fluorescent molecules or **fluorophores** can absorb high-energy radiation and emit this radiation in the form of light at a lower energy. When high-energy radiation is absorbed, the fluorescent molecule or atom moves from a low energy level to a high energy level. When it returns to a resting ground state, some of this absorbed energy is released in the form of light or low-energy **photons**. This process is called fluorescence and it is these low energy photons that are observed in fluorescence assays. Each fluorophore has a specific **excitation maximum** (also called an absorption maximum), the optimal wavelength required to excite the molecule, and a specific **emission maximum**, the optimal wavelength at which photons are released. For example, fluorescein (FITC), a common fluorophore used in fluorescence microscopy, absorbs light energy at a wavelength maximum of 494nm (blue light) and emits longer-wavelength light at a maximum of ~520nm (green light). The difference along the spectrum between the excitation wavelength and the emission wavelength is termed the **Stokes shift**. The Stokes shift of fluorophores conventionally used in fluorescence microscopy is often between 20 and 40nm. However, this can vary significantly depending on the fluorophore or **fluorescent protein** (e.g. GFP) one is visualizing. A list of fluorophores commonly used in fluorescence microscopy, including fluorescent proteins, is given in Table 17.1. It can be seen that a wide range of 'colours' are available and that many of them can be separated along the spectrum, enabling us to label multiple structures and proteins within cells simultaneously. An example of a multicolour fluorescence image is shown in Chapter 18. We can commonly label with up to four colours in the same sample, and with some spectral-based imaging systems there is the potential to image with between eight and ten colours (see section 17.1.4).

See also Plate 14 (p.8)

Molecules only remain in their excited high-energy state transiently (time-scale of nanoseconds) before returning to their resting ground state. This relaxation of a fluorescence molecule to its ground state can be measured using techniques such as fluorescence lifetime imaging microscopy (FLIM). The lifetime of a fluorescent molecule varies greatly and depends on a number of factors including the fluorophore type and environmental factors such as temperature, pH, and viscosity. Also, of importance to fluorescently labelled specimens, this excitation and relaxation event will only occur a defined number of times before **photobleaching** occurs and the molecule will no longer fluoresce. Photobleaching can be utilized to study protein dynamics in living cells in techniques such as fluorescence recovery after photobleaching (FRAP) and fluorescence loss in photobleaching (FLIP) (Lippincott-Schwartz et al. 2001).

17.1.2 How does fluorescence microscopy work?

A fluorescence light microscope enables detection and localization of fluorescence within cells where antibodies conjugated to fluorescent markers have been utilized to detect specific proteins of interest, or reporter genes such as GFP have been utilized to highlight the expression of genes of interest.

Table 17.1 Fluorophores and fluorescent proteins commonly used in fluorescent light microscopy

Fluorophore	Ex. max. (nm)	Em. max. (nm)	Applications	Colour emitted
Hoechst 33342	352	455	Used to stain nuclear DNA in live cells	Blue
DAPI	359	461	Used to stain DNA in fixed cells	Blue
FITC (fluorescein)	494	520	Antibody conjugate for localization by microscopy	Green
EGFP (enhanced green fluorescent protein)	498	516	Used as reporter gene to localize proteins of interest inside live cells	Green
Rhodamine	550	573	Antibody conjugate for localization by microscopy	Red
DsRed (*Discosoma* red fluorescent protein)	556	583	Used as a reporter gene to localize proteins of interest inside live cells	Red
Cy3	554	566	Antibody conjugate for localization by microscopy	Red
Cy5	649	666	Antibody conjugate for localization by microscopy	Infra-red

Ex. max., excitation maximum; Em. max., emission maximum.

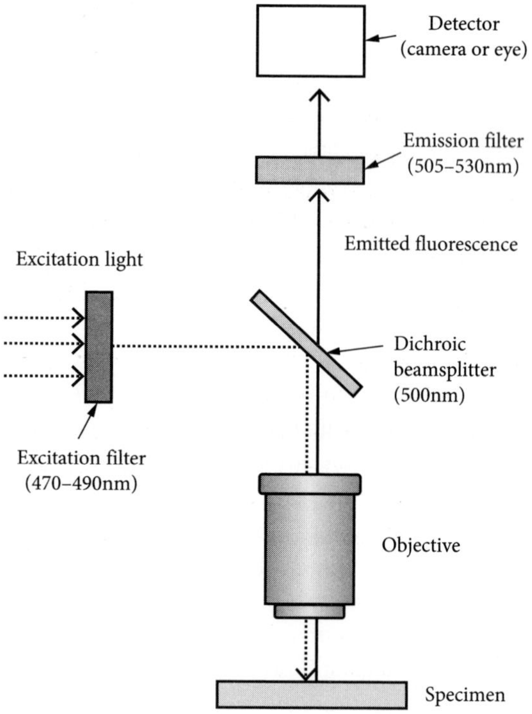

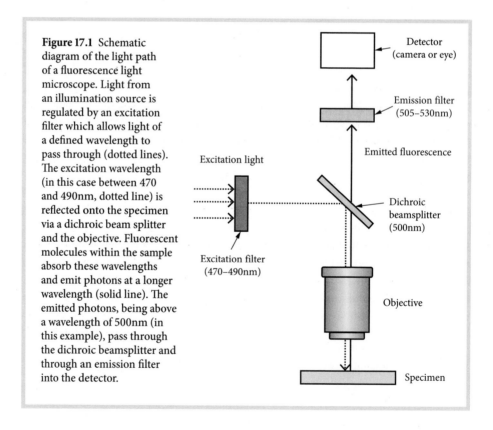

Figure 17.1 Schematic diagram of the light path of a fluorescence light microscope. Light from an illumination source is regulated by an excitation filter which allows light of a defined wavelength to pass through (dotted lines). The excitation wavelength (in this case between 470 and 490nm, dotted line) is reflected onto the specimen via a dichroic beam splitter and the objective. Fluorescent molecules within the sample absorb these wavelengths and emit photons at a longer wavelength (solid line). The emitted photons, being above a wavelength of 500nm (in this example), pass through the dichroic beamsplitter and through an emission filter into the detector.

A fluorescence light microscope comprises some common basic components: a light source, excitation filters, dichroics and mirrors, an emission filter, a series of lenses, and a detection device (Figure 17.1). These components collectively modulate how specific wavelengths of light pass through to the sample and how photons emitted by the sample

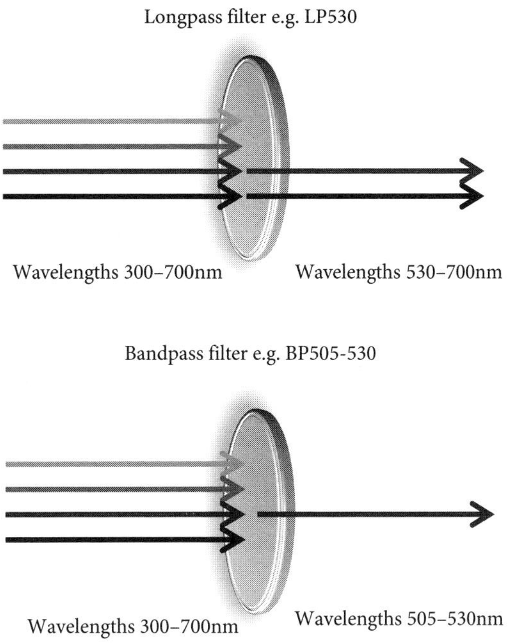

Figure 17.2 Schematic diagram of light transmission through a longpass and a bandpass filter, two of the most common types of filter used to modulate light wavelengths through a fluorescent microscope. The shaded lines represent light of different wavelengths as indicated by the text. In the longpass filter example all light above a wavelength of 530nm is able to transmit, whilst light below 530nm is blocked. The bandpass filter is designed to allow a limited wavelength of light to transmit; in the example shown only light between 505 and 530nm can be transmitted.

are collected for detection. The initial component in a fluorescent light microscope is the illumination source. Common types include mercury and xenon arcs, metal halides, LEDs (light emitting diodes), and laser sources. The wavelengths of the photons that enter the optical path from the illumination source are regulated using excitation filters. These are bandpass filters which allow light of a defined wavelength range to pass through (Figure 17.2). Light that passes through is then reflected onto the sample via a dichroic beam splitter. The latter is a coated glass element specifically designed to reflect selective wavelengths of light and allow the transmission of longer wavelengths, such as that emitted from the fluorescently labelled sample. In many fluorescent light microscopes both excited and emitted light travel through the objective lens, a method referred to as epi-fluorescence. In epi-fluorescence light reflected from the specimen is collected rather than transmitted through the specimen, which enables the collection of low-emission fluorescent signal without the presence of high-intensity excitation light. Emitted light then passes back through the dichroic beam splitter to the detector. However, prior to detection, it is filtered through an emission filter. Emission filters are often one of two types: bandpass or longpass (Figure 17.2). Bandpass filters, as with the emission filter, allow light of a defined wavelength range to pass through and therefore prevent background signal (noise) from being detected. Longpass filters allow all the light above a specific wavelength to pass through and are useful for collecting emission from fluorophores that have a wide emission profile or where the signal is limited and therefore as much light as possible needs to be collected to generate an image.

The production of sufficient photons from the sample will result in a signal that can be detected and recorded as an image. Signal detection (image capture) in light microscopy is performed using one of two technologies—camera or photomultiplier tubes (PMTs). Both these technologies are designed to detect the relatively low numbers of photons typically emitted by samples in fluorescence microscopy. Highly sensitive digital cameras, termed charged-coupled devices or CCD cameras, are commonly used. These cameras enable low signal collection which is achieved by cooling the collection chip to temperatures below –30°C. Recent advances in CCD technology include the development of the electron-multiplying charge-coupled device (EMCCD). This device allows the signal to be increased without increasing background noise, and hence enables much higher sensitivity to be achieved, resulting in a clearer image of the sample.

A PMT amplifies the detected photons emitted from the sample and converts them into electrons. This amplification process can be controlled by regulating the voltage on the PMT, so that as the voltage is increased the resultant image becomes brighter. However, increasing the voltage can also cause undesirable background noise and lead to reduced image contrast (quality). In many microscopy systems this is overcome by imaging the sample a number of times and utilizing software to generate an 'average' image. The averaging of a series of images results in the reduction of background noise because of its random location, and improves the signal which is consistent in location.

17.1.3 Improving the resolution of a light microscope

Resolution in light microscopy refers to the ability to differentiate two closely positioned structures on a specimen as separate entities. Images can be resolved in the 2D plane (the lateral axis, also called the *xy* plane) or the 3D plane (axial, also called the *z* plane). The resolution of an image is directly limited by the emission wavelength of the fluorophore being imaged and the numerical aperture (the light-collecting ability) of the objective.

Lateral (xy plane) resolution in light microscopy

In light microscopy we are limited to separating structures that are no closer together than 200–300nm. The resolving power of an imaging system in the *xy* plane (lateral) can be calculated using Abbe's law:

$$R = 0.6\lambda/\mathrm{NA} \quad (17.1)$$

where R is the resolving power in nanometres, λ is the emission wavelength, and NA is the numerical aperture of the objective. The use of a shorter excitation wavelength (in the UV region) will improve image resolution, but this often results in extensive sample damage (especially to live cells) and poor image quality, and is expensive. Therefore wavelengths above 400nm are generally used.

Certain characteristics of the lens can also cause loss of resolution in images, in particular the depth of field (also referred to as depth of focus). Depth of field relates to the 3D range (axial plane) over which a lens can produce a sharp in-focus image. Beyond this point an out-of-focus signal is collected, which appears as blurriness in a 2D image and can reduce the lateral (*xy* axis) resolution of the structures being imaged. As a result techniques such as confocal microscopy and wide-field deconvolution microscopy (see section 17.1.4) have been developed which enable researchers to view only the signal at the focal plane and reduce the contribution to the final image of an out-of-focus signal.

Axial (z-plane) resolution in light microscopy

Axial resolution or the degree of detail we can see in neighbouring focal planes is limited by the following equation:

$$R = 2\lambda\eta/\mathrm{NA}^2 \qquad 17.2$$

where λ is the emission wavelength of the fluorophore, η is the refractive index of the sample, and NA is the numerical aperture of the objective. Using this equation, imaging the fluorophore FITC in live cells with a refractive index of 1.33 through a 63x/1.4NA objective will give an axial resolution of ~700nm. Confocal and deconvolution microscopy techniques improve this resolution to ~350nm, but this is still insufficient if we want to study structures such as endosomes or clusters of secretory vesicles which can have diameters <100nm.

Advanced imaging techniques such as structured illumination microscopy (SIM) and stimulated emission depletion microscopy (STED) overcome these limitations, but they are very new technologies and are currently expensive to purchase (Huang et al. 2010).

An imaging technology that enables improved axial resolution in living cells is **total internal reflection fluorescence (TIRF) microscopy** (Millis 2012). A laser is used to illuminate the sample at an angle, instead of directly, so that light is reflected from the bottom of the sample. This causes a shallow 'wave' of excitation which only excites molecules within ~100nm of the bottom surface of the sample. Utilized predominantly in the study of events at the cell surface, TIRF microscopy ensures that only molecules within this 100nm region are excited, thus improving the axial resolution by a factor of almost 10. However, TIRF microscopy does not improve the lateral resolution of a standard imaging system. TIRF can also be used on sectioned tissue to improve the axial resolution of structures of interest and has been utilized in super-resolution imaging techniques such as photo-activatable localization microscopy (PALM) (Betzig et al. 2006).

17.1.4 Confocal microscopy

Confocal microscopes allow imaging of a number of different fluorophores in both two and three dimensions with a high degree of image resolution. They can also be used to perform live cell imaging at high speed (Pawley 2006; Wilson 2011).

Confocal microscopes work by blocking out-of-focus signals from passing to the detector. Emitted light from the sample is passed though a small hole, or pinhole aperture, in the light path before it is collected by the system's detectors (Figure 17.3). The pinhole aperture is located between the detector and the sample and is in a common focal position to the objective lens—hence the term 'confocal' ('common focus'). As the aperture has a small diameter, any photons emanating from beyond the focal plane are blocked and hence only the signal from the focal plane is imaged, resulting in improved image resolution.

Detection of signal through a confocal microscope requires very sensitive detection methods utilizing either cooled CCD digital cameras or PMTs. This is because, although the entire volume of the cell is illuminated with a powerful excitation source (often a laser), only a relatively small fraction of the total signal that is produced will pass through the pinhole aperture to the detectors. For example, if a 15μm thick HeLa cell was being imaged by a confocal microscope and the pinhole aperture was set to enable

Figure 17.3 Schematic representation of a confocal microscope. Object A is in focus through the objective, and as a result light emitted from this structure (solid black line) passes through the pinhole and is collected by the detector (PMT). However, object B is below the plane of focus and therefore is out of focus from the pinhole. Therefore the light from object B (dotted line) is blocked, resulting in the imaging of object A alone. *Z* represents the axial plane.

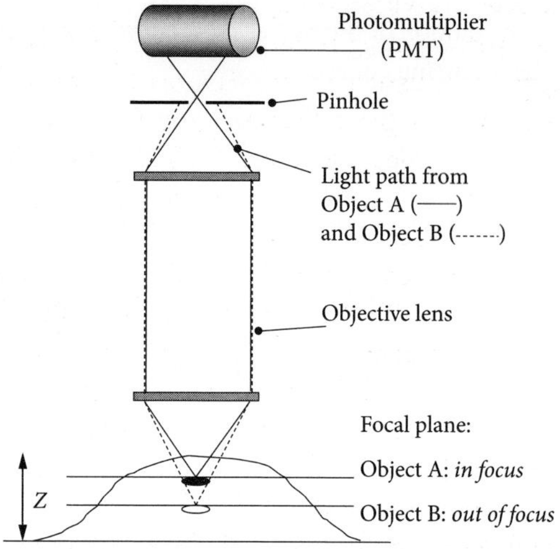

the collection of signal from 1µm thick optical sections, based on the diameter of the pinhole and the emission characteristics of the fluorophore, we would essentially collect only ~6% of the total signal generated by the cell.

3D image datasets of the specimen can also be generated using confocal microscopes. With reference to the 15µm thick HeLa cell discussed above, we could image a series of 1µm thick optical sections through the entirety of the cell and study the spatial relationships between the proteins in 3D. The number of optical sections we would need to image, or sample, is determined by the Nyquist theorem, which states that optimal sampling is achieved by using units two times smaller than the smallest feature you expect to see in your specimen. The reason for this is to prevent 'oversampling' or the collection of data that will not provide any additional information. Based on this theorem, we would need to image every 0.5µm through our 15µm HeLa cell, resulting in 30 confocal images.

An increasing number of 'spectral' confocal microscopes are coming onto the market which do not utilize the 'classic' emission filter to control signal onto a single PMT but use a series of PMT detectors assembled in an array which collects light along the entire light spectrum. These systems are more flexible than traditional systems with fixed emission filters, as the region of the spectrum required to obtain the optimal signal from a fluorophore or fluorescent protein can be customized to within 10–20nm using a slider within the software of the system. Whilst conventional 'non-spectral' imaging systems could potentially utilize custom-prepared emission filters with very narrow emission ranges, a wide range of filters would be required and hence the system would lack the convenience of a spectral system.

Laser scanning confocal microscopy

Laser scanning confocal microscopy is one of the most commonly used high-resolution imaging platforms in modern biological research as it enables 2D and 3D imaging of samples ranging from single cells to tissue sections, and even whole organisms. Modern systems incorporate a number of features which enable multicolour imaging of fixed and live cells, time lapse, and dynamic studies such as protein diffusion and protein–protein interactions (Wilson 2011).

In laser scanning confocal microscopy a laser point with a diameter less than the resolving power of the system is focused on the sample and moved linearly across it. This has the effect of generating a signal that is theoretically smaller than the resolving power of the microscope and thus at the limits of resolution, with a resultant improvement of resolution by a factor of ~1.4. The pinholes are adjustable, allowing users to generate the best possible images of their samples. For example, in cases where the signal is low, a large pinhole setting can be used to increase signal collection but this will compromise the resolution. With a bright sample, a smaller pinhole setting can be used, achieving maximal resolution from the system.

One drawback of laser scanning confocal microscopy is the effect of photobleaching (see section 17.1.1) which occurs as the result of exposure of the fluorophore to a high-power light source (laser). This will commonly occur if the sample being imaged has a low fluorescence signal and it is necessary to expose the fluorophore to higher amounts of laser energy to generate sufficient signal beyond the background noise.

Spinning disc confocal microscopy

Spinning disc confocal microscopes are utilized in imaging samples where both high lateral resolution and high speed are required (Nakano 2002), such as imaging the movements of intracellular structures in live cells in 3D or over short time-scales (often of the order of tens of images a second) such as calcium sparks in the response to an agonist, or the movement of vesicles between membrane-bound compartments such as endosomes or the Golgi apparatus. As with laser scanning confocal microscopes, lasers are used as excitation sources but, instead of scanning individual points within the sample, they illuminate the entire field of view. This has the effect of instantly producing a full image at the same resolution as a laser scanning confocal. Illumination of the entire field of view is achieved using two sets of pinholes placed on a disc in a concentric array and spun at high speed. The disc, originally developed by the German physicist Paul Nipkow (Nipkow disc), has a micro-lens in each pinhole which simultaneously focuses the laser beam onto the sample in an array of excitation points, in effect generating thousands of excitation points at once. Emitted light is then collected through a second pinhole to remove any out-of-focus emission, separated by a dichroic mirror, and collected onto a CCD camera. 3D confocal images of thick samples (up to 100μm) can be achieved quickly and accurately using spinning disc confocal microscopes.

One drawback of spinning discs is that the pinholes are fixed and therefore cannot be adjusted if samples exhibit low fluorescence signal, as can be done with a laser scanning system. This means that samples need to emit a good fluorescence signal to be effectively imaged by this technique.

Multiphoton microscopy

Multiphoton microscopes (Ustione and Piston 2011) are classed as confocals as they enable the observation of fluorescence signal that is only present at the focal plane. However, multiphoton microscopes do not have pinholes within the optical pathway as conventional confocal systems do. Instead, they utilize excitation sources that emit in the infra-red range and excite fluorophores through two- or three-photon excitation, an event which only occurs at the focal plane because of the focusing effect of the objective lens. Multiphoton microscopy is often utilized in areas of research where greater depth of imaging is required, as greater penetration of tissue is possible with infra-red photons. Conventional confocal imaging is limited to a depth of approximately 100μm into tissue; beyond this, light scatter is too great to resolve any detail. With multiphoton microscopy, the imaging depth can be extended to 500μm, and near to 1000μm in some systems, and makes imaging fluorescent dyes and proteins in thick tissue or whole animals more feasible.

17.1.5 Deconvolution microscopy

An alternative method of improving the resolution of an image is to process the raw image using mathematical algorithms, a process termed **deconvolution** (Wallace et al. 2001; Swedlow and Platani 2002; Biggs 2010).

In light microscopy, light scatter and out-of-focus blur are introduced because of factors such as refractive index changes and the passage of photons through optical components (e.g. objectives and filters). To overcome this we can simulate the optical path and generate a mathematical model of the 'convolution' that can occur, i.e. the aberrations that are introduced to a photon as it passes through the optical path of the microscope. This mathematical model can then be utilized to improve the quality and resolution of the images collected from the microscope (Figure 17.4). The simulated model of the system is referred to as the **point spread function** (PSF) and describes the 'shape' of the light that is emitted from the sample using fluorescent beads and particles that are smaller than the resolving potential of the microscope. Deconvolution is commonly performed on a fluorescence light microscope, where the images collected contain both in-focus and out-of-focus signals. An important addition to this type of system is its ability to collect image stacks through the sample of interest. This often involves the use of a motorized focus drive that allows accurate stepping through the sample whilst image collection occurs. The image stacks are then collected and analysed to determine which model can be used for subsequent image analysis.

Deconvolution algorithms fall into two classes: deblurring and image restoration. Deblurring algorithms, also referred to as nearest-neighbour or no-neighbour algorithms, are used predominantly on small-volume images (e.g. five to ten images in a 3D dataset) or single 2D images. Deblurring algorithms introduce a blurring effect on the image which is subsequently removed from the original data. This type of deconvolution is quick and does not require large amounts of computer processing capacity. However, it must be used with care as the processing *removes* data from the image and consequently may cause artefacts to occur in image intensity and morphological changes to structures within the picture.

Image restoration algorithms are significantly more reliable, and utilize the data set as a 3D volume. They identify out-of-focus signal (pixels) within the volume and restore these pixels to their 'origin' based on the information obtained from the calculated PSF. Algorithms such as linear least squares, Tikhonov–Miller regularization, or Wiener deconvolution are image restoration procedures. Restorative algorithms work on the basis that as the raw data is produced in a convolved manner, deconvolution must restore this data to a theoretical model based on the information obtained from the PSF.

Both restorative and deblurring deconvolution algorithms function iteratively, with round after round of blurring (convolving) and deblurring (deconvolving) until the difference

Figure 17.4 Schematic diagram of deconvolution processing. When viewed through a microscope, 3D samples contain both in-focus and out-of-focus signals. (a) When focused on the black boxed X at position 0 we can simultaneously detect out of focus structures above (+1) and below (–1) the focal position. The resultant 2D image will contain all this information, causing a loss in image resolution. (b) By collecting a series of images through the 3D volume of our sample we can ascertain where each structure is in focus (boxed X), and remove the out-of-focus structures by utilizing a deconvolution processing step. (c) An example of deconvolution in human epithelial cells displaying fluorescent mitochondria. The top panel is a raw image which contains both in-focus and out-of-focus structures. Following deconvolution, greater detail can be observed, and an improvement in the image quality is evident in the lower panel.

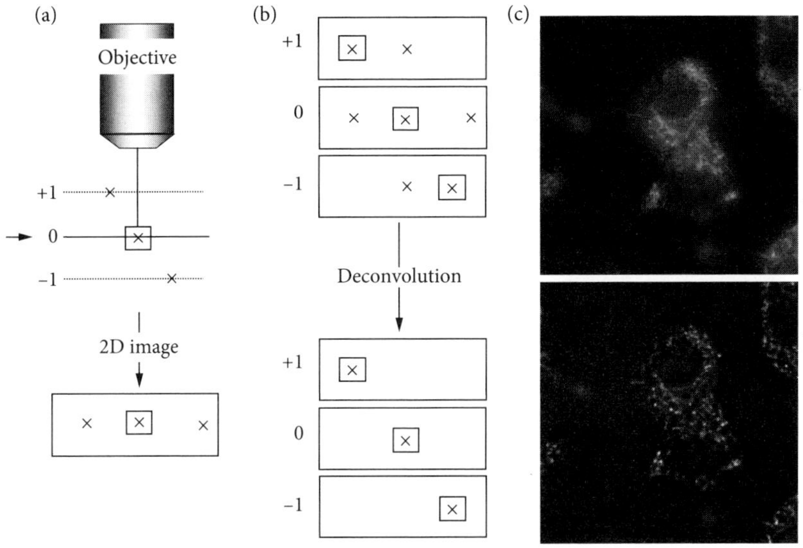

between the raw dataset and the processed dataset is minimized. In commercial software packages this can usually be controlled, but is often set to ten or 15 iterations. This method of repeated deconvolution steps is referred to as constrained iterative deconvolution.

17.1.6 Sample preparation for fluorescence microscopy

Optimal sample preparation is necessary for imaging cells by fluorescence light microscopy. An overview of sample preparation for both live and fixed samples is shown in Figure 17.5. Sample preparation for live cell imaging is relatively straightforward and involves seeding adherent cells onto glass-based culture dishes or similar suitable substrates. An experiment may require a transfection step where exogenous genes, fused to reporter genes such as GFP, are constructed, introduced, and expressed in the cells. Imaging is often performed 24–72 hours after transfection and may require buffered culture medium to maintain healthy cells.

See Chapter 18

Imaging specific antigens within fixed cells and tissue requires more sample preparation but can be achieved quickly with very reproducible results. A technique called immunofluorescence (IF) is used in which the antigen of interest is detected with an antigen-specific antibody followed by localization of the primary antibody with a secondary antibody conjugated to a fluorescent tag.

Figure 17.5 Flow diagram of sample preparation for fluorescence light microscopy of either live or fixed cells.

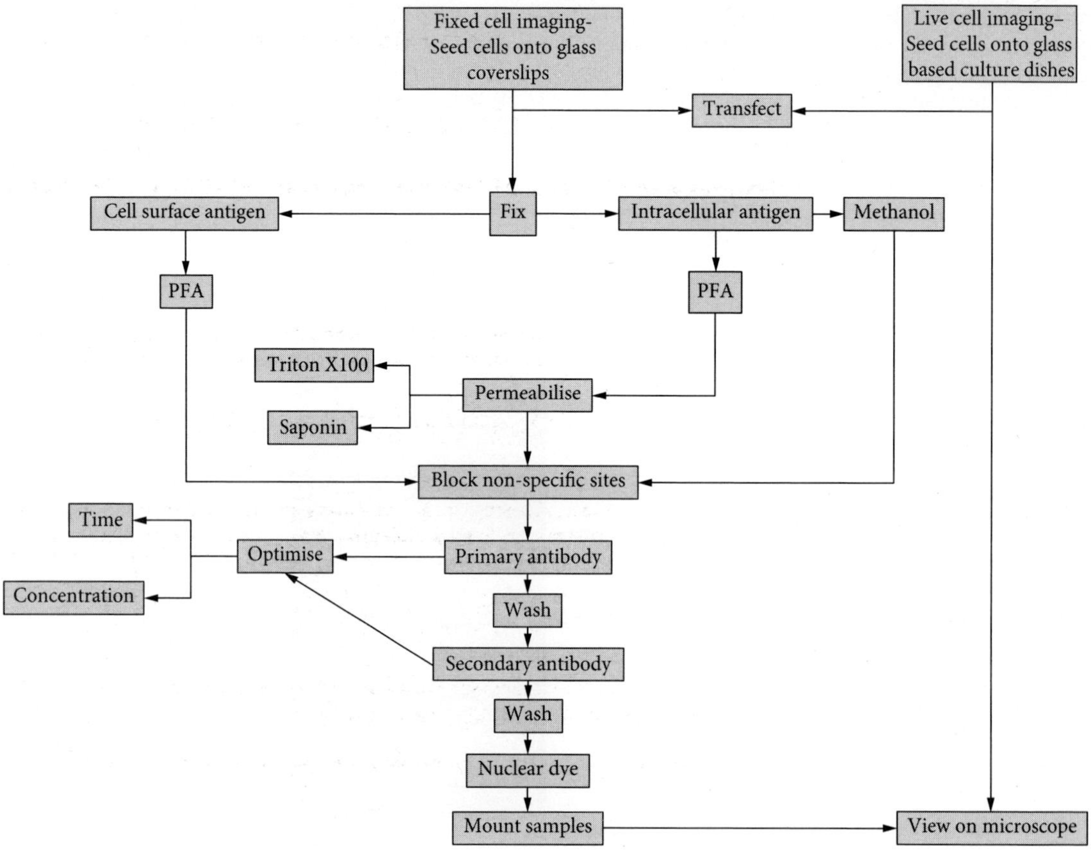

Cells are initially fixed chemically using cross-linkers, such as aldehydes (e.g. paraformaldehyde (PFA)) or alcohols (e.g. methanol). The fixation step can be followed by a cell permeabilization step if intracellular antigens are being localized as it will enable antibodies to access intracellular antigens. Following permeabilization, the sample is incubated with primary antibody targeting the antigen of interest. After a few washes to remove any excess unbound primary antibody, the sample is incubated with secondary fluorescently conjugated antibody directed against the primary antibody. This step is followed by the addition of a dye that will label nuclear DNA (e.g. DAPI). The sample is then mounted to preserve the fluorescent signal and finally visualized on the microscope.

CASE STUDY 17.1

Using fluorescence light microscopy to study the role of cell surface receptors in cell invasion

A researcher is investigating the trafficking of a cell surface receptor and its role in cell invasion. He wants to observe the dynamics of the receptor as it cycles between the cell surface and intracellular membrane bound compartments in live cells. He would also like to study the intracellular localization of this receptor relative to structures such as endosomes and lysosomes. Finally, he would like to study the behaviour and localization of this receptor in invasion assays where cells are plated onto collagen matrices.

? QUESTION
Which light microscopy technologies would be best suited to performing these studies?

The questions could potentially be addressed using two systems: laser scanning confocal microscopy and total internal reflection fluorescence (TIRF) microscopy.

Let us consider the live cell studies first. In these experiments the researcher wants to observe the receptor at the cell membrane. The axial (z) resolution of a light microscope is in the region of 700–1000nm. This level of resolution would not discriminate between events that occur at the cell membrane from events that occur just inside the cell. Based on the improved axial resolution offered by TIRF microscopy, this type of study would be achievable as only receptors in the plasma membrane would be excited by the evanescent wavefront which extends approximately 100nm into the cell.

The localization studies and cell invasion assays could be performed on a laser scanning confocal. By utilizing antibodies specific to proteins known to reside in organelles such as endosomes and lysosomes, and to the receptor, the researcher could perform multicolour imaging studies on fixed cells. Alternatively, the studies could be performed using an epi-fluorescent microscope and applying deconvolution algorithms to improve the resolution of the data. Deconvolution could also be used on the confocal data.

The migration (invasion) assays could also be performed on a laser scanning confocal microscope utilizing 3D imaging, as long as the cells are seeded on the surface of the collagen. If the cells were to migrate further than 50–100mm into the collagen, it may be necessary to perform the studies using multiphoton microscopy. Multiphoton microscopy would allow imaging at high resolution up to approximately 500μm into the sample and would enable the imaging of large groups of cells. It would also enable imaging of cells implanted in the collagen.

17.2 Transmission electron microscopy

17.2.1 What is transmission electron microscopy?

Transmission electron microscopes (TEMs) were originally developed to overcome the resolution limitations of the light microscope (summarized in Table 17.2). They enable detailed views to be obtained of cellular ultrastructure, the organization of membranous subcompartments, (such as cytoskeletal filaments, the nuclear envelope, the cell wall, the Golgi apparatus, mitochondria, and chloroplasts), and individual macromolecular complexes such as the ribosome at resolutions ranging from 0.3 to 10nm. Such high-resolution structural images can provide valuable information about the functions of macromolecular complexes within the cell.

A schematic representation of a TEM is shown in Figure 17.6. A beam of electrons from the electron source (or gun) situated at the top of an evacuated optical column is transmitted down the optical axis of the column. The electrons pass through the specimen and are focused and magnified by a series of electromagnetic lenses to produce an image which can be viewed on a viewing screen, or alternatively captured by a recording device (the detector) and then viewed on a computer screen.

The electron gun is composed of an electron source (the cathode) and an anode responsible for extracting and accelerating electrons. The source may be either field

Table 17.2 Fluorescence and electron microscopy technologies: applications and resolving power

Microscope technology	Applications	Resolving power (*xy* plane, *z* plane)
Epi-fluorescence microscope	Imaging fluorescently labelled live or fixed samples	200–300nm, 700nm
Line-scanning confocal microscope	High-resolution imaging of fluorescent samples in 3D Samples are often fixed but can also be live	150–200nm, 350nm
Spinning disc confocal microscope	High-speed high-resolution imaging of fluorescently labelled live or fixed samples in 3D	150–200nm, 350nm
TIRF Microscope	Imaging the surface of cells, e.g. trafficking of cell surface receptors or channels in living cells	200–300nm, 100nm
Multiphoton microscope	Imaging thick biological samples (>100 μm) (either live or fixed samples)	200–300nm, 700nm
Deconvolution processing of epi-fluorescent microscopy images	Imaging fluorescently labelled live or fixed samples	150–200nm, 350nm
TEM of chemically fixed plastic-embedded sections	Imaging of thin sections of cells prepared by microtomy/ ultramicrotomy	5–75nm, N/A
TEM of freeze-substituted and embedded sections	Imaging of thin sections of cells prepared by microtomy/ ultramicrotomy	3–5 nm, N/A
Transmission electron cryo-tomography (cryo-ET)	3D structure determination of cryo-immobilized regions of cells or pleiomorphic viruses in the near-native state	4–10nm, 6–15 nm
Single-particle negative-stain TEM	Structure probing of small macromolecular assemblies (<250kDa) which have been covered in a layer of heavy metals; 2D averaging or 3D reconstruction may be used	2–4nm, 2–4nm
Single-particle transmission electron cryo-microscopy	3D imaging of macromolecular assemblies (>250kDa) and viruses cryo-immobilized in a thin vitreous layer in near-native state.	0.3–1nm, 0.3–1nm
Serial surface view imaging (scanning electron microscopy)	3D imaging of large cells or tissues	1–10nm, 10–30nm

Figure 17.6 Schematic diagram of a transmission electron microscope (TEM).

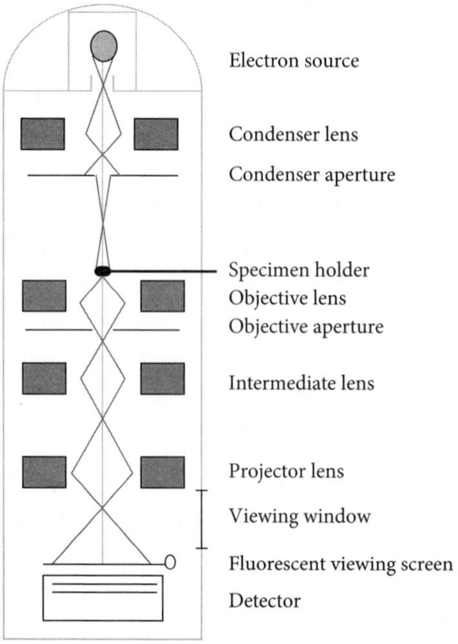

emission or thermionic depending on the microscope type. Field emission guns (FEG-TEMs) are operated at relatively high extraction voltages (200–300kV) which allow the electrons to penetrate the specimen better and therefore are ideal for the study of thicker specimens (150–300nm deep) such as sections of cells or tissues. Modern thermionic TEMs are usually operated between 80 and 120kV, and can produce similar image quality to a FEG if specimen thickness is kept to between 50 and 100nm. The microscope column is maintained at high vacuum using a combination of different vacuum pumps. This is necessary as collisions between the electrons and any gaseous molecules would make it impossible to generate a clear image.

Three types of electromagnetic lens are used in TEMs. The condenser lens focuses the beam onto the specimen. The objective lens recombines electrons scattered from the sample and focuses them onto the image plane to form an image. A series of intermediate and projector lenses then magnify the image. This array of lenses allows TEMs to operate at overall magnifications in the range of 20x–300 000x.

Electrons are viewed indirectly using either a fluorescent viewing screen or a recording device. A recorded exposure is known as an **electron micrograph** and is captured by either film or digital camera (CCD). Film is generally considered a superior recording medium as a greater area can be recorded and it encodes information to a higher resolution. However, as film must be digitized (or scanned), which is a time-consuming process, a CCD camera is commonly used as it allows direct readout and image inspection.

Transmission of electrons through a specimen produces an image which corresponds to a projection, or 'shadow', of the specimen. The shadow comprises dark and light areas that provide information about the structure of the specimen. This contrast arises as a result of scattering of electrons as they pass through the specimen. Heavy atoms (e.g. heavy metals such as lead or uranium) will scatter electrons more strongly than light atoms (e.g. carbon, oxygen, and nitrogen). Hence, depending on the atomic composition of the specimen, contrast may arise either by exclusion of strongly scattered electrons by the objective aperture of the microscope (known as amplitude contrast), or by interference of weakly scattered electrons with the unscattered beam (known as phase contrast). The image produced in a TEM is a 2D projection of the specimen. It is possible to reconstruct a 3D representation of the specimen computationally; this is discussed further in Section 17.2.3.

Although TEMs can achieve relatively high resolving powers (0.1–0.3nm), they can only transmit electrons through thin specimens (<300nm). Furthermore, the areas visualized are limited in size, typically to an area of 1–2μm^2, and can be difficult to interpret, requiring time-consuming data collection and image processing steps. However, in combination with other sources of information, such as fluorescence light microscopy, meaningful high-resolution visualization of cells and tissues is possible.

17.2.2 Sample preparation and sectioning

An integral component of successful electron imaging is sample preparation which is required as the harsh environment of the electron microscope (i.e. the vacuum and the ionizing effect of the electron beam) can damage the specimen under investigation. Three sample preparation techniques are described in this section: negative staining, plastic embedding, and cryogenic fixation. The approach used to prepare the sample will depend on the sample type (e.g. whether it is a purified macromolecular complex or a tissue sample), the features you wish to observe, and how well you require the sample to be preserved. Figure 17.7 provides an overview of the approaches that can be used to prepare different samples.

A key part of sample preparation is fixation—the aim is to preserve the structure of the specimen while introducing as few artefacts and changes as possible. There are two main fixation methods: chemical-fixation and **cryogenic** fixation (also referred to as immobilization). Chemical-fixation is generally the most prone to specimen alteration, but is still commonly applied as it is simple and quick to use while still providing useful information. Structural information is better preserved using cryogenic fixation, but is technically more challenging. Cryogenic samples are visualized at temperatures of −180°C; this is known as electron cryo-microscopy. Chemically fixed specimens are generally observed at room temperature (23°C).

Negative staining

A purified solution of macromolecular complexes, viruses, or bacterial cells can be imaged quickly at room temperature using negative staining. In this technique, an

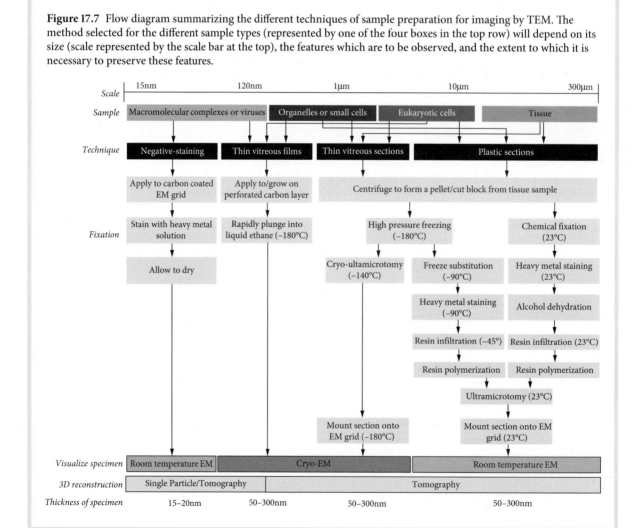

Figure 17.7 Flow diagram summarizing the different techniques of sample preparation for imaging by TEM. The method selected for the different sample types (represented by one of the four boxes in the top row) will depend on its size (scale represented by the scale bar at the top), the features which are to be observed, and the extent to which it is necessary to preserve these features.

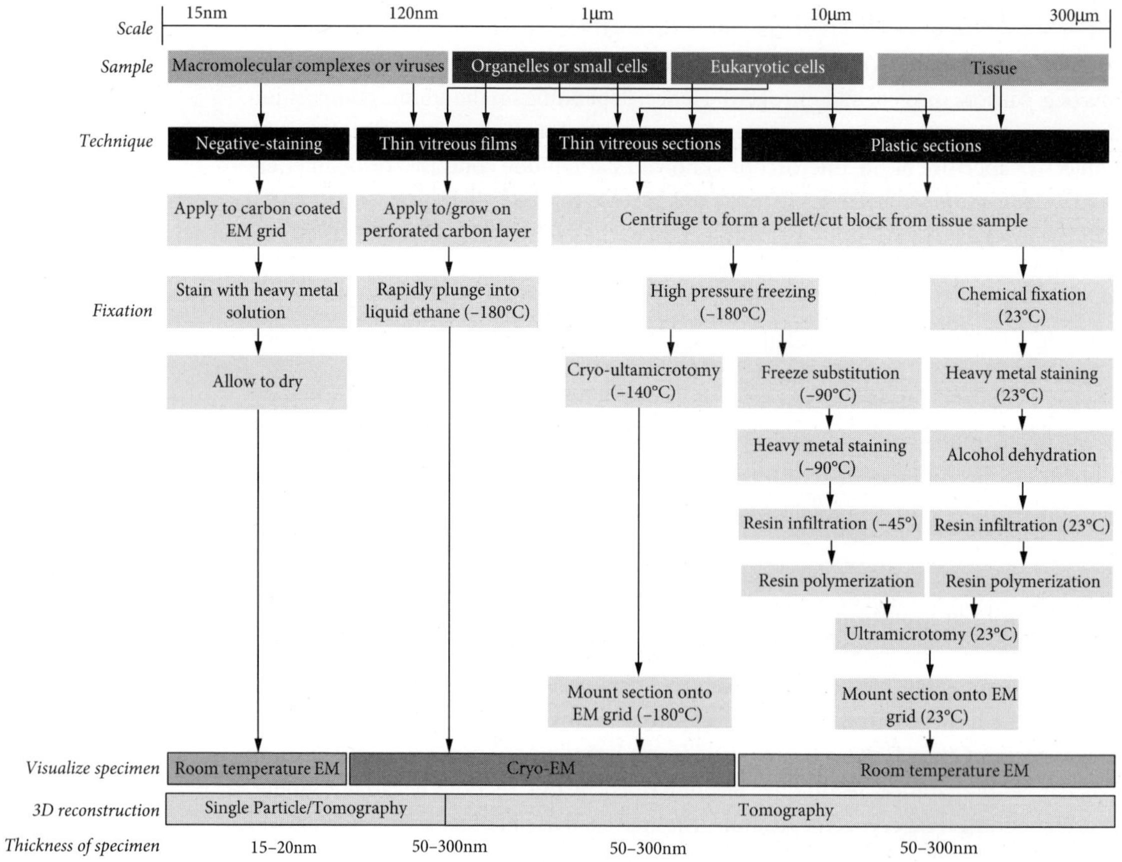

aliquot of the aqueous sample is applied to an EM grid which has been covered in a thin layer (20nm) of material, such as evaporated carbon or formvar plastic, to provide planar support. Once the particles in solution have adsorbed onto this layer, the grid is stained and fixed using a heavy metal stain such as uranyl acetate and then dried. This procedure will produce an 'envelope' of stain around the macromolecules or cells. The stained regions scatter electrons strongly, which increases contrast and hence improves the ability to differentiate features within a sample.

Plastic embedded and thin sectioning of cells

In this method, the cells or tissue are first fixed chemically with an aldehyde (usually glutaraldehyde) to cross-link the proteins, followed by a secondary fixation step using osmium tetroxide. Osmium tetroxide interacts with, and hence stains, the lipids of cellular membranes. The sample is then dehydrated by serial treatments with increasing concentrations of either alcohol or acetone (30–100%) and embedded in epoxy resin. To stabilize the specimen, the resin is polymerized by either exposure to UV light or heating and then sectioned using a **microtome**. The slices are placed onto 3mm metal mesh discs and stained using uranyl acetate (or sometimes lead citrate). As with negative staining, the heavy metal atoms surround (envelope) the biological structures and improve image contrast. The procedure is the most prone to artefacts because of the harshness of each step. However, historically it provided the foundation of our understanding of cellular ultrastructure and can still be used at the early stages of a project for quick evaluation of specimens. An example of a TEM image of a sample prepared by plastic embedding and sectioning is shown in Figure 17.8(a).

Cryogenic fixation

In cryogenic fixation the sample is frozen rapidly (cooling rate ~10 000°C/sec) so that the water in and around the sample is **vitrified**, or preserved in an amorphous state. During this process cells and their components are immobilized in their functional states. Preparations of macromolecular complexes, virions, organelles, or thin or small cells (e.g. *Dictyostelium discoidium*) can be vitrified as a thin layer over a perforated layer of carbon film (30–40nm thick) which has been overlayed onto a standard EM specimen grid by rapidly plunging it into liquid ethane maintained at about −180°C. This technique is known as cryo-EM of vitreous thin films. In some cases the membrane boundary or 'edge' of a eukaryotic cell grown in a monolayer may also be thin enough (200–500nm) to prepare in this way (Medalia et al. 2002). An example of a TEM image of a vitreous thin-film preparation is shown in Figure 17.8(b).

In specimens more than 1000–2000nm thick, at least some ice crystals will form which can obscure the sample features during imaging. Thicker specimens (2–200μm) can be vitrified by raising the pressure around the specimen (to up to 2000 bar) whilst simultaneously lowering the temperature to cryogenic temperatures using a device called a high-pressure freezer.

Following cryo-fixation, the frozen sample may be sectioned directly using a technique called cryo-electron microscopy of thin vitreous sections (CEMOVIS). However, this is extremely challenging as all sample manipulations must be performed at cryogenic temperatures. For example, sections 50–100nm thick must be cut by a diamond knife using a cryo-ultramicrotome (Al-Amoudi et al. 2004).

Figure 17.8 TEM images of *Ostreococcus tauri*, the smallest known unicellular eukaryote, prepared by two different methods. (a) A TEM image of *O.tauri* which has been chemically fixed, plastic embedded, and sectioned at room temperature. (b) A single unstained cell has been prepared by vitrification in a thin layer of water inside a carbon hole (2μm in diameter and visible at the outer edges). Identifiable cellular compartments include the cell nucleus (n), mitochondria (m), chloroplast (c), endoplasmic reticulum (er), the Golgi apparatus (g), and granules (gr). Cryo-EM not only preserves membranous structures in a state close to native, but also allows additional detail in the form of macromolecular structures such as the ribosome (dense dots distributed through the cytoplasm) to be visualized. The scale bar represents 200nm.
Reproduced from Henderson, G.P., et al. (2007) *PLoS One* **2**: e749. Adapted and printed by permission under the Creative Commons Attribution 2.5 license.

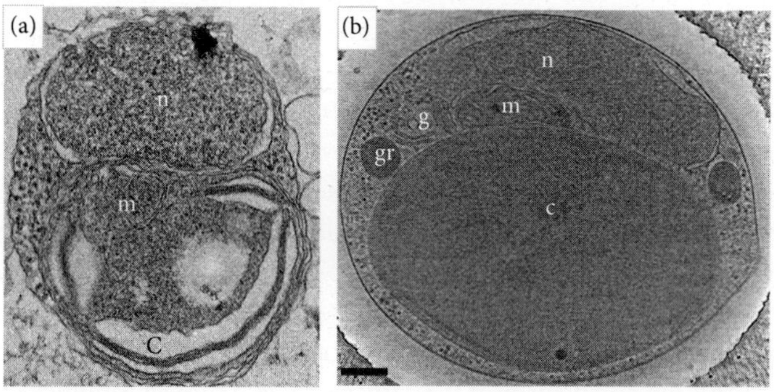

An alternative to direct sectioning of frozen material is freeze substitution and plastic embedding. In freeze substitution the vitrified water is replaced with a solvent such as acetone at a low temperature (−90°C). The sample is then fixed and stained using uranyl acetate, for example, and the specimen is infiltrated with a resin, again at a low temperature (−45°C). The resin is polymerized (−25°C) and the temperature of the specimen is increased to room temperature for sectioning (Sosinsky et al. 2008).

17.2.3 Three-dimensional reconstruction of TEM images

As described in section 17.2.1, the image produced in a TEM is a 2D projection of the specimen. From this 2D projection it is possible to reconstruct computationally a 3D representation of the specimen. This involves recording multiple 2D projections (images) of the specimen in different orientations and then 'back-projecting' them to re-create what the specimen really looks like in 3D. This approach also allows us to incorporate structural information gathered from complementary techniques, such as X-ray crystallography, into our analysis, enabling a more complete picture of a 3D object to be built up.

Three 3D reconstruction methods are available: **single-particle reconstruction** (Orlova and Saibil 2011) and **electron crystallography** for the study of macromolecular complexes, and **tomographic reconstruction** for the study of cells (Leis et al. 2009). These methods are now typically applied to cryo-EM images, but were all originally developed and applied to stained specimens. Descriptions of the reconstruction process are beyond the scope of this book, but readers may wish to consult the references provided here.

17.2.4 Scanning electron microscopy (SEM)

Unlike TEMs, which view a projection or 'shadow' of thin specimens, SEMs are able to provide excellent views of surface topography of large areas (up to 50μm^2) of sample, much larger than is possible by TEM. A resolution of 3nm can be achieved by modern SEMs and are capable of magnifying the specimen between 20x and 30 000x.

As with TEM, SEM comprises an optical column which houses an electron gun and lenses responsible for accelerating and focusing electrons into a specimen chamber (Figure 17.9). However, unlike TEM, the specimen is then scanned in a raster pattern (i.e. in a sequence of horizontal lines) and images are built up gradually by detecting secondary electrons that are ejected from each point of the specimen. SEM beams are not expected to penetrate the sample, and consequently relatively low acceleration voltages are used (5–30kV). Samples are placed on a specimen stage, and the stage is moved and tilted so that a clear image can be recorded. The position of the stage (and sample) is monitored by the user using an infra-red camera and TV screen.

Sample preparation for SEM involves fixing the cells or tissues chemically, dehydrating the sample by substituting water with increasing concentrations of alcohol or acetone, and substituting again with liquid carbon dioxide. The carbon dioxide is then evaporated in a process known as **critical point drying**. Once the carbon dioxide has evaporated, a sputter coater is used to coat the specimen with metal atoms (usually gold or platinum).

SEM tomography

Although SEM is used to study the surface features of specimens, tomographic imaging can be performed by a process called serial surface view imaging to achieve a detailed

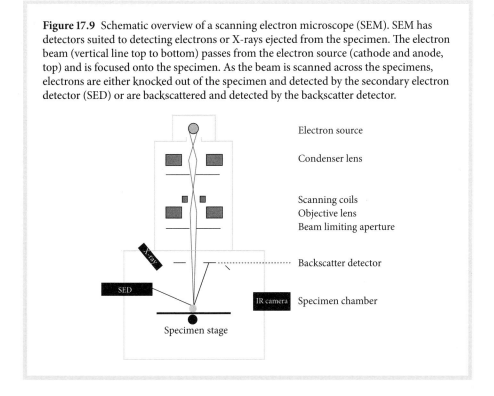

Figure 17.9 Schematic overview of a scanning electron microscope (SEM). SEM has detectors suited to detecting electrons or X-rays ejected from the specimen. The electron beam (vertical line top to bottom) passes from the electron source (cathode and anode, top) and is focused onto the specimen. As the beam is scanned across the specimens, electrons are either knocked out of the specimen and detected by the secondary electron detector (SED) or are backscattered and detected by the backscatter detector.

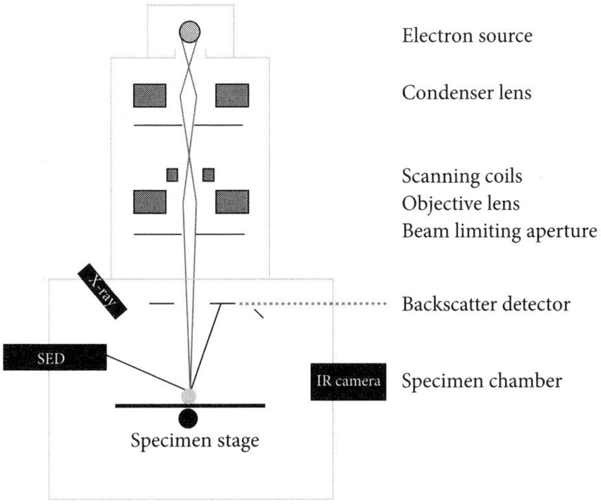

view of the 3D spatial organization of cells or tissues (Bennett et al. 2009). To carry out serial surface imaging, the surface of the specimen is imaged after consecutive layers of the specimen have been physically cut away using either a diamond knife or an ion beam. A 3D volume is then produced by computationally 'stacking' the images which represent physical slices through the specimen. The resolution of the resultant image in the *xy* plane is typically 1–10nm. However, the resolution in the *z* plane will be determined by the thickness of the slices. Typical volumes studied are 10 × 5 × 10μm. Using TEM to study an equivalent volume is a considerable undertaking, and therefore SEM is an attractive technique for such studies, capable of achieving resolutions very close to that of TEM (10–20nm).

17.2.5 Macromolecular localization and structure determination

Although electron microscopy can generate 3D snapshots of cells, it is often impossible to identify specific proteins of interest from raw images or **tomograms** alone unless the protein is part of a particularly large complex, such as a ribosome, the 26S proteasome, or the nuclear pore complex. This is because of macromolecular crowding, low contrast, and low signal-to-noise ratio. To assist in identification, two approaches are available. One is a computational strategy called **template matching** which is used when the structure of a protein complex is known. It involves assigning protein identity to the observed shapes by identifying similarities in shape and size to predetermined protein structures from X-ray crystallography or single-particle cryo-EM studies (Beck et al. 2011). The second approach is to use **electron-dense labels** to identify the protein of interest and is commonly used to identify smaller proteins or when the structure has not yet been determined.

Electron-dense protein tags and immuno-electron microscopy

Immuno-electron microscopy uses conjugated antibodies to localize proteins of interest in plastic-embedded sections, similar to immunofluorescence light microscopy. Antibodies conjugated to electron-dense tags such as colloidal gold particles, usually 5–20nm in size, are used to probe for specific antigens on the surface of a section. The section is incubated with a droplet of the appropriate primary antibody, followed by incubation with the secondary antibody which is gold conjugated and then viewed using TEM. Cryo-sections cannot be probed using this method. Instead, metal-binding proteins can be used as clonable tags in the same way that fluorescent proteins such as GFP are used in fluorescence microscopy.

Metallothionein (MT) is a small (~6kDa) cysteine-rich heavy-metal-binding protein able to bind up to 40 gold atoms. Any MT when expressed as a fusion product to a protein of interest can result in metal particles about 1nm in size which can be observed by TEM (Diestra et al. 2009). This marker is preferable to immuno-gold labelling as the MT protein is immediately adjacent to the protein of interest, whereas the gold attached to an antibody may be displaced by as much as 10–15nm from the target because of the intervening antibody. MT tags are most successfully used with unstained freeze-substituted resin-embedded specimens. In frozen hydrated samples the water layer will obscure the small protein by matching its contrast. This limitation can be overcome by fusing multiple MTs together to increase the size of the heavy atom particle tagged to the protein of interest.

Figure 17.10 Electron-visible protein tags for electron microscopy and tomography. In this example the coding sequences for ferritin (FtnA) and green fluorescent protein (GFP) were fused to a membrane targeting sequence (mts-GFP-FtnA). (a) Electron cryo-tomography demonstrated the accumulation of electron dense granules at the cell poles after the *E. coli* cells had been grown in the presence of 1mM $FeSO_4$ for 3 hours. The inset shows an overview of the cell which was segmented to produce a 3D model of the ferritin localization as shown in (b). The outer membrane is coloured in yellow, the inner membrane in cyan, the membrane vesicles in green, and mts-FtnA in pink. (c) Fluorescence microscopy (monitoring GFP) was able to confirm the localization of the expression construct to the cell poles. Scale bars: (a) 50nm; (b) 100nm; (c) 2000nm). For a colour reproduction of this figure, please see Plate 15. Adapted from Wang,Q., et al. (2011) *Structure* **19**: 147–54. Reprinted with permission from Elsevier.

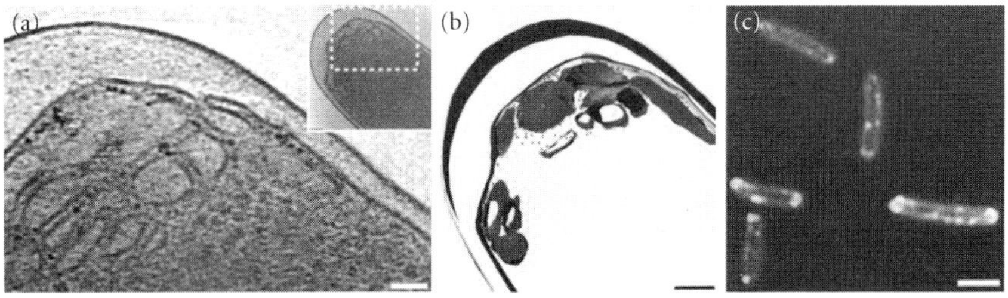

Another example of a metal-binding protein which has been developed for cryo-EM is ferritin (Wang et al. 2011). This 19kDa protein can assemble into spherical multimers, 12nm in diameter. In the presence of sufficient metal ions, the core of the ferritin multimer will fill to produce a 7.5nm electron-dense particle which is sufficiently large for most cryo-EM applications. A benefit of these clonable tags is that double-fusion proteins can be created, comprising GFP and a heavy-metal-binding protein. With such constructs, samples can be visualized using both fluorescence light microscopy and electron microscopy. An example of this is shown in Figure 17.10.

17.3 Correlative light electron microscopy

Although electron microscopy can provide detailed structural information on organelle, membrane, and macromolecular complexes, it cannot be used with live cells to track cellular events. In contrast, fluorescence microscopy can be used on live cells to track dynamic localization and/or function of specific molecules, but cannot resolve images to the same level as electron microscopy. Correlative light electron microscopy (CLEM) is a technique that combines fluorescence microscopy and transmission electron microscopy to produce high-resolution images that include localized fluorescence information specific to the protein of interest. In CLEM, regions of interest are first identified by fluorescence microscopy and are then studied at high resolution using electron microscopy. Examples of events that have been studied using this approach are endocytic vesicle formation (Kukulski et al. 2011), virus entry and budding (Jun et al. 2011), and organelle function (van Driel et al. 2009).

To implement CLEM, we must have a way of imaging the same region of interest in the electron microscope as identified in the fluorescence microscope. This is achieved using 'finder grids'. Finder grids are marked with an array of letters that allow specific regions to be easily identified in both the light and electron microscope. Once images or tomograms are recorded in the electron microscope, the images from both fluorescence microscopy and electron microscopy are merged. This means developing a way of 'correlating' images recorded by two different instruments by matching their coordinate systems, scaling (to account for differences in magnification), and then rotating the images so that they can be overlayed one on top of the other. The precision, or accuracy with which the two images can be overlaid, is of utmost importance as this will ultimately determine the validity of their interpretation.

CLEM studies can be undertaken on either plastic embedded sections (produced by freeze substitution) or frozen-hydrated specimens (either intact cells grown on electron microscopy grids or thin vitreous sections). For either approach both fluorescence microscopy and electron microscopy must be performed after the sample has been fixed so that identical time points are studied in both microscopes. This is more challenging for frozen-hydrated specimens as the sample must be maintained at cryogenic temperatures during fluorescence microscopy. Therefore cryo-stages for light microscopy have been developed (van Driel et al. 2009; Jun et al. 2011).

CASE STUDY 17.2

Studying the influence of virus infection on host cellular ultrastructure using electron microscopy and correlative light electron microscopy

Confocal microscopy studies of virus-infected cells have revealed interesting changes in the morphology of the cell membrane leading up to and at the time of virus budding. The virus is known to bud from the host membrane 24 hours post-infection. However, the effect of virus factors on the host membrane is not clearly understood. GFP-labelled virus factor 1 (VF1) has been shown to associate with the host plasma membrane at the time of budding. Additionally, these experiments suggest that VF1 is likely to co-localize with the virus genetic material (in this case a ribonucleoprotein complex (RNP)) and that the cytoskeleton is also associated with the viral RNPs. We are interested in learning more about the mechanism by which VP1 facilitates virus budding. Specifically, we would like to know if any actin/microtubule rearrangements take place at the sites of budding and if these components interact with the virus RNP. Additionally, we would like to determine the 3D structure of the viral RNP in association with any interacting complexes so that we are able to identify interacting surfaces on the virus nucleocapsid (protein) by fitting the X-ray crystal structure.

QUESTION

How should the cells be prepared so that the above questions can be studied by transmission electron microscopy?

These questions require that we locate a small specific region of the cell and identify an event (the virus budding) which takes place over a relatively short time. For these reasons we must use cryo-fixation to preserve the structures of interest rapidly—chemical-fixation is simply too slow. Our next choice for sample preparation is influenced by the thickness of the specimen. If the host cells are thin enough at their edges (where budding may take place), the cells can be cultured on EM grids and prepared for cryo-EM by plunge freezing into liquid ethane. Using a confocal cryo-stage these EM grids can be inspected at cryogenic temperatures and GFP-labelled VF1 budding sites identified. The EM grids can then be transferred into the electron microscope for high-resolution inspection. To determine the 3D structure of budding complexes we would use cryo-tomography to generate a 3D reconstruction of the region of the cell we are interested in.

However, if the cells are too thick to be imaged directly, we will need to cryo-immobilize them and subsequently section them for inspection. If we are interested in the structure of membranous regions of the budding site, we may conveniently use freeze substitution and plastic-embedded sectioning. The fluorescent signal is preserved in these sections and hence the same correlative approach as described in the previous paragraph can be used, but at room temperature. However, if we are interested in the organization of cytoskeletal elements as well as the structure of viral RNPs (macromolecular details), we must use CEMOVIS, i.e. direct sectioning of the high-pressure frozen and unstained material followed by cryo-microtomy and cryo-tomography.

17.4 Chapter summary

- Fluorescence molecules can be used to localize a number of proteins and structures within live and fixed cells using fluorescence microscopy.
- Fluorescence microscopy has an inherent resolution limit imposed by the optics of the microscope. The resolution of images can be improved by using techniques such as confocal microscopy.
- A number of different confocal techniques are available which enable different types of sample imaging. For example, laser scanning confocal microscopy enables 3D imaging of cells, spinning disc confocal microscopy can be used to image movements of intracellular structures in living cells, and TIRF microscopy can be used to image the surface of cells (e.g. trafficking of cell surface receptors).
- Deconvolution is an image-processing step in which raw images are processed computationally using mathematical algorithms to improve image resolution.
- The resolving power of the electron microscope is greater than that of fluorescence microscopes and can provide detailed structural information on organelles, membranes, and macromolecular complexes.
- A number of procedures are available for preparing a specimen for electron microscopy including negative staining, cryo-fixation, and plastic embedding and thin sectioning. It is important to be aware of the benefits and drawbacks of each.
- 2D images from electron microscopy can be reconstructed computationally to create 3D models which can be analysed in combination with information from other structural techniques such as X-ray crystallography.
- In order to identify specific proteins in electron microscopy images, heavy metal labels can be used in a similar way to fluorescent molecules in light microscopy.
- Correlative light electron microscopy (CLEM) combines fluorescence microscopy and transmission electron microscopy to produce high-resolution images that include

localized fluorescence information specific to the protein of interest. In CLEM, regions of interest are first identified by fluorescence and then studied at high resolution using electron microscopy.

✓ References

Al-Amoudi, A., Chang, J.J., Leforestier, A., et al. (2004) Cryo-electron microscopy of vitreous sections. *EMBO J* **23**: 3583–8.

Beck, M., Topf, M., Frazier, Z., et al. (2011) Exploring the spatial and temporal organization of a cell's proteome. *J Struct Biol* **173**: 483–96.

Bennett, A.E., Narayan, K., Shi, D., et al. (2009) Ion-abrasion scanning electron microscopy reveals surface-connected tubular conduits in HIV-infected macrophages. *PLoS Pathoge* **5**: e1000591.

Betzig, E., Patterson, G.H., Sougrat, R., et al. (2006) Imaging intracellular fluorescent proteins at nanometer resolution. *Science* **313**: 1642–5.

Biggs, D.S.C. (2010) 3D deconvolution microscopy. *Curr Protoc Cytom* **52**: 12.19.1–20.

Diestra, E., Fontana, J., Guichard, P., Marco, S., and Risco, C. (2009) Visualization of proteins in intact cells with a clonable tag for electron microscopy. *J Struct Biol* **165**, 157–68.

Henderson, G.P., Gan, L., and Jensen, G.J. (2007) 3-D ultrastructure of O. tauri: electron cryotomography of an entire eukaryotic cell. *PloS One* **2**: e749.

Huang, B., Babcock, H., and Zhuang, X. (2010) Breaking the diffraction barrier: super-resolution imaging of cells. *Cell* **143**: 1047–58.

Jun, S., Ke, D., Debiec, K., et al. (2011) Direct visualization of HIV-1 with correlative live-cell microscopy. *Structure* **19**: 1573–81.

Kukulski, W., Schorb, M., Welsch, S., Picco, A., Kaksonen, M., and Briggs, J.A. (2011) Correlated fluorescence and 3D electron microscopy with high sensitivity and spatial precision. *J Cell Biol* **192**: 111–19.

Leis, A., Rockel, B., Andrees, L., and Baumeister, W. (2009) Visualizing cells at the nanoscale. *Trends Biochem Sci* **34**: 60–70.

Lippincott-Schwartz, J., Snapp, E., and Kenworthy, A. (2001) Studying protein dynamics in living cells. *Nat Rev Mol Cell Biol* **2**: 444–56.

Medalia, O., Weber, I., Frangakis, A.S., Nicastro, D., and Baumeister, W. (2002) Macromolecular architecture in eukaryotic cells visualized by cryoelectron tomography. *Science* **298**: 1209–13.

Millis, B.A. (2012) Evanescent-wave field imaging: an introduction to total internal reflection fluorescence microscopy. *Methods Mol Biol* **823**: 295–309.

Nakano, A. (2002) Spinning-disk confocal microscopy: a cutting-edge tool for imaging of membrane traffic. *Cell Struct Funct* **27**: 349–55.

Orlova, E.V., and Saibil, H.R. (2011) Structural analysis of macromolecular assemblies by electron microscopy. *Chem Rev* **111**: 7710–48.

Pawley, J.B. (ed) (2006) *Handbook of Biological Confocal Microscopy* (3rd edn). New York: Springer.

Sosinsky, G.E., Crum, J., Jones, Y.Z., et al. (2008) The combination of chemical fixation procedures with high pressure freezing and freeze substitution preserves highly labile tissue ultrastructure for electron tomography applications. *J Struct Biol* **161**: 359–71.

Swedlow, J.R. and Platani, M. (2002) Live cell imaging using wide-field microscopy and deconvolution. *Cell Struct Funct* **27**: 335–41.

Ustione, A. and Piston, D.W. (2011) A simple introduction to multiphoton microscopy. *J Microsc* **243**: 221–6.

van Driel, L.F., Valentijn, J. A., Valentijn, K.M., Koning, R.I., and Koster, A.J. (2009) Tools for correlative cryo-fluorescence microscopy and cryo-electron tomography applied to whole mitochondria in human endothelial cells. *Eur J Cell Biol* **88**: 669–84.

Wallace, W, Schaefer, L.H., and Swedlow, J.R. (2001) A working person's guide to deconvolution in light microscopy. *BioTechniques* **31**: 1076–8.

Wang, Q., Mercogliano, C.P., and Löwe, J. (2011) A ferritin-based label for cellular electron cryotomography. *Structure* **19**: 147–54.

Wilson, T. (2011) Resolution and optical sectioning in the confocal microscope. *J Microsc* **244**: 113–21.

18 Histopathology in biomolecular research

Noelyn Hung and Tania Slatter

Chapter overview

Study of actual tissue, diseased or normal, is fundamental to biomedical research. Tissue can be examined macroscopically (i.e. by the naked eye) or using a microscope (microscopically). This area of study is called histology or histo-**pathology** if the specimen under investigation is affected by disease. This chapter will provide practical insight into the techniques and skills necessary to handle and investigate human pathology samples successfully. We will look at the different ways in which tissues can be obtained from a patient and how they are preserved. We will then summarize the macroscopic features that are characteristic of pathological samples, focusing particularly on tumour tissues. Additionally, techniques that are frequently used to visualize abnormal and normal tissues and cells microscopically will be covered: histological staining, immunohistochemistry, and *in situ* hybridization. As we progress through the chapter, we will highlight how data obtained using these techniques is important in a range of settings including clinical and diagnostic laboratories as well as research laboratories. Although the focus of this chapter is human tissue, much of it applies equally well to studies on murine or other animal tissues.

LEARNING OBJECTIVES

This chapter will enable the reader to:

- identify ethical issues underpinning work with human tissue;
- summarize the various types of tissue-sampling techniques available and identify macroscopic features that characterize normal and pathological samples (primarily a range of different tumours);
- outline the steps involved in preparing tissues for microscopic analysis;
- describe the techniques of histological staining, immunohistochemistry, and *in situ* hybridization and select the most appropriate method to answer a defined research question;
- appreciate the value of studying tissues and how it can complement *in vitro* research as well as assist in the diagnosis and monitoring of disease.

18.1 Introduction

Studying tissues improves our understanding of disease states. For example, normal tissue can be distinguished from diseased tissue by features such as changes in cell morphology, expression and localization of particular proteins, or changes in cell proliferation and cell death.

In the laboratory, cells and compounds are extracted for simplicity of manipulation and interpretation. Therefore analysis of tissue samples is vitally important for validating *in vitro* findings. Although cell cultures contribute enormously to our understanding of cellular processes, they are limited by their inability to examine a system in the context of the surrounding **stromal** cells, stromal matrix, neighbouring cellular interactions, or immune reactions. Examination of the tissue as a whole allows for the 'natural environment' and the response within that environment to be assessed. The importance of these interactions in promoting our understanding of cellular processes is highlighted by the recent findings of the role mast cells play in carcinogenesis (Gilfillan and Beaven 2011) and regulatory T cells play in tumour tolerance and growth (Whiteside 2012). Environmental conditions also alter **epigenetic** and epigenomic status and add to our understanding of the complexity of disease pathogenesis (Herceg and Vaissiere 2011; Martin-Subero 2011). Studying tissues enables a more complete picture to be attained, and thus is an essential aspect of biomedical research

But do you know what to do with the generous donation of a 'piece of tissue or tumour'? In this chapter we will describe the pathological features associated with benign and malignant cells and describe three methods by which tissue samples can be analysed:

- histological stains to visualize particular cell or tissue types or cell components;
- immunohistochemistry to detect the protein levels and localization of particular proteins in a tissue sample using an antibody directed against the protein of interest;
- *in situ* hybridization to identify specific DNA or RNA sequences in a tissue section using probes complementary to the target nucleic acid.

Biomedical research data obtained in a laboratory setting is often translated into a clinical environment to develop new diagnostic approaches and therapies. Laboratory investigation of tissue samples, often using the histological techniques listed above, is integral to this process. For example, large-scale genomic projects have been instrumental in identifying new tumour markers, an example of which is the discovery of isocitrate dehydrogenase 1 mutations (IDH1) in brain tumours (Parsons et al. 2008; Yan et al. 2009). The next step is the adoption of these markers in a clinical setting. Indeed, some diagnostic laboratories now screen for IDH1 mutations using immunohistochemistry techniques (Capper et al. 2009).

Research with human tissue samples can involve the following stages:

(i) obtaining ethical approval from a research ethics committee;

(ii) obtaining the tissue sample;

(iii) macroscopic analysis of the tissue;

(iv) preparing the tissue sample to enable microscopic analysis;

Each stage is described in turn in the following sections.

18.2 Ethical approval and consent

The use of human tissue requires ethical approval and the informed consent of the donor prior to experimentation. Any research that uses human tissue samples must be conducted in accordance with the statutory regulations that govern such work. In the UK, the Human Tissue Act 2004 governs work with human tissue. In New Zealand the Human Tissue Act 2008 and in Australia the Human Tissue Act 1982, with amendments in 2006, serve the same purpose. Each country has its own regulatory framework, but typically requires obtaining written approval from an ethics committee to undertake the project before any work, or consent process, begins. The committee is charged with the responsibility of ensuring that the tissue donors will be given adequate information about the project for them to provide consent. The committee will also check that your methods of obtaining consent and the experimental design are appropriate for answering the research question. For example, are enough cases available for analysis in order for a statistically significant outcome to be reached? This compliance process may take months, so plenty of forward thinking is required.

Ethics committees require the data to be stored securely in an anonymized way, but often in a way that allows for de-anonymization back to the donor if necessary. Access to the tissues should be controlled so that they are only used for the approved study and by the approved study investigators. Ethical questions arise if the tissues are wasted because of inexperienced handling or processing, and need to be carefully addressed in, for example, a teaching facility. Signed declarations by study staff agreeing to keep any information gained (inadvertently or otherwise) in a confidential manner are also useful to emphasize this privilege. It is unacceptable to retain human tissues without the express consent of the donor or next of kin.

18.3 Tissue organization and data storage

When building a cohort of cases it is wise to invest time in organizing an electronic database and suitable physical storage space. Many file-making and database programs are available. Thought should be given to how the samples will be anonymized in an ordering system that could be de-anonymized in the future. In our experience a basic Excel spreadsheet will fulfill most of the requirements. Time and date of collection, tissue fixation, type, and length, and tissue location from the body are all examples of valuable information that can easily be lost as the study progresses and therefore should be carefully noted. You should also consider the clinical data (e.g. date of diagnosis, stage at diagnosis, radiographic findings, treatment given, or date of death) that might be required for the final analysis; who will hold this information, and who will update it as cases are added.

18.4 Obtaining tissue samples for research

There are a number of ways in which tissue samples can be obtained, including autopsy specimens, surgical tissue resection, excision **biopsy**, and exfoliative biopsy. The methods outlined here relate primarily to human studies, but are equally applicable to animal studies.

18.4.1 The autopsy

The terms autopsy, necropsy, and post-mortem refer to the process of methodically examining, interpreting, and diagnosing the pathology of a cadaver. The post-mortem emerged from the detailed anatomical examinations of the Renaissance, and when it became obvious to observers such as William Osler that clinical observations made during life had significant anatomical foundations, the discipline of morbid anatomy developed. Later, the term 'pathology' or the study of 'suffering/disease' was applied to most departments of morbid anatomy and morbid anatomy references.

Today's post-mortem examinations are performed in suites for which the name 'mortuary' is rather outdated (except to refer to a storage facility), as they have more in common with the modern surgical theatre. The internal examination of the chest, abdominal, and cranial contents comprises the 'full' post-mortem. In addition, other tissues such as joints/cavities, bones, skeletal muscles, spinal canal, and peripheral nerves make a significant contribution to the diagnoses. As part of the post-mortem, body weight, measurements, and external appearances are recorded. Skin incisions ensure no disfigurement of exposed skin, and usually comprise a longitudinal abdominal midline incision that separates at the xiphisternum to each shoulder tip. This allows exposure and removal of a front chest 'shield' by incising the ribs on both sides, which can be repositioned following the post-mortem to reconstruct the chest cage. Each organ is weighed for comparison with tables of mean weights and sampled for light microscopy examination. Depending on the findings, a small piece of tissue which appears normal and any abnormal areas are sampled. These samples are then processed for histology (see section 18.6) and used for further tests such as PCR or immunohistochemistry (see section 18.7). Microbiological, virological, forensic, and toxicological samples can also be taken as appropriate.

See Chapter 2

18.4.2 Surgical tissue resections

Resection is the removal of part or all of a tissue or organ during surgery and is a valuable source of tissue material. The largest resection specimens are amputations and whole or partial organ excisions. Many limbs amputated for diabetic ulcers, neuropathies, or tumours also contain many other tissues, such as fat, muscle, blood vessels, and skin, which, with ethical approval, can serve as control tissues. Non-neoplastic chronic disease specimens are often sent by the surgeon directly for disposal and do not undergo examination by the pathologist, and therefore are an under-utilized resource for research tissues.

Smaller samples from colonoscopy, bronchoscopy, or needle/cores of tissue are less useful as a tissue source, because of their small size. Nonetheless, careful serial sectioning can produce many useful sections.

You will need to orientate these smaller tissue samples properly to aid recognition. In many epithelial, layered, or covering tissue samples (e.g. intestine and skin), the area of interest can be quickly cut through (extinguished) during sectioning. Embedding of samples for sectioning onto glass slides, electron microscopy chucks, or frozen section chucks should reflect the required orientation. Chuck is the technical term for what the tissue is mounted on. In some samples, practice recognizing the bovine mucosa and muscularis mucosa of the intestine, for example, will aid in the orientation of a precious small transmural or mucosal biopsy before trimming and sectioning.

Other cellular material

Blood samples have long been used in the analysis of haematological malignancies, but other fluid samples are often overlooked. For example, urine, cerebrospinal fluid, and aspirated fluid from ascites can contain abundant cellular material suitable for the research laboratory, and collection is easy and often less traumatic for the patient. The cellular material can be prepared by filter, cytospin, or cell block, and readily used for immunohistochemical techniques. For the best cellular preservation ask for samples that have freshly accumulated.

Imprints or touch preparations (made by touching a cut lymph node or tumour onto a glass slide) can also provide adequate samples for diagnosis or multiple immunohistochemical or histological stains, even when there is only a small amount of specimen. These preparations are also useful for confirming the types of cells submitted for test tube analysis, as the exact tissue piece submitted can be first imprinted on a slide without the loss of significant tissue for analysis.

18.5 Recognizing pathology

Many good experiments have been hampered by the inadvertent investigation of, for example, normal tissue elements or necrotic parts of a tumour. Hence, correct selection and identification of any tissue for an experiment is critical. The following sections provide some guidelines on how to recognize macroscopic features of pathology in common tissue samples. Of course, examination of a microscopic section is the best way to confirm a tissue sample/cell type. However, some generalizations for recognition at the time of collection are possible, as each tissue and organ has its own characteristics. The **hyperplastic** or **hypertrophied** tissue is usually recognized by tissue enlargement, organ expansion (by radiological imaging), organ weight, or histological appearance. For example, a parathyroid or the thyroid gland (goitre) will be enlarged compared with a normal gland. The hyperplastic synovium of rheumatoid arthritis produces an inflamed pannus (thick inflamed tissue) that extends around joint tissues, destroying joint architecture and stability as it grows. Enlarged lymph nodes (lymphadenopathy) reflect hyperplasia of an immune response or neoplastic process.

The classical features of acute inflammation (heat, redness, swelling, pain, and loss of function) are difficult to recognize in the devascularized pathological specimen. However, infection produces areas of abscess formation, diffuse infiltration, softening, and/or colour change in a tissue which sometimes only microscopic examination will confirm.

The specific tissue and chronicity will determine the pathological features of infarction. The classical description of a triangular lesion with the base to the outside of the lesion, such as renal, pulmonary, brain, or splenic infarction, indicates the occluded or critical blood supply distribution. Newer radiological imaging techniques have greatly enhanced the early detection of infarction in the brain, for example, where a vague softness to palpation was often the only macroscopic sign.

18.5.1 Neoplasia

Neoplasia can be generally categorized as benign or **malignant** based on several characteristics, which are routinely used by the examining clinician and radiologist. Understanding the basics of neoplasia is important, as tumours are often the base material for new molecular studies from which the next generation of cancer hallmarks are established (Hanahan and Weinberg 2011).

Benign tumours

Benign tumours usually grow slowly and produce a smooth circumscribed border. In many instances a fibrous capsule, or a pseudo-capsule of surrounding stretched and compressed tissue, surrounds the border microscopically.

Thus when examined by the clinician (by **palpation**) the benign tumour has a smooth edge and is detached from the surrounding tissues, hence imparting tissue mobility. The tumour will be of uniform appearance with no areas of necrosis or haemorrhage (e.g. the common uterine fibroid (correctly termed a leiomyoma) or the benign breast lesion, (fibro-**adenoma**)). Microscopically the tumour cells bear a close resemblance to normal cells with good differentiation (specialization of a cell) and few indicators of cell proliferation (i.e. mitotic figures, variation in cell size, or immunohistochemical markers of proliferation such as proliferating cell nuclear antigen (PCNA) or Ki67).

Malignant tumours

A malignant neoplasm invades surrounding tissues, grows across and destroys tissue planes, and produces an irregular infiltrative margin. For the palpating clinician this presents as an irregular mass that is non-mobile and fixed to surrounding tissues. Radiological imaging reveals a poorly circumscribed edge indicative of destruction and invasion into surrounding tissues is observed. The radiological image may contain variable-density areas reflecting tissue necrosis or haemorrhage. On a cut section the malignant tumour has an irregular poorly demarcated boundary that invades and fixes surrounding tissues, and is variable because of fibrosis, necrosis, and/or haemorrhage. Necrotic tumour is soft and pale, and is easily smeared or broken apart when handled. Characteristics of a selection of neoplasias (breast, colon, prostate, renal, and dermal) are as follows.

Breast

The benign fibro-adenoma presents as a well-circumscribed firm pale tumour that easily shells out from the surrounding tissue. Frequently, fibro-adenomas are at least a centimetre in size and are excised for cosmetic reasons. Only one or two sections will be required for diagnosis, and hence the spare tissue can be used for research purposes.

A breast specimen excised for malignancy is orientated, weighed, measured, described, occasionally X-rayed (to determine the presence of microcalcification), and/or photographed. Individual surgical margins are identified in the fresh state using a paint that withstands tissue processing. As uniform and swift fixation is important, parallel incisions through the tissue at 1–2cm intervals are made prior to immersion in **fixative**;

volumetric ratio of at least ten parts fixative to one part tissue is used (see section 18.6.1). Fresh tumour tissue can be extracted when the parallel incisions are made—diagnostic orientation and integrity must be maintained at this stage. At least 24 hours of fixation is required, and should not be compromised for research samples.

The ducts, lobules, and connective tissue of normal breast tissue are pale and soft or rubbery. Breast cysts contain clear, opaque, or blue fluid, and are common in breast tissue from older patients. Cancerous breast tissue is recognizable by its pale firm pear-like **scirrhous** consistency, traversing tissue planes and invading adjacent tissues. A needle inserted into malignant tumour will often impart a firm gritty consistency. The edge of the tumour will irregularly penetrate the adjacent yellow fat. Sampling necrotic tumour (soft and pale or slightly yellow) or the scirrhous (fibrotic) centre, where malignant cells can be sparse, should be avoided. The advancing periphery of a breast cancer is usually the most cellular and hence the most suitable for sampling.

Cancerous breast ducts are about 1mm in diameter. They are identifiable by their passage through fat surrounded by a thin layer of connective tissue and often calcified contents. **Carcinoma** *in situ* (especially of the comedo type) is obvious macroscopically because of its protrusion from ducts, often like soft toothpaste from a tube.

Lumpectomy specimens are small and hence there is only a minimal amount of spare tissue. Similarly, diagnostic needle biopsy and fine-needle aspirate also lack spare tissue. In these instances spare histological sections or imprints (see below) may have to suffice for research purposes. However, a tissue block made from the aspirate may yield abundant material and sections.

Needle biopsies provide a small amount of tissue, but occasionally spare sections. If such scarce tissue is necessary for your study it is best to develop a collaborative relationship with the surgical pathologist in order to obtain these precious materials.

An invasive malignancy is generally readily identified microscopically because there is a proliferation of invading mitotic, and often pleomorphic, cells. *In situ* neoplasia is defined by a basement membrane that can be highlighted with special stains such as PAS (Periodic acid-Schiff) or smooth muscle actin if a myoepithelial layer is also present.

Bowel

For a malignant colon specimen, the tissue is similarly orientated, weighed, measured, described, and occasionally photographed. Usually an incision is made along the antimesenteric border to expose the lumen and the contents are carefully washed away. Fresh tumour tissue may be obtained before the specimen is immersed in fixative. Immersion in enough fixative to obtain a volumetric ratio of at least 10 parts fixative to one part tissue will ensure good fixation. The specimen may also require placement and pin fixation to ensure appropriate orientation of later sections. This fixation will prevent tissues from twisting or retracting from each other, and inappropriately orientating in a tissue section. Care is needed during sampling as the centre of a colonic cancer is typically ulcerated and at least superficially necrotic (see Case Study 18.1). The periphery of a colon cancer will often bear remnant adenoma with varying grades of **dysplasia**, and is sometimes recognizable grossly by a rolling raised edge. On cut section the tumour is pale and firm with an irregular invading edge surrounded by thickened desmoplastic (fibrotic) stroma.

Prostate

The prostate gland presents a distinct challenge when trying to identify tumour macroscopically. Histological confirmation is frequently required as benign prostatic hypertrophy/hyperplasia is diffusely mixed throughout the gland, and a carcinoma will usually diffusely infiltrate without demarcated margins or areas of necrosis or haemorrhage. Thus both the benign and the malignant gland will present as an enlarged irregular organ with a cut section in which accurate discrimination between benign and malignant is difficult. The most secure diagnosis of malignant areas is established from microscopic examination.

Kidney

A renal tumour is perhaps one of the most interesting tumours to examine macroscopically. One of the most common of these tumours is the clear cell carcinoma, which is a yellow to orange circumscribed mass, with tongues of tumour that may invade the surrounding capsule, veins, or renal pelvis. Areas of haemorrhage or necrosis are also readily seen, making sampling of viable tumour tissue less challenging than for most other tumours.

Skin

Skin biopsies come in a variety of forms, either incisional (curette, punch) or excisional. Most incisional biopsies will yield small amounts of tissue that are ideally vertically orientated to show dermis, epidermis, and hypodermis. Excisional biopsies maintain a margin of normal tissue. The most common types of skin cancer, such as basal cell carcinoma and squamous cell carcinoma, can deeply invade surrounding tissues with poor circumscription. While the basal cell carcinoma frequently becomes centrally cystic and collapses, the squamous cell carcinoma is likely to develop central areas of necrotic keratinous material. This soft pale necrotic material is also evident in metastatic lymph nodes and should be avoided during sampling for research.

The biological progression of malignant melanoma from radial to vertical growth phase can often be recognized macroscopically, as the radial growth phase invades squamous epithelium (usually the epidermis) and the vertical growth phase invades the dermis or deeper. The latter produces a palpable raised nodule within the lesion that reflects vertical growth proliferation in the radial growth phase background. As new clones of malignant cells develop, the border expands irregularly and the amount of pigment made by the tumour varies, producing an irregularly circumscribed and irregularly coloured lesion.

18.6 Tissue preparation

Once the required tissue has been collected and identified, it is prepared so that it can be presented as a tissue section on a glass slide for analysis by light or electron microscopy. The typical preparative steps include fixing the tissue chemically, sample processing followed by embedding in a solid medium, and sectioning. Figure 18.1 outlines the steps involved in creating a paraffin-embedded block. An alternative to chemical-fixation is frozen section fixation. Both these processes are described here.

18.6.1 Fixation

To prevent post-mortem decay surgically removed tissue must be preserved, i.e. fixed, as quickly as possible. The type of fixative chosen will depend on the type of tissue and

Figure 18.1 Steps involved in creating a paraffin-embedded block. Fixation: tissue collected for analysis is fixed first, commonly using 10% neutral buffered formalin (NBF). Fixed tissue is placed into a processing cassette. Care must be taken to ensure that the cassette is labelled clearly with a solvent-resistant pen or in pencil. To obtain sections of bone the tissue is first decalcified in acid (such as 10% formic acid) or a less harsh chelating agent (e.g. 10% EDTA) prior to processing. **Processing**: fixed tissue is prepared for paraffin infiltration. During processing the tissue is dehydrated using a series of alcohols to remove the water content, the alcohol is removed (often using xylene), and paraffin wax is added. **Embedding**: the paraffin-immersed tissue is placed in a mould and additional paraffin wax is added to completely cover the tissue. The wax is solidified and a paraffin block ready for sectioning is made. **Sectioning**: paraffin-embedded blocks are clamped into a **microtome** and sections of the tissue are made. The sections are placed onto a flotation bath of warm water to smooth out wrinkles, placed onto glass slides, and dried.

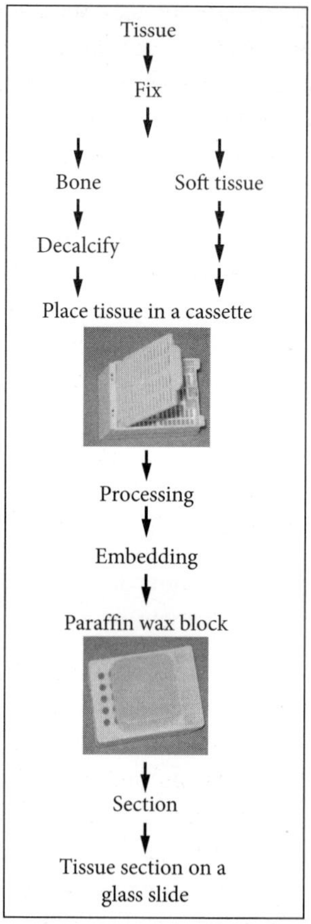

the downstream applications. The most widely used fixative is formaldehyde, a type of aldehyde that is used as a 3.7–4% buffered solution which is also referred to as 10% neutral buffered formalin (NBF) or simply as 'formalin' (Bancroft and Gamble 2008). Formaldehyde fixation is ideal for a number of routine histological applications such as **haematoxylin and eosin** (H&E) staining and immunohistochemistry. Optimal tissue fixation requires the following.

- The tissue is placed into the fixative as quickly as possible after removal. Fixation inactivates enzymes which would otherwise destroy tissue (autolysis), and prevents bacterial decomposition (putrefaction).
- The fixative penetrates the tissue. The smaller the piece of tissue, the better the fixation will be. Ideally, tissue should be cut up into pieces less than 5mm thick prior to the addition of the fixative.
- Enough fixative is used. The volume of fixative should be at least 10 times and ideally up to 15–30 times greater than the volume of tissue.
- The tissue is left in the fixative for a sufficient time. Tissue pieces less than 2mm thick will require 4–8 hours to be fixed. Larger tissue pieces will require longer fixation (e.g. overnight). For downstream immunohistochemistry protocols, tissue should be removed from formalin fixation as soon as possible.

Formalin is toxic and a human carcinogen (Liteplo 2002). Formalin-free fixatives that are ethanol based and hence carry a lower health risk are available (Dotti et al. 2011). However, the tissue morphology can be different with these alternatives.

Fixatives other than formalin may be more appropriate for some applications. Alternative methods of fixation include heat, glutaraldehyde, glyoxal, osmium tetroxide, potassium dichromate, mercuric chloride, picric acid, zinc salts, cupric acid, alcohol (methanol and ethanol), acetone, and acetic acid. Specific examples where alternatives to formalin are more suited include glutaraldehyde and osmium tetroxide for electron microscopy, because of their superior ability to preserve cellular morphology, osmium tetroxide for analysis of fat tissue, alcohol for cytology, and zinc salts for improved immunoreactivity in immunohistochemistry. Extra care must be taken with handling tissues fixed with the formalin-free alternatives, as viruses may still be active.

Different fixatives act by cross-linking proteins to different extents. Although this process preserves the structural integrity of the cell and cellular components, it can alter or mask antigens. This is a factor for consideration when choosing a fixative, and when considering how long to leave a tissue to fix.

Frozen tissue

Chemically fixed and processed tissue is not appropriate for some applications, and instead tissue is snap frozen in liquid nitrogen after collection. Applications that use frozen tissue include special stains for fat and urates, tests that require tissue enzymes to be active, and immunohistochemistry protocols that require much improved immunoreactivity towards primary antibodies. Snap freezing in liquid nitrogen is a useful way of preserving DNA and RNA, and is also adequate for limited morphological assessment such as the distinction between malignant and non-malignant cells. Prior to freezing, tissue can be immersed in a cryoprotectant solution such as sucrose to avoid ice crystal formation. Before or after freezing OCT (optimal cutting temperature) medium is added to the tissue to create a mould for cutting thin sections on a **cryostat**.

18.6.2 Tissue processing

During tissue processing for light microscopy, water is extracted from the fixed tissue and replaced with paraffin wax so that thin sections (3–5μm) can be cut. Samples are dehydrated by graded ethanol washes from 0% to 100% (to prevent tissue distortion), followed by xylene washes to allow efficient impregnation of the hydrophobic paraffin wax which is added last. This step is usually automated. It is important that the tissue is fixed throughout, otherwise even more distortion will be created in the unfixed middle parts of the tissue block.

Any water- or alcohol-soluble products are lost from the tissue during tissue processing. Therefore a frozen section may be more suitable if it is necessary to preserve these substances. A good example is fats within the liver, which will be readily lost during processing but are demonstrable with frozen sections.

Although we have only discussed tissue orientation and embedding of single tissue pieces, tissue microarrays can be an efficient way of examining many different tumours, or tumours of the same type, by having up to 30 pieces of each tumour on one slide (van de Rijn and Gilks 2004). Tissue microarrays are not limited to tumours, and many different tissues of interest can be included on the same slide. In-house tissue microarrays (sometimes referred to as 'sausages') can also be an efficient use of tissue, and later reagents, by including a positive and negative control and several test tissues on one slide.

18.6.3 Tissue embedding

Once the tissue is fixed, it is placed in a solid medium (block) so that thin sections can be cut. The most common type of block for formalin-fixed tissue is paraffin wax. In some applications other materials, such as epoxy resin and water-soluble waxes, are used to create the solid tissue support. The former is used with very thin sections or sections of hard tissues (uncalcified bone), and the latter with adipose tissue.

The tissue should be orientated in the cassette block or on the cryostat chuck to ensure that particular areas of interest will cut onto the slide foremost. Although routinely used cassettes will accommodate tissue pieces as large as 25 × 20 × 5mm, it is important that fixation is not hampered by thick tissue. The thinner or smaller the tissue, the quicker the fixative will act and the better the fixation will be.

18.7 Microscopic analysis of tissue samples

18.7.1 Histology stains

In a light (and electron) microscope it is difficult to identify particular cell types or cell components unless the specimen is stained to highlight these features. The most widely used stain for histology is the haematoxylin and eosin (H&E) stain. Haematoxylin is a basic dye used to stain the cell nucleus blue, and eosin is an acidic dye that stains the cytoplasm and connective tissue varying shades of pink. Special stains can be used to identify specific tissue components. In most cases these can be used with formalin-fixed tissue sections. However, some staining methods are not compatible with this fixation method. A selection of special stains, their use, and the commonly used fixation method is given in Table 18.1. Many detailed protocols are available for each staining method and are described in histology textbooks (e.g. Bancroft and Gamble 2008).

18.7.2 Immunohistochemistry stains

Immunohistochemistry (IHC) is used to detect a specific protein in a tissue section. IHC can tell you how much of a protein is present and where that protein is located in the tissue. Antibodies are used to bind to the protein of interest, the antibody is detected with a coloured stain, and hence the protein of interest is detected for analysis using light microscopy. Two widely used stains are horseradish peroxidase (HRP), which highlights the protein of interest with a brown colour, and alkaline phosphatase (AP), which produces a red colour. The steps involved in the IHC method are outlined in Figure 18.2,

Table 18.1 Special stains commonly used in histology

Stain	Tissue component	Colour	Type of tissue fixation required
Alcian blue	Acetic mucins	Blue	Formalin-fixed
Congo red	Amyloid deposit	Red–orange	Formalin-fixed
Giemsa	Distinguishes haematopoietic cell types	Blue	Formalin- or alcohol-fixed
Gordon & Sweet's	Reticular fibres	Black	Formalin-fixed
Gram stain	Bacteria	Gram-negative: red Gram-positive: blue/black	Formalin-fixed
Luxol Fast Blue MBS	Myelin	Blue	Formalin-fixed
Martius/scarlet/blue (MSB)	Connective tissue	Collagen: blue Muscle and fibrin: red	Formalin-fixed
Masson's Trichrome	Connective tissue and muscle	Collagen: blue Muscle fibres: red	Formalin-fixed
Methenamine silver	Urates	Black	Frozen tissue
Oil Red O	Fat	Red	Frozen tissue
Periodic acid-Schiff (PAS)	Carbohydrate/fungi	Purple	Formalin-fixed
Perls' Prussian blue	Haemosiderin	Blue	Formalin-fixed
Toluidine blue	Mast cells	Red–purple	Formalin-fixed
von Kossa	Calcium	Brown–black	Formalin-fixed

and an outline for the enzymatic detection of HRP and AP labels is given in Figure 18.3. One application of immunohistochemisty in the diagnostic laboratory is to aid cancer diagnosis by identifying specific tumour antigens (proteins). For example, in glioma, an antibody to the most common IDH1 mutation, R132H, is applied to tissue sections. The antibody binds to proteins with the R132H mutation and is detected by addition of a secondary labelled antibody or polymer. Once the label is developed, the tumour is scored positive for mutant IDH1 upon detection of the coloured cells by light microscopy.

There are a number of variables in the IHC method and therefore any new application is optimized first on a control tissue known to be positive for the protein of interest. Once it is confirmed that the antibody binds to the correct part of the tissue/cell and not non-specifically (background staining), the optimized conditions are ready to be used on experimental tissues.

Key steps that can be modified to provide successful IHC results include the antigen retrieval method used to unmask the epitopes in formalin-fixed tissue, the primary antibody dilution, and the length of time the tissue is incubated with the antibody. The choice of primary antibody is important. Ideally, you would choose an antibody that has already been shown to work on formalin-fixed tissue. Two groups of antibodies are commonly used in IHC, monoclonal and polyclonal antibodies. Polyclonal antibodies recognize multiple epitopes of a protein of interest, whilst monoclonal antibodies recognize a single epitope. Both types of antibody have advantages and disadvantages. With monoclonal antibodies there is less chance for cross-reactivity with other proteins (less background staining), and polyclonal antibodies have a greater chance of working on formalin-fixed tissue. Antibodies may detect other proteins, leading to non-specificity

See Chapter 9

Figure 18.2 The steps involved in immunohistochemistry (IHC), immunofluorescence (IF), fluorescence *in situ* hybridization (FISH), and chromogenic *in situ* hybridization (CISH). Detailed descriptions of each method are provided in Box 18.1. HRP, horseradish peroxidase; AP, alkaline phosphatase.

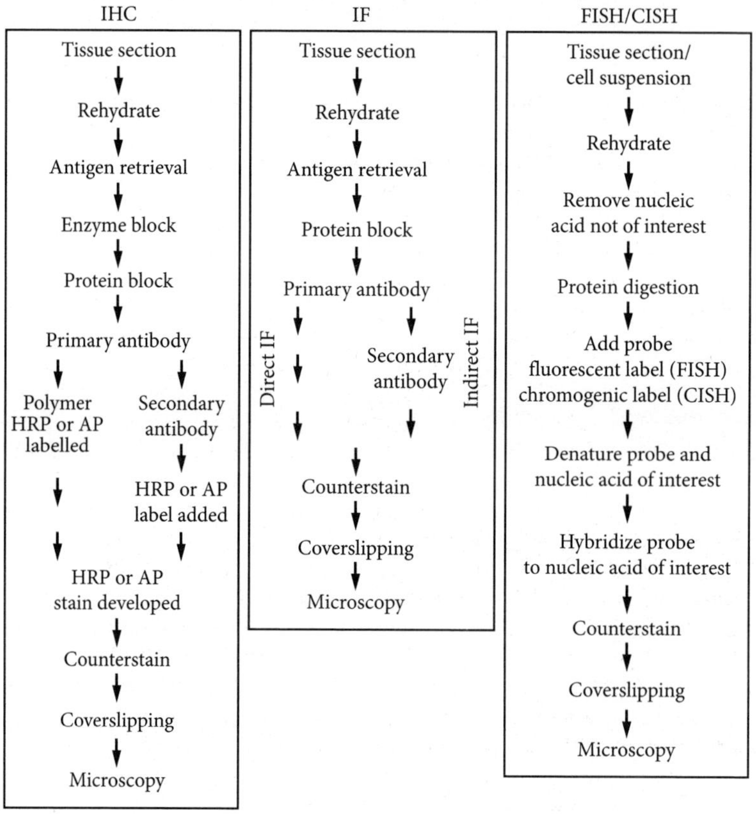

Figure 18.3 Enzymatic detection of horseradish peroxidase (HRP) and alkaline phosphatase (AP) labels. In immunohistochemistry the detection of a protein of interest requires a coloured precipitate to be produced where the primary antibody is bound. The DAB chromogen is often used for a HRP detection system. DAB is mixed with a substrate buffer containing hydrogen peroxide and added to the tissue section towards the end of the IHC procedure; a brown precipitate is formed upon reaction with the HRP label already present on the slide. The Fast Red chromogen is often used for the AP detection system. Fast Red is mixed with naphthol and added to the tissue section; a red precipitate is formed upon reaction with the AP label already present on the slide.

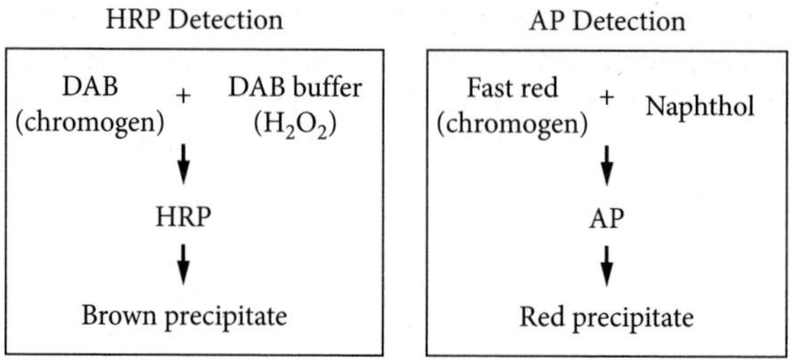

The varied uses of tissue sections

BOX 18.1

The immunohistochemistry (IHC), immunofluorescence (IF), fluorescence *in situ* hybridization (FISH), and chromogenic *in situ* hybridization (CISH) procedures begin with tissue sections cut onto special glass slides (coated with an adhesive or positive charge which helps to keep the tissue on the slide). Slides are baked at ~60°C to further aid tissue adhesion to the slide.

Immunohistochemistry (IHC) – The paraffin wax is removed (often with xylene), the slides are rehydrated (using a series of alcohol solutions with reducing alcohol content), and antigen retrieval is performed. *Antigen retrieval* reveals the antigens (required for immunoreactivity) that were masked during formalin fixation, and can be performed using heat (microwave in citrate buffer at pH 6, or in tris-EDTA buffer at pH 9) or enzymatic (proteinase K digestion) methods. *Blocking* steps are used to prevent non-specific background staining. A protein block containing normal serum (3–10%) is used to prevent the antibody binding non-specifically to the tissue and slide. Additional tissue components can be blocked if required, for example to block endogenous mouse IgG when mouse monoclonal antibodies are used on mouse tissue. *Primary antibody* is diluted and added to each slide. The incubation time is variable, typically from 1 hour to overnight. There is an inverse relationship between the dilution of antibody and the time required for adequate staining, i.e. less time is required if a more concentrated antibody is used. After antibody incubation, excess antibody is removed by washing. Detection of primary antibody uses polymers conjugated with HRP or AP, or secondary antibodies labelled with biotin. Following polymer incubation, excess polymer is removed by washing, and the peroxidase labels are developed (HRP is developed using DAB, and AP is developed using Fast Red). If secondary antibodies are used, slides are washed after secondary antibody incubation and an additional incubation period with streptavidin labelled with HRP or AP is required before the stain is developed. If the DAB signal is weak the staining can be enhanced with 'DAB enhancer' solutions. A *counterstain* is added, commonly haematoxylin to stain the cell nucleus. A light blue counterstain is preferable for investigating nuclear-bound antibodies as this allows the DAB or Fast Red stain to be seen. For long-term storage stained slides are cover-slipped. DAB-stained slides are dehydrated and cover-slipped using a hydrophobic mountant. The Fast Red stain is incompatible with a hydrophobic mountant, and so these slides are cover-slipped in an aqueous mountant.

Immunofluorescence (IF) – The initial part of the IF procedure is similar to that for IHC except that an enzyme block step is not required and the primary antibody is labelled with a fluorescent dye (direct IF), or the secondary antibody is labelled with the fluorescent dye (indirect IF). Instead of haematoxylin, DAPI is used to counterstain the nucleus. The results of IF staining are viewed using a fluorescent microscope.

Fluorescence/chromogenic *in situ* hybridization (FISH/CISH) – Paraffin-embedded tissue sections are deparaffinized and hydrated (not required if fixed cell suspensions dried onto slides are used). If DNA is to be detected, RNA is removed by incubation with RNAse I. Protein is digested with proteases to aid the access of the probe to the target sequence. The probe (fluorescently labelled for FISH and with a chromogenic label (HRP or AP) for CISH) is added to the slide, the probe and tissue are denatured to convert DNA to single-stranded copies, and the probe is hybridized to the complementary sequence of the DNA sequence of interest. The counterstains are DAPI for FISH and haematoxylin for CISH. Staining results are viewed with a light microscope (CISH) or a fluorescent microscope (FISH).

See Chapter 9

or cross-reactivity. Checking the Western blot results that are commonly provided by the manufacturer of the antibodies will determine if this is an issue. As Western blotting separates proteins on the basis of size, ideally the antibody should detect a single band corresponding to the size of the protein of interest.

18.7.3 Immunofluorescence stains

Immunofluorescence (IF) is similar to IHC, but fluorescent signals are used instead of peroxidases to identify proteins. The steps involved in the IF procedure are outlined in Figure 18.2. Advantages of IF include identifying multiple proteins in the same tissue section by using different coloured fluorescent dyes and compatibility with higher-resolution imaging (e.g. confocal microscopy). A major issue with IF is fading of the fluorescent signal. This can be prevented by using high-quality fluorescent dyes that are much more stable, preventing exposure of the slide to light during the staining procedure, and cover-slipping with an 'anti-fade' mountant.

See Chapter 17

18.7.4 *In situ* hybridization stains

In situ hybridization is used to detect specific DNA and RNA sequences in a tissue section. Instead of antibodies, probes (short sequences of oligonucleotides) with a complementary sequence to the DNA or RNA sequence of interest are used. If fluorescently labelled probes are used, the technique is referred to as FISH (fluorescent *in situ* hybridization). If probes labelled with peroxidase (HRP or AP) are used, the technique is referred to as CISH (chromogenic *in situ* hybridization). This technique is applied if protein detection is not possible, for example to visualize telomeres in a cell. It is also commonly used in cancer research to investigate changes in gene dosage, such as loss of one or both copies of a particular gene. FISH/CISH is usually performed using paraffin-embedded tissue sections, frozen tissue sections post-fixed in 4% paraformaldehyde vapour, or methanol/acetic acid-fixed cell suspensions dried onto glass slides. The steps involved in the FISH and CISH procedures are outlined in Figure 18.2.

The key to successful *in situ* hybridization is the use of good quality probes. Probes consist of 20–50 oligonucleotides. Specific modifications can be made to create probes that can bind tighter and with greater specificity to the target nucleic acid, that is, PNA (protein nucleic acid) probes, in which the oligonucleoides are added to a peptide backbone. FISH, like IF, requires care to prevent fading of the fluorescent label. Again this can be achieved by using stable fluorescent labels, protecting the slide during staining from light, and coverslipping slides in mountant that prevents fading.

18.7.5 Staining of multiple targets

The IHC, IF, FISH, and CISH techniques are all compatible with identifying more than one target at a time. The use of fluorescence in IF and FISH allows up to four targets to be identified in the same tissue section relatively easily. The fluorescence labels must be selected carefully to avoid using labels with overlapping emission spectra. Multiple staining is easier if primary antibodies are labelled directly. If indirect IF is used, multiple staining is easier if the primary antibodies are raised in different species, thus allowing the multiple primary antibodies, and subsequently the multiple secondary antibodies, to be added at the same time. However, if, for example, two mouse primary antibodies are used, more incubation steps are required: (i) the antibody to the first protein of interest; (ii) the secondary antibody to the first primary antibody; (iii) the antibody to the second protein of interest; (iv) the secondary antibody to the second primary antibody with a different fluorescent label to that used in incubation step (ii).

The IHC and CISH methods can easily detect two targets; one target can be stained using DAB and the second target stained using Fast Red. Unlike fluorescence, chromogenic multiple stains are viewed at the same time under the microscope. Therefore to distinguish the stains it is preferable for the target stained with DAB to be in a different part of the cell to the target stained with Fast Red.

If you use sections from archived blocks that are several years old remember that the age of the block can have detrimental effects on antigen preservation and retrieval. For example, older tissue processors or manual processing techniques were often temperature variable, and fixation periods could have been longer. Check with the source of your blocks to find out when they were processed and what processing technique was used.

Mr Smith's colectomy

CASE STUDY 18.1

Ethical approval is granted, Mr Smith is delighted that his diseased tissues will benefit medical research, he has had enough time to read and understand the study information, he has asked a few questions, and he has signed the consent form. The surgeon has removed his colonic cancer which has now just been delivered from theatre to the pathology laboratory or 'cut up room'.

QUESTION 1
What will the pathologist need to consider when fixing this specimen?

The colon needs to be opened (along the antimesenteric border or away from the cancer) and cleaned to allow the fixative good access to the mucosa. The thickness of the tissues should also be considered. If the fixative needs to diffuse into a lot of thick fatty tissue, it is often best to open or cut the tissue before fixation. Care should be taken not to ruin any pathological or anatomical relationships (such as resection margins).

Often, the colon will curl and the muscle layers will contract when fixing, so it will be opened and pinned out to maintain good tissue orientation for subsequent sectioning. A large specimen like a colonic resection will require at least 24 hours to fix properly and at least 10 times its volume of formalin.

For the pathological report, the specimen will be described and measured, and the features of the cancer such as size, extent of spread, necrosis, and relationship to other tissues will be documented in detail (Figure 18.4). Tissue samples are selected from strategic parts of the specimen to illustrate the pathological features microscopically. A search will be made through the mesentery noting location, size, appearance, and number of mesenteric lymph nodes to document tumour spread.

The samples will be carefully logged and placed in the labelled tissue cassettes for processing. Usually the tissue side of interest will be placed face down in the cassette, and later embedded, facing outwards for trimming and sectioning. Small samples for light

Figure 18.4 Freshly excised colonic cancer measuring about 20mm in diameter. The arrow indicates the necrotic centre.

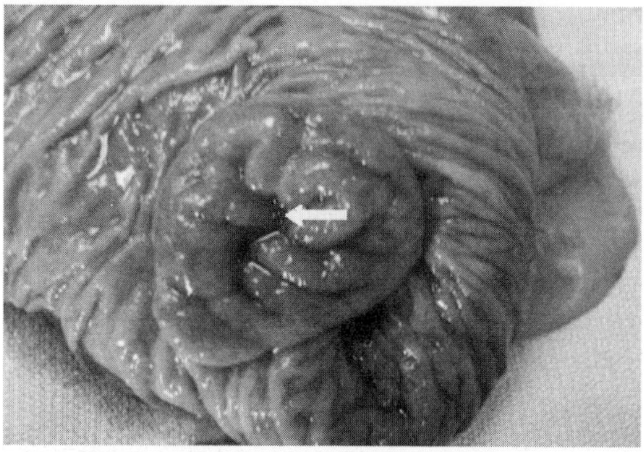

microscopy will fix more quickly, but remember that some fixatives take longer to diffuse into tissues than 10% formalin. The tissue is best cut into 2–3mm cubes for transmission electron microscopy when fixed in glutaraldehyde. Processing is also best achieved with smaller samples to allow for diffusion though the tissue.

QUESTION 2

You are collecting tissue samples, and have asked the pathologist for pieces of tissue from the centre and periphery of the cancer. You wish to determine what sort of immune response exists at the infiltrating edge of the cancer. The pathologist provides you with two samples as requested, one from the centre and one from the periphery. What are the considerations pertaining to the tissue collection location?

The centres of colonic cancers are often necrotic, and indeed in the centre of this specimen (Figure 18.4) you can just see the start of central collapse and ulceration (white arrow). Make sure that the tissue you are studying is not necrotic by doing a H&E section. The periphery of the cancer may also contain precursor benign adenoma or non-invasive cancer. If the tissue is going to be frozen for analysis, a frozen section or even touch preparation can confirm the nature of the tissue.

The inflammatory infiltrate will also change according to the pathology present. A necrotic sample will contain dead and dying acute inflammatory cells, epithelial or stromal cells, and probably bacteria. At the outer edge of the cancer there are areas where an inflammatory infiltrate is sparse. Therefore IHC or *in situ* hybridization studies will be the most informative to illustrate exactly which inflammatory cells are present and where. Flow cytometry is also useful, but remember to check the tissue sample for an inflammatory infiltrate in the first instance, or a false-negative result is possible. Furthermore, you may need to determine that the infiltrate you are examining is reactive to the cancer, and not normal or adenomatous tissue.

See Chapter 16

QUESTION 3

How will you ensure that invasive epithelial cancer has been sampled?

A hallmark of cancer is the ability of malignant cells to invade surrounding tissues, and therefore one should not find epithelial cells away from their usual intramucosal

Figure 18.5 H&E stain of a section of colon cancer. Low-power magnification H&E section showing the full thickness of the bowel wall from lumen to peritoneal cavity. To the left of the image is an invading cancer. The invading edge has an abrupt transition, surrounded by a few patches of chronic inflammatory cells (yellow arrows). There is a progression from mucosa (yellow star) to remnant adenoma (blue star), which is difficult to identify macroscopically, to invasive cancer (white star). The white arrow indicates the muscularis mucosa which separates intramucosal cancer from cancer invading the submucosa below. The blue dots on the left, within the cancer, indicate lakes of extracellular mucin produced by the cancer. For a colour reproduction of this figure, please see Plate 16.

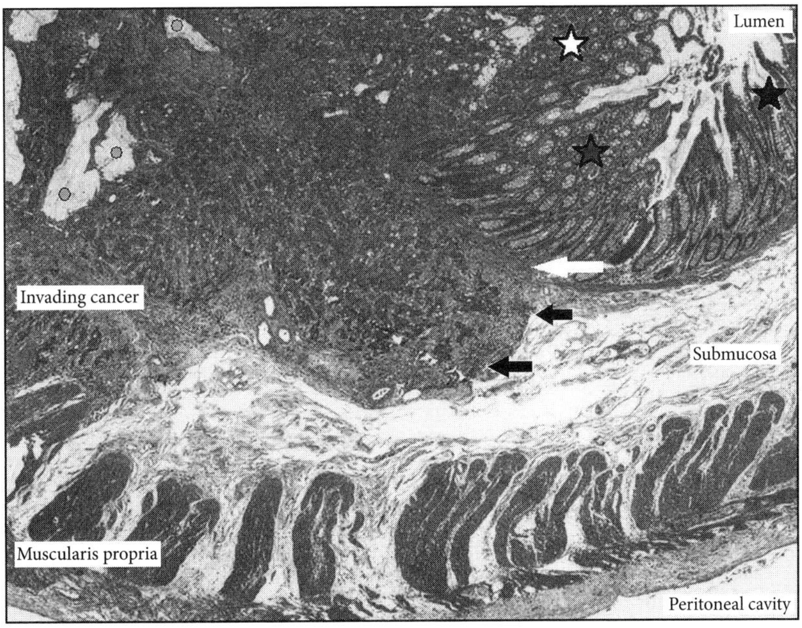

residence below the muscularis mucosa (with minor exceptions). Being able to recognize the muscularis mucosa means that anything beyond this is more than likely invasive cancer (Figure 18.5). This can be especially helpful in well-differentiated cancers where glands are well formed and cytology is uniform. Hence, embedding a sample to demonstrate the mucosa, muscularis mucosa, and deeper layers can be very helpful.

? QUESTION 4
Identifying lymphocyte types: How do you separate B cells from T cells?

Historically, the identification of the B and T lymphocytes in H&E or Giemsa-stained sections, for example, was impossible. On rare occasions a cerebriform nuclear morphology in the right pathological lesion (e.g. a malignant T-cell lymphoma of the skin known as mycosis fungoides) highlighted a possible T-cell lymphocyte, but most often immunohistochemistry is used. The challenge now is to subtype lymphocytes, and while flow cytometry offers a fast and multi-antibody approach, the number of antibodies that can be applied to a single section hinders studies on histological samples. Recent immunological advances now necessitate T-helper 1, 2, and 17 phenotypic analysis, for example, and require double staining. Ensuring that you have optimized both antibodies is essential. Immunofluoresence can be a useful tool in these situations (see Figure 18.6).

Figure 18.6 Identifying lymphocyte types using immunofluorescence to highlight the T-helper 17 response: CD4 cells highlighted in green fluorescence, interleukin 17 expression in red fluorescence, and nuclei by DAPI staining in blue. The combined result, illustrating co-localization, is illustrated in the bottom right panel (white arrow). Note the autofluorescent blood vessel wall in the top right panel (blue star). For a colour reproduction of this figure, please see Plate 17.

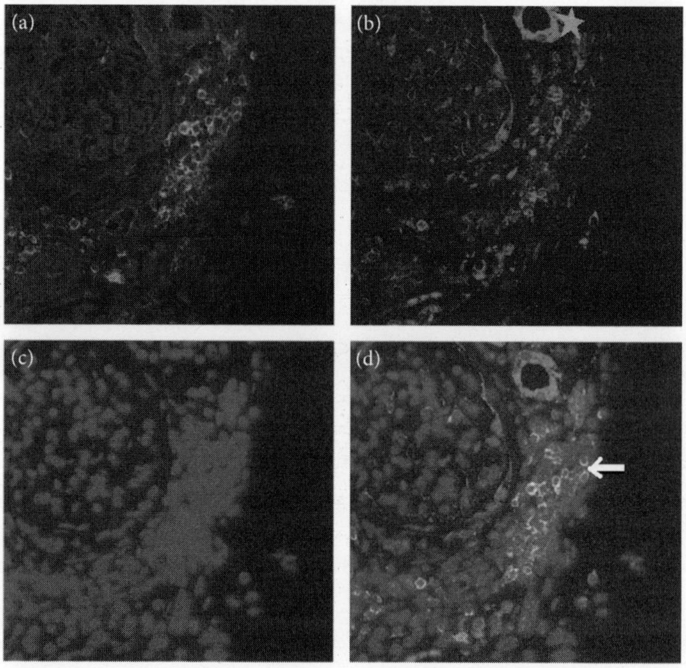

18.8 Chapter summary

- This chapter provides an overview of the powerful results that can be derived from the study of tissues. While it is not possible to cover every aspect or avenue of investigation, the basic requirements of ethics, collection, processing and tools of investigation have been outlined.

- Many good experiments have been hampered by the inadvertent investigation of normal tissue elements or necrotic parts of a tumour. Hence correct selection and identification of any tissue for an experiment is critical.

- Ensuring appropriate tissue preservation and processing is also critical to obtaining reliable investigative results.

- The use of stains can aid identification of different tissue elements. A number of stains are available: histology stains, immunohistochemistry stains, immunofluorescence stains, and *in situ* hybridization stains.

- **Finally, as a non-medically qualified scientist, do not forget to consult surgeons and pathologists who are familiar with tissue retrieval and can often provide valuable assistance.**

References

Bancroft, J. D. and Gamble, M. (2008) *Theory and practice of histological techniques*. Edinburgh: Churchill Livingstone.

Capper, D., Zentgraf, H., Balss, J., Hartmann, C., and von Deimling, A. (2009) Monoclonal antibody specific for IDH1 R132H mutation. *Acta Neuropathol* **118**: 599–601.

Dotti, I., Bonin, S., Basili, G., and Faoro, V. (2011) Formalin-free fixatives. In: Stanta, G. (ed), *Guidelines for Molecular Analysis in Archive Tissues.* New York: Springer.

Gilfillan, A.M. and Beaven, M.A. (2011) Regulation of mast cell responses in health and disease. *Crit Rev Immunol* **31**: 475–529.

Hanahan, D. and Weinberg, R.A. (2011) Hallmarks of cancer: the next generation. *Cell* **144**: 646–74.

Herceg, Z. and Vaissiere, T. (2011) Epigenetic mechanisms and cancer: an interface between the environment and the genome. *Epigenetics* **6**: 804–19.

Liteplo, R. (2002) *Formaldehyde*. Geneva: World Health Organization.

Martin-Subero, J.I. (2011) How epigenomics brings phenotype into being. *Pediatr Endocrinol Rev* **9** (Suppl 1): 506–10.

Parsons, D.W., Jones, S., Zhang, X., et al. (2008) An integrated genomic analysis of human glioblastoma multiforme. *Science* **321**: 1807–12.

van de Rijn, M. and Gilks, C.B. (2004) Applications of microarrays to histopathology. *Histopathology* **44**: 97–108.

Whiteside, T.L. (2012) What are regulatory T cells (Treg) regulating in cancer and why? *Semin Cancer Biol* **22**: 327–34.

Yan, H., Parsons, D.W., Jin, G., et al. (2009) IDH1 and IDH2 mutations in gliomas. *N Engl J Med* **360**: 765–73.

4

Working with models in the biomolecular sciences

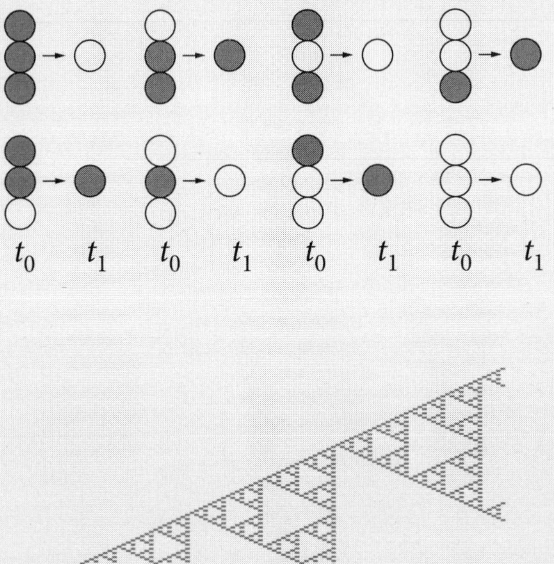

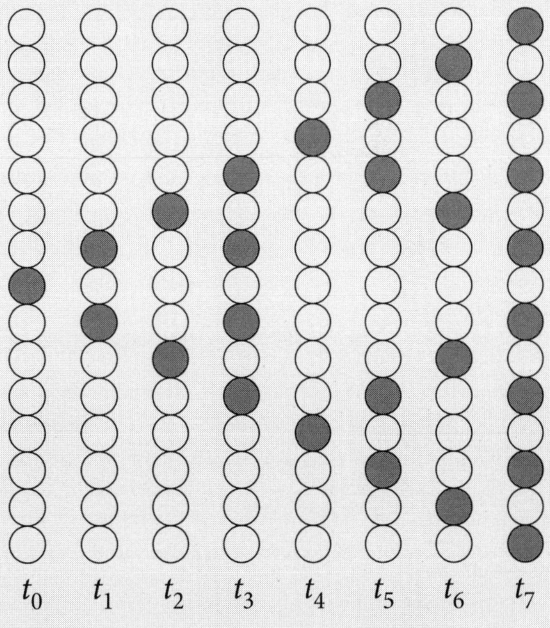

Mouse models in bioscience research

19

Brian Corbett and Jeannie Chin

Chapter overview

Animal models can be crucial for answering important questions in biological and biomedical research. Whereas biochemical or cell biological methodologies provide mechanistic insight into the regulation and role of a specific element (such as a protein or drug) at the molecular or cellular level, animal models are critical for understanding the role of those elements at an organismal level. A number of model organisms are commonly used in biological research, including worm, fly, zebrafish, mouse, rat, and non-human primates. However, in this chapter we will focus specifically on mouse models. We will cover different types of genetic manipulations in mice, how they are generated, and examples of their applications.

LEARNING OBJECTIVES

This chapter will enable the reader to:

- describe the different types of genetically engineered mice used in biomedical research;
- differentiate between the genetic manipulations used to create each type of mouse model;
- outline the basic steps for generating each of the mouse models discussed in this chapter;
- assess the advantages and disadvantages of each type of mouse model;
- determine which mouse models are best suited to address different types of experimental questions.

19.1 Introduction

Research involving animals has played a critical role in advancing our understanding of both the normal functions of the body and the mechanistic basis for various diseases. Such vital information has allowed, and continues to enable, the biomedical research community to develop novel therapeutics for human diseases and conditions that would otherwise have remained untreatable. Although, *in vitro* and *in silico* studies play an important role in investigating the cellular or molecular roles of particular proteins or genes, an understanding of how these proteins and genes contribute to systems level functioning in an organism as a whole requires the analysis of their functions in animal models.

As animal models are often deemed the most appropriate for addressing particular scientific questions, nearly all countries have developed governmental regulatory programmes to ensure that researchers are adequately trained to work with animals. It is mandatory that the minimum number of animals are used for each study, that they are properly cared for, and that every effort is made to avoid pain or distress to research animals. All studies also have to be scientifically validated. Thus, within this context of regulated animal research, the benefits in the form of information gained can be achieved while ensuring that all studies minimize the use of, and the distress to, animals in research.

Although a number of model organisms are commonly used in biological research, each with their specific applications, advantages, and limitations (summarized in Table 19.1), in this chapter we will focus specifically on mouse models. For a number of reasons mouse models are particularly useful for probing the function of genes and proteins *in vivo*. First, mice are mammalian and have a fundamental anatomy that is comparable to that of humans, making research in mouse models more easily translatable to human conditions. In addition, 99% of human genes have identifiable mouse **homologues** (genes of similar sequence, structure, and function), and vice versa (Waterston et al. 2002). The anatomical and genetic similarities between humans and mice allow the generation of disease and treatment models that have been very useful for understanding human disease. Furthermore, compared with other mammalian models, such as non-human primates, mice are relatively affordable to maintain in large numbers. Moreover, colonies of mice can be bred fairly rapidly. Mice have a gestation period of about three weeks, with litter sizes of six to 12 pups depending on the background **strain** of the mouse. Mice reach maturity at around 3 months of age and have a lifespan of approximately 2 years. Finally, the mouse **genome** has been fully sequenced and is simple to experimentally manipulate, making genetic alterations an extremely powerful tool.

The genetically altered mouse models we will discuss in this chapter are outlined in Table 19.2 and described below.

- **Transgenic mice** Mice in which exogenous DNA, i.e. a **transgene**, is integrated into the mouse genome at a random site, is inherited in Mendelian fashion, and is expressed under the control of an experimentally specified **promoter**.
- **Knockin mice** Mice in which a transgene replaces the endogenous gene at a specific genetic **locus**, allowing expression of the transgene to be controlled by an endogenous promoter.
- **Knockout mice** Mice in which a specific endogenous gene is targeted and deleted from the locus, or rendered non-functional either by replacement with or addition of irrelevant DNA.
- **Conditional knockout mice** Mice in which the knockout of a specific endogenous gene depends on the presence of an activator transgene driven by an experimentally chosen promoter.
- **Mice with inducible or repressible gene expression** Mice in which expression of a specific gene can be experimentally induced or repressed in a reversible manner via administration of a drug.

Table 19.1 Model organisms used in biomedical and biological research

Animal	Major uses	Advantages	Disadvantages	Lifespan
Worm *Caenorhabditis elegans*	Genetic screening and manipulation, anatomical studies	Easily manipulable genome, low anatomic variability, short inter-generational interval allows rapid expansion of colony, easy and affordable to maintain in large numbers	Less translational than mammalian models, small size restricts experiments	2–3 weeks
Fly *Drosophila melanogaster*	Genetic screening and manipulation	Most genetically manipulable organism for research, easy to produce in large numbers for screening, short inter-generational interval allows rapid expansion of colony, affordable to maintain in large numbers	Less translational than mammalian models, small size restricts experiments	3–4 weeks
Zebrafish *Danio rerio*	Genetic screening, embryology and vertebrate development studies	Easily manipulable genome, easy to produce in large numbers for screening, transparent embryos allow easy visualization of internal organs or fluorescently labelled cells	Less translational than mammalian models, small size restricts experiments	2–3 years
Mouse *Mus musculus*	Genetic engineering, behavioural studies, modelling human disease	Manipulable genome, ability to reduce variability between subjects by controlling genetic background and experimental conditions, exhibit many behaviours with human equivalents, inter-generational interval allows rapid expansion of colony	Large colonies expensive to maintain (relative to fly or worm models)	2–3 years
Rat *Rattus norvegicus*	Pharmacology, behavioural studies	Intermediate physical size facilitates studies of mammalian physiology while reducing costs associated with larger animals, exhibit many behaviours with human equivalents	Less manipulable genome	2.5–3.5 years
Non-human primate Various species	Studies of higher-order cognitive and behavioural function, translational assessments for biomedical research	Highly evolved nature allows study of higher-order cognitive and behavioural functions, highly translational	Expensive to maintain and care for	20–50 years, depending on species

19.2 Transgenic mice

Transgenic mice are useful tools for investigating the role of a gene at the organismal *in vivo* level. The term 'transgenic' mouse refers to a mouse with exogenous DNA, or a transgene, introduced into its genome. The transgene may be **homologous**, meaning that its origin is from the same species (in this case mouse), or **heterologous**, meaning that its origin is from a different species (e.g. human). Both homologous and heterologous transgenic mice typically over-express the transgene, i.e. they express the transgene at levels higher than that gene is naturally expressed. It is important to note that the term 'transgenic' is sometimes used to describe a number of different types of genetically altered mice with various genomic manipulations, such as knockout or knockin mice.

Table 19.2 Genetically altered mouse models used in biomedical and biological research

Model	Characteristic	Major uses	Advantages	Disadvantages
Transgenic	Over-expresses exogenous gene in specific tissues or cell types	Assess how over-expression of a heterologous or homologous transgene will affect the mouse	Does not disturb mouse genome	Random transgene integration (no control over copy number or expression pattern)
Knockin	Expresses exogenous gene in place of endogenous mouse gene	Assess effects of a gene with altered function	Gene is expressed under control of natural promoter.	Not possible to experimentally regulate levels of gene expression
Knockout	Lacks the gene of interest	Determine the consequences of the loss of function of a gene	Gene is absent from entire genome	Gene is absent throughout development and may elicit unknown compensatory effects
Cre–*lox* system	Conditionally knocks out or expresses a gene of interest in specific cells	Assess gene function in a specific tissue or cell type	Much higher spatial resolution and control of gene expression	Integration of Cre or floxed elements may occur incorrectly
Tet-On and Tet-Off systems	Conditionally and reversibly turn on/off gene expression by adminstration of tetracycline or doxycycline	Assess effects of transient expression of gene of interest, or assess effects of gene expression only at a certain time period	Possibility to avoid developmental effects of gene manipulation, and provides ability to examine before and after effects of gene regulation	Gene expression can sometimes be 'leaky' and not necessarily strictly under control of tetracycline or doxycycline

However, since we will discuss the latter types of genomic manipulations in different sections of this chapter, we will specifically refer to transgenic mice as having exogenous DNA introduced into their genome.

19.2.1 Application of transgenic mice in research

Transgenic mice are used when a researcher wants to assess the role of novel or increased expression of a gene of interest *in vivo*. The primary questions a transgenic mouse model can answer are:

- How will the presence of a heterologous transgene affect the organism?
- How will over-expression of a homologous transgene affect the organism?

Transgenes that express human genes are common as they allow researchers to investigate the role of these genes, particularly those carrying disease-related mutations, in disease pathogenesis. Transgenic mice are useful because, with the exception of transgene introduction, the genome is left untouched (although see section 19.2.2 for a discussion of random transgene integration). This allows any results to be attributed to the transgene. If another model was employed, for example a knockin mouse, it would be unclear whether the results obtained were due directly to the transgene or to the manipulation of the mouse's natural wild-type gene. The use of such transgenic mice to study disease is exemplified by the many transgenic mouse models of cancer and Alzheimer's disease (Chin 2011). Two applications of transgenic mice are presented below.

Heterologous transgenic mice: Alzheimer's disease transgenic mice

Several mutations in the human gene that encodes the amyloid precursor protein (APP) have been linked to autosomal dominant forms of Alzheimer's disease. These mutations result in increased levels of aggregation-prone amyloid beta (Aβ) peptides that are released when APP is cleaved by proteases. Aβ is the primary constituent of amyloid plaques, a major pathological hallmark of the disease. Although Aβ appears to play a critical role in Alzheimer's disease pathogenesis, the exact mechanisms by which it does so are unclear. For this reason, a number of transgenic mice have been created that express human APP with one or more disease-linked mutations. These mice have provided great insight into the mechanistic basis of Alzheimer's disease (Chin 2011). Notably, while mice do express a murine form of APP from which Aβ peptides are produced, mouse Aβ differs from human Aβ by three amino acids. This three amino acid difference renders mouse Aβ much less likely to aggregate into neurotoxic forms, which may explain why mouse APP/Aβ does not induce the cognitive deficits associated with Alzheimer's disease. Thus, in order to investigate how human APP/Aβ affects neuronal function and contributes to disease pathogenesis, *heterologous* transgenic mice must be employed. It is important to keep in mind that the endogenous mouse APP gene is not manipulated in any way in this type of transgenic mouse model. This is important for maintaining the normal functions of murine APP during development and lifespan of the mouse, and for allowing any observed effects to be attributed to the toxic gain of function consequences induced by the human APP gene and the resulting aggregation-prone Aβ. Thus if your scientific question calls for the introduction of a heterologous mutation-carrying gene in a mouse, without disrupting endogenous gene function, transgenic models are the most appropriate. Any observed effects can then be attributed exclusively to the introduction of the transgene, as the endogenous mouse gene is left unaltered and retains normal function.

Homologous transgenic mice: BDNF transgenic mice

Sometimes an investigator may not be interested in expressing a heterologous gene into a mouse, but rather in over-expressing a particular mouse gene in order to examine its consequences. Transgenic mice also offer a means for accomplishing this. One example is the use of transgenic mice that over-express murine brain-derived neurotrophic factor (BDNF). BDNF has been proposed to mediate the antidepressant effects of the selective serotonin reuptake inhibitor (SSRI) class of antidepressants for the successful treatment of depression. As patients typically require several weeks of treatment with SSRIs before improvement is noticeable, it is unlikely that their effectiveness as antidepressants is mediated by their acute modulation of serotonin levels and related neurotransmission. However, prolonged treatment with SSRIs also increases BDNF levels in the brain. As exogenous BDNF has antidepressant actions when applied to key areas of the brain, and its upregulation by SSRIs occurs over a time course of weeks, it has been proposed that upregulation of BDNF is the mechanism by which SSRIs exert antidepressant actions (Shirayama et al. 2002). This hypothesis implies that increased levels of BDNF may be sufficient to attenuate depression-like symptoms. Therefore if a researcher is interested in investigating the role of increased levels of BDNF in a mouse model of depression, the transgenic model is appropriate because it allows for over-expression of BDNF. Indeed, BDNF-transgenic mice have been created to examine the role of BDNF in depression (Govindarajan et al. 2006). The endogenous mouse BDNF gene is left unaltered, but transgenic mice have additional copies of the BDNF gene at novel loci in their genome. Expression of BDNF transgenes can be targeted to the forebrain areas most likely to play roles in depression, for example by using the calcium-calmodulin protein kinase II (CaMKII) promoter to drive BDNF expression. The choice and use of promoters to drive transgene expression is of critical importance, and is discussed more extensively in section 19.2.3.

19.2.2 General considerations for transgenic mice

While transgenic mice have very useful applications, there are also some limitations that must be kept in mind. First, the genomic locus or loci into which a transgene integrates is random. Although there are some regions of the genome that appear to be more susceptible to integration, where the transgene will integrate cannot be controlled unless specific gene-targeting techniques are used (see section 19.3.1). Thus position effects that influence the pattern and level of transgene expression may occur. The novel locus into which the transgene integrates may be in a region of the genome primarily comprised of heterochromatin, or tightly packed transcriptionally silent DNA. Additionally, the transgene may be inserted into an endogenous gene, thereby disrupting the function of that gene, or be subject to the regulatory elements that control that gene's expression. Furthermore, additional silencing positions lead to mosaicisms in which gene expression may not be expressed in all cells of a specified tissue (Sauer 1998). Finally, the number of transgene copies that integrate into the genome is also random and can alter transgene expression pattern and levels. Therefore a number of transgenic organisms expressing the same transgene are typically created, and each one is evaluated to determine expression stability, level, and patterns using standard techniques such as *in situ* hybridization, immunohistochemistry, and Western blot analyses. The **founder**, i.e. the transgenic mouse that best matches the research goal, is identified from these analyses. The transgenic line is then propagated from that founder (see section 19.2.3).

See Chapters 9 and 18

It is also important to bear in mind the response of cells to over-expression of the transgene. When a gene or protein is over-expressed, compensatory mechanisms may be induced in the cells in which expression occurs. Compensatory mechanisms refer to reciprocal responses of other components in the cell in response to the presence of high levels of the over-expressed gene product. For example, if the expression of a receptor ligand is increased, the expression of its cognate receptor might be decreased in an attempt to maintain normal levels of signalling in that pathway. While this is an oversimplified example, it serves to highlight the fact that the effect of transgene expression may be unpredictable.

Finally, transgenic mice typically express the transgene from embryonic development. Such early expression may not be ideal for some scientific questions, such as when the aim is to understand the role of the gene/protein in an adult organism. Expression of the transgene during embryogenesis may induce undesired, and perhaps unrelated, effects that might hinder the evaluation of the gene of interest. Such effects can be minimized by temporally and/or spatially regulating gene expression using regulatable gene expression models (see sections 19.4 and 19.5).

19.2.3 Generation of transgenic mice

An overview of how transgenic mice are generated is shown diagrammatically in Figure 19.1 and described below. Transgenic mouse generation begins by designing the genetic **construct** to be introduced into the host mouse genome. The genetic construct

Figure 19.1 Generation of transgenic mice. (a) First, the transgene construct is developed, which contains both the coding region of the gene to be expressed and the promoter that will be used to drive gene expression. The transgene construct is ligated into a vector and cloned in a bacterial host. Restriction enzymes are then used to cut the circular gene targeting vector to linearize the construct, which is then purified. (b) The double-stranded DNA constructs are injected into the male pronucleus of a fertilized egg removed from a recently mated female. Successful integration of the DNA construct is enhanced if the eggs have been recently fertilized (within 3 hours). (c) The transgenic fertilized eggs are injected into the oviducts of a pseudo-pregnant female mouse. (d) When the offspring are born, their genotypes are assessed to identify the transgenic mice. Only about a third of the mice will be transgenic. (e) The transgenic mice are bred with non-transgenic wild-type mice to produce offspring that will be analysed to assess the levels and patterns of transgene expression. (f) The transgenic mouse line with the most ideal expression level and pattern will be further bred to propagate the colony of transgenic mice.

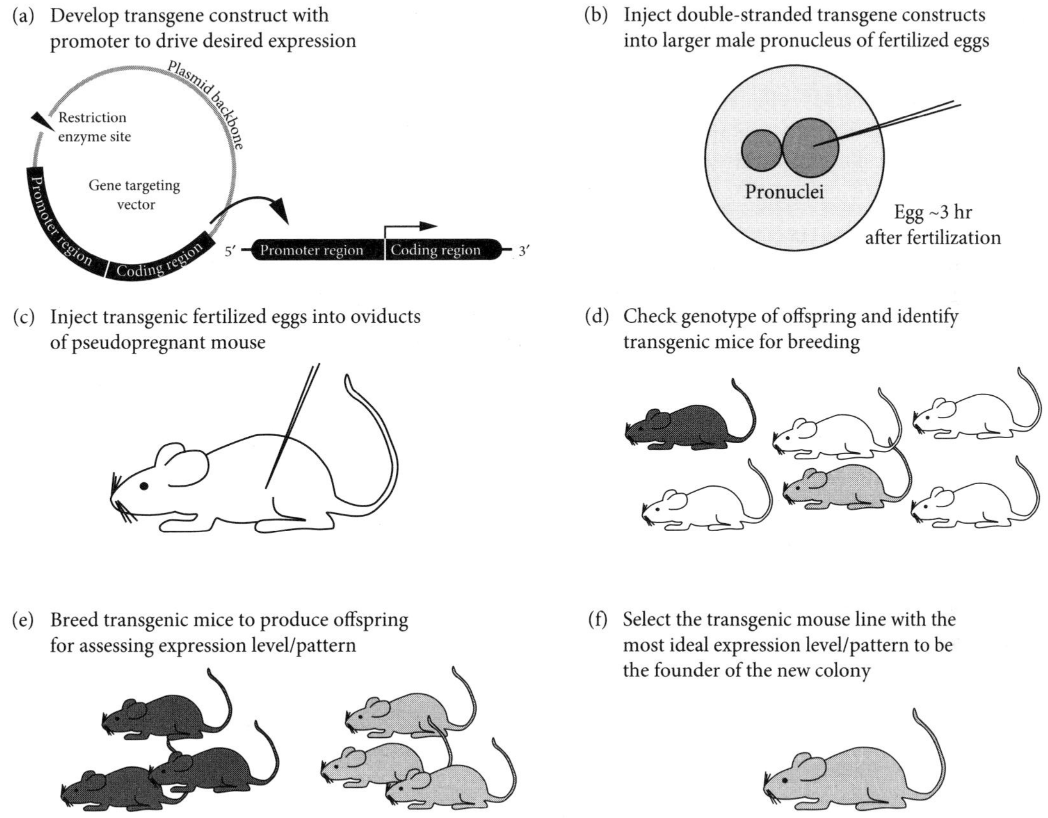
(a) Develop transgene construct with promoter to drive desired expression
Plasmid backbone
Restriction enzyme site
Gene targeting vector
Promoter region
Coding region
5′
Promoter region
Coding region
3′
(b) Inject double-stranded transgene constructs into larger male pronucleus of fertilized eggs
Pronuclei
Egg ~3 hr after fertilization
(c) Inject transgenic fertilized eggs into oviducts of pseudopregnant mouse
(d) Check genotype of offspring and identify transgenic mice for breeding
(e) Breed transgenic mice to produce offspring for assessing expression level/pattern
(f) Select the transgenic mouse line with the most ideal expression level/pattern to be the founder of the new colony

comprises two key components: the DNA coding region (cDNA) of the transgene, and the promoter region that drives gene expression. Promoters are genetic sequences, usually located upstream of the 5′ end of a gene, that regulate the transcription of that gene. Promoters control when and where the transgene will be expressed, and are selected on the basis of the aim of the experiment. If expression of the transgene is to be restricted to a particular cell type, then a cell-type-specific promoter is used. For example, if the experiment calls for transgene expression to occur only in glial cells of the brain, then the transgene would be placed under control of the glial fibrillary acidic protein (GFAP) promoter. GFAP is only expressed in glia, as the combination of proteins required to bind to the GFAP promoter sequence and promote its transcription are only expressed in this cell type. Other cell-type-specific promoters can be used to drive expression, thus providing many options for regulating transgene expression in most cell types.

Transgene constructs typically consist of exons, the coding regions of DNA. However, if the researcher is interested in splice variants of the gene of interest, then introns, the non-coding regions of DNA that separate exons into discrete units and allow alternative splicing events to take place, can be included. If the transgene is heterologous, splicing might not necessarily be predictable as the gene is no longer subject to its endogenous splicing machinery.

See Chapter 1

Following its design, the transgene construct is ligated into a vector. A common type of vector is a **plasmid**, a naturally occurring circle of DNA that can easily be replicated by bacteria to produce many copies of the plasmid containing the transgene construct. Once numerous copies are made in bacteria, the circular plasmids are linearized using restriction enzymes to cut the plasmid and isolate the transgene construct from the plasmid backbone. The linear DNA is purified and is then ready to be inserted into the genome of a host mouse. This is typically accomplished by microinjecting the linear double-stranded DNA of the transgene construct into the male pronucleus (a nucleus containing haploid DNA) of fertilized eggs collected from recently mated female mice. This refers to a 3–5 hour window in which an ovum has become fertilized, but fusion between sperm and ovum nuclei has not yet taken place. The microinjected DNA constructs randomly integrate into the genome and a zygote is then formed following nuclear fusion. The zygotes are in turn transferred into the oviducts of a pseudo-pregnant mouse—a female mouse that has been mated with a sterile male mouse. This confers the appropriate hormonal response required for gestation, although the sterile male mouse did not actually impregnate the surrogate mother. If the zygote transfer is successful, embryos will be brought to term and transgenic mice will be generated (Carter and Shieh 2010; Miller 2011).

The success of transgenic mouse generation relies on a number of factors. First, the microinjected transgene must be incorporated into the host genome of the pronucleus. There is some variability in this step, due in part to the randomness of double-stranded breaks that occur naturally in the host genomic DNA. Transgenes integrate at the sites where double-stranded breaks occur because the naturally occurring ligation repair processes incorporate the double-stranded linear transgene contruct into the host genome. Multiple copies of the transgene may be incorporated if there are multiple double-stranded DNA breaks in the genome. This random integration can be overcome, if desired, by directing the transgene to specific sites in the genome. This method of genomic incorporation is similar to knockin and knockout mouse generation (see section 19.3.7).

See Chapter 2

Once the transgenic mice are born (about six to ten mice per litter), their transgenic genotype is confirmed. This is typically achieved by taking a small tail biopsy from which genomic DNA is extracted and analysed by the polymerase chain reaction (PCR) for the presence of the transgene. Only about a third of the offspring will successfully express the transgene. Transgenic mice with confirmed genotypes are called founders, and are mated with non-transgenic wild-type mice to produce offspring. Following Mendelian genetics, mating a transgenic mouse with a non-transgenic mouse will produce litters in which 50% of offspring are non-transgenic, and 50% are transgenic and have transgene expression levels and patterns identical to the transgenic parent. Some of these transgenic offspring will be sacrificed and analysed to assess transgene expression patterns and levels in different tissues. This will include using Northern blot analysis and *in situ* hybridization to examine RNA levels and patterns, and Western blot analysis and immunohistochemistry to examine protein levels and distribution. As the number and pattern of integration sites can vary, different founders will have different expression patterns

See Chapters 9 and 18

and levels. Only the founder whose offspring exhibit the most ideal expression patterns and levels will be kept and continually bred to propagate the transgenic line, while the other founders are typically discarded. Occasionally, multiple founders are kept if they exhibit different expression levels or patterns of experimental interest.

19.3 Gene targeted mice

Transgenic models add an exogenous gene to the genome, but endogenous genes may also be *targeted* and manipulated. Such *gene-targeted* approaches include knockin mice, as discussed in section 19.3.1, and knockout mice, discussed in section 19.3.4. In knockin mice, endogenous genes are replaced with exogenous DNA at the same locus as the wild-type gene, allowing the expression of this newly introduced gene to be controlled by the same cellular and transcriptional machinery as the naturally occurring gene. This makes the **gene expression** pattern of the exogenous gene identical to that of the natural gene, with respect to cell-type specificity, temporal patterning, expression levels, and any regulation-dependent mechanisms. In knockout mice, endogenous genes are typically replaced with non-functional DNA (which does not encode protein), effectively *knocking out* the gene of interest. Thus knockout mice are ideally suited for investigating the consequences of the loss of function of a particular gene. Knockin mice, on the other hand, are particularly useful for studying the effects of a gene with altered function, whether it is a gain or a partial loss of function. This section will focus on the applications, limitations, and generation of both knockin and knockout mice.

19.3.1 Knockin mice

Transgenic mice are useful for assessing the effects of a heterologous gene or the over-expression of a homologous gene, typically under the control of highly active promoters. However, they are not optimal for all research questions. As transgene expression is driven by an experimentally chosen promoter, expression levels are typically higher than what would be found endogenously. Moreover, transgene expression is not subject to the cellular mechanisms that normally regulate that gene's expression. Knockin mice are unique in that they allow gene expression levels and patterns to remain under the control of the naturally occurring promoter and any genomic locus-specific effects.

19.3.2 Applications of knockin mice in research

Disease-linked point mutations in voltage-gated sodium channels

Assume that a researcher is interested in a particular point mutation in the gene encoding a voltage-gated sodium channel, because genetic screening and analysis has linked this mutation to epilepsy in patients. *In vitro* experiments in cell lines demonstrate that this particular mutation results in impaired inactivation of the channel, resulting in increased sodium currents (Rhodes et al. 2005). This result suggests that this point mutation could underlie seizures and an epilepsy phenotype *in vivo*. However, to determine whether this is true requires an animal model. A transgenic model would not be appropriate as simply over-expressing the mutated version of the gene in a transgenic mouse that also expresses the normal endogenous mouse gene may not result in any phenotype. Therefore, to model this disease appropriately it is necessary to *replace* the endogenous gene with the mutated gene, rather than simply *add* a copy of the mutant gene to the genome. Thus, knockin mice are particularly useful for examining the natural consequences of altered protein function since the exogenous protein is expressed in the same pattern and at the same levels as the endogenous protein is expressed in a wild-type mouse. This feature makes knockin mice useful in the investigation of many diseases related to altered protein function resulting from a mutation in the coding region of its gene.

Disease-linked point mutations causing partial loss of function

Another example in which a knockin mouse is the most suitable for modelling a condition is when genetic mutations result in a partial loss of protein function. Let us assume that a researcher is studying a disease in which a mutation in the protein-binding domain of protein XYZ is believed to play an important role in this disease. The mutation does not completely prevent binding of protein XYZ with its partners, but it significantly decreases its binding ability. Humans afflicted with the disease express normal levels of this protein, but its function is altered. Therefore a knockin mouse that replaces the endogenous gene with a mutant copy is most appropriate for modelling this disease. A transgenic mouse would simply add a mutant copy of the gene to the genome, resulting in expression of the mutant protein in addition to the normal protein. Disease may not even develop in such transgenic mice because expression of the endogenous gene is left unaltered and may fulfill its functions normally. Furthermore, a knockout mouse (see section 19.3.4), in which the gene is completely ablated, is also inappropriate because the mutation associated with this disease does not result in a complete loss of function.

19.3.3 General considerations for knockin mice

A key consideration in the use of a knockin mouse is the construct to be used as the knockin gene. Typically, the knockin gene is a mouse gene that contains mutations designed to recapitulate mutations observed in humans. Therefore a thorough understanding of the amino acid sequences in both the human and mouse genes, which are often not completely homologous, is necessary to faithfully replicate the human mutation. If the human and mouse genes share common amino acid sequences, the decision as to which amino acid(s) to mutate in the mouse gene may be fairly simple. However, if this is not the case, the creation of a knockin mouse becomes more complex. For example, let us consider a case in which the human mutation is R311A—arginine replaced by alanine at position 311 amino acids away from the N-terminus of the protein. If the mouse and human genes are not highly homologous, it is unlikely that the 311th amino acid of the mouse protein is also arginine. However, by examining regions of homology between the two proteins, one may find, for example, that there is a homologous region in the mouse gene that codes for an arginine residue 295 amino acids from the N-terminus and is likely to be homologous to the human arginine 311 residue. Therefore an R295A mutation may be chosen for the mouse gene. It is also important to ensure that mutating this residue in the mouse gene does in fact replicate the effects of the corresponding mutation in the human gene. For example, for the situation involving altered sodium-channel function described earlier, it may be appropriate to transfect a construct containing the mutant mouse gene into a cell line and confirm that sodium-channel currents are increased similarly to the mutant human gene.

19.3.4 Knockout mice

Knockout mice replace a gene of interest with non-functional DNA, essentially 'knocking out' or ablating that gene. The generation of knockin and knockout mice is similar and is described in section 19.3.7. Both genetic manipulations target a specific locus, the only difference being the sequence of DNA used to replace the endogenous gene. The knockin replacement encodes a gene of altered function, while the knockout encodes non-functional DNA. The primary application of a knockout mouse is to allow a researcher to determine which cellular processes a particular gene is necessary for. By comparing the biochemical, cellular, and/or behavioural **phenotypes** of knockout mice with non-transgenic wild-type mice, researchers can determine which processes are affected by the ablation of the targeted gene. This is a powerful technique for assessing gene function.

19.3.5 Applications of knockout mice in research

Knockout mice have been used in many studies to assess the roles of genes in various biological disciplines, emphasizing their utility in biomedical research. Gene knockout can also be combined with transgene over-expression to obtain even greater insight into gene function. An example of this is presented here.

Use of knockout mice to study the role of Fyn tyrosine kinase in brain function

The activity of various tyrosine kinases modulates brain function both during development and in adult organisms, where they augment neuronal function during memory formation. However, in the early 1990s it was unclear which tyrosine kinases performed which roles. One study examined four different lines of knockout mice in which a different tyrosine kinase had been knocked out in each mouse line (Grant et al. 1992). In studying each of these lines of knockout mice, the researchers discovered that knocking out the tyrosine kinase Fyn created subtle abnormalities in brain development, although mice were able to survive, mature normally, and reproduce. Indeed, *fyn* knockout mice were relatively normal with two significant exceptions—they had reduced ability to modulate neuronal function, and they were impaired with respect to learning and memory. Therefore this study demonstrated that Fyn played an important role in neuronal function and learning and memory. However, as the knockout of Fyn also created developmental abnormalities, it was unclear whether those abnormalities (although subtle) or the continued absence of Fyn in adult mice were responsible for the later impairment in learning and memory exhibited by adult mice. To address this question, the same group later created a transgenic mouse that expressed Fyn under the control of the calcium-calmodulin protein kinase II (CaMKII) promoter mentioned in section 19.2.1, which only becomes active in primarily pyramidal neurons two weeks postnatally (Kojima et al.

1997). By **cross-breeding** *fyn* knockout mice with Fyn transgenic mice, the researchers created mice that had no Fyn during development, but expressed Fyn beginning two weeks after birth. They found that the developmental abnormalities were still present in the cross-bred mice, but that learning and memory were restored. Therefore the absence of Fyn in adult mice, and not the initial developmental abnormalities, was responsible for the learning and memory impairments.

19.3.6 General considerations for knockout mice

A few considerations must be taken into account when using knockout mice. The first is whether or not the gene is **haplosufficient**. Expression of only one copy of a haplosufficient gene is sufficient to maintain wild-type function, whereas expression of both copies of a **haploinsufficient** gene is necessary to maintain wild-type function. This is relevant, as a heterozygous mouse model may not display any phenotype differences if the knocked-out gene is haplosufficient. In this instance, a homozygous knockout must be generated, which can easily be accomplished by breeding together two heterozygous knockouts. Researchers often compare the biochemical and behavioural phenotypes of wild-type heterozygous knockout mice and homozygous knockout mice to assess differences between mice with normal levels of gene expression, levels of approximately 50% of normal gene expression, or no gene expression, respectively, to understand that gene's function.

Even if the ablated gene is responsible for an important biological process, there may be no obviously altered phenotype. There are two main reasons for this: the presence of genes coding for proteins with overlapping functions, or the induction of compensatory mechanisms triggered by the lack of expression of the ablated gene. Genes with overlapping function are quite common throughout the genome. These genes are usually **paralogues**, meaning they are homologous but distinct genes within a species genome coding for distinct proteins. An example of homologous but distinct proteins is the Ras (RAt Sarcoma) family of proteins. When bound to guanosine triphosphate (GTP), Ras is activated and promotes activation of its downstream target proteins. Hydrolysis of GTP to guanosine diphosphate (GDP) inactivates Ras proteins and hence their downstream targets. Owing to similarities in structure and function, the ablation of certain Ras proteins does not affect normal growth and development, which is indicative of overlapping function with other members of the Ras family (Nakamura et al. 2008).

Furthermore, the ablation of a gene and lack of its expression may trigger the induction of compensatory mechanisms that result in partial or even complete amelioration of the consequence(s) of lost gene expression. Such amelioration can occur through compensatory increases in the expression of other proteins involved in the same pathway in an effort of the system to normalize. Additionally, a genetic knockout can sometimes also result in alterations in the expression of genes not specifically related to its physiological role. For example, knockout of the D4 dopamine receptor affects not only dopamine receptor expression and function, but also results in alterations in glutamate receptor expression (Gan et al. 2004), a neurotransmitter system that is independent of the dopamine receptor system. Thus knockout models can provide a great deal of information regarding a gene's role, but care must be taken to consider all the parameters that might be affected by gene ablation in order to interpret the role of the gene of interest correctly. Moreover, because of possible alterations in proteins not specifically involved in the role of the knocked-out gene, biological processes altered in the knockout should not be assumed to be alterations directly regulated by the knocked-out protein, but rather processes dependent on expression of the the knocked-out protein.

A final point to consider is that the gene of interest may be an essential gene, meaning that its expression is vital for the growth and development of the organism. A gene whose ablation results in the inability of the mouse to survive past the early developmental stages is referred to as **embryonic lethal** or **neonatal lethal** depending on the developmental stage at which the mice typically die. This becomes an important issue if a homozygous knockout is necessary to address the scientific question at hand (e.g. due to haplosufficiency) *and* the knockout gene is an essential gene. In such cases, regulatable gene expression models that knock out a gene only in specific cell types or at specific time-points during development (Cre–*lox*) or at a time determined by the researcher (Tet) should be utilized. These models have other important applications as well and will be the focus of section 19.4.

19.3.7 Generation of gene targeted mice

Unlike transgenic mice, in which exogenous DNA is incorporated into the genome at random locations and regulated by experimentally chosen promoters, both knockin and knockout mice incorporate exogenous DNA at a *specific* locus. These models require a slightly more complex system for generation. Analogous to the transgenic construct used to introduce exogenous DNA into the mouse genome in transgenic mice, an **expression cassette** is first designed in gene targeted models. The expression cassette contains the sequence of DNA targeted to replace the endogenous gene, a positive drug selector (NEO^R, explained in detail later), a negative drug selector (thymidine kinase), and two regions of homologous DNA (homology arms) that flank both the replacement sequence and the *neo*-cassette but *not* the thymidine kinase negative drug selector region (Figure 19.2) (Carter and Shieh 2010; Doyle et al. 2012). These homology arms are homologous to the genomic sequences that flank the endogenous gene targeted for replacement, and thus allow direct integration of the expression cassette to this specific locus via homologous recombination mediated by endogenous cellular machinery. The exogenous gene is essentially 'swapped' for the endogenous gene, and therefore comes under the control of the endogenous promoter of that locus. The replacement sequence for a knockin mouse is a gene with an altered function, whereas for a knockout mouse the replacement sequence consists of non-functional DNA. With this exception, the remainder of the process for generating genetically targeted models is identical.

Once an expression cassette is designed, it is introduced into mouse embryonic stem cells, typically by **transfection**. The transfected cells are subjected to two drug selectivity treatments. The first involves exposure to neomycin, an antibiotic that is toxic to cells unless they express NEO^R, a bacterial gene that confers resistance to neomycin in the cells expressing it. Therefore any cells that are resistant to neomycin treatment will have the cassette integrated into their genome. However, this does not indicate whether or not the cells have integrated the cassette to the targeted locus. To determine if integration is locus-specific, the cells are treated with ganciclovir, which is toxic to cells expressing thymidine kinase. As the thymidine kinase region is located outside the homology arms, it will not be expressed if the cassette integrated at the target locus, and thus the cells will be insensitive to ganciclovir treatment. If the thymidine kinase gene is expressed,

Figure 19.2 Generation of genetically targeted mice. (a) The first step in creating a genetically targeted mouse is to develop the gene targeting construct and vector, containing homology arms flanking the replacement sequence and the *neo* cassette. The thymidine kinase region (tk) is placed outside the homology arms. Once vector construction is complete, a unique restriction enzyme is used to linearize the construct. (b) Linearized vector is either injected or electroporated into embryonic stem cells. (c) Cells are treated with neomycin. Only cells containing the *neo* cassette, and thus successfully transfected, are resistant to neomycin and will survive. (d) Cells are then treated with ganciclovir, which is toxic to cells expressing thymidine kinase (tk). As the tk region is located outside the homology arms, only cells that underwent locus-specific integration will not have the tk region and will be insensitive to ganciclovir and hence survive. (e) The surviving cells are injected into the inner cell mass of a 3.5-day-old blastocyst. (f) The blastocyst is injected into a pseudo-pregnant female mouse. (g) The offspring are genotyped to determine which have had the gene of interest successfully knocked in/out. (h) The knockin/out mice are bred to homozygosity before (i) breeding is expanded.

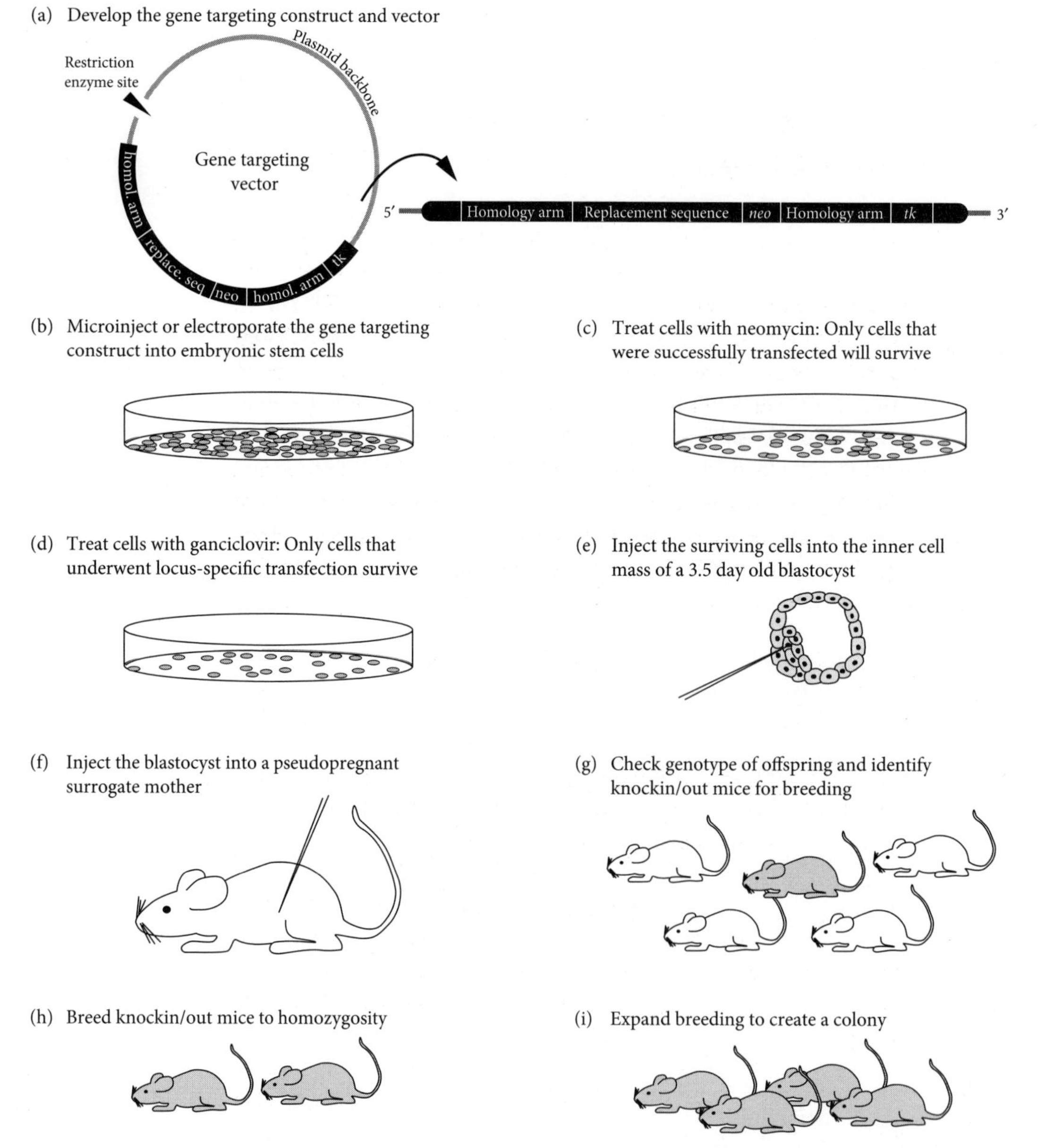
(a) Develop the gene targeting construct and vector
Restriction enzyme site
Plasmid backbone
Gene targeting vector
homol. arm
replace. seq
neo
homol. arm
tk
5′
Homology arm
Replacement sequence
neo
Homology arm
tk
3′
(b) Microinject or electroporate the gene targeting construct into embryonic stem cells
(c) Treat cells with neomycin: Only cells that were successfully transfected will survive
(d) Treat cells with ganciclovir: Only cells that underwent locus-specific transfection survive
(e) Inject the surviving cells into the inner cell mass of a 3.5 day old blastocyst
(f) Inject the blastocyst into a pseudopregnant surrogate mother
(g) Check genotype of offspring and identify knockin/out mice for breeding
(h) Breed knockin/out mice to homozygosity
(i) Expand breeding to create a colony

this indicates that the integration of the expression cassette was not restricted to the target locus, and hence the cells will die when treated with ganciclovir. Thus cell survival following neomycin treatment indicates that integration has taken place, whereas subsequent cell survival following ganciclovir treatment indicates that integration occurred at the target locus via the homology arms. The selected **embryonic stem cells** are then transferred to the inner cell mass of a 3.5-day-old blastocyst, a stage early in development that consists of 64 or 128 cells. The blastocyst is then transferred to a pseudo-pregnant surrogate mother where the embryos gestate. The embryonic stem cells then become a germ cell lineage in the embryo, and will eventually develop and reproduce to give rise to offspring that express the knockin (or knockout) gene in all cell types in which its promoter is active (Carter and Shieh 2010; Doyle et al. 2012).

19.4 Regulatable gene expression systems (Cre–*lox* models)

As discussed, genes of interest can be expressed in specific cell types in transgenic mice by using cell-type-specific promoters. Additionally, regulatable gene expression systems can also be used to restrict expression of specific genes in certain cell types as well as turn gene expression on or off at specified times. Specifically, we will focus on the Cre–*lox* system which allows the *conditional knockout* of a gene in specific cell types or regions, providing *spatial* control of gene regulation. In section 19.5 we will discuss the tetracycline-controlled transcriptional systems which allow genes to be induced (Tet-On) or repressed (Tet-Off) in mice by introducing the antibiotic chemical tetracycline into the mouse's diet. Since gene expression can be turned on and off by administering or removing tetracycline, these systems provide *temporal* control of gene regulation. Here we use the term 'regulatable gene expression system' to refer collectively to conditional knockout (Cre–*lox*), inducible (Tet-On), and repressible (Tet-Off) models.

19.4.1 Cre–*lox* conditional knockout mice

The Cre–*lox* model provides a way of knocking out genes in specific cell types or tissues (conditional gene knockout). This system was adapted from the P1 bacteriophage, in which the Cre recombinase protein (a site-specific DNA recombinase) binds to *loxP* sites in the bacteriophage DNA, excising the DNA between two *loxP* sites via recombination. Therefore, when cells that have *loxP* sites in their genome express Cre, a site-specific recombination event can occur between the *loxP* sites and the double-stranded DNA is cut at both *loxP* sites by the Cre protein. Thus the targeted gene is excised, and DNA ligase then rejoins the cut strands. Therefore conditional knockouts can be used to excise, or knock out, a gene of interest in the cell types of interest. This is accomplished by breeding together mice with one of two specific genetic components: a transgenic *cre* gene under the control of a specific-cell-type promoter that drives Cre expression only in the cells of interest, or a knocked-in gene of interest (GOI) that is flanked by *loxP* sites. A GOI flanked by *loxP* sites is said to be 'floxed'. Cre recognizes and binds

Figure 19.3 Regulatable mice: Cre–*lox* system. (a) To create a mouse in which gene expression can be conditionally knocked out in specific cell types, a transgenic mouse that expresses Cre under the control of a cell-type-specific promoter (CTSP) is cross-bred with a knockin mouse that expresses the gene of interest (GOI) flanked by two *loxP* sites (the GOI is 'floxed'). As the floxed gene has been knocked in, its expression is under the control of the endogenous promoter. (b) Left panel: Cells of the targeted tissues have naturally occurring transcription factors that recognize the CTSP and bind to it, initiating transcription of Cre. Cre protein is produced, which recognizes and binds to the *loxP* sites of the floxed GOI, and excises the DNA between the two *loxP* sites via circularization recombination. In this manner, all cells of the specified type will express Cre and thus have the GOI knocked out. Right panel: Cells that are not of the specified type do not have the transcription factors that are necessary to initiate transcription of Cre. Therefore no Cre protein that can bind the *loxP* sites that flank the GOI is produced. The GOI will not be knocked out in these cells, and expression is maintained under the control of the endogenous promoter.

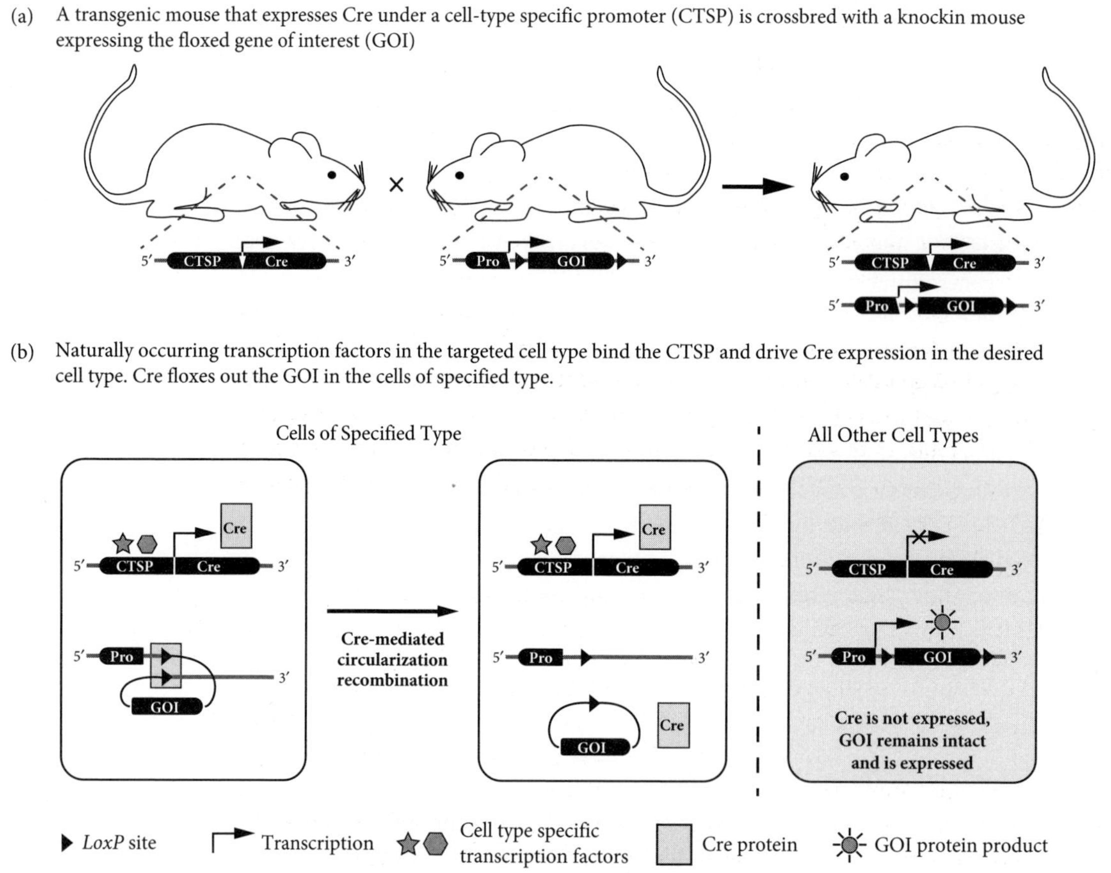

to the *loxP* sites of the floxed GOI and effectively excises any genetic material between the *loxP* sites (Sauer 1998), as illustrated in Figure 19.3. Therefore the knockout is conditional because it occurs only under the condition that cells express both Cre and the floxed GOI. As *loxP* sites do not naturally exist in the mammalian genome, floxing a GOI ensures that only the GOI is excised from the genome. Moreover, floxed GOIs are removed only in the presence of Cre. Thus, if Cre expression is placed under the control of a cell-type-specific promoter (CTSP), the excision of a floxed GOI can be restricted to specific cell types.

19.4.2 Applications of and considerations for Cre–*lox* conditional knockout mice in research

By conditionally knocking out genes, Cre–*lox* mice allow a researcher to restrict the knock out of genes to specific cell types as well as knocking out genes that would not be able to be knocked out by traditional methods because of embryonic or neonatal lethality issues. Some examples that illustrate the use of conditional knockouts to answer particular research questions are provided here.

Brain-region-specific knockout of a neuronal protein for the study of Alzheimer's disease

Let us assume that a researcher studies a mouse model of Alzheimer's disease, and discovers that the expression of a particular protein is significantly decreased exclusively in the hippocampus, a brain area important for learning and memory that is vulnerable in Alzheimer's disease. As the mice exhibit memory deficits that correlate with the decreased expression of this protein, the researcher hypothesizes that loss of this protein in the hippocampus is responsible for memory deficits. To test this hypothesis, the researcher wants to demonstrate that removing this protein exclusively from the hippocampus is sufficient to induce cognitive deficits. A traditional knockout is not ideal in this situation because it would ablate gene expression in the entire organism, which would not recapitulate what is observed in this mouse model. However, a Cre–*lox* conditional knockout is ideal. By crossing a transgenic mouse that expresses Cre under the control of a hippocampus-specific promoter with a knockin mouse that expresses a floxed version of the gene of interest, the researcher can excise and knock out the gene of interest exclusively in the hippocampus.

*Use of the Cre–*lox *system to knock out a gene that is embryonic or neonatal lethal*

Brain-derived neurotrophic factor (BDNF) is a growth factor that plays an essential role in brain development as well as in brain function in adult organisms. As BDNF plays a role in early brain development, traditionally generated **homozygous** BDNF knockout mice die within the first few weeks of life. To circumvent this problem, a Cre–*lox* model can be generated that expresses a floxed BDNF gene and Cre under the control of the calcium-calmodulin-dependent protein kinase II (CamKII) promoter (Monteggia et al. 2007). CamKII is primarily expressed in the forebrain, a region of interest in BDNF-related research. More importantly, CamKII expression begins post-natally, approximately 2 weeks after birth, and therefore BDNF can be conditionally knocked out in the forebrain after the period of brain development has been completed. This advance allowed researchers both to generate a viable homozygous knockout and to restrict the knockout of BDNF to a brain region of interest. While repressible and inducible models (see section 19.5) provide more precise temporal control of expression than the Cre–*lox* system, certain promoters drive expression at particular stages of development and thus are naturally regulated in a temporal manner. Therefore these promoters can be exploited to drive expression at desired phases of development using a Cre–*lox* model.

Cre–lox technology for the creation of a conditional knockin model

Although appropriate promoter selection or Tet-On inducible models (see section 19.5) is typically used to induce gene expression in a cell-type-specific or time-controlled manner, this can also be accomplished with a Cre–*lox* model. A floxed STOP codon can be inserted between a promoter and the coding region of a gene of interest. This codon will prevent translation of the protein product of that gene in any cell that does not also express Cre. However, cells that express Cre will have the STOP codon excised and removed, subsequently allowing normal expression of the gene product. The primary advantage of this method is that gene expression is limited to the expression pattern of two promoters: the endogenous promoter of the gene with the floxed STOP codon (assuming that its promoter was not manipulated) and the promoter controlling Cre expression.

Conditional knockouts are a powerful tool for establishing tissue-specific expression. However, as with other models, data must be interpreted carefully. Models must always be verified to ensure that their expected/proposed knockout is indeed functioning properly and models the intended phenomenon. In addition, integration of the Cre and floxed genetic elements can sometimes occur incorrectly, or in the wrong direction, preventing expression of the desired genetic element.

19.4.3 Generation of Cre–*lox* conditional knockout mice

As the Cre–*lox* system is a commonly used model, there are many established lines of mice that exist with knocked-in floxed genes of interest and transgenic mice that express Cre under the control of a vast number of different promoters. A number of Cre-expressing mouse lines have been produced with specific expression of Cre in a variety of specific cell types in tissues such as brain, muscle, cardiac tissue, endothelium, immune system, adipose tissues, bone, and more, and thus are applicable for research in a number of scientific disciplines (Wang 2009). Many can be purchased directly from mouse repositories and distributors such as the Jackson Laboratories (http://cre.jax.org/strainlist.html), and others can be obtained from the researchers in academia and the pharmaceutical industry who created them and/or maintain databases to track the many available Cre mouse lines (http://www.mshri.on.ca/nagy/).

The generation of a Cre–*lox* mouse to manipulate a gene of interest can be as simple as cross-breeding a mouse that expresses Cre under the promoter of choice with a mouse that has the floxed gene of interest. If there is not an established Cre or floxed gene of interest, the researcher must generate his/her own mouse line by developing a transgenic mouse with Cre under the control of a cell-type/temporal pattern-specific promoter and/or a knocked-in floxed gene intended for Cre-dependent knockout. Transgenic Cre mice can be created using standard transgenic expression techniques (see section 19.2.3), and knockin mice expressing a floxed gene of interest can be generated using standard transgenic and gene-targeted model techniques (see section 19.3.7).

Most Cre-expressing mouse lines are characterized by their producers, but their expression patterns can be verified by crossing the Cre-expressing mouse with a Cre reporter mouse. A Cre reporter mouse is a knockin mouse that expresses a floxed reporter gene of interest, often an enzyme or fluorescent protein. A list of readily available Cre reporter strains and comparison between them can be found on the Jackson Laboratories website (http://cre.jax.org/crereporters.html). A common Cre reporter mouse is the Rosa 26 reporter line, which has a floxed STOP signal placed upstream of the *lacZ* gene to prevent *lacZ* expression in the absence of Cre. The *lacZ* gene encodes β-galactosidase, an enzyme whose expression can be easily detected with standard histochemical techniques. If a fluorescent protein such as green fluorescent protein (GFP) is used as the reporter, Cre expression will trigger the floxing of the STOP signal upstream of the GFP sequence. In this case, all Cre-expressing cells will fluoresce green when visualized with epi-fluorescent microscopy. Therefore, Cre reporter mouse lines are useful for verifying Cre expression patterns.

See Chapter 17

19.5 Regulatable gene expression systems (Tet-Off/Tet-On mice)

We have previously discussed how the choice of promoter and the use of Cre–*lox* models can confer cell type and tissue specificity as well as developmentally regulated temporal specificity. However, more precise temporal control over gene expression can be achieved using a tetracycline-controlled transcriptional activation model (Tet model). Tet models are of two types, Tet-On and Tet-Off, and are used to turn gene expression on and off reversibly. Gene expression in a Tet model is controlled by tetracycline (or its analogue doxycycline) provided in the mouse's diet, which serves as a trigger for initiating or repressing gene regulation. This system differs from the Cre–*lox* system in which gene knock-out is irreversible once site-specific recombination has occurred. In addition, controlling the timing of gene expression using Tet models can overcome limitations such as embryonic lethality, as gene expression can be turned on or off after birth, or after a certain age. Tet models are also advantageous for experiments in which only transient gene expression is desired.

Like Cre–*lox* models, Tet models involve two main components—an activator and a responder. In Tet-Off systems (in which tetracycline represses gene expression), the activator is tTA (tetracycline-responsive transcriptional activator), which is expressed as a transgene driven by the promoter of choice. When tTA is expressed, it binds to a specific DNA sequence (the responder) called the TRE (transcriptional response element) and initiates expression of the GOI in cells that contain both tTA- and TRE-responsive genes. However, when tetracycline (or its analogue doxycycline) is present, it binds to the tTA and prevents it from binding to the TRE, thereby turning GOI expression off (Figure 19.4). The level of expression of the TRE-regulated GOI can be increased or decreased by varying the amount of tetracycline in the mouse's diet. Therefore expression of the GOI can be turned off by administering tetracycline to mice (usually in their food), and turned back on by removing tetracycline from the diet. Tetracycline controls the temporal pattern of expression, and the promoter used to drive the tTA expression controls which cells/tissues express the GOI, providing the ability to control gene expression both spatially and temporally.

A variant of the Tet-Off model is the Tet-On model, in which tetracycline turns gene expression on (Figure 19.4). In the Tet-On model, rtTA (reverse tTA) binds TRE sequences only in the *presence* of tetracycline (or doxycycline), and thus tetracycline (or doxycycline) turns TRE-regulated gene expression on. Both tTA and rtTA expression can be driven by a selected promoter of choice, so that expression can be ubiquitous

Figure 19.4 Regulatable mice: Tet systems. (a) To create a mouse in which gene expression can be reversibly turned on or off, a transgenic mouse that expresses either tTA (tetracycline transactivator) or rtTA (reverse tetracycline transactivator) under the control of a cell-type-specific promoter (CTSP) is cross-bred with a transgenic or knockin mouse that expresses the gene of interest (GOI) under control of a TRE (tetracycline responsive element). (b) Left panel, Tet-Off system: In the absence of tetracycline or doxycycline (DOX), the cells of specified type have naturally occurring transcription factors that bind to the cell-type-specific promoter (CTSP) and initiate transcription of tTA. The expressed tTA protein binds to the TRE element upstream of the GOI, allowing transcription of the GOI and production of the GOI protein product (upper left box). No other cell types have the naturally occurring transcription factors necessary to bind the CTSP; thus no tTA is produced, and the GOI cannot be expressed (upper right box). When DOX is administered to the mice in the cells of specified type, DOX binds to the expressed tTA and induces a conformational change in its structure such that it is no longer able to bind the TRE upstream of the GOI. Therefore, in the presence of DOX, no GOI expression occurs in the Tet-Off system (bottom left box). DOX does not affect other cell types since no tTA is expressed in the first place (bottom right box). Right panel, Tet-On system: In the absence of DOX, the cells of specified type produce rtTA because they have the naturally occurring transcription factors necessary to initiate rtTA expression. However, in the absence of DOX, the conformation of rtTA prevents it from binding to the TRE upstream of the GOI. No GOI expression occurs (top left box). No other cell types have the correct transcription factors necessary to drive rtTA expression, and so no GOI expression occurs (top right box). In the presence of DOX, DOX binds the rtTA that is expressed in cells of specified type and changes rtTA conformation such that it can bind the TRE upstream of the GOI. GOI expression occurs (bottom left box). DOX does not affect other cell types since no rtTA is expressed in the first place (bottom right box).

(a) A transgenic mouse carrying tTA or rtTA under a cell-type specific promoter (CTSP) is crossbred with a transgenic mouse carrying a TRE upstream of the gene of interest (GOI).

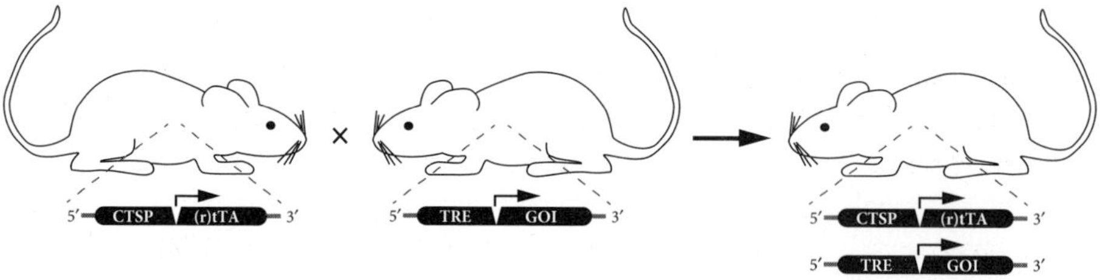

(b) Naturally occurring transcription factors in the targeted cell type bind the CTSP and drive tTA or rtTA expression in the desired cell type. Doxycycline (DOX) administered in the diet binds rtTA or tTA the specified cell type and induces a conformational change that inhibits tTA from binding the TRE (Tet-Off system) or allows rtTA to bind the TRE (Tet-On system).

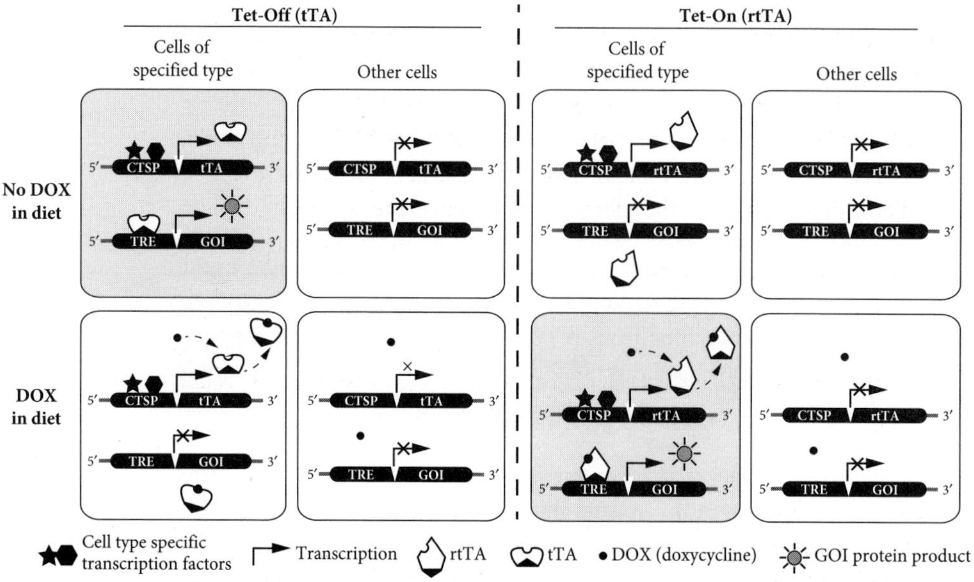

or cell type/region specific. The researcher controls where the GOI will be expressed by driving (r)tTA expression with cell-type/region-specific promoters. Moreover, the researcher chooses which target GOI to regulate by creating a knockin mouse that has a TRE sequence upstream of the knocked in GOI (Sun et al. 2007). The choice of Tet-Off or Tet-On models is influenced by the scientific question to be addressed, as each has both benefits and disadvantages that will be discussed in the following sections.

19.5.1 Applications of Tet-Off/Tet-On mice in research

Tet-Off models are often used when the aim of the experiment is to maintain gene expression (ON) for most of the time, and restrict it in certain circumstances only or after a certain developmental time period. In such cases, mice will be administered tetracycline or doxycycline when gene expression needs to be repressed. Tet-On systems are used in cases where expression of transgenes is required only for a specific period of time, typically after the mouse has matured. In these cases, tetracycline or doxycycline will be administered when gene expression needs to be turned on. An example of an application of the Tet-Off system is given here.

Use of a Tet-Off model to assess the role of tau in Alzheimer's disease

Tau is a microtubule binding protein that becomes hyperphosphorylated in Alzheimer's disease, leading to the formation of neurofibrillary tangles, one of the pathological hallmarks of Alzheimer's disease. However, it is unclear whether hyperphosphorylated tau itself or the neurofibrillary tangles it forms are more detrimental in this disease. To address this question, transgenic mice have been created using the Tet-Off system to reversibly express mutant forms of tau that are prone to hyperphosphorylation and tangle formation (Santacruz et al. 2005). In this system, the responder transgene consists of mutant tau under the control of a TRE. An activator transgene in a second mouse line is used to drive tTA expression regulated by the CamKII promoter, which drives expression in the forebrain region. Mice carrying the responder or activator transgenes are cross-bred to produce mice that contain both transgenes and express mutant tau in forebrain structures. At various ages associated with different stages of tau pathology, tau expression can be repressed by adding tetracycline or doxycycline to the mouse food. Tetracycline or doxycycline then binds to tTA, rendering it incapable of binding to the TRE driving mutant tau expression. Tau expression remains repressed as long as tetracycline or doxycycline is included in the diet. Using this system, the effects of mutant tau expression on brain function can be assessed by conducting various behavioural and/or electrophysiological tests in the presence of hyperphosphorylated tau, tangles, or both. Tau expression can then be repressed to determine whether the observed phenotypes are eliminated. As the Tet-Off system allows gene expression to be turned on and off reversibly, it allows researchers to demonstrate that the expression of a particular gene causes a particular phenotype, and that eliminating expression of that gene normalizes the phenotype.

19.5.2 General considerations for Tet-Off/Tet-On mice

There are some key factors to consider when using Tet-Off or Tet-On mice. When possible, choosing the Tet model that necessitates the shortest duration of tetracycline/

doxycycline administration is prudent, because long-term administration of these drugs can be associated with some side effects. Moreover, the regulation of gene expression can sometimes be 'leaky'. Leakiness refers to some level of GOI expression that may occur in cells that do not express (r)tTA (and therefore should not express the GOI) because of unwanted residual activity of the TRE-driven GOI. Leakiness can also result from the fact that rtTA has some affinity for TRE sequences even in the absence of tetracycline or doxycycline. In both cases, 'leaky' expression cannot be controlled by tetracycline or doxycycline administration. Leaky expression levels are often low, or negligible, and are sometimes acceptable. However, if leaky expression levels are too high, regulating the expression levels of the GOI with tetracycline or doxycycline is not possible because it is already being expressed in the absence of the drugs. In this case, re-derivation of the mice may be necessary. Often several lines of mice are created, and the line with the least leakiness and the best experimental control of expression is propagated.

19.5.3 Generation of Tet-Off/Tet-On mice

Tet models are binary models that require two components to drive regulated gene expression. The first component is the tetracycline-controlled transcriptional activator (tTA or rtTA), which is typically produced in a transgenic mouse that expresses (r)tTA under control of a promoter that drives expression in desired cell types or regions. The second component is the gene of interest (GOI) under the regulation of a tetracycline-responsive element (TRE), which is typically integrated into the genome of a separate transgenic or knockin mouse. Cross-breeding the (r)tTA mouse with the TRE mouse results in offspring that express (r)tTA in the cells of interest in TRE mice, thus allowing for tetracycycline/doxycycline regulation of gene expression (Figure 19.4). A large number of transgenic mouse lines which express (r)tTA under the control of promoters that drive expression in various cell types and regions have already been created. Such mice can be obtained from mouse repositories and distributors such as the Jackson Laboratories (http://jaxmice.jax.org/), and others can be obtained from the researchers in academia and the pharmaceutical industry who created them. If the (r)tTA mice and TRE mice that appropriately address the question at hand already exist, Tet model generation may be as simple as obtaining and cross-breeding the two mouse lines. However, if the desired mouse lines have not been developed for one or both of the necessary genetic components, transgenic or knockin mice must be created using the methods described in sections 19.2.3 and 19.3.7.

19.6 Chapter summary

- **Mouse models are extremely useful for probing the role of genes *in vivo*.**
- **Mouse models can be used to study the normal roles of genes as well as the mechanisms by which certain mutations in genes give rise to human disease.**

- Several types of genetically engineered mice can be generated to address many types of research question. The choice of mouse model depends on the question being investigated.
- Transgenic mice over-express an exogenous gene of interest in specific tissues or cell types. Such mice are particularly useful for addressing questions such as: How will the presence of a heterologous transgene affect the organism? How will over-expression of a homologous transgene affect the organism?
- Knockin mice carry an exogenous gene in place of the endogenous murine gene, and expression of the knocked in gene is under control of the promoter of the endogenous mouse gene. Knockin mice are particularly useful for studying the effects of a gene with altered function.
- Knockout mice lack the gene of interest, which has been targeted and deleted from its genome. Knockout mice are ideally suited to investigating the consequences of the loss of function of a particular gene.
- The expression of genes can be controlled spatially and/or temporally using the Cre–lox system or the Tet-On/Off system.

Probing the role of a genetic mutation associated with cancer

CASE STUDY 19.1

You are researching the role of a specific genetic mutation in a rare form of cancer. Your *in vitro* experiments using cell lines suggest that the genetic mutation causes a loss of function in the gene protein product, which may be responsible for increasing the probability of developing cancer. However, this hypothesis has yet to be examined in a mouse model of disease to determine whether such a loss of function is sufficient to cause the type of cancer associated with this mutation in human patients. The mouse and human genes have high homology.

QUESTION

Which type of genetically altered mouse model would be the most appropriate for testing whether the genetic mutation is sufficient to cause cancer?

The most appropriate model is a knockin mouse that expresses a version of the mouse gene that has been mutated to mimic the point mutations associated with the human disease. This model will most closely mimic the mutation in human cancer patients, since the gene and its protein product will be expressed and present in appropriate cell types, albeit with a loss of function. Since the genetic mutation causes a loss of function, a knockout mouse model might also be appropriate. However, in the knockout mouse, the gene and its protein product will be completely absent from the genome, which does not truly mimic the human condition. Such absence may preclude your ability to determine whether the presence of the mutant protein has other effects apart from or in addition to loss-of-function effects. A transgenic mouse that expresses the mutant human gene is also unsuitable as the endogenous mouse gene will still be expressed and serve its normal function, and hence will mask the effect of the mutant human gene.

CASE STUDY 19.2 Assessing the role of a novel gene in brain development

See Chapter 6

Your research suggests that the gene you have been studying may have a role in the normal development of the brain. Neuronal cell lines in which you have knocked down expression of your gene (using RNAi techniques) demonstrate that expression of lower than normal levels of your gene inhibits the growth and elaboration of complex structures in these cells.

QUESTION

Which type of genetically altered mouse model would be the most appropriate for testing whether your gene plays a critical role in brain development?

A knockout mouse in which your gene has been ablated is the most appropriate model to test your hypothesis. As your experiments demonstrate that decreasing levels of gene expression strongly affect neuronal growth and structural integrity, a knockout model will simulate this manipulation *in vivo*. Moreover, since you are interested in its role in brain development, it is appropriate for gene expression to be completely absent even throughout developmental stages. A knockout model will allow you to determine whether your gene is necessary for normal brain development. A conditional knockout mouse is not necessary in this case, since ablation of the gene throughout development is satisfactory. However, if you decide to examine which stage of brain development the function of your gene is necessary for, you may wish to create a conditional knockout mouse so that you can suppress gene expression only at specific stages of brain development to assess at which stage gene expression is critical for normal brain development. You may also wish to complement studies in knockout mice with studies of a transgenic mouse that over-expresses your gene of interest to assess whether over-expression of your gene results in effects that are opposite to the results observed in knockout mice.

References

Carter, M. and Shieh, J. (2010) *Guide to Research Techniques in Neuroscience*. San Diego, CA: Academic Press.

Chin, J. (2011) Selecting a mouse model of Alzheimer's disease. *Methods Mol Biol* **670**: 169–89.

Doyle, A., McGarry, M.P., Lee, N.A., and Lee, J.J. (2012) The construction of transgenic and gene knockout/knockin mouse models of human disease. *Transgenic Res* **21**: 327–49.

Gan, L., Falzone, T.L., Zhang, K., Rubinstein, M., Baldessarini, R.J., and Tarazi, F.I. (2004) Enhanced expression of dopamine D(1) and glutamate NMDA receptors in dopamine D(4) receptor knockout mice. *J Mol Neurosci* **22**: 167–78.

Govindarajan, A., Rao, B.S., Nair, D., et al. (2006) Transgenic brain-derived neurotrophic factor expression causes both anxiogenic and antidepressant effects. *Proc Natl Acad Sci USA* **103**: 13208–13.

Grant, S.G., O'Dell, T.J., Karl, K.A., et al. (1992) Impaired long-term potentiation, spatial learning, and hippocampal development in fyn mutant mice. *Science* **258**: 1903–10.

Kojima, N., Wang, J., Mansuy, I.M., Grant, S.G., Mayford, M., and Kandel, E.R. (1997) Rescuing impairment of long-term potentiation in fyn-deficient mice by introducing Fyn transgene. *Proc Natl Acad Sci USA* **94**: 4761–5.

Miller, R.L. (2011) Transgenic mice: beyond the knockout. *Am J Physiol Renal Physiol* **300**: F291–300.

Monteggia, L.M., Luikart, B., Barrot, M., et al. (2007) Brain-derived neurotrophic factor conditional knockouts show gender differences in depression-related behaviors. *Biol Psychiatry* **61**: 187–97.

Nakamura, K., Ichise, H., Nakao, K., et al. (2008) Partial functional overlap of the three ras genes in mouse embryonic development. *Oncogene* **27**: 2961–8.

Rhodes, T.H., Vanoye, C.G., Ohmori, I., Ogiwara, I., Yamakawa, K., and George, A.L., Jr. (2005) Sodium channel dysfunction in intractable childhood epilepsy with generalized tonic–clonic seizures. *J Physiol* **569**: 433–45.

Santacruz, K., Lewis, J., Spires, T., et al. (2005) Tau suppression in a neurodegenerative mouse model improves memory function. *Science* **309**: 476–81.

Sauer, B. (1998) Inducible gene targeting in mice using the Cre/lox system. *Methods* **14**: 381–92.

Shirayama, Y., Chen, A.C., Nakagawa, S., Russell, D., and Duman, R.S. (2002) Brain-derived neurotrophic factor produces antidepressant effects in behavioral models of depression. *J Neurosci* **22**: 3251–61.

Sun, Y., Chen, X., and Xiao, D. (2007) Tetracycline-inducible expression systems: new strategies and practices in the transgenic mouse modelling. *Acta Biochim Biophys Sin (Shanghai)* **39**: 235–46.

Wang, X. (2009) Cre transgenic mouse lines. *Methods Mol Biol* **561**: 265–73.

Waterston, R.H., Lindblad-Toh, K., Birney, E., et al. (2002) Initial sequencing and comparative analysis of the mouse genome. *Nature* **420**: 520–62.

20 Mathematical models in biomolecular sciences

Josie A. Athens

Chapter overview

In this chapter we will look at methods that allow us to perform experiments *in silico*, i.e. on the computer. While the classic *in vivo* and *in vitro* experiments usually provide better models for understanding biomolecular-related problems, computer simulations based on mathematical models allow us, among other things, to:

- test new hypotheses;
- explore different theoretical scenarios and predict potential outcomes based on a particular set of assumptions;
- estimate biological **parameters** from experimental data.

We can formally express the relationship between biological variables such as genes, molecules, or cells using an equation or a system of equations. These equations represent a mathematical model of our problem of interest. We can analyse the model using computer simulations in which we investigate the **dynamics** (change) of our variables in time, in space, or in both time and space.

LEARNING OBJECTIVES

This chapter will enable the reader to:

- summarize the major components of mathematical models used in biomolecular sciences;
- differentiate between discrete and continuous models used in biomolecular sciences;
- recognize when to consider a mathematical model as part of the research design;
- interpret results from computer simulations;
- run simple mathematical models to simulate biological problems.

20.1 Introduction

In biomolecular sciences we have the opportunity to answer any question by using different models. A model is neither true nor false; rather, it is useful or not useful. For the purposes of this book, we can define a model as a cognitive construct that allows us to better understand a biological phenomenon, such as signal transduction or disease, within a theoretical framework.

Sometimes we use another species to study a particular human disease; experiments performed in an animal model are described as *in vivo*. To have more control over the variables involved we sometimes use cell lines and, in this case, the experiments are performed *in vitro*. Sometimes we need a more theoretical approach to test a particular hypothesis or to help us design a new experiment. In this theoretical approach we can use mathematical models, and the experiments usually are performed *in silico*, i.e. using a computer.

The use of mathematical models in biology has been described in a number of ways, such as mathematical biology, biometrics, bioinformatics, theoretical biology, and computational biology. When more traditional experiments are also involved in the study (so-called 'wet' experiments), the discipline is known as systems biology (Hood et al. 2004; Kitano 2002).

20.1.1 Why use a mathematical model?

The main roles of mathematical models are associated with *prediction* and *understanding* (Keeling and Rohani 2008). We can use a mathematical model to:

- test a hypothesis;
- predict potential outcomes;
- estimate parameters from experimental data.

A mathematical model will allow us to simulate experiments at a relatively low cost in a relatively short time. In some instances performing a real experiment may not be feasible because of ethical, financial, or design constraints. For example, a mathematical model can be proposed to simulate the evolution of phenomena that can take too much time to observe, such as the development of cancer or AIDS.

In molecular biology, mathematical models can help us to understand the relationship between biological variables. In some instances we can use a feedback process in which a mathematical model is proposed on the basis of experimental observations and then used to guide new experiments to extend the model and thus further our understanding of the research question.

20.1.2 The modelling approach

We can think of mathematical modelling as a tool or technique to answer scientific questions. As with other techniques, when we want to use mathematical models, we use the scientific method. Therefore, the experimental design can be divided into three general phases:

- conceptualization;
- implementation;
- analysis.

In the conceptualization phase, the relationships between the variables of interest are identified and formalized in a mathematical equation or set of equations. In most cases, the equations have to be translated to a computer program that can be interpreted by specialized software. In the implementation phase the program is run on a computer and analyses performed to evaluate the model. For example, in a sensitivity analysis, we run the model using different parameter values and assess whether model predictions are consistent with experimental observations. Once we have performed the initial analysis, we can use the model to answer our original research question. When experimental data are available, parameter values are fitted using optimization techniques similar to those used in linear regression (least-squares). Finally, model predictions from the simulations are further analysed within the current theoretical framework and may suggest the development of new experiments. Ideally, the model and the experiments will provide mutual feedback in an iterative process which allows the scientist to better understand potential and observed outcomes generated under relevant biological scenarios.

In mathematical models the main independent variable is time or space (and sometimes both). For example, in signal transduction models we can use kinetic rates to simulate a metabolic pathway, and in metastasis models we may be interested in both time (kinetics) and space (where the tumour cells are located). For simplicity, in this chapter we will consider only the case of time. In other words, we are interested in the dynamics of a system (Kaplon and Glass 1995). This being the case, the scientific problems that can be approached with this kind of mathematical model must be able to provide information about the variables, such as concentration, status, or relationship, at different times.

A mathematical model has two types of components: variables and parameters. During a simulation, variables (e.g. concentration) are usually dynamic and change with time, whereas parameters (e.g. mutation rate, proliferation rate, probability of being active) remain constant. However, it is also possible to have a dynamic parameter as a function of time.

In mathematical models there is a trade-off between prediction and understanding (Keeling and Rohani 2008). A simple mathematical model requires fewer variables and/or parameters than a complex model, but may only provide general predictions. A recommended approach is to start with a simple mathematical model and then extend it in steps by adding more variables and parameters so as to build up a more detailed understanding of the problem.

Once the main variables have been identified, it is helpful to draw a picture in which we can identify the relationships and general characteristics of the components of our system. In the case of biological networks, a standard graphical notation has been proposed (Kitano 2003; Kitano et al. 2005). From these diagrams the relationships are formally established in an equation or set of equations.

20.1.3 Equations

We can define an experiment as the manipulation of an independent variable in order to assess its effect on a dependent variable under controlled conditions. For example, a group of randomly selected individuals receive a treatment, while another group who do not receive treatment are used as controls, and we measure a relevant outcome such as change in body temperature. In mathematics the relationship between one or more variables is established formally with an equation. A general statement could simply specify that our dependent variable y (e.g. change in body temperature) is a function (f) of the variable x (e.g.treatment or control group):

$$y = f(x) \tag{20.1}$$

There are several examples of the use of equations in biomolecular sciences. It is likely that readers are familiar with the Michaelis–Menten equation used to explain enzyme kinetics, or with the Scatchard equation to study antibody affinity. In the following sections we introduce some approaches that have been used in mathematical modelling. We start with **difference equations**, as they are the easiest to explain and are the foundation of more complicated tools, and finish with **ordinary differential equations (ODEs)**.

Mathematical biology has grown substantially as a new discipline, and in this chapter we give a brief introduction to the topic with some applications for molecular biology. For a more comprehensive coverage of mathematical biology in general you may wish to consult books by Edelstein-Keshet (2005) and Vries (2006), and for a particular emphasis on molecular biology Fall et al. (2002) and Istrail et al. (2003).

20.2 Difference equations

Difference equations are mathematical expressions that relate variable values at discrete times. With a simple difference equation we compute the value that a variable of interest (e.g. the concentration of a transcription factor) has at each unit of time (e.g. each minute), using an updating rule that establishes how the variable value at a given time relates to its value at a previous time:

A simple case is when we consider, for example, that the expression level of a particular transcriptional factor y at time t is a function of its expression level at a previous time $t-1$:

$$y_t = f(y_{t-1}) \tag{20.2}$$

If we assume that our unit of time is minutes and that the current time t is 10min, equation (20.2) tells us that the concentration of gene y at 10min is a direct consequence of the concentration of gene y at 9min.

To visualize the dynamics of y we can plot its values at different times (i.e. a plot of y versus time). If we are interested in understanding the updating rule, we can plot the values of y_t versus the values of y_{t-1}; this plot is known as the return map.

Difference equations can be linear or non-linear. In the following sections we introduce both types and present some applications to biology.

20.2.1 Linear difference equations

The simplest example of a difference equation is when the outcome of the return map is a straight line, i.e. when we have a linear function. For example, we can assume that at time t we will have β times what was expressed at time $t-1$:

$$y_t = \beta y_{t-1} \tag{20.3}$$

In discrete models the step between each time is constant (e.g. 1min). In the current example let us assume that our time step is 1min. To simulate equation (20.3) we will need:

- a starting expression level (**initial condition**) of y, for example 0.1mg/mL;

- the value of the parameter β which can be estimated from either experimental data or theoretical assumptions;
- a final time t_n when we will stop the simulation, for example 20min.

Equation (20.3) is so simple that we can run simulations without even using a computer. Interestingly, we can perform experiments by changing the value of our parameter. In these experiments we could analyse the effect of different values of β on the dynamics of y. These experiments are known as **sensitivity analysis**.

For our first experiment let us asume that $\beta = 1$. As a consequence, the expression level of y does not change with time, i.e. it is in a steady state equal to the initial condition. The first 20min in Figure 20.1 represent the steady state.

In the next experiments we assume that gene y is in a steady state until we introduce a **perturbation** (change) to the system. Assume that at time $t = 20$min the gene is activated and its expression level is increased by 20% (i.e. $\beta = 1.2$). Figure 20.1(a) shows the result of the simulation. The system is in the steady state for the first 20min and then grows exponentially by 20% at time 21min. What is the underlying assumption? In this simple example we assume that the effect of the activation is immediate, and so no delay is observed.

If the value of our parameter was <1 then we would be simulating inhibition. For example, Figure 20.1(b) shows the effect of 20% inhibition ($\beta = 0.8$). Again, the inhibition starts immediately 20min after the start of the experiment. It can clearly be seen that at 21min the expression level of y has dropped from 0.1 to 0.08mg/mL.

20.2.2 Non-linear difference equations: the logistic equation

When the outcome of the return map is a curve instead of a straight line, we have a non-linear relationship. In molecular biology there are many complicated interactions within the variables of interest, including several control mechanisms such as feedback loops. These interactions and controls make linear relationships very unlikely in biology and, although linear models are a good starting point, we need to consider non-linear models as a better approximation to biological data.

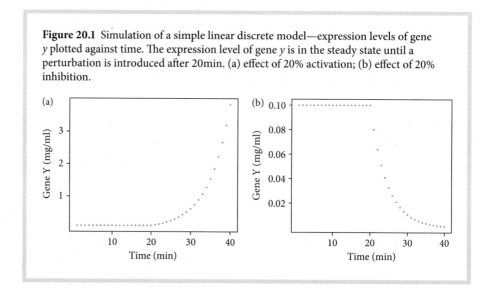

Figure 20.1 Simulation of a simple linear discrete model—expression levels of gene y plotted against time. The expression level of gene y is in the steady state until a perturbation is introduced after 20min. (a) effect of 20% activation; (b) effect of 20% inhibition.

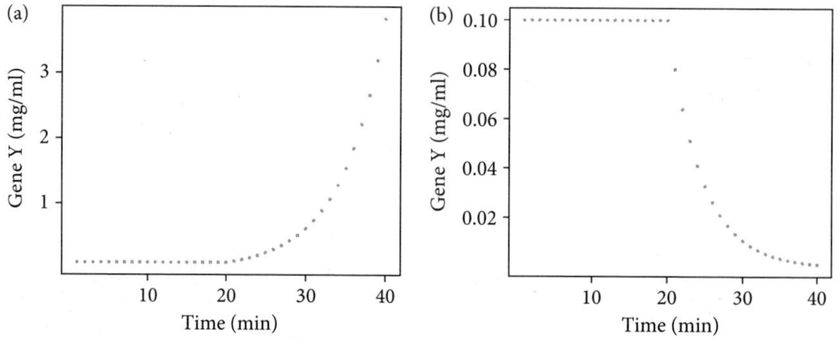

Let us start by trying to find a good example of the simulation shown in Figure 20.1(a): the case of the cytokine interleukin 2 (IL-2). In a subset of T cells, IL-2 has an autocrine effect, i.e. IL-2 activates (among other genes) the expression of IL-2. As in any 'normal' condition we would expect some kind of feedback mechanism that will control the expression and avoid an unlimited exponential growth of cells. To keep things simple, we do not want to involve other genes in our equation yet, so we can model this control as a negative feedback loop in which we stop the activation at higher concentrations. A simple equation would be:

$$y_t = \beta y_{t-1}(1 - y_{t-1}) \tag{20.4}$$

Continuing with our cytokine example, in equation (20.4) y_t is the proportion of active IL-2 at time t (e.g. 3min), y_{t-1} is the proportion of active IL-2 at the previous time (e.g. 2min), and the parameter β is the autostimulation rate. Equation (20.4) is known as the logistic equation or quadratic map because the outcome of the return map is a parabola. At very small values of y (e.g. 0.001), $(1 - y_{t-1}) \approx 1$ and equation (20.4) behaves like equation (20.3) and we observe exponential growth for $\beta > 1$. As y increases, the second term becomes more important and the expression levels increase non-exponentially until a plateau is reached which represents saturation or maximum expression level. In Figure 20.2(a) we simulate the scenario for an activation effect of 20%. This kind of dynamics is very common in biology because of the saturation effect. In this scenario, the system reaches **equilibrium** (shown by the plateau on the graph) approximately 20min after introducing the perturbation (t = 40min).

It is well known that the logistic equation is able to produce complicated dynamics with small changes in its parameter (May 1976). For example, if we increase the value of the activation effect to $\beta = 3.3$ (activation of 230%), instead of reaching a fixed point at equilibrium, the system shows cyclic behaviour with period 2, i.e. it oscillates between two values (Figure 20.2(b)). An oscillation can be the result of a negative feedback loop

Figure 20.2 Simulation of the logistic equation—expression levels of gene y plotted against time. Gene y stays in the steady state for 20min until activation starts. (a) Dynamics of an activation effect of 20%; the system reaches equilibrium after 40min. (b) Dynamics of an activation effect of 3.3 (230%); the system undergoes periodic oscillations at equilibrium. In this example, the periodic oscillations at equilibrium represent the two possible expression levels that gene y can have as a result of, for example, the presence of feedback loops.

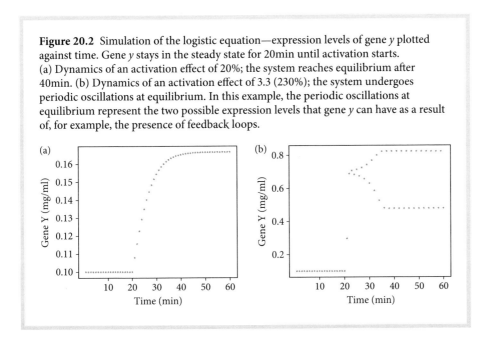

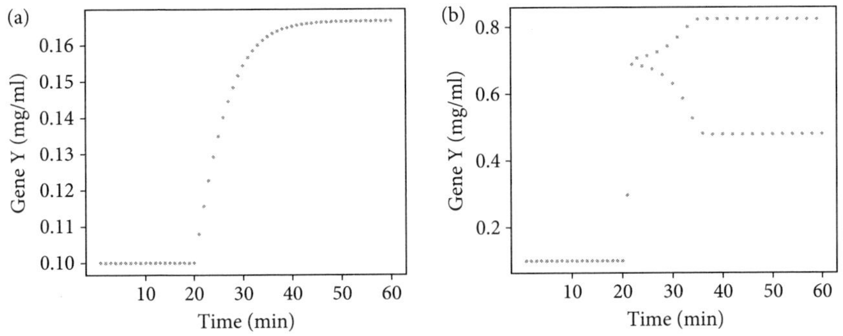

and indicates that, for example, a nuclear factor is active at one discrete time and is inactive at the next discrete time. One would expect these transitions to occur continuously, but they may be observed or simulated at discrete times. Oscillations similar to those shown in Figure 20.2(b) have been reported for the nuclear factor kappa-B (NF-κB) (Hoffmann et al. 2002).

Equation 20.4 can be used when the outcome has a value between 0 and 1, as in proportions or probabilities. However, if the dependent variable y can take values > 1, the equation needs to include the cap on y, i.e. the maximum value that y can reach. With this generalization, equation 20.4 becomes:

$$y_t = \beta y_{t-1}\left(1 - \frac{y_{t-1}}{y_{max}}\right) \qquad (20.5)$$

where y_{max} is the saturation point or maximum value of y. If we are considering proportions, the maximum value is 1 and we would have equation (20.4) again. Note that equation (20.5) is very similar to the Michaelis–Menten equation.

CASE STUDY 20.1

Time to become a palpable tumour

The aim of the experiment is to estimate the time required for a breast cancer cell to become a palpable tumour of at least 10^9 cells. The approximate doubling time of a breast tumour is 1.8–47.5 days.

QUESTION

Which is the most appropriate mathematical model for estimating this time?

Consider the options. We can assume that the doubling of the tumour always takes the same amount of time and therefore is discrete. As we are not interested in other aspects, such as nutrient and oxygen supply (e.g. angiogenesis), the interaction with the immune response, or the presence of chemotherapy, we can propose a simple discrete model such as the one described by equation (20.5) in which only one variable is involved. To solve this equation we need to know the value of two parameters: the proliferation constant and the maximum number of cells that a tumour can have (its maximum size or lethal burden cap). As we are modelling doubling, the value of β is 2. The time step is constant and we assume that the doubling time is 10 days, so each step represents 10 days. We can assume a lethal burden tumour cap of 10^{13} and run the simulations starting with one tumour cell until we reach a size of 10^9 cells. At the end of the simulation, remember to multiply the number of steps by the assumed doubling time (in our example, 10 days). For further information, see Spencer et al. (2004).

20.3 Cellular automata

A **cellular automaton** is an extension of difference equations in which we are able to follow the dynamics of one or more variables or elements which reside in a lattice or grid. The grid is an ensemble of contained spaces known as 'cells'. In the simplest case, each cell is occupied by a single binary element. For example, the elements may represent genes or proteins that could be either active or inactive. The status of each element is defined by the updating rules that usually take into account the status that a particular element had at the previous discrete time, as well as the status of its close neighbours at the same previous discrete time.

For example, we can extend the linear discrete model described by equation (20.2) by considering that the expression level of gene y now depends not only on its value at a previous time (e.g. 1min previously) but also on the expression level of genes x and z at the same previous time:

$$y_t = f(x_{t-1}, y_{t-1}, z_{t-1}) \tag{20.6}$$

For simplicity, instead of measuring expression levels, we can consider only two possible values for all the variables involved, i.e. whether the genes are expressed ('on') or not expressed ('off'). If we arrange our genes in a grid in which each 'cell' (member of the grid) is occupied by one gene and define the relationships between them (updating rules), we will generate a cellular automaton.

In a simple case, the grid has only one dimension and the expression state of one of the genes at time t depends on its expression state at the previous time, as well as on the expression state of its two neighbours at the same previous time. Given that three genes are involved in the updating rule and that all of them can have only two states ('on' or 'off'), we have eight possible combinations. For each triplet combination, the updating rule establishes an outcome. For example, the rule could establish that if the three genes were active at time $t = 0$, the one in the middle would be inactive at time $t = 1$ (see left upper corner in Figure 20.3(a)). If we represent an active element by 1 and an inactive element by 0, we can define the updating rule mathematically by a binary number containing eight elements (the ordered possible outcomes). This binary number is used to name the updating rule. For example, the outcome: 0 1 0 1 1 0 1 0 corresponds to 'rule 90'.

Figure 20.3(a) shows the updating rules for 'rule 90'. In Figure 20.3, filled circles represent a gene that is expressed ('on' state) and empty circles represent a gene that is not expressed ('off' state). The eight updating rules apply to the middle gene in each case.

To run a simulation, we need to start with an initial condition and then follow the rules shown in Figure 20.3(a). It happens that the first and the last member of the grid only have one neighbour. One option for applying the rules to these members is to assume that they are connected, so that instead of having a line we have a ring (in geometry, a 'torus'). Figure 20.3(b) shows the dynamics of running rule 90 for up to seven times starting with just one expressed gene. The simulation runs from left to right (time is shown along the bottom). When we run the simulation 100 times, a **fractal** pattern is observed as shown in Figure 20.3(c). In a fractal one can observe the same pattern at different scales. Biological examples are snowflakes, leaf borders, and animal stripes.

Let us assume that we have a protein which when activated makes the cell generate a particular pigment, and that its status (active or inactive) can be simulated with a one-dimensional cellular automaton like the one shown in Figure 20.3. In this example, the final pigmentation would be similar to the pattern shown in Figure 20.3(c). This pattern has been found in some seashells and has been used to simulate morphogenesis.

Cellular automata can be extended to include several variables interacting with each other and with the environment. These models, known as agent-based models (ABMs), allow analysis in time and space as well as the presence of both deterministic and random (stochastic) events. ABMs are sufficiently powerful and flexible to be considered a

Figure 20.3 Simulation of a one-dimensional cellular automaton. (a) Rule 90. The updating rule of a cell is given by its previous state as well as the previous state of its two neighbours. Filled circles represent an 'on' state and empty circles an 'off' state. In each of the eight possible rules, we follow the outcome of the gene in the middle circle. (b) Simulation of rule 90 for seven time steps. After the first time steps have been simulated, a characteristic triangular pattern emerges. (c) Simulation of rule 90 for 100 time steps. For larger simulations, the initial pattern that was observed in (b) can be detected at different scales, generating a fractal. A fractal similar to (c) can be observed in some seashells.

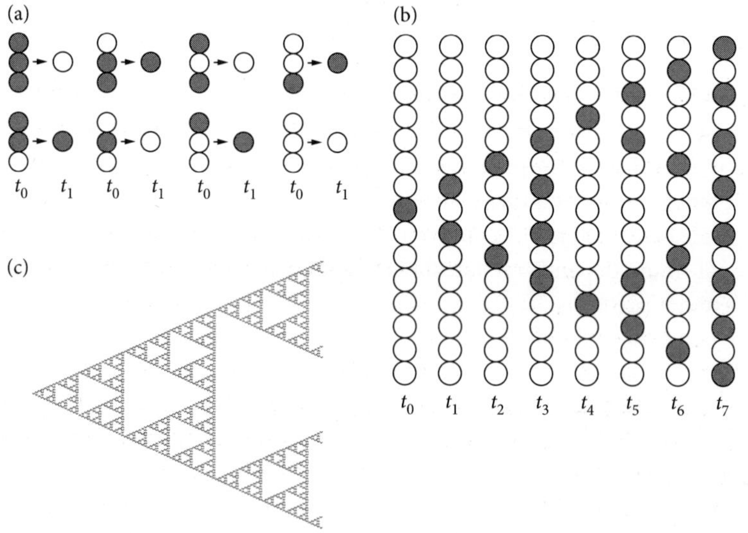

good example of synthetic life. ABMs can be simulated with specialized software such as NetLogo (http://ccl.northwestern.edu/netlogo) or SWARM (http://www.swarm.org).

As an example of how a discrete model can develop, an initial cellular automaton developed to study antigen presentation (Seiden and Celada 1992) was extended to an ABM able to simulate various immune-associated phenomena such as vaccination (Kohler et al. 2000) and autoimmunity (Calcagno et al. 2011).

20.4 Networks

One way of describing the relationship between variables graphically is to use **networks** or graphs. In these diagrams each variable represents a node or vertex, which may be connected to other members with a link known as an edge. In molecular biology, network theory has been applied to the study of gene regulatory networks and metabolic networks (Barabási and Oltvai 2004; Schlitt and Brazma 2007).

One way of inferring gene networks is to analyse data in which cells are cultured in the presence or absence of a particular stimulus and **microarrays** are taken at different times. In these experiments, we are interested in analysing how this perturbation (stimulus) of the system affects the dynamics of the gene expression levels under investigation.

For example, we can extend equation (23.3) to consider that the expression level of gene y at time t is a linear combination of all the expression levels at the previous time (Dewey and Galas 2001; Wu and Dewey 2006). A measurement is represented as $a_i(t)$, indicating the mRNA level of the ith gene at time t where i ranges from 1 to M (the number of genes), and t ranges from 1 to N (the number of times). The corresponding model is:

$$a_i(t) = \sum_{j=1}^{M} \lambda_{i,j} a_j(t-1) \quad (20.7)$$

Equation (20.7) relates the mRNA expression level at time t to the linear combination of all other mRNA expression levels at the previous time $t - 1$. The transition coefficients $\lambda_{i,j}$ are the elements of an $M \times M$ transition matrix and are the parameters that are calculated from the data. These coefficients are unitless and show how strongly weighted each contribution from the previous time will be to the production of the ith gene. These parameters are calculated using a generalized matrix inversion technique known as singular value decomposition and give the phenomenological influence of the expression level of the jth gene on the production of the ith gene.

In this case we are interested not in predicting expression levels, as these are part of our data, but in understanding the relationship between all the genes analysed and, more specifically, trying to identify those genes that are most affected by the perturbation.

To plot a network we need to know how the nodes (gene expression levels in our example) are connected. We can record the information used to plot the network in an $M \times M$ matrix, called the **adjacency matrix**, in which each row and each column represents a node. For example, in Figure 20.4 we have four genes: A, B, C, and D. Let us assume that we used the discrete model shown in equation (20.7) to analyse a hypothetical experiment in which we quantified the expression levels of these four genes using microarrays. A transition matrix is estimated from the model (see the example in Figure 20.4). In the left panel of Figure 20.4, each number represents the strength of the predicted connection between the row genes and the column genes. For example, we can see that the strongest connection (weighted connection of 92) is between gene B and gene C, and the weakest connection is between gene D and gene A (weighted connection of 1). When we compare these two extreme cases, we can assume that the former represents a real phenomenological association, whereas the latter is not a real phenomenological connection. In an adjacency matrix, connection is identified by 1 and no connection is identified by 0. Thus, to construct an adjacency matrix we need to define an arbitrary threshold identifying a potential biological connection.

In the example shown in the left panel of Figure 20.4 all values greater than an arbitrary threshold of 50 have been highlighted. Those values that are greater than or equal to this threshold in the transition matrix (Figure 20.4, left panel) are given a value of 1 in the adjacency matrix (Figure 20.4, middle panel); otherwise they are given a value of 0 in the adjacency matrix. For example, the adjacency matrix in the middle panel of Figure 20.4 shows that gene B activates or inhibits gene C, but not the other way around (the row for C has a zero in the column for B). This directed graph is a representation of the phenomenological network being studied. It is called a phenomenological network because we have only found mathematical associations that may or may not represent biological relationships.

Figure 20.4 Rationale for generating networks from the transition matrix. Starting with the absolute estimated values in the transition matrix (left panel) an arbitrary threshold is chosen to generate the adjacency matrix (central panel). In this example, all values ≥50 have been highlighted and transformed to 1 in the adjacency matrix. The adjacency matrix provides the information needed to draw a network (right panel). In this example the hypothetical gene D has been highlighted because this is the one with the most connections ('hub'). In the right panel, genes A, B, C and D are called nodes. The influence of one gene on the expression level of either another gene or itself is represented by arrows known as edges or links.

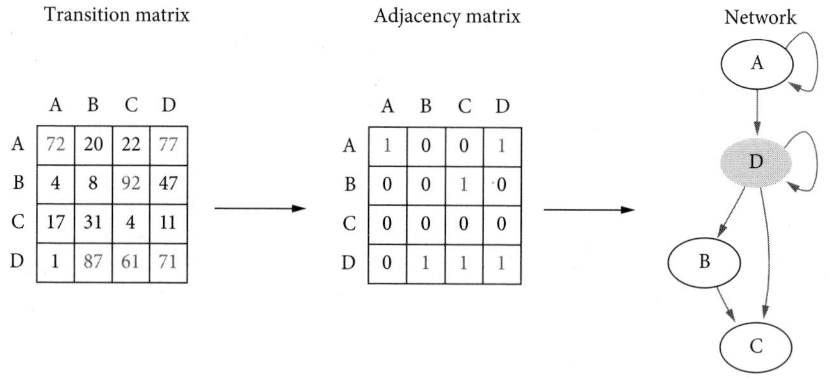

In the trivial example shown in the right panel of Figure 20.4, arrows represent potential biological associations or influences, but we do not know whether the association is positive (activation) or negative (inhibition). The model suggests that the expression level of gene A will influence the expression levels of genes A and D. Gene D has an influence on the expression level of genes D, B, and C, and finally gene B has an influence on the expression level of gene C. In this network, gene D represents a hub as it is the most connected gene.

This approach was used to identify the pro-inflammatory cytokine interleukin-1 (IL-1) as a major contributor of the phenomenological network of the breast cancer cell line SUM-149 (Streicher et al. 2007).

There are alternative methods of inferring networks from gene expression datasets. For example, the correlations among gene expression levels over time can be used to predict potential biological networks (Opgen-Rhein and Strimmer 2007). A review of different modelling approaches to the study of gene regulatory networks can be found in Schlitt and Brazma (2007). More detailed information on network theory can be found in Barabási (2002), Alon (2007) and Bornholdt and Schuster (2003).

20.5 Markov chains

In all the examples previously discussed in this chapter, we have used deterministic models—knowing the values of the parameters and initial conditions, we are able to predict the value of our independent variables. Some events, like mutation, are random. Models which introduce random events are known as probabilistic or stochastic models.

A **Markov chain** is an example of a stochastic discrete model in which a variable undergoes transitions from different states with a given probability. The transitions depend only on the current state of the variable and the outcome of the random event. For example, let us assume that a gene *y* can transit from the steady state *WT* (wild type) to a mutant *Mut* with a probability of 10^{-6} on each division cycle, this means that if a cell undergoes a million division cycles, on average we would expect gene *y* to mutate once.

Stochastic models are generally more time consuming than deterministic models as several simulations (e.g. 100 simulations) should be run to report the average dynamics of the system. Nevertheless, they can provide more realistic results when it is known that random events like motion or mutation may be involved. For example, stochastic Markov chains have been used as an alternative way of studying the process of phototransduction (Houillon et al. 2010).

20.6 Petri nets

Petri nets (PNs) are an extension of directed graphs (like the one shown in Figure 20.4) that have been used to study metabolic networks using a discrete approach (Hardy and Robillard 2004; Moore et al. 2005). In PNs there are two types of nodes: places and transitions. In metabolic networks, places can represent molecular species and transitions can represent reactions (Goss and Peccoud 1998). Places only have connections with transitions, and transitions only have connections with places; these connections are known as arcs, and can represent kinetic constants. The other elements in PNs are 'tokens', which can be present in places and represent the number of molecules. Following a synapsis analogue, a transition is 'fired up' when it receives a given number of tokens.

Figure 20.5 shows an example of a Petri net representing the interaction of the epidermal growth factor (EGF) with its receptor (EGFR). In the initial condition we have one molecule of EGF and one molecule of EGFR (both places, i.e. EGF and EGFR, have one token each). To become active, a transition needs to receive at least one token for each place with an input to the transition. In our initial condition, the only transition that is currently active is T0 because there is one token in the place EGF and one token in the place EGFR. If we run a simulation, in the next step T0 becomes inactive and releases one token to the complex place (EGF:EGFR). In this scenario either transition T2 or transition T1 can become active. In a stochastic simulation, we give probabilities to the corresponding arcs; in the simplest scenario they have the same probability, so on average T2 and T1 are fired the same number of times. If T1 is fired up, the complex dissociates; this step represents the reversibility of the reaction. On the other hand, if T2 is fired up, the complex is internalized. Figure 20.5 represents just the initial step of the EGFR pathway (Oliveira et al. 2004).

PNs have the advantage of not requiring detailed information about reaction rates that may not be known or hard to obtain (Schlitt and Brazma 2007). A list of PN software can be found at the PNs world web site (http://www.informatik.uni-hamburg.de/TGI/PetriNets/).

Figure 20.5 Petri net representing the interaction of the epidermal growth factor (EGF) with its receptor (EGFR): empty circles (known as places) represent molecular species; rectangles (known as transitions) represent reactions or biological process like internalization; small filled circles (known as tokens) indicate the number of available molecular species at a given time. The formation of the complex EGF:EGFR is reversible and controlled by reactions T0 and T1. The internalization (int) of the complex is controlled by reactions T2 and T3. The Petri net shows the initial condition when reaction T0 is fired up by the input of the two tokens coming from EGF and EGFR (both with one token each).

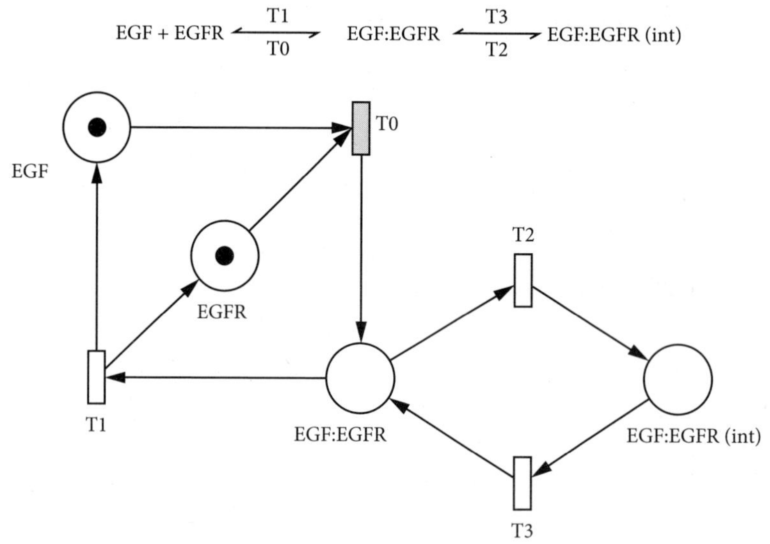

20.7 Ordinary differential equations (ODEs)

Ordinary differential equations (ODEs) are the most widely used mathematical models in mathematical biology. Their main advantage is that they are usually able to provide more accurate predictions than other models. In ODEs time is continuous, so we are looking at instantaneous rates. Although some ODEs can be solved analytically, we generally use a numerical approach to simulate the system.

To introduce ODEs we will use a simple mathematical model that was proposed for studying the negative feedback signalling module between nuclear factor κB (NF-κB) and its inhibitor I-κB (Hoffmann et al. 2002). Figure 20.6(a) shows a diagram representing the model. In this simple model, NF-κB is in the cytosol inhibited by I-κB at rate β; in the presence of IKK, NF-κB is activated and translocates to the nucleus where it activates the expression of I-κB at rate γ. Nuclear I-κB binds nuclear NF-κB, restoring its inhibition. The parameters α and δ are self-regulation rates of NF-κB and I-κB, respectively.

The model can be formally expressed by the following set of ODEs:

$$\frac{dNF\kappa B}{dt} = IKK - \alpha(NF\kappa B) - \beta(I\kappa B) \tag{20.8}$$

Figure 20.6 Simple model of the negative feedback loop between the nuclear factor κB (NF-κB) and its inhibitor I-κB. (a) Diagram of the model. When NF-κB is activated by IKK, I-κB is degraded and NF-κB is activated. NF-κB activates the expression of I-κB that in turn is able to inhibit NF-κB again. The dynamics of this system depends on the amount of feedback regulation, controlled by parameters β and γ, and the amount of self-regulation, controlled by parameters α and δ. (b) Dynamics of I-κB under different scenarios. The dashed curve corresponds to continuous oscillations with high feedback efficiency. The solid curve corresponds to damping oscillations with intermediate feedback. The dotted curve corresponds to a system that reaches a steady state after approximately 2 hours with low feedback efficiency. From Hoffmann, A. et al. (2002) Science 298: 1241. Reprinted with permission from AAAS.

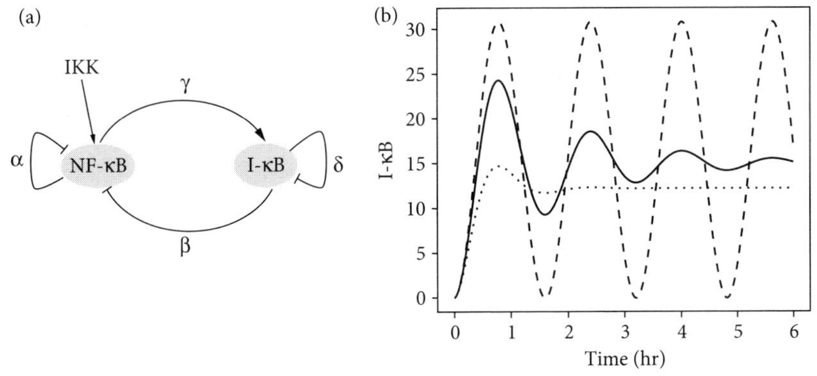

$$\frac{dI\kappa B}{dt} = \gamma(NF\kappa B) - \delta(I\kappa B) \tag{20.9}$$

Equation (20.8) represents the instantaneous change in the concentration of NF-κB over time. In equation (20.8) the concentration of NF-κB is increased by the concentration of IKK and is decreased by both the concentration of I-κB (at rate β) and its own concentration (at rate α). In equation (20.9) the concentration of I-κB is increased by the concentration of NF-κB (at rate γ) and decreased by its own concentration (at rate δ).

Specialized software is needed to solve this system of ODEs numerically. We can choose commercial software such as Maple (http://www.maplesoft.com), Mathematica (http://www.wolfram.com), MATLAB (http://www.mathworks.com), Berkeley Madonna (http://www.berkeleymadonna.com), or STELLA (http://www.iseesystems.com), or free software such as R (http://cran.r-project.org), SciPy (http://www.scipy.org), or XPPAUT (http://www.math.pitt.edu/~bard/xpp/xpp.html). The following information is needed to run a simulation:

- the set of equations;
- the values of the initial conditions;
- the values of all the parameters;
- the final time when the simulation will stop;
- the time step to solve the equations numerically (this time is usually relatively small and does not need to be fixed throughout the simulation);
- the integration method (a classic integration method is Runge–Kutta (RK4)).

Figure 20.6(b) shows the dynamics of I-κB under different sets of parameters that represent potential biological scenarios. The dashed curve represents persistent oscillations of I-κB when the NF-κB pathway has been activated; the solid curve corresponds to damped oscillations and the dotted curve represents the case of gradual rising to a plateau level. This simple mathematical model was extended to study the effect of different mutants on the dynamics of NF-κB and various I-κB subunits (Hoffmann et al. 2002).

In molecular biology, we can highlight examples of the contribution of ODEs to the understanding of the cell cycle (Tyson and Novak 2008; He et al. 2011) and signal transduction (Kholodenko et al. 1999; Brown et al. 2004).

The **systems biology mark-up language** (SBML) was developed to facilitate the exchange of mathematical models used in biochemistry among different software applications (Hucka et al. 2003). SBML has become the standard format for reporting biochemical network models and is usually required in scientific papers published in open access journals. Most of the mathematical software previously mentioned has the ability to importing and export to SBML format. Furthermore, free software specifically designed to work with SBML, such as COPASI (http://www.copasi.org), is also available. This software enables the user to express equations in terms of the biochemical reactions instead of classical ODEs (which can also be visualized using the software).

CASE STUDY 20.2 A dose-response induction of autoimmune disease

The aim of the study is to analyse the effect of different concentrations of specific T cells on the development of autoimmune diseases.

QUESTION
Which is the most appropriate mathematical model for studying this process?

Consider the options. In our study the outcome is having or not having the autoimmune disease, and the independent variables are the number or concentration of relevant cell populations such as effector cells and regulatory cells. Although it is possible to study changes in cell numbers at discrete times, it makes more sense to consider the change as a continuous variable, and therefore we propose a time-continuous model. If we have used a discrete approach, for example an agent-based model, we will need to define all the updating rules (equations) that allow us to simulate the relationships between the cells and the emergence of disease. When we use a continuous model we only need to estimate cell proliferation rates, and we can define a set of two ODEs, one for the effector cells and one for the regulatory cells, to simulate the experiments. Once we have run our simulations, we can test our theoretical predictions with laboratory experiments. For further information, see Segel et al. (1995).

20.8 Chapter summary

- Mathematical models complement laboratory experiments and are an alternative set of tools that allow us to understand biological phenomena better.
- They can be used to help in the interpretation of high-throughput technologies, including microarray analysis and proteomics, or to analyse/predict the effect that a

set of independent variables have on a particular outcome (dependent variable) under relevant theoretical scenarios. Those cases for which specific experiments cannot be performed because of ethical, economical, or time limitations, but can be simulated using a theoretical framework, are of particular interest.

- Systems biology is the integration of different disciplines, including mathematics, to study biological phenomena from a global perspective.
- Mathematical models can be *continuous*, when they consider instantaneous changes over time, or *discrete*, when the system evolves at equal integer time steps.
- Mathematical models can be *deterministic* when a single outcome can be predicted based upon the values of the initial conditions and the parameters, or *stochastic* when events occur at random.
- The predictive power of a mathematical model depends mainly on the accuracy of its parameters, the underlying assumptions, its complexity, and its capability to fit experimental data.
- The same biological problem can be approached using more than one class of mathematical model, each with different levels of complexity. Choosing a particular model requires good communication between scientists from different disciplines (e.g. biology and mathematics).

References

Alon, U. (2007) *An Introduction to Systems Biology: Design Principles of Biological Circuits.* Boca Raton, FL: Chapman & Hall/CRC.

Barabási, A.L. (2002) Statistical mechanics of complex networks. *Rev Mod Phys* **74**: 47–97.

Barabási, A.L., and Oltvai, Z.N. (2004) Network biology: understanding the cell's functional organization. *Nat Rev Genet* **5**: 101–13.

Bornholdt, S. and Schuster, H.G. (eds) (2003) *Handbook of Graphs and Networks: From the Genome to the Internet.* Weinheim: Wiley–VCH.

Brown, K.S., Hill, C.C., Calero, G.A., et al. (2004) The statistical mechanics of complex signaling networks: nerve growth factor signaling. *Phys Biol* **1**: 184–95.

Calcagno, C., Puzone, R., Pearson, Y.E., et al. (2011) Computer simulations of heterologous immunity: highlights of an interdisciplinary cooperation. *Autoimmunity* **44**: 304–14.

Dewey, T.G. and Galas, D.J. (2001) Dynamic models of gene expression and classification. *Funct Integr Genomics* **1**: 269–78.

Edelstein-Keshet, L. (2005) *Mathematical Models in Biology.* Philadelphia, PA: Society for Industrial and Applied Mathematics.

Fall, C.P., Marland, E., Wagner, J., and Tyson, J.J. (2002) *Computational Cell Biology.* New York: Springer.

Goss, P.J. and Peccoud, J. (1998) Quantitative modeling of stochastic systems in molecular biology by using stochastic Petri nets. *Proc Natl Acad Sci USA* **95**: 6750–5.

Hardy, S. and Robillard, P.N. (2004) Modeling and simulation of molecular biology systems using petri nets: modeling goals of various approaches. *J Bioinform Comput Biol* **2**: 595–613.

He, E., Kapuy, O., Oliveira, R.A., Uhlmann, F., Tyson, J.J., and Novák, B. (2011) System-level feedbacks make the anaphase switch irreversible. *Proc Natl Acad Sci USA* **108**: 10016–21.

Hoffmann, A., Levchenko, A., Scott, M.L., and Baltimore, D. (2002) The IκB–NF-κB signaling module: temporal control and selective gene activation. *Science* **298**: 1241–5.

Hood, L., Heath, J.R., Phelps, M.E., and Lin, B. (2004) Systems biology and new technologies enable predictive and preventative medicine. *Science* **306**: 640–3.

Houillon, A., Bessière, P., and Droulez, J. (2010) The probabilistic cell: implementation of a probabilistic inference by the biochemical mechanisms of phototransduction. *Acta Biotheor* **58**: 103–20.

Hucka, M., Finney, A., Sauro, H.M., et al. (2003) The systems biology markup language (SBML): a medium for representation and exchange of biochemical network models. *Bioinformatics* **19**: 524–31.

Istrail, S., Pevzner, P., and Shamir, R. (2003) *Computational Molecular Biology* Amsterdam: Elsevier.

Kaplan, D. and Glass, L. (1995) *Understanding Nonlinear Dynamics*. New York: Springer.

Keeling, M.J. and Rohani, P. (2008) *Modeling Infectious Diseases in Humans and Animals*. Princeton, NJ: Princeton University Press.

Kholodenko, B.N., Demin, O.V., Moehren, G., and Hoek, J.B. (1999) Quantification of short term signaling by the epidermal growth factor receptor. *J Biol Chem* **274**: 30169–81.

Kitano, H. (2002) Computational systems biology. *Nature* **420**: 206–10.

Kitano, H. (2003) A graphical notation for biochemical networks. *Biosilico* **1**: 169–76.

Kitano, H., Funahashi, A., Matsuoka, Y., and Oda, K. (2005) Using process diagrams for the graphical representation of biological networks. *Nat Biotech* **23**: 961–6.

Kohler, B., Puzone, R., Seiden, P.E., and Celada, F. (2000) A systematic approach to vaccine complexity using an automaton model of the cellular and humoral immune system. I. Viral characteristics and polarized responses. *Vaccine* **19**: 862–76.

May, R.M. (1976) Simple mathematical models with very complicated dynamics. *Nature* **261**: 459–67.

Moore, J.H., Boczko, E.M., and Summar, M.L. (2005) Connecting the dots between genes, biochemistry, and disease susceptibility: systems biology modeling in human genetics. *Mol Genet Metab* **84**: 104–11.

Oliveira, J.S., Jones-Oliveira, J.B., Dixon, D.A., Bailey, C.G., and Gull, D.W. (2004) Hyperdigraph-theoretic analysis of the EGFR signaling network: initial steps leading to GTP:Ras complex formation. *J Comp Biol* **11**: 812–42.

Opgen-Rhein, R. and Strimmer, K. (2007) From correlation to causation networks: a simple approximate learning algorithm and its application to high-dimensional plant gene expression data. *BMC Syst Biol* **1**: 37.

Schlitt, T., and Brazma, A. (2007) Current approaches to gene regulatory network modelling. *BMC Bioinformatics* **8** (Suppl 6): S9.

Segel, L.A., Jäger, E., Elias, D., and Cohen, I.R. (1995) A quantitative model of autoimmune disease and T-cell vaccination. Does more mean less? *Immunol Today* **16**: 80–4.

Seiden, P.E. and Celada, F. (1992) A model for simulating cognate recognition and response in the immune system. *J Theor Biol* **158**: 329–57.

Spencer, S.L., Berryman, M.J., García, J.A., and Abbott, D. (2004) An ordinary differential equation model for the multistep transformation to cancer. *J Theor Biol* **231**: 515–24.

Streicher, K.L., Willmarth, N.E., García, J., Boerner, J.L., Dewey, T.G., and Ethier, S.P. (2007) Activation of a nuclear factor κB/interleukin-1 positive feedback loop by amphiregulin in human breast cancer cells. *Mol Cancer Res* **5**: 847–61.

Tyson, J.J. and Novak, B. (2008) Temporal organization of the cell cycle. *Curr Biol* **18**: R759–68.

Vries, G.D. (2006) *A Course in Mathematical Biology : Quantitative Modeling with Mathematical and Computational Methods*. Philadelphia, PA: Society for Industrial and Applied Mathematics.

Wu, X. and Dewey, T.G. (2006) From microarray to biological networks: analysis of gene expression profiles. *Methods Mol Biol* **316**: 35–48.

Glossary

3D reconstruction – A computational image-processing technique used to generate 3D representations of a specimen from 2D information recorded when using an electron microscope.

Adapter – A short piece of chemically synthesized DNA used to bind to the end of nucleic acid fragments (in next-generation sequencing used to create a library).

Adenoma – A benign epithelial tumour arising in glands or forming a glandular pattern.

Adjacency matrix – A matrix representation of a graph in which connections between row elements with column elements are given the value 1.

Adult stem cell – An unspecialized stem cell found inside tissues or organs after embryonic development. An adult stem cell remains in a non-dividing (quiescent) state until activated for tissue or organ development.

Affinity – A measure of the preference for two macromolecules to interact.

Affinity chromatography – Purification of proteins using the specificity of a target protein for an immobilized binding ligand.

Affinity tag – An additional protein or short peptide sequence, fused to a protein of interest, that has known affinity for a molecule or ion.

Agarose gel electrophoresis – An electric current is used to move negatively charged nucleic acid (DNA or RNA) molecules across a solid agarose matrix towards the positive electrode. Molecules are separated according to molecular weight.

Amelogenin – A protein found in teeth, the gene for which is located on the X and Y chromosomes in humans.

Ammonium sulphate precipitation – Use of high concentrations of ammonium sulphate to differentially reduce the solubility of proteins, which can then be removed by centrifugation.

Amplicon – Formal name for new DNA sequences synthesized during PCR.

Amplified DNA – DNA sequence copied during PCR, for example.

Analyte – Any substance or material that is being analysed.

Annealing – Binding of an oligonucleotide to single-stranded DNA molecules. In the case of PCR, binding of oligonucleotide primers to template DNA molecules.

Antigen – A substance that is capable of generating a specific immune response and reacting with the products of that response (i.e. antigen is recognized by an antibody).

Antisense strand (DNA) – The complementary DNA strand to the **s**ense DNA strand and therefore also to the mRNA strand. The antisense strand is the template strand used during mRNA transcription. Within RNAi pathways, the strand of the RNA duplex which is antisense to the mRNA transcript becomes the guide strand, and its complementarity with the mRNA strand is used to bind and silence the mRNA.

Apo state – The state of a protein when it is not bound to a ligand. See **holo state** for the opposite.

Apoptosis – The process of controlled cell death, a sequence of events resulting in the elimination of cells from a tissue or organism.

Argonaute – Catalytic functioning protein of the RISC protein complex. RISCs containing the AG02 Argonaute cleave mRNA transcripts. RISCS containing the AG01, AG03, or AG04 Argonautes suppress mRNA translation or cause decay through de-adenylation.

ARMS PCR – **A**mplification **R**efractory **M**utation **S**ystem PCR: a means of typing alleles using a panel of reactions, each targeting different polymorphic sequences of a locus.

Aseptic – Free from pathogens; a technique carried out in sterile conditions to reduce the risk of microbial contamination.

Aspect ratio – The ratio of the height of an object to its width.

Autofluorescence – Natural emission of light (photons) by a cell or particle when excited by a high-energy source (laser).

Auxotrophic selection – Selection of bacterial or fungal cells based on the loss of their ability to synthesize certain molecules required for growth.

Axial resolution – The smallest distance that two objects can be separated in three dimensions.

Bacterial artificial chromosome (BAC) – Recombinant plasmid designed to clone large fragments of DNA and allow propagation in a bacterial host.

Bacteriophage – A virus capable of infecting a bacterial cell which uses the host's transcriptional and translational apparatus to replicate and express its genome. The phage may be lytic (replicated phage destroy the host cell) or non-lytic.

Bacteriophage M13 – A virus that specifically infects Gram-negative bacteria such as *E. coli.* The phage particles are extruded from the bacterial cell and can be isolated from the culture medium. They contain a single-stranded copy of the bacteriophage genome, including any gene that has been cloned into the vector.

Bait – Something used to attract an interacting partner that has affinity for it. In protein binding studies, the bait is typically the known protein.

Base peak (in mass spectrometry) – The most intense peak within the mass spectrum which is normalized to 100% abundance.

B factor – A measure of the mobility of an atom. Various kinds of disorder can 'spread' an atom's electron density. B factors inform about local resolution; the higher the B factor the lower is the certainty about the atom's position.

Binding capacity (in chromatography) – The amount of a protein (e.g. in milligrams) which can be bound per millilitre of chromatography matrix. Influences the choice of column volume for the amount of protein you require.

Binding curve – A set of data points plotted as a curve quantitating the binding or association of two biomolecules (e.g. DNA and protein).

Bioinformatics – A scientific field that applies computer science and statistics to investigate molecular biology.

Biopsy – A procedure that removes a sample of cells or tissues for histological examination by a pathologist.

Biotinylated – Description of a biological molecule to which biotin has been covalently attached.

Bipotent – The potential of a stem cell to differentiate into two cell types of an organism.

BLAST – **B**asic **L**ocal **A**lignment **S**earch **T**ool, hosted by the National Center for Biotechnology Information, US Library of Medicine. DNA sequences can be probed for homologous regions using this tool.

Buffer – Reagent intended to maintain the pH of a solution within a certain range.

Buffer exchange – Changing protein solution composition using multiple cycles of ultrafiltration, concentration, and dilution into buffer of choice.

Calorimeter – Instrument for measuring the amount of heat produced in any system.

Capillary electrophoresis – Precise method of separating biomolecules on the basis of their charge and molecular weight, frequently using fluorescent-based detection systems.

Capture step (in protein purification) – Initial chromatography step which will work reliably with cell extract samples. A cheap matrix that will bind a large amount of protein is often used.

Carbon nanotubes – Very small tubes composed of one or more layers of graphene rolled into a tube.

Carcinoma – A malignant tumour derived from epithelial tissue.

Cassette (expression) – Part of a DNA vector that directs the cell's machinery to make RNA and protein. Comprises the gene coding sequences and the sequences that regulate its expression.

Cassette (mutagenesis) – A section of DNA that would normally be generated by annealing synthetic oligonucelotides containing the desired mutations with overlapping ends to create complementary sequences for cloning into a suitable vector to replace the corresponding region of the wild-type gene. A cassette can also be a larger DNA fragment produced by PCR mutagenesis, for example, that is then digested with appropriate restriction enzymes before replacing the corresponding wild-type region.

Catabolite repression – The process in bacteria of maintaining transcriptional inactivity of genes encoding proteins for metabolism of alternative carbon sources when another source is abundant. For example, glucose exerts catabolite repression on the *lac* operon for lactose utilization.

cDNA – Single-stranded complementary DNA synthesized from an mRNA template using the enzyme reverse transcriptase, primer and dNTPs. Second-strand DNA synthesis is by DNA polymerase using the first synthesized strand as the template.

cDNA library – A collection of DNA sequences generated from mRNA sequences, usually from a single organism.

Cell density – The number of viable cells per volume (either per cm^2 or per ml) which is useful to know when calculating how many cells to seed into fresh tissue culture vessels for experiments.

Cell lines – An established culture of cells. A primary culture that has undergone subculture becomes a cell line. May be immortal or non-immortal.

Cell sorter – A flow cytometer capable of physically separating a heterogenous mixture of cells into single-cell types and depositing them into tubes, 96-well plates, and other collection devices. Separation is based on classifying each cell type by virtue of a unique marker expression.

Cell surface marker – A protein on the outer membrane of a cell used as a unique classifier of a cell type; a binding receptor for a ligand, or an antibody/antigen target.

Cell viability – An assessment of the functional state of a cell: the ability to undergo cell division; being metabolically active; not undergoing cell death.

Cellular automaton (CA) – A discrete model in which the elements from a grid can change its status (value) according to a set of rules that usually take into account the status of other elements (including itself) at a previous time.

Chemical shift – Change in NMR transition energies, and peak positions, due to molecular structure that allows signals from atoms of the same chemical to be separated. Measured in ppm.

Chemical shift assignment – The connection of chemical shifts in NMR spectra to the originating atom in a molecular structure or sequence.

Chemical shift mapping – The use of changes in chemical shift upon ligand binding to determine where the ligand binds to its target.

ChIP – A technique used to determine whether a protein is bound to a particular DNA region in a chromatin context.

ChIP-on-chip – A chromatin immunoprecipitation reaction followed by hybridization of the resultant DNA on a chip to determine all DNA sequences bound to the protein of interest.

ChIP-sequencing – A chromatin immunoprecipitation reaction followed by sequencing of resultant DNA to determine all DNA sequences bound to the protein of interest.

Chiral molecules – Molecules whose structure is asymmetric around one or more atoms. Most biomolecules are chiral.

Chromatin – Eukaryotic DNA is packed in the nucleus in the form of chromatin. Chromatin consists of DNA with histones and other proteins such as **transcription factors** associated with it.

Chromatography – The process of purifying molecules by sieving or differential binding and **elution** using a solid matrix.

Chromophore – A chemical group capable of selective light absorption, also potentially resulting in emission of energy at a certain wavelength, which may be in a visible or non-visible part of the electromagnetic spectrum. A **fluorophore** is a type of chromophore.

Circular dichroism (CD) spectroscopy – A method of measuring the interaction of **chiral molecules** with circularly polarized light.

Cloning – Method of copying a DNA sequence *in vivo* or *in vitro* (molecular cloning). Cells or organisms can also be copied and cloned.

Cluster differentiation – A term used in antibody/antigen classification based on structural differences in the receptor or antigen protein complex.

Coat protein – Any of the proteins that make up the capsid (outer coat) of a virus (or **bacteriophage**).

Codon usage – The frequency with which particular codons for amino acids are used in the genome of an organism.

Colorimetric assay – An assay that entails a change in the colour of a reagent when a reaction (or interaction) occurs.

Competent (cells) – Cells that have been treated to enhance their ability to take up DNA.

Combinatorial cloning – A process that allows transfer of a cDNA from a single donor vector to multiple destination vectors via a shared recombination mechanism.

Complementation assay – An experiment designed to test whether the loss of function caused by mutation or deletion of a chromosomal gene X can be recovered by transforming the cell with a plasmid carrying a homologous gene.

Complementarity determining regions (CDRs) – The **hypervariable regions** within an antibody that forms the **paratope** or binding site for the antigen.

Confluency – The coverage of the cell culture vessel by adherent cells. 100% confluency means that there is no room left for the cells to grow and leads to contact inhibition.

Constant neutral loss scanning – An MS/MS experiment used to monitor a particular loss of mass from precursor ions.

Constitutive expression – A gene that is transcribed constantly.

Construct (DNA) – A functional unit of DNA that contains the information necessary for the transfer and expression of a target gene.

Contact inhibition – Inhibition of cell growth that occurs when two cells come into contact with each other. **Apoptosis** of the cells can also be naturally induced.

Contamination (in PCR) – Intrusive and unwanted DNA source liable to be amplified in a PCR.

Coomassie blue – Commonly used blue dye for staining proteins following gel-based separations.

Copy number variation – A structural variation that results in an abnormal number of copies (due to either a duplication or a deletion) of a section or sections of DNA. They are generally large, ranging from 1kb to several megabases of DNA.

Correlation spectrum – Spectrum which connects chemical shifts between two atoms, typically via scalar couplings. Correlation spectra must be 2D or higher.

Correlative light electron microscopy – A method which involves imaging the same region of a cell using both fluorescent light microscopy and transmission electron microscopy. The images are then correlated and the information obtained from one considered in the context of the other.

Cosmid – A hybrid between a plasmid and the cos sites of the lambda bacteriophage, into which large DNA fragments can be cloned to allow propagation.

Critical point drying – A method of dehydrating a specimen for scanning electron microscopy. Water is first displaced with acetone, and then with liquid carbon dioxide. The pressure and temperature are adjusted to enable liquid carbon dioxide to vaporize without changing density.

Cross-breeding – The process of breeding together two pure-bred animals with the intent of producing offspring which share the traits of both parental lineages.

Cryogenic – Refers to very low temperatures. In the context of specimen preparation for microscopy the specimen is kept immobilized at very low temperatures.

Cryostat – Device used to cut frozen tissue into thin sections.

Crystal – An ordered array of molecules.

Damped (in AFM) – A damped oscillator will experience a reduction in the amplitude of its oscillations over time. This is usually the result of friction and therefore the degree of damping is proportional to the velocity of the oscillator.

Data system (in mass spectrometry) – A computer that controls all aspects of the mass spectrometer and which processes the amplified signals received from the detector to produce a mass spectrum.

Deconvolution (in fluorescence microscopy) – An image analysis process which improves the resolution within a fluorescence image by correcting for aberrations occurring during image acquisition.

Deep sequencing – Where next-generation DNA sequencing (sequence-by-synthesis or sequence-by-ligation methods) is done in a massively parallel fashion and sequence information is captured by a computer. Such techniques can achieve sequence reads of >80Mb/hour.

Deflection (in AFM) – A measure of an object's or a material's displacement due to applied load (force). It can be either a distance or an angular measurement.

Deformation – Change in the shape or size of an object due to an applied force. See also **Elastic**.

Degenerate (in NMR) – When two chemical shifts and their peaks in the NMR spectrum overlap so much that they cannot be separated.

Degenerate primers – Primers lacking stringency for a unique target sequence.

Deletion (DNA) – The removal of one or more nucleotides from a DNA sequence.

Denaturant – A chemical that destabilizes and eventually unfolds proteins. Common examples are urea and guanidine hydrochloride.

Desalting – Removal of low molecular weight solutes from macromolecules using a gel filtration matrix to change solution conditions between steps.

Detector (in mass spectrometry) – Where the separation of the ions achieved in the mass analysis is measured electronically.

Deterministic – An event whose outcome can be accurately predicted given enough information from the system.

Dialysate – The solution used for **dialysis**, after dialysis has been performed.

Dialysis – Changing solution conditions by diffusion across a semi-permeable membrane.

DICER – A member of the RNAse III protein family. Its main function is to cleave longer dsRNAs into the short 21–23nt siRNA and pre-miRNA to miRNA.

Dielectric constant – Required to estimate the strength of certain charge-affected interactions (such as electrostatic interactions); it accounts for the effects of the intervening medium. In biological systems, this medium is most often water, which has a dielectric constant D of 80. It is a 'relative permittivity' measurement.

Difference equation – An equation in which the time is discrete. Expresses the dynamics of a system in terms of events that occurred at a previous time.

Differentiation – The process by which a non-specialized cell becomes specialized which is brought about by an **epigenetic** change. This results in a shift in the physical characteristics (morphology, metabolism, membrane potential, etc.)

Diffraction pattern – The raw data of an X-ray diffraction experiment consisting of an array of intensity peaks on a detector generated by the diffracted X-rays.

Diodes (in surface plasmon resonance) – In this context, the term refers to the twin-terminal structure of a *light-emitting* device.

Dissociation (in cell culture) – Breaking by enzymatic or mechanical means of bonds between extracellular proteins which anchor the cells to the culture vessel to obtain a cell suspension for subculturing or cell seeding.

Divalent cations (in PCR) – Inorganic molecules important in the PCR reaction.

DNA binding domain – Protein domain that has the ability to bind to a particular sequence of DNA.

DNA ligase – An enzyme which joins a 5′-phosphate end to a 3′-hydroxyl end of DNA to create a new high-energy phosphodiester bond. The enzyme requires the presence of ATP as an energy source. It is used, for example, to join an insert DNA fragment to a cloning vector and to seal phosphodiester backbone gaps created during certain mutagenesis experiments such as multisite mutagenesis.

DNA polymerase – An enzyme that catalyses the template-directed primer-dependent addition of deoxynucleotide units, using dNTPs, to the 3′ end of a DNA chain.

DNA probe – See **probe**.

DNA sequencing – Method of establishing the base sequence of a DNA molecule.

DNA shuffling – An approach to the random exchange of mutations or natural sequence variations between homologous genes. This generates a library of variant genes that can be screened to identify new functional properties. It is normally used to introduce variation into proteins.

DpnI – A methylation-dependent restriction enzyme that will introduce a phosphodiester bond cleavage at the nucleotide sequence GATC where the A is methylated. It will digest either double strands of methylated DNA or a single strand of methylated DNA within a heteroduplex molecule, such as those produced during the **QuikChange® mutagenesis** process.

Droplet PCR – Microfluidic method of carrying out PCR in aqueous droplets.

Dynamic range (in mass spectrometry) – Concentration range over which the increase in signal is proportional to the increase in concentration of the analyte.

Dynamics – The study of how things change with time. In biochemistry the most common synonymn is kinetics.

Dysplasia – Disordered growth. This may occur because of abnormal differentiation or maturation of tissue, but also refers to pre-neoplastic changes in the epithelium.

Elastic – In materials science, a material is elastic if it can be deformed reversibly. See also **viscoelastic**.

Electron crystallography – A crystallographic approach to 3D structure determination which involves visualizing thin crystalline layers by electron microscopy. Diffraction patterns and images are recorded, providing both amplitude and phase information about the structure of the crystallized macromolecule.

Electron-dense label – A clonable protein tag which includes a metal-binding domain. Upon incubation with a metal salt, the domain will bind multiple metal atoms to form a cluster which is visible in the electron microscope.

Electron density map (X-ray crystallography) – A 3D map showing the density of electrons across the structure. As electrons are concentrated around atoms and bonds, the map shows the location of these atoms and bonds allowing a model of the structure to be built.

Electron micrograph – Magnified image recorded from an electron microscope.

Electroporation – A technique in which a brief electrical field is applied to the cell membrane, causing a brief (and reversible) increase in cell permeability.

Electrospray ionization (ESI) – An ionization technique where analyte molecules in solution are sprayed through a capillary held at a high electric potential to produce highly charged solution droplets that subsequently shrink (through solvent evaporation) to leave charged analyte ions in the gas phase.

Electrostatic deflection (in cell sorting) – Generation of a high-voltage potential either side of the sort stream. Attracts charged droplets from the main stream so that they form 'side streams' which are directed into the sort collection device.

Ellipticity (in CD) – The extent to which a sample interacts with circularly polarized light.

Elution – The process of washing bound proteins from a chromatography matrix.

Elution volume – The volume of buffer required to wash a protein of interest from a chromatography column.

Embryonic lethal – Describes a manipulation that results in the inability of an organism to survive past the embryonic developmental stages.

Embryonic stem cell – A **totipotent** cell derived from the inner mass of a blastocyte. Embryonic stem cells are capable of differentiating into all three germ layers (endoderm, mesoderm, and ectoderm), which in turn give rise to all of a mammal's tissue and organs through organogenesis.

Emission maximum – Refers to the wavelength at which maximum intensity of electromagnetic rediation emission is observed.

EMSA (**E**lectrophoretic **M**obility **S**hift **A**ssay) – A method of detecting DNA–protein interactions via reduced electrophoretic mobility of a protein–nucleic acid complex compared with the free nucleic acid in a native polyacrylamide or agarose gel.

Emulsion PCR – A type of PCR where the PCR template and reagents are present in an oil emulsion. Used with some next-generation sequencing platforms.

Endonuclease – An enzyme that catalyses the hydrolysis of a phosphodiester bonds within a nucleic acid molecule (i.e. at internal sites).

Endpoint (in PCR) – Point at the end of a PCR reaction at which PCR products can be quantified in order, permitting cross-comparison between reactions.

Enthalpy – A thermodynamic measurement which represents the difference between the energy consumed on reaction/interaction with that released, often observed as heat.

Entropy – A thermodynamic measure of the state of disorder of a system. Together with **enthalpy**, it contributes to the overall free energy of a system or process.

Epigenetic – Literal meaning is 'above' or 'over' genetics. It is the study of heritable changes in gene expression or cellular phenotype caused by mechanisms other than changes in the underlying DNA sequence.

Episomal DNA –An autonomously replicating DNA molecule that is separate from, and independent of, the cell's chromosomal DNA.

Epitope – A localized region on the surface of an antigen that is capable of combining with a specific antibody, i.e. the region that is recognized (**paratope**). The term 'epitope' can also be applied to an interaction partner for proteins other than antibodies.

Epitope tags – Using recombinant DNA techniques, genetic sequences coding for epitopes that are recognized by common antibodies are fused to the gene. This is a strategy used to bypass production of protein-specific antibodies, enabling localization, purification, and further molecular characterization of the protein of interest with antibodies specific for the epitope tag. See also **affinity tags**.

Equilibrium – The state in which the elements of a system do not change over time.

Equilibrium constant – The ratio of concentrations of products to substrates at equilibrium.

Euchromatin – A lightly packed form of **chromatin** which is transcriptionally active.

EVCs – **E**xternally **V**isible **C**haracteristics: observable phenotypic traits such as eye colour and hair colour.

Excitation maximum – Wavelength of light which causes a fluorescent molecule to fluoresce maximally.

Exonuclease – An enzyme that catalyses the hydrolysis of a phosphodiester bond sequentially from the end of a nucleic acid molecule (i.e. removes one nucleotide at a time).

Experiments *in silico* – Experiments performed using computer software.

Expression cassette – See **cassette (expression)**.

Expression vector – Plasmid-based or engineered virus which combine the features required for gene insertion with features that enable transcription and translation of the cloned gene inside a prokaryotic or eukaryotic host cell.

Extension (in PCR) – Phase of the PCR cycle in which a new strand of DNA is synthesized.

Fab region – Fragment of an antibody generated by digestion with papain, containing one **paratope**.

Familial searching – Means of forensic DNA database searching intended to locate possible relatives via a suspect DNA profile.

Fc region – **F**ragment **c**ryatallizable region; part of an antibody that reacts with cell surface receptors and proteins such as protein A and protein G.

Feedback loop – Takes information from the past and uses it to affect or influence the same system in the future. As such, a feedback loop uses the history of a system to feed back into itself and has an associated response time.

Fingerprint (spectrum) (in NMR) – A spectrum which is distinctive to a particular molecule because of a unique distribution of peaks.

Fixative – An agent used to preserve tissue for gross or microscopic analysis.

Flow cell – A device fitted in a detection instrument that allows solution to pass through and within which a certain property (usually absorbance or fluorescence of electromagnetic radiation) of the solution can be monitored.

Flow rate (in chromatography) – Speed at which solutions can be washed through a column.

Flow-through – Solution of molecules which washes through a chromatography **matrix** during sample loading (usually the unwanted impurities).

Fluorescence – The property of absorbing light (electromagnetic radiation) of short wavelength and emitting light of longer wavelength.

Fluorescence emission maxima – See **emission maximum**.

Fluorescence-based detection methods – Approach to DNA detection that rely on the incorporation or release of **fluorophores**.

Fluorescent conjugate – Chemical attachment of a fluorescent molecule to a protein.

Fluorescent protein – A protein which is intrinsically fluorescent.

Fluorophore – A molecule or part of a molecule that emits fluorescence following excitation with light.

Footprint (DNA) – A series of fragments of DNA obtained by digestion of DNA by DNase I. If a protein is bound to the DNA, some DNA fragments will be missing as they will be protected by the protein.

Forensic DNA database – A computer database of DNA profiles collected in the interests of criminal justice.

Forensic DNA profiling – Method of establishing the genotype of an individual at highly variable alleles to assist in human identification in the interests of criminal justice.

Forward light scatter (in flow cytometry) – Light scatter detected forward of the laser source; indicative of particle size.

Founder – A mouse that has been created from a genetically manipulated egg or embryo, and that is used to breed and produce additional transgenic mice carrying the same genetic manipulation.

Fourier transform–ion cyclotron resonance (FT-ICR) – A type of mass analyser where ions are trapped inside a magnetic field. The ions are induced to oscillate in a circular motion perpendicular to the plane of the magnetic field. The frequency of oscillation of the ions is dependent on m/z.

Fractal – An object with self-similar patterns exhibited at different scales.

Fractions (in chromatography) – Small volumes of liquid collected during chromatography for protein purification.

Fragment library – A collection of nucleic acid fragments flanked by **adapters** for addition to a next-generation sequencing platform.

Free energy – A thermodynamic quantity that determines whether a reaction or process will occur spontaneously (energy released from a system) or not.

Fusion protein – A translated protein product comprising two different polypeptide sequences joined by a short peptide linker into one protein chain.

Fusion tag – See **affinity tag**.

Gel filtration – Chromatography method separating molecules based on their different sizes and shapes.

Gene cloning – Method of copying a DNA sequence *in vivo* or *in vitro*. Involves the insertion of the gene (sequence) of interest into a cloning vector and subsequent propagation of the **recombinant DNA** molecule in a host organism.

Gene expression – The process by which a gene is regulated and its product is synthesized; usually involves multiple steps.

Genetic drift – A change in the frequency of alleles in a population, causing gene variants to disappear and homogeneity to occur.

Genome – The entirety of an organism's hereditary information, usually encoded in DNA.

Genomic DNA library – A collection of clones that comprise the full genome of a particular organism.

Glioblastoma – Most common and most malignant primary brain tumour (grade IV); arises from glial cells.

Gradient elution – Use of gradually changing solution conditions to differentially elute molecules from a chromatography **matrix**, e.g. increasing salt concentration in **ion-exchange chromatography**.

Growth curve – An empirical model of the growth of a population of cells over time; usually includes lag, log, plateau, and death phases.

Guide strand (in RNAi) – RNAi is mediated by dsRNA duplexes. The guide strand is complementary in sequence with the target mRNA sequence to be silenced and thus is able to initiate target-sequence cleavage mediated by the **RISC** complex.

Haematoxylin and eosin – A commonly used stain in histology. Haematoxylin is used to stain cell nuclei blue, and eosin stains the cytoplasm and connective tissue shades of red, pink, and orange.

Hanging drop – Method devised by Harrison in 1907 that is frequently employed for culturing stem cells. Involves culturing cells in droplets suspended from a cover slip.

Haploinsufficient – A gene that requires the presence of both alleles to be functional or to produce the wild-type phenotype.

Haplosufficient – A gene that fulfills normal function even when one copy of the gene is mutant or deleted.

Hapten – A small molecule that can illicit an immune response only when coupled with a carrier protein.

Hayflick limit – Maximum number of times a cell can replicate, i.e. when the telomeres become too short to allow further cell division.

HeLa – Cell line commonly used in research. Generated from cervical cancer cells taken from Henrietta Lacks in 1951 and cultured without permission. It is thought that a fraction of other established cell lines may be contaminated with HeLa.

Heteroduplex – Double-strand DNA molecule in which the two complementary strands differ in sequence composition.

Heterologous – Derived from a different species.

Heterozygous – Having two different alleles for a single gene or trait.

High-efficiency mutagenesis Approaches to primer-directed mutagenesis that include a step, such as digestion with *Dpn*I, to destroy the wild-type template DNA, thus preventing it from generating a background of transformants carrying wild-type plasmid DNA rather than mutant plasmid.

High-temperature-activated DNA polymerase – A DNA polymerase inactive at low temperatures, which can be activated on heating, reducing the possibility for non-specific strand elongation.

High-throughput – A method optimized to generate a large amount of data in a short time.

Holo state – The state of a protein when it is bound to a ligand.

Homologous – Derived from the same species.

Homologue – A gene or protein of similar sequence, structure, and function.

Homozygous – Having two identical alleles for a single gene or trait.

HSQC – **H**eteronuclear **S**ingle **Q**uantum **C**oherence (spectrum). A 2D spectrum that correlates two different types of nuclei (e.g. ^{1}H and ^{15}N, or ^{1}H and ^{13}C).

Hydrodynamic flow – The flow or movement of a liquid.

Hydrodynamic focusing (in flow cytometry) – The use of fluidic flow (hydro) to channel or focus a sample into the central core of the stream.

Hydrophobic – Literally means 'fear of water', and is used to refer to parts of a protein structure that are non-polar and therefore do not interact with water.

Hydrophobic interaction chromatography (HIC) – Separation of proteins based on their different hydrophobic properties.

Hyperplastic (hyperplasia) – Enlargement of an organ due to increased number of cells. Can only occur in tissues containing cells that are capable of undergoing mitotic division.

Hypertrophy – Enlargement in an organ or body part due to an increase in cell size; usually seen in muscle, but it can occur in other organs such as the kidney.

Hypervariable regions – Regions found in both the heavy and light chains of an immunoglobulin molecule that make up the unique antigen binding site. See also **complementarity determining regions**.

Hypoxia – Deprivation of an adequate oxygen supply (oxygen concentration of ~5%) which can promote specific cell behaviour and is often used to replicate the environment of stem cells *in vitro*.

Immortal cell lines – The absence of cell senescence, usually due to upregulation of telomerase which abolishes the **Hayflick limit**, thus allowing the cells to continue replicating.

Immunophenotyping – Classification of cell types by the use of antigen-specific antibodies (immunoglobulins).

Immunoprecipitation – Purification of protein from a complex sample in solution using specific antibodies to precipitate a particular protein.

***In vitro* fertilization** – Commonly referred to as IVF. The process of fertilizing a woman's egg with a man's sperm in a laboratory. The term *in vitro* refers to 'outside the body'.

Inclusion bodies – An insoluble form in which recombinant proteins are often expressed in *E. coli*, usually considered undesirable as protein must be solubilized and refolded for **chromatography**.

Induced-fit – A mechanism of binding between two molecules that may alter their conformation slightly for best fit at the interaction interface.

Inducible expression – Expression of a gene that is regulated by the presence of a substance (an inducer) in the environment.

In-frame fusion – Joining together DNA molecules encoding two separate proteins so that their reading frames are continuous and unbroken.

Inhibitors – Molecules that block/interfere with the function of another.

Initial condition – The starting value of a variable.

Initialization phase (in PCR) – High-temperature phase at the start of a PCR reaction intended to prevent mispriming.

Insertion (DNA) – Introduction of one or more nucleotides in a DNA sequence.

Insertion vector – A lambda vector that is constructed by deleting some of the non-essential genes of the lambda genome.

Intermediate purification step – A second chromatography step, exploiting another property of the protein, to remove some remaining impurities.

Interphase – The phase of the cell cycle in which preparation for cell division takes place.

Intrinsic fluorescence – Refers to where a protein possesses fluorescent properties within its natural sequence (the protein does not require modification/labelling to become fluorescent).

Intrusive DNA – Unwanted contaminating sequences liable to be amplified during PCR.

Ion-exchange chromatography (IEX) – Separation of molecules using differences in the charged properties of molecules in a sample mixture.

Ionic strength – A measure of the concentration of salts in solution, an important property influencing protein solubility and strength of protein binding to many matrices.

Ionization source (in mass spectrometry) – Region of the mass spectrometer where the analyte molecules are converted to ions prior to analysis.

Ionized – Addition of a positive or negative charge to a molecule.

ISEL (*In Situ* End Labelling) – An assay used to measure DNA fragmentation.

Isocratic elution – Running chromatographic purification using a single buffer, most commonly used in size exclusion chromatography.

Isoelectric focusing – Separation of proteins on a pH gradient gel such that proteins migrate to the point in the gel where their overall charge is zero (isoelectric point).

Isoelectric point – The pH at which a protein has no overall charge; an important factor in ion-exchange chromatography and affecting protein solubility.

Isotonic saline solution – Salt solution with the same osmotic properties as the cytoplasm of a cell to prevent cell swelling or shrinkage. Used in cell sorters as the drop has to hold an electric charge which water will not do.

J coupling – See **scalar coupling**.

K_a – Equilibrium association constant: a measure of the degree of association between a protein and its ligand.

K_d – Equilibrium dissociation constant: a measure of the degree of dissociation between a protein and its ligand. Inversely related to K_a.

Klenow fragment – Large fragment of DNA polymerase I. It exhibits 5′ to 3′ polymerase activity and 3′ to 5′ exonuclease activity. It is used to label by fill-in 5′-overhangs of double-stranded DNA.

Lag phase – Period of growth in which cells adapt to their environment and do not actively divide, represented by a non-existent or very shallow gradient on a growth curve.

Laser (**L**ight **A**mplification by the **S**timulated **E**mission of **R**adiation) – a pure light wavelength generating device.

Lateral resolution – The smallest distance that two objects can be separated in two dimensions.

Libraries of variants – A large number of mutated genes usually generated as a result of saturation or random mutagenesis. Often the gene encodes a protein and therefore the library will express a range of mutated proteins from which new versions with desirable properties can be selected.

Linear amplification – Production of one new copy of a DNA molecule per cycle as DNA molecules produced during previous cycles cannot be used as templates for further DNA synthesis.

Linear flow rate – Number used to describe the recommended **flow rate** for different matrices when making self-packed columns.

Lipofection – Transfer of genetic material into a cell using **liposomes** which are vesicles that package the genetic material, fuse with the lipid bilayer of the host cell membrane, and release the material once inside the cell.

Liposome – An artificial vesicle composed of a lipid bilayer that can be used to transport negatively charged nucleic acids across the lipid bilayer of the cell membrane.

Liquid chromatography – Separation of analytes by passing them through a column, with compounds eluting at different times depending on their affinity for the column material under the buffer conditions being used.

List mode data file (in flow cytometry) – Type of electronic file used to save the measurements detectors make as a cell passes through the interrogation laser(s) with results stored in chronological order for each cell in the sample.

Locus – The specific location of a gene or DNA sequence on a chromosome.

Log phase – The period of growth in which cells **proliferate** exponentially, represented by a steep gradient on a growth curve.

Loops (in PCR) – Artefacts formed by partial self-hybridization of oligonucleotide primers.

Low-template DNA analysis – PCR amplification of low copy numbers of template molecules.

Lumpectomy – A surgical procedure in which only the tumour and some surrounding tissue is removed.

Lyse – Cause or produce disintegration of a compound, substance, or cell (lysed cells release their cytosolic contents).

Major groove (DNA) – 22Å wide spacing in the 3D structure of the double-stranded DNA helix. The two DNA strands are not directly opposite each other, and produce two grooves of different sizes.

Malignancy – Commonly called cancer. A tumour with an aggressive behaviour including invasion and destruction of adjacent tissues, and the capacity to spread to distant sites.

Markov chain – Discrete model in which elements transit to different status with a given probability based upon the information on the previous status.

Mass accuracy – Difference between measured and theoretical molecular mass.

Mass analyser (in mass spectrometry) – Region of the mass spectrometer where ions are separated according to their m/z ratio.

Mass range limit – Maximum *m/z* range over which an analyser can measure.

Mass spectrometry (MS) – An instrumental technique used to determine the molecular mass of a chemical compound.

Mass spectrum – Graphical representation of the raw data showing *m/z* versus ion intensity.

Mass-to-charge ratio (*m/z*) – Quantity measured by mass spectrometry.

Mate pair library – Collection of pairs of nucleic acid fragments separated by several kilobases in the genome that are ligated either side of an chemically synthesized internal **adapter** sequence to maximize sequence coverage across a genome.

Materials Transfer Agreement (MTA) – A legal agreement that controls the conditions of use of materials exchanged between collaborating parties.

Matrix (in chromatography) – Solid support used to bind proteins during chromatography. **Resin** is another term, often used interchangeably.

Matrix-assisted laser desorption ionization (MALDI) – Ionization technique where analyte molecules in solution are mixed with a matrix compound and left to crystallize on a target plate. The co-crystallized sample is inserted into the mass spectrometer and then irradiated with a laser which leads to matrix and analyte molecules 'sputtering' from the surface into the gas phase where the analyte ions are formed.

Mean residue ellipticity (MRE) – Ellipticity of a protein corrected for its concentration and the number of peptide bonds. A measure of how much of the protein has secondary structure.

Melting temperature (T_m**)**– Temperature at which an oligonucleotide primer seperates from its template.

Metaphase – Stage of mitosis in the eukaryotic cell cycle in which condensed and highly coiled chromosomes align in the middle of the cell before being separated into each of the two daughter cells.

Methylation – Attachment of methyl groups to a molecule; an element involved in gene regulation resulting in the conversion of cytosine in a CpG dinucleotide to 5-methyl cytosine.

Microarray – Also known as a DNA chip; a collection of microscopic DNA molecules organized and spotted onto a solid surface. Enables rapid analysis of the expression of many genes simultaneously.

Microfluidic – Description of a device that allows containment and manipulation of tiny volumes (nanolitres or less) of fluid by active (micro) components, such as micropumps or microvalves.

Microinjection – Method of introduction of new DNA directly into the nucleus of a cell.

Microtome – Instrument used to cut thin slices (sections) from a biological specimen.

Minor groove (DNA) – 12Å wide spacing in the 3D structure of the double-stranded DNA helix. The narrowness of the minor groove means that the edges of the bases are more accessible in the major groove.

miRNA (microRNA) – Short non-protein-coding post-transcriptional regulatory RNA molecules also called mishybridization.

Misannealing – Annealing of an oligonucleotide primer to a non-intended target.

Mis-incorporation (in PCR) – Incorrect incorporation of bases into a newly synthesized DNA strand.

Model – In X-ray crystallography, the structural model that an X-ray crystallographer builds into an electron density map. The model includes coordinates for all the atoms as well as their occupancy and B factor.

Molecular cloning – See **cloning**.

Molecular tumbling – The random tumbling of molecules in solution due to Brownian motion. Larger molecules tumble more slowly due to their larger mass.

Monoclonal antibody – Immunoglobin molecule produced by one particular B-lymphocyte cell line. As such all copies of the monoclonal antibody have the same primary sequence and react with the same affinity to the same **epitope** structure.

Monolayer (cells) – Single layer of cells adhered to the surface of the culture vessel.

Multipotent – Potential of a stem cell to differentiate into multiple cell types of an organism, such as haematopoietic cells.

Mutagenesis – Process of introducing changes into a DNA sequence.

Mutated protein – Protein encoded by a gene that differs from the wild-type gene as a result of one of more mutations that lead to a change in the amino acid sequence of the protein.

Mycoplasma – A very small (<1μm) bacterium that can contaminate cell cultures, leading to problems with growth, adherence and genetic aberrations

Neonatal lethal – Manipulation that results in the inability of the mouse to survive past the first few days after birth.

Neoplasia (neoplasm) – A 'new growth' of tissue. Commonly known as a tumour, either benign or malignant.

Network (in mathematical modelling) – A graphic representation of relationships between different elements of a system. A network or graph has two kinds of elements: nodes (or vertices) and edges (or arcs). Nodes connect to other nodes with edges.

Non-immortal cell lines – Cells with a **Hayflick limit,** or a finite number of population doublings, due to continuously shortened telomeres. Cells reaching their limit can no longer proliferate.

NTA – Nitrilotriacetic acid (used with a bound metal ion, e.g. Ni^{2+}, Co^{2+}) for His-tagged protein purification.

Nucleosome – Basic unit of chromatin composed of DNA wrapped around histone proteins. Each nucleosome has dimers of H3, H4, H2A, and H2B histones and 146bp of DNA

Nullipotency – Cell which displays no differentiation potential, i.e. is terminally differentiated, such as a red blood cell.

Obligate – For protein interaction; refers to where the protein interaction can essentially be regarded as permanent since it is a member of a functional complex (a subunit).

Off-target non-specific (in RNAi) – Unintentional phenotypic effects which arise as a result of RNAi treatment but which do not relate to the specific RNA duplex employed in the assay.

Off-target specific (in RNAi) – Unintentional phenotypic effects which arise as a result of RNAi treatment and are related to the specific RNA duplex employed in the assay.

Oligonucleotide – A short sequence, normally 15–20nt, that is produced chemically. Oligonucleotides of up to around 100nt can be produced. Normally oligonucleotides are ordered from specialist suppliers, and are now quite a cheap and affordable reagent custom-made by specialist manufacturers.

Oligopotency – Potential of a stem cell to differentiate into a few cell types of an organism.

Optical and magnetic tweezers – Method of trapping either dielectric (optical) or magnetic particles in a harmonic potential. These trapped particles can then be used for a variety of applications.

Optical lever – Allows measurement of very small degrees of deflection to be performed by detecting the positional displacement of reflected light from the back of a lever.

Optimization (in PCR) – Configuration of a PCR protocol according to the competing demands of fidelity, efficiency, and yield.

Orbitrap – Type of mass analyser where ions are trapped in an electrostatic field and induced to oscillate along a spindle electrode. The longitudinal frequency of oscillation along this electrode is indicative of the ion's m/z.

Ordinary differential equation (ODE) – Equation in which we analyse the infinitesimal change (derivative) of the dependent variable when the independent variable has an infinitesimal change in its value. For mathematical models used in molecular biology, the standard independent variable is time or space.

Palpation – Part of a physical examination used by a healthcare practitioner to feel an object/lesion in order to determine its physical characteristics.

Panning – Affinity selection technique used to screen for 'active' molecules amongst a mixture.

Paralogues – Genes that derive from the same ancestral gene but currently reside at different locations within the same genome.

Parameter – Measured element characteristic of a system needed to understand the relationship between variables.

Paratope – Part of an antibody that specifically binds with an **epitope**.

Partial products (in PCR) – Incompletely synthesized strands of DNA.

Passaging – See **subculturing**.

Passenger strand (in RNAi) – RNAi is mediated by dsRNA duplexes. The passenger strand has the same sequence as the target mRNA sequence to be silenced and thus this strand is discarded during **RISC** loading.

Paternity testing – Method of establishing the male parent of a child by DNA analysis.

Pathology – Science and study of disease, especially the causes and development of abnormal conditions, both gross and microscopic.

PCR additive – Any molecule intended to affect PCR.

Periplasm – Region between the inner cytoplasmic membrane and the external outer membrane of Gram-negative bacteria.

Perturbation – Change in at least one condition on a particular system.

Petri net (PN) – Dynamic network that evolves at a discrete time. A PN has four elements: places, transitions, arcs, and tokens.

Phage display – A particular protein display method, in which expressed proteins are displayed on the surface of bacteriophage which assemble in an *E. coli* host.

Phase (in X-ray crystallography) – The phase, or more properly the phase shift, describes the offset (in radians or degrees) of the scattered wave relative to the incident wave before it was diffracted by the crystal.

Phase shift – In terms of sinusoidal functions (or waves) a phase shift or phase difference is an angular measurement of the offset between two waves.

Phenotype – The physical, behavioural, and/or biochemical characteristics of an organism.

Phosphatase – An enzyme that catalyses the removal of a phosphate group from a substrate.

Photobleaching – Irreversible loss of **fluorescence** by a fluorescent molecule following repeated rounds of excitation.

Photon – Particle of light.

piRNA (PIWI-interacting RNAs) – Known to be involved in silencing of retrotransposons in germ line cells. Their interference pathways do not appear to be conserved with those of siRNAs and miRNAs.

Plasmids – Circular duplex DNA molecules that replicate independently of the chromosomal DNA. Occur naturally in bacteria and, in very limited cases, eukaryotic cells (e.g. 2μm plasmid in *Saccharomyces cerevisiae*). Used extensively as vectors in molecular biology.

Plate reader – Laboratory instrument designed to detect biological, chemical, or physical events in samples contained in microtitre plates, usually via optical absorbance or fluorescence.

Platform (in next-generation sequencing) – Apparatus used for performing next-generation sequencing.

Pluripotent – Potential of a stem cell to differentiate into all cells of an organism with the exception of the embryo, i.e. cultured embryonic stem cells.

Point mutation – Single base-pair change within a DNA sequence.

Point spread function – Diffraction pattern formed by a point of light as it passes through the optical path of a light microscope.

Polar – Term applied to a molecule or chemical group that has unequal distribution of charge (where electrons favour one atom over another in the group). The resulting separation of charge results in the ability of the group to interact with water and other similarly polar entities.

Polishing purification step – Final chromatography step, often size exclusion, used to remove aggregated protein when the sample may appear pure by **SDS-PAGE**.

Polyclonal antibodies – An antibody preparation isolated from the sera of an immunized animal. Polyclonal antibodies are able to recognise multiple **epitopes** on the same antigen. Polyclonal antibodies may also recognise non-related target antigens within a sample.

Polyacrylamide gel electrophoresis (PAGE) – Gel-based method of separating biomolecules according to their molecular weight.

Polymerase chain reaction (in PCR) – Rapid and powerful technique to allow the selective amplification of a targeted region of DNA or cDNA.

Potency – Potency of a cell signifies its potential to differentiate into various cell types of an organism.

Precursor ion scanning – MS/MS experiment used to detect all the precursor ions which dissociate to produce a fragment ion of a particular *m/z*.

pre-miRNA (precursor miRNA) – Endogenous regulatory RNA intermediate. Generated in the nucleus by Drosha and exported into the cytoplasm for processing by **DICER**.

pre-shRNA (precursor shRNA) – Short hairpin RNA designed to mimic the structure and function of a pre-miRNA.

Primer – DNA polymerases require an existing 3′ end of a nucleotide sequence to extend during the DNA synthesis reaction. A primer is usually a single-stranded DNA sequence that anneals to the template strand to provide the initiation site for DNA synthesis. In practice primers are usually synthetically produced oligonucleotides, although longer DNA sequences can also act in this regard.

Primer dimers – Artefacts caused by self-hybridization between oligonucleotide primer pairs.

pri-miRNA (primary miRNA) – Endogenous regulatory RNA intermediate. Transcribed from the nucleus and processed by Drosha into a pre-miRNA.

pri-shRNA (primary shRNA) – Short hairpin RNA designed to mimic the structure and function of a pri-miRNA.

Probe – Small length of chemically synthesized nucleic acid used to bind to its complementary sequence on a target nucleic acid fragment. The probe is labelled with biotin, a fluorophore, or a radioisotope at its 3′ or 5′ end.

Product ion scanning – MS/MS experiment used to detect all the fragment ions produced by a particular precursor.

Proliferate – Cells growing and increasing in number due to cell division.

Promoter – Specific sequence of DNA, usually just upstream of the coding sequence of the gene that controls the transcription of that particular gene.

Proof-reading DNA polymerase – DNA polymerase that contains a 3′ to 5′ DNA exonuclease activity that facilitates the removal of a wrongly incorporated nucleotide so that the correct nucleotide can then be inserted by the 5′ to 3′ DNA synthesis activity of the enzyme. The proof-reading activity of a DNA polymerase is important in ensuring that DNA is replicated in a high-fidelity manner with an error rate of around 1 in 10^6–10^8nt incorporated. Such DNA polymerases are used in site-directed mutagenesis procedures. A DNA polymerase that does not contain this proof-reading activity will have a higher error rate during DNA synthesis, perhaps one error in 10^3–10^4nt incorporated. Such an enzyme is used in certain **random mutagenesis** methods.

Protein display – A molecular biology method that results in expression of recombinant genes such that the gene products are accessible to interacting partners, i.e. the proteins are displayed on a surface (such as a cell surface).

Protein interaction modules – Small defined regions of globular protein structure (domains) known to mediate protein–protein interaction, usually by recognition of consensus target sequences.

Proteome – Total protein complement of the cell or organism.

Proteomics – Analysis of the **proteome**.

Proteotypic peptides Specific peptide sequence which are unique to a particular protein (or isoform), normally associated with their use as surrogates for protein quantification by MS.

psi – Measurement of pressure (pounds per square inch (of mercury)).

Pull-down assay – An assay for measuring interaction that relies on the 'pulling' or extraction of a protein out of a mixture by virtue of its interaction with another protein that is on a solid support or precipitated from solution.

Quadripotent – Potential of a stem cell to differentiate into four cell types of an organism.

Quadrupole – Component of a mass spectrometer which acts as an 'filter', only allowing ions of a certain mass-to-charge ratio through. A quadrupole consists of four parallel rods, with diametrically opposed rods connected electronically. Ions travelling down the centre of the quadrupole will have a stable or unstable trajectory depending on the amplitude and frequency of oscillation of the electronic potential applied to each pair of rods.

Qualitative – Non-numerical descriptions based on some quality, characteristic, or observation rather than on some quantity or measured value.

Quantization (in NMR)– Requirement from quantum mechanics that very small objects can only take up particular discrete energies.

Quantitative – Where a reaction or interaction can be described and measured in numerical terms.

Quiescent – State of the cell when it is not actively dividing. Cells are quiescent in the G_0 phase of the cell cycle.

QuikChange® mutagenesis – Method for high-efficiency mutagenesis commercially produced by Stratagene, now owned by Agilent Technologies. The principle of the approach can be duplicated without using a commercial kit.

Radius of curvature – In geometry the radius of curvature refers to a curve that can be described as the arc of a circle whose radius is the radius of curvature.

Ramachandran plot – Scatter plot of backbone torsion angles (phi and psi) for a protein main chain. Some combinations of phi and psi are forbidden as they result in steric clashes between carbonyl oxygens and side chains. Certain angle combinations are indicative of secondary structure. In a good structure the majority of residues should be within 'allowed' regions of the Ramachandran plot.

Random match probability – Probability that a given forensic DNA profile will be encountered in an individual in the general population.

Random mutagenesis – Process of introducing mutations, usually point mutations, into a population of genes to generate a library of mutated genes that might encode a library of variant proteins.

Recombinant DNA – Artificial DNA molecule constructed by splicing together component sequences.

Recombinant protein – Protein from a non-native source, made by introducing the gene into a different host organism (using recombinant DNA techniques) where it is expressed *in situ*. A protein produced by transcription/translation of a recombinant DNA molecule.

Reducing agent – Compounds used to maintain cysteine in the sulphydryl (–SH) form, reducing protein aggregation and retaining activity (e.g. mercaptoethanol, dithiothreitol).

Refinement (in X-ray crystallography) – Tells you how often, on average, each data point was measured in an X-ray diffraction experiment.

Refractive index – Property of a material that changes the speed of light, computed as the ratio of the speed of light in a vacuum to the speed of light through the material.

Replacement vector – Lambda vector that is constructed so that non-essential genes (the central stuffer region) are replaced by the gene of interest.

Reporter gene – Gene whose product is easily detected and not ordinarily present in an organism or cell type under study that is expressed as part of a DNA construct introduced experimentally.

Repressor – Protein that blocks transcription of an operon by binding to its operator region.

Resolution – (in mass spectrometry) – Measure of the ability of an instrument to distinguish two analytes of slightly different *m/z*.

Resolution (in X-ray crystallography) – Measure of the ability to distinguish between neighbouring features in an electron density map.

Restriction endonuclease – Enzyme that cuts DNA molecules at a limited number of specific nucleotide sequences (recognition sites) to generate a double-stranded DNA cut.

Reverse transcriptase – Enzyme that synthesizes a cDNA strand from an mRNA molecule, using a primer and dNTPs.

R factor – This statistic reports on how well your X-ray model predicts the experimental data.

R_{free} – This statistic is a cross-validation that tests how well your X-ray model predicts an independent set of experimental data.

RISC (**R**NA **I**nduced **S**ilencing **C**omplex) – Trimeric complex involving one of the four **Argonaute** proteins, **DICER**, and the cofactor TRBP. The **guide strand** (antisense to the mRNA transcript) of the RNAi duplex is incorporated into the RISC, and RNA silencing occurs.

RNAi (RNA interference) – Ability of small (~23nt) double-stranded RNA to produce sequence specific gene-silencing effects in a cell.

Salting in/out – Using different salt concentrations to change protein solubility (see also **ammonium sulphate precipitation**).

Sample introduction (in mass spectrometry) – Method of introducing the sample into the mass spectrometer.

Scaffold – Platform on which to build or modify a protein or protein complex.

Scalar coupling – Small changes in NMR transition energies caused by neighbouring nuclei influencing each other via electrons in chemical bonds. Used to transfer information between connecting signals in HSQC spectra. J-couplings are measured in hertz (Hz) and labelled $^{X}J_{Y}$ where x is the number of bonds the coupling is over and Y denotes the connected atoms. For example, $^{2}J_{HN-H\alpha}$ is a coupling across two bonds between an HN and an Hα atom.

Scanning probe – Scanning probe microscopy is a branch of microscopy where a probe is scanned across a sample in order to interrogate it. It is used to build up a picture.

ScFv (single-chain variable fragment) – **Fusion protein** (genetically engineered) of the hypervariable regions of the heavy and light chains of an immunoglobulin. It contains one paratope.

Scirrhous – Firm dense consistency in a tumour.

SDS-PAGE (**s**odium **d**odecyl **s**ulphate **p**olyacrylamide **g**el **e**lectrophoresis) – A method for analytical (usually) separation of macromolecules based on their size (or molecular mass). SDS is an anionic detergent used to ensure that all proteins in a mixture are negatively charged and migrate towards the cathode, separating through the pores in the acrylamide-based gel on the basis of relative molecular mass.

Secondary antibody – An antibody that has been raised to the **Fc region** of a primary antibody. Manufacturers conjugate secondary antibodies to fluorescent tags for use in immunofluorescence microscopy, or HRP for use in **Western blotting**.

Seeding – Diluting a cell suspension to the required concentration and distributing amongst new culture vessels after **subculture** for use in experiments.

Seed region – Portion of the **guide strand** that initiates target sequence binding.

Selected/multiple reaction monitoring (SRM/MRM) – Mass spectrometric technique which will record a signal only if a particular analyte mass gives rise to a particular fragment mass upon fragmentation. It is highly sensitive and selective, and is commonly used for the quantification of proteolytic peptides.

Self-renewal – Ability of a cell to undergo several cell cycles of division whilst maintaining an undifferentiated state.

Sense strand – DNA strand which has the same sequence as the mRNA strand following mRNA transcription. Within RNAi pathways, the strand of the RNA duplex which is sense to the mRNA transcript is the passenger strand and thus is discarded.

Sensitivity analysis – Mathematical technique to evaluate the effect of different parameter values on the dynamics of a system.

Sepharose – Type of chromatography matrix used in SEC or which is modified to produce matrices for IEX, HIC, or affinity chromatography.

Sequence motif – Nucleotide or amino acid sequence pattern that is widespread and has, or is conjectured to have, a biological significance, perhaps in mediating interactions with other molecules.

Sequencing chemistry – Method used to generate the sequence, i.e. cycle sequencing.

SH3 domain – The SRC homology 3 domain is a small protein domain of about 60 amino acids that recognizes a proline-rich **epitope** on particular protein surfaces.

shRNA (short hairpin RNA) – A 19–29nt RNA sequence which folds and self-anneals to form a duplex containing a stem–loop at the location of the fold.

Shuttle vector – A vector that is constructed to enable replication in two different host organisms (e.g. bacteria and yeast) or mammalian cells.

Side scatter (in flow cytometry) – Wide-angle light scatter of the laser light source when a cell or particle passes through the beam. Measurements indicate cellular granularity and cytoplasmic complexity.

Single molecule – Single-molecule experiments, in contrast with ensemble experiments, are able to measure the behaviour of one molecule at a time. This has the advantage that rare but significant events are not hidden beneath the average behaviour of the system, as can be the case in ensemble measurements.

Single-nucleotide polymorphism – Any type of genetic variation at a single nucleotide.

Single-particle reconstruction – Method of calculating a 3D reconstruction of a macromolecular complex from many 'single particles' (macromolecular complexes) imaged in different orientations by electron microscopy. The assumption is that all the complexes have identical structures.

siRNA (small interfering RNA) – A 21–23nt RNA duplex with 5′ monophosphate groups and 3′ dinucleotide overhangs. It guides the cleavage and degradation of its complimentary mRNA sequence.

Site-directed mutagenesis – Process of introducing point mutations, insertions, or deletions into a DNA sequence at specific sites.

Size-exclusion chromatography (SEC) – Synonymous with **gel filtration**.

SNP – See **single nucleotide polymorphism**.

Solid phase PCR – PCR reaction where the primers are covalently linked to a solid support. Used with some next-generation sequencing platforms.

Specificity – Quality of being specific; having a preference for reacting or interacting with a particular entity from a population.

Sphaeroplast – Viable yeast cell from which the cell wall has been removed enzymically.

Spin (of atomic nuclei) – Quantum mechanical phenomenon which makes some types of atomic nuclei behave as if they spin like a gyroscope. As nuclei are charged, this leads to the generation of a magnetic field.

Stable isotope dilution – Quantification of an analyte by comparing the amount of endogenous analyte with a known amount of synthetic version which is tagged with a stable 'heavy' isotope.

Stable isotope labelling – (in NMR) – Replacement of naturally abundant NMR-inactive nuclei (e.g. ^{12}C) in a sample by rarer NMR-active nuclei (e.g. ^{13}C). Only isotopes which are non-radioactive or 'stable' are used.

Stable knockdown – Reduction or elimination of the expression of a gene of interest as a result of a **stable transfection**.

Stable transfection – Integration of an exogenous genetic sequence into the genome of a cell which is then present in the genome of all subsequent daughter cells.

Stem cell – Undifferentiated, unspecialized cell with the ability to proliferate, self-renew, and differentiate into specialized cells of a particular tissue or organ. See also **Adult stem cell** and **Embryonic stem cell**.

Stiffness – Extent to which a material can resist deformation or deflection under an applied force.

Stochastic – A random event, i.e. with a given probability.

Stokes shift – Difference in wavelength between the maximum absorption (excitation) wavelength and emission wavelength of a flurophore.

Stopped-flow fluidics – Method of rapidly mixing two reactants together in order to measure reaction kinetics on a millisecond timescale.

Strain (mice) – Laboratory mice that have been bred in isolation for generations, producing a group of mice that are nearly identical and may carry certain physiological traits of interest.

Strand separation (in PCR) – Phase of the PCR cycle in which the strands of a DNA duplex molecule are separated by heating.

Stringency – Specificity of an oligonucleotide primer for its template target.

Stroma – Connective tissue and blood vessel framework that supports an organ or other structure such as a tumour.

STRs (short tandem repeats) – highly polymorphic genetic loci.

Structure factor – Resultant wave scattered by the sample. It has both a magnitude (or amplitude) and phase.

Sub-clone – Process of moving a cDNA from its original vector to a new destination vector.

Subculturing – Preferably carried out in the log growth phase. Involves dissociating a cell culture from the culture vessel and dividing between new vessels and fresh media to continue proliferation and expand the cell line.

Substitution (DNA) – Exchange of one nucleotide(s) for another, resulting in a variation in the DNA sequence.

Supershift – If an antibody against the protein of interest is added in an **EMSA** reaction, the antibody–protein–DNA complex will migrate even more slowly than the DNA–protein, resulting in a supershift of labelled DNA in the native gel.

System biology mark-up language (SBML) – Standard representation format used to express mathematical models from biological phenomena.

Tag – A stretch of amino acids added to an expressed protein to aid purification or immunological detection of the protein. See **Affinity tag**.

Tandem mass spectrometry (MS/MS) – Mass spectrometry method for peptide identification, where an intact peptide is isolated by a quadrupole (MS) and fragmented, and the fragment masses are measured (MS/MS). The pattern of fragment ions along with the mass of the intact peptide is usually sufficient to identify the amino acid sequence of the peptide.

***Taq* polymerase** – *Thermus aquaticus* DNA polymerase, an enzyme derived from a thermophilic bacterium capable of synthesizing DNA at high temperatures.

Template – Single-strand DNA sequence, often the wild-type sequence, to which a primer anneals and which is used by a DNA polymerase to generate the newly synthesized DNA strand based on complementary Watson–Crick base pairing.

Template matching (in electron microscopy)– Way of identifying macromolecular complexes within a tomogram by identifying similarities in shape and size with predetermined structures identified by another structural technique (e.g. single-particle reconstruction or X-ray crystallography).

Thermal cycler – Instrument in which PCR can be undertaken.

Time-of-flight (TOF) analyser – Type of mass analyser where the analyte ions are all given the same kinetic energy such that their velocity is proportional to their m/z ratio.

Tip–sample interaction (in AFM) – Forces acting on the tip of the scanning probe due to its interaction with a sample.

Tomogram – 3D reconstruction of a unique specimen calculated using methods for tomographic reconstruction.

Tomographic reconstruction – 3D reconstruction of the entire area (e.g. a 2000 nm^2 field-of-view). The specimen is tilted and imaged over a range of angles (e.g. ±65°) to produce a tilt series of images. These images are aligned computationally to a common coordinate system and reconstructed into 3D.

Topography – Relief or landscape profile across the surface of an object. A topographical map of a sample will comprise a set of x, y, and z coordinates.

Total internal reflection fluorescence (TIRF) microscopy – Microscopy technique that enables the imaging of fluorescent molecules present within 100nm of an interface. For example, in cells that adhere to a glass substrate only fluorescence at the base, or bottom, of the cell will be imaged.

Totipotent – Potential of a stem cell to differentiate into all cell types of an organism, i.e. the zygote.

Transcription factor – A protein that binds to a specific DNA sequence (through its DNA binding domain) and regulates gene expression by promoting or blocking the transcription of that gene.

Transduction – Transfer of genetic material to a mammalian cell using viral particles.

Transfection – Transfer of genetic material to a mammalian cell using non-viral methods (e.g. **plasmids** or **electroporation**).

Transformant – A bacterial cell that has taken up a plasmid or recombinant plasmid and which is normally identified initially because it gives rise to a colony of cells on a selective agar plate containing an antibiotic for which the plasmid carries a resistance gene.

Transformation – Introduction of plasmid or recombinant plasmid into a bacterial host cell. Also the change in a eukaryotic cell caused by infection with a cancer-causing virus.

Transgene – A gene or genetic material that is transferred into an organism using genetic engineering approaches.

Transient (for protein interaction) – where the complex that is formed is not permanent, such as an enzyme–inhibitor or receptor–ligand interaction.

Transient knockdown – Reduction or elimination of the expression of a gene of interest as a result of **transient transfection**.

Transient transfection – Transfection in which the the newly introduced genetic material does not integrate into the host genome, and therefore is lost as cells undergo mitosis.

Translational repression – Interference with translational initiation pathways. One method by which expression of mRNA is silenced.

Transmission (in mass spectrometry) – Ratio of the number of ions reaching the detector to the number entering the mass spectrometer.

Tripotent – Potential of a stem cell to differentiate into three cell types of an organism.

TUNEL (**T**erminal Transferase d**U**TP **N**ick **E**nd **L**abelling) – DNA end-labelling assay used to measure fragmentation such as that seen during the process of apoptosis.

Two-dimensional electrophoresis (2-DE) – Sequential in-gel separation of proteins based on their overall charge followed by molecular weight.

Ultrafiltration – Centrifugation of a protein solution over a membrane with specific molecular weight cut-offs to separate small molecules from macromolecules.

Unipotent – Potential of a stem cell to differentiate into one cell type of an organism.

Uracil-*N*-glycosylase – By error during DNA replication the DNA polymerase can incorporate occasional dUTP rather than dTTP. In forming the phosphodiester bond the dUTP is converted into dUMP in the DNA. Uracil-*N*-glycosylase is an enzyme encoded by the *ung* gene in *E. coli* that removes these dUMPs before the DNA is repaired by the introduction of the correct dTMP.

Vacuum system (in mass spectrometry) – A set of vacuum pumps used to maintain a high vacuum within the mass spectrometer to ensure that the analyte ions are not lost through collisions with background gas molecules.

Vector – Vehicle for transporting foreign genetic material into a cell (see also **plasmids**).

Viscoelastic – Material with both liquid-like (viscous) and solid-like (elastic) behaviour. The type of behaviour observed is typically dependent on the time-scale over which it is observed.

Vitrification – Means of immobilizing an aqueous specimen by reducing its temperature so rapidly that crystalline water configurations (i.e. ice crystals) cannot form.

Void volume – Minimum volume in which a molecule can be eluted from a column, by passing only through the channels around the beads.

Western blotting – Proteins are removed from a gel by a process of transfer (or blotting) on to a nitrocellulose membrane (high-protein affinity) and subsequently identified by specific interaction with particular antibodies.

Wild type – Naturally occurring DNA sequence of a gene.

WW domain – Small motif characterized by two conserved tryptophan residues (spaced ~20–23 amino acid residues apart) which recognizes proteins that contain specific proline-rich sequence motifs.

Yeast artificial chromosome (YAC) – Cloning vector that allows the combination of large foreign DNA fragments with the small elements necessary for replication in yeast cells.

Yield – Amount of protein produced at the end of the purification process, often expressed in milligrams or as a percentage of the total in the starting material.

Young's modulus – Measure of **stiffness** of elastic materials.

Y-STR (**s**hort **t**andem **r**epeat) – Highly polymorphic genetic loci located on the Y chromosome.

Index

F

G

N

O

P

S

T

U

V

W

X

Y

Plate 1

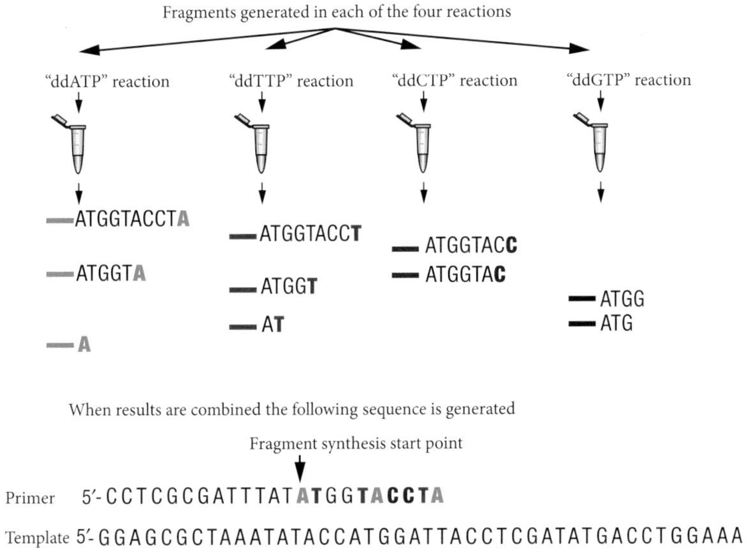

Plate 2

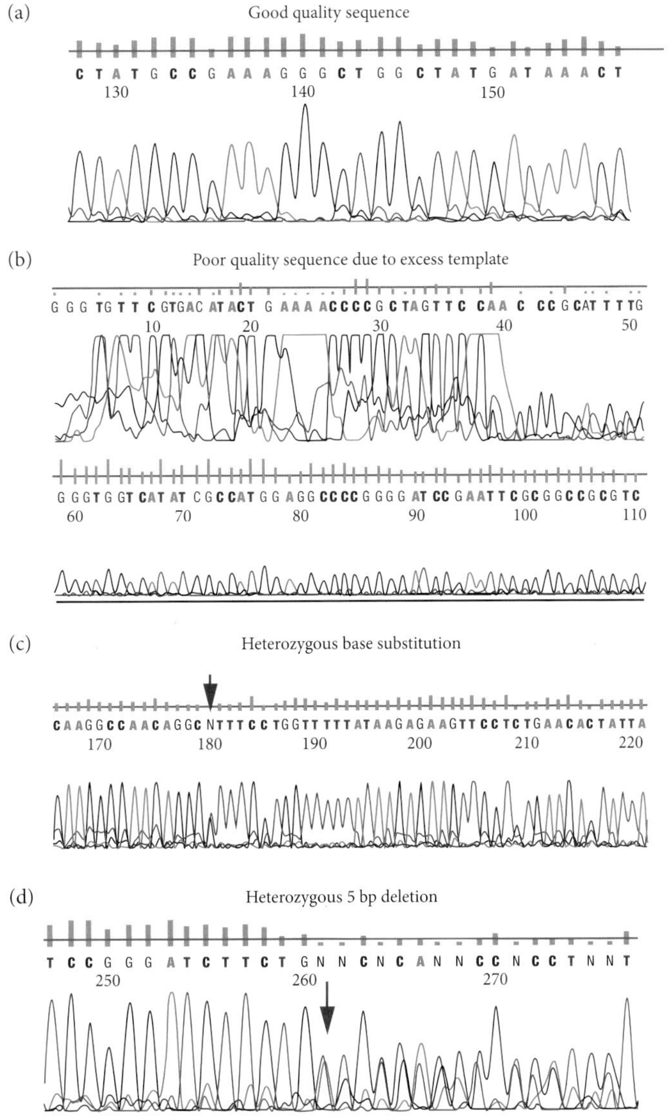

Plate 3

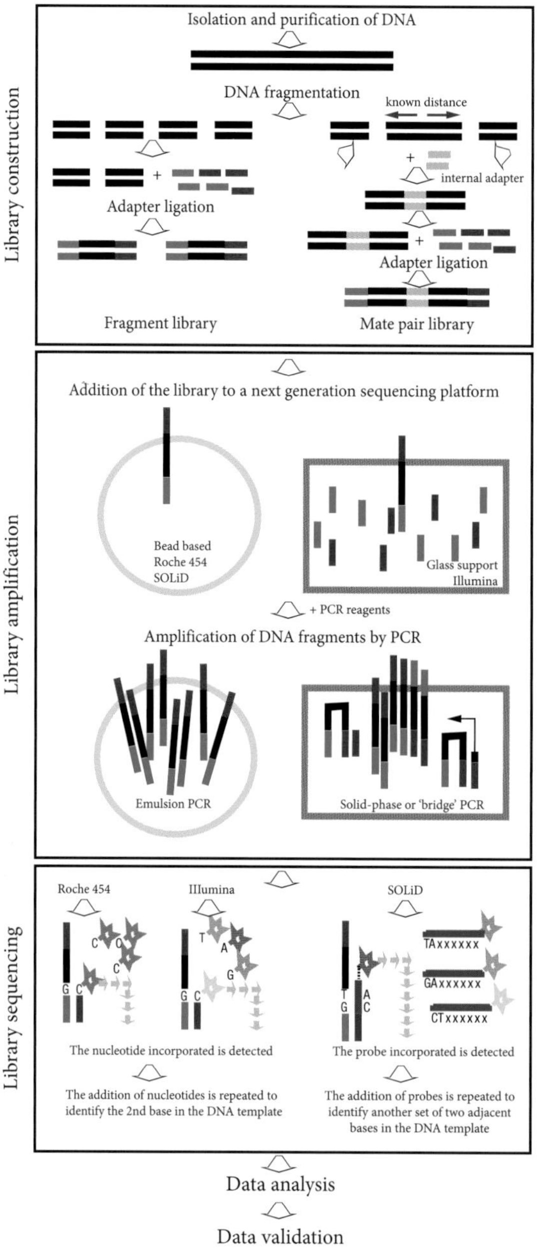

Library construction
Isolation and purification of DNA
DNA fragmentation
known distance
internal adapter
Adapter ligation
Adapter ligation
Fragment library
Mate pair library
Library amplification
Addition of the library to a next generation sequencing platform
Bead based
Roche 454
SOLiD
Glass support
Illumina
+ PCR reagents
Amplification of DNA fragments by PCR
Emulsion PCR
Solid-phase or 'bridge' PCR
Library sequencing
Roche 454
IIlumina
SOLiD
TAxxxxxx
GAxxxxxx
CTxxxxxx
The nucleotide incorporated is detected
The probe incorporated is detected
The addition of nucleotides is repeated to identify the 2nd base in the DNA template
The addition of probes is repeated to identify another set of two adjacent bases in the DNA template
Data analysis
Data validation

Plate 4

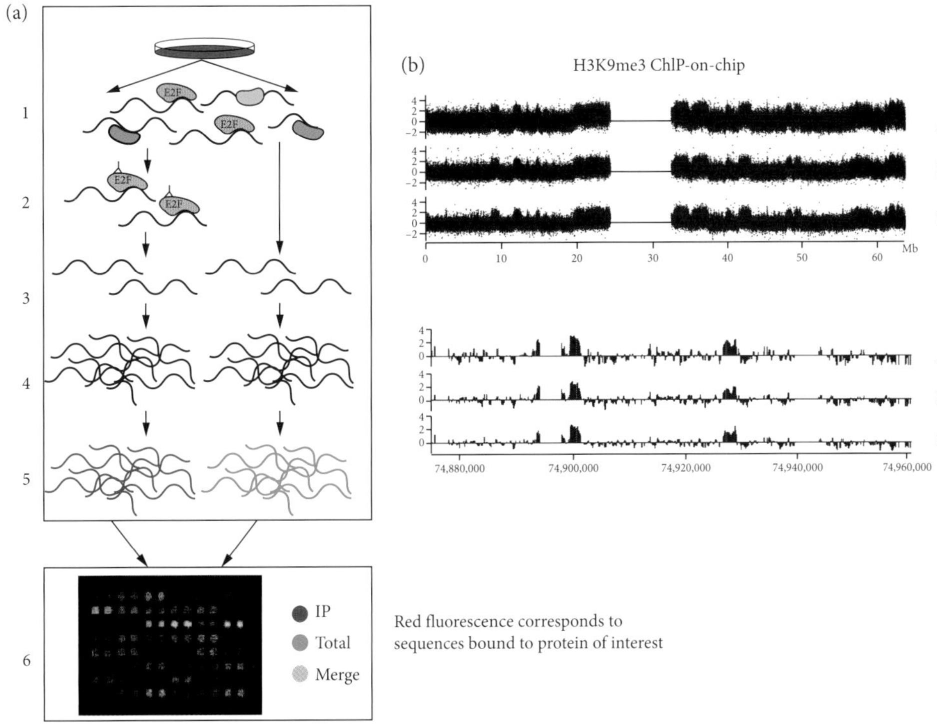

Plate 5

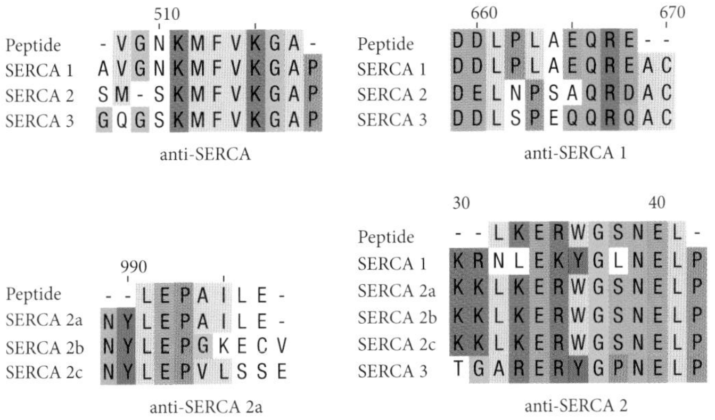

Plate 6

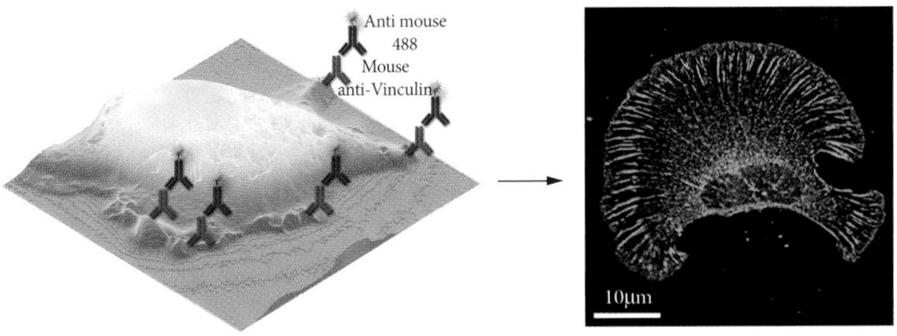

Fixed cell stained with mouse vinculin primary Ab and counter stained with anti-mouse Alexa fluor 488 secondary Ab

Cellular localisation of Vinculin visualized using fluorescence microscopy

Plate 7

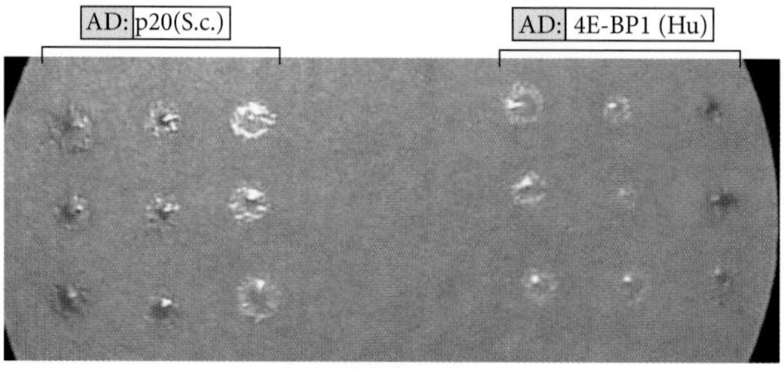

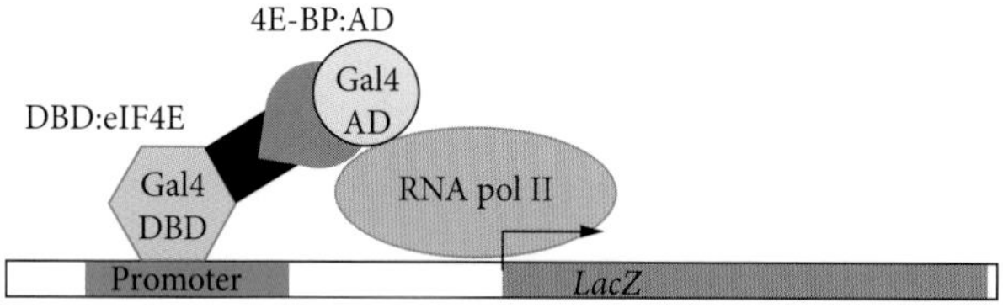

Plate 8

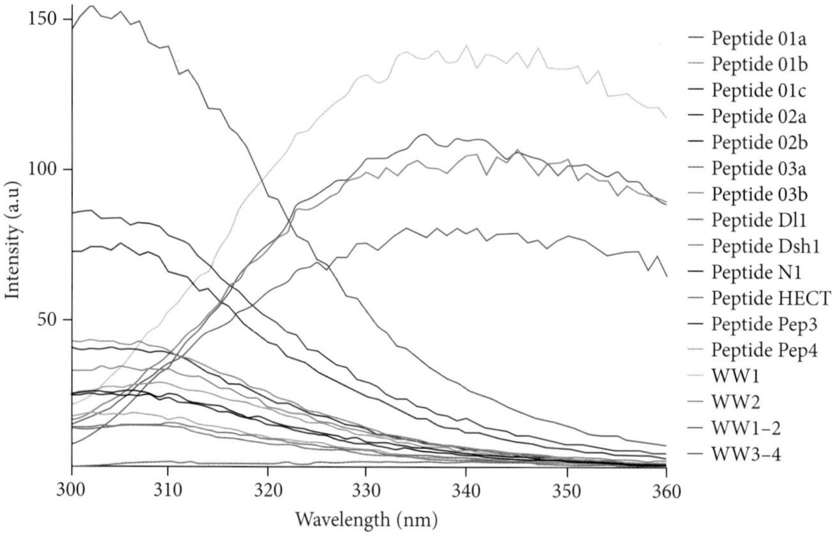

Plate 9

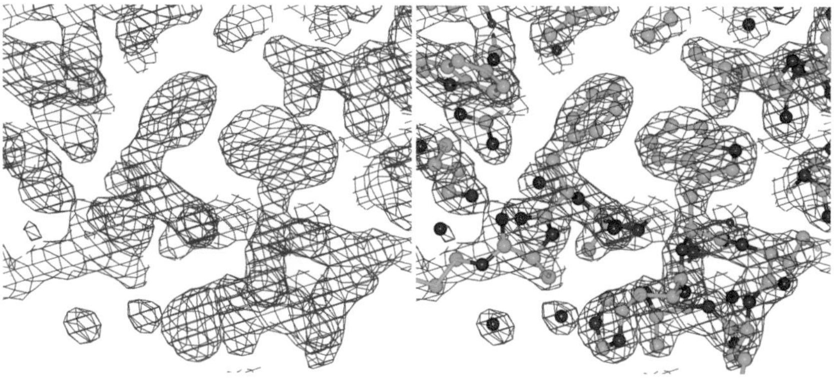

Plate 10

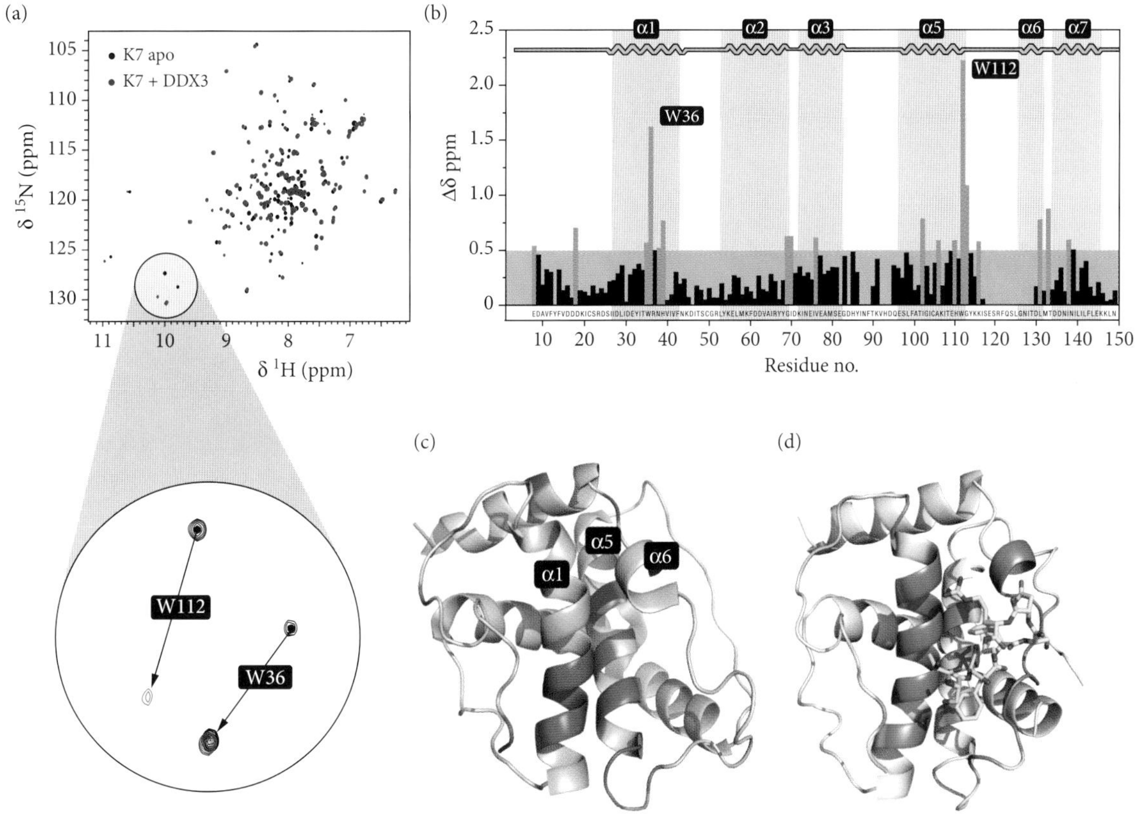

Plate 11

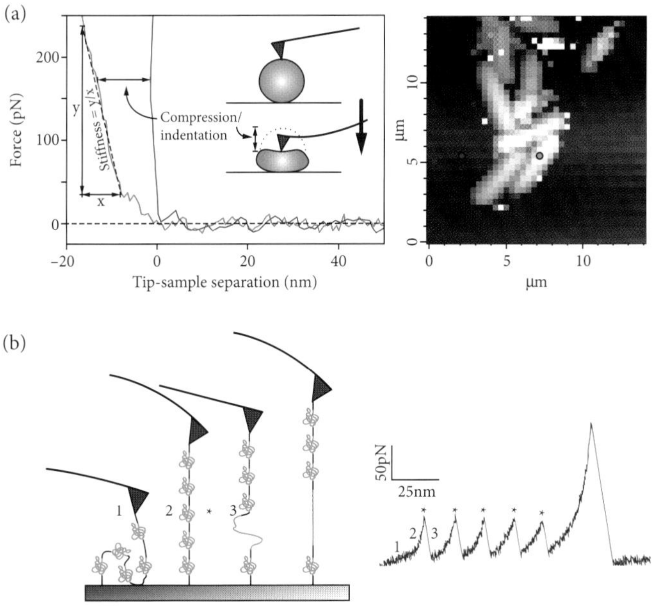

Plate 12

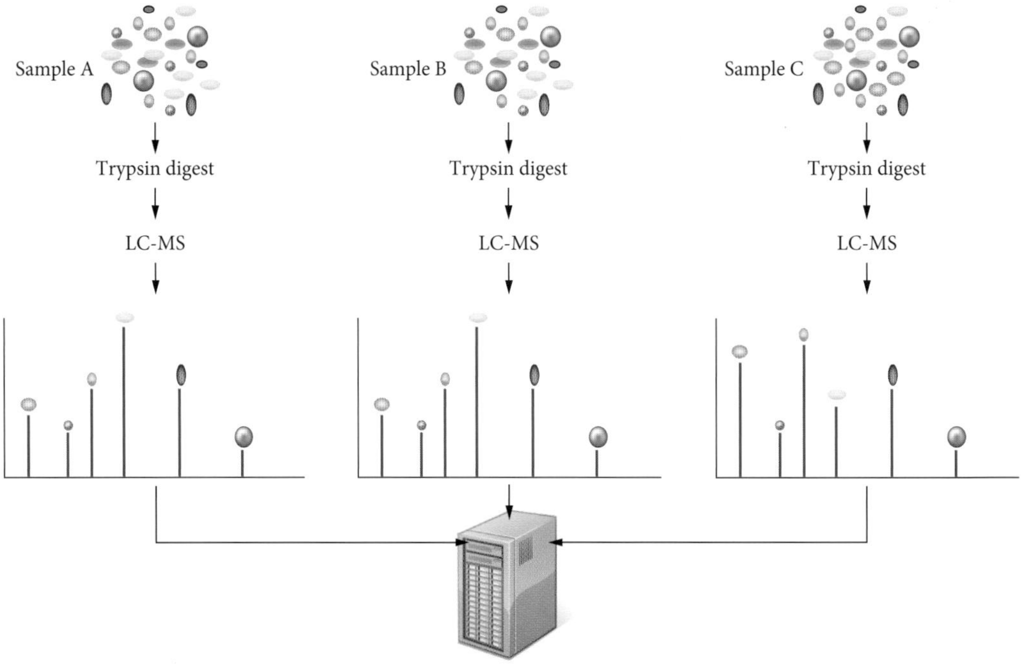

m/z	Retention	Sample B vs A	Sample C vs A	Sample D vs A
.....	.....	.....	.....	.....
.....	.....	.....	.....	.....
.....	.....	.....	.....	.....

Plate 13

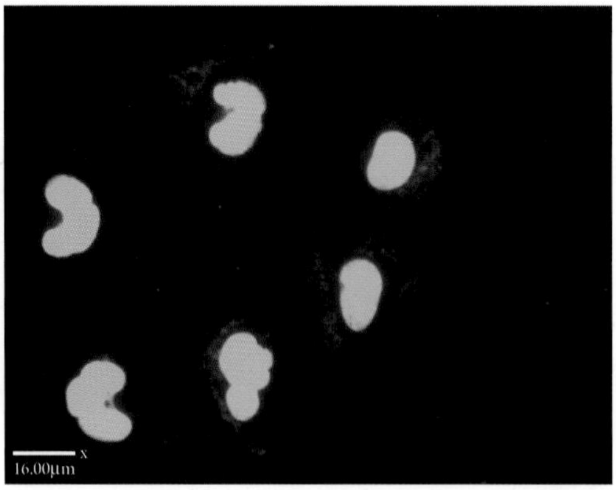

Plate 14

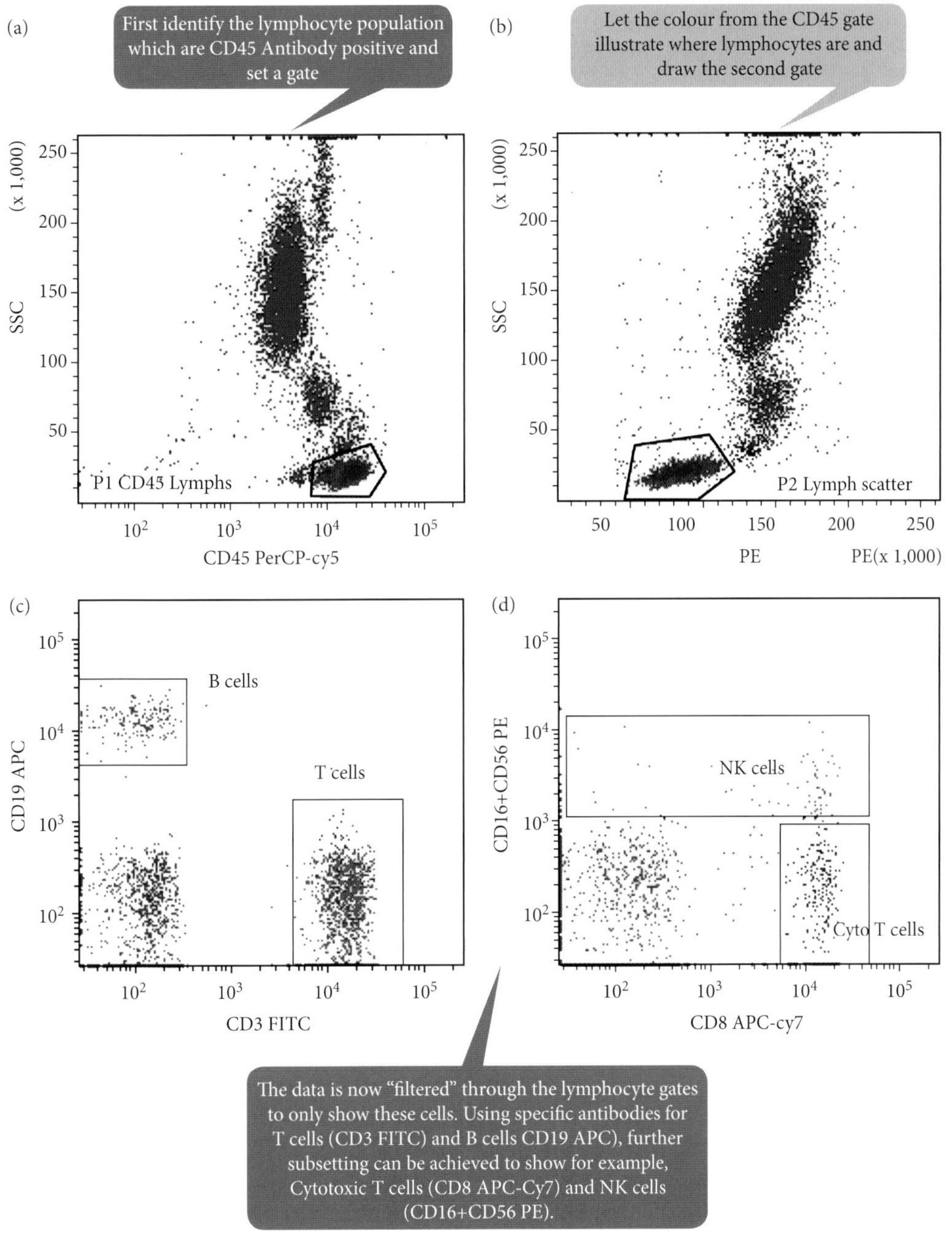

Population	#Events	%Parent	%Total
All events	10,000	####	100.0
P1 CD45 Lymphs	3,324	33.2	33.2
P2 Lymph scatter	3,260	98.1	32.6
B cells	195	6.0	2.0
T cells	1,841	56.5	18.4
NK cells	66	3.6	0.7
Cyto T cell	325	17.7	3.2

Finally, the sequence of gates used are summarised as a hierarchy and presented as a table. Basic statistics are also included below.

Plate 15

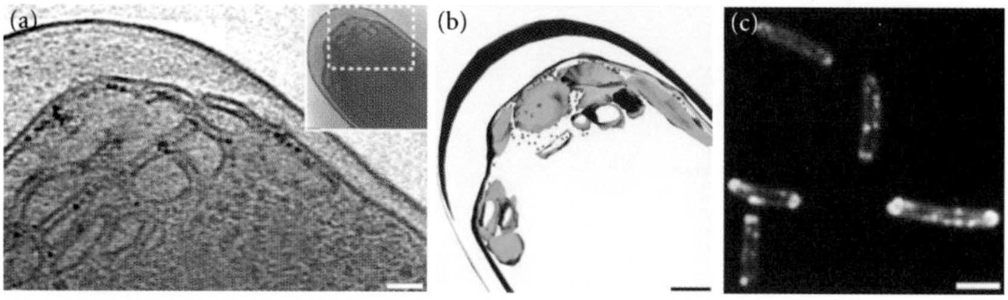

Plate 16

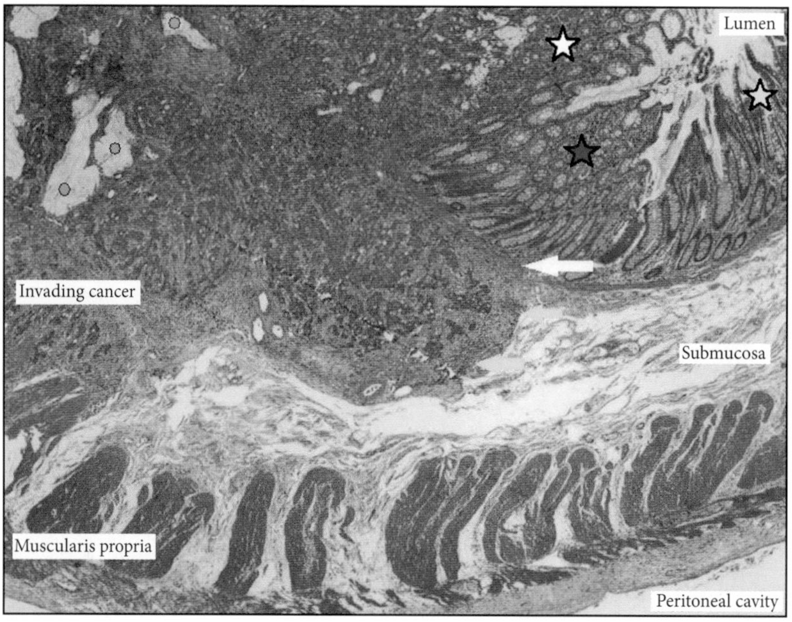

Plate 17

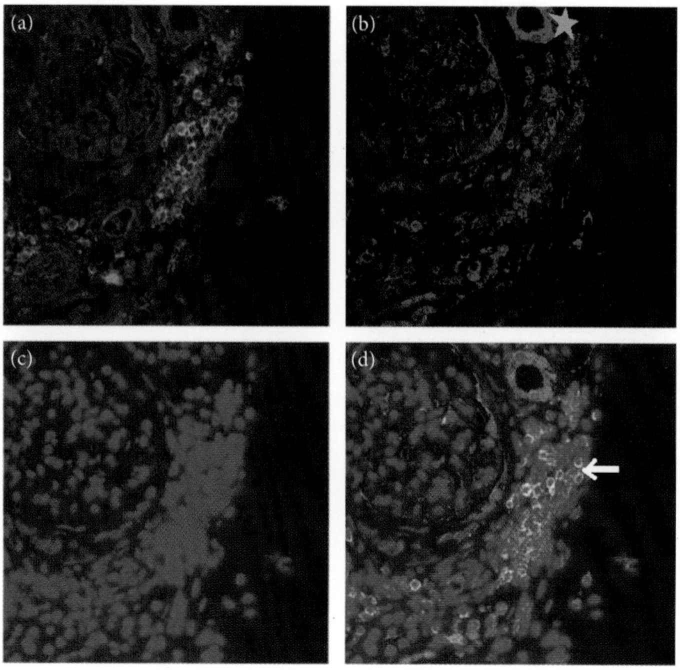